Complexity and Nonlinearity in Cardiovascular Signals

Riccardo Barbieri • Enzo Pasquale Scilingo
Gaetano Valenza
Editors

Complexity and Nonlinearity in Cardiovascular Signals

Springer

Editors
Riccardo Barbieri
Department of Electronics
Informatics and Bioengineering
Politecnico di Milano
Milan, Italy

Gaetano Valenza
Department of Information Engineering
& Bioengineering and Robotics
Research Center "E. Piaggio"
University of Pisa, School of Engineering
Pisa, Italy

Enzo Pasquale Scilingo
Department of Information Engineering
& Bioengineering and Robotics
Research Center "E. Piaggio"
University of Pisa, School of Engineering
Pisa, Italy

ISBN 978-3-319-58708-0 ISBN 978-3-319-58709-7 (eBook)
DOI 10.1007/978-3-319-58709-7

Library of Congress Control Number: 2017946615

Printed on acid-free paper

This Springer imprint is published by Springer Nature
The registered company is Springer International Publishing AG
The registered company address is: Gewerbestrasse 11, 6330 Cham, Switzerland

Foreword

Efficient and accurate cardiovascular monitoring in clinical practice constitutes a public health stake of paramount importance. The automated and reproducible analysis of cardiovascular signals has thus received significant research efforts over the last two decades. Such research, conducted by teams from all over the world, gathering renowned medical doctors and experts in statistical signal processing, led to a very diverse body of science, both in the broad range of pathophysiological issues they have addressed and in the richness of the formal concepts and corresponding technical tools they have invented and developed. Advances in sensor technologies as well as in computer facilities, together with progress in medicine and physiology, are continuously and further enlarging the potential benefits of cardiovascular monitoring, hence further motivating the need for accurate cardiovascular signal analysis tools and methodologies. The timing is thus excellent for the publication of a volume focused on recent advances on the quantitative assessment of complexity and nonlinearity in the cardiovascular system able to reflect on, organize together, compare, and survey the current knowledge and know-how in the field, aiming to tame the diversity of available conclusions and achieved results and to help frame future research and development.

The book, proposed by R. Barbieri, E. Pasquale Scilingo, and G. Valenza, world-renowned scientists who have significantly and consistently contributed to the field and have a rich view of its various and different practical and methodological aspects, is offering state-of-the-art contributions in cardiovascular signal analysis. The volume opens with a chapter, proposed by editors and collaborators, reviewing the physiological mechanisms underlying the nature of cardiovascular signals, and is organized in two sections, dedicated respectively to methodological developments and applications.

The Methodology Section will review and compare the use of several concepts, stemming from information theory and advanced statistical signal processing, as well as scale-free dynamics and nonlinear or nonstationarity signal processing methods.

The complexity of cardiovascular signals has indeed been probed using concepts ranging from chaos (Schulz et al., Faes et al.) and entropies (Hu et al., Valenza et al.) to symbolic complexity (Schulz et al.). The modeling of cardiovascular signals has required extensive use of nonstandard signal processing tools: time-varying analysis (Valenza et al.), time-frequency analysis (Orini et al.), multiscale analysis (Vanello et al.), and fractal analysis (Castiglioni et al., Paradisi et al.).

The Application Section offers a broad overview of the potential interests of cardiovascular signal analysis for the investigations of several different physiological issues and for the assessment of pathologies, thus paving the way toward clinical practice of these modern tools: heart failure (Liu et al.), diabetic cardiac neuropathy (Jelinek et al.), depression and bipolar disorder (Valenza et al.), epilepsy (Romigi et al.), atrial fibrillation (Alcaraz et al.), fetal heart rate monitoring (Signorini et al.), acute brain injury (Almeida et al.), and sleep apnea (Varon et al.). The Application Section also contains two review articles, one dedicated to the relations of age and gender with complexity measures (Schroeder et al.) and another one offering a historical perspective on the impacts of academic research efforts on clinical practice, as well as on the relations between these two worlds (Sassi et al.).

In sum, R. Barbieri, E. Pasquale Scilingo, and G. Valenza have assembled a remarkable collection of contributions, whose reading is both instructive and pleasant, offering an up-to-date overview of the field, its history, as well as its possible future.

CNRS Senior Researcher, Ecole Normale Supérieure de Lyon Patrice Abry
Lyon, France
April 15, 2017

Contents

Part I
Introduction

Chapter 1
Introduction to Complex Cardiovascular Physiology

Michele Orini, Riccardo Barbieri, Mimma Nardelli, Enzo Pasquale Scilingo, and Gaetano Valenza

Abstract This chapter aims at providing a brief overview of the main aspects in cardiovascular physiology that have encouraged and justified the use of advanced nonlinear signal processing methodologies for the study of the cardiovascular system. This system, in fact, constantly adapts to changes in internal and external conditions to maintain blood pressure homeostasis through complex and dynamic feedback mechanisms that simultaneously affect several processes such as heart rate, cardiac output, blood pressure, respiration, peripheral resistance etc. Therefore, there is a need for nonlinear, non-stationary, and multivariate approaches to assess cardiovascular interactions and their causal structure in health and disease.

This chapter aims at providing a brief overview of the main aspects in cardiovascular physiology that have encouraged and justified the use of advanced nonlinear signal processing methodologies for the study of the cardiovascular system. This system, in fact, constantly adapts to changes in internal and external conditions to maintain blood pressure homeostasis through complex and dynamic feedback mechanisms that simultaneously affect several processes such as heart rate, cardiac output, blood pressure, respiration, peripheral resistance etc. Therefore, there is a need for nonlinear, non-stationary, and multivariate approaches to assess cardiovascular interactions and their causal structure in health and disease. This introduction focuses on short-term regulation of the cardiovascular function, whose primary

---

M. Orini (✉)
Institute of Cardiovascular Science, University College of London, London, UK
e-mail: m.orini@ucl.ac.uk

R. Barbieri
Department of Electronics, Informatics and Bioengineering, Politecnico di Milano, Milan, Italy

M. Nardelli
Bioengineering and Robotics Research Centre "E. Piaggio" and Department of Information Engineering, University of Pisa, Pisa, Italy

E.P. Scilingo • G. Valenza
Department of Information Engineering & Bioengineering and Robotics Research Center "E. Piaggio", University of Pisa, School of Engineering, Pisa, Italy

R. Barbieri et al. (eds.), *Complexity and Nonlinearity in Cardiovascular Signals*,
DOI 10.1007/978-3-319-58709-7_1

Table 1.1 Important variables of the cardiovascular system

Measure	Typical value	Normal range
End-diastolic volume (EDV)	120 mL	65–240 mL
End-systolic volume (ESV)	50 mL	16–143 mL
kStroke volume (SV)	70 mL	55–100 mL
Ejection fraction (Ef)	58%	55–70%
Cardiac output (CO)	4.9 L/min	4.0–8.0 L/min
Systolic pressure		90–119 mmHg
Diastolic pressure		60–79 mmHg
Heart rate (HR)	70 bpm	60–100 bpm

aim is to ensure that blood pressure allows oxygen and nutrients supply to match body's needs. The most important parameters that determine the cardiovascular function as well as their normal range are reported in Table 1.1. Mean arterial pressure (MAP) is determined by the interaction between cardiac output (CO), total peripheral resistance (TPR), heart rate (HR), and stroke volume (SV) [1]:

$$MAP = CO \times TPR = (HR \times SV) \times TPR \tag{1.1}$$

Indeed, the autonomic nervous system is able to continuously and dynamically regulate blood pressure by adjusting these parameters. Details on neural control of the cardiovascular system, as well as on gender differences in the autonomic nervous system regulation, baroreflex, heart rate variability, and respiratory sinus arrhythmia follow below.

1.1 Neural Control of the Cardiovascular System

The autonomic nervous system plays a primarily role to maintain homeostasis through the regulation of arterial pressure and all major cardiovascular variables.

The efferent autonomic signals are transmitted to the various organs of the body through two major subdivisions called the sympathetic nervous system and the parasympathetic nervous system [2]. The sympathetic and parasympathetic nerve fibers secrete mainly one or the other of two synaptic transmitter substances: acetylcholine or norepinephrine. Fibers that secrete acetylcholine are called cholinergic, while those that secrete norepinephrine are called adrenergic, a term derived from adrenalin, which is an alternate name for epinephrine.

Table 1.2 shows the effects on different visceral functions of the body caused by stimulating either the parasympathetic nerves or the sympathetic nerves. From this table, it can be seen that sympathetic stimulation causes excitatory effects in some organs but inhibitory effects in others. Likewise, parasympathetic stimulation causes excitation in some but inhibition in others. Also, when sympathetic stimulation excites a particular organ, parasympathetic stimulation sometimes inhibits it, demonstrating that the two systems occasionally act reciprocally to each other. However, most organs are dominantly controlled by one or the other of the two systems.

Table 1.2 Effect of sympathetic and parasympathetic activation

Organ	Sympathetic stimulation	Parasympathetic stimulation
Heart	Increased rate	Slowed rate
Heart	Increased force of contraction	Decreased force of contraction
Coronaries	Constricted (α), dilated (β_2)	Dilated
Vessel	Constricted	No effect
Arterioles	Constricted	No effect
Adrenal medullae	Release of (nor) epinephrine	No effect
Arterial pressure	Short-term increase	Short-term decrease

In contrast with skeletal nervous system regulation, activation of autonomic effectors is enabled by low frequency stimulation. In general, only one nerve impulse every few seconds suffices to maintain normal sympathetic or parasympathetic effect. Moreover, the sympathetic and parasympathetic systems are continuously active, and the mean basal rates of activity are known, respectively, as sympathetic tone and parasympathetic tone [2]. While the tonic influences of a system are usually associated with an average modulation, variation over the tone is usually called phasic stimulation.

The activity of the autonomic nerves that regulate the cardiovascular function is determined by a network of neurons located in the medulla oblongata that receive inputs from other central structures including the hypothalamus, cerebral cortex, and medullary chemoreceptors; and from peripheral reflexes arising from baroreceptor, chemoreceptor, mechanoreceptor, thermoreceptor, and nociceptor afferents located in the blood vessels, heart, lungs, skeletal muscles, skin, and viscera [3].

The sympathetic nervous system can rapidly increase MAP by constricting arterioles, and therefore peripheral resistance, or by increasing heart rate or stroke volume, whereas the predominant effect of parasympathetic activation is a rapid decrease of heart rate, which influences cardiac output [1].

1.1.1 The Parasympathetic Nervous System

About 75% of all parasympathetic nerve fibers are in the vagus nerves (cranial nerve X), passing to the entire thoracic and abdominal regions of the body. The vagus nerves supply parasympathetic nerves to the heart, lungs, esophagus, stomach, entire small intestine, proximal half of the colon, liver, gallbladder, pancreas, kidneys, and upper portions of the ureters [2].

Parasympathetic division consists of long preganglionic fibers that synapse on short postganglionic fibers arising from ganglia located close to the effector targets. Therefore, parasympathetic discharge causes fairly localized responses [3].

Either all or almost all of the postganglionic neurons of the parasympathetic system are cholinergic. Thus virtually all the terminal nerve endings of the parasympathetic system secrete acetylcholine. Acetylcholine activates mainly two types of receptors, called muscarinic and nicotinic receptors.

1.1.1.1 Effect of the Parasympathetic Nervous System

- **Effects on the heart:**
  Tonic parasympathetic activation predominates over sympathetic tone at rest [4]. Parasympathetic stimulation on the heart causes mainly opposite effects than sympathetic stimulation: decreased heart rate and strength of contraction. Heart rate reduction is mediated by inhibition of the sympathetic nervous system and by direct hyperpolarization of sinus nodal cells. The membrane hyperpolarization decreases the spontaneous firing rate of the sinoatrial node and slows conduction in the atrioventricular node, thereby slowing the intrinsic heart rate [4].

  The effects of parasympathetic regulation of ventricular activation and repolarization have not been extensively studied, and parasympathetic innervation of the ventricular myocardium was considered minimal for many years [5]. However, direct vagal nerve stimulation has been recently shown to decreased heart rate, left ventricular pressure and contractility, and to prolong ventricular action potential duration in an in-vivo porcine model [5].
- **Effects on blood vessels:**
  Parasympathetic stimulation has almost no effects on most blood vessels except to dilate vessels in certain restricted areas, such as in the blush area of the face [2].
- **Effects on arterial pressure:**
  Parasympathetic stimulation can provoke a decrease in arterial pressure, because it decreases pumping by the heart but has virtually no effect on vascular peripheral resistance [2]. Very strong vagal parasympathetic stimulation can almost stop or occasionally actually stop the heart entirely for a few seconds and cause temporary loss of all or most arterial pressure.

Clinical Implications While elevated sympathetic activity is associated with an adverse prognosis, a high level of parasympathetic activation confers cardio-protection by several potential mechanisms [4]. Epidemiological data indicate that the resting heart rate predicts mortality. The higher the vagus nerve activity is, the slower the heart rate is, the greater the increase in the parasympathetic component of heart rate variability is, and the better the outcome is [4]. Moreover, parasympathetic activation and its physiological effects are attenuated in heart failure. And either direct or indirect vagus nerve stimulation could have direct beneficial effects on remodeling and clinical outcomes. Vagal nerve stimulation has also been shown to decrease infarct size [6], reduce the ventricular fibrillation threshold and decrease the incidence of ventricular arrhythmias and mortality during ischemia [5]. Due to its cardio-protective effect, mainly mediated by antagonizing sympathetic overdrive, vagal nerve stimulation devices have been proposed to treat heart failure [7, 8]. Many approaches are under development to modulate autonomic activity, that include vagus nerve stimulation (VNS) [9], spinal cord stimulation [10], and baroreceptor activation [11]. However, despite great expectations, recent clinical trials have shown that VNS does not reduce the rate of death or heart failure events in chronic heart failure patients [12].

1.1.2 Respiratory Sinus Arrhythmia

Cardiorespiratory interactions are mediated by both mechanical effects related to hemodynamics, and neural control mainly related to vagal (parasympathetic) activity.

Respiratory sinus arrhythmia (RSA) is the variation in heart rate that occurs during a breathing cycle, or heart rate variability in synchrony with respiration. In short-term HRV, RSA is the most prominent and consistent component. It appears in power spectrum of RR interval as a peak within the so-called HF band (0.15–0.4 Hz) or, more appropriately, as a peak at respiratory frequency. Because the magnitude of RSA is attenuating with progressive suppression of cardiac vagal activity and abolished by complete vagal blockage with atropine, RSA has been proposed and widely used as a quantitative index of cardiac vagal function [13–15].

The magnitude of RSA increases with rest and decreases with strain or tension. RSA is increased in the supine position and decreased in the upright position [16]. It becomes greatest during sleep and is greater during slow wave sleep than REM sleep. It also increases with relaxation and decreases with physical and mental stresses [13]. Moreover, RSA decreases with advancing age and severity of cardiac diseases. In particular, in patients with coronary artery disease, RSA at rest decreases progressively with advancing severity of coronary artery disease [13].

Although RSA was already studied at the beginning of the twentieth century, it is still far from being completely understood, and it is consider a complex phenomenon whose physiological role is still a matter of debate [17].

The importance of RSA is motivated by the following reasons: (i) It is frequently employed as an index of cardiac vagal tone or even believed to be a direct measure of vagal tone. For this reason RSA is classically described as a vagally mediated increase and decrease in heart rate concurrent with inspiration and expiration, respectively. (ii) It is used to index disease risk or severity [18]; (iii) It is a central point in the evolution theory of neural control of cardiorespiratory interactions [19]. However, there are many concerns regarding the interpretation of RSA, which are described in the following section.

Interpretation of RSA The simple magnitude of RSA is often assumed to be a valid index of cardiac parasympathetic control. However, as explained in [19], there are many significant caveats regarding vagal tone interpretation. The main concerns are:

(i) Respiratory parameters of rate and volume can confound relations between RSA and cardiac vagal tone. Alterations in respiration rate and tidal volume had profound effects upon RSA magnitude: RSA magnitude under steady-state conditions is inversely related to respiration rate and directly related to tidal volume. However, these effect have been reported to be unrelated to directly determined levels of cardiac vagal tone [18]. A number of laboratory studies have documented that changes in respiratory rate and tidal volume can seriously confound the association of RSA and cardiac vagal tone. Respiratory influences upon RSA amplitude become a problem for assessment of cardiac vagal tone whenever respiratory rate and/or tidal volume substantially differ between groups or conditions, and whenever RSA, respiratory parameters and cardiac vagal tone do not covary with each other [19]. The most likely explanation for the RSA variations due to changes in respiratory rate and volume, is that they are related to phasic patterning of vagal effects upon heart rate, owning to respiratory gating [20]. Therefore, in this case RSA cannot be used to quantitatively measure vagal tone.

To exclude respiratory confounding effects from RSA, subjects are sometimes asked to breath at a paced rate [18]. Another technique is to adjust for the influence of tidal volume. The measure is simply RSA divided by tidal volume when time domain-measures are used [18], or the transfer function from cross-spectral analysis of RR interval and respiratory time series when spectral analysis is employed [21]. In both cases, this characterizes the amount of RSA amplitude change per tidal volume. Due to the reciprocal relation between rate and tidal volume, this adjustment may reduce or eliminate the RSA dependency upon respiratory rate [19]. Finally, experts suggest that the only way to determine whether RSA differences are related to respiratory parameters in a particular study is to measure both respiration and RSA, as well as to examine relations between the two [19].

(ii) Although within-subject relations between RSA and cardiac vagal control are often strong, between-subject associations may be relatively weak [19]. Furthermore, it is not clear what exactly individual differences in RSA represent.

(iii) RSA measurement is strongly influenced by concurrent levels of momentary physical activity, which can bias estimation of individual differences in vagal tone. Therefore, it is plausible that even small laboratory differences in movement during baseline measurement may produce effects upon RSA that could be wrongly inferred as evidence of constitutional differences in autonomic control [19].

(iv) RSA amplitude is affected by sympathetic tone and may not be a pure vagal measure. RSA was shown to be sensitive to variations in cardiac vagal tone when cardiac sympathetic tone was absent, or was relatively low and stable. However, there is not confirmation about the parasympathetic specificity of RSA, i.e. that RSA always specifically reflects cardiac vagal control independently from how cardiac sympathetic activity changes [19]. It has been shown that RR interval correction of RSA can reduce or eliminate the influence of basal levels of cardiac sympathetic tone. Concerning possible sympathetic effects on RSA, it is worth mention that a pronounced decrease in RSA magnitude may signify true reduction of vagal outflow from brain to heart, or a primary increase in sympathetic tone that leads to an interaction with vagal activity, or both.

(v) RSA and cardiac vagal tone may dissociate under certain circumstances. Dissociation has been observed in experiments in which phenylephrine was used to activate vagal baroreflex response. The relation between RSA and vagal tone was relatively proportional and linear until heart rate slows down to very low levels. As heart rate further decreases, so does magnitude of RSA [22]. Some results suggest that the relation between RSA and vagal tone is quadratic across the entire range of vagally mediated heart change [22].

Another instance of dissociation between cardiac vagal tone and RSA appears to occur during stimulation of the carbon dioxide chemoreceptors [19].

Recently, it was suggested that the apparent associations between RSA and cardiac vagal tone can be explained as indirect consequences; i.e., whenever the cardiac vagal tone changes in response to the resting level of the cardiopulmonary system, RSA appears to change parallel to it [13]. This hypothesis, which supports the idea that the degree of respiratory modulation of cardiac vagal outflow and cardiac vagal tone may be regulated separately and independently of each other, can explain the possible dissociation between the respiratory cardiac vagal modulation and cardiac vagal tone under certain conditions. According to this hypothesis, RSA should be considered as an intrinsic resting functionality within the cardiopulmonary system [13].

Origin of RSA Three main hypothesis exist to explain the origin of RSA. According to them, RSA may be due to (i) A central mechanism [23], (ii) The baroreflex [24–26], (iii) The mechanical stretching of the sinoatrial node [27, 28]; (iv) A mixture of these mechanisms [29].

Those who support the central origin for RSA, pointed out that vagal-cardiac motoneuron membrane potentials fluctuate at respiratory frequencies, modulate responsiveness of vagal motoneurons to arterial baroreceptor inputs, and impose a respiratory rhythm on vagal-cardiac nerve traffic and heart periods. Therefore, central respiratory gating of vagal motoneuron responsiveness [20] is sufficient to explain respiratory sinus arrhythmia [23]. According to this hypothesis, the strong correlation between systolic pressure and RR intervals at respiratory frequencies reflects the influence of respiration on these two measures, rather than arterial baroreflex physiology [20, 30]. A central mechanism is thought to be more likely than baroreflex also due to the observation that the latencies between a change in the respiratory component of the arterial pressure and a related change in the respiratory component of th RR (i.e. RSA) reported in the literature is too short to be due to baroreflex mechanisms. Moreover, RSA is still present in subjects with baroreflex failure [29], and causal analysis also support the hypothesis that baroreflex cannot be considered as the unique origin of RSA [29, 31–33].

The hypothesis that RSA is mainly a reflex phenomenon, driven by incoming information from baroreceptors, relies on a model [24, 34]. According to this hypothesis the RSA is due to respiratory induced blood pressure oscillations (mainly due to variations in the stroke volume following intrathoracic pressure changes) that are translated into heart rate by the baroreflex [25]. Consequently, RSA would reduce diastolic pressure variability. This hypothesis does not refute animal experiments that show respiration to modulate centrally the blood pressure to heart period reflex, but it suggests that in awake humans this phenomenon is insufficient to explain respiration-to-heart rate relations [25].

A third hypothesis suggests that, given the complexity of human cardio-respiratory physiology, RSA should be the results of many causes. Among others, there are the feedback from the lungs and, possibly, atrial stretch receptors [26].

This debate points out the need of improved methodologies that undertake the limitation of classical temporal or spectral analysis [35].

Physiological Role of RSA Respiratory sinus arrhythmia is a physiologic phenomenon reflecting respiratory-circulatory interactions universally observed among vertebrates. Some studies have shown that the efficiency of pulmonary gas exchange is improved by RSA, suggesting that RSA may play an active physiologic role [36, 37]. The matched timing of alveolar ventilation and its perfusion with RSA within each respiratory cycle could save energy expenditure by suppressing unnecessary heartbeats during expiration and ineffective ventilation during the ebb of perfusion. Thus RSA may have a positive influence on gas exchange at the level of the lung via efficient ventilation/perfusion matching [13, 37]. This hypothesis is in agreement with the definition of the RSA as an "intrinsic resting function of cardiopulmonary system" previously mentioned [13].

However, recent results [17] have been presented that do not support the hypothesis that RSA optimizes pulmonary gas exchange efficiency via clustering of heart beats in inspiration. In that study [17], contrary to assumptions in the literature, no significant clustering of heart beats was observed, even with high levels of RSA enhanced by slow breathing [17]. It is suggested that some process other than the clustering of heart beats may be responsible for previously reported associations between RSA and indexes of pulmonary gas exchange efficiency. Authors speculate that the enhancement of gas exchange may be mediated by mechanical effects. The falling intra thoracic pressure associated with spontaneous inspiratory effort lowers right atrial pressure and facilitates blood flow to the right ventricle because the rate of venous return changes inversely with right atrial pressure. This increased venous return during inhalation may be transmitted to the pulmonary artery on subsequent beats, resulting in a matched increase in both pulmonary capillary blood flow and alveolar ventilation. This thoracic pump effect, which is distinct from RSA, may be greater during slower breathing and may confound the apparent relationship between RSA and pulmonary gas exchange efficiency [17].

1.1.3 The Sympathetic Nervous System

The sympathetic nervous system innervates the heart, blood vessels, kidneys and adrenal medulla, and has a predominant role in cardiovascular regulation.

Each sympathetic pathway from the cord to the stimulated tissue is comprised of two neurons, a short preganglionic neuron and a long postganglionic neuron. Such an arrangement allows sympathetic discharge to cause diffuse responses involving multiple regional effectors [3]. All preganglionic sympathetic neurons are cholinergic, while most of the postganglionic sympathetic neurons are adrenergic. Ordinarily, the norepinephrine secreted directly into a tissue remains active for only a few seconds, demonstrating that its reuptake and diffusion away from the tissue are rapid. Epinephrine and norepinephrine are almost always released by the adrenal medullae at the same time as the different organs are stimulated directly by generalized sympathetic activation. Therefore, the organs are actually stimulated in two ways: Directly by the sympathetic nerves and indirectly by the adrenal medullary hormones. The two means of stimulation support each other, and either can in most instances substitute for the other. However, some exceptions exist. For example, recent work has shown that sympathetic nerve stimulation, not circulating norepinephrine, modulates cardiac repolarization dispersion [38]. When secreted into the blood by the adrenal medullae, both norepinephrine and epinephrine remain very active for 10–30 s; but their activity declines to extinction over 1 to several minutes [2].

Before a transmitter substance secreted at an autonomic nerve ending can stimulate an effector organ, it must first bind with specific receptors on the effector cells. There are two major types of adrenergic receptors, alpha receptors and beta receptors. Norepinephrine and epinephrine have slightly different effects in exciting the alpha and beta receptors. Norepinephrine excites mainly alpha receptors but excites the beta receptors to a lesser extent as well. Conversely, epinephrine excites both types of receptors approximately equally. Therefore, the relative effects of norepinephrine and epinephrine on different effector organs are determined by the types of receptors in the organs.

Sympathetic nerves are continuously active so all innervated blood vessels remain under some degree of continuous constriction. Direct observation of the sympathetic nerve activity shows that (i) Discharges occur in a synchronized fashion, with many of the nerves in the bundle being active at approximately the same time, and (ii) That discharges generally occur in a highly rhythmical fashion. Postganglionic sympathetic nerves are comprised of hundreds to thousands of unmyelinated fibers that fire action potentials at almost the same time (synchronization) to give discharges of summed spikes. Two features seem to characterize sympathetic discharges: their firing frequency and amplitude. Some researches have made the hypothesis of a differential control over the amplitude and frequency of sympathetic activity [39]. This hypothesis suggests that the network of cells involving the rostral ventrolateral medulla (at brainstem) provide the basal level of nerve recruitment and determine the firing frequency based on the intrinsic rhythmicity and phasic input from arterial baroreceptors, while inputs from cell groups with direct projections to the spinal cord provide an extra level of gain/recruitment of fibers [39].

Assessment of the Sympathetic Nerve Activity Because it is difficult to directly record sympathetic outflow in humans, plasma concentrations of norepinephrine are often used as a surrogate measure of postganglionic sympathetic nerve activity. Caution is warranted, however, as blood levels will be affected by changes not only in the release of noreprinephrine but also in its reuptake and metabolism (clearance from the blood). Furthermore, plasma norepinephrine provides limited insight into regional differences in sympathetic nerve activity as it reflects the contribution of norepinephrine spillover from all potential sources [3].

Radiotracer technology has been used extensively for studying norepinephrine kinetics in humans and has now become a gold standard for assessing sympathetic nerve activity in humans [39]. Norepinephrine in the plasma reflects the transmitter released by sympathetic nerves that has spilled over into the circulation. Norepinephrine spillover rate gives the rate at which norepinephrine released enters plasma and provides information about the activity of sympathetic nerves in regions that are not accessible by percutaneous microneurography, such as the heart and kidney [40]. While this technique offers good estimations of regional sympathetic nerve activity, limitations are related to its low repeatability, to the fact that it does not allow for continuous recording and to the evidence of non linear relationship between actual sympathetic activity and norepinephrine spillover [39].

In humans, sympathetic activity is often measured at the peroneal nerve, and the most common measurement is muscle sympathetic neural activity, often called MSNA. The technique involves the percutaneous insertion of a high-impedance tungsten microelectrode (the tip of which is only a few microns in diameter). Most sympathetic neural recordings involve multiunit recordings, but single-neuron recordings are also possible. The level of muscle sympathetic neural activity did not correlate to the resting heart rate or blood pressure (within normal range) but was found to relate to cardiac output and thus total peripheral resistance in males [39]. Muscle sympathetic neural activity is characterized by a striking inter-individual variability in resting humans, but it is very reproducible in a given person [40]. Recently, it has been suggested to be a good index of whole-body "net" sympathetic vasoconstrictor tone in healthy young men [40].

Researchers have also attempted to use non-invasive marker to track sympathetic nerve activity. Heart rate variability was shown to provide valuable information about sympathetic tone [41–43]. Although heart rate variability analysis has the great advantage of being the non-invasive and relatively simple, there are serious limitations to this technique. Specifically, while the low frequency variability in heart rate is influenced by the sympathetic nervous system, examples where increases in sympathetic activity are not associated with changes in low frequency variability have been reported [39, 44].

Effect of the Sympathetic Nervous Activity As also shown in Table 1.2, the sympathetic nerves have a dominant role in cardiovascular control due to their effects to increase cardiac rate and contractility, cause constriction of arteries and veins, cause release of adrenal catecholamines, and activate the reninangiotensin-aldosterone system [3].

- **Effects on the heart:**
  Cardiac sympathetic innervation of the heart includes innervation of the sinoatrial node, which allows sympathetic nerves to increase heart rate by increasing the slope of diastolic depolarization during the spontaneous sinoatrial node action potential [40]. Sympathetic nerves also innervate the myocardium; increases in sympathetic activity increase myocardial contractility and, therefore, increase

stroke volume. Thus, the sympathetic nervous system has both chronotropy and inotropy effects on the heart [3]. Given the ability to modulate both cardiac rate and stroke volume, the autonomic nerves provide an important remote mechanism to rapidly adjust cardiac output to meet short-term changes in the body's needs. In humans, there is a good deal of tonic vagal discharge and a moderate amount of tonic sympathetic discharge. The interplay of these tonic activities results in a resting heart rate that is $\approx$ 30% lower than the intrinsic heart rate of 90–100 beats/min and a cardiac output that is $\approx$ 30% higher than in the absence of sympathetic discharge [39]. Additional sympathetic discharge can increase heart rate and stroke volume and increase cardiac output. Conversely, withdrawal of tonic vagal or sympathetic discharge has opposing effects to increase or decrease cardiac output, respectively [39].

Sympathetic activation affects the action potential duration of cardiac myocyte by modulating calcium and potassium ionic currents. Recently, experimental and theoretical studies have linked sympathetic activity to low-frequency oscillations in ventricular repolarization [45–47]. Beat-to-beat repolarization variability, often estimated as QT interval variability from the ECG also increases with tonic sympathetic activation [48, 49].

- **Effects on blood vessels:** Sympathetic innervation of the peripheral vasculature causes vasoconstriction primarily through the action of norepinephrine at post-synaptic α-adrenergic receptors [40].

  Vascular capacitance, mainly driven by venous capacitance, is strongly influenced by sympathetic nervous activity. It is often overlooked that the venous circulation receives considerable sympathetic innervation, and with 70% of the blood volume can play a significant role in the acute cardiovascular responses to sympathetic activation. Venoconstriction in the splanchnic circulation results in a significant shift of blood towards the heart, increasing diastolic filling, and thus increasing cardiac output. Blood that is forced out of the veins returns to the heart, increasing end-diastolic volume and, via the Frank-Starling mechanism, increasing stroke volume and cardiac output. As 20% of blood volume is located in the veins of the splanchnic circulation, translocation of blood from this venous reservoir due to sympathetic venoconstriction is a particularly effective way to quickly redistribute blood from the venous side to the arterial side of the circulation. Recently, it has been suggested that increases in venomotor tone driven by sympathetic activity may be important mediators in cardiovascular disease development [39].
- **Effects on blood pressure:** Sympathetic stimulation increases both propulsion by the heart and resistance to flow, which usually causes a marked acute increase in arterial pressure [2]. To understand how sympathetic activity can control blood pressure, it is important to remind relationship reported in (1.1), which shows that blood pressure is determined by the product between cardiac output $\times$ total peripheral resistance. Flow varies directly (and resistance inversely) with the fourth power of the vessel radius. As a result, even small changes in vessel caliber can have relatively large effects on vascular resistance and blood flow. Sympathetic neural control of arteriolar resistance therefore offers a powerful mechanism to regulate regional blood flows to individual organs and tissues. As

the arterioles are the major contributors to total peripheral resistance, sympathetic control plays a principal role in the regulation of systemic blood pressure [3]. Sympathetic control of blood vessels is mediated by the baroreflex, which is one of the most important mechanisms for the short-term maintaining of the blood pressure.

Pathophysiology What distinguishes the sympathetic nervous system is the emerging evidence that overactivity is strongly associated with a variety of cardiovascular diseases [39, 50, 51]. Whether this increased sympathetic activity acts as a driver of the disease progression or whether it is merely a follower is currently matter of debate. However, beta blockers that treat and inhibit excess sympathetic stimulation are currently widely used to treat most forms of cardiac diseases and reduce cardiac risk.

In the following, we revised some of the conditions that are known to be associated to chronic sympatho-excitation [39, 40].

- **Hypertension:** The study of the mechanisms underlying hypertension is clinically relevant because hypertension is a causative factor in the development of heart failure, renal failure, and stroke. The causes of sympatho-excitation in association with hypertension are unclear but may involve increases in chemoreflex sensitivity to hypoxia or hypercapnia [40]. However, it is important to consider that in the vast majority of cardiovascular diseases, there is a disproportionate increase in renal sympathetic activity compared with sympathetic activity to the muscle [39]. Chronic changes in blood pressure may be related with sympathetic action on the kidney.
- **Heart failure:** Among patients with heart failure, muscular sympathetic activity is strikingly increased. In terms of the occurrence of sympathetic bursts, a healthy person may experience 30–50 bursts per 100 heart beats, whereas patients with heart failure can experience as many as 90–100 bursts per 100 heart beats. This extreme sympathoexcitation has been shown to be a predictor of mortality for patients with heart failure [40, 52]. Rapid increases in cardiac sympathetic activity are associated with ventricular arrhythmias, coronary occlusion, and damage to myocytes associated with the resulting high norepinephrine levels. Moreover, the degree of sympatho-activation appears to be a good indicator of long-term prognosis [39]. The fundamental processes underlying the sympathetic activation in heart failure remain uncertain.
- **Obesity:** Muscle sympathetic neural activity is increased markedly among obese patients and decreases with weight loss induced by exercise and diet [40]. Obesity is often related with hypertension, with up to 70% of newly diagnosed hypertensive cases are attributable to obesity [39]. Obese hypertensive subjects present high level of norepinephrine spillover from the kidneys. The mechanism(s) by which weight gain elicits sympathetic neural activation remains unclear [39].
- **Sleep apnea:** Sleep-related breathing disorders play an important pathophysiological role in cardiovascular disease, and sympathetic activation is thought to be a key mechanism linking sleep apnea to cardiovascular disease. For instance, in

patients with obstructive sleep apnea, muscular sympathetic activity is increased, even during the awake state without apnea [40]. Over time, periodic nocturnal sympathetic activation related to apnea appears to evolve into a rise in the mean daytime level of sympathetic activity.

- **Mental stress:** Large-scale studies link hypertension development with chronic mental stress in the workplace. Blood pressure has been shown to be elevated soon after migration, presumably due to stress [39]. Stress is recognized as a cardiac risk factor, with epidemiological data showing that chronic stress is associated with coronary heart disease and short-term emotional stress can act as a trigger of cardiac events among individuals with advanced atherosclerosis [53]. A stress-specific coronary syndrome, called Takotsubo cardiomyopathy, also exists. The physiological response to stress and its association with cardiac risk are thought to be mediated, at list in part, by a modulation of the sympathetic nerve activity [53]. However, the role of the sympathetic nervous system in these events need further studies [54–57].

1.1.4 Baroreflex

The baroreflex is a negative feedback system that buffers short-term fluctuations in arterial pressure [1] by modifying the cardiovascular variables which, as shown in (1.1), determine arterial pressure. Under a variety of environmental and physiological circumstances, arterial blood pressure is regulated around a narrow range and arterial baroreflexes are critically important for the beat-to-beat regulation of blood pressure [1]. The importance of the baroreflex has been demonstrated in several experimental data in which it has been observed that, in many animal species, arterial baroreceptor denervation results in an increase of the variability of blood pressure [52].

The arterial baroreflex senses changes in blood pressure via baroreceptors, which are sensory afferent nerve endings mainly located in the carotid sinus and the aortic arch. The baroreceptors respond to stretching of the vessel wall. In general, increases in this stretching as the result of a short-term increase in blood pressure lead to an increase in afferent input into central autonomic nuclei (notably the nucleus tractus solitarius). This increase in afferent input results in a reflex decrease in sympathetic neural outflow (sympathetic baroreflex), which in turn decreases vasoconstrictor tone, myocardial contractility (to decrease stroke volume), and heart rate. These sympathetic influences work in conjunction with parasympathetic influences on the sinoatrial node to decrease heart rate (vagal cardiac baroreflex). During a short-term decrease in blood pressure, the opposite occurs, and the autonomic nervous system acts to increase vasoconstriction, increase stroke volume, and increase heart rate [40].

Baroreflex Control of Sympathetic and Parasympathetic Outflow At normal arterial pressures there is activity from all neural components of the arterial baroreflex: both afferent and efferent limbs. Both resistance and capacitance vessels are partially constricted. With regard to determinants of heart rate, the sinoatrial and atrioventricular nodes receive sympathetic and parasympathetic innervation, and both efferent outflows are active at rest. Baseline heart rate is determined by the balance between these two opposing influences and the relative contribution of the sympathetic and parasympathetic nervous systems varies among species. In conscious humans, basal parasympathetic nerve activity to the heart is the major determinant of baseline heart rate, whereas basal sympathetic nerve activity has small or negligible effects. In addition, variation in resting heart rate among individuals is caused mainly by differences in basal parasympathetic tone. There are significant differences in the time delay of the response mediated by parasympathetic and sympathetic efferents. Following a rapid rise in arterial pressure, parasympathetic activation produces an immediate reaction (between 200 and 600 ms) [52]. On the contrary, the reaction to cardiac and vasomotor sympathetic activation occurs with a 2–3 s delay and reaches maximal effect more slowly. An even more sluggish response has been observed in the baroreflex control of venous return. Therefore, the ability of the baroreflex to control heart rate on a beat-to-beat basis is exerted through vagal but not sympathetic activity [52]. However, even for fast maneuvers, like standing up from the supine position, the ability to increase heart rate quickly is not of primary importance. Heart-transplant patients provide dramatic proof of this contention since they can only change their heart rate by way of circulating hormones, like norepinephrine, which takes at least a minute to have effect on heart rate after standing up. Still, these individuals can stand up without any problem [58]. However, when the sympathetic innervation to the vasculature is lost, as it may occur in autonomic failure, then the patient can no longer stand up since blood pressure starts to drop immediately and, after some 5 s, consciousness is lost due to cerebral hypoxia [58]. Therefore, the sympathetic efferent part of the baroreflex is of much more importance, since it induces vasoconstriction in response to a drop in blood pressure [58, 59]. Thus, for the control of blood pressure sympathetic outflow to the heart and vasculature is far more important than what the vagus nerve does to heart rate [59].

Although baroreflex has been widely studied, many aspects of the cardiovascular control are still unclear. For instance, in a recent debate [23, 25, 29], the involvement of the baroreflex in the respiratory sinus arrhythmia (oscillations in heart period synchronous with respiration, described in the following, has been discussed. Some authors [25], advocated that baroreflex should be considered as the main responsible for respiratory sinus arrhythmia. Following this hypothesis, the first explanation of respiratory sinus arrhythmia is that it is due to respiratory induced blood pressure oscillations that are translated into heart rate oscillations by the baroreflex. Others suggested that respiratory sinus arrhythmia has a central origin [20, 23, 29] and is not baroreflex mediated [60]. This debate clearly shows that cardiovascular control, including baroreflex, is far from being completely understood and that much work is still needed.

In a recent study, orthostatic hypotension was observed in patients with pure autonomic failure despite the increase in leg vascular resistance [61]. The results shown in [61] demonstrate that leg vasoconstriction during orthostatic challenges in patients with pure autonomic failure is not abolished. This indicates that the sympathetic nervous system is not the sole or pivotal mechanism inducing leg vasoconstriction during orthostatic challenges. Additional vasoconstrictor mechanisms may compensate for the loss in sympathetic nervous system control.

Baroreceptors In the normal operating range of arterial pressure, around 100 mmHg, even a slight change in pressure causes a strong change in the baroreflex signal to readjust arterial pressure back toward normal. The baroreceptors respond extremely rapidly to changes in arterial pressure; in fact, the rate of impulse firing increases in the fraction of a second during each systole and decreases again during diastole. Furthermore, the baroreceptors respond much more to a rapidly changing pressure than to a stationary pressure. That is, if the mean arterial pressure is 150 mmHg but at that moment is rising rapidly, the rate of impulse transmission may be as much as twice that when the pressure is stationary at 150 mmHg [62]. The arterial baroreflex also responds to the normal small variations in blood pressure that are continually induced by the respiratory cycle and by changes in posture in healthy, resting humans.

Although the arterial baroreceptors provide powerful moment-to-moment control of arterial pressure, their importance in long-term blood pressure regulation has been controversial [62]. Some physiologists consider the baroreflex relatively unimportant in chronic regulation of arterial pressure because baroreceptors tend to reset in one to two days to the pressure level to which they are exposed. This resetting of the baroreceptors may attenuate their potency as a control system for correcting disturbances that tend to change arterial pressure for longer than a few days at a time. Experimental studies, however, have suggested that the baroreceptors do not completely reset and may therefore contribute to long-term blood pressure regulation, especially by influencing sympathetic nerve activity of the kidneys [62]. Recently, it has been suggested that arterial baroreceptors play a role in long-term regulation of arterial pressure under conditions of increased dietary salt intake [39].

Measurement of Baroreflex Sensitivity Traditionally, baroreflex sensitivity (BRS) is measured as the induced heart period lengthening (in ms) divided by the causative change in previous systolic pressures (in mmHg) [58, 63]. The baroreflex appears to buffer (unwanted) blood pressure variability by modulating heart rate variability. When the sympathetic baroreflex is less sensitive, the response of the sympathetic nerves to a given change in arterial pressure will be less pronounced and may be less able to return the pressure to baseline levels [40].

Originally, a pharmacological tool was used to quickly increase or decrease blood pressure, i.e. an intravenous bolus injection of phenylephrine or nitroprusside, which increases and decreases blood pressure, respectively. While vasoconstrictor drugs, as phenylephrine, allows one to mainly explore the vagal component of the baroreceptor control of heart rate, the excitation of the sinus node that accompanies a reduction in arterial pressure caused by the administration of vasodilators, as nitroglycerin, is partly mediated through sympathetic mechanisms. Therefore vasodilators have been used to obtain information on the sympathetic limb of heart rate control [52].

The lack of selectivity in the response has been claimed as one of the major limitations of the use of vasoactive drugs. Indeed, the pressure stimulus causes a simultaneous activation of multiple reflexogenic areas, particularly cardiopulmonary receptors, which may interfere with or even counteract the arterial baroreceptor reflex. Moreover, vasoactive drugs may directly affect the transduction properties of baroreceptors, the central nervous system part of the reflex arc and the response of the sinus node [52].

Later, it was shown that this intervention could be replaced by the measurement of spontaneous blood pressure variations and the ensuing heart period variations [64, 65]. Specifically, the variations around the "Eigen frequency" of the baroreflex at 0.1 Hz are useful, while the variations around respiratory frequency are controversial [23, 25, 26, 29], for this purpose. Resonances in low frequency are induced mainly by the latencies in the sympathetic efferent arm of the baroreflex, excited by spontaneous variations in blood pressure. This explanation for the genesis of so-called low-frequency blood pressure oscillations received its theoretical basis by the modeling work of De Boer et al. [24, 58].

Two basic approaches have been established for spontaneous BRS quantification: one based on time domain and the other on frequency domain measurements. In time-domain, the sequence method [66] is based on the identification of three or more consecutive beats in which progressive increases/decreases in systolic blood pressure are followed by progressive lengthening/shortening in RR interval. The sensitivity of the reflex is obtained by computing the slope of the regression line relating changes in systolic pressure to changes in RR interval [52]. Evaluation of BRS by spectral methods is based on the concept that each spontaneous oscillation in blood pressure elicits an oscillation at the same frequency in RR interval by the effect of arterial baroreflex activity [52]. There are some differences in the computational algorithms of spectral indexes of BRS [63], which can be divided in autoregressive spectral methods [67, 68] and transfer function based methods [65]. More recently, advanced time-varying methods to assess changes in the baroreflex sensitivity have been also proposed [69–71].

However, these methodologies provide only a partial characterization of the baroreflex, and probably not even the most important component. The baroreflex sensitivity describes the ability of the baroreflex to change heart period (or heart rate) in response to a change in blood pressure [58], while it gives no information about the changes in the blood vessels operated by sympathetic nerves. Given that the changes in heart period are principally related to the vagal efferent activity to the heart, the BRS is sometime called vagal BRS or cardiac BRS [72].

An index of baroreflex control of sympathetic outflow (sympathetic BRS) is provided by the relationship between muscular sympathetic nervous activity and diastolic blood pressure during the drug boluses [72]. Diastolic blood pressure is used because muscular sympathetic nervous activity correlates more closely with diastolic blood pressure than with systolic pressure. A recent study showed that, within individuals, there is a low correlation between the sensitivities of the cardiac and sympathetic efferent arms of the baroreflex mechanism, i.e. between cardiac and sympathetic baroreflex sensitivity [72]. This result emphasizes the distinct pathways associated with cardiac and sympathetic baroreflex control and the need to use caution when drawing conclusions about the baroreflex as a whole if only one or the other arm of the reflex is studied [72].

Clinical Relevance A quantitative description of BRS can provide a useful index of neural regulation of the sinus atrial node. This information has clinical and prognostic value in a variety of cardiovascular diseases, including myocardial infarction and heart failure [50, 52]. The notion that BRS may be altered in patients with cardiovascular disease is not new [73]. Cardiovascular diseases are often accompanied by an impairment of baroreflex mechanisms, with a reduction of inhibitory activity and an imbalance in the physiological sympathetic-vagal outflow to the heart, thus resulting in a chronic adrenergic activation [52].

Age and blood pressure have been demonstrated to be the most important correlates of BRS. Although the exact mechanism is not known, loss of arterial distensibility is generally regarded to be the main mechanism responsible for reduction of BRS in older subjects [52]. The clinical relevance of spontaneous BRS analysis [64] is shown by its ability to detect early impairment of autonomic function [74] and to provide information of prognostic value, as in patients after stroke [75] or myocardial infarction [76], or in the diagnosis of brain death [77]. In particular, the study described in [76] enrolled almost 1300 patients under 80 years and showed that impaired vagal reflexes, expressed by a depressed BRS (< 3 ms/mmHg), was a significant predictor of total cardiac mortality, independently of well-established risk factors such as depressed left ventricular function and the number of ectopic beats/hour [52].

Patients with low heart-rate variability or BRS have a reduced capability to antagonise sympathetic activation through vagal mechanisms [76]. However, a low BRS added predictive value to that of low heart-rate variability [76].

A number of papers have supported the pathophysiological and clinical relevance of spontaneous BRS estimates. Indeed, the information provided by laboratory and spontaneous methods appears to be complementary, when exploring the complexity inherent in baroreflex cardiovascular modulation [78].

1.2 Sympatho-Vagal Nonlinear Interactions

Sympathetic and parasympathetic systems activities were traditionally considered to produce opposite effects in the autonomic control. When this dynamic balance changes into a static imbalance, for example, under environmental stress, the organism becomes vulnerable to pathology.

The classical model of sympatho-vagal balance was proposed by William Cannon in the beginning of twentieth century. He suggested the existence of a mutual influence, which conducts to a decrease of the parasympathetic activity in case of increased sympathetic activity and vice versa, with a linear-continuum model of activity of the autonomic nervous system in its two subdivisions. The model of Cannon was based on a negative feedback mechanism, regulated by the hypothalamus [79].

Cannon's pioneer studies about emotional physiology, highlights the role of the sympathetic nervous system in the coordination and mediation with the neuroendocrine and cardiovascular modifications, which follow an emotional stimulus. He called this response "emergency reaction", which is known as *"flight or fight response"* of the sympathetic nervous system [80]. In the same theory, the parasympathetic system was indicated to be the accountable to the *"rest and digest"* functions, preserving the basal conditions of the organism during the resting state, in terms of heart rate, respiration and metabolism.

In the 1930s, Rosenblueth and Simeone observed that when a vagal stimulus was given to anesthetized cats, their heart rate decreased and the effect was more evident under tonic sympathetic stimulation [81]. This was successively confirmed by Adli Samaan [82].

A review about this interpretation of the interaction between sympathetic and parasympathetic activities was written by Levy in 1971 [83], who introduced the definition of "accentuated antagonism", as the phenomenon where the cardiac response to neural activity in one autonomic division depends on the level of activity in the other autonomic division. Following this definition, it was suggested that the effects of combined sympathetic and vagal stimulation on the heart rate are not simply additive. Levy studied the consequence of a vagal stimulation on the heart, and he found in his experiments that the effects of the accentuated antagonism are modulated in accordance with the cardiac effector tissue [84]. This concept is detectable in the activity of sinoatrial nodal tissue, where the action of sympathetic stimulation produces a less relevant effect when the vagal activity rises. This kind of interaction is not appreciable in the neural control of atrioventricular conduction. When the parasympathetic stimulation of the heart is not accompanied by a considerable sympathetic activity, the only response is a feeble negative inotropic effect on the ventricular myocardium.

The accentuated antagonism was found to be influenced by time factor in the sequence of excitation of vagal and sympathetic nervous systems [85, 86]. Furukawa et al. found that when a simultaneous stimulation of sympathetic and parasympathetic systems is produced, the inhibitory effect of vagal stimulation prevails on the effect of the contemporary sympathetic stimulation. However the autonomic response tends to slide back towards the level before the stimulation and the variation of the activity as a function of the time is more relevant when the frequency of the excitation sequence is high.

Yang and Levy showed that a prolonged stimulation of sympathetic system leads to a reduction of the effects of vagal nervous system activity [87], in a sort of sympathetic inhibition of vagus nerve. This means that the nature and the magnitude of sympatho-vagal interaction can be influenced by previous consistent stimulations, which can provoke a mitigation of this phenomenon. From the biochemical point of view, this outcome is the consequence of the accumulation of neuropeptide Y

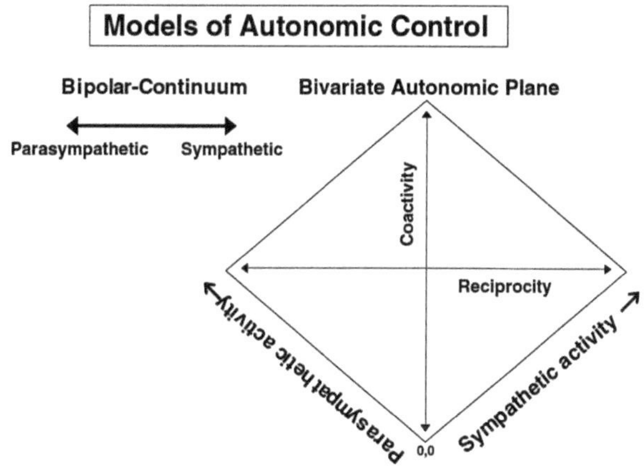

Fig. 1.1 Model of autonomic nervous system of Bernston et al. [95]

which inhibits the vagal nervous termination at the junctional level [88]. Many other studies in the literature highlighted the inhibition of chronotropic responses to vagal stimulation, after a previous intense sympathetic activity [89–92].

At the end of twentieth century, Bernston et al. [93] proposed a non-linear model to describe these interactions, based on a bidirectional modulation of vagal control and supported by the study of Porter et al. [94]. This model is characterized by coactive or independent changes in the autonomic divisions. Both the subsystems in this model are considered to be tonically active and to cooperate in normal and emergency conditions. Even if one or the other system can prevail in the control, it is the total balance of activity of the two to maintain the homeostasis of the organism. This bivariate representation of autonomic nervous system activity is shown in Fig. 1.1. The psychophysiological responses are depicted, in this autonomic space model, as vectors on an autonomic plane bounded by orthogonal sympathetic and parasympathetic axes. Bernston's model incorporates the classical reciprocal modes of autonomic control (sympathetic activation/parasympathetic withdrawal; parasympathetic activation/sympathetic withdrawal), and also adds the broader range of autonomic modes (coactivation, coinhibition, and uncoupled sympathetic or parasympathetic changes). These four modes correspond to the four quadrants of the bivariate sympathetic-parasympathetic autonomic plane, two of them are the branches where the systems activities are negatively correlated (sympathetic reciprocal and parasympathetic reciprocal modes), and two where the branches are positively correlated (coactivation and coinhibition). In contrast with the linear model of Cannon, the Bernston model assumes that rostral neural networks action, following particular psychological and physical states, can produce reciprocal, autonomous or also coactive changes in sympathetic and parasympathetic activity.
The nonlinear aspects of the dynamic interaction between sympathetic and parasympathetic nervous systems were studied by Sunagawa and Kawada [96, 97]. In their research, they estimated the transfer function from nerve stimulation to heart rate. This model highlighted the crucial role of the dynamical interaction of the autonomic nervous system subdivisions in the regulation of heart rate,

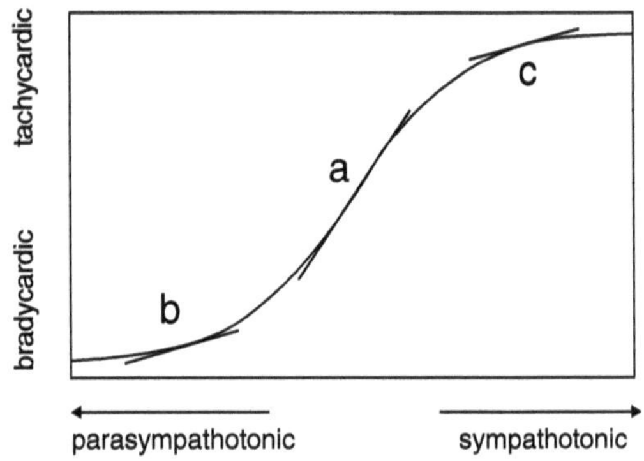

Fig. 1.2 Scheme of the mechanism explaining the nonlinear transfer function between autonomic stimulation and heart rate [96]

considering the continuous changes in sympathetic and vagal activities due to external stimuli. The authors considered a system where the input was the vagal or the sympathetic stimulation and the output was the effect on the heart rate. Given the intrinsic nonlinearity of biological systems, the Gaussian white noise approach was applied, in order to be able to calculate the linear transfer function also in presence of nonlinear dynamics [98]. The experiment was addressed to examine the tonic vagal stimulation effects on the dynamic heart rate response to sympathetic stimulation and vice versa in α-chloralose-anesthetized rabbits. Confirming the theory of Levy, Sunagawa et al. studied that stimulation of either the sympathetic and parasympathetic nervous systems increased dynamic response of heart rate to the concurrent stimulation of its counterpart. This phenomenon was called *bidirectional augmentation* and was explained by considering a sigmoidal static relationship between autonomic nervous system and heart rate. The scheme of this reciprocal excitation is shown in Fig. 1.2. From the graph it could be seen that when the sympathetic nervous system is stimulated alone, the mean heart rate is high and (c) can be considered the operating point. If a simultaneous vagal stimulation is produced, the operating point of the system becomes (a). The slope of the tangential lines in the operating points depends on the gain of the stimulation. For example, the slope in (a) is larger than in (c) because the concomitant vagal stimulation increased the gain of the dynamic heart rate. When the stimulation involves only the parasympathetic nervous system, the operating point of heart rate is in (b) and the dynamic gain is also low.

This model has several limitations, e.g. the absence of precise information about the operating points and the inability to express the real nonlinear properties of input-output relationship of the system [97]. Regarding the last observation, Borger et al. demonstrated that the gain of the transfer function decreased with the increase of the mean stimulation frequency of Gaussian white noise [99]. This property was not explained through the sigmoid curve of Fig. 1.2. In order to overcome these limitations, Kuwada et al. applied a neural network analysis to obtain further information about the static and dynamic regulation of heart rate. This approach was used to analyze data recorded during three experimental protocols in a rabbit model. In the first two protocols either the parasympathetic or the sympathetic systems

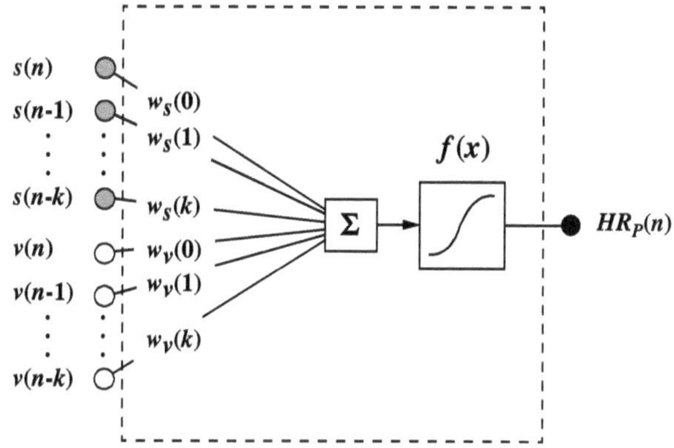

Fig. 1.3 Two-layered neural networks used in [97] for the identification of heart rate characteristics

were stimulated, while in the third protocol both system were simultaneously stimulated. They used a simple two layers neural network for the identification of the characteristics of heart rate (see Fig. 1.3). The first layer received as input the time series of command signals for sympathetic stimulation $(s(n), \ldots, s(n-k))$ and for vagal elicitation $(v(n), \ldots, v(n-k))$. The weights related to the response of the two systems are shown in the figure with the symbols w_s and w_v. The output of the neural network is given by the following equation:

$$HR_P(n) = f\left\{\sum_{i=0}^{k}[w_s(i) \times s(n-i) + w_v(i) \times (n-i)]\right\} \quad (1.2)$$

The function f, defined in the second layer, is:

$$f(x) = a \times x + b \quad (1.3)$$

The authors chose a linear function in order to plot the scattergrams of the measured responses versus the linearly predicted outputs. The scattergrams highlighted a sigmoidal dependence, as shown in Fig. 1.4, on the left.

Then the authors used a nonlinear sigmoidal function:

$$f(x) = \frac{P_1}{1 + \exp[P_2 \times (P_3 - x)]} + P_4 \quad (1.4)$$

where P_1 is the range of heart rate response, P_2 is the slope coefficient, P_3 is the midpoint of the operating range and P_4 is the minimum of heart rate. In Fig. 1.4, on the right, it is easy to see the result of the application of this nonlinear neural network, with the linear dependence between the linearly predicted values of heart rate and the measured heart rate. In the same study, the authors simulated static heart rate response to combined stimulation of vagal and sympathetic systems. The results are displayed in Fig. 1.5, which shows that in absence of sympathetic stimulation, the heart rate decreases along the parasympathetic axis (curve a–b) and when only the sympathetic system is stimulated, it increases along its axis (curve a–c). The

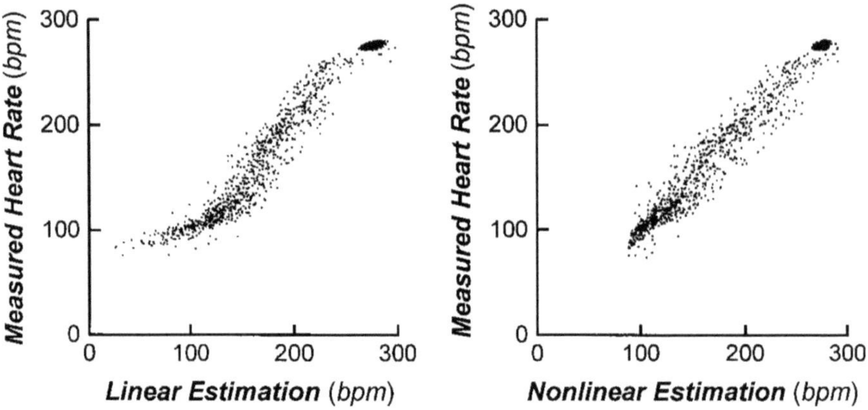

Fig. 1.4 Scattergrams of measured heart rate values as a function of the output of the linear neural network (*on the left*) and of the nonlinear neural network (*on the right*) [97]

Fig. 1.5 Graph of the heart rate response following combined sympathetic and vagal stimulation [97]

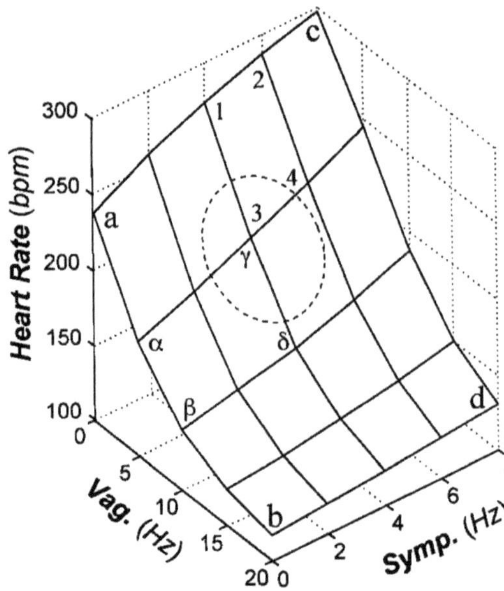

slopes of this curve change when the stimulation is combined, in the first case it rises when the sympathetic system is stimulated (curve c–d) and in the second is reduced by the activation of vagal nervous system (curve b–d).

There is a region, circumscribed by a dashed line in Fig. 1.5, where the combined action of the two system produces steeper slopes along the two directions (lines 3–4 and γ–δ). This suggests that, under a balanced condition of the stimulation, a bidirectional augmentation can be produced.

The hypothesis of nonlinearity in heart rate dynamics was supported also by other studies in the literature. Long range power-law correlations in beat-to-beat fluctuations were discussed in several works, and this condition is typical in physical systems near a critical point [100].

Tulppo et al. tested the hypothesis of fractal organization of autonomic dynamics in heart rate regulation [101]. The fractal properties of heart rate consist in the self-similarity fluctuations which are invariant over a large range of scales and are supposed to depend on the characteristics derived from deterministic chaos in the control of autonomic nervous system [102]. The loss of the fractal properties of heart rate dynamics has been identified in presence of pathological diseases, e.g. congestive heart failure [100], and during aging [103]. This interpretation apparently conflicts with the classical theory of homeostasis, opening to possible chaotic characteristics of autonomic nervous system functioning.

1.3 Heart Rate Variability

Heart rate varies over very short periods of time, even from beat to beat [104]. These variations do not follow strictly regular patterns, but repeating phenomena may be observed. Best known is the respiratory sinus arrhythmia: quickening of the heartbeat at inspiration, slowing at expiration [104]. This is entirely vagally mediated. But this is not the only rhythm that can be discerned. Slow variations, with repetition periods of approximately 10 s, not related to respiration, are present most of the time [105]. These may be very prominent in the upright posture [104, 106]. Even slower rhythms can be observed over the course of the day if sufficient time is allowed for recording, with the slowest one being the circadian pattern-speeding heart rate up during the waking day and slowing it down during sleep.

In 1981, Akselrod et al. introduced power spectral analysis of these heart rate fluctuations to quantitatively evaluate beat-to-beat cardiovascular control [14, 41]. Since then, heart rate variability analysis has received great attention from physiologists, clinicians and bioengineers, because it is considered as a useful tool to assess sympathetic and parasympathetic outflows [42, 107, 108]. Frequency-domain analyses contributed to the understanding of the autonomic background of RR interval fluctuations in the heart rate record [41]. However, as discussed in the following paragraph, the usefulness of HRV indices to evaluate nerve traffic is controversial [43, 44, 64, 104, 109].

The clinical importance of HRV became apparent in the late 1980s when it was confirmed that HRV was a strong and independent predictor of mortality following an acute myocardial infarction [41].

Nowadays, heart rate variability analysis is considered as a non invasive methodology of substantial utility to evaluate autonomic control mechanisms and to identify patients with an increased cardiac mortality [50, 51, 110], but it is accepted that it does not represent a quantitative measure of absolute autonomic outflow [64, 109]. However, when associated to other components of cardiovascular variability (blood pressure variability, respiration, baroreflex etc.), heart rate variability provides important information about the autonomic control of circulation, in normal and diseased conditions [64, 109].

Indices of HRV Many indices have been proposed to quantify HRV properties. Most of them were included in the Task Force of 1996 [41], and are reported in the Table 1.3. They can divided in two groups:

- Time-domain indices:
  These indices, reported in the upper part of Table 1.3, are the most simple. In spite of their simplicity, a number of reports have confirmed, also in prospective studies, their prognostic value as predictors of arrhythmic and total cardiac mortality not only after myocardial infarction but also in patients with heart failure [41, 110, 111]. In a continuous electrocardiographic (ECG) record, each QRS complex is detected, and the so-called normal-to-normal (NN) intervals (that is all intervals between adjacent QRS complexes resulting from sinus node

Table 1.3 Indices of HRV, from [41]

Variable (Units)	Description
Time-domain	
SDNN (ms)	SD of all NN
SDANN (ms)	SD of the averages of NN in all 5-min segments of the entire recording
RMSSD (ms)	The square root of the mean of the sum of the squares of differences between adjacent NN
SDNN index (ms)	Mean of the SD of all NN for all 5-min segments of the entire recording
SDSD (ms)	SD of differences between adjacent NN
NN50 count	Number of pairs of adjacent NN differing by more than 50 ms in the entire recording
pNN50 (%)	NN50 count divided by the total number of all NN
HRV triangular index	Total number of all NN divided by the height of the histogram of all NN measured on a discrete scale with bins of 7.8125 ms (1/128 s)
TINN (ms)	Baseline width of the minimum square difference triangular interpolation of the highest peak of the histogram of all NN
Differential index (ms)	Difference between the widths of the histogram of differences between adjacent NN measured at selected heights
Logarithmic index	Coefficient ϕ of the negative exponential curve $ke^{-\phi t}$, which is the best approximation of the histogram of absolute differences between adjacent NN
Frequency-domain	
Power (ms^2)	The variance of NN over 5 min
VLF (ms^2)	Power in VLF range ($f \leq 0.04$ Hz)
LF (ms^2)	Power in LF range ($f \in [0.04–0.15]$ Hz)
LF norm (nu)	LF power in norm units LF/(total power-VLF) × 100
HF (ms^2)	Power in HF range ($f \in [0.15–0.4]$ Hz)
HF norm (nu)	HF power in norm units HF/(total power-VLF) × 100
LF/HF	Ratio LF(ms^2)/HF(ms^2)

NN NN intervals corresponding to heart period during sinus rhythm, *SD* Standard deviation depolarizations), are determined and mathematically processed [41, 112]. Time-domain indices can be divided in two subgroups: statistical indices, mainly obtained form mean and standard deviations of heart period, and geometric indices, mainly obtained from the probability density function of the NN.

- Frequency domain indices:
  Interest for frequency domain parameters derives from the possibility of identifying periodic oscillations in HRV and to correlate these oscillation patterns to neural discharge. In fact, vagal and sympathetic cardiac controls operate on heart rate in different frequency bands [64]. Low frequency (LF; 0.04–0.15 Hz) fluctuations in HR are affected by electrical stimulation of both vagal and sympathetic cardiac nerves in animals. Similarly, in humans, LF powers are reduced by either parasympathetic or sympathetic blockade; parasympathetic blockade by atropine eliminates most HR fluctuations at high frequency (HF; 0.15–0.4 Hz); thus, HRV at HF is a satisfactory, although partly incomplete, measure of vagal cardiac control, whereas LF components reflect both sympathetic and parasympathetic modulation, without excluding a role of humoral factors, gender and age. The LF/HF ratio was proposed as measure of the sympatho-vagal balance [42, 108] and was also effective to study autonomic control in those clinical conditions associated with an increase sympathetic and reduced vagal modulation of sinus node such as, for example, the acute phase of myocardial infarction or the initial phases of heart failure [110]. Since the beginning, however, it was evident that the physiological interpretation of LF and HF components and of LF/HF ratio was progressively more problematic when moving from short-term to 24 hour recordings, and recently it has been suggested that spectral analysis of HRV must be restricted to short term recordings under controlled conditions in order to measure more correctly and to interpret more safely LF and HF components [110].

  The origin and the interpretation of the LF rhythm of heart rate variability, is by far the most controversial. It is known to be related to Mayer's wave, oscillations of arterial pressure lower than respiration with a period of approximately 10 s [105]. Several data suggest that heart rate oscillations buffer Mayer's wave [105]. The function as well as physiological determinants of Mayer's wave frequency and amplitude are still largely unknown [105].
- Advanced indices: Several methodologies have been proposed to improve the assessment of time-domain and spectral indices or to overcome their limitations. These include time-varying autoregressive models [113, 114], time-frequency methods [115–117], hybrid parametric and non-parametric methods [108, 118], point process models [119] and non-linear methodologies [120].

Clinical Relevance The clinical relevance of HRV analysis is related to the well-established link between autonomic cardiac control and cardiovascular mortality, including sudden cardiac death, with HRV being a key marker of such a relationship. In fact, reduced HRV is associated with increased mortality after myocardial infarction and increased risk of sudden arrhythmic death [64]. One of the basic assumptions used to explain the negative predictive value of reduced HRV was the concept that overall HRV was largely dependent on vagal mechanisms and that a reduction in HRV could reflect an increased sympathetic and a reduced vagal modulation of sinus node; i.e., an autonomic imbalance favouring cardiac electrical instability [110]. Changes in HRV have been also shown to identify favorable changes in cardiac autonomic control after cardiac resynchronization therapy in patients with severely symptomatic heart failure [64]. These observations strongly suggest that HRV, in addition to representing a research tool, should become a more widely employed clinical parameter [64].

Criticisms The reliability of HRV as autonomic marker has been seriously criticized. Criticisms do not deny the clinical relevance of HRV, but they question its physiological interpretation. In 1997, Dr Ekberg wrote a critical review [121] that support the opinion that calculations of sympathovagal balance, proposed ten years before [42], may obscure rather than illuminate human physiology and pathophysiology. In the paper, Dr Ekberg claimed that the ratio of the LF power to HF power is not suitable to quantitatively evaluate autonomic nervous activity owing to the following reasons [121]:

- Vagal contributions to baseline LF RR-interval fluctuations are great, and evidence that baseline LF RR-interval spectral power is related quantitatively to sympathetic-cardiac nerve traffic is nonexistent.
- Most evidence refutes the notion that LF RR-interval spectral power tracks baroreflex-mediated changes of sympathetic nerve activity.
- Baseline respiratory-frequency RR-interval fluctuations are related significantly but imperfectly to the level of human vagal-cardiac nerve traffic.
- Moderate changes of arterial pressure, which alter vagal-cardiac nerve activity, do not change HF RR-interval fluctuations, and changes of breathing frequency and depth, which profoundly alter HF RR-interval fluctuations, may not change vagal-cardiac nerve activity at all.

Finally he questioned the hypothesis that sympathetic and parasympathetic nervous systems continuously interact [44], suggesting that the construct of sympatho-vagal balance imposes attributes on physiological regulatory mechanisms that they do not possess [121].

This critical appraisal caused many of the members of the task force [41] to answer to these criticisms to support the usefulness of HRV indices in cardiovascular control analysis [43].

Ten years later, a series of editorial papers about whether cardiovascular variability is or is not an index of autonomic control of circulation [64, 109] refocused the debate about the role of HRV indices in physiology. All authors agreed that neither low frequency (< 0.15 Hz) nor high frequency (> 0.15 Hz) indices can be used as exclusive markers of sympathetic and parasympathetic activity, respectively [109]. And some of them suggested that heart rate variability provides only a qualitative marker of cardiac parasympathetic regulation through respiratory sinus arrhythmia (RSA). In the debate, on one side, Taylor and Studinger [64, 109] argued that quantification of autonomic activity has not been achieved, and suggested that efforts should be focused on establishing more direct links to underlying physiology. For these authors, the point is that there is more to be learned from cardiovascular periodicities than what they may or not say regarding baseline levels of autonomic nerve traffic [64, 109]. On the other side, Parati et al. [64, 109] claimed that HRV analysis has already established a basis for autonomic interpretation of results and that future modeling will expose underlying physiology.

These debates and discussions, which should include the more recent debate about the origin of respiratory sinus arrhythmia [23, 25] and Mayer's Wave [105], show that research in cardiovascular variability is far from being over and much effort is needed in multi-signal analysis.

1.3.1 Representations of Heart Rate Variability

In the literature, several methodologies to mathematically describe heart rate variability have been proposed.

- Interpolation: Heart rate variability can be estimated just by evenly resampling the inverse interval function, i.e. the reciprocal of the heart period. In the interpolation process, the n-th heart period should be arbitrarily assign to a given temporal instant, which usually is the time of occurrence of one of the two QRS complexes which determine the beginning and the end of the heart period. This procedure is not based on a physiological model. The representation of a discrete unevenly sampled time series as a continuous signal is often a requirement of many signal processing techniques, among which there is time-frequency analysis.
- Autoregressive model: Autoregressive models as well as ARMA and ARMAX models are widely used to describe heart rate variability and the interactions of cardiovascular signals [114, 122, 123]. The use of these models is rarely motivated by cardiovascular physiology, and is principally due to the possibility of using very powerful methodologies of time series spectral and cross spectral analysis. Although the use of these models is rarely based on cardiovascular physiology, it offers the possibility of using very powerful methodologies of time series spectral and cross spectral analysis.
- IPFM: The integral pulse frequency modulation (IPFM) model has been used to explain the mechanisms of control of the autonomic system over the heart rate [124–126]. The IPFM model is based on the hypothesis that the sympathetic and parasympathetic influences on the sino-atrial node can be represented by a single modulating signal, and the beat trigger impulse is generated when the integral of this function reaches a threshold. The heart timing signal, which assume the IPFM model, has been demonstrated to provide an unbiased estimation of the ANS modulation, even in the presence of isolated ectopic beats [124, 125]. Recently, a time-varying threshold IPFM model was proposed to be used in situations in which the mean heart period is time varying, as during exercise stress testing [127].

- Point process: Recently, a point process model has been also used to study heart rate and heart rate variability [119, 128]. A point process is a random process for which any realization consists of a set of isolated points in time. In this framework, the waiting time until the next R-wave event or, equivalently, the length of the next R-R interval, obeys an history-dependent inverse Gaussian process probability density. The time-varying parameters of this model are estimated by local maximum likelihood and assessed using a model goodness-of-fit by Kolmogorov-Smirnov tests based on the time-rescaling theorem [128, 129]. The mean and the standard deviation of the probability density function are used to give instantaneous estimates of heart rate and heart rate variability. Recently, this point process framework has been combined with bi-variate autoregressive models to include the influence of respiration and to assess cardiovascular interactions [21, 70].

1.4 Gender/Age Differences in the Autonomic Nervous System Regulation

Gender differences in the autonomic nervous system control have been described in previous studies. Studying rats, some discrepancies are detectable in pre-natal phases, as the distribution of neuropeptides in the brain. Fewer ganglionic neurones are present in females and higher acetylcholinesterase activity is observable in males [130–132].

The male or female sex hormones, especially estrogen, plays a crucial role in the gender difference for the autonomic activity [133]. Estrogen modulates the autonomic nervous system activity, suppressing sympathetic activity and elevating the parasympathetic tone. Some experimental studies in rats suggested that the inhibition of sympathetic system is more appreciable in females, for whom the release of noradrenaline is lower [134]. In addition, estrogen is associated with an increase of acetylcholine release, whereas the testosterone increases the noradrenaline synthesis [135].

All these aspects support the hypothesis that the parasympathetic activity is greater in females and the sympathetic tone is more relevant in males. Other recent studies suggested that estrogen level also influences the nervous density and nervous growth factor. The anti-apoptopic effect of this hormone can explain how the aging influence on the autonomic activity is more evident in males than in females [135].

Although female sex hormones have been implicated in protecting females against several cardiovascular diseases, the mechanism by which this occurs is still unknown [136]. During resting conditions, healthy men and women have similar blood pressure but heart rate tends to be higher in women. Some studied suggested that gender differences in baroreceptor reflex regulation of sympathetic nerve activity may provide a mechanism by which females are protected against the development of hypertension [136]. A previous study about response to postural changes indicated that, when changing from sitting to standing positions, the increases in heart rate and total peripheral resistance were greater in women than men [137]. A possible mechanism of protection against hypertension in females, if the form of hypertension is initiated by retention of sodium and water, can be detected in the reflex-mediated inhibition of sympathetic nerve activity which increases blood volume and activates cardiopulmonary receptors. Under these conditions, females may inhibit sympathetic nerve activity more efficiently, leading to a greater excretion of sodium and water and delaying the onset of hypertension.

Gender differences had been also found in response to adrenergic nerve stimulation and it was supposed the presence of differences in the clearance of released noradrenaline. An increase in adrenaline clearance in women was studied by Lenders et al. [138]. Also the sympathetic activation of the adrenal medulla resulted to be attenuated in females and this finding could explain the supposed gender differences in cardiovascular responses to stress [139].

The pathophysiological states of cardiovascular system present many differences due to gender in the incidence and clinical course. For example, gender-based differences exist in the prognosis after myocardial infarction [140, 141] and non-ischaemic cardiomyopathy [142, 143].

Experimental and clinical studies suggested the presence of sex-specific differences in cardiac hypertrophy and in patients who are more than 60 years old, with better left ventricular remodelling in women than in men. The improvement in the systolic function and left ventricular adaptation were more pronounced before menopause and tended to disappear thereafter [143]. However, it remains to be proven if the difference in behavior of the cardiac autonomic nervous system in men and women contributes to the large gender differential in morbidity and mortality of heart disease [144, 145]. In Table 1.4 the sex-related differences in the pathophysiological incidence in the two genders, given the results in the literature, are shown.

Table 1.4 Gender-related differences in autonomic nervous system activity during physiological and disease states (modified from [135])

Physiological or disease state	Gender difference
Disease state	
Prevalence of Raynaud's syndrome	$F > M$
Prevalence of presyncope/syncope	$F > M$
Survival of non-CAD heart failure	$F > M$
Survival after myocardial infarction	$M > F$
Risk of ventricular arrhythmias	$M > F$
Central obesity	$M > F$
Physiological state	
Systolic blood pressure rise on aerobic exercise	$M > F$
Haemodynamic response to isometric exercise	$M > F$
Tolerance to cold temperature	$M > F$
Tolerance to repeated hypoglycaemia	$M > F$

CAD coronary artery disease

The quantification of heartbeat complex dynamics has been proven to provide relevant information on psychophysiological and pathological states [146] being modulated by aging [146, 147]. Previous research highlighted a strong relationship between the aging process and cardiovascular dynamics, as well as respiration, physical activity and electrodermal response.

The analysis of breathing patterns in elderly patients, using both standard and nonlinear analyses, showed several differences in the autonomic regulation due to aging [148, 149]. Studying physical activity in elderly, the reduction in circadian activity rhythm amplitude has been linked to the development of dementia in older women [150]. Concerning cardiovascular variability, many HRV correlates have been related to age using both standard and nonlinear methods. Specifically, a significant decrease of the standard parameters, defined in the time and frequency domains, was found in elderly people, with major influence on men [151, 152]. Ryan et al. showed a significant loss of HRV high frequency power (0.15–0.40 Hz) and of the ratio between the low frequency and the high frequency spectral power, in elderly men and women [153]. Nonlinear scaling properties in HRV series from elderlies were studied in 24 h recordings using Detrended Fluctuation Analysis (short- and long-term correlations indexes), Lyapunov exponent, and Approximate entropy [154]. A day-night modulation was found in all nonlinear HRV indices, which significantly correlated with age.

Recently, short-term HRV series have been studied as related to age. Particularly, Voss et al. studied HRV standard and nonlinear features extracted within 5 min series gathered from 1906 subjects, as divided into two groups: age 25–49 years, and 50–74 years [155]. Results revealed that compression entropy, Detrended Fluctuation Analysis, Poincaré plot analysis, Segmented Poincaré plot analysis, Irreversibility analysis, and symbolic analysis were statistically different between the two groups. This results confirmed previous evidences [156], highlighted that cardiovascular complexity changes according to age.

Extensive further details on gender/age differences in ANS dynamics, with special focus on cardiovascular complexity, can be found in the chapter authored by Schroeder et al.

References

1. Heesch, C.M.: Reflexes that control cardiovascular function. Adv. Physiol. Educ. **277**(6), S234–S243 (1999). [Online]. Available: http://advan.physiology.org/content/277/6/S234.short
2. Guyton, A., Hall, J.: The autonomic nervous system and the adrenal medulla. In: Guyton, A.C., Hall, J.E. (eds.) Textbook of Medical Physiology, pp. 748–760. Elsevier and Saunders, Philadelphia (2006)
3. Thomas, G.D.: Neural control of the circulation. Adv. Physiol. Educ. **35**(1), 28–32 (2011). [Online]. Available: http://advan.physiology.org/content/35/1/28.abstract
4. Olshansky, B., Sabbah, H.N., Hauptman, P.J., Colucci, W.S.: Parasympathetic nervous system and heart failure. Circulation **118**(8), 863–871 (2008). [Online]. Available: http://circ.ahajournals.org/content/118/8/863.short

5. Yamakawa, K., So, E.L., Rajendran, P.S., Hoang, J.D., Makkar, N., Mahajan, A., Shivkumar, K., Vaseghi, M.: Electrophysiological effects of right and left vagal nerve stimulation on the ventricular myocardium. Am. J. Physiol. Heart Circ. Physiol. **307**(5), H722–H731, 2014. [Online]. Available: http://www.pubmedcentral.nih.gov/articlerender.fcgi?artid=4187397&tool=pmcentrez&rendertype=abstract
6. Shinlapawittayatorn, K., Chinda, K., Palee, S., Surinkaew, S., Thunsiri, K., Weerateerangkul, P., Chattipakorn, S., KenKnight, B.H., Chattipakorn, N.: Low-amplitude, left vagus nerve stimulation significantly attenuates ventricular dysfunction and infarct size through prevention of mitochondrial dysfunction during acute ischemia-reperfusion injury. Heart Rhythm **10**(11), 1700–1707 (2013). [Online]. Available: http://www.sciencedirect.com/science/article/pii/S1547527113008473
7. De Ferrari, G.M., Crijns, H.J., Borggrefe, M., Milasinovic, G., Smid, J., Zabel, M., Gavazzi, A., Sanzo, A., Dennert, R., Kuschyk, J., Raspopovic, S., Klein, H., Swedberg, K., Schwartz, P.J.: Chronic vagus nerve stimulation: a new and promising therapeutic approach for chronic heart failure. Eur. Heart J. **32**(7), 847–855 (2011). [Online]. Available: http://eurheartj.oxfordjournals.org/content/32/7/847.abstract
8. Lopshire, J., Zipes, D.: Device therapy to modulate the autonomic nervous system to treat heart failure. Curr. Cardiol. Rep. **14**(5), 593–600 (2012). [Online]. Available: http://www.scopus.com/inward/record.url?eid=2-s2.0-84870586959&partnerID=40&md5=2e4e9453f0b6a546744573944f78f6ec
9. Zannad, F., De Ferrari, G.M., Tuinenburg, A.E., Wright, D., Brugada, J., Butter, C., Klein, H., Stolen, C., Meyer, S., Stein, K.M., Ramuzat, A., Schubert, B., Daum, D., Neuzil, P., Botman, C., Castel, M.A., D'Onofrio, A., Solomon, S.D., Wold, N., Ruble, S.B.: Chronic vagal stimulation for the treatment of low ejection fraction heart failure: results of the NEural Cardiac TherApy foR Heart Failure (NECTAR-HF) randomized controlled trial. Eur. Heart J. **36**(7), 425–433 (2015). [Online]. Available: http://eurheartj.oxfordjournals.org/content/36/7/425.abstract
10. Premchand, R.K., Sharma, K., Mittal, S., Monteiro, R., Dixit, S., Libbus, I., DiCarlo, L.A., Ardell, J.L., Rector, T.S., Amurthur, B., KenKnight, B.H., Anand, I.S.: Autonomic regulation therapy via left or right cervical vagus nerve stimulation in patients with chronic heart failure: results of the ANTHEM-HF Trial. J. Card. Fail. **20**(11), 808–816 (2014). [Online]. Available: http://www.sciencedirect.com/science/article/pii/S1071916414011890
11. Abraham, W.T., Zile, M.R., Weaver, F.A., Butter, C., Ducharme, A., Halbach, M., Klug, D., Lovett, E.G., Müller-Ehmsen, J., Schafer, J.E., Senni, M., Swarup, V., Wachter, R., Little, W.C.: Baroreflex activation therapy for the treatment of heart failure with a reduced ejection fraction. JACC Heart Fail. **3**(6), 487–496 (2015). [Online]. Available: http://dx.doi.org/10.1016/j.jchf.2015.02.006
12. Gold, M.R., Van Veldhuisen, D.J., Hauptman, P.J., Borggrefe, M., Kubo, S.H., Lieberman, R.A., Milasinovic, G., Berman, B.J., Djordjevic, S., Neelagaru, S., Schwartz, P.J., Starling, R.C., Mann, D.L.: Vagus nerve stimulation for the treatment of heart failure. J. Am. Coll. Cardiol. **68**(2), 149–158 (2016). [Online]. Available: http://linkinghub.elsevier.com/retrieve/pii/S0735109716324044
13. Hayano, J., Yasuma, F.: Hypothesis: respiratory sinus arrhythmia is an intrinsic resting function of cardiopulmonary system. Cardiovasc. Res. **58**(1), 1–9 (2003). [Online]. Available: http://cardiovascres.oxfordjournals.org/content/58/1/1.abstract
14. Akselrod, S., Gordon, D., Ubel, F.A., Shannon, D.C., Barger, A.C., Cohen, R.J.: Power spectrum analysis of heart rate fluctuations: a quantitative probe of beat-to-beat cardiovascular control. Science **213**, 220–222 (1981)
15. Saul, J.P., Berger, R.D., Albrecht, P., Stein, S.P., Chen, M.H., Cohen, R.J.: Transfer function analysis of the circulation: unique insights into cardiovascular regulation. Am. J. Physiol. Heart Circ. Physiol. **261**(4), H1231–H1245 (1991). [Online]. Available: http://ajpheart.physiology.org/content/261/4/H1231.abstract
16. Pagani, M., Lombardi, F., Guzzetti, S., Rimoldi, O., Furlan, R., Pizzinelli, P., Sandrone, G., Malfatto, G., Dell'Orto, S., Piccaluga, F., Turiel, M., Baselli, G., Cerutti, S., Malliani,

A.: Power spectral analysis of heart rate and arterial pressure variabilities as a marker of sympatho-vagal interaction in man and conscious dogs. Circ. Res. **59**, 178–193 (1986)
17. Tzeng, Y.C., Sin, P.Y.W., Galletly, D.C.: Human sinus arrhythmia: inconsistencies of a teleological hypothesis. Am. J. Physiol. Heart Circ. Physiol. **296**(1), H65–H70 (2009). [Online]. Available: http://ajpheart.physiology.org/content/296/1/H65.abstract
18. Grossman, P., Wilhelm, F.H., Spoerle, M.: Respiratory sinus arrhythmia, cardiac vagal control, and daily activity. Am. J. Physiol. Heart Circ. Physiol. **287**(2), H728–H734 (2004). [Online]. Available: http://ajpheart.physiology.org/content/287/2/H728.abstract
19. Grossman, P., Taylor, E.W.: Toward understanding respiratory sinus arrhythmia: relations to cardiac vagal tone, evolution and biobehavioral functions. Biol. Psychol. **74**(2), 263–285 (2007). [Online]. Available: http://www.sciencedirect.com/science/article/B6T4T-4M7VFNK-1/2/8584deabf148494e1491 1e8fed4cf52c
20. Eckberg, D.L.: The human respiratory gate. J. Physiol. **548**(2), 339–352 (2003). [Online]. Available: http://jp.physoc.org/content/548/2/339.abstract
21. Chen, Z., Brown, E.N., Barbieri, R.: Assessment of autonomic control and respiratory sinus arrhythmia using point process models of human heart beat dynamics. IEEE Trans. Biomed. Eng. **56**(7), 1791–1802 (2009). [Online]. Available: http://dx.doi.org/10.1109/TBME.2009. 2016349
22. Goldberger, J.J., Challapalli, S., Tung, R., Parker, M.A., Kadish, A.H.: Relationship of heart rate variability to parasympathetic effect. Circulation **103**(15), 1977–1983 (2001). [Online]. Available: http://circ.ahajournals.org/content/103/15/1977.abstract
23. Eckberg, D.L.: Point:counterpoint: respiratory sinus arrhythmia is due to a central mechanism vs. respiratory sinus arrhythmia is due to the baroreflex mechanism. J. Appl. Physiol. **106**(5), 1740–1742 (2009); discussion 1744. [Online]. Available: http://dx.doi.org/10.1152/ japplphysiol.91107.2008
24. deBoer, R.W., Karemaker, J.M., Strackee, J.: Hemodynamic fluctuations and baroreflex sensitivity in humans: a beat-to-beat model. Am. J. Physiol. **253**(3 Pt 2), H680–H689 (1987)
25. Karemaker, J.M.: Counterpoint: respiratory sinus arrhythmia is due to the baroreflex mechanism. J. Appl. Physiol. **106**(5), 1742–1743 (2009); discussion 1744. [Online]. Available: http://dx.doi.org/10.1152/japplphysiol.91107.2008a
26. Karemaker, J.M.: Last word on point:counterpoint: respiratory sinus arrhythmia is due to a central mechanism vs. respiratory sinus arrhythmia is due to the baroreflex mechanism. J. Appl. Physiol. **106**(5), 1750 (2009). [Online]. Available: http://dx.doi.org/10.1152/ japplphysiol.00225.2009
27. Blain, G., Meste, O., Bermon, S.: Influences of breathing patterns on respiratory sinus arrhythmia in humans during exercise. Am. J. Physiol. Heart Circ. Physiol. **288**(2), H887–H895 (2005)
28. Meste, O., Khaddoumi, B., Blain, G., Bermon, S.: Time-varying analysis methods and models for the respiratory and cardiac system coupling in graded exercise. IEEE Trans. Biomed. Eng. **52**(11), 1921–1930 (2005)
29. Julien, C., Parkes, M.J., Tzeng, S.Y.C., Sin, P.Y.W., Ainslie, P.N., van de Borne, P., Fortrat, J.-O., Custaud, M.-A., Gharib, C., Porta, A., Vallais, F., Baselli, G., Pagani, M., Lucini, D., Hughson, R.L., Taylor, J.A., Tan, C.O., Baekey, D.M., Dick, T.E., Paton, J.F.R., Taha, B.: Comments on point:counterpoint: respiratory sinus arrhythmia is due to a central mechanism vs. respiratory sinus arrhythmia is due to the baroreflex mechanism. J. Appl. Physiol. **106**(5), 1745–1749 (2009). [Online]. Available: http://dx.doi.org/10.1152/japplphysiol.00196.2009
30. Badra, L.J., Cooke, W.H., Hoag, J.B., Crossman, A.A., Kuusela, T.A., Tahvanainen, K.U.O., Eckberg, D.L.: Respiratory modulation of human autonomic rhythms. Am. J. Physiol. Heart Circ. Physiol. **280**(6), H2674–H2688 (2001). [Online]. Available: http://ajpheart.physiology. org/content/280/6/H2674.abstract
31. Porta, A., Baselli, G., Rimoldi, O., Malliani, A., Pagani, M.: Assessing baroreflex gain from spontaneous variability in conscious dogs: role of causality and respiration. Am. J. Physiol. Heart Circ. Physiol. **279**(5), H2558–H2567 (2000)

32. Vallais, F., Baselli, G., Lucini, D., Pagani, M., Porta, A.: Spontaneous baroreflex sensitivity estimates during graded bicycle exercise: a comparative study. Physiol. Meas. **30**(2), 201–213 (2009). [Online]. Available: http://dx.doi.org/10.1088/0967-3334/30/2/007
33. Porta, A., Catai, A.M., Takahashi, A.C.M., Magagnin, V., Bassani, T., Tobaldini, E., van de Borne, P., Montano, N.: Causal relationships between heart period and systolic arterial pressure during graded head-up tilt. Am. J. Physiol. Regul. Integr. Comp. Physiol. **300**(2), R378–R386 (2011). [Online]. Available: http://ajpregu.physiology.org/content/300/2/R378.abstract
34. de Boer, R., Karemaker, J., Strackee, J.: Relationships between short-term blood-pressure fluctuations and heart-rate variability in resting subjects II: a simple model. Med. Biol. Eng. Comput. **23**, 359–364 (1985). [Online]. Available: http://dx.doi.org/10.1007/BF02441590
35. Sin, P.Y.W., Galletly, D.C., Tzeng, Y.C.: Influence of breathing frequency on the pattern of respiratory sinus arrhythmia and blood pressure: old questions revisited. Am. J. Physiol. Heart Circ. Physiol. **298**(5), H1588–H1599 (2010). [Online]. Available: http://ajpheart.physiology.org/content/298/5/H1588.abstract
36. Hayano, J., Yasuma, F., Okada, A., Mukai, S., Fujinami, T.: Respiratory sinus arrhythmia: a phenomenon improving pulmonary gas exchange and circulatory efficiency. Circulation **94**(4), 842–847 (1996). [Online]. Available: http://circ.ahajournals.org/content/94/4/842.abstract
37. Yasuma, F., Hayano, J.-I.: Respiratory sinus arrhythmia* : Why does the heartbeat synchronize with respiratory rhythm? Chest **125**(2), 683–690 (2004). [Online]. Available: http://chestjournal.chestpubs.org/content/125/2/683.abstract
38. Yagishita, D., Chui, R.W., Yamakawa, K., Rajendran, P.S., Ajijola, O.A., Nakamura, K., So, E.L., Mahajan, A., Shivkumar, K., Vaseghi, M.: Sympathetic nerve stimulation, not circulating norepinephrine, modulates t-peak to t-end interval by increasing global dispersion of repolarization. Circ. Arrhythm. Electrophysiol. **8**(1), 174–185 (2015). [Online]. Available: http://circep.ahajournals.org/content/8/1/174.abstract
39. Malpas, S.C.: Sympathetic nervous system overactivity and its role in the development of cardiovascular disease. Physiol. Rev. **90**(2), 513–557 (2010). [Online]. Available: http://physrev.physiology.org/content/90/2/513.abstract
40. Charkoudian, N., Rabbitts, J.A.: Sympathetic neural mechanisms in human cardiovascular health and disease. Mayo Clin. Proc. **84**(9), 822–830 (2009). [Online]. Available: http://www.mayoclinicproceedings.com/content/84/9/822.abstract
41. Task Force of the European Society of Cardiology the North American Society of Pacing: Heart rate variability : standards of measurement, physiological interpretation, and clinical use. Circulation **93**(5), 1043–1065 (1996). [Online]. Available: http://circ.ahajournals.org/content/93/5/1043.short
42. Pagani, M., Lombardi, F., Guzzetti, S., Rimoldi, O., Furlan, R., Pizzinelli, P., Sandrone, G., Malfatto, G., Dell'Orto, S., Piccaluga, E.: Power spectral analysis of heart rate and arterial pressure variabilities as a marker of sympatho-vagal interaction in man and conscious dog. Circ. Res. **59**(2), 178–193 (1986). [Online]. Available: http://circres.ahajournals.org/content/59/2/178.abstract
43. Malliani, A., Pagani, M., Montano, N., Mela, G.S.: Sympathovagal balance: a reappraisal. Circulation **98**(23), 2640–2643 (1998). [Online]. Available: http://circ.ahajournals.org/content/98/23/2640.2.short
44. Eckberg, D.L.: Sympathovagal balance: a critical appraisal. Circulation **96**(9), 3224–3232 (1997). [Online]. Available: http://circ.ahajournals.org/content/96/9/3224.short
45. Pueyo, E., Orini, M., Rodríguez, J.F., Taggart, P.: Interactive effect of beta-adrenergic stimulation and mechanical stretch on low-frequency oscillations of ventricular action potential duration in humans. J. Mol. Cell. Cardiol. **97**, 93–105 (2016). [Online]. Available: http://linkinghub.elsevier.com/retrieve/pii/S0022282816301146
46. Hanson, B., Child, N., Van Duijvenboden, S., Orini, M., Chen, Z., Coronel, R., Rinaldi, C.A., Gill, J.S., Gill, J.S., Taggart, P.: Oscillatory behavior of ventricular action potential duration in heart failure patients at respiratory rate and low frequency. Front. Physiol. **5**,

414 (2014). [Online]. Available: http://www.pubmedcentral.nih.gov/articlerender.fcgi?artid=4211392&tool=pmcentrez&rendertype=abstract
47. Rizas, K.D., Nieminen, T., Barthel, P., Zürn, C.S., Kähönen, M., Viik, J., Lehtimäki, T., Nikus, K., Eick, C., Greiner, T.O., Wendel, H.P., Seizer, P., Schreieck, J., Gawaz, M., Schmidt, G., Bauer, A.: Sympathetic activity-associated periodic repolarization dynamics predict mortality following myocardial infarction. J. Clin. Invest. **124**(4), 1770–1780 (2014). [Online]. Available: http://www.scopus.com/inward/record.url?eid=2-s2.0-84902156467&partnerID=40&md5=f08dade5cb8e62aa61ead95613c55139
48. Baumert, M., Porta, A., Vos, M.A., Malik, M., Couderc, J.-P., Laguna, P., Piccirillo, G., Smith, G.L., Tereshchenko, L.G., Volders, P.G.A., QT interval variability in body surface ECG: measurement, physiological basis, and clinical value : position statement and consensus guidance endorsed by the European Heart Rhythm Association jointly with the ESC Working Group on Cardiac Cellular Electrophysiology. Europace **18**(6), 925–944 (2016). [Online]. Available: http://europace.oxfordjournals.org/content/early/2016/01/27/europace.euv405.abstract
49. Baumert, M., Schlaich, M.P., Nalivaiko, E., Lambert, E., Sari, C.I., Kaye, D.M., Elser, M.D., Sanders, P., Lambert, G.: Relation between QT interval variability and cardiac sympathetic activity in hypertension. Am. J. Physiol. Heart Circ. Physiol. **300**(4), H1412–H1417 (2011). [Online]. Available: http://ajpheart.physiology.org/content/300/4/H1412.abstract
50. Shen, M.J., Zipes, D.P.: Role of the autonomic nervous system in modulating cardiac arrhythmias. Circ. Res. **114**(6), 1004–1021 (2014). [Online]. Available: http://circres.ahajournals.org/content/114/6/1004.abstract
51. Fukuda, K., Kanazawa, H., Aizawa, Y., Ardell, J.L., Shivkumar, K.: Cardiac innervation and sudden cardiac death. Circ. Res. **116**(12), 2005–2019 (2015). [Online]. Available: http://circres.ahajournals.org/cgi/doi/10.1161/CIRCRESAHA.116.304679
52. La Rovere, M.T., Pinna, G.D., Raczak, G.: Baroreflex sensitivity: measurement and clinical implications. Ann. Noninvasive Electrocardiol. **13**(2), 191–207 (2008). [Online]. Available: http://dx.doi.org/10.1111/j.1542-474X.2008.00219.x
53. Steptoe, A., Kivimaki, M.: Stress and cardiovascular disease. Nat. Rev. Cardiol. **9**(6), 360–370 (2012). [Online]. Available: http://dx.doi.org/10.1038/nrcardio.2012.45
54. Hopf, H.-B., Skyschally, A., Heusch, G., Peters, J.: Low-frequency spectral power of heart rate variability is not a specific marker of cardiac sympathetic modulation. J. Am. Soc. Anesthesiol. **82**(3), 609–619 (1995)
55. Taylor, J.A., Carr, D.L., Myers, C.W., Eckberg, D.L.: Mechanisms underlying very-low-frequency rr-interval oscillations in humans. Circulation **98**(6), 547–555 (1998)
56. Goldstein, D.S., Bentho, O., Park, M.-Y., Sharabi, Y.: Low-frequency power of heart rate variability is not a measure of cardiac sympathetic tone but may be a measure of modulation of cardiac autonomic outflows by baroreflexes. Exp. Physiol. **96**(12), 1255–1261 (2011)
57. Reyes del Paso, G.A., Langewitz, W., Mulder, L.J., Roon, A., Duschek, S.: The utility of low frequency heart rate variability as an index of sympathetic cardiac tone: a review with emphasis on a reanalysis of previous studies. Psychophysiology **50**(5), 477–487 (2013)
58. Karemaker, J., Wesseling, K.: Variability in cardiovascular control: the baroreflex reconsidered. Cardiovasc. Eng. **8**, 23–29 (2008). [Online]. Available: http://dx.doi.org/10.1007/s10558-007-9046-4
59. Karemaker, J.M.: Why do we measure baroreflex sensitivity the way we do? Clin. Auton. Res. **12**(6), 427–428 (2002)
60. Taylor, J.A., Eckberg, D.L.: Fundamental relations between short-term rr interval and arterial pressure oscillations in humans. Circulation **93**(8), 1527–1532 (1996). [Online]. Available: http://circ.ahajournals.org/cgi/content/abstract/93/8/1527
61. Groothuis, J.T., Thijssen, D.H.J., Lenders, J.W.M., Deinum, J., Hopman, M.T.E.: Leg vasoconstriction during head-up tilt in patients with autonomic failure is not abolished. J. Appl. Physiol. **110**(2), 416–422 (2011). [Online]. Available: http://jap.physiology.org/content/110/2/416.abstract
62. Guyton, A., Hall, J. (eds.): Textbook on Medical Physiology. Elsevier, Philadelphia (2005)

63. Laude, D., Elghozi, J.-L., Girard, A., Bellard, E., Bouhaddi, M., Castiglioni, P., Cerutti, C., Cividjian, A., Rienzo, M.D., Fortrat, J.-O., Janssen, B., Karemaker, J.M., Lefthériotis, G., Parati, G., Persson, P.B., Porta, A., Quintin, L., Regnard, J., Rüdiger, H., Stauss, H.M.: Comparison of various techniques used to estimate spontaneous baroreflex sensitivity (the eurobavar study). Am. J. Physiol. Regul. Integr. Comp. Physiol. **286**(1), R226–R231 (2004). [Online]. Available: http://dx.doi.org/10.1152/ajpregu.00709.2002
64. Parati, G., Mancia, G., Rienzo, M.D., Castiglioni, P., Taylor, J.A., Studinger, P.: Point:counterpoint: cardiovascular variability is/is not an index of autonomic control of circulation. J. Appl. Physiol. **101**(2), 676–682 (2006). [Online]. Available: http://jap.physiology.org/content/101/2/676.abstract
65. Robbe, H.W., Mulder, L.J., Rüddel, H., Langewitz, W.A., Veldman, J.B., Mulder, G.: Assessment of baroreceptor reflex sensitivity by means of spectral analysis. Hypertension **10**(5), 538–543 (1987)
66. Parati, G., Di Rienzo, M., Bertinieri, G., Pomidossi, G., Casadei, R., Groppelli, A., Pedotti, A., Zanchetti, A., Mancia, G.: Evaluation of the baroreceptor-heart rate reflex by 24-hour intra- arterial blood pressure monitoring in humans. Hypertension **12**(2), 214–222 (1988). [Online]. Available: http://hyper.ahajournals.org/content/12/2/214.abstract
67. Pagani, M., Somers, V., Furlan, R., Dell'Orto, S., Conway, J., Baselli, G., Cerutti, S., Sleight, P., Malliani, A.: Changes in autonomic regulation induced by physical training in mild hypertension. Hypertension **12**(6), 600–610 (1988)
68. Mainardi, L.T., Bianchi, A.M., Furlan, R., Piazza, S., Barbieri, R., di Virgilio, V., Malliani, A., Cerutti, S.: Multivariate time-variant identification of cardiovascular variability signals: a beat-to-beat spectral parameter estimation in vasovagal syncope. IEEE Trans. Biomed. Eng. **44**(10), 978–989 (1997)
69. Orini, M., Laguna, P., Mainardi, L.T., Bail6n, R.: Assessment of the dynamic interactions between heart rate and arterial pressure by the cross time-frequency analysis. Physiol. Meas. **33**(3), 315–331 (2012). [Online]. Available: http://dx.doi.org/10.1088/0967-3334/33/3/315
70. Chen, Z., Purdon, P., Harrell, G., Pierce, E., Walsh, J., Brown, E., Barbieri, R.: Dynamic assessment of baroreflex control of heart rate during induction of propofol anesthesia using a point process method. Ann. Biomed. Eng. **39**, 260–276 (2011). [Online]. Available: http://dx.doi.org/10.1007/s10439-010-0179-z
71. Keissar, K., Maestri, R., Pinna, G.D., Rovere, M.T.L., Gilad, O.: Non-invasive baroreflex sensitivity assessment using wavelet transfer function-based time-frequency analysis. Physiol. Meas. **31**(7), 1021–1036 (2010). [Online]. Available: http://dx.doi.org/10.1088/0967-3334/31/7/011
72. Dutoit, A.P., Hart, E.C., Charkoudian, N., Wallin, B.G., Curry, T.B., Joyner, M.J.: Cardiac baroreflex sensitivity is not correlated to sympathetic baroreflex sensitivity within healthy, young humans. Hypertension **56**(6), 1118–1123 (2010). [Online]. Available: http://hyper.ahajournals.org/content/56/6/1118.abstract
73. Eckberg, D.L., Drabinsky, M., Braunwald, E.: Defective cardiac parasympathetic control in patients with heart disease. N. Engl. J. Med. **285**(16), 877–883 (1971). [Online]. Available: http://www.nejm.org/doi/full/10.1056/NEJM197110142851602
74. Frattola, A., Parati, G., Gamba, P., Paleari, F., Mauri, G., Di Rienzo, M., Castiglioni, P., Mancia, G.: Time and frequency domain estimates of spontaneous baroreflex sensitivity provide early detection of autonomic dysfunction in diabetes mellitus. Diabetologia **40**, 1470–1475 (1997). [Online]. Available: http://dx.doi.org/10.1007/s001250050851
75. Robinson, T.G., Dawson, S.L., Eames, P.J., Panerai, R.B., Potter, J.F.: Cardiac baroreceptor sensitivity predicts long-term outcome after acute ischemic stroke. Stroke **34**(3), 705–712 (2003). [Online]. Available: http://stroke.ahajournals.org/content/34/3/705.abstract
76. La Rovere, M.T., Bigger, J.T., Marcus, F.I., Mortara, A., Schwartz, P.J.: Baroreflex sensitivity and heart-rate variability in prediction of total cardiac mortality after myocardial infarction. Lancet **351**(9101), 478–484 (1998). [Online]. Available: http://www.sciencedirect.com/science/article/pii/S0140673697111448

77. Conci, F., Di Rienzo, M., Castiglioni, P.: Blood pressure and heart rate variability and baroreflex sensitivity before and after brain death. J. Neurol. Neurosurg. Psychiatry **71**(5), 621–631 (2001). [Online]. Available: http://jnnp.bmj.com/content/71/5/621.abstract
78. Parati, G.: Arterial baroreflex control of heart rate: determining factors and methods to assess its spontaneous modulation. J. Physiol. **565**(3), 706–707 (2005). [Online]. Available: http://jp.physoc.org/content/565/3/706.short
79. Cannon, W.B.: The Wisdom of the Body. W.W. Norton, New York (1932)
80. Cannon, W.B.: Bodily Changes in Pain, Fear, Hunger, and Rage. Appleton, New York (1929)
81. Rosenblueth, A., Simeone, F.A.: The interrelations of vagal and accelerator effects on the cardiac rate. Am. J. Physiol. Legacy Content **110**(1), 42–55 (1934)
82. Samaan, A.: The effect of adrenalin, atropine and ether anaesthesia on the heart rate of normal dogs and the animal deprived of different parts of the autonomic nervous system. Arch. Int. Pharmacodyn. Thér. **50**, 101–27 (1935)
83. Levy, M.N.: Brief reviews sympathetic-parasympathetic interactions in the heart. Circ. Res. **29**(5), 437–445 (1971)
84. Levy, M.N.: Cardiac sympathetic-parasympathetic interactions. Fed. Proc. **43**(11), 2598–2602 (1984)
85. Levy, M.N., Blattberg, B.: Progressive reduction in norepinephrine overflow during cardiac sympathetic nerve stimulation in the anaesthetized dog. Cardiovasc. Res. **10**(5), 549–555 (1976)
86. Furukawa, Y., Levy, M.N.: Temporal changes in he sympathetic-parasympathetic interactions that occur in the perfused canine atrium. Circ. Res. **55**(6), 835–841 (1984)
87. Yang, T., Levy, M.N.: Sequence of excitation as a factor in sympathetic-parasympathetic interactions in the heart. Circ. Res. **71**(4), 898–905 (1992)
88. Warner, M.R., Levy, M.N.: Neuropeptide y as a putative modulator of the vagal effects on heart rate. Circ. Res. **64**(5), 882–889 (1989)
89. Potter, E.K.: Prolonged non-adrenergic inhibition of cardiac vagal action following sympathetic stimulation: neuromodulation by neuropeptide Y? Neurosci. Lett. **54**(2), 117–121 (1985)
90. Potter, E.: Presynaptic inhibition of cardiac vagal postganglionic nerves by neuropeptide y. Neurosci. Lett. **83**(1–2), 101–106 (1987)
91. Warner, M.R., Levy, M.N.: Inhibition of cardiac vagal effects by neurally released and exogenous neuropeptide y. Circ. Res. **65**(6), 1536–1546 (1989)
92. Smith-White, M.A., Herzog, H., Potter, E.K.: Role of neuropeptide yy 2 receptors in modulation of cardiac parasympathetic neurotransmission. Regul. Pept. **103**(2), 105–111 (2002)
93. Berntson, G.G., Cacioppo, J.T., Fieldstone, A.: Illusions, arithmetic, and the bidirectional modulation of vagal control of the heart. Biol. Psychol. **44**(1), 1–17 (1996)
94. Porter, T.R., Eckberg, D.L., Fritsch, J.M., Rea, R.F., Beightol, L.A., Schmedtje, J.F. Jr, Mohanty, P.K.: Autonomic pathophysiology in heart failure patients. Sympathetic-cholinergic interrelations. J. Clin. Invest. **85**(5), 1362 (1990)
95. Berntson, G.G., Cacioppo, J.T.: Heart rate variability: stress and psychiatric conditions. Dyn. Electrocardiogr. 57–64 (2004)
96. Sunagawa, K., Kawada, T., Nakahara, T.: Dynamic nonlinear vago-sympathetic interaction in regulating heart rate. Heart Vessels **13**(4), 157–174 (1998)
97. Kawada, T., Sugimachi, M., Shishido, T., Miyano, H., Sato, T., Yoshimura, R., Miyashita, H., Nakahara, T., Alexander, J., Sunagawa, K.: Simultaneous identification of static and dynamic vagosympathetic interactions in regulating heart rate. Am. J. Physiol. Regul. Integr. Comp. Physiol. **276**(3), R782–R789 (1999)
98. Marmarelis, V.: Analysis of Physiological Systems: The White-Noise Approach. Springer Science and Business Media, New York (2012)
99. Berger, R.D., Saul, J.P., Cohen, R.J.: Transfer function analysis of autonomic regulation. I. canine atrial rate response. Am. J. Physiol. Heart Circ. Physiol. **256**(1), H142–H152 (1989)

100. Peng, C.-K., Havlin, S., Stanley, H.E., Goldberger, A.L.: Quantification of scaling exponents and crossover phenomena in nonstationary heartbeat time series. Chaos Interdisciplinary J. Nonlinear Sci. **5**(1), 82–87 (1995)
101. Tulppo, M.P., Kiviniemi, A.M., Hautala, A.J., Kallio, M., Seppänen, T., Mäkikallio, T.H., Huikuri, H.V.: Physiological background of the loss of fractal heart rate dynamics. Circulation **112**(3), 314–319 (2005)
102. Goldberger, A.L.: Fractal mechanisms in the electrophysiology of the heart. IEEE Eng. Med. Biol. Mag. **11**(2), 47–52 (1992)
103. Iyengar, N., Peng, C., Morin, R., Goldberger, A.L., Lipsitz, L.A.: Age-related alterations in the fractal scaling of cardiac interbeat interval dynamics. Am. J. Physiol. Regul. Integr. Comp. Physiol. **271**(4), R1078–R1084 (1996)
104. Karemaker, J.: The riddles of heart rate variability. Clin. Auton. Res. **11**, 65–66 (2001). [Online]. Available: http://dx.doi.org/10.1007/BF02322048
105. Julien, C.: The enigma of mayer waves: Facts and models. Cardiovasc. Res. **70**(1), 12–21 (2006). [Online]. Available: http://dx.doi.org/10.1016/j.cardiores.2005.11.008
106. Cooke, W.H., Hoag, J.B., Crossman, A.A., Kuusela, T.A., Tahvanainen, K.U.O., Eckberg, D.L.: Human responses to upright tilt: a window on central autonomic integration. J. Physiol. **517**(2), 617–628 (1999). [Online]. Available: http://jp.physoc.org/content/517/2/617.abstract
107. Malliani, A., Pagani, M., Lombardi, F., Cerutti, S.: Cardiovascular neural regulation explored in the frequency domain. Circulation **84**, 482–492 (1991)
108. Montano, N., Ruscone, T.G., Porta, A., Lombardi, F., Pagani, M., Malliani, A.: Power spectrum analysis of heart rate variability to assess the changes in sympathovagal balance during graded orthostatic tilt. Circulation **90**, 1826–1831 (1994)
109. Malliani, A., Julien, C., Billman, G.E., Cerutti, S., Piepoli, M.F., Bernardi, L., Sleight, P., Cohen, M.A., Tan, C.O., Laude, D., Elstad, M., Toska, K., Evans, J.M., Eckberg, D.L.: Cardiovascular variability is/is not an index of autonomic control of circulation. J. Appl. Physiol. **101**(2), 684–688 (2006). [Online]. Available: http://jap.physiology.org/content/101/2/684.short
110. Lombardi, F.: Clinical implications of present physiological understanding of hrv components. Card. Electrophysiol. Rev. **6**, 245–249 (2002). [Online]. Available: http://dx.doi.org/10.1023/A:1016329008921
111. La Rovere, M.T., Pinna, G.D., Maestri, R., Mortara, A., Capomolla, S., Febo, O., Ferrari, R., Franchini, M., Gnemmi, M., Opasich, C., Riccardi, P.G., Traversi, E., Cobelli, F.: Short-term heart rate variability strongly predicts sudden cardiac death in chronic heart failure patients. Circulation **107**(4), 565–570 (2003). [Online]. Available: http://circ.ahajournals.org/content/107/4/565.abstract
112. Rajendra Acharya, U., Paul Joseph, K., Kannathal, N., Lim, C., Suri, J.: Heart rate variability: a review. Med. Biol. Eng. Comput. **44**(12), 1031–1051 (2006). [Online]. Available: http://dx.doi.org/10.1007/s11517-006-0119-0
113. Mainardi, L.T., Bianchi, A.M., Baselli, G., Cerutti, S.: Pole-tracking algorithms for the extraction of time-variant heart rate variability spectral parameters. IEEE Trans. Biomed. Eng. **42**(3), 250–259 (1995); ID: 1
114. Mainardi, L.T.: On the quantification of heart rate variability spectral parameters using time-frequency and time-varying methods. Philos. Trans. A Math. Phys. Eng. Sci. **367**(1887), 255–275 (2009)
115. Novak, P., Novak, V.: Time/frequency mapping of the heart rate, blood pressure and respiratory signals. Med. Biol. Eng. Comput. **31**(2), 103–110 (1993)
116. Pola, S., Macerata, A., Emdin, M., Marchesi, C.: Estimation of the power spectral density in nonstationary cardiovascular time series: assessing the role of the time-frequency representations (tfr). IEEE Trans. Biomed. Eng. **43**(1), 46 (1996)

117. Orini, M., Bailon, R., Laguna, P., Mainardi, L.T.: Modeling and estimation of time-varying heart rate variability during stress test by parametric and non parametric analysis. In: Proc. Computers in Cardiology, 30 September–3 October 2007, pp. 29–32. [Online]. Available: http://ieeexplore.ieee.org/ielx5/4729058/4745405/04745413.pdf?tp=
118. Orini, M., Bailón, R., Enk, R., Koelsch, S., Mainardi, L., Laguna, P.: A method for continuously assessing the autonomic response to music-induced emotions through hrv analysis. Med. Biol. Eng. Comput. **48**(5), 423–433 (2010). [Online]. Available: http://dx.doi.org/10.1007/s11517-010-0592-3
119. Barbieri, R., Matten, E.C., Alabi, A.A., Brown, E.N.: A point-process model of human heartbeat intervals: new definitions of heart rate and heart rate variability. Am. J. Physiol. Heart Circ. Physiol. **288**(1), H424–H435 (2005). [Online]. Available: http://ajpheart.physiology.org/content/288/1/H424.abstract
120. Sassi, R., Cerutti, S., Lombardi, F., Malik, M., Huikuri, H.V., Peng, C.-K., Schmidt, G., Yamamoto, Y., Gorenek, B., Lip, G.Y.H., Grassi, G., Kudaiberdieva, G., Fisher, J.P., Zabel, M., Macfadyen, R.: Advances in heart rate variability signal analysis: joint position statement by the e-Cardiology ESC Working Group and the European Heart Rhythm Association co-endorsed by the Asia Pacific Heart Rhythm Society. Europace **17**(9), 1341–1353 (2015). [Online]. Available: http://europace.oxfordjournals.org/content/17/9/1341.abstract
121. Eckberg, D.L.: Temporal response patterns of the human sinus node to brief carotid baroreceptor stimuli. J. Physiol. **258**(3), 769–782 (1976). [Online]. Available: http://jp.physoc.org/content/258/3/769.abstract
122. Mainardi, L.T., Bianchi, A.M., Cerutti, S.: Time-frequency and time-varying analysis for assessing the dynamic responses of cardiovascular control. Crit. Rev. Biomed. Eng. **30**(1–3), 175–217 (2002); lR: 20051116; PUBM: Print; JID: 8208627; RF: 112; ppublish
123. Baselli, G., Caiani, E., Porta, A., Montano, N., Signorini, M.G., Cerutti, S.: Biomedical signal processing and modeling in cardiovascular systems. Crit. Rev. Biomed. Eng. **30**(1–3), 55–84 (2002)
124. Mateo, J., Laguna, P.: Analysis of heart rate variability in the presence of ectopic beats using the heart timing signal. IEEE Trans. Biomed. Eng. **50**(3), 334–343 (2003)
125. Mateo, J., Laguna, P.: Improved heart rate variability signal analysis from the beat occurrence times according to the IPFM model. IEEE Trans. Biomed. Eng. **47**(8), 985–996 (2000)
126. Pyetan, E., Akselrod, S.: Do the high-frequency indexes of hrv provide a faithful assessment of cardiac vagal tone? a critical theoretical evaluation. IEEE Trans. Biomed. Eng. **50**(6), 777–783 (2003)
127. Bailón, R., Laouini, G., Grao, C., Orini, M., Laguna, P., Meste, O.: The integral pulse frequency modulation model with time-varying threshold: application to heart rate variability analysis during exercise stress testing. IEEE Trans. Biomed. Eng. **58**(3), 642–652 (2011)
128. Barbieri, R., Brown, E.: Analysis of heartbeat dynamics by point process adaptive filtering. IEEE Trans. Biomed. Eng. **53**(1), 4–12 (2006)
129. Brown, E.N., Barbieri, R., Ventura, V., Kass, R.E., Frank, L.M.: The time-rescaling theorem and its application to neural spike train data analysis. Neural Comput. **14**(2), 325–346 (2002). [Online]. Available: http://dx.doi.org/10.1162/08997660252741149
130. Rugarn, O., Hammar, M., Theodorsson, A., Theodorsson, E., Stenfors, C.: Sex differences in neuropeptide distribution in the rat brain. Peptides **20**(1), 81–86 (1999)
131. Beaston-Wimmer, P., Smolen, A.J.: Gender differences in neurotransmitter expression in the rat superior cervical ganglion. Dev. Brain Res. **58**(1), 123–128 (1991)
132. Loy, R., Sheldon, R.A.: Sexually dimorphic development of cholinergic enzymes in the rat septohippocampal system. Dev. Brain Res. **34**(1), 156–160 (1987)
133. Du, X.-J., Riemersma, R.A., Dart, A.M.: Cardiovascular protection by oestrogen is partly mediated through modulation of autonomic nervous function. Cardiovasc. Res. **30**(2), 161–165 (1995)

134. Du, X.-J., Dart, A.M., Riemersma, R.A., Oliver, M.F.: Sex difference in presynaptic adrenergic inhibition of norepinephrine release during normoxia and ischemia in the rat heart. Circ. Res. **68**(3), 827–835 (1991)
135. Dart, A.M., Du, X.-J., Kingwell, B.A.: Gender, sex hormones and autonomic nervous control of the cardiovascular system. Cardiovasc. Res. **53**(3), 678–687 (2002)
136. Hinojosa-Laborde, C., Chapa, I., Lange, D., Haywood, J.R.: Gender differences in sympathetic nervous system regulation. Clin. Exp. Pharmacol. Physiol. **26**(2), 122–126 (1999)
137. Frey, M.A.B., Tomaselli, C.M., Hoffler, W.G.: Cardiovascular responses to postural changes: differences with age for women and men. J. Clin. Pharmacol. **34**(5), 394–402 (1994)
138. Lenders, J.W., De Boo, T., Lemmens, W.A., Reuenga, J., Willemsen, J.J., Thien, T.: Comparison of blood pressure response to exogenous epinephrine in hypertensive men and women. Am. J. Cardiol. **61**(15), 1288–1291 (1988)
139. Frankenhaeuser, M., Dunne, E., Lundberg, U.: Sex differences in sympathetic-adrenal medullary reactions induced by different stressors. Psychopharmacology **47**(1), 1–5 (1976)
140. Marrugat, J., Antó, J.M., Sala, J., Masiá, R., R. Investigators, et al.: Influence of gender in acute and long-term cardiac mortality after a first myocardial infarction. J. Clin. Epidemiol. **47**(2), 111–118 (1994)
141. Tunstall-Pedoe, H., Morrison, C., Woodward, M., Fitzpatrick, B., Watt, G.: Sex differences in myocardial infarction and coronary deaths in the scottish monica population of glasgow 1985 to 1991 presentation, diagnosis, treatment, and 28-day case fatality of 3991 events in men and 1551 events in women. Circulation **93**(11), 1981–1992 (1996)
142. Adams, K.F., Sueta, C.A., Gheorghiade, M., O'Connor, C.M., Schwartz, T.A., Koch, G.G., Uretsky, B., Swedberg, K., McKenna, W., Soler-Soler, J., et al.: Gender differences in survival in advanced heart failure insights from the first study. Circulation **99**(14), 1816–1821 (1999)
143. Simon, T., Mary-Krause, M., Funck-Brentano, C., Jaillon, P., C.I. Investigators, et al.: Sex differences in the prognosis of congestive heart failure results from the Cardiac Insufficiency Bisoprolol Study (CIBIS II). Circulation **103**(3), 375–380 (2001)
144. Ramaekers, D., Ector, H., Aubert, A., Rubens, A., Van de Werf, F.: Heart rate variability and heart rate in healthy volunteers. is the female autonomic nervous system cardioprotective? Eur. Heart J. **19**(9), 1334–1341 (1998)
145. Eaker, E., Packard, B., Thom, T.: Epidemiology and risk factors for coronary heart disease in women. Cardiovasc. Clin. **19**(3), 129–145 (1988)
146. Marmarelis, V.Z.: Nonlinear Dynamic Modeling of Physiological Systems, vol. 10. Wiley, New York (2004)
147. Kirsch, M.R., Monahan, K., Weng, J., Redline, S., Loparo, K.A.: Entropy-based measures for quantifying sleep-stage transition dynamics: relationship to sleep fragmentation and daytime sleepiness. IEEE Trans. Biomed. Eng. **59**(3), 787–796 (2012)
148. Téllez, J.P., Herrera, S., Benito, S., Giraldo, B.F.: Analysis of the breathing pattern in elderly patients using the hurst exponent applied to the respiratory flow signal. In: 2014 36th Annual International Conference of the IEEE Engineering in Medicine and Biology Society (EMBC), pp. 3422–3425. IEEE, New York (2014)
149. Giraldo, B.F., Téllez, J.P., Herrera, S., Benito, S.: Study of the oscillatory breathing pattern in elderly patients. In: 2013 35th Annual International Conference of the IEEE Engineering in Medicine and Biology Society (EMBC), pp. 5228–5231. IEEE, New York (2013)
150. Tranah, G.J., Blackwell, T., Stone, K.L., Ancoli-Israel, S., Paudel, M.L., Ensrud, K.E., Cauley, J.A., Redline, S., Hillier, T.A., Cummings, S.R., et al.: Circadian activity rhythms and risk of incident dementia and mild cognitive impairment in older women. Ann. Neurol. **70**(5), 722–732 (2011)
151. Umetani, K., Singer, D.H., McCraty, R., Atkinson, M.: Twenty-four hour time domain heart rate variability and heart rate: relations to age and gender over nine decades. J. Am. Coll. Cardiol. **31**(3), 593–601 (1998)

152. Antelmi, I., De Paula, R.S., Shinzato, A.R., Peres, C.A., Mansur, A.J., Grupi, C.J.: Influence of age, gender, body mass index, and functional capacity on heart rate variability in a cohort of subjects without heart disease. Am. J. Cardiol. **93**(3), 381–385 (2004)
153. Ryan, S.M., Goldberger, A.L., Pincus, S.M., Mietus, J., Lipsitz, L.A.: Gender-and age-related differences in heart rate dynamics: are women more complex than men? J. Am. Coll. Cardiol. **24**(7), 1700–1707 (1994)
154. Beckers, F., Verheyden, B., Aubert, A.E.: Aging and nonlinear heart rate control in a healthy population. Am. J. Physiol. Heart Circ. Physiol. **290**(6), H2560–H2570 (2006)
155. Voss, A., Heitmann, A., Schroeder, R., Peters, A., Perz, S.: Short-term heart rate variability – age dependence in healthy subjects. Physiol. Meas. **33**(8), 1289 (2012)
156. Voss, A., Schulz, S., Schroeder, R., Baumert, M., Caminal, P.: Methods derived from nonlinear dynamics for analysing heart rate variability. Philos. Trans. R. Soc. Lond. A Math. Phys. Eng. Sci. **367**(1887), 277–296 (2009)

Part II
Methods

Chapter 2
Symbolic Dynamics, Poincaré Plot Analysis and Compression Entropy Estimate Complexity in Biological Time Series

Steffen Schulz and Andreas Voss

Abstract Methods from nonlinear dynamics have shown new insights into alterations of the cardiovascular system under various physiological and pathological conditions, and thus providing additional prognostic information. In this chapter prominent complexity methods of non-linear dynamics as symbolic dynamics, Poincaré plot analyses, and compression entropy are introduced and their algorithmic implementations and application examples in clinical trials are provided. Especially, we will give their basic theoretical background, their main features and demonstrate their usefulness in different applications in the field of cardiovascular and cardiorespiratory time series analyses.

2.1 Introduction

Linear time and frequency domain measures are often not sufficient enough to quantify the complex dynamics of physiological systems and their related time series. Therefore, various efforts have been made to apply nonlinear complexity measures to analyze, e.g. the heart rate variability (HRV) [1]. These approaches differ from the traditional time- and frequency domain HRV analyses because they quantify the signal properties instead of assessing only the magnitude or the frequency power of the heart rate time series. They assess the self-affinity of heartbeat fluctuations over multiple time scales (fractal measures); the regularity/irregularity or randomness of heartbeat fluctuations (entropy measures); the coarse-grained dynamics of HR fluctuations based on symbols (symbolic dynamics) and the heartbeat dynamics based on a simplified phase-space embedding [1].

Symbolic dynamics is based on a coarse graining of the dynamics of a signal. The time series (in our cases the ECG or the noninvasively recorded blood pressure curve) are transformed into symbol sequences with symbols of a given alphabet. Some detail information is lost but the coarse dynamic behavior retains and

S. Schulz • A. Voss (✉)
Institute of Innovative Health Technologies IGHT, Ernst-Abbe-Hochschule, Jena, Germany
e-mail: andreas.voss@eah-jena.de

R. Barbieri et al. (eds.), *Complexity and Nonlinearity in Cardiovascular Signals*,
DOI 10.1007/978-3-319-58709-7_2

can be analyzed. Depending on the time series we have to define the type and number of symbols. Long time series allow a higher number of symbols (a higher resolution of the dynamics) than short time series. Combining successive symbols into usually called 'words', enable the extraction of patterns that represent different system states. In the literature different applications of symbolic dynamics can be found in respect to biological time series analyses, ranging from cardiovascular physiology to the field of neuroscience. In cardiovascular physiology, symbolic dynamics was applied, e.g. for the determination of the sympathovagal balance towards a sympathetic activation and vagal withdrawal during graded orthostatic challenge [2], for monitoring the complexity of the cardiac control [3], to evaluate the maternal baroreflex regulation during gestation [4], to investigate autonomic regulation (cardiorespiratory system) during acute psychosis in patients suffering from paranoid schizophrenia and their healthy first-degree relatives [5], to detect pathological states and improve the risk stratification in cardiology [6]. In the field of neuroscience, symbolic dynamics was applied, e.g. to characterized brain microstates [7], to study the recurrence of microstates during stimulus [8], to detect determinism in periictal intracranial electroencephalographic signals [9], to predict epileptic seizures [10], to assess causal relations between brain signals under different levels of consciousness [11], to detect the development of epileptic seizure [12, 13], and to investigate the central–autonomic coupling in mental disorder [14, 15].

Poincaré plot is a popular HRV analysis tool among clinicians due to its ability to visually represent nonlinear dynamics of HRV. It is a geometrical representation of a time series in a Cartesian plane [16]. Poincaré plot analysis (PPA) represents a nonlinear quantitative technique of phase-space characterization, whereby the shape of the plot can be categorized into functional classes, as suggested by Kamen et al. [17]. PPA allows calculating of heart rate dynamics with trends [17, 18]. The Poincaré plots are two- or three-dimensional graphical representation (scatter plots) of each NN interval or in the time series plotted against subsequent NN intervals. Babloyantz et al. [19] qualitatively and quantitatively analyzed electrocardiograms with Poincaré sections starting in 1988. Typically, PPA shows an elongated cloud of points oriented along the line of identity. Only for graphical illustration an ellipse characterizing the shape of the cloud of points can be drawn in the plot whereas the center of the ellipse is the mean NN value. In general, the Poincaré plot can be evaluated quantitatively through the computation of the SD indexes of the plot [20]. Three indices be calculated: the standard deviation of the instantaneous NN interval variability (minor axis of the ellipse–SD1), the standard deviation of the long term NN interval variability (major axis of the ellipse–SD2) and the axes ratio (SD1/SD2) [21, 22]. Thereby, the Poincaré plot provides information about the beat-by-beat behavior of the heart [21]. Analysis of Poincaré plots revealed increased randomness in beat-to-beat heart rate behavior demonstrated by an increase in the ratio between short-term and long-term HRV suggested that more random short-term heart rate behavior may be associated with a complicated clinical course [23]. This measure has not been used much for risk stratification, but has proved useful for detecting editing problems that significantly influence the calculation of HRV variables [24].

However, it has been shown that the indices SD1 and SD2 of Poincaré plot analysis represent more or less only linear features of the heart rate dynamics [25]. Therefore, new methods of analyzing the Poincaré plots were developed to retain the nonlinear features [26, 27]. Poincaré plot is commonly applied to assess the dynamics of HRV [28–30] and in risk stratification, e.g. for sudden death [26], in patient suffering from dilated cardiomyopathy [31], in cardiac arrhythmia subjects [32, 33].

Among the various approaches that have made progress in the quantification of HRV complexity, entropy measures have gained a significant interest. Shannon entropy, conditional entropy, approximate entropy and sample entropy, respectively are some of the most applied approaches of entropy estimation [34, 35]. Complexity analysis can be performed through the evaluation of entropy and entropy rate. Entropy (e.g., Shannon entropy) calculates the degree of complexity of the distribution of the samples of a signal [34]. For HRV analysis, entropy is not calculated directly over the samples of the biological time series but over patterns of length L (i.e., ordered sequences of L samples). In this case, entropy measures the complexity of the pattern distribution as a function of L [34]. In 1977 Ziv and Lempel [36] developed a universal algorithm for lossless data compression (LZ77) using string-matching on a sliding window. Lossless compression ensures that the original information can be exactly reproduced from the compressed data. The LZ77 algorithm is widely used and implemented in compression utilities such as GIF image compression and WinZip®. LZ77 was modified by Baumert et al. in 2004 [37] to analysis heart rate time series called the compression entropy (H_c). in the field of spontaneous fluctuations of cardiovascular oscillations entropy based methods were applied to investigate fetal development [38], to determine age effects on the autonomic system [39–41], in differentiating pathological states from healthy states [42–46], for monitoring cardiac autonomic function [34, 47], in typifying the effects of a pharmacological treatment [48–51], and in predicting risk [37, 52].

In this chapter, some of the prominent complexity measures and their enhancements as symbolic dynamics, Poincaré plot analyses and compression entropy are introduced and their algorithmic implementations and applications in clinical trials are discussed. Especially, we will give their basic theoretical background, their main features and demonstrate their usefulness in different applications in the field of cardiovascular and cardiorespiratory time series analyses.

2.2 Methods

2.2.1 Symbolic Dynamics

Symbolic dynamics did start with Hadamard's ideas about complex systems in 1898 [53]. The most important result of his work was the simplified description of sequences that can arise in geodesic flows on surfaces of negative curvature. He introduced a finite set of forbidden symbol pairs and defined possible sequences as those that do not contain any forbidden symbol pair. Later on, Morse and Hedlund [54] used this method in 1938 to prove the existence of periodic and other dynamics in different classical dynamical systems. They showed that in many circumstances such a finite description of a systems dynamics is possible and performed their investigations by finding interesting sequences satisfying the constraints defined by the corresponding symbolic dynamical system.

In 1983, Aizawa [55] reported that symbolic dynamics could be systematically constructed near the onset point of chaos by taking into account the topological similarity of the Lorenz map and Hao [56] investigated symbolic dynamics applying sequences of two, three and more letters. Although symbolic dynamics was originally developed as a method to study basic structures and behavior of dynamical systems, the methods and the technology have found significant application in various fields of sciences. Some of the major fields for applications of symbolic dynamics are linear algebra, data storage (coding), data transmission (information theory), data analysis and processing as well as life sciences. One of the first life science applications were from Paulus et al. [57] who applied symbolic dynamics in animal experiments.

The research groups of Kurths and Voss were the first who introduced symbolic dynamics into heart rate variability analysis in 1993 [58] and later 1995 [59] and 1996 [6] developing special optimized measures for analysis of the heart rate dynamics. The introduced measures were finally validated in risk stratification based on data from 572 patients after myocardial infarction [60].

The analysis of symbolic dynamics has been proven to be sufficient for the investigation of complex systems and describes dynamic aspects within time series [6, 61, 62]. The concept of symbolic dynamics is based on a coarse-graining of the dynamics. To classify dynamic changes within, e.g. heart rate time series $RR_1, RR_2, \ldots, RR_n$ they were transformed into a symbol sequence $s_1, \ldots s_n, s_i \in A$ with four symbols from the alphabet $A = \{0, 1, 2, 3\}$ (Fig. 2.1). There, according to the transformation rules in equation μ is the mean beat-to-beat interval (RR), a is a special scaling parameter equal to 0.1 and RR_n is the beat-to-beat interval at the time point n.

$$0: \mu < RR_i \leq (1+a) * \mu$$

$$1: (1+a) * \mu < RR_i < \infty$$

$$2: (1-a) * \mu < RR_i \leq \mu$$

$$3: 0 < RR_i \leq (1-a) * \mu$$

where $i = 1,2,3 \ldots$.

In this way some detailed information is lost but the more general dynamic behavior can be analyzed [59]. It is important to mention that small changes

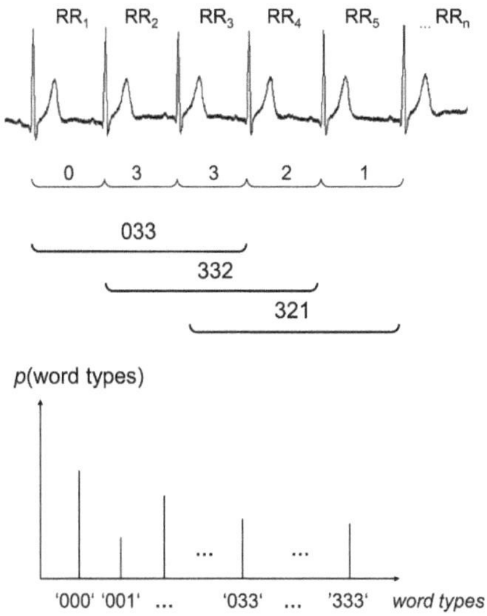

Fig. 2.1 Basic principle of symbolic dynamics analyses based on RR-intervals

of the threshold parameters a and μ do not significantly influence the analysis results. There are several quantities that characterize such symbol strings. We obtain 64 different word types (000,001, ..., 333) using three successive symbols from the alphabet A to characterize symbol strings. The Shannon [63] and Rényi [64] entropies calculated from the distribution of word types p(word types) are suitable measures to quantify the dynamic behavior of heart beat time series and their complexity. High values of these entropies refer to higher complexity in the investigated heart beat time series, vice versa.

An example of word type distributions (histograms) obtained from BBI time series of white noise, a healthy subject, a patient suffering from coronary heart disease and a patient suffering from atrial fibrillation are shown in Fig. 2.2.

The Shannon entropy (H_{Shannon}) is the classical measure of information and is defined as [59]

$$H_{Shannon} = -\sum\nolimits_{i=1}^{k} p_i \log_2 p_i$$

where p is the probability distribution of every single word type and k (=64) is the total number of word types. The Rényi entropy generalizes the Shannon entropy and is estimated from α-weighted probabilities distributions (for HRV analysis often an α value of 0.25 or 4 is in use) by [59]

$$H_{\text{Renyia}} = -\frac{1}{3} log_2 \sum_{i=1}^{k} p_i^a.$$

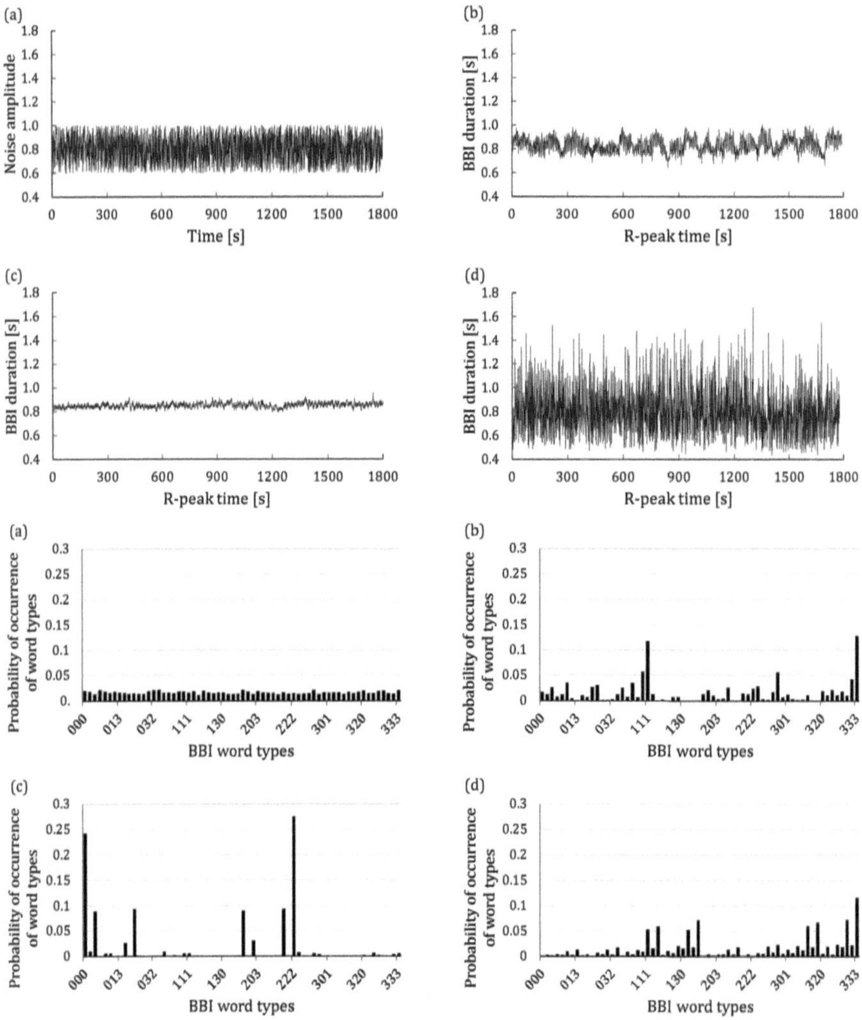

Fig. 2.2 Symbolic dynamics: Tachograms and word distributions (histograms) obtained from BBI time series applying symbolic dynamic analysis of (**a**) white noise, (**b**) a healthy subject, (**c**) a patient suffering from coronary heart disease and (**d**) a patient suffering from atrial fibrillation. (BBI = beat-to-beat intervals)

The word type distribution p differs between short (30 min) and long term (24 h) measures in healthy subjects as well as patient with ischemic cardiomyopathy.

Another procedure which based on word distribution is to count the number of words with length 3 that seldom or never occur so called 'forbidden words'

$$p_{WTi} = \frac{n_{WTi}}{N} < 0.001$$

with p_{WTi} as probability of word types i occurrence, N is the number of all words and n_{WTi} is the number words in type i defined by $n_{WTi} = \sum_j W_{ij}$ with W_{ij} as single word j of word type i.

A high number of forbidden words reflect a reduced dynamic behavior in the heart beat time series. If the heart beat time series is highly complex in the Shannonian sense, only a few forbidden words are found. Words with a probability less than 0.1% were counted. The parameter 'wsdvar' is a measure of heart rate time series variability depending on a word sequence. The resulting word sequence $\{w_1, w_2, w_3, \ldots\}$ is transformed into a sequence $\left\{\overline{S}_1, \overline{S}_2, \overline{S}_3, \ldots\right\}$, where 'wsdvar' is defined as the standard deviation of this sequence $\overline{S}_i$. Further, a high percentage of word, which consist only of symbols '0' and '2' define the parameter 'wpsum02' that is a good measure of decreased HRV, conversely a measure of increased HRV is given by 'wpsum13' (symbols '1' and '3').

In 1999 Wessel et al. [65] applied a new method of long-term symbolic dynamics in patients with dilated cardiomyopathy. They investigated 24 h heart beat time series which allow a higher number of symbols than short-term time series. For the transformation of the time series they used five symbols from the alphabet $A = \{0, 1, 2, 3, 4, 5\}$ instead of four symbols. Thereby, a higher resolution of the analysis was given. The time series are subdivided into short word sequences (bins) of word length k. The number of word types depends on the number of RR-intervals within the time series and is given by word types = (number of symbols)$^{\text{word length}}$.

For example for a 24 h recording (mean heart rate: 80 bpm, number of symbols = 6, word length = 4, word types = 1296) the time series has a length N of 115,200 RR-intervals in the tachogram, so that there are about 89 words in each bin.

$$p(w_k) = \frac{N}{wordtypes} = \frac{115199}{6^4} = 89$$

The accuracy of the word distribution will be reduced by too few words per bin. Voss et al. [6] defined a heuristic basis of 20 as the averaged minimal number of words per bin.

Calculating the percentage of words consisting only of a unique type of symbols is other approach to find epochs of low or high variability. In this way 6 successive symbols of a simplified alphabet [6], consisting only of symbols '0' or '1', $A = \{0, 1\}$ were observed. Here the symbol '0' stands for a difference between two successive beats lower than a special limit (5, 10, 20, 50, and 100 msec) whereas '1' represents those cases where the difference between two successive beats exceeds this special limit:

$$'1' : |RR_n - RR_{n-1}| \leq limit$$

$$'0' : |RR_n - RR_{n-1}| < limit$$

Words consisting only of a unique type of symbol (either all '0' or all '1') were counted. For low variability ('plvar') of sequences containing six consecutive '0' were assessed, whereas for high variability ('phvar') six consecutive '1' were evaluated. There, an increase of '000000' (plvar5, plvar10, plvar20) sequences resulting in increased values of 'plvar' and a decrease in '111111' (plvar20, plvar50, plvar100) sequences leading to reduced values of 'phvar' indicate reduced system complexity.

In 2001 Porta et al. [34] introduced a modified procedure of symbolic dynamics. Here the length of the RR-intervals was limited to 300 beats. The full range of the sequences was uniformly spread on 6 levels (0 to 5), and patterns of length $L = 3$ were constructed [66]. All patterns with $L = 3$ were grouped without any loss into four families. These were: (1) patterns with 0 variations—0 V, (2) patterns with 1 variation—1 V, (3) patterns with 2 like variations—2LV and (4) patterns with 2 unlike variations—2UV. The rates of occurrence of these patterns will be indicated as 0 V%, 1 V%, 2 LV% and 2 UV% [67].

To obtain more detailed information about the dynamics of heart rate some new pattern families were introduced [44, 68]: 'ramp'/'ASC' (three successive symbols form an ascending ramp), 'decline'/'DESC' (three successive symbols form a descending ramp), 'PEAK' (second symbol is larger than the other two symbols forming a peak) and 'VAL' (second symbol is smaller than the other two symbols forming a valley).

A further extension of the classical symbolic dynamic approach represents the segmented short term symbolic dynamic. The segmented short-term symbolic dynamics (SSD) was introduced in order to describe nonlinear aspects within long-term RR time series applying a 24-hour segmentation algorithm in an enhanced way [45, 69]. Therefore, the related time series were segmented in both 1 min overlapping and non-overlapping time windows of 15 min, 30 min or 60 min duration. Within each of these windows a symbol- and word transformation were performed in accordance to the classical symbolic dynamics approach with the scaling parameter $a = 0.1$ [58]. For each segment ($s = 1 \ldots S$, S–number of segments) from the 24-hour interval time series, several parameters based on the word distribution were estimated:

- $pW000_S$ to $pW333_S$ – Probability of occurrence of each word type (000,001, ... ,333) within the interval time series;
- $pTH1_S$ to $pTH20_S$–Number of words with a probability higher than a threshold pTH (1–20%);
- m_pW000 to m_pW333 as the mean values and s_pW000 to s_pW333 as the standard deviations of the parameters $pW000_S$ to $pW333_S$ and $pTH1_S$ to $pTH20_S$;
- *Shannon_pW*xxx—The Shannon entropy calculated from the distribution of each single word type 'xxx' over all windows were estimated as a suitable measure to quantify the dynamic behavior and the complexity of the word type occurrence in the windowed time series.

Cysarz et al. [3, 70] investigated different approaches for the transformation of the original time series to the symbolic time series. They investigated three different transformation methods as: (1) symbolization according to the deviation from the average time series (σ-method), (2) symbolization according to several equidistant levels between the minimum and maximum of the time series (max-min-method), (3) binary symbolization of the first derivative of the time series (binary Δ-coding-method). Each method was applied to cardiac interbeat interval series *RRi* and its difference ΔRR_I, and the occurrence of short sequences ("words") of length $k = 3$ were analyzed. The sequences of length k were categorized according to their amount of variations between successive symbols as proposed by Porta et al. [34]. They could show that all methods are capable of reflecting changes of the cardiac autonomic nervous system, here during head-up tilt. In addition, they demonstrated that changes of cardiovascular regulation during pharmacological challenges can be assessed by the analysis of symbolic dynamics derived from the RR interval series independently of the specific symbolic transformation rule. Finally, they concluded that the 'standard' setting of the different parameters of the investigated methods yield reasonable results and can be used in future clinical studies. However, more refined results could be obtained by an optimized setting of the parameters. Especially, the impact of the parameters should be investigated in more details.

Recently methods have been developed to analyze couplings in dynamic systems. In the field of medical analysis of complex physiological system is a growing interest in how insight may be gained into the interaction between regulatory mechanisms in healthy and diseased persons. For the analyses of the physiological regulatory systems (cardiovascular, cardiorespiratory) as well as the quantification of their interactions, a variety of linear as well as nonlinear methods ranging from parametric models in the time- or in the frequency domain to model-free approaches in the information domain have been proposed [71]. Linear methods are partly insufficient to quantify nonlinear structures and the complexity of physiological systems (time series)—the joint symbolic dynamics (JSD) [4, 72] and high resolution joint symbolic dynamics (HRJSD) [73] overcomes the limitations.

2.2.1.1 Joint Symbolic Dynamics – (JSD)

In 2002, Baumert et al. [4] introduced the method of joint symbolic dynamics (JSD) to investigate the nonlinear interaction between blood pressure and heart beat time series in a more complex way. This method based on the analysis of dynamic processes by means of symbols [6] and allows a simplified quantification of the dynamics of two time series. Therefore, both time series were transformed into symbol sequences of different words. The symbol sequences consist only of '0' and '1', increasing values are coded as '1' and decreasing and unchanged values are coded as '0', respectively, and words of length 3 were formed using a shift of one symbol in time series. Following indices were estimated:

- Normalized probabilities of all single word type ($k = 64$) occurrences were computed as (JSD1-JSD64) using an 8x8 word distribution density matrix;

- SumSym—Symmetric word types where the pattern in both time series are equal (for cardiovascular time series this index reflecting baroreflex-like response patterns);
- SumDiam—Diametric word types, where the response of the first time series to second time series changes is asymmetric (for cardiovascular time series this index reflecting the opposite of baroreflex; patterns);
- $\text{JSD}_{\text{Shannon}}$—The Shannon entropy computed over all 64 word types in the word distribution matrix.

With the JSD method a rough assessment of the overall short-term bivariate interactions can be obtained. In 2011, Kabir et al. [74] proposed a tertiary symbolization scheme and quantified the relative frequency of word types, capturing RSA patterns and thereby adding physiological a-priori knowledge to the analysis. To address the issue of different frequencies between cardiac and respiratory oscillators, Hilbert transformation was introduced to obtain the instantaneous respiratory phase (RP) sampled at the R peak in ECG, yielding beat-to-beat symbol sequences of changes in RR interval and respiratory phase. [75].

2.2.1.2 High Resolution Joint Symbolic Dynamics – (HRJSD)

HRJSD was introduced to quantify the effect of antipsychotics on cardiovascular couplings in patients suffering from acute schizophrenia [73]. The idea of HRJSD is to classify frequent deterministic patterns lasting three beats (symbols). The HRJSD approach enables the classification and characterization of short-term cardiovascular regulatory bivariate coupling patterns which are dominating the interaction generated by the autonomic nervous system. In contrast to other coupling approaches HRJSD emphasizes a clear characterization how the couplings are composed by the different regulatory aspects of the autonomic nervous system. HRJSD based on the classical JSD [4] analyzing dynamic processes by means of symbols. HRJSD has been successful applied to investigate cardiovascular, cardiorespiratory and central-autonomic couplings [5, 14, 76, 77].

In short, HRJSD works in the way that both time series (e.g. heart rate (BBI) and respiratory frequency (RESP)) were transformed into symbol sequences (word length of 3) applying 3 symbols $A = \{0,1,2\}$ and a threshold level l_{BBI} and l_{RESP} for symbol (Fig. 2.3).

Threshold levels for symbol transformation are: no threshold (similar to JSD); the spontaneous baroreflex sensitivity based on the sequence technique [78, 79] and 25% of the standard deviation of the investigated time series as an adapted threshold to the individual physiological dynamic variability.

Thereby, X represents a bivariate sample vector with x_{BBI} and x_{RESP} as the n beat-to-beat values of BBI and RESP, respectively.

$$X = \left\{ \left[x_n^{BBI}, x_n^{RESP}\right]^T \right\}_{n=0,1,\ldots} x \in R$$

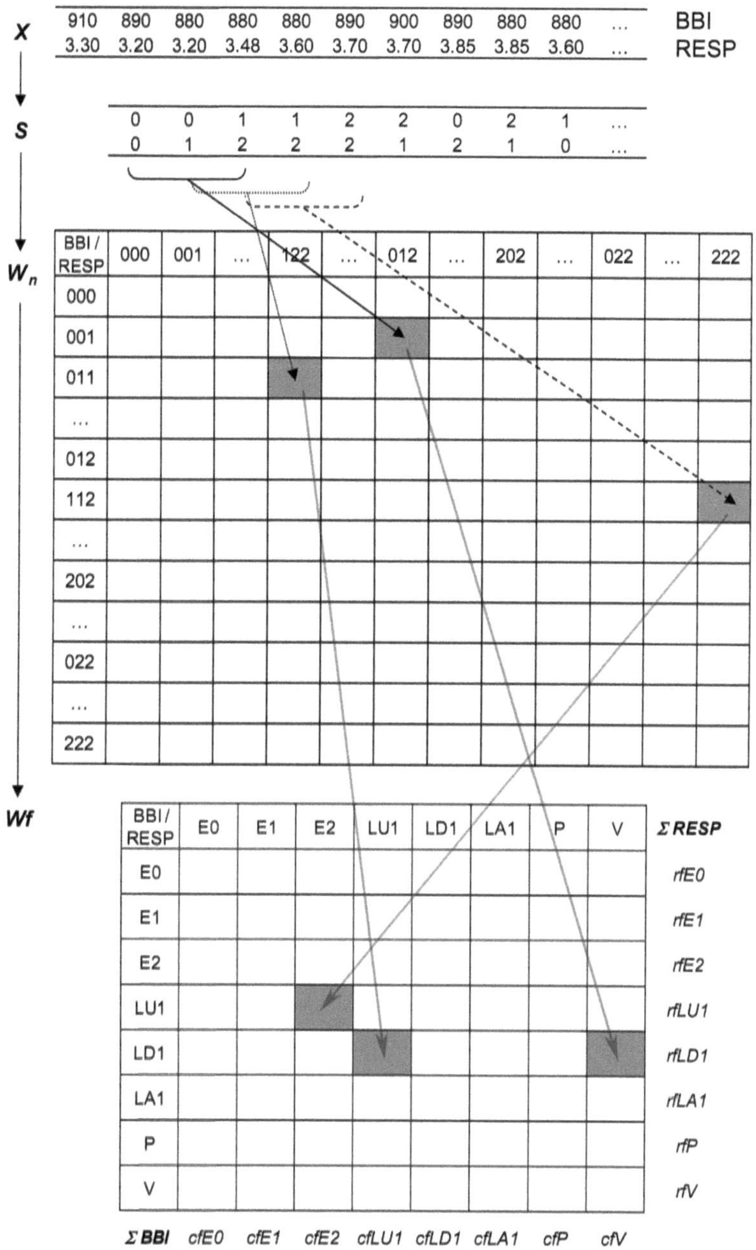

Fig. 2.3 Basic principle of High Resolution Joint Symbolic Dynamics (HRJSD) analysis: (top) Transformation of the bivariate sample vector X (BBI = beat-to-beat intervals [ms]; RESP = respiratory frequency [s]) into the bivariate symbol vector S ('0': decreasing values; '1': equal values and '2': increasing values); (middle): Word distribution density matrix W_n (27x27); (bottom): Word pattern family distribution density matrix Wf (8x8) with 8 pattern families wf, rf_{BBI} row sum of specific word family and cf_{RESP} column sum of specific word family

For the symbol transformation X is afterwards transformed into a bivariate symbol vector S defined as

$$S = \left\{ \left[s_n^{BBI}, s_n^{RESP} \right]^T \right\}_{n=0,1,\ldots} \quad s \in 0, 1, 2$$

with the following definitions

$$s_n^{BBI} = \begin{cases} 0: \left(x_{n+1}^{BBI} - x_n^{BBI}\right) < -l^{BBI} \\ 1: -l^{BBI} \le \left(x_{n+1}^{BBI} - x_n^{BBI}\right) \le l^{BBI} \\ 2: \left(x_{n+1}^{BBI} - x_n^{BBI}\right) > l^{BBI} \end{cases}$$

$$s_n^{RESP} = \begin{cases} 0: \left(x_{n+1}^{RESP} - x_n^{RESP}\right) < -l^{RESP} \\ 1: -l^{RESP} \le \left(x_{n+1}^{RESP} - x_n^{RESP}\right) \le l^{RESP} \\ 2: \left(x_{n+1}^{RESP} - x_n^{RESP}\right) > l^{RESP} \end{cases}.$$

Symbol sequences S was subdivided into short words (bins) w_k of length $k = 3$. Three symbols led to 27 different word types for BBI (w_{BBI}) and RESP (w_{RESP}) (word types ranging from: 000,001,…,221,222). Symbol sequences with increasing values were coded as "2", decreasing values were coded as '0' and unchanging (no variability) values were coded as '1'. All single word types $w_{\mathrm{BBI,RESP}}$ (total number of all word type combinations $27 \times 27 = 729$) were afterwards grouped into eight pattern families wf whereby the sum of probabilities of all single word families' occurrences $p(wf)$ was normalized to 1 (Fig. 2.3).. The 8 pattern families (E0, E1, E2, LU1, LD1, LA1, P, V) represent different aspects of autonomic modulation (strong and weak increase/ decrease, no variability, alternations) and were sorted into an 8x8 pattern family density matrix Wf resulting in 64 cardiorespiratory coupling patterns. The pattern definition (Fig. 2.4) is as follows:

- E0, E1 and E2—No variation within the word consisting of three symbols of type '0', '1' and '2', respectively.
- LU1 and LD1—One variation within the word consisting of two different symbols with low increasing behavior (LU1) and low decreasing behavior (LD1) of BBI and SYS.
- LA1—One variation within the word consisting of two different alternating symbols of type '0' and '2' with an increasing-decreasing behavior of BBI and SYS.
- P and V—Three variations within the word consisting of three different symbols with peak-like behavior (P) and with valley-like behavior (V) of BBI and SYS.

Besides the 64 coupling patterns, 8 pattern families for BBI and RESP the sum of each ($n = 8$) column cf_{RESP} ($cfE0$, $cfE1$, $cfE2$, $cfLU1$, $cfLD1$, $cfLA1$, cfP, cfV), the

Word family	Definition	Effect on HR	Effect on SYS/RESP	Effect on EEG
E0	no variation within the word consisting of three symbols of type '0' (decreasing BBI, SYS & EEG behavior; '000')	↑	↓	↓
E1	no variation within the word consisting of three symbols of type '1' (unchanging BBI, SYS & EEG behavior; '111')	→	→	→
E2	no variation within the word consisting of three symbols of type '2' (increasing BBI, SYS & EEG behavior; '222')	↓	↑	↑
LU1	one variation within the word consisting of two different symbols with low increasing behavior of BBI, SYS & EEG ('122', '022', '112', '221', '220', '211', '121', '212')	↘	↗	↗
LD1	one variation within the word consisting of two different symbols with low decreasing behavior of BBI, SYS & EEG ('011', '001', '002', '110', '100', '200', '010', '101')	↗	↘	↘
LA1	one variation within the word consisting of two different alternating symbols of type '0' and '2' with an increasing-decreasing behavior of BBI, SYS & EEG ('020', '202')	↗↘↗ ↘↗↘	↘↗↘ ↗↘↗	↘↗↘ ↗↘↗
P	three variations within the word consisting of three different symbols with peak-like behavior of BBI, SYS & EEG ('120', '201', '210')	↘↗	↗↘	↗↘
V	three variations within the word consisting of three different symbols with valley-like behavior of BBI, SYS & EEG ('021', '102', '012')	↗↘	↘↗	↘↗

Fig. 2.4 Definition of 8 pattern families of High Resolution Joint Symbolic Dynamics (HRJSD) (HR = heart rate; BBI = beat-to-beat intervals; SYS = systolic blood pressure; RESP = respiratory frequency; EEG: mean power EEG time intervals in relation to BBI)

sum of each ($n = 8$) row rf_{BBI} ($rfE0$, $rfE1$, $rfE2$, $rfLU1$, $rfLD1$, $rfLA1$, rfP, rfV) and the Shannon entropy ($\mathrm{HRJSD_{Shannon}}$) of Wf as a measure of the overall complexity of the coupling (can be determined.

$$\mathrm{HRJSD_{Shannon}} = -\sum_{i,j=1}^{8}\left[p\left(wf_{i,j}\right) log_2 p\left(wf_{i,j}\right)\right]$$

Being a unique feature of the HRJSD approach in contrast to other coupling approaches is the identification of different physiological regulatory patterns generated by the interplay of the involved physiological regulatory systems. From the aspect of biomedical signal processing based on symbolic analysis, the HRJSD approach, based on a redundancy reduction strategy and grouping of single-word types into eight pattern families, enables a detailed description and quantification of bivariate couplings. As a further unique feature in contrast to the classical JSD approach and other coupling approaches [15, 80] HRJSD emphasizes a clear characterization of how the couplings are composed, with regard to the different regulatory aspects of the interacting systems. To summarize, HRJSD approach creates a bridge between univariate and bivariate symbolic analyses, allowing the quantification and classification of deterministic regulatory bivariate coupling patterns depending on the experimental conditions at hand [73].

As a further enhancement on joint symbolic dynamics analyses we have recently extended the HRJSD approach to analyses three signals for multivariate analyses (mHRJSD). Thereby, X represents a trivariate sample vector with e.g. x_{BBI}, y_{SYS} and z_{RESP} as the n beat-to-beat values of BBI, SYS (systolic blood pressure) and RESP, respectively.

$$X = \left\{ \left[x_n^{BBI}, y_n^{SYS}, z_n^{RESP}\right]^T \right\}_{n=0,1,\dots} x \in R$$

In mHRJSD X is afterwards transformed into a trivariate symbol vector S defined as

$$S = \left\{ \left[s_n^{BBI}, s_n^{SYS}, s_n^{RESP}\right]^T \right\}_{n=0,1,\dots} s \in 0, 1, 2.$$

8 pattern families (E0, E1, E2, LU1, LD1, LA1, P, V) were derived from each of the three time series and sorted into an 8x8x8 pattern family density matrix *Wf* leading to multivariate coupling pattern. As an example, the pattern family 'E0' from BBI time series is coupled with the eight pattern families from SYS and 8 from RESP as: BBI-E0/SYS-E0/RESP-E0, BBI-E0/SYS-E0/RESP-E1, BBI-E0/SYS-E0/RESP-E2, BBI-E0/SYS-E0/RESP-LU1, ..., BBI-E0/SYS-V/RESP-V. Thus, the pattern family 'E0' (BBI-E0/SYS-E0/RESP-E0) contains word types that consist only of the '0' symbol. On one hand, this means that BBI decreases over three values and which were therefore coded by '0' three times (representing an increase of the mean heart rate over three values) whereas on the other hand, SYS values increases and RESP values decrease over three values.

2.2.1.3 Joint Conditional Symbolic Analysis – (JCSA)

In 2015, Porta et al. [81] extended the short-term symbolic dynamics procedure [34] to a bivariate and multivariate ones—the joint symbolic analysis (JSA) and the joint conditional symbolic analysis (JCSA). In principle, they created for two time series (x, y) for each time series separately the pattern 0 V, 1 V, 2LV and 2UV. Afterwards, joint patterns by associating one pattern of x and one of y were formed. Thereby, coordinated activity between x and y was investigated by considered introducing two categories: (i) coordinated (C) joint Scheme (0 V–0 V, 1 V–1 V, 2LV–2LV and 2UV–2UV) and (ii) uncoordinated (UNC) joint scheme with different classes (e.g. 1 V and 2UV). The percentage of 0 V–0 V, 1 V–1 V, 2LV–2LV and 2UV–2UV patterns (i.e. 0 V–0 V%, 1 V–1 V%, 2LV–2LV% and 2UV–2UV%) was computed by dividing their amount by the number of C ones and, then, by multiplying the result by 100. In the case of JCSA, x and y were conditioned on a third signal s (e.g. the respiratory phase: inspiration (INSP) and expiration (EXP) phases, the transitions from INSP to EXP (INSP–EXP) and vice versa (EXP–INSP) over s). Thus, e.g. the coordinated pattern 0 V–0 V, 1 V–1 V, 2LV–2LV and 2UV–2UV whose symbols are associated with events all occurring in the inspiratory (INSP) phase are classified as 0 V–0 V$|_{INSP}$, 1 V–1 V$|_{INSP}$, 2LV–2LV$|_{INSP}$ and 2UV–2UV$|_{INSP}$. Analogously, they defined 0 V–0 V$|_{EXP}$, 1 V–1 V$|_{EXP}$, 2LV–2LV$|_{EXP}$ and 2UV–2UV$|_{EXP}$ as the joint patterns whose values are all linkable to the expiratory (EXP) phase. The percentage of the 0 V–0 V, 1 V–1 V, 2LV–2LV and 2UV–2UV patterns occurring in the INSP phase (i.e. 0 V–0 V%$|_{INSP}$, 1 V–1 V%$_{INSP}$, 2LV–2LV%$_{INSP}$ and 2UV–2UV%$|_{INSP}$) were computed by dividing the number of the 0 V–0 V, 1 V–1 V, 2LV–2LV and 2UV–2UV patterns occurring in the INSP phase by the number of coordinated patterns (C) occurring in the INSP phase and, then, by multiplying the results by 100 [81, 82].

2.2.1.4 Symbolic Coupling Traces – (SCT)

A further JSD extension represents the symbolic coupling traces (SCT) introduced by Wessel et al. [83]. SCT based on the analysis of structural patterns and enables the detection of the direction (bidirectional) of time-delayed couplings in short-term bivariate time series. Using the classical JSD algorithm two time series $x(t)$ and $y(t)$ were transformed into symbol sequences $s_x(t)$ and $s_y(t)$ using also the alphabet $A = \{0,1\}$ and afterwards series of word $w_x(t)$ and $w_y(t)$ of length $l = 3$ were formed. In contrast to JSD a delay-time probability matrix $\Pi(\tau) = (p_{ij}(\tau))$ was estimated describing how word W_i would occur in w_x at time t and W_j would occur in w_y at time $(t + \tau)$ with p_{ij} as the joint probabilities of the words. For the quantification of SCT used only the symmetric and diametric traces of the bivariate word distribution matrix (BWD), thereby excluding random effects and including only significant coupling information whereby following measures can be calculated:

The trace T of the matrix $\Pi(\tau)$ defined as

$T(\tau) = \sum_{i=j} p_{ij}(\tau)$ represents the fraction of both time series, which are structurally equivalent to each other at lag τ;

The trace $\overline{T}(\tau)$ of the matrix $\Pi(\tau)$ defined as

$\overline{T}(\tau) = \sum_{i=1,\ldots,d,j=d+1-i} p_{ij}(\tau)$ describing the fraction of both signals, which are structurally diametric at lag τ (d is the number of different patterns). Both parameters vary from 0 to 1 and comprise the diagonals of the BWD only. Finally, the difference $\Delta T = T - \overline{T}$ can be calculated to determine the exact detection of lags (delayed couplings) between two time series. Thereby, T only captures influences which preserve the structure of the transmitted pattern of dynamics (symmetrical influences) and $\overline{T}$ only quantifies influences which inverts the dynamical structure of the driver (diametrical influence) [84]. The lags τ should limited to $20 \leq \tau \leq 20$ (sampling units) in order to focus on short time-delayed dependencies only. The main advantages of SCT are its ability to detect delayed coupling (time lags), its applicability to moderately noisy time series (< 10 dB) and its insensitivity to non-stationarities. [71, 85].

2.2.1.5 Specifics and Restrictions

SD: Detailed information will be lost; outliers (ectopic beats and noise) influence symbol strings; the accuracy of the word distribution will be reduced by too few words per bin (heuristic basis of 20 as the averaged minimal number of words per bin are needed [6]; no stationarity required; short- and long term time series analyses; only short-term fluctuations can be investigated depending on the word length (3–6 symbols).

JSD/HRJSD: Coarse-graining without any information about information flow; based on a bivariate redundancy reduction strategy to group single word types into eight pattern families; allowing a detailed quantification and classification of deterministic regulatory bivariate coupling patterns depending on experimental conditions; no stationarity required; short- and long term time series analyses; only short-term fluctuations can be investigated depending on the word length (3 symbols).

SCT: High noise levels (>10 dB) influence detection of delayed coupling; critical significance level has to calculated; no stationarity required; short- and long term time series analyses; only short-term fluctuations can be investigated depending on the word length (3 symbols).

2.2.2 *Poincaré Plot Analyses*

The Poincaré plot analysis (PPA) is a method that allows calculating of heart beat dynamics with trends [17, 18, 86]. The Poincaré plots are two-dimensional graphical representation (scatter plots) of each RR-interval or in the time series plotted against the subsequent RR-interval.

It provides a visual and quantitative analysis of RR-interval sequences. Babloyantz et al. [19] qualitatively and quantitatively analyzed electrocardiograms with Poincaré sections in 1988.

Thereby, the shape of the plot, that is assumed to be influenced by changes in the vagal and sympathetic modulation, can be used to classify the sequence into one of several classes. The plots provide detailed beat-to-beat information on the behavior of the heart [17]. Typically, PPA shows an elongated cloud of points oriented along the line of identity. Only for graphical illustration an ellipse characterizing the shape of the cloud of points can be drawn in the plot whereas the center of the ellipse is the mean RR value. In general, three indices are calculated from the Poincaré plots: the standard deviation of the instantaneous BBI variability (minor axis of the ellipse—$SD1$), the standard deviation of the long term BBI variability (major axis of the ellipse—$SD2$) and the axes ratio ($SD1/SD2$) [21, 22, 86].

The level of short-term variability can be quantified by $SD1$ and the level of long-term variability is given by $SD2$. $SD1$ and $SD2$ are calculated using:

$$x_n = (x_1, x_2, \ldots, x_{N-1}) \text{ and } x_{n+1} = (x_2, x_3, \ldots, x_N)$$

$$SD1 = \sqrt{VAR\frac{x_n - x_{n+1}}{\sqrt{2}}}$$

$$SD2 = \sqrt{VAR\frac{x_n + x_{n+1}}{\sqrt{2}}},$$

where *VAR* is the variance, x_n is a time series with $n = 1, \ldots, N - 1$ (N–length of time series) and x_{n+1} is the same time series shifted by one.

Laitio et al. [23] showed that an increased *SD*1/*SD*2 ratio was the most powerful predictor of postoperative ischemia. Stein et al. [87] also demonstrated that an increased *SD*1/*SD*2 ratio had the strongest association with mortality. Further, Mäkikallio et al. [88] found for healthy subjects $SD1/SD2 \approx 0.2$, for post infarction patients with ventricular tachyarrhythmia $SD1/SD2 \approx 0.3$ and for patients with a previous myocardial infarction who had experienced ventricular fibrillation $SD1/SD2 \approx 0.6$. Voss et al. [61] found a significant reduced value for $SD1 \approx 15$ in patient with dilated cardiomyopathy in comparison to healthy subjects $SD1 \approx 24$. Furthermore, Voss et al. showed [89] that the cloud of the Poincaré plot are characterized by high variability for healthy subjects; reduced to a narrow distribution of the cloud for patients at high risk; and arrhythmias producing separate clouds in the Poincaré plot. Arrhythmias can be easily visually detected connecting each single point of the cloud by trajectories. Moreover, they presented first attempts of lagged- and three-dimensional Poincaré plots.

An example of standard Poincaré plots obtained from BBI time series of white noise, a healthy subject, a patient suffering from coronary heart disease and a patient suffering from atrial fibrillation are shown in Fig. 2.5.

2.2.2.1 Segmented Poincaré Plot Analysis – (SPPA)

The SPPA approach was introduced by Voss et al. [26] as a nonlinear approach of phase-space characterization for the nonlinear quantification of NN time series based on the traditional Poincaré plot analysis. Thereby, the cloud of points is segmented into 12 × 12 equal rectangles whose size depends on the standard deviations *SD*1 (height) and *SD*2 (width) of BBI, SYS and DIA NN time series of the Poincaré plot.

Thereby, the cloud of points presented by PPA is then rotated $\alpha = 45$ degrees around the main focus of the plot allowing for a simplified *SD*1/*SD*2 adapted probability-estimating procedure (Fig. 2.6). Starting from the main focus of the plot, a grid of 12 × 12 rectangles is drawn into the plot [27]. For the estimation of the single probabilities (p_{ij}) within each rectangle the number of points within every rectangle is counted and normalized by the total number of all points. Based on these single probabilities all row (i) and column (j) probabilities are calculated by summation of the related single probabilities as:

- SPPA_r_i = single probability of each row with $i = 1$–12,
- SPPA_c_j = single probability of each column with $j = 1$–12.

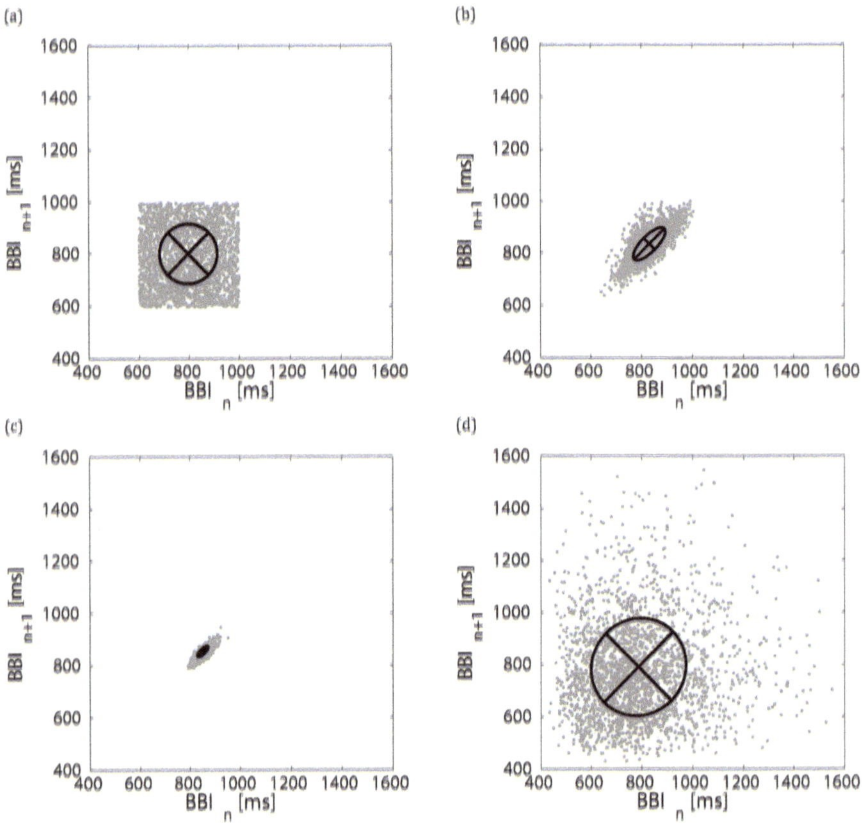

Fig. 2.5 Poincaré plot analysis: Standard Poincaré plots of BBI time series of (**a**) white noise, (**b**) a healthy subject, (**c**) a patient suffering from coronary heart disease and (**d**) a patient suffering from atrial fibrillation. (BBI = beat-to-beat intervals)

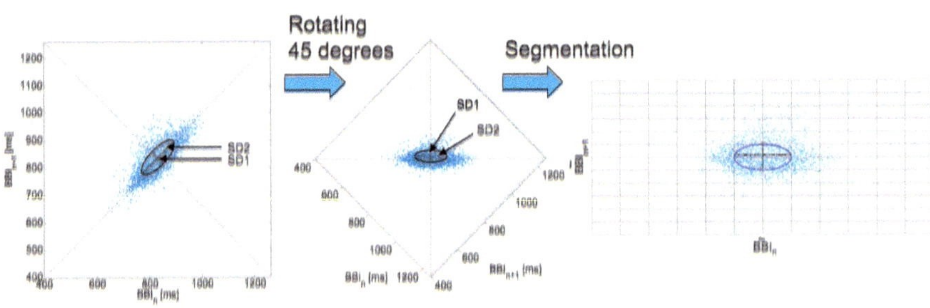

Fig. 2.6 Basic principle of segmented Poincaré plot analysis (SPPA); (left) standard Poincaré plot with *SD*1 and *SD*2; (middle) the rotated Poincaré plot (45 degrees) and (right) the segmented Poincaré plot for SPPA. (BBI = beat-to-beat intervals; $\widetilde{BBI}$ = transformed BBI after segmentation; *SD*1 = minor axis of the cloud (short-term variability); *SD*2 = major axis of the cloud (long-term variability))

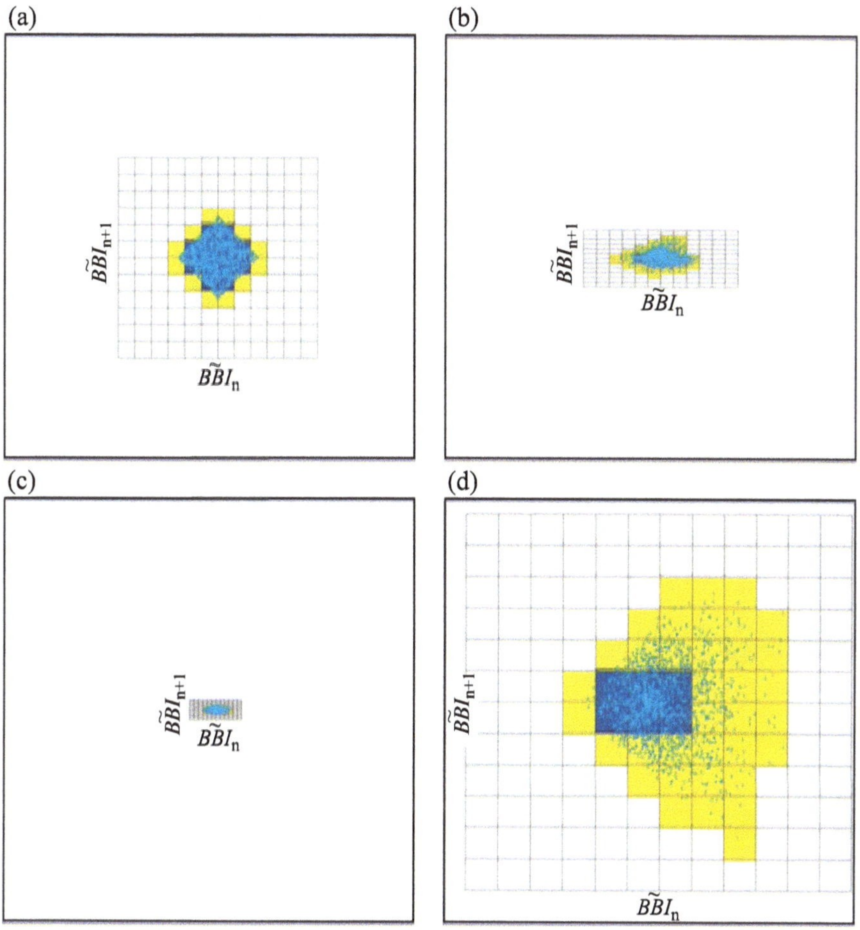

Fig. 2.7 Segmented Poincaré plot analysis: Segmented Poincaré plots of BBI time series of (**a**) white noise, (**b**) a healthy subject, (**c**) a patient suffering from coronary heart disease and (**d**) a patient suffering from atrial fibrillation. (BBI = beat-to-beat intervals)

Additionally, the Shannon entropy [bit] of the 12×12 probability matrix quantifying its disorder or uncertainty will be estimated.

An example of segmented Poincaré plots obtained from BBI time series of white noise, a healthy subject, a patient suffering from coronary heart disease and a patient suffering from atrial fibrillation are shown in Fig. 2.7.

2.2.2.2 Bivariate Segmented Poincaré Plot Analysis – (2DSPPA)

2DSPPA [90] based on the univariate SPPA [26] and works in the way that BBI time series were plotted over systolic or diastolic blood pressure (SYS, DIA) NN time series. In this study, the interactions between BBI and SYS, BBI and DIA as well as between SYS and DIA were investigated. Based upon the single probabilities (see SPPA), all row (i) and column (j) probabilities were calculated by summation of the related single probabilities:

- 2DSPPA_r_i = single probability of each row with $i = 1$–12, (e.g. BBI-SYS_r_1 = the single probability of the BBI time series plotted over the systolic blood pressure (SYS) NN time series in row 1)
- 2DSPPA_c_j = single probability of each column with $j = 1$–12.

2.2.2.3 Trivariate Segmented Poincaré Plot Analysis – (3DSPPA)

Three-dimensional SPPA (3DSPPA) uses three shifted signals from a time series (univariate) or three different signals (multivariate: e.g. BBI, SBP, DIA and RESP) plotted in several box models. 3DSPPA based on specific subdivisions of a cubic box model which is similar to SPPA [16, 91] and is able to investigate multivariate couplings between subsystems, e.g. of cardiovascular and cardiorespiratory autonomic regulation. Thereby, the 3DSPPA considers two varying positions of the cloud of points:

- Non-rotated version
- Rotated version [91].

According to the size of the boxes within the 3D cubes plot the cubic box model is subdivided into 12 × 12 × 12 equal dimensioned cubelets for a total number of $N = 1728$ cubelets. In addition, 3DSPPA considers two approaches differing in the dimension of the cubelets calculated for both the rotated and non-rotated position of the cloud of points:

- Adapted 3DSPPA (calculation of the standard deviations $SD1$ and $SD2$ of the time series; cubelet size is adapted to the calculated SD with regard to the axis; 3D cubic box model consists of 12 × 12 × 12 equal cubelets whereby the center of the cubic box represents the main focus of the cloud of points)
- Predefined 3DSPPA (3D cubic box model represents the basic model with regard to all investigated subjects).

In accordance to SPPA for each cubelet the probability of occurrence ($p_{r,c,d}$) of data points within the cubelet is calculated where the indices r represents the row, c the column and d the depth of the specific cubelet (1–12) [16, 91].

2.2.2.4 Lagged Segmented Poincaré Plot Analysis – (LSPPA)

The SPPA does not provide information about physiological changes, and especially, their relation to frequency bands corresponding to high-frequency (HF), low-frequency (LF) and very low-frequency (VLF) bands. To overcome this limitation,

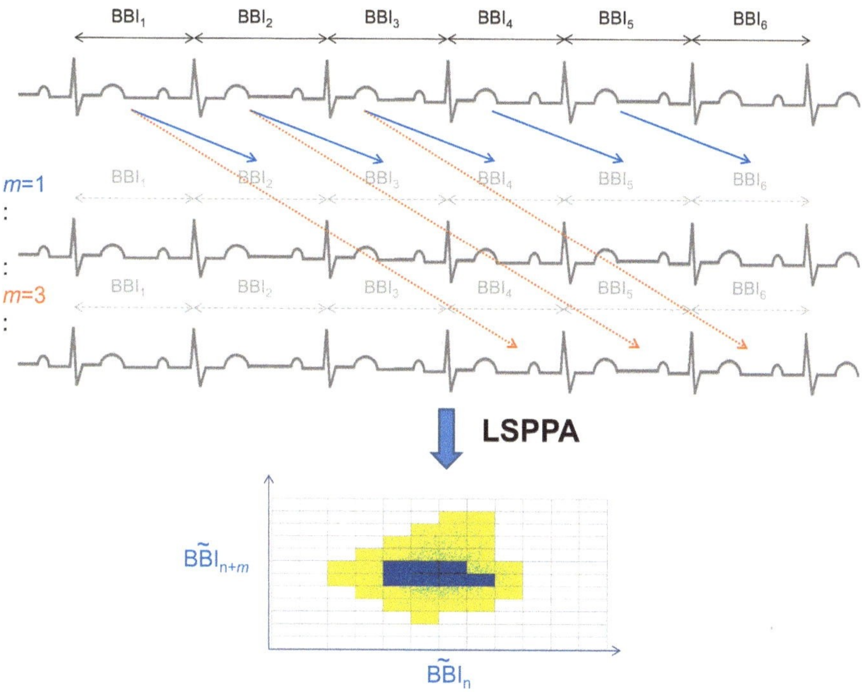

Fig. 2.8 Basic principal of time correlation analysis applying lagged segmented Poincaré plot analysis (LSPPA) for BBI time series (lags: $m = 1, m = 3$). (BBI = beat-to-beat intervals; $\widetilde{BBI}$ = transformed BBI after segmentation)

the lagged segmented Poincaré plot analysis (LPPA) was introduced (Fig. 2.8) [27]. LSPPA based on SPPA considering NN_n as a function of its subsequent NN_{n+m} by a lag of one ($m = 1$). Thereby, the lag is defined as the distance between the investigated *NN*. Assuming that LSPPA can describe time correlations, we have investigated the patterns of LSPPA, applying lags from $m = 1$–100. To make the calculated results comparable, we used the same time series length for each lag by cutting an offset (specific number of *NN*; *offset* = 100-*m*) at the end of each time series leading to:

$$NN_n = \left(NN_1, NN_2, \ldots, NN_{N-m-offset}\right)$$

and

$$NN_{n+m} = \left(NN_2, NN_3, \ldots, NN_{N-offset}\right)$$

For each lag, the normalized probability of point occurrence in rows and columns related to the 12×12 rectangle grid is evaluated. According to SPPA, the size

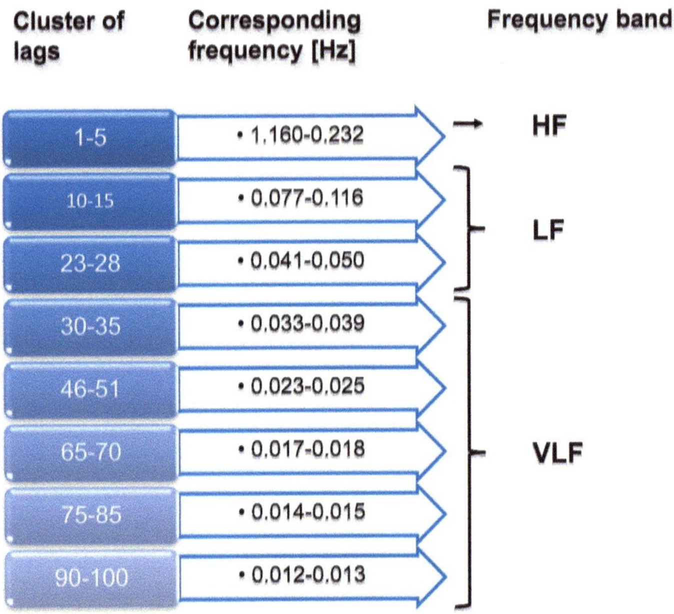

Fig. 2.9 Declaration of clusters of lags including their corresponding frequencies and the related frequency bands for lagged segmented Poincaré plot analysis

of the rectangles is adapted on $SD1$ and $SD2$ and depends on m in the lagged version of SPPA where the corresponding frequency f is calculated as follows: $f \;\; 1/m^{*} meanNN$. For a mean heart rate of 60 beats/min ($meanNN = 1000$ msec) the lowest frequency resolution will be 0.01 Hz. Afterwards, the lags are combined into eight different clusters covering the most significant frequencies (Fig. 2.9). The defined clusters include 5–11 lags independently to usually considered frequency bands. Each cluster consists of at least 75% lags with highly significant indices. Lower clusters include 5–6 lags, whereas the upper ones include 10–11 lags, according to the degree of differences between the related frequencies [27, 91].

The clusters correspond to the standardized frequency bands as HF-band (0.15–0.4 Hz) corresponds to cluster I, LF-band (0.04–0.15 Hz) corresponds to (clusters II and III), and VLF-band (0.003–0.04 Hz) corresponds to (clusters IV–VIII). Thus, LSPPA is able to provide additional information about various oscillations of the underlying physiological mechanisms of autonomic regulation as respiratory (0.145–0.6 Hz), intrinsic myogenic (0.052–0.145 Hz), neurogenic/sympathetic (0.021–0.052 Hz) and endothelial (0.0095–0.021 Hz) activities [27] (Fig. 2.9).

2.2.2.5 Complex Correlation Measure – (CCM)

Karmakar et al. [92] introduced the Complex Correlation Measure (CCM) quantifying temporal aspect of the Poincaré plot and is a function of multiple lag correlation of a investigated time series and incorporating point-to-point variation of the signal in contrast to the standard *SD*1 and *SD*2 PPA indices.

CCM is computed in a windowed manner embedding the temporal information of the investigated time series. Here, a moving window of three consecutive points from the Poincaré plot is considered and the area of the triangle formed by these three points is computed that measures the temporal variation of the points in the window. If three points are aligned on a line then the area is zero, which represents the linear alignment of the points. Moreover, since the individual measure involves three points of the two dimensional plot, it is comprised of at least four different points of the time series for lag $m = 1$ and at most six points in case of lag $m \geq 3$. Hence the measure conveys information about four different lag correlation of the signal. [16, 92] CCM may be used to study the lag response behavior of any pathological condition in comparison with normal subjects [16].

2.2.2.6 Specifics and Restrictions

PPA /SPPA/LSPPA*:* *SD*1 and *SD*2 dependent on other time-domain measures; short- and long term time series analyses; identifies outlier (ectopic beat or artefact); multi-lagged Poincare´ plot analyses provide more information than any measure from single lagged Poincare´ plot analyses [93]; different lag plots (e.g. $m = 3$) better reveal the behavior of the signal than the single lag plot; Poincaré plot at any lag *m* is more of a generalized scenario, where other levels of temporal variation of the dynamic system are hidden [16]; SPPA retains nonlinear features of the investigated time series, therefore overcoming some limitations of traditional PPA.

CCM is not only related to the *SD*1 and *SD*2, but it also provides temporal information, which can be used to quantify the temporal dynamics of the system [16].

2.2.3 Compression Entropy

An approach to describe the entropy of a text was introduced in the framework of algorithmic information theory. Here, the entropy (Kolmogorov-Chaitin complexity) of a given text is defined as the smallest algorithm that is capable of generating the text [94]. Although it is theoretically impossible to develop such an algorithm data compression techniques might be a sufficient approximation. Ziv and Lempel [36] introduced in 1977 an universal algorithm for lossless data compression (LZ77) using string-matching on a sliding window. Today, this algorithm is widely used and implemented in compression utilities such as GIF image compression and

WinZip®. The algorithm is briefly described here that a sequence $x = x_1, x_2, \ldots x_n$ of symbols of length L of a given alphabet Θ of size $\Phi = |\Theta|$ has to be compressed. Subsequences $x_m, x_{m+1}, \ldots, x_n$ of x will be denoted as x_m^n. The algorithm keeps the most w recently encoded symbols in a sliding window. The not-yet-encoded sequence of symbols is stored in the look-ahead buffer of size b. The compressor positioned at p the first string in the look-ahead buffer looks for the longest string match of length n between the not-yet-encoded string and the strings x_p^{p+n-1} and the already-coded string in the sliding window $x_{p-w+v}^{p-w+v+n-1}$ beginning at position v. Therefore, the matching string is fully described by n and v. In short, the LZ77 algorithm operates in the following steps:

1. Encode the first w symbols without compression,
2. Set the pointer $p = w + 1$,
3. Find for some v in the range of $1 \leq v \leq w$ the largest n in the range of $1 \leq n \leq b$ such that $x_p^{p+n-1} = x_{p-w+v}^{p-w+v+n-1}$.
4. Encode the integers n and v into unary-binary code and the symbol $x_{p+n} \in \Theta$ without compression,
5. Set the pointer p to $p = n + 1$ and go to step 3 (iterate).

Assuming that the source is an ergodic process and the length of the compressed text is large ($L \to \infty$) the entropy of the compressed string H_c is determined as the length M of the compressed string divided by the length L of the original data series [37, 95].

$$H_c = M/L$$

This algorithm can be applied in a modified way for analysis of heart beat time series [37]. Here, the compression entropy H_c of heart beat time series is affected by the sample rate s, the window length w and the look-ahead buffer size b and hence denoted as $H_c^{s,w,b}$. In summary, $H_c^{s,w,b}$ indicates in which extent data from heart beat time series or each other time series can be compressed using the detection of repetitive sequences. H_c decreases with increasing window length w but the influence of the look-ahead buffer size b on H_c is only marginal. The parameters w and buffer b were optimized to $H_c^{7,3}$ ($w = 7, b = 3$) [96]. The dependency of $H_c^{s,w,b}$on w and b suggests a sensitive of $H_c^{s,w,b}$to vagally rather than sympathetically mediated components of HRV. Assuming that the compressibility of a time series is a measure of its nonlinear complexity, the complexity of heart rate in high-risk patients is reduced and, therefore, compression entropy decreases with increasing risk [96].

Examples of compression entropy (compressibility) obtained from BBI time series of white noise, a healthy subject, a patient suffering from coronary heart disease and a patient suffering from atrial fibrillation are shown in Fig. 2.10.

2013, Baumert et al. [35] investigated the compressibility of heart rate time series on multiple time scales, using a coarse-graining procedure applying the H_c algorithm ten times. Thereby, scale one corresponds to the original time series, higher scales (2–10) were obtained by coarse-graining procedure, using the procedure proposed by Costa et al. [97] multiscale entropy algorithm. Here, an one-dimensional discrete time series, $\{x_1, \ldots, x_i, \ldots, x_N\}$, the coarse-grained

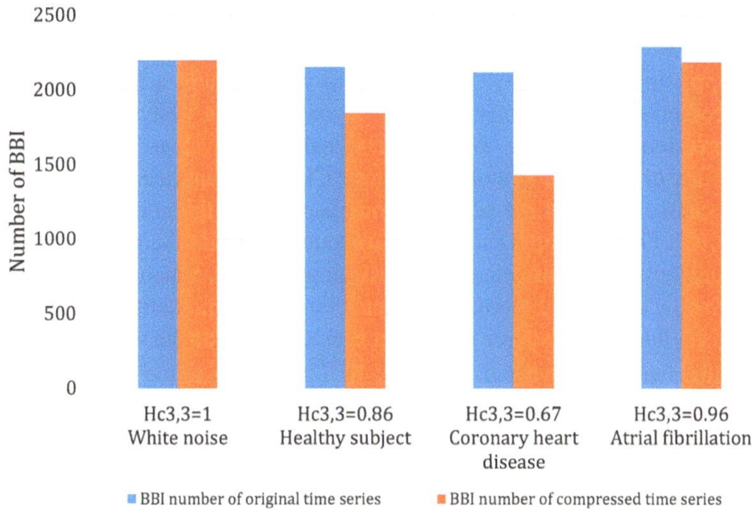

Fig. 2.10 Compression entropy: Compressibility (window length, $w = 3$, buffer length, $b = 3$) for BBI time series. Illustrative comparison of the length of the original (*blue*) vs. compressed (*orange*) BBI time series of white noise, healthy subjects, patients suffering from coronary heart disease and atrial fibrillation. (BBI = beat-to-beat intervals)

time series $\{x(\tau)\}$ determined by the scale factor τ is constructed by $x_j^{(\tau)} = 1/\tau \sum_{i=(j-1)\tau+1}^{j\tau} x_i$ where τ represents the scale factor and $1 \leq j \leq L/\tau$. In other words, coarse-grained time series for scale τ were obtained by taking arithmetic mean of τ neighboring original values without overlapping. For scale 1, the coarse grained time series is simply the original time series. They concluded that multiscale investigation of compressibility may provide an alternative method for entropy assessment of biomedical signals over different time scales and be particularly useful for heart rate complexity analysis.

A limitation of all univariate methods is that they are not able to quantify the direct interrelationships between different time series. Therefore, they have limited power to reveal the underlying physiological mechanisms responsible for changes in time series complexity.

Recently, Li and Vitányi [94] compared several entropy and entropy rate based methods. They investigated the hypothesis that the linear model-based (MB) approach for the estimation of conditional entropy (CE) can be utilized to assess the complexity of the cardiac control in healthy individuals in comparison to nonlinear model-free (MF) methods such as corrected ApEn, SampEn, corrected CE, two k-nearest-neighbor CE procedures and permutation CE. They found, that the MB approach can be utilized to monitor the changes of the complexity of the cardiac control, thus speeding up dramatically the CE calculation. They concluded that, due to that the remarkable performance of the MB approach challenges the notion, generally assumed in cardiac control complexity analysis based on CE, about the need of MF techniques and could allow real time applications.

2017, Schumann et al. [98] introduced the cross-compression entropy (CH_c). CH_c combines the data compression technique based on symbol transformation and the principle of causality. It estimates to which extent a time series (target) can be compressed by another (source) beyond its own past. The improvement of compression considering the source signal is equivalent to an increase of predictability of the target's future. Therefore, CH_c represents a measure of causality between coarse-grained time series in compliance with the notion of Wiener and Granger [99]. First, the input time series are transformed into sequences of symbols [77]. In analogy to the univariate compression procedure, a memory window and a buffer window are shifted along the target series [37, 96]. Additionally, a memory window is defined in the source series that is used to encode the target symbols. The target buffer window covers B_y symbols $Y_p^{p+B_y}$, starting at the current data point Y_p (coding position). These target symbols are encoded using the symbols of the source memory window $X_{p-M_x}^{p-1}$ with length M_x. The longest subseries $X_{p-M_x+v-1}^{p-M_x+v+n-2}$, lasting n source symbols starting at element v, that matches the target sequence Y_p^{p+n-1} is extracted. Instead of encoding the whole target string, the starting point v and the length n of its equivalent in the source memory and the successor X_{p+n} is stored. Hence, n target symbols can be passed and the new coding position is X_{p+n+1}.

If there is a redundant substring, of the same length or longer, included in the target memory, compressibility is not improved by taking the source signal into account. Therefor matching source symbols are ignored, if a substring $Y_{p-M_y+k-1}^{p-M_y+k+l-2}$ with a length of $l \geq n$ in the target memory matches Y_p^{p+l-1}.

$CH_{cX \rightarrow Y}$ is defined as the proportion of iterations that can be saved compressing Y_i by X_i with respect to the original length of Y_i. Assuming the input Y_i of length N_0 is compressed by X_i in N_{com} iterations, CH_c is calculated by the equation below.

$$CH_{cX \rightarrow Y} = \frac{N_0 - N_{com}}{N_0}$$

Because redundant symbol patterns at different positions in the source window (start v) can contribute to the compression, CH_c allows the delay of coupling to vary during the time of acquisition. The interaction across various time scales can be covered by CH_c due to a variable length of redundant symbol substrings n. Although CH_c is very flexible, it is easy to interpret and has the natural boundaries of zero and one. The compression algorithm is well known and easy to implement. Symbolization is robust to outliers and does not require stationarity of the data.

2.2.3.1 Specifics and Restrictions

Dependency on sampling rate, the window length, the look-ahead buffer size and a threshold; implementation has to consider integer numbers; relatively insensitive to artefacts and non-stationarity which have only a locally limited influence; short-term method where longer lasting patterns reflecting especially sympathetically mediated influences are not or only partially considered for compression; short- and long term time series analyses.

2.3 Clinical Applications

2.3.1 Schizophrenia: Cardiorespiratory System

Schizophrenia is a severe mental disorder associated with a significantly increased cardiovascular mortality rate. However, the underlying mechanisms leading to cardiovascular disease (CVD) are not fully known. For the quantitative analyses of the cardiorespiratory system in univariate and bivariate ways, several linear and nonlinear time series analysis approaches were developed. Studies indeed showed that the coupling between the cardiovascular system and respiration is strongly nonlinear [100]. Therefore, linear methods seem to be inappropriate and not able to fully address physiological regulatory mechanisms within the cardiovascular system. Methods based on entropies have the common feature that they analyze a putative information transfer between time series and address either the uncertainty or predictability of time series. Complexity analysis can be performed by evaluating the entropy and entropy rate. Entropy (e.g., Shannon or Renyi) calculates the degree of complexity of a signal's sample distribution. However, a limitation of all univariate nonlinear methods is that they are not able to quantify the direct interrelationships such as the nonlinear influence of respiration on heart rate. Therefore, they have limited power to reveal the underlying physiological mechanisms responsible for changes in cardiorespiratory complexity [76].

We investigated [76, 101, 102] non-medicated 23 patients with schizophrenia (SZ; 12 male; 30.4 $\pm$ 10.3 years) and 23 healthy subjects (CO, 13 male; 30.3 $\pm$ 9.5 years) matched regarding to age and sex (Table 2.1) to quantify heart rate- and respiratory variability (HRV, RESPV) and their dynamics as well as the cardiorespiratory coupling (CRC).

Patients were included only when they had not taken any medication for at least 8 weeks. Diagnosis of paranoid schizophrenia was established when patients fulfilled DSM-IV criteria (Diagnostic and statistical manual of mental disorders, 4th edition. Psychotic symptoms were quantified using the Positive and Negative Syndrome Scale (PANSS). A careful interview and clinical investigation was performed for all controls to exclude any potential psychiatric or other disease as well as interfering medication. The Structured Clinical Interview SCID II and a personality inventory (Freiburger Persönlichkeitsinventar, FPI) were additionally applied for controls to detect personality traits or disorders which might influence autonomic function. This study complied with the Declaration of Helsinki. All participants gave written informed consent to a protocol approved by the local Ethics

Table 2.1 Clinical and demographic data of enrolled study participants

Data	Healthy subjects (CO)x	Schizophrenic patients (SZ)
Number of participants	23	23
Gender (male/female)	13/10	12/11
Age (mean ± std. in years)	30.3 ± 9.5	30.4 ± 10.3
PANSS, mean (min-max)	n.a.	85.7 (43–124)
SANS, mean (min-max)	n.a.	49.6 (14–81)
SAPS, mean (min-max)	n.a.	60.9 (6–108)

Psychotic symptoms for acute schizophrenia were quantified using the scale for the assessment of positive symptoms (SAPS) and negative symptoms (SANS) and positive and negative syndrome scales (PANSS); n.*A. means* not applicable

Committee of the University Hospital Jena. Patients were advised that the refusal of participating in this study would not affect future treatment.

From all subjects a high resolution short-term ECG (1000 Hz sampling frequency) and synchronized calibrated respiratory inductive plethysmography signal (LifeShirt®, Vivometrics, Inc., Ventura, CA, U.S.A.) were recorded for 30 min. Investigations were performed between 3 and 6 p.m. in a quiet room which was kept comfortably warm (22–24 °C) and began after subjects had rested in supine position for 10 min. Subjects were asked to relax and to breathe normally to avoid hyperventilation. No further instruction for breathing was given. Subjects were asked explicit not to talk during the recording. Following time series were automatically extracted from the raw data records as:

- Time series of heart rate consisting of successive beat-to-beat intervals (BBI) and
- Time series of respiratory frequency (RESP) as the time intervals between consecutive breathing cycles.

Quantification of HRV, RESPV and CRC was done by applying several complexity approaches as symbolic dynamics (SD), compression entropy (H_C), sample entropy (SampEn), Poincaré plot analysis (PPA), Higuchi fractal dimension (HD), high resolution joint symbolic dynamics (HRJSD), cross conditional entropy (CCE), cross multiscale entropy (CMSE) and normalized short time partial directed coherence (NSTPDC) [71, 76, 103].

The nonparametric exact two-tailed Mann-Whitney U-test (SPPS 21.0) was performed to non-normally distributed indices (significant Kolmogorov-Smirnov test) to evaluate continuous baseline variables as well as differences in autonomic indices between SZ and CO. The significance level was set to $^{**}p < 0.01$ (Bonferroni-Holm adjustment: $^{***}p < 0.001$). In addition, all results were presented as mean ± standard deviation.

We found (Table 2.2, Figs. 2.11, 2.12, and 2.13) significant difference in HRV, RESPV and their dynamics as well as in CRC in SZ in comparison to CO ($p < 0.001$). Especially, we demonstrated by different complexity based measures (univariate, bivariate) that HRV was characterized by reduced complexity

Table 2.2 Univariate statistical analysis results of heart rate- and respiratory variability in the complexity domain to discriminate between patients suffering from schizophrenia (SZ) and healthy subjects (CO)

			CON	SZ
	Index	p	Mean ± Std	Mean ± Std
SD	$SD_{Renyi025\_BBI}$	**	3.74 ± 0.37	3.47 ± 0.37
	$Forbword_{\_BBI}$	**	15.04 ± 13.25	25.91 ± 12.35
	$SD_{Renyi025\_RESP}$	***	3.23 ± 0.15	3.47 ± 0.19
	$Forbword_{\_RESP}$	**	31.61 ± 4.53	23.65 ± 6.41
H_C	H_{C_BBI}	***	0.82 ± 0.10	0.69 ± 0.10
	H_{C_RESP}	n.s.	0.59 ± 0.08	0.59 ± 0.12
SampEn	$SampEn_{\_BBI}$	**	2.29 ± 0.30	1.96 ± 0.47
	$SampEn_{\_RESP}$	n.s.	1.32 ± 0.37	1.49 ± 0.50
PPA	$SD1/SD2_{\_BBI}$	***	0.46 ± 0.16	0.29 ± 0.11
	$SD1/SD2_{\_RESP}$	***	0.11 ± 0.03	0.18 ± 0.06
HD	HD_{RSA}	**	1.14 ± 0.03	1.19 ± 0.05
HRJSD	$HRJSD_{Renyi025}$	***	4.06 ± 0.11	4.37 ± 0.15
CCE	$CCEuf_{\_BBI\text{-}RESP}$	**	0.12 ± 0.07	0.16 ± 0.08
CMSE	$CMSE_{scale1}$	***	2.13 ± 0.67	1.43 ± 0.42
NSTPDC	$H_{BBI \rightarrow RESP}$	**	0.78 ± 0.07	0.71 ± 0.09
	$H_{RESP \rightarrow BBI}$	***	0.76 ± 0.05	0.71 ± 0.05

BBI beat-to-beat intervals, *RESP* time intervals between consecutive breathing cycles, *SD* symbolic dynamics, H_C compression entropy, *SampEn* sample entropy, *PPA* Poincaré plot analysis, *HD* Higuchi fractal dimension (*RSA* respiratory sinus arrhythmia), *HRJSD* high resolution joint symbolic dynamics, *CCE* cross conditional entropy, *CMSE* cross multiscale entropy; *NSTPDC* normalized short time partial directed coherence, mean value ± standard deviation, *p* univariate significance (**$p < 0.01$, ***$p < 0.001$, *n.s.* not significant)

(Figs. 2.11, and 2.12), whereas RESPV was characterized by increased complexity and CRC was reduced accompanied by reduced complexity in schizophrenia.

These results suggest a parasympathetic withdrawal and an ongoing sympathetic predominant activation in cardiac autonomic regulation. Bär et al. [43] suggested that the reduction in heart rate complexity indicates that heart rate cannot adapt to different requirements arising from posture or exertion and that the heart is at higher risk of developing arrhythmias in those patients. Voss et al. [104] found also reduced HRV (linear, nonlinear) in schizophrenic patients and their relatives, indicating an impaired autonomic function, widely believed to be caused by a reduced vagal tone. They further stated that at this moment the association of a reduced HRV and higher cardiac risk (high mortality rates) for those patients has not been satisfactorily proven. In general, the reduction in cardiac complexity supports the thesis of a changed sympathetic/parasympathetic heart rate control in schizophrenic patients [1]. A reduction of cardiac complexity (the increase of cardiac regularity) is considered as a marker of pathology.

Considering respiration and respiratory variability (Fig. 2.13) as well as their complexity we found, in accordance to previous findings [102, 105, 106]. Homma

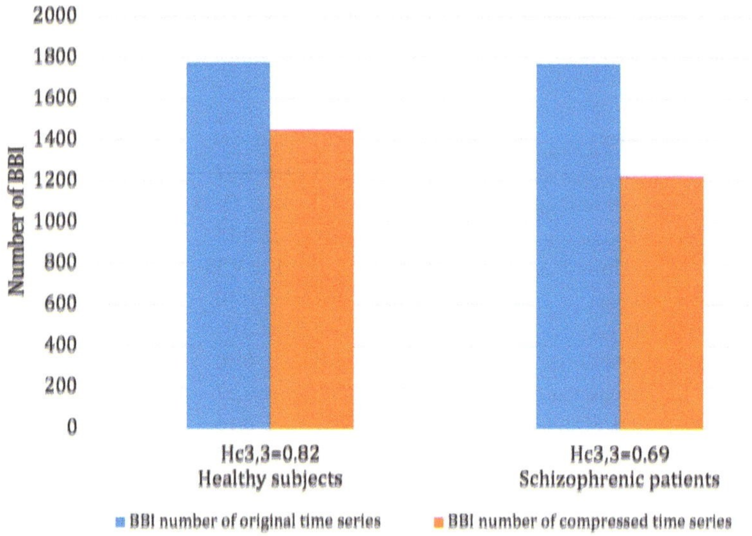

Fig. 2.11 Compression entropy: Compressibility (window length, $w = 3$, buffer length, $b = 3$) for BBI time series. Illustrative comparison of the length of the original (*blue*) vs. compressed (*orange*) BBI time series for patients suffering from (*right*) schizophrenia ($H_c^{3,3} = 0.69$) and (*left*) healthy subjects ($H_c^{3,3} = 0.82$). (BBI = beat-to-beat intervals)

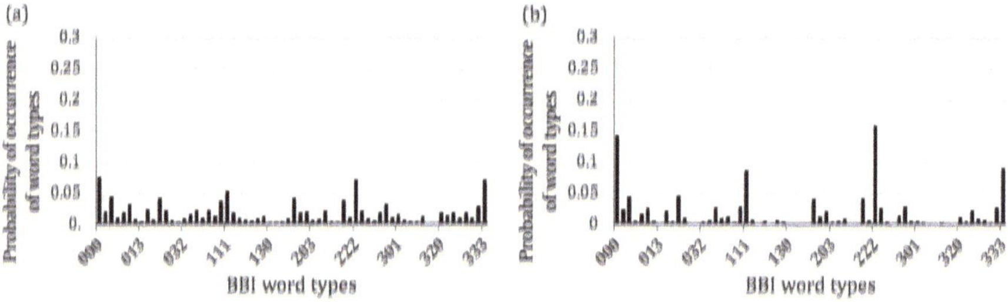

Fig. 2.12 Symbolic dynamics: Word distributions (histograms) obtained from BBI time series applying symbolic dynamic analysis of (**a**) healthy subjects and (**b**) patients suffering from schizophrenia. (BBI = beat-to-beat intervals)

et al. [107] stated that the final respiratory output involves a complex interaction between the brainstem and higher centers, including the limbic system and cortical structures. Respiration is primarily regulated for metabolic and homeostatic purposes in the brainstem and also changes in response to changes in emotions, such as sadness, happiness, anxiety or fear. Williams et al. [108] could show a functional disconnection in autonomic and central systems for processing threat-related signals in patients with paranoid schizophrenia and hypothesized that paranoid cognition may reflect an internally generated cycle of misattribution regarding incoming fear

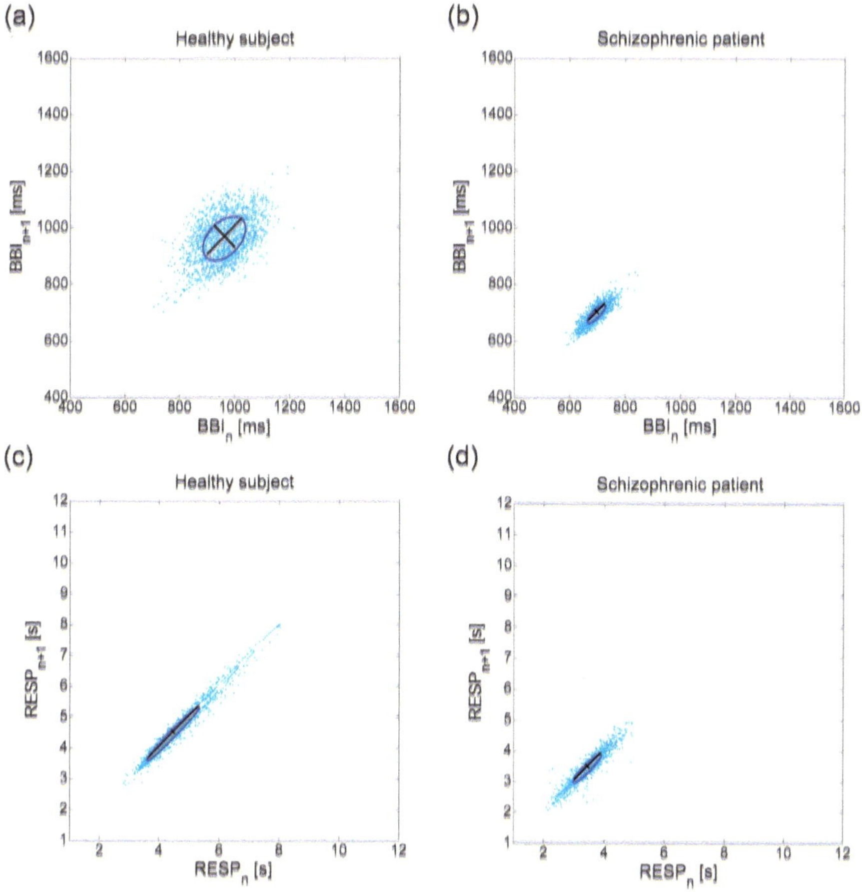

Fig. 2.13 Poincaré plot analysis: Standard Poincaré plots of BBI time series of (**a**) a healthy subject and (**b**) a patient suffering from schizophrenia, and of respiratory frequency intervals (RESP) of (**c**) a healthy subject and (**d**) a patient suffering from schizophrenia. (BBI = beat-to-beat intervals; RESP = respiratory frequency)

signals due to a breakdown in the regulation of these systems. Boiten et al. [109] found that respiration patterns reflect the general dimensions of emotional response that are linked to response requirements of the emotional situations. Furthermore, the ratio of inspiration time to expiration time is closely related to emotions, but however, are inconsistent discussed [109]. The found alterations in RESPV likely can be explained that a dysregulation of arousal, as suggested in paranoid schizophrenia in amygdalae prefrontal circuits, might contribute to the correlation of psychopathology and breathing alterations [102]. In contrast to HRV respiration was characterized by increased entropy indices ($SD_{Renyi025\_RESP}$, $SampEn_{\_RESP}$, and HDRSA) describing the complexity and randomness of the respiratory time series. HD_{RSA} indicating that fractal characteristics (morphological structure) of the RSA signal were increased in SZ indicating that the underlying rhythm of the RSA signal is randomly fluctuating. These findings pointing to increased irregularity of respiration. It could be shown that a small change in respiratory functioning may lead to background symptoms of panic and anxiety in any disorder as a results of the link between the central nervous system and respiration [110]. It is well proven that schizophrenia is related to panic attacks [111, 112], further supporting the altered CRC in SZ might be at least partly related to panic attacks in the acute psychotic state. It was shown that changed complexity indices, resulting from depressed organ function, a loss of interaction among subsystems, an overwhelming action of a subsystem over others and an impairment of regulatory processes, is a clear hallmark of a pathological situation [113]. We believe that the found alterations in CRC might reflect arousal and permeant stress situation in acutely ill SZ patients. This can be is confirmed by comparable physiological changes within the cardiorespiratory system in healthy subjects during stress conditions [102].

In conclusion, we showed that complexity indices from HRV; RESPV and CRC analyses could contribute to enhanced risk stratification in schizophrenic patients and possible will identify in the near future those patients at higher risk cardiovascular disease. At the moment we are just at the beginning to understand the interrelationship between the cardiorespiratory system in psychotic states and the related brainstem neural networks and control mechanisms.

2.3.2 *Schizophrenia: Blood Flow*

We investigated the impairment of microcirculation in schizophrenic patients by means of spectral analysis of blood flow signals and to determine if microcirculation is unequally altered in different tissue depths. Furthermore, the impact of gender and age on the spectral parameters of the Laser Doppler Flowmetry (LDF) signal in healthy and diseased microcirculation were analyzed. The Segmented Spectral Analysis (SSA) algorithm was applied to LDF signals of a provoked post-ischemic stage and compared to the traditional total spectral analysis (TSA). 15 healthy subjects (CON, mean age 32.4 years) and 15 patients (PAT, mean age 33.0 years) were enrolled. Spectral analysis was performed on two LDF signals at a depth of 2 mm and 6–8 mm. Features in five frequency subintervals were determined. Our results indicate that microcirculation is strongly impaired in patients. SSA of blood flow revealed differences between CON and PAT in all three frequency intervals referring to local vasomotion (endothelial $p = 0.03$; sympathetic $p = 0.02$, myogenic $p = 0.03$) as well as the respiratory ($p = 0.02$) and cardiac ($p = 0.006$) bands in the deeper tissue. In contrast, in the near-surface tissue only the endothelial ($p = 0.006$) and cardiac ($p = 0.006$) components were altered. Furthermore, SSA determined a gender- and age dependency regarding blood flow. In conclusion, we could demonstrate that microcirculation in schizophrenic patients is significantly impaired and that blood flow in the near-surface skin and in the superficial muscle tissue is affected differently, depending on its location in the near-surface skin or in the superficial muscle tissue. These alterations of microcirculation are more pronounced in the deeper tissue depth of about 6–8 mm and are influenced by gender and age. [114].

2.3.3 Idiopathic Dilated Cardiomyopathy

Identifying idiopathic dilated cardiomyopathy (IDC) patients who are at risk of sudden death is still unsolved. The presence of autonomic imbalance in patients with IDC might predict sudden death and tachyarrhythmic events. The aim of this study was to analyze the suitability of blood pressure variability (BPV) compared to heart rate variability (HRV) for non-invasive risk stratification in IDC patients. Therefore, continuous non-invasive blood pressure and high-resolution electrocardiogram were recorded from 91 IDC patients for 30 min. During a median follow-up period of 28 months (range: [17–38] months), 14 patients died due to sudden death or necessary resuscitation due to a life-threatening arrhythmia. HRV and BPV analyses were performed in time domain, frequency domain, and nonlinear dynamics. We found that dynamics of blood pressure regulation was significantly changed in high-risk patients, indicating an increased BPV. BPV indexes from nonlinear symbolic dynamics revealed significant univariate (sensitivity: 85.7%; specificity 77.9%; area under receiver-operator characteristics [ROC] curve: 87.8%) differences. In an optimum multivariate set consisting of two clinical indexes (left ventricular end-diastolic diameter, New York Heart Association) and one nonlinear index (symbolic dynamics), highly significant differences between low- and high-risk IDC groups were estimated (sensitivity of 92.9%, specificity of 86.5%, and area under ROC curve of 95.3%). In sum, diastolic BPV indexes, especially those from symbolic dynamics, appear to be useful for risk stratification of sudden death in patients with IDC. [115].

2.3.4 Depression

Due to that major depressive disorders (MDD) are associated with an increased risk for cardiovascular morbidity and mortality and it is known that MDD are accompanied with an autonomic dysfunction with increased sympathetic and/or reduced parasympathetic activity to data only limited information are available about the degree and complexity of cardiovascular regulation. We investigated 57 non-medicated depressed patients in comparison to 57 healthy subjects matched with respect to age, gender and group size and the influence of MDD on autonomous nervous system by means of linear and nonlinear indices from heart rate- and blood pressure variability (HRV, BPV). Complexity indices from nonlinear dynamics demonstrated considerable changes in autonomous regulation due to MDD. For the first time we could show that non-medicated depressed patients who were matched with respect to age, gender and group size reveal a significant changed short-term as well as long-term complexity of cardiovascular regulation. Complexity indices from nonlinear dynamics showed considerable changes in autonomous regulation caused by the disease depression. This finding was supported by linear indices only from BPV. These results suggest substantial changes in autonomic control probably due to a change of interactions between different physiological control loops in MDD. [44].

2.3.5 *Pregnancy*

Hypertensive pregnancy disorders affect 6–8% of gestations representing the most common complication of pregnancy for both mother and fetus. In this study, 10 healthy non-pregnant women, 66 healthy pregnant women and 56 hypertensive pregnant women (chronic hypertension, pregnancy induced hypertension and PE) were investigated applying the three-dimensional segmented Poincaré plot analyses (3DSPPA) to detect hypertensive pregnancy disorders and especially pre-eclampsia (PE). From all subjects 30 min of beat-to-beat intervals (BBI), respiration (RESP), non-invasive systolic (SBP) and diastolic blood pressure (DBP) were continuously recorded and analyzed. Non-rotated adapted 3DSPPA discriminated best between hypertensive pregnancy disorders and PE concerning coupling analysis of 2 or 3 different systems (BBI, DBP, RESP and BBI, SBP, DBP) reaching an accuracy of up to 82.9%. This could be increased to an accuracy of up to 91.2% applying multivariate analysis differentiating between all pregnant women and PE. In conclusion, 3DSPPA could be a useful method for enhanced risk stratification in pregnant women [91].

2.4 Conclusion

It becomes apparent that cardiovascular regulation is one of the most complex systems in human due to the fact, that a variety of factors influencing the cardiovascular system (e.g. heart rate) [103]. The application of nonlinear dynamics is motivated by the fact that the control systems of the cardiovascular system have been shown to be nonlinear because of its high complexity and the nonlinear interactions between the physiological subsystems [103, 116]. Several of nonlinear indices have been proven to be of diagnostic relevance or have contributed to risk stratification. In this chapter we focused on the nonlinear methods symbolic dynamics, Poincaré plot analysis, and compression entropy and their recent enhanced versions. The introduced complexity methods revealed new insights into the changed autonomic nervous system under various physiological and pathophysiological conditions, provide additional prognostic information and definitely complement traditional time and frequency domain analyze. These methods have already been proven to be of diagnostic relevance or have contributed to risk stratification in different diseases.

Several metrics assessing the system's complexity and information flow among its components are based on symbolization techniques. Indices derived from symbolic dynamics might provide a reliable alternative to markers of complexity and causality that can be derived from model-based multiple linear regression approaches in time and frequency domains [117–120]. In the medical field, it can be expected that symbolic dynamics will have a powerful impact in the coming years by playing a significant role in tailoring individual treatments, improving diagnostics and therapy, managing patient data and reducing the cost of healthcare systems via a more precise risk stratification [15].

Research on HRV has proven that Poincaré plot analysis is a powerful tool to evaluate short term and long term HRV. Researchers have investigated a number of methods: converting the two- or three-dimensional Poincaré plot into various one-dimensional views; segmented the plot to retain nonlinear properties of the underlying system; multiple lagged plot analysis to investigate time correlation of the system; the fitting of an ellipse to the plot shape; and measuring the correlation coefficient of the plot. Poincaré plot analyses play a more and more important role and—as an advantage—are easier to understand and interpret [16].

Methods based on entropies have in common that they analyze a putative information transfer within one or between time series and address the uncertainty or predictability of time series. Compression entropy represents only one complexity measure within the big group of methods calculating entropy and/or entropy rates. This method is easy to apply, working time efficiently and allows straightforward physiological interpretations.

Applications of these introduced methods in future studies are very promising and should (depending on the complexity of the investigated system) be performed in a multivariate manner complemented with linear and further successful non-linear approaches.

References

1. Voss, A., Schulz, S., Schroeder, R., Baumert, M., Caminal, P.: Methods derived from nonlinear dynamics for analysing heart rate variability. Philos. Transact. A Math. Phys. Eng. Sci. **367**, 277–296 (2009)
2. Porta, A., Tobaldini, E., Guzzetti, S., Furlan, R., Montano, N., Gnecchi-Ruscone, T.: Assessment of cardiac autonomic modulation during graded head-up tilt by symbolic analysis of heart rate variability. Am. J. Physiol. Heart Circ. Physiol. **293**, H702–H708 (2007)
3. Cysarz, D., Porta, A., Montano, N., Leeuwen, P.V., Kurths, J., Wessel, N.: Quantifying heart rate dynamics using different approaches of symbolic dynamics. Eur. Phys. J. Spec. Top. **222**, 487–500 (2013)
4. Baumert, M., Walther, T., Hopfe, J., Stepan, H., Faber, R., Voss, A.: Joint symbolic dynamic analysis of beat-to-beat interactions of heart rate and systolic blood pressure in normal pregnancy. Med. Biol. Eng. Comput. **40**, 241–245 (2002)

5. Schulz, S., Haueisen, J., Bar, K.J., Andreas, V.: High-resolution joint symbolic analysis to enhance classification of the cardiorespiratory system in patients with schizophrenia and their relatives. Philos. Trans. A Math. Phys. Eng. Sci. **373**, 20140098 (2015)
6. Voss, A., Kurths, J., Kleiner, H.J., Witt, A., Wessel, N., Saparin, P., et al.: The application of methods of non-linear dynamics for the improved and predictive recognition of patients threatened by sudden cardiac death. Cardiovasc. Res. **31**, 419–433 (1996)
7. Wackermann, J., Lehmann, D., Michel, C.M., Strik, W.K.: Adaptive segmentation of spontaneous EEG map series into spatially defined microstates. Int. J. Psychophysiol. **14**, 269–283 (1993)
8. Beim Graben, P., Hutt, A.: Detecting recurrence domains of dynamical systems by symbolic dynamics. Phys. Rev. Lett. **110**, 154101 (2013)
9. Schindler, K., Gast, H., Stieglitz, L., Stibal, A., Hauf, M., Wiest, R., et al.: Forbidden ordinal patterns of periictal intracranial EEG indicate deterministic dynamics in human epileptic seizures. Epilepsia. **52**, 1771–1780 (2011)
10. Hively, L.M., Protopopescu, V.A., Munro, N.B.: Enhancements in epilepsy forewarning via phase-space dissimilarity. J. Clin. Neurophysiol. **22**, 402–409 (2005)
11. Mashour, G.A.: Cognitive unbinding: a neuroscientific paradigm of general anesthesia and related states of unconsciousness. Neurosci. Biobehav. Rev. **37**, 2751–2759 (2013)
12. Staniek, M., Lehnertz, K.: Symbolic transfer entropy: inferring directionality in biosignals. Biomed. Tech. (Berl). **54**, 323–328 (2009)
13. Voss, A., Schulz, S., Schroeder, R.: Monitoring in cardiovascular disease patients by nonlinear biomedical signal processing. Conf. Proc. IEEE Eng. Med. Biol. Soc. **2011**, 6564–6567 (2011)
14. Schulz, S., Bolz, M., Bar, K.J., Voss, A.: Central- and autonomic nervous system coupling in schizophrenia. Philos. Trans. A Math. Phys. Eng. Sci. **374**, 20150178 (2016)
15. Porta, A., Baumert, M., Cysarz, D., Wessel, N.: Enhancing dynamical signatures of complex systems through symbolic computation. Philos. Trans. A Math. Phys. Eng. Sci. **373**, 20140099 (2015)
16. Khandoker, A.H., Karmakar, C., Brennan, M., Palaniswami, M., Voss, A.: Poincaré plot methods for heart rate variability analysis. Springer, London (2013)
17. Kamen, P.W., Krum, H., Tonkin, A.M.: Poincare plot of heart rate variability allows quantitative display of parasympathetic nervous activity in humans. Clin. Sci. (Lond.) **91**, 201–208 (1996)
18. Weiss, J.N., Garfinkel, A., Spano, M.L., Ditto, W.L.: Chaos and chaos control in biology. J. Clin. Invest. **93**, 1355–1360 (1994)
19. Babloyantz, A., Destexhe, A.: Is the normal heart a periodic oscillator? Biol. Cybern. **58**, 203–211 (1988)
20. Woo, M.A., Stevenson, W.G., Moser, D.K., Trelease, R.B., Harper, R.M.: Patterns of beat-to-beat heart rate variability in advanced heart failure. Am. Heart J. **123**, 704–710 (1992)
21. Brennan, M., Palaniswami, M., Kamen, P.: Poincare plot interpretation using a physiological model of HRV based on a network of oscillators. Am. J. Physiol. Heart Circ. Physiol. **283**, H1873–H1886 (2002)
22. Kamen, P.W., Tonkin, A.M.: Application of the Poincare plot to heart rate variability: a new measure of functional status in heart failure. Aust. NZ J. Med. **25**, 18–26 (1995)
23. Laitio, T.T., Huikuri, H.V., Kentala, E.S., Makikallio, T.H., Jalonen, J.R., Helenius, H., et al.: Correlation properties and complexity of perioperative RR-interval dynamics in coronary artery bypass surgery patients. Anesthesiology. **93**, 69–80 (2000)
24. Kleiger, R.E., Stein, P.K., Bigger Jr., J.T.: Heart rate variability: measurement and clinical utility. Ann. Noninvasive Electrocardiol. **10**, 88–101 (2005)
25. Brennan, M., Palaniswami, M., Kamen, P.: Do existing measures of Poincare plot geometry reflect nonlinear features of heart rate variability? I.E.E.E. Trans. Biomed. Eng. **48**, 1342–1347 (2001)

26. Voss, A., Fischer, C., Schroeder, R., Figulla, H.R., Goernig, M.: Segmented Poincare plot analysis for risk stratification in patients with dilated cardiomyopathy. Methods Inf. Med. **49**, 511–515 (2010)
27. Voss, A., Fischer, C., Schroeder, R., Figulla, H.R., Goernig, M.: Lagged segmented Poincare plot analysis for risk stratification in patients with dilated cardiomyopathy. Med. Biol. Eng. Comput. **50**, 727–736 (2012)
28. Tulppo, M.P., Makikallio, T.H., Takala, T.E., Seppanen, T., Huikuri, H.V.: Quantitative beat-to-beat analysis of heart rate dynamics during exercise. Am. J. Phys. **271**, H244–H252 (1996)
29. Rajendra Acharya, U., Joseph, K.P., Kannathal, N., Lim, C.M., Suri, J.S.: Heart rate variability: a review. Med. Biol. Eng. Comput. **44**, 1031–1051 (2006)
30. Tulppo, M.P., Makikallio, T.H., Seppanen, T., Airaksinen, J.K., Huikuri, H.V.: Heart rate dynamics during accentuated sympathovagal interaction. Am. J. Phys. **274**, H810–H816 (1998)
31. Fischer, C., Seeck, A., Schroeder, R., Goernig, M., Schirdewan, A., Figulla, H.R., et al.: QT variability improves risk stratification in patients with dilated cardiomyopathy. Physiol. Meas. **36**, 699–713 (2015)
32. Stein, P.K., Reddy, A.: Non-linear heart rate variability and risk stratification in cardiovascular disease. Ind. Pacing Electrophysiol. J. **5**, 210–220 (2005)
33. Rydberg, A., Karlsson, M., Hornsten, R., Wiklund, U.: Can analysis of heart rate variability predict arrhythmia in children with Fontan circulation? Pediatr. Cardiol. **29**, 50–55 (2008)
34. Porta, A., Guzzetti, S., Montano, N., Furlan, R., Pagani, M., Malliani, A., et al.: Entropy, entropy rate, and pattern classification as tools to typify complexity in short heart period variability series. I.E.E.E. Trans. Biomed. Eng. **48**, 1282–1291 (Nov 2001)
35. Baumert, M., Voss, A., Javorka, M.: Compression based entropy estimation of heart rate variability on multiple time scales. Conf. Proc. IEEE Eng. Med. Biol. Soc. **2013**, 5037–5040 (2013)
36. Ziv, J., Lempel, A.: Universal algorithm for sequential data compression. IEEE Trans. Inf. Ther. **23**, 337–343 (1977)
37. Baumert, M., Baier, V., Haueisen, J., Wessel, N., Meyerfeldt, U., Schirdewan, A., et al.: Forecasting of life threatening arrhythmias using the compression entropy of heart rate. Methods Inf. Med. **43**, 202–206 (2004)
38. Van Leeuwen, P., Cysarz, D., Edelhauser, F., Gronemeyer, D.: Heart rate variability in the individual fetus. Auton. Neurosci. **178**, 24–28 (2013)
39. Boettger, M.K., Schulz, S., Berger, S., Tancer, M., Yeragani, V.K., Voss, A., et al.: Influence of age on linear and nonlinear measures of autonomic cardiovascular modulation. Ann. Noninvasive Electrocardiol. **15**, 165–174 (2010)
40. Voss, A., Schroeder, R., Heitmann, A., Peters, A., Perz, S.: Short-term heart rate variability–influence of gender and age in healthy subjects. PLoS One. **10**, e0118308 (2015)
41. Porta, A., Faes, L., Bari, V., Marchi, A., Bassani, T., Nollo, G., et al.: Effect of age on complexity and causality of the cardiovascular control: comparison between model-based and model-free approaches. PLoS One. **9**, e89463 (2014)
42. Javorka, M., Trunkvalterova, Z., Tonhajzerova, I., Javorkova, J., Javorka, K., Baumert, M.: Short-term heart rate complexity is reduced in patients with type 1 diabetes mellitus. Clin. Neurophysiol. **119**, 1071–1081 (2008)
43. Bär, K.J., Boettger, M.K., Koschke, M., Schulz, S., Chokka, P., Yeragani, V.K., et al.: Non-linear complexity measures of heart rate variability in acute schizophrenia. Clin. Neurophysiol. **118**, 2009–2015 (2007)
44. Schulz, S., Koschke, M., Bär, K.J., Voss, A.: The altered complexity of cardiovascular regulation in depressed patients. Physiol. Meas. **31**, 303–321 (2010)
45. Schulz, S., Ritter, J., Oertel, K., Witt, K., Bär, K.J., Guntinas-Lichius, O., et al.: Altered autonomic regulation as a cardiovascular risk marker for patients with sudden sensorineural hearing loss. Otol. Neurotol. **35**, 1720–1729 (2014)

46. Bari, V., Valencia, J.F., Vallverdu, M., Girardengo, G., Marchi, A., Bassani, T., et al.: Multiscale complexity analysis of the cardiac control identifies asymptomatic and symptomatic patients in long QT syndrome type 1. PLoS One. **9**, e93808 (2014)
47. Baumert, M., Javorka, M., Seeck, A., Faber, R., Sanders, P., Voss, A.: Multiscale entropy and detrended fluctuation analysis of QT interval and heart rate variability during normal pregnancy. Comput. Biol. Med. **42**, 347–352 (2012)
48. Bär, K.J., Koschke, M., Berger, S., Schulz, S., Tancer, M., Voss, A., et al.: Influence of olanzapine on QT variability and complexity measures of heart rate in patients with schizophrenia. J. Clin. Psychopharmacol. **28**, 694–698 (2008)
49. Porta, A., Guzzetti, S., Montano, N., Pagani, M., Somers, V., Malliani, A., et al.: Information domain analysis of cardiovascular variability signals: evaluation of regularity, synchronisation and co-ordination. Med. Biol. Eng. Comput. **38**, 180–188 (2000)
50. Bär, K.J., Schuhmacher, A., Hofels, S., Schulz, S., Voss, A., Yeragani, V.K., et al.: Reduced cardio-respiratory coupling after treatment with nortriptyline in contrast to S-citalopram. J. Affect. Disord. **127**, 266–273 (2010)
51. Ebert, A., Jochum, T., Ritter, J., Boettger, M.K., Schulz, S., Voss, A., et al.: Does parasympathetic modulation prior to ECT treatment influence therapeutic outcome? Prog. Neuro-Psychopharmacol. Biol. Psychiatry. **34**, 1174–1180 (2010)
52. Faes, L., Porta, A., Rossato, G., Adami, A., Tonon, D., Corica, A., et al.: Investigating the mechanisms of cardiovascular and cerebrovascular regulation in orthostatic syncope through an information decomposition strategy. Auton. Neurosci. **178**, 76–82 (2013)
53. Hadamard, J.: Les surfaces à courbures opposées et leurs lignes géodésiques. J. Math. Pures Appl. **Ser. V**, 27–73 (1898)
54. Morsem, M., Hedlund, G.A.: Symbolic dynamics. Amer. J. Math. **60**, 815–866 (1938)
55. Aizawa, Y.: Symbolic dynamics approach to intermittent chaos. Prog. Theor. Phys. **70**, 1249–1263 (1983)
56. Hao, B.L.: Elementary symbolic dynamics and chaos in dissipative systems. World Scientific Publishing, Singapore (1990)
57. Paulus, M.P., Geyer, M.A., Gold, L.H., Mandell, A.J.: Application of entropy measures derived from the ergodic theory of dynamical systems to rat locomotor behavior. Proc. Natl. Acad. Sci. U. S. A. **87**, 723–727 (1990)
58. Voss, A., Dietz, R., Fiehring, H., Kleiner, H.J., Kurths, J., Saparin, P., et al.: High resolution ECG, heart rate variability and nonlinear dynamics: tools for high risk stratification. Comput. Cardiol. 261–264 (1993)
59. Kurths, J., Voss, A., Saparin, P., Witt, A., Kleiner, H.J., Wessel, N.: Quantitative analysis of heart rate variability. Chaos. **5**, 88–94 (1995)
60. Voss, A., Hnatkova, K., Wessel, N., Kurths, J., Sander, A., Schirdewan, A., et al.: Multiparametric analysis of heart rate variability used for risk stratification among survivors of acute myocardial infarction. Pacing Clin. Electrophysiol. **21**, 186–192 (1998)
61. Voss, A., Schroeder, R., Truebner, S., Goernig, M., Figulla, H.R., Schirdewan, A.: Comparison of nonlinear methods symbolic dynamics, detrended fluctuation, and Poincare plot analysis in risk stratification in patients with dilated cardiomyopathy. Chaos. **17**, 015120 (2007)
62. Żebrowski, J.J., Poplawska, W., Baranowski, R., Buchner, T.: Symbolic dynamics and complexity in a physiological time series. Chaos, Solitons Fractals. **11**, 1061–1075. %@ 0960–0779 (2000)
63. Shannon, C.E.: A mathematical theory of communication. Bell Syst. Tech. J. **27**, 379–423 and 623–656 (1948)
64. Rényi, A.: On measures of entropy and information. Proc. Fourth Berkeley Symp. Math. Stat. Probability. **1**, 547–561 (1961)
65. Wessel, N., Brückner, C., Malberg, H., Schumann, A., Reinsperger, F., Osterziel, K., et al.: Long-term symbolic dynamics for heart rate variability analysis in patients with dilated cardiomyopathy. Cardiomyopathy. Computers in Cardiology. **26**, 253–256 (1999)

66. Guzzetti, S., Borroni, E., Garbelli, P.E., Ceriani, E., Della Bella, P., Montano, N., et al.: Symbolic dynamics of heart rate variability: a probe to investigate cardiac autonomic modulation. Circulation. **112**, 465–470 (2005)
67. Porta, A., Faes, L., Mase, M., D'Addio, G., Pinna, G.D., Maestri, R., et al.: An integrated approach based on uniform quantization for the evaluation of complexity of short-term heart period variability: application to 24 h Holter recordings in healthy and heart failure humans. Chaos. **17**, 015117 (2007)
68. Heitmann, A., Huebner, T., Schroeder, R., Perz, S., Voss, A.: Multivariate short-term heart rate variability: a pre-diagnostic tool for screening heart disease. Med. Biol. Eng. Comput. **49**, 41–50 (2011)
69. Voss, A., Schroeder, R., Caminal, P., Vallverdú, M., Brunel, H., Cygankiewicz, I., et al.: Segmented symbolic dynamics for risk stratification in patients with ischemic heart failure. Cardiovasc. Eng. Technol. **1**, 290–298 (2010)
70. Cysarz, D., Van Leeuwen, P., Edelhauser, F., Montano, N., Somers, V.K., Porta, A.: Symbolic transformations of heart rate variability preserve information about cardiac autonomic control. Physiol. Meas. **36**, 643–657 (2015)
71. Schulz, S., Adochiei, F.C., Edu, I.R., Schroeder, R., Costin, H., Bär, K.J., et al.: Cardiovascular and cardiorespiratory coupling analyses: a review. Philos. Trans. A Math. Phys. Eng. Sci. **371**, 20120191 (2013)
72. Baumert, M., Baier, V., Truebner, S., Schirdewan, A., Voss, A.: Short- and long-term joint symbolic dynamics of heart rate and blood pressure in dilated cardiomyopathy. I.E.E.E. Trans. Biomed. Eng. **52**, 2112–2115 (2005)
73. Schulz, S., Tupaika, N., Berger, S., Haueisen, J., Bär, K.J., Voss, A.: Cardiovascular coupling analysis with high-resolution joint symbolic dynamics in patients suffering from acute schizophrenia. Physiol. Meas. **34**, 883–901 (2013)
74. Kabir, M.M., Saint, D.A., Nalivaiko, E., Abbott, D., Voss, A., Baumert, M.: Quantification of cardiorespiratory interactions based on joint symbolic dynamics. Ann. Biomed. Eng. **39**, 2604–2614 (Oct 2011)
75. Baumert, M., Javorka, M., Kabir, M.: Joint symbolic dynamics for the assessment of cardiovascular and cardiorespiratory interactions. Philos. Trans. A Math. Phys. Eng. Sci. **373**, 20140097 (2015)
76. Schulz, S., Bär, K.J., Voss, A.: Analyses of heart rate, respiration and cardiorespiratory coupling in patients with schizophrenia. Entropy. **17**, 483–501 (2015)
77. Schulz, S., Haueisen, J., Bär, K.J., Voss, A.: Quantification of cardiorespiratory coupling in acute schizophrenia applying high resolution joint symbolic dynamics. In: Computing in Cardiology Conference (CinC), pp. 101–104. Zaragoza, Spain (2013)
78. Bertinieri, G., Di Rienzo, M., Cavallazzi, A., Ferrari, A.U., Pedotti, A., Mancia, G.: Evaluation of baroreceptor reflex by blood pressure monitoring in unanesthetized cats. Am. J. Phys. **254**, H377–H383 (1988)
79. Laude, D., Elghozi, J.L., Girard, A., Bellard, E., Bouhaddi, M., Castiglioni, P., et al.: Comparison of various techniques used to estimate spontaneous baroreflex sensitivity (the EuroBaVar study). Am. J. Phys. Regul. Integr. Comp. Phys. **286**, R226–R231 (2004)
80. Schulz, S., Adochiei, F.C., Edu, I.R., Schroeder, R., Costin, H., Bar, K.J., et al.: Cardiovascular and cardiorespiratory coupling analyses: a review. Philos. Trans. A Math. Phys. Eng. Sci. **371**, 20120191 (2013)
81. Porta, A., Marchi, A., Bari, V., Heusser, K., Tank, J., Jordan, J., et al.: Conditional symbolic analysis detects nonlinear influences of respiration on cardiovascular control in humans. Philos. Trans. A Math. Phys. Eng. Sci. **373**, 20140096 (2015)
82. Bari, V., Marchi, A., De Maria, B., Rossato, G., Nollo, G., Faes, L., et al.: Nonlinear effects of respiration on the crosstalk between cardiovascular and cerebrovascular control systems. Philos. Trans. A Math. Phys. Eng. Sci. **374**, 20150179 (2016)
83. Wessel, N., Suhrbier, A., Riedl, M., Marwan, N., Malberg, H., Bretthauer, G., et al.: Detection of time-delayed interactions in biosignals using symbolic coupling traces. EPL. **87**, 10004 (2009)

84. Suhrbier, A., Riedl, M., Malberg, H., Penzel, T., Bretthauer, G., Kurths, J., et al.: Cardiovascular regulation during sleep quantified by symbolic coupling traces. Chaos. **20**, 045124 (2010)
85. Schulz, S., Voss, A.: Cardiovascular and cardiorespiratory coupling analysis—State of the art and future perspectives, In: 2014 8th Conference of the European Study Group on Cardiovascular Oscillations (ESGCO), Trento, pp. 25–26. (2014).
86. Mourot, L., Bouhaddi, M., Perrey, S., Rouillon, J.D., Regnard, J.: Quantitative Poincare plot analysis of heart rate variability: effect of endurance training. Eur. J. Appl. Physiol. **91**, 79–87 (2004)
87. Stein, P.K., Domitrovich, P.P., Huikuri, H.V., Kleiger, R.E.: Traditional and nonlinear heart rate variability are each independently associated with mortality after myocardial infarction. J. Cardiovasc. Electrophysiol. **16**, 13–20 (2005)
88. T. Mäkikallio, Analysis of heart rate dynamics by methods derived from nonlinear mathematics: clinical applicability and prognostic significance: Oulun Yliopisto, 1998.
89. Voss, A., Kurths, J., Kleiner, H.J., Witt, A., Saparin, P., Dietz, R., et al.: New methods for the detection of high risk patients in cardiology. Biomed. Tech. (Berl). **39**, 134–143 (1994)
90. Seeck, A., Baumert, M., Fischer, C., Khandoker, A., Faber, R., Voss, A.: Advanced Poincare plot analysis differentiates between hypertensive pregnancy disorders. Physiol. Meas. **32**, 1611–1622 (2011)
91. Fischer, C., Voss, A.: Three-dimensional segmented Poincare plot analyses SPPA3 investigates cardiovascular and cardiorespiratory couplings in hypertensive pregnancy disorders. Front. Bioeng. Biotechnol. **2**, 51 (2014)
92. Karmakar, C.K., Khandoker, A.H., Gubbi, J., Palaniswami, M.: Complex correlation measure: a novel descriptor for Poincare plot. Biomed. Eng. Online. **8**, 17 (2009)
93. Lerma, C., Infante, O., Perez-Grovas, H., Jose, M.V.: Poincare plot indexes of heart rate variability capture dynamic adaptations after haemodialysis in chronic renal failure patients. Clin. Physiol. Funct. Imaging. **23**, 72–80 (2003)
94. Li, M., Vitányi, P.M.B.: An introduction to kolmogorov complexity and its applications. Springer Publishing Company, Inc., Heidelberg (1997)
95. Baumert, M., Baier, V., Voss, A., Brechtel, L., Haueisen, J.: Estimating the complexity of heart rate fluctuations–an approach based on compression entropy. Fluct. Noise Lett. **5**, L557–L563 (Dec 2005)
96. Truebner, S., Cygankiewicz, I., Schroeder, R., Baumert, M., Vallverdu, M., Caminal, P., et al.: Compression entropy contributes to risk stratification in patients with cardiomyopathy. Biomed. Tech. (Berl). **51**, 77–82 (2006)
97. Costa, M., Goldberger, A.L., Peng, C.K.: Multiscale entropy analysis of complex physiologic time series. Phys. Rev. Lett. **89**, 068102 (2002)
98. Schumann, A., Schulz, S., Voss, A., Scharbrodt, S., Baumert, M., Bär, K.-J.: Baroreflex coupling assessed by cross-compression entropy. Front. Physiol. **8**, 282 (2017)
99. Granger, C.W.J.: Investigating causal relations by econometric models and cross-spectral methods. Econometrica. **37**, 424–438 (1969)
100. Novak, V., Novak, P., de Champlain, J., Le Blanc, A.R., Martin, R., Nadeau, R.: Influence of respiration on heart rate and blood pressure fluctuations. J. Appl. Physiol. (1985). **74**, 617–626 (1993)
101. Schulz, S., Bär, K.J., Voss, A.: Respiratory variability and cardiorespiratory coupling analyses in patients suffering from schizophrenia and their healthy first-degree relatives. Biomed. Tech. (Berl). **57**(Suppl. 1), 1044 (2012)
102. Bär, K.J., Rachow, T., Schulz, S., Bassarab, K., Haufe, S., Berger, S., et al.: The phrenic component of acute schizophrenia–a name and its physiological reality. PLoS One. **7**, e33459 (2012)
103. Voss, A., Schulz, S., Schroeder, R., Baumert, M., Caminal, P.: Methods derived from nonlinear dynamics for analysing heart rate variability. Philos. Trans. A Math. Phys. Eng. Sci. **367**, 277–296 (2009)

104. Voss, A., Schulz, S., Schroder, R.: Analysis of cardiovascular oscillations using nonlinear dynamics methods for an enhanced diagnosis of heart and neurological diseases and for risk stratification. In: E-Health and Bioengineering Conference (EHB), vol. 2011, pp. 1–6 (2011)
105. Peupelmann, J., Boettger, M.K., Ruhland, C., Berger, S., Ramachandraiah, C.T., Yeragani, V.K., et al.: Cardio-respiratory coupling indicates suppression of vagal activity in acute schizophrenia. Schizophr. Res. **112**, 153–157 (2009)
106. Schulz, S., Bar, K.J., Voss, A.: Respiratory variability and cardiorespiratory coupling analyses in patients suffering from schizophrenia and their healthy first-degree relatives. Biomed. Tech. (Berl). **57**, 1044 (2012)
107. Homma, I., Masaoka, Y.: Breathing rhythms and emotions. Exp. Physiol. **93**, 1011–1021 (2008)
108. Williams, L.M., Das, P., Harris, A.W., Liddell, B.B., Brammer, M.J., Olivieri, G., et al.: Dysregulation of arousal and amygdala-prefrontal systems in paranoid schizophrenia. Am. J. Psychiatry. **161**, 480–489 (2004)
109. Boiten, F.A., Frijda, N.H., Wientjes, C.J.: Emotions and respiratory patterns: review and critical analysis. Int. J. Psychophysiol. **17**, 103–128 (1994)
110. Weiden, P.J., Weiden, M.: Schizophrenia and respiratory symptoms: a serious, but overlooked, comorbidity. CNS Spectr. **15**, 10–13 (2010)
111. Goodwin, R., Lyons, J.S., McNally, R.J.: Panic attacks in schizophrenia. Schizophr. Res. **58**, 213–220 (Dec 1 2002)
112. Buckley, P.F., Miller, B.J., Lehrer, D.S., Castle, D.J.: Psychiatric comorbidities and schizophrenia. Schizophr. Bull. **35**, 383–402 (2009)
113. Porta, A., Di Rienzo, M., Wessel, N., Kurths, J.: Addressing the complexity of cardiovascular regulation. Philos. Transact. A Math. Phys. Eng. Sci. **367**, 1215–1218 (2009)
114. Voss, A., Seeck, A., Israel, A.K., Bar, K.J.: Enhanced spectral analysis of blood flow during post-occlusive reactive hyperaemia test in different tissue depths. Auton. Neurosci. **178**, 15–23 (2013)
115. Voss, A., Goernig, M., Schroeder, R., Truebner, S., Schirdewan, A., Figulla, H.R.: Blood pressure variability as sign of autonomic imbalance in patients with idiopathic dilated cardiomyopathy. Pacing Clin. Electrophysiol. **35**, 471–479 (2012)
116. Glass, L.: Chaos and heart rate variability. J. Cardiovasc. Electrophysiol. **10**, 1358–1360 (1999)
117. Porta, A., Castiglioni, P., Di Rienzo, M., Bari, V., Bassani, T., Marchi, A., et al.: Short-term complexity indexes of heart period and systolic arterial pressure variabilities provide complementary information. J. Appl. Physiol. **113**, 1810–1820 (2012)
118. Porta, A., Castiglioni, P., Di Rienzo, M., Bassani, T., Bari, V., Faes, L., et al.: Cardiovascular control and time domain Granger causality: insights from selective autonomic blockade. Philos. Trans. A Math. Phys. Eng. Sci. **371**, 20120161 (2013)
119. Baccala, L.A., de Brito, C.S., Takahashi, D.Y., Sameshima, K.: Unified asymptotic theory for all partial directed coherence forms. Philos. Trans. A Math. Phys. Eng. Sci. **371**, 20120158 (2013)
120. Blinowska, K.J., Kaminski, M., Brzezicka, A., Kaminski, J.: Application of directed transfer function and network formalism for the assessment of functional connectivity in working memory task. Philos. Trans. A Math. Phys. Eng. Sci. **371**, 20110614 (2013)

Chapter 3
Information Decomposition: A Tool to Dissect Cardiovascular and Cardiorespiratory Complexity

Luca Faes, Giandomenico Nollo, and Alberto Porta

Abstract This chapter reports some recent developments of information-theoretic concepts applied to the description of coupled dynamical systems, which allow to decompose the entropy of an assigned target system into components reflecting the information stored in the system and the information transferred to it from the other systems, as well as the nature (synergistic or redundant) of the information transferred to the target. The decomposition leads to well-defined measures of information dynamics which in the chapter will be defined theoretically, computed in simulations of linear Gaussian systems and implemented in practice through the application to heart period, arterial pressure and respiratory time series. The application leads to decompose the information carried by heart rate variability into amounts reflecting cardiac dynamics, vascular and respiratory effects on these dynamics, as well as the interaction between cardiovascular and cardiorespiratory effects. The analysis of head-up and head-down tilt test protocols demonstrates the relevance of information decomposition in dissecting cardiovascular control mechanisms and accept or reject physiological hypotheses about their activity.

3.1 Introduction

The short-term autonomic regulation in humans is accomplished through the combined activity of several physiological systems, including the cardiac, vascular and pulmonary systems, which have their own internal regulation but also interact

L. Faes (✉) • G. Nollo
Bruno Kessler Foundation, Trento, Italy

BIOtech, Department of Industrial Engineering, University of Trento, Trento, Italy
e-mail: feas.luca@gmail.com

A. Porta
Department of Biomedical Sciences for Health, University of Milan, Milan, Italy

Department of Cardiothoracic, Vascular Anesthesia and Intensive Care, IRCCS Policlinico San Donato, Milan, Italy

R. Barbieri et al. (eds.), *Complexity and Nonlinearity in Cardiovascular Signals*,
DOI 10.1007/978-3-319-58709-7_3

with each other in order to preserve the homeostatic function [1]. The effects of this complex regulation are manifested in the time course of the output physiological variables of the various organ systems. For instance, it is well known that the beat-to-beat variability of the heart period (HP) results from the activity of important physiological regulatory mechanisms, including the cardiovascular effects from systolic arterial pressure (SAP) to HP manifested through the so-called baroreflex, the cardiorespiratory effects from respiration (RESP) to HP known as respiratory sinus arrhythmia, as well as central neuroautonomic commands that affect the heart rate dynamics independently of the cardiac baroreflex and cardiopulmonary pathways [2]. An important but poorly addressed aspect of physiological regulation is the fact that all these mechanisms do not operate independently of each other, but rather constitute intertwined elements of the overall control of the heart rate. A clear example of the interrelation among cardiovascular and cardiorespiratory mechanisms is the coexistence of the central respiratory drive modulating cardiac vagal motor neuron responsiveness (effect RESP→HP) with the mechanical vasculo-pulmonary influences transmitted to the sinus node through the cardiac baroreflex (effect RESP→SAP→HP) [3]. These mechanisms are often simultaneously active together with the cardiac baroreflex mediating the effects of SAP on HP unrelated to respiration, and a variety of mechanisms modulating HP dynamics independently of RESP and SAP, such as the central commands independent of afferent inputs which originate from respiratory and vasomotor centers in the brainstem [4].

In order to dissect these physiological mechanisms and clarify their relative contribution to the complexity of the cardiac dynamics, there is the need of devising methods able to quantify the predictable portion of these dynamics, to investigate how this portion arises from internal HP dynamics and from causal interactions driven by SAP and RESP, and to explore how collective causal interactions (e.g., {RF,SAP}→HP) arise from different effects (e.g., SAP→HP, RF→HP, RF→SAP→HP). In the present work, this issue is approached in the novel framework of information dynamics [5, 6], considering three nested information decomposition strategies fully dissecting the statistical dependencies within a network of dynamical systems composed of a target system and two connected source systems. The three decompositions allow to separate: (i) the *predictive information* about the target system into amounts quantifying the *information storage* in the system and the *joint information transfer* from the source systems [7]; (ii) the information storage into the *internal information* ascribed exclusively to the target dynamics and three *interaction storage* terms accounting for the modification of the information shared between the target dynamics and the sources [8, 9]; and (iii) the joint information transfer into amounts of information transferred individually from each source when the other is assigned (*conditional information transfer*) and a term accounting for the modification of the information transferred due to cooperation between the sources (*interaction information transfer*) [10]. With this approach, we first define several entropy-based measures of information dynamics, stating their properties and reciprocal relations. Then, we provide a data-efficient approach for the computation of these measures, which yields their exact values in the case of stationary Gaussian systems. Even though these hypotheses might appear to be quite restrictive at the first sight, they are largely exploited in multivariate modeling of cardiovascular variability interactions [2, 11, 12], and have often been found to be satisfied in well-controlled experimental protocols for which computations based on Gaussianity have yielded comparable results than model-free nonlinear approaches [8, 13]. Here we exploit the theoretical framework and the proposed computation approach, applied to short term HP, SAP and RESP time series measured from healthy subjects, in order to elucidate the mechanisms of cardiac, cardiovascular and cardiorespiratory regulation in the resting supine position, and to assess the reciprocal modification of these mechanisms induced by two stimulations known to evoke adjustments in the cardiovascular control related to opposite effects on central circulatory volemia, i.e. head-up tilt (HUT) [9, 14] and head-down tilt (HDT) [10].

3.2 Basic Concepts of Probability and Information Theory

We start recalling some basic concepts of probability and information theory [15], which will be used respectively in Sect. 3.3.1 to provide a mathematical framework for the description of physiological networks, and in Sect. 3.3.2 to define the measures of information dynamics descriptive of the dynamical structure of these networks.

A *random variable* is a mathematical variable whose value is subject to variations due to chance. In this work we will deal with continuous random variables, i.e. variables that can take values inside an infinite-dimensional set usually denoted as the domain. The generic scalar random variable V with domain D_V is characterized by its distribution function, intended as a function which assigns a probability to each measurable subset of D_V. Formally, the probability for the variable V of taking values within the interval $[a,b]\subseteq D_V$ is determined by the integral $\Pr\{a \leq V \leq b\} = \int_a^b f_V(v)dv = F_V(b) - F_V(a)$, where f_V is the probability density function of the variable and F_V is its cumulative distribution function. The cumulative distribution quantifies the probability that the variable V has v as its upper bound, $F_V(v) = \Pr\{V \leq v\}$, while the probability density is mathematically defined as the derivative of the cumulative distribution, in a way such that $F_V(v) = \int_{-\infty}^{v} f_V(u)du$. These definitions extend in a straightforward way to the generic k-dimensional variable $\mathbf{W} = [W_1, \ldots, W_k]$ by defining the joint probability density $f_{\mathbf{W}}(w_1, \ldots, w_k)$ and performing multiple integration over the domain of each scalar component to get the cumulative distribution. Moreover, the conditional probability density of V given $\mathbf{W}$ expresses the probability of observing the value v for V given that the values $\mathbf{w} = w_1, \ldots, w_k$ have been observed for $W_1, \ldots, W_k$: $f_{V|\mathbf{W}}(v|\mathbf{w}) = f_{V,\mathbf{W}}(v, w_1, \ldots, w_k) / f_{\mathbf{W}}(w_1, \ldots, w_k)$.

Probability distributions are represented in a convenient way in the framework of information theory. The central quantity in this framework is the *entropy* of a random variable, intended as a measure of the uncertainty about the outcomes of this variable. Specifically, the (differential) entropy of the continuous variable V measures the average amount of information contained in V or, equivalently, the average amount of uncertainty about V that is resolved upon revealing its outcomes $v \in D_V$:

$$H(V) = -\int_{D_V} f_V(v)\log f_V(v)dv. \tag{3.1}$$

The definition in Eq. (3.1) extends easily to a generic k-dimensional variable $\boldsymbol{W}$ by performing multiple integration over the domain of each scalar component. Then, the *conditional entropy* of V given $\boldsymbol{W}$ is defined using the conditional density function as follows

$$H(V|\boldsymbol{W}) = -\iint_{D_V, D_W} f_{V,\mathbf{W}}(v,\boldsymbol{w})\log f_{V|\mathbf{W}}(v|\boldsymbol{w})\,dvd\boldsymbol{w}, \tag{3.2}$$

measuring the residual information contained in V when $\boldsymbol{W}$ is assigned, intended as the average uncertainty that remains about the outcomes of V when the outcomes of $\boldsymbol{W}$ are known.

Entropy and conditional entropy are related to each other by the equation $H(V)=H(V|\boldsymbol{W})+I(V;\boldsymbol{W})$, where $I(V;\boldsymbol{W})$ is the *mutual information* (MI) between V and $\boldsymbol{W}$ defined as

$$I(V;\boldsymbol{W}) = \iint_{D_V, D_W} f_{V,\mathbf{W}}(v,\boldsymbol{w})\log\frac{f_{V|\mathbf{W}}(v|\boldsymbol{w})}{f_V(v)}dvd\boldsymbol{w}, \tag{3.3}$$

measuring the amount of information shared between V and $\boldsymbol{W}$ intended as the average reduction in uncertainty about the outcomes of V obtained when the outcomes of $\boldsymbol{W}$ are known. Moreover, the *conditional mutual information* (CMI) between V and $\boldsymbol{W}$ given a third variable $\boldsymbol{U}$, $I(V;\boldsymbol{W}|\boldsymbol{U})$, quantifies the information shared between V and $\boldsymbol{W}$ which is not shared with $\boldsymbol{U}$, intended as the average reduction in uncertainty about the outcomes of V provided by the knowledge of the outcomes of $\boldsymbol{W}$ that is not explained by the outcomes of $\boldsymbol{U}$:

$$\begin{aligned} I(V;\boldsymbol{W}|\boldsymbol{U}) &= H(V|\boldsymbol{U}) - H(V|\boldsymbol{W},\boldsymbol{U}) \\ &= I(V;\boldsymbol{U},\boldsymbol{W}) - I(V;\boldsymbol{U}) \end{aligned}, \tag{3.4}$$

where each conditional entropy or mutual information term is computed according to (3.2) and (3.3), respectively.

Another interesting information-theoretic quantity is the *interaction information*, which is a measure of the amount of information that a target variable V shares with two source variables $\boldsymbol{W}$ and $\boldsymbol{U}$ when they are taken individually but not when they are taken together:

$$I(V;\boldsymbol{W};\boldsymbol{U}) = I(V;\boldsymbol{U}) + I(V;\boldsymbol{W}) - I(V;\boldsymbol{U},\boldsymbol{W}). \tag{3.5}$$

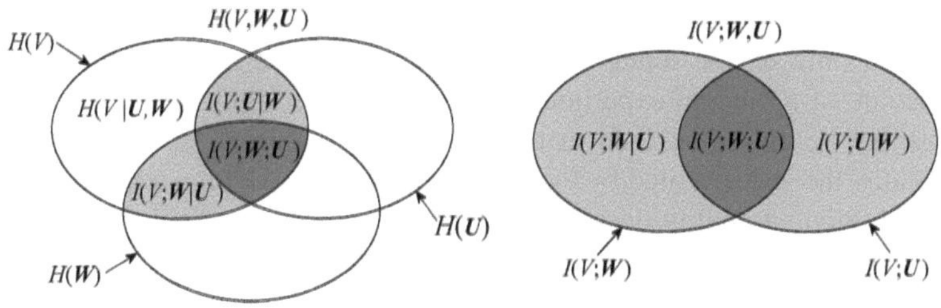

Fig. 3.1 Entropy diagram (*left*) and mutual information diagram (*right*) depicting the relations between the basic information-theoretic measures defined for three random variables V,$\boldsymbol{W}$,$\boldsymbol{U}$: the entropy $H(.)$, the conditional entropy $H(.|.)$, the mutual information $I(.|.)$, the conditional mutual information $I(.;.|.)$, and the interaction information $I(.;.;.)$

Alternatively, the interaction information can be intended as the amount of information bound up in the set of variables $\{V,\boldsymbol{W},\boldsymbol{U}\}$ beyond that which is present in the individual subsets $\{V,\boldsymbol{W}\}$ and $\{V,\boldsymbol{U}\}$. Contrary to all other information measures defined in (3.1, 3.2, 3.3, and 3.4) which are never negative, the interaction information defined in (3.5) can take on both positive and negative values, with positive values indicating *redundancy* (i.e., $I(V;\boldsymbol{U},\boldsymbol{W})<I(V;\boldsymbol{U})+I(V;\boldsymbol{W})$) and negative values indicating *synergy* (i.e., $I(V;\boldsymbol{U},\boldsymbol{W})>I(V;\boldsymbol{U})+I(V;\boldsymbol{W})$) between the two sources $\boldsymbol{U}$ and $\boldsymbol{W}$ that share information with the target V.

Mnemonic Venn diagrams of the information measures recalled above, showing how these measures quantify the amounts of information contained in a set of variables and shared between variables, are shown in Fig. 3.1. The several rules that relate the different measures with each other can be inferred from the figure; for instance, the chain rule for entropy decompose the information contained in the target variable V as $H(V)=I(V;\boldsymbol{U},\boldsymbol{W})+H(V|\boldsymbol{U},\boldsymbol{W})$, the chain rule for mutual information decomposes the information shared between the target V and the two sources $\boldsymbol{W}$ and $\boldsymbol{U}$ as $I(V;\boldsymbol{U},\boldsymbol{W})=I(V;\boldsymbol{U})+I(V;\boldsymbol{W}|\boldsymbol{U})=I(V;\boldsymbol{W})+I(V;\boldsymbol{U}|\boldsymbol{W})$, and the interaction information between V, $\boldsymbol{W}$ and $\boldsymbol{U}$ results as $I(V;\boldsymbol{U};\boldsymbol{W})=I(V;\boldsymbol{U})-I(V;\boldsymbol{U}|\boldsymbol{W})=I(V;\boldsymbol{W})-I(V;\boldsymbol{W}|\boldsymbol{U})$.

Note that, since the natural logarithm is commonly used in the computation of entropy and mutual information for continuous variables, all the measures defined above are expressed in natural units, or *nats*.

3.3 Information Decomposition in Networks of Interacting Physiological Dynamic Systems

3.3.1 Dynamic Systems, Stochastic Processes and Their Probabilistic Description

In this study, we consider the heart, the respiratory system and the vascular system as dynamical systems that assume diverse states at different instants of time. These systems are also assumed to be stochastic, meaning that their current state does not depend only on the inputs to the system and on its initial state but also on the outcome of a random experiment. Therefore, the evolution over time of these systems can be only described in probabilistic terms using stochastic processes. Specifically, the states visited by the cardiac system X over time are represented by the heart period variability described as a stochastic process $X=\{X_n\}$, $n=,1,2,\ldots,$ where the random variable X_n describes the n-th heart period, i.e. the state assumed by X at the n-th time step. Using the same notation, we denote the stochastic processes descriptive of respiration and systolic arterial pressure variability as $Y=\{Y_n\}$ and $Z=\{Z_n\}$, $n=,1,2,\ldots,$ which represent the dynamic states visited by the respiratory and vascular systems, respectively. The three systems taken together constitute the physiological network **S**={X,Y,Z}, whose activity is described by the vector stochastic process $\boldsymbol{S}=\{\boldsymbol{S}_n\}$, where the vector random variable $\boldsymbol{S}_n=[X_n\ Y_n\ Z_n]^{\mathrm{T}}$ quantifies in probabilistic terms the state of the overall observed system at the n-th time step. Then, a realization of the stochastic process X is the time series $x=[x_1\ x_2 \ldots x_N]$, containing the values of heart period collected over N consecutive heartbeats. The same holds for the sequences of the consecutively measured SAP values and respiration samples, indicated respectively with the time series $y=[y_1\ y_2 \ldots y_N]$, and $z=[z_1\ z_2 \ldots z_N]$. The three synchronous sequences are collected in the multivariate time series $\boldsymbol{s}=[\boldsymbol{s}_1\boldsymbol{s}_2 \ldots \boldsymbol{s}_N]$, $\boldsymbol{s}_n=[x_n\ y_n\ z_n]^{\mathrm{T}}$, $n=1,\ldots,N$, which takes the form of an $M\times N$ data matrix representing a realization of the vector process $\boldsymbol{S}$ ($M=3$ in this case). An overall description of the physiological network and related processes and time series is presented in Fig. 3.2.

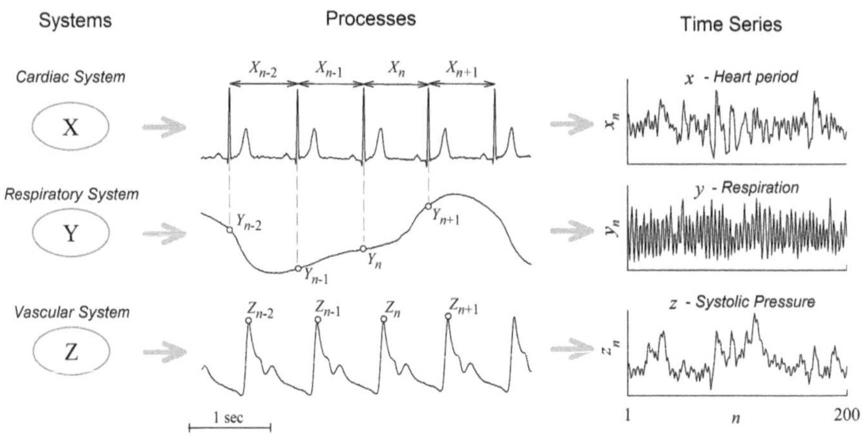

Fig. 3.2 Representation of a physiological network based on the theory of dynamical systems. The cardiac, respiratory and vascular systems are seen as the dynamical systems X, Y and Z, which visit different states at each heartbeat. The evolution of the network across time is quantified by the stochastic processes X, Y and Z: the state of the network at time n is described by the random variables of the present heart period X_n, respiratory amplitude Y_n, and systolic arterial pressure Z_n. A realization of these processes is the multivariate time series $\{x,y,z\}$

When considered individually, each stochastic process describing the dynamic activity of a node of the observed physiological network is characterized by a family of probability density functions, or equivalently by the corresponding family of cumulative distribution functions. Accordingly, the temporal statistical structure of the process X is described by the probability density functions $f_{X_n}(x_n)$ or by the cumulative distribution functions $F_{X_n}(x_n)$, $n =, 1, 2, \ldots, N$. The probability density is then defined in a straightforward way for the other processes, $f_{Y_n}(y_n)$, $f_{Z_n}(z_n)$, for joint variables taken from the same process, $f_{X_{n1} \cdots X_{nk}}(x_{n1}, \ldots, x_{nk})$, and for sets of vector variables taken from the overall vector process $\boldsymbol{S}=\{X\ Y\ Z\}$, $f_{S_{n1} \cdots S_{nk}}(\boldsymbol{s}_{n1}, \ldots, \boldsymbol{s}_{nk})$, $\forall n_1, \ldots, n_k, \forall k \geq 1$.

Setting a temporal reference frame in which n represents the present time, we denote as X_n the random variable describing the present of X, and as $X_n^-=[X_1\ X_2 \cdots X_{n-2}\ X_{n-1}]^T$ the (n-1)-dimensional vector variable describing the past of X. The same notation applies for Y and Z, and extends to $\boldsymbol{S}_n^- = \left[\boldsymbol{S}_1^T\ \boldsymbol{S}_2^T \cdots \boldsymbol{S}_{n-2}^T\ \boldsymbol{S}_{n-1}^T\right]^T$ to denote the 3(n-1)-dimensional vector variable describing the past of $\boldsymbol{S}$. In general, the operation of separating the present from the past allows to consider the flow of time and to study the causal interactions within and between processes by looking at the statistical dependencies among these variables. In fact, the dynamic properties of a system are studied in the information domain introducing the concept of *transition probability*, which is the probability associated with the transition of the system from its past states to its present state. Thus, the state transition of the network S relevant to the system X is described by the conditional probability density $f_{X_n|S_n^-}\left(x_n|\boldsymbol{s}_n^-\right)$, which reduces to any restricted conditional density in particular cases (e.g., when only the individual dynamics of the process X are of interest, the conditional density $f_{X_n|X_n^-}\left(x_n|x_n^-\right)$ is considered).

A useful property of stochastic processes is stationarity, which defines the time-invariance of any joint probability density taken from the process, i.e., $f_{S_{n1} \ldots S_{nk}}(\boldsymbol{S}_{n1}, \ldots, \boldsymbol{S}_{nk}) = f_{S_{n1} \ldots S_{nk+m}}(\boldsymbol{S}_{n1}, \ldots, \boldsymbol{S}_{nk})\ \forall n_1, \ldots, n_k, \forall m, k \geq 1$. For stationary processes, we use the shorthand notation $f_S(\boldsymbol{S}_{n1}, \ldots, \boldsymbol{S}_{nk})$ for the probability density. When the process is stationary, the fact that the probability density is the same at all times allows to pool together the observations measured across time order to estimate the densities, thus enabling the estimation of probabilities from individual realizations. For a stationary stochastic process, also the transition probabilities are time-independent, i.e. $f_{X_n|S_n^-}\left(x_n|\boldsymbol{S}_n^-\right) = f_{X_{n+m}|S_{n+m}^-}\left(x_{n+m}|\boldsymbol{S}_{n+m}^-\right) \equiv f_S\left(x_n|\boldsymbol{S}_n^-\right)\ \forall n, m \geq 1$. An important class of dynamic processes is that of *Markov processes*, which are processes for which the present depends on the past only through a finite number of time steps. Specifically, the process X embedded in the network $\boldsymbol{S}$ is a Markov process of order m if its transition probability function satisfies the condition $f_S(x_n|\,\boldsymbol{s}_n^-)=f_S(x_n|\boldsymbol{s}_{n-m}, \ldots, \boldsymbol{s}_{n-1})$. The Markov property allows to investigate the state transitions relevant to the considered target process by restricting the analysis to the consideration of the past m states visited by the network. Stationarity constitutes the basis for the theoretical computation of information-theoretic quantities in the simulations of Sect. 3.4, and of their estimation on cardiovascular time series described in Sect. 3.5.

3.3.2 Information Decomposition in Multivariate Stochastic Processes

In the following we present a framework to dissect the information contained in the target system of a network composed by several interacting dynamic systems into amounts reflecting basic elements of information processing, like the new information generated at each time step, the information stored in the system, the information transferred to it from the other connected systems, and the redundant or synergistic modification of the information transferred from two source systems towards the target. Here, 'information' and its constituent elements are quantified through the utilization of well-known standard information theoretic measures like those defined in Sect. 3.2, applied by taking as arguments for these measures proper combinations of the past and present states of the considered stochastic processes.

Let us consider a 'target' system in our network, say X, and interpret the remaining systems Y and Z as 'source' systems. The 'information' contained in the target X varies at each time step in a way such that, when the network transits from past states to the present state, new information is produced about X in addition to the information which is carried from the past states of the whole network **S**={X,Y,Z}. In the framework of information dynamics [5, 8], this process is described by defining entropy-based functionals which characterize the dynamical structure of the target process X by relating the present state X_n to the past states of the overall process $\boldsymbol{S}_n^-$. Here, assuming that $\boldsymbol{S}$ is a Markov process of order m, the whole past $\boldsymbol{S}_n^-$ is truncated at lag m, i.e., $\boldsymbol{S}_n^- \cong \boldsymbol{S}_n^m = [S_{n-1}^T \cdots S_{n-m}^T]^{\mathrm{T}}$ (the same holds for the past of the individual processes: $X_n^m = [X_{n\text{-}1} \cdots X_{n\text{-}m}]^{\mathrm{T}}$, $Y_n^m = [Y_{n\text{-}1} \cdots Y_{n\text{-}m}]^{\mathrm{T}}$, $Z_n^m = [Z_{n\text{-}1} \cdots Z_{n\text{-}m}]^{\mathrm{T}}$). The amount of information contained in the target process X at time n is defined as the entropy of the variable X_n:

$$H_X = H(X_n), \tag{3.6}$$

where, under the assumption of stationarity of the process X, dependence on the time index n is omitted in the formulation of the entropy H_X. Then, exploiting the chain rule for entropy, the target information can be decomposed as

$$H_X = P_X + N_X = I\left(X_n; \boldsymbol{S}_n^m\right) + H\left(X_n | \boldsymbol{S}_n^m\right) \tag{3.7}$$

where $P_X = I\left(X_n; \boldsymbol{S}_n^m\right)$ is the *prediction entropy* of the target X, which quantifies the predictive information of X as the information shared between the present X_n and the past of the whole network $\boldsymbol{S}_n^m$, and $N_X = H\left(X_n | \boldsymbol{S}_n^m\right)$ is the newly generated information that appears in the target process X after the transition from the past states to the present state, measured as the conditional entropy of the present of X given the past of the whole network $\boldsymbol{S}_n^m$. The decomposition in Eq. (3.7) evidences how the information carried by the target of a network of interacting processes can be dissected into an amount that can be predicted from the past states of the network, which is thus related to the concept of information stored in the target node the network, and an amount that is not predictable from the history of any observed system, which is thus related to the concept of new information produced at the target node.

The predictive information quantifies how much the uncertainty about the current state of the target system is reduced by the knowledge of the past states visited by the overall multivariate system. To understand the contribution of the different parts of the multivariate system to this reduction in uncertainty, the predictive information can be decomposed into amounts related to the concepts of *information storage* and *information transfer*. Specifically, the prediction entropy of the target process X can be expanded as

$$P_X = S_X + T_{YZ \to X} = I\left(X_n; X_n^m\right) + I\left(X_n; Y_n^m \oplus Z_n^m | X_n^m\right), \tag{3.8}$$

where $\oplus$ denotes vector concatenation $\left(Y_n^m \oplus Z_n^m = \left[Y_n^{m \cdot T} Z_n^{m \cdot T}\right]^T\right)$. In Eq. (3.8), $S_X = I\left(X_n; X_n^m\right)$ is the *self entropy* of X, which quantifies the information storage in the target as the amount of information shared between the present of the target X_n and its past X_n^m, and $T_{YZ \to X} = I\left(X_n; Y_n^m \oplus Z_n^m | X_n^m\right)$ is the *joint transfer entropy* from (Y, Z) to X, which quantifies the information jointly transferred from the two sources Y and Z to the target X as the amount of information contained in the past of the sources $Y_n^m \oplus Z_n^m$ that can be used to predict the present of the target X_n above and beyond the information contained in the past of the target X_n^m. This second decomposition resulting from Eq. (3.8) is useful to dissect the whole information that is contained in the past of the observed network and is available to predict the future states of the target into a part that is specifically stored in the target itself, and another part that is exclusively transferred to the target from the sources.

In turn, the information storage can be further expanded into further levels of information decomposition that evidence how the past of the various processes interact with each other in determining the information stored in the target. In particular, the self entropy of X is expanded as

$$S_X = S_{X|Y,Z} + I_{X;Y,Z}^X = I\left(X_n; X_n^m | Y_n^m \oplus Z_n^m\right) + I\left(X_n; X_n^m; Y_n^m \oplus Z_n^m\right), \tag{3.9}$$

where $S_{X|Y,Z} = I\left(X_n; X_n^m | Y_n^m \oplus Z_n^m\right)$ is the *conditional self entropy* of X given Y and Z, which quantifies the so-called internal information in the target intended as the amount of information contained in the past of the target X_n^m that can be used to predict the present X_n above and beyond the information contained in the past of the sources $Y_n^m \oplus Z_n^m$, and $I_{X;Y,Z}^X = I\left(X_n; X_n^m; Y_n^m \oplus Z_n^m\right)$ is the *interaction self entropy* of X in the context of the network process $\{X,Y,Z\}$, which quantifies the so-called interaction information storage measured as the interaction information of the present of the target X_n, its past X_n^m, and the past of the sources $Y_n^m \oplus Z_n^m$. In turn, the interaction self entropy can be expanded as

$$I_{X;Y,Z}^X = I_{X;Y}^X + I_{X;Z}^X - I_{X;Y;Z}^X = I_{X;Y|Z}^X + I_{X;Z|Y}^X + I_{X;Y;Z}^X, \tag{3.10}$$

where $I^X_{X;Y} = I\left(X_n; X_n^m; Y_n^m\right)$ and $I^X_{X;Z} = I\left(X_n; X_n^m; Z_n^m\right)$ are the interaction self entropies of X in the context of the bivariate processes $\{X,Y\}$ and $\{X,Z\}$, $I^X_{X;Y|Z} = I\left(X_n; X_n^m; Y_n^m|Z_n^m\right)$ and $I^X_{X;Z|Y} = I\left(X_n; X_n^m; Z_n^m|Y_n^m\right)$ are the conditional interaction self entropies of X in the context of the full processes $\{X,Y,Z\}$, and $I^X_{X;Y;Z} = I\left(X_n; X_n^m; Y_n^m; Z_n^m\right)$ is the *multivariate interaction entropy* between X, Y and Z in the context of the network process $\{X,Y,Z\}$, which quantifies the so-called multivariate interaction information of the present of the target X_n, its past X_n^m, the past of one source Y_n^m, and the past of the other source Z_n^m. This last expansion puts in evidence basic atoms of information about the target, which quantify respectively the interaction information of the present and the past of the target with one of the two sources taken individually, and the interaction information of the present of the target and the past of all processes. This last term expresses the information contained in the union of the four variables (X_n, X_n^m, Y_n^m, Z_n^m), but not in any subset of these four variables.

Finally, as a last decomposition of the information transfer, we show how the joint transfer entropy can be expanded to evidence how the past of the sources interact with each other in determining the information transferred to the target. Specifically, the joint transfer entropy from (Y,Z) to X can be expanded as

$$T_{YZ\to X} = T_{Y\to X} + T_{Z\to X} - I^X_{Y;Z|X} = T_{Y\to X|Z} + T_{Z\to X|Y} + I^X_{Y;Z|X}, \tag{3.11}$$

where $T_{Y\to X} = I\left(X_n; Y_n^m|X_n^m\right)$ and $T_{Z\to X} = I\left(X_n; Z_n^m|X_n^m\right)$ are the transfer entropies from each individual source to the target in the context of the bivariate processes $\{X,Y\}$ and $\{X,Z\}$, $T_{Y\to X|Z} = I\left(X_n; Y_n^m|X_n^m \oplus Z_n^m\right)$ and $T_{Z\to X|Y} = I\left(X_n; Z_n^m|X_n^m \oplus Y_n^m\right)$ are *conditional transfer entropies* quantifying the information transfer from one source to the target conditioned to the other source in the context of the whole network process $\{X,Y,Z\}$, and $I^X_{Y;Z|X} = I\left(X_n; Y_n^m; Z_n^m|X_n^m\right) = I\left(X_n; Y_n^m|X_n^m\right) - I\left(X_n; Y_n^m|X_n^m \oplus Z_n^m\right)$ is the *interaction transfer entropy* between Y and Z to X in the context of the network process $\{X,Y,Z\}$, which quantifies the so-called interaction information transfer measured as the interaction information of the present of the target X_n, and the past of the two sources Y_n^m and Z_n^m, conditioned to the past of the target X_n^m.

The Venn diagram depicting all the decompositions defined above is shown in Fig. 3.3. The diagram evidences how the information contained in the target process X at any time step (all non-white areas) splits in a part that can be explained from the past of the whole network (predictive information, coloured areas) and in a part which is not explained by the past (new information, gray). The predictable part is the sum of a portion explained only by the target (information storage, red shades) and a portion explained by the sources (information transfer, blue shades). In turn, the information storage is in part due exclusively to the target dynamics (internal information, $S_{X|Y,Z}$) and in part to the interaction of the dynamics of the target and the two sources (interaction information storage, $I^X_{X;Y;Z}$, which is the sum of the interaction storage of the source and each target plus the multivariate interaction information). Similarly, the information transfer

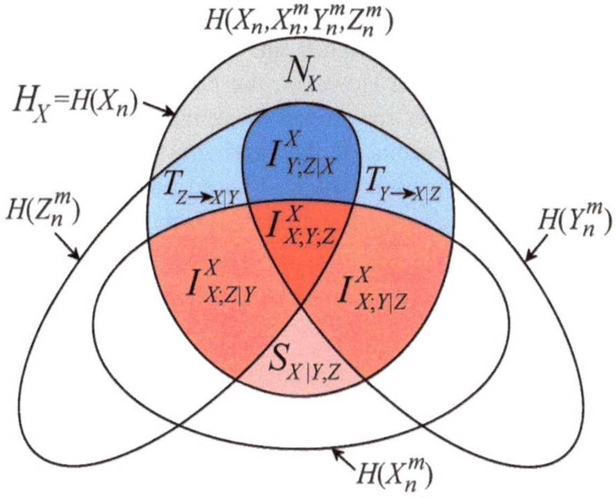

Fig. 3.3 Graphical representation of the information theoretic quantities resulting from the decomposition of the information carried by the target X of a network of interacting stationary processes $S=\{X,Y,Z\}$. The entropy decomposition dissects the entropy of the process (H_X) as the sum of the new information (N_X, gray) and the predictive information (P_X, *colored areas*). The predictive information decomposition dissects P_X as the sum of the information storage (S_X, *red shades*) and the information transfer ($T_{ZY\rightarrow X}$, *blue shades*). The information storage decomposition dissects S_X as the sum of the internal information ($S_{X|YZ}$), conditional interaction terms ($I^X_{X;Z|Y}$ and $I^X_{X;Y|Z}$) and multivariate interaction ($I^X_{X;Y;Z}$). The information transfer decomposition dissects $T_{ZY\rightarrow X}$ as the sum of conditional information transfer terms ($T_{Z\rightarrow X|Y}$ and $T_{Y\rightarrow X|Z}$) and interaction information transfer ($I^X_{Y;Z|X}$). Figure redrawn from [10]

can be ascribed to an individual source when the other is assigned (conditional information transfer, $T_{Y\rightarrow X|Z}$, $T_{Z\rightarrow X|Y}$) or to the interaction between the two sources (interaction information transfer, $I^X_{Y;Z|X}$), Note that all interaction terms can take on either positive values, reflecting redundant cooperation between the past states of the processes involved in the measures while they are used to predict the present of the target, or negative values, reflecting synergetic cooperation; since the interaction terms reflect how the interaction between source variables may lead to the elimination of information in the case of redundancy or to the creation of new information in the case of synergy, they quantify the concept of *information modification*. This concept and those of information storage and information transfer constitute the basic elements to dissect the more general notion of information processing in networks of interacting dynamical systems [6].

3.4 Computation of Information Dynamics

The practical computation of the measures appearing in the information decompositions presented in Sect. 3.3 presupposes to provide estimates of the entropy, MI and conditional MI for the vector variables enclosing the present and the past of the observed processes. In the most general case, and when nonlinear effects are relevant, non-parametric approaches are recommended to yield model-free estimates of entropy and MI [7]. However, the necessity to estimate entropies of variables of very high dimension may impair the reliability of model-free estimators, especially when short realizations of the processes are available. In this study we present an estimation approach which is based on assuming a joint Gaussian probability density for the observed stationary Markov processes and exploiting exact derivations that hold in this case [8].

3.4.1 Linear Gaussian Computation

In this section we provide a derivation of the exact values of information dynamics under the assumption that the observed dynamical network **S**={X,Y,Z} is composed by Gaussian systems. Specifically, we assume that the overall vector process $\boldsymbol{S}$=[X Y Z]^{T} has a joint Gaussian distribution, which means that any vector variable $\boldsymbol{S}_{nk} = \left[\boldsymbol{s}_{n_1}^T \ldots \boldsymbol{s}_{n_k}{}^T\right]^T$ obtained sampling the process at times $n_1,\ldots,n_k$, $\forall n_1,\ldots,n_k,\forall k\geq 1$, has the known probability density

$$f_S\left(\boldsymbol{S}_{nk}\right) = \frac{1}{\sqrt{(2\pi)^d\left|\boldsymbol{\Sigma}_{nk}\right|}} e^{-\frac{1}{2}\boldsymbol{S}_{nk}^{\mathrm{T}}\boldsymbol{\Sigma}_{nk}^{-1}\boldsymbol{S}_{nk}}, \tag{3.12}$$

where d is the dimension of $\boldsymbol{S}_{nk}$, $\boldsymbol{\Sigma}_{nk} = \mathrm{E}\left[\boldsymbol{S}_{nk}\boldsymbol{S}_{nk}^{\mathrm{T}}\right]$ is the covariance matrix of $\boldsymbol{S}_{nk}$ and $|\bullet|$ is the matrix determinant. Particularizing the density (3.12) to the variables representing the present of the target process and the past of the target and source processes in the network, and exploiting the basic information-theoretic relations (3.1, 3.2, and 3.3), the measures of information dynamics for joint Gaussian processes can be obtained as follows.

The entropy of the target process X evaluated at time n is given by the known Eq. [15]

$$H_X = H\left(X_n\right) = \frac{1}{2}\ln\left(2\pi e\sigma^2\left(X_n\right)\right), \tag{3.13}$$

where $\sigma(X_n)$=E[$X_n{}^2$] is the variance of the target process. Moreover, exploiting a central result published in [16], we derive that the conditional entropy of the present of the target given the past m points of the whole network process $\boldsymbol{S}$ is given by

$$N_X = H\left(X_n|\boldsymbol{S}_n^m\right) = \frac{1}{2}\ln\left(2\pi e\sigma^2\left(X_n|\boldsymbol{S}_n^m\right)\right), \tag{3.14}$$

where $\sigma^2(X_n|\boldsymbol{S}_n^m)$ is the *partial variance* of X_n given $\boldsymbol{S}_n^m$, quantified as the variance of the prediction error of a linear regression of X_n on $\boldsymbol{S}_n^m$. The same formulation holds for any 'restricted' linear regression where one or more of the processes composing $S=[X,Y,Z]$ are excluded from the set of regressors. Thus, the regression of X_n performed respectively on X_n^m, $X_n^m \oplus Y_n^m$, $X_n^m \oplus Z_n^m$, $Y_n^m \oplus Z_n^m$, and $X_n^m \oplus Y_n^m \oplus Z_n^m$, yields the partial variances $\sigma^2(X_n|X_n^m)$, $\sigma^2(X_n|X_n^m \oplus Y_n^m)$, $\sigma^2(X_n|X_n^m \oplus Z_n^m)$, $\sigma^2(X_n|Y_n^m \oplus Z_n^m)$, and $\sigma^2(X_n|X_n^m \oplus Y_n^m \oplus Z_n^m)$. Then, the variance of X and the different partial variances can be combined as follows to estimate predictive information, information storage and joint information transfer,

$$\begin{aligned} P_X &= \frac{1}{2} \ln \frac{\sigma^2(X_n)}{\sigma^2(X_n|S_n^m)} \\ S_X &= \frac{1}{2} \ln \frac{\sigma^2(X_n)}{\sigma^2(X_n|X_n^m)}, \; T_{YZ \to X} = \frac{1}{2} \ln \frac{\sigma^2(X_n|X_n^m)}{\sigma^2(X_n|X_n^m \oplus Y_n^m \oplus Z_n^m)} \end{aligned}, \tag{3.15a}$$

bivariate and full internal information

$$\begin{aligned} S_{X|Y} &= \frac{1}{2} \ln \frac{\sigma^2(X_n|Y_n^m)}{\sigma^2(X_n|X_n^m \oplus Y_n^m)}, S_{X|Z} = \frac{1}{2} \ln \frac{\sigma^2(X_n|Z_n^m)}{\sigma^2(X_n|X_n^m \oplus Z_n^m)} \\ S_{X|Y,Z} &= \frac{1}{2} \ln \frac{\sigma^2(X_n|Y_n^m \oplus Z_n^m)}{\sigma^2(X_n|X_n^m \oplus Y_n^m \oplus Z_n^m)} \end{aligned}, \tag{3.15b}$$

and bivariate and conditional information transfer

$$\begin{aligned} T_{Y \to X} &= \frac{1}{2} \ln \frac{\sigma^2(X_n|X_n^m)}{\sigma^2(X_n|X_n^m \oplus Y_n^m)}, T_{Z \to X} = \frac{1}{2} \ln \frac{\sigma^2(X_n|X_n^m)}{\sigma^2(X_n|X_n^m \oplus Z_n^m)} \\ T_{Y \to X|Z} &= \frac{1}{2} \ln \frac{\sigma^2(X_n|X_n^m \oplus Z_n^m)}{\sigma^2(X_n|X_n^m \oplus Y_n^m \oplus Z_n^m)}, T_{Z \to X|Y} = \frac{1}{2} \ln \frac{\sigma^2(X_n|X_n^m \oplus Y_n^m)}{\sigma^2(X_n|X_n^m \oplus Y_n^m \oplus Z_n^m)} \end{aligned}. \tag{3.15c}$$

The interaction terms of the various decompositions are then obtained directly from the quantities in Eqs. (3.15a, 3.15b, and 3.15c): the interaction information storage terms are $I^X_{X;Y} = S_X - S_{X|Y}$ and $I^X_{X;Z} = S_X - S_{X|Z}$ for the bivariate processes, and $I^X_{X;Y,Z} = S_X - S_{X|Y,Z}$ for the whole network process; the conditional interaction storage terms are $I^X_{X;Y|Z} = S_{X|Z} - S_{X|Y,Z}$ and $I^X_{X;Z|Y} = S_{X|Y} - S_{X|Y,Z}$; the interaction information transfer is $I^X_{Y;Z|X} = T_{Y \to X} + T_{Z \to X} - T_{YZ \to X}$; and the multivariate interaction information is $I^X_{X;Y;Z} = S_X - S_{X|Y} - S_{X|Z} + S_{X|Y,Z}$.

Thus, the computation of all measures of information dynamics for jointly Gaussian processes is straightforward once the partial variance of the present of X given the various combinations of the past of X, Y and Z are obtained. The partial variance of X_n given the generic d-dimensional vector variable $\boldsymbol{V}$, which in our case can be any of X_n^m, $X_n^m \oplus Y_n^m$, $X_n^m \oplus Z_n^m$, $Y_n^m \oplus Z_n^m$, and $X_n^m \oplus Y_n^m \oplus Z_n^m$, is given by [8, 16]

$$\sigma^2(X_n|\boldsymbol{V}) = \sigma^2(X_n) - \Sigma(X_n;\boldsymbol{V})\,\Sigma(\boldsymbol{V})^{-1}\Sigma(X_n;\boldsymbol{V})^{\mathrm{T}}. \tag{3.16}$$

where $\Sigma(X_n;\boldsymbol{V})$ is an $1 \times d$ vector containing the covariances between X_n and each scalar element of $\boldsymbol{V}$, and $\Sigma(\boldsymbol{V})$ is the $d \times d$ covariance matrix of $\boldsymbol{V}$. Therefore, the computation of the partial variances amounts to evaluating the relevant covariance and cross-covariance matrices between the present and past variables of the various processes, to be inserted in (3.16). In general, these matrices contain as scalar elements the covariance between two time-lagged variables taken from the processes *X*, *Y*, and *Z*, which in turn appear as elements of the autocovariance of the whole observed process $\boldsymbol{S}=[X\ Y\ Z]^{\mathrm{T}}$, defined at each lag $k\geq0$ as $\boldsymbol{\Gamma}_k=\mathrm{E}[\boldsymbol{S}_n\ \boldsymbol{S}^{\mathrm{T}}_{n\text{-}k}]$. Now we turn to show how this autocovariance matrix can be computed from the parameters of the vector autoregressive (VAR) formulation of the process $\boldsymbol{S}=[X\ Y\ Z]^{\mathrm{T}}$:

$$\boldsymbol{S}_n=\sum_{k=1}^{m}\mathbf{A}_k\boldsymbol{S}_{n-k}+\boldsymbol{U}_n, \tag{3.17}$$

where m is the order of the Markov process, $\mathbf{A}_k$ are $M\times M$ coefficient matrices and $\boldsymbol{U}_n$ is a zero mean Gaussian white noise process with diagonal covariance matrix $\boldsymbol{\Lambda}$. The autocovariance of the process (3.17) is related to the VAR parameters via the well known Yule-Walker equations:

$$\boldsymbol{\Gamma}_k=\sum_{l=1}^{m}\mathbf{A}_l\boldsymbol{\Gamma}_{k-l}+\delta_{k0}\boldsymbol{\Lambda}, \tag{3.18}$$

where δ_{k0} is the Kronecher product. In order to solve Eq. (3.17) for $\boldsymbol{\Gamma}_k$, with $k=0, 1,\ldots, m-1$, we first express Eq. (3.16) in a compact form as $\boldsymbol{\phi}_n=\mathbf{A}\boldsymbol{\phi}_{n-1}+\boldsymbol{E}_n$, where

$$\boldsymbol{\phi}_n=\left[\boldsymbol{S}_n^{\mathrm{T}}\boldsymbol{S}_{n-1}^{\mathrm{T}}\cdots\boldsymbol{S}_{n-m+1}^{\mathrm{T}}\right]^{\mathrm{T}},\ \mathbf{A}=\begin{bmatrix}\mathbf{A}_1 & \cdots & \mathbf{A}_{m-1} & \mathbf{A}_m\\ \mathbf{I}_M & \cdots & \mathbf{0}_M & \mathbf{0}_M\\ \vdots & \ddots & \vdots & \vdots\\ \mathbf{0}_M & \cdots & \mathbf{I}_M & \mathbf{0}_M\end{bmatrix},\ \boldsymbol{E}_n=\left[\boldsymbol{U}_n^{\mathrm{T}}\mathbf{0}_{1\times M(m-1)}\right]^{\mathrm{T}}. \tag{3.19}$$

Then, the covariance matrix of $\boldsymbol{\phi}_n$, which has the form:

$$\boldsymbol{\Psi}=\mathrm{E}\left[\boldsymbol{\phi}_n\boldsymbol{\phi}_n^{\mathrm{T}}\right]=\begin{bmatrix}\boldsymbol{\Gamma}_0 & \boldsymbol{\Gamma}_1 & \cdots & \boldsymbol{\Gamma}_{m-1}\\ \boldsymbol{\Gamma}_1^{\mathrm{T}} & \boldsymbol{\Gamma}_0 & \cdots & \boldsymbol{\Gamma}_{m-2}\\ \vdots & \vdots & \ddots & \vdots\\ \boldsymbol{\Gamma}_{m-1}^{\mathrm{T}} & \boldsymbol{\Gamma}_{m-2}^{\mathrm{T}} & \cdots & \boldsymbol{\Gamma}_0\end{bmatrix}, \tag{3.20}$$

can be expressed as $\boldsymbol{\Psi}=\mathbf{A}\boldsymbol{\Psi}\,\mathbf{A}^{\mathrm{T}}+\boldsymbol{\Xi}$, where $\boldsymbol{\Xi}=\mathrm{E}\left[\boldsymbol{E}_n\boldsymbol{E}_n^{\mathrm{T}}\right]$ is the covariance of $\boldsymbol{E}_n$. This last equation is a discrete-time Lyapunov equation, which can be solved for $\boldsymbol{\Psi}$ yielding the autocovariance matrices $\boldsymbol{\Gamma}_0,\ldots,\boldsymbol{\Gamma}_{m\text{-}1}$. Finally, the autocovariance can be calculated recursively for any lag $k\geq m$ by repeatedly applying Eq. (3.18). This shows how the autocovariance sequence can be computed up to arbitrarily high lags starting from the parameters of VAR representation of the observed Gaussian process.

3.4.2 *Theoretical Example*

In this Section we show the computation of the terms appearing in the information decompositions defined in Sect. 3.3 using simulations. In order to make the interpretation free of issues related to practical estimation of the measures, we simulate Gaussian VAR processes and exploit the procedure described in Sect. 3.4.1 to quantify all measures of information dynamics. Specifically, we consider a stationary multivariate Gaussian process $\boldsymbol{S}=\{X,Y,Z\}$, with temporal statistical structure described by the following VAR model:

$$\begin{aligned} X_n &= cY_{n-1} + Z_{n-1} + U_n \\ Y_n &= V_n \\ Z_n &= -0.8Z_{n-2} + (1-c)\,Y_{n-1} + W_n \end{aligned}, \tag{3.21}$$

where U, V and W are univariate Gaussian processes of zero mean and unit variance, uncorrelated with each other and without temporal structure. The process (3.21) is a continuous Markov process of order $m=2$, for which the present variable $\boldsymbol{S}_n$ has a multivariate Gaussian probability distribution of the form (3.12). The temporal structure of the process is depicted by the graphical representations of Fig. 3.4a, where interactions within and between processes are depicted specifying all time-lagged influences (left), focusing on the influences from the past to the present of the processes (middle), and evidencing the influences among processes in a compact representation (right). The process X has no internal dynamics, but it exhibits temporal structure arising from the causal interactions originating from Y and Z, both occurring with lag 1. The process Z has autodependency effects at lag 2, and is the target of a causal interaction originating in Y and occurring with lag 1. The process Y has no temporal structure at all. The simulation parameter c sets the strength of the causal effects originating from the process Y in a way such that higher values of c determine stronger coupling from Y to X and weaker coupling from Y to Z.

The results of information decomposition performed assuming the process X as the target, and the processes Y and Z as sources, are reported in Fig. 3.4b. The values of the measures resulting from the decomposition of the predictive information defined in (3.8) (left), the decomposition of the information storage defined in (3.9, and 3.10) (middle), and the decomposition of information transfer defined in (3.11) (right), computed as a function of the coupling parameter c, are obtained according to the procedure described in Sect. 3.4.1 starting from the known values of the model parameters. The behaviour exhibited by the prediction entropy P_X at increasing the

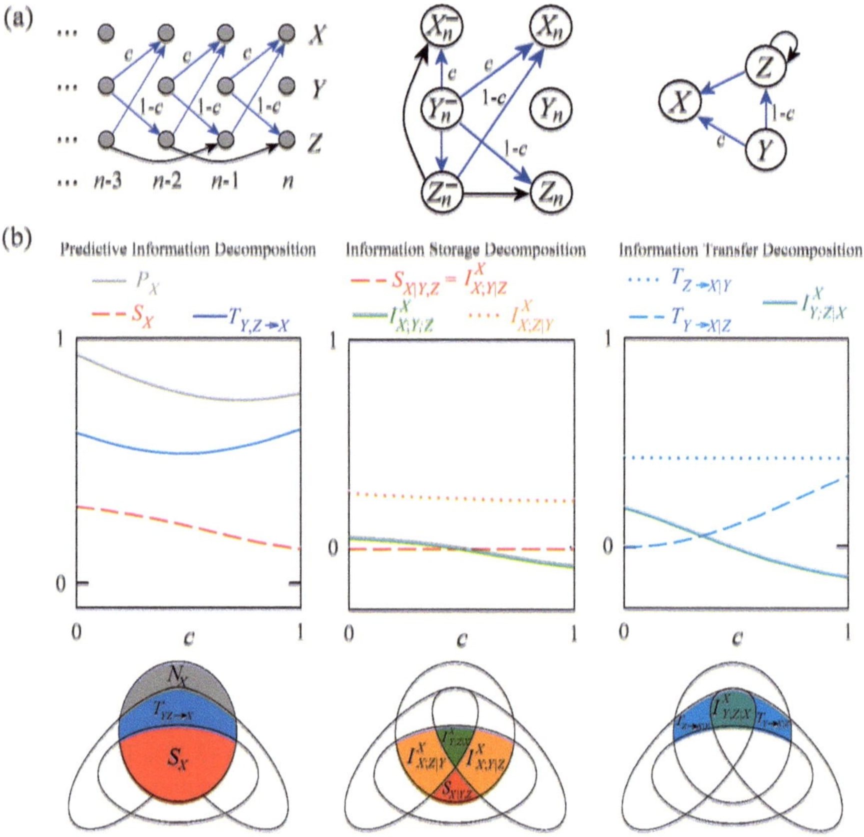

Fig. 3.4 Information decomposition for a stationary Gaussian trivariate process $\{X,Y,Z\}$ of order two, with internal dynamics in Z and causal interactions from Y to X and from Z to X. (**a**) microscopic (*left*), mesoscopic (*middle*) and macroscopic (*right*) graph representations of the process. (**b**) Decompositions of predictive information, $P_X=S_X+T_{ZY\rightarrow X}$, information storage, $S_X=S_{X|YZ}+I^X{}_{X;Z|Y}+I^X{}_{X;Y|Z}+I^X{}_{X;Z;Y}$, and information transfer, $T_{ZY\rightarrow X}=T_{Z\rightarrow X|Y}+T_{Y\rightarrow X|Z}+I^X{}_{Y;Z|X}$, performed as a function of the coupling parameter c

parameter c from 0 to 1 is the result of a decreasing trend of the self entropy S_X and a non-monotonic variation of the joint transfer entropy $T_{YZ\rightarrow X}$. In turn, the decrease of the information storage with the parameter c is the result of a decrease of the multivariate interaction information $I^X_{X;Z;Y}$, that evidences a shift from redundancy to synergy, and of a slight decrease of the conditional interaction self entropy $I^X_{X;Z|Y}$; note that both the conditional self entropy $S_{X|YZ}$ and the conditional interaction self entropy $I^X_{X;Y|Z}$ are always zero regardless of the values of c. The trend of the joint information transfer results from the increase of the conditional transfer entropy $T_{Y\rightarrow X|Z}$ and the contemporaneous decrease of the interaction transfer entropy $I^X_{Y;Z|X}$ in the presence of a stable conditional transfer entropy $T_{Z\rightarrow X|Y}$. The trends exhibited by the measures are descriptive of the causal statistical structure displayed by the network for the different parameter configurations. Indeed, the absence of internal information reflects the absence of internal dynamics in the target system. Of note, the absence of internal dynamics does not prevent the target system from storing information; in this case, the whole information storage is due to causal interactions from the sources to the target. The increase of the information transfer from Y to X given Z, and the constant values of the information transfer from Z to X given Y, reflect respectively the strong dependence of the coupling $Y \rightarrow X$, and the independence of the coupling $Z \rightarrow X$, on the parameter c. Finally, the decrease of the interaction terms denote the occurrence of a shift from redundant to synergistic contributions of the sources to the information storage in the target, as well as of the joint contribution of the two sources to the information transfer: with $c=0$, the interactions are entirely redundant because all causal effects from Y to X are mediated by Z; with $c=1$, the interactions are maximally synergetic because Y and Z provide information to X according to fully separate mechanisms.

3.5 Application to Physiological Networks

This Section is relevant to the practical computation of the information-theoretic measures defined in Sect. 3.3, estimated as described in Sect. 3.4, in physiological networks where the assigned target process is the heart period and the sources are systolic arterial pressure and respiration. The analysis is focused on the decomposition of information carried by the cardiac system into an amount of new information produced by the system at each heartbeat and an amount that can be inferred by the memory of the whole observed physiological network. The latter term is further decomposed into the amount of information stored in the cardiac system and transferred to it from the vascular and respiratory systems. Moreover, information storage and transfer are further split to evidence contributions ascribed to the individual source systems as well as interaction terms denoting information modification due to redundant or synergistic source cooperation. This chain of decompositions allows to dissect the variability of the heart rate into contributions which are related to the multiple physiological mechanisms (e.g., the baroreflex and the respiratory sinus arrhythmia among others) involved into its generation.

The relevance of the measures of information dynamics in reflecting these mechanisms and their reciprocal activation/deactivation is proved considering two experimental protocols which are known to produce adjustments of the cardiovascular control evoked by opposite effects on the central circulatory volemia: head-up tilt (HUT) is associated with a reduction of venous return and consequent sinoaortic and carotid unloading of the baroreceptors that lead to vagal withdrawal and sympathetic activation [17]; head-down tilt (HDT) induces an increase in venous return, central blood volume and central venous pressure that lead to cardiopulmonary loading and sympathetic inhibition [18].

3.5.1 Experimental Protocols and Data Analysis

The analysis exploited two sets of recordings collected during experimental protocols planned to study the effect on the cardiovascular control of postural challenges associated respectively with unloading (i.e., HUT) and loading (i.e., HDT) of the baroreceptors. In the HUT protocol [8], we studied 15 young healthy subjects (seven females, aging from 22 to 32years, median 25years) in whom, after 15min of adaptation in the resting supine position, the surface electrocardiogram (lead II), the noninvasive finger blood pressure (Finapres, Ohmeda) and the spontaneous respiratory flow (based on nasal thermistor) were measured for 10min in the supine position (REST) and for additional 10min in the 60° upright body position that was reached passively using a motorized table. The recordings in the HUT condition started after two minutes from the tilt transition. In the HDT protocol [10], we studied 13 healthy men aged from 41 to 71years (median: 59years) in whom the electrocardiogram (lead II) and noninvasive finger blood pressure (Nexfin, BMEYE, Amsterdam, The Netherlands) were recorded with a sampling rate of 400Hz. Moreover, a respiratory signal was derived from the respiratory-related amplitude modulations of the ECG. As in the previous protocol, 15min were allowed in the resting baseline position for subjects' stabilization before starting the measurements. Then, each recording session consisted in 10min of baseline recording at REST followed by 10min of recording during HDT with a table inclination of −25°. The recordings of the HDT session started 5min after tilting the table. During the protocol, the subjects breathed according to a metronome at 16breaths/min to prevent modifications in the magnitude of respiratory sinus arrhythmia due to alterations of the breathing rate.

After detecting the R-wave on the ECG and locating the R-wave peak using parabolic interpolation, the temporal distance between two consecutive R-wave apexes was computed and utilized as an approximation of HP. The maximum of the arterial pressure inside the n-th HP (i.e. HP_n) was taken as the n-th SAP (i.e. SAP_n). In the HUT protocol, the n-th value of the respiratory time series (i.e. $RESP_n$) was obtained sampling the respiratory nasal flow signal at the time of the R-wave peak of the ECG denoting the beginning of the n-th HP. In the HDT protocol, the amplitude of the first QRS complex delimiting HP_n, taken as the difference between the ECG at the R-wave apex and the isoelectric line, was taken as the n-th respiratory value (i.e. $RESP_n$). In all conditions, the occurrences of R-wave and SAP peaks were carefully checked to avoid erroneous detections or missed beats, and if isolated ectopic beats affected HP and SAP values, these measures were linearly interpolated using the closest values unaffected by ectopic beats.

HP, SAP and R sequences of consecutive synchronous values were chosen inside the REST and HUT periods for the first protocol, and inside the REST and HDT periods for the second protocol. The length of the analyzed time series was 300 beats for the HUT protocol, and 256 beats for the HDT protocol, thus allowing to focus on short-term cardiovascular regulatory mechanisms [19]. All time series were linearly detrended before the analysis. Stationarity of the analyzes time series was tested using a test that checks the steadiness over time of mean and variance of the selected sequences [20]. Then, for each subject and experimental condition, realizations of the observed process $S=\{X,Y,Z\}$ were obtained by normalizing the measured multivariate time series $\{\mathrm{HP}_n, \mathrm{RESP}_n, \mathrm{SAP}_n\}$, i.e. subtracting the mean from each series and dividing the result by the standard deviation. The resulting time series $\{X_n, Y_n, Y_n\}$ was fitted with a VAR model in the form of Eq. (3.16), where model identification was performed using the standard vector least squares method and the model order was optimized according to the Akaike Information Criterion [21]. The estimated model coefficients were exploited to derive the covariance matrix of the vector process, and the covariances between the present and the past of the processes were used as in Eq. (3.16) to estimate all the partial variances needed to compute the measures of information dynamics. In all computations, to account for fast vagal reflexes capable to modify HP in response to within-beat changes of RESP and SAP, the present RESP and SAP variables were included into the vectors representing their past values; this choice, which is in agreement with the measurement convention that allows zero-lag effects from RESP and SAP to HP (see Fig. 3.2), corresponds to use $[Y_n\ Y_n^m]$ and $[Z_n\ Z_n^m]$, in place of Y_n^m and Z_n^m, in the computation of the partial variances performed using Eq. (3.16).

3.5.2 Results and Discussion

Figure 3.5 reports the results of information decomposition applied to the variability of the normalized HP process X during the HUT protocol. The entropy decomposition shown in Fig. 3.5a documents that the new information produced by the process at each heartbeat was significantly lower in the upright body position than in the supine position. Since entropies are computed for normalized series with unit variance (H_X=0.5ln2πe in all cases), the decrease of the new information N_X corresponds to a statistically significant increase of the predictive information $P_X=H_X-N_X$ (see Eq. (3.7)). The significant increase of the prediction entropy with HUT was the result of an increase of the self entropy of HP accompanied by unchanged values of the joint transfer entropy from RESP and SAP to HP (increase of S_X and stable $T_{YZ\to X}$, Fig. 3.5a). In turn, the significant increase of the self entropy was due to the observation of significantly higher values of the conditional self entropy of HP given RESP and SAP and of the interaction self entropy between HP and SAP given RESP (increase of $S_{X|Y,Z}$ and $I^X_{X;Z|Y}$, Fig. 3.5b); the interaction self entropy between HP and RESP given SAP, $I^X_{X;Y|Z}$, and the multivariate interaction self entropy, $I^X_{X;Y;Z}$, were slightly negative, indicating a small prevalence of synergistic interaction, and unaffected by HUT. Finally, the unvaried joint transfer entropy in the upright position resulted from a significant increase of the conditional transfer entropy from RESP to HP given SAP compensated by the decrease of the conditional transfer entropy from SAP to HP given RESP (increase of $T_{Z\to X|Y}$ and decrease of $T_{Y\to X|Z}$, Fig. 3.5c), in the presence of unaltered

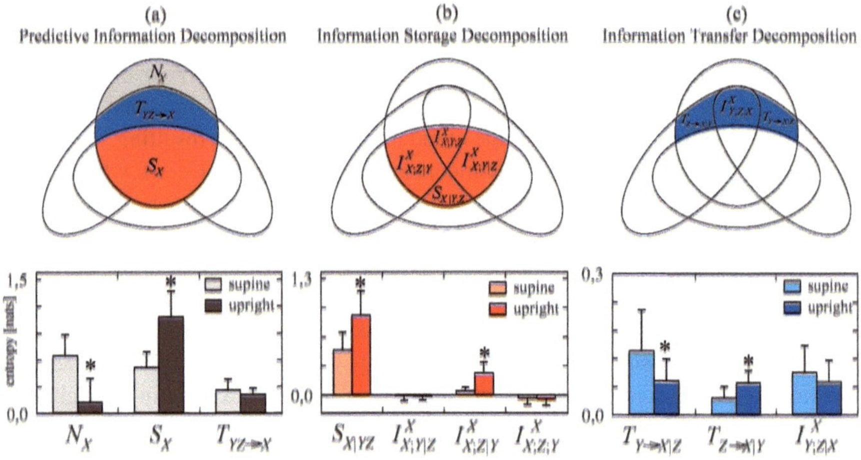

Fig. 3.5 Information decomposition during head-up tilt (HUT). Heart period is measured as the target process X of the physiological network including also respiration and systolic arterial pressure as source processes Y and Z. (**a**) predictive information decomposition, $P_X=H_X-N_X=S_X+T_{ZY\to X}$. (**b**) Information storage decomposition, $S_X=S_{X|YZ}+I^X_{X;Z|Y}+I^X_{X;Y|Z}+I^X_{X;Z;Y}$. (**c**) Information transfer decomposition, $T_{ZY\to X}=T_{Z\to X|Y}+T_{Y\to X|Z}+I^X_{Y;Z|X}$. All measures are computed as mean+std.dev. over 15 subjects in the resting supine position and in the upright position assumed during HUT. Asterisks denote statistically significant difference ($p<0.05$) between supine and upright, assessed by the Wilcoxon signed rank test. Figure modified from [14]

redundant transfer entropy $I^X_{Y;Z|X}$. Overall, these results indicate that the transition from the resting supine position to the orthostatic HUT position is associated with a higher predictability of the HP node of the physiological network (higher predictive information), that results from higher information stored in the node and unchanged information transferred to it. The larger storage is due to larger internal information (stronger internal dynamics of HP) and larger interaction information between HP and SAP (higher redundant cardiovascular cooperation); the unchanged transfer is the balance between an increased baroreflex coupling (higher information transfer from SAP to HP) and a decreased cardiorespiratory coupling (lower information transfer from RESP to HP), in the presence of an unchanged redundant interaction between SAP and RESP while they transfer information to HP.

Figure 3.6 reports the results of information decomposition applied to the variability of the normalized HP process X during the HDT protocol. The entropy decomposition of Fig. 3.6a documents that the new information about HP was significantly higher during HDT than in the baseline resting condition, indicating a decrease in the predictive information. The lower prediction entropy was reflected by a lower self entropy (Fig. 3.6a), which in turn was reflected by a lower conditional self entropy (Fig. 3.6b). All other terms of the information decompositions, including the joint transfer entropy ($T_{YZ\to X}$ in Fig. 3.6a) and its constituent factors ($T_{Y\to X|Z}$, $T_{Z\to X|Y}$ and $I^X_{Y;Z|X}$ in Fig. 3.6c), as well as the constituent factors of the

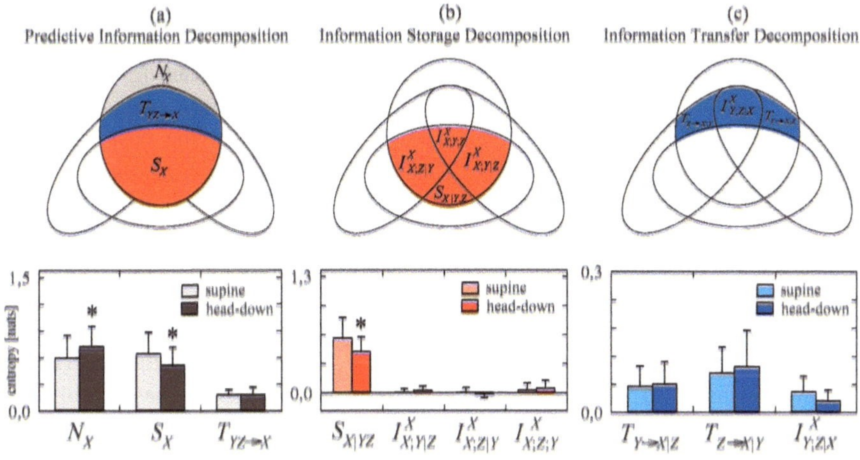

Fig. 3.6 Information decomposition during head-down tilt (HDT). Heart period is measured as the target process X of the physiological network including also respiration and systolic arterial pressure as source processes Y and Z. (**a**) predictive information decomposition, $P_X=H_X-N_X=S_X+T_{ZY\to X}$. (**b**) Information storage decomposition, $S_X=S_{X|YZ}+I^X{}_{X;Z|Y}+I^X{}_{X;Y|Z}+I^X{}_{X;Z;Y}$. (**c**) Information transfer decomposition, $T_{ZY\to X}=T_{Z\to X|Y}+T_{Y\to X|Z}+I^X{}_{Y;Z|X}$. All measures are computed as mean+std.dev. over 13 subjects in the resting supine position and in the head-down position assumed during HDT. Asterisks denote statistically significant difference ($p<0.05$) between supine and head-down, assessed by the Wilcoxon signed rank test. Figure modified from [10]

interaction self entropy $I^X_{X;Y,Z}$ (i.e., $I^X_{X;Z|Y}$, $I^X_{X;Y|Z}$ and $I^X_{X;Y;Z}$ in Fig. 3.6b), did not change significantly during HDT in comparison with the baseline resting condition. These results indicate that, contrary to what happens for HUT, HDT reduces the predictive information about the HP node of the physiological network, and this increased complexity is exclusively related to a weakening of the self-predictable dynamics of the HP (decreased information storage and unchanged information transfer). In turn, the lower information storage was entirely ascribed to a decrease in the internal information of the cardiac node, with unchanged interaction terms, and the stable information transfer was the result of unchanged amounts of conditional and redundant interaction transfer.

The results obtained for the two studied protocols clearly document the involvement of the cardiovascular control in regulating the dynamics of the heart period during head-up and head-down solicitation. This involvement is demonstrated by the changes exhibited by the considered information-theoretic measures in comparison with the resting supine condition. Interestingly, the main changes documented by the predictive information decomposition were of opposite sign: the new information generated by the cardiac system decreased (and the predictive information increased) during HUT, while the new information increased (and the predictive information decreased) during HDT. Moreover, since the joint information transfer did not change during HUT and HDT compared to the baseline condition, the variations of the predictive information were reflected in variations of the same sign in the information storage: the information stored in the heart period dynamics increased significantly during HUT, and decreased significantly during HDT. These results point out that the complexity of the cardiac dynamics, intended as the predictability of the current HP on its past values and on the past values of SAP and RESP, is subjected to changes that depend on the type of orthostatic stimulation: positive orthostatism leads to reduction in HP complexity, while negative orthostatism increases the HP complexity. Since complexity measures are under the influence of the autonomic nervous system, with higher or lower complexity reflecting respectively a shift of the sympatho-vagal balance towards parasympathetic or sympathetic activation [22], we conclude that the changes of predictive information and information storage reflect an increase of the sympathetic activity during HUT, and of the vagal activity during HDT. These results are compatible with physiological mechanisms that lead to vagal withdrawal and sympathetic activation during HUT, like the reduction of venous return and the baroreceptor unloading [23, 24], and to a sympathetic inhibition during HDT, like the cardiopulmonary loading and the acute central circulatory hypervoleamia [18, 25].

The information storage decomposition documented another common trait of the cardiac dynamics in the two protocols, i.e. the fact that variations in the information storage are largely determined by variations in the internal information. Indeed, the increase of the self entropy during HUT, as well as its decrease during HDT, were accompanied by statistically significant variations of the same sign in the conditional self entropy. These results suggest the importance of the internal dynamics of the cardiac systems, i.e. of predictable HP dynamics that cannot be explained by the effects of SAP and RESP on HP, in the generation of the information stored in the system. The internal information reflects mechanisms of regulation of the cardiac dynamics which are fully unrelated to RESP and SAP, possibly including direct sympathetic influences on the sinus node which are unmediated by the activation of baroreceptors and/or low pressure receptors, and central commands originating from respiratory centers in the brainstem that are independent of afferent inputs [4, 26]. Thus, the higher internal information found during HUT might reflect the stronger sympathetic drive directed to the sinus node during the postural stress. On the other hand, the lower internal information found during HDT might be due to the activation of a central mechanism, independent on the cardiac baroreflex and cardiopulmonary stimulation, that makes the cardiac dynamics more complex and limits their self-predictability.

In both HUT and HDT, the information transferred jointly from RESP and SAP to HP did not change in comparison with the baseline condition. However, this similar behaviour was the result of different trends of the information quantities composing the joint transfer entropy: during HUT, the conditional information transfer decreased from RESP to HP and increased from SAP to HP; during HDT, none of the terms composing the information transfer changed significantly. The opposite trends of cardiovascular and cardiorespiratory information transfer during HUT document respectively a higher importance of the baroreflex pathway and a lower relevance of the cardiopulmonary pathway in contributing to heart rate variability during positive orthostatism. These results are in agreement with the increase of the baroreflex coupling and weakening of the respiratory sinus arrhythmia which are known effects of the postural stress induced by HUT [27, 28]. On the other hand, the invariance of the amounts of joint and conditional information transfer from RESP and SAP to HP observed moving from rest to HUT suggests to exclude the cardiac baroreflex control and cardiopulmonary reflexes from the physiological mechanisms altered by negative orthostatism. These last results, together with the reduction of the cardiac internal information, support the central drive hypothesis for the increase of the respiratory sinus arrhythmia which is known to occur during HDT [26]: indeed, the lower self entropy of HP given SAP and RESP observed in the absence of alterations in the cardiac baroreflex and in the cardiopulmonary reflexes could be explained by an enhanced activity of respiratory centers affecting directly the sinus node, resulting in an increase of respiratory sinus arrhythmia and in the complexity of the cardiac dynamics.

The analysis of the interaction information transfer documented that SAP and RESP contribute redundantly to the information transferred to HP in the resting supine condition as well as during HUT and HDT. Remarkably, the common information held by the vascular and respiratory systems about the cardiac dynamics constitutes an important amount (more than 20% in each condition) of the information transferred jointly from SAP and RESP to HP. The redundant nature of the contribution of SAP and R to HP finds explanation in the coexistence of several mechanisms of cardiovascular and cardiorespiratory regulation, including the direct effects of RESP on SAP—due to respiratory-related changes of the intrathoracic pressure—which are then translated to HP through the baroreflex [29], and more complex interactions and integrations between vasomotor and respiratory centers occurring at the brainstem level. This amount of redundancy might accomplish a principle of fault tolerance and harmonization of neural responses which, given its prevalence over synergy, seems to be more valuable than the reactivity and flexibility which might derive from synergetic interaction. A similar prevalence of redundancy over synergy was not documented for the interaction terms involving the past of HP, i.e. the interaction self entropies between HP and RESP given SAP or between HP and SAP given RESP, and the multivariate interaction self entropy between HP, SAP and RESP. These quantities were indeed rather small and often equally distributed between negative and positive values, indicating the lack of a prevalence of net synergetic or redundant contributions to the cardiac information storage. Moreover, these quantities did not change significantly between conditions, with the only exception of the increase of the interaction self entropy between HP and SAP given RESP during HUT that can be ascribed to the solicitation of baroreflex effects in this condition [8]. In general, these results do not exclude the absence of synergistic interactions in the cardiovascular control, because the approach presented here estimates the net balance between redundancy and synergy, and because it tends to favour mathematically the detection of redundancy over synergy [30]. Thus, future studies assessing redundancy and synergy as independent quantities [14, 30], and exploring the balance between redundancy and synergy using different estimation approaches [7], are needed to understand whether and to what extent synergy plays a role in the neural control of cardiovascular and cardiorespiratory physiological networks.

Finally, we remark some limitations of the experimental protocols adopted in the HUT and HDT studies [8, 10] which may have an impact on the comparison of the information-theoretic measures performed here. First, the two groups had small sample size of (15 and 13 subjects) and were not homogeneous in terms of age and gender, as the HUT and HDT protocols involved respectively young subjects balanced by gender and older male subjects. Since age and gender are known to have an impact on cardiovascular variability [31], it is possible that some of the differences observed in the information measures between the two protocols are due these factors. Moreover, the different modalities of measurement for respiration (nasal flow and ECG-derived) are likely determining different phases in the recorded respiratory time series. Since the information measures computed in this study are based on VAR models encompassing several interaction lags, we expect a low impact of the respiration phase bias on our results. However, effects on the measures of information transfer cannot be excluded, and future works are needed to explore the effect of different modalities to record respiratory activity on the results of cardiorespiratory coupling analyses.

3.6 Conclusions

This work introduced a framework for the full quantitative description of the information carried by the target of a network of interacting dynamical systems, first evidencing the amounts of new (unpredictable) information produced by the system at each time step and of information that can be predicted by the past states of the network, and then dissecting this predictive information through different decomposition strategies. The decompositions are relevant to separation of the target predictive information, separation of the information stored in the target system, and separation of the information transferred to the target from two source systems. These decompositions yield terms that elucidate the specific contributions of the individual systems to the target dynamics, as well as terms describing the balance between redundancy and synergy of the two sources while they contribute to information storage and information transfer.

A viable approach for the exact computation of these measures has been presented and applied to simulated networks, allowing us to relate modifications of the various measures of information dynamics to the causal statistical properties of the network links. Moreover, we illustrated how information storage, transfer and modification interact with each other to give rise to the predictive information of a target dynamical system connected to multiple source systems. In fact, though confirming that different measures like storage, transfer and modification reflect different aspects of information processing (respectively, regularity, causality and synergy/redundancy), we have shown that these measured can undergo concurrent modifications in response to specific system alterations. Therefore, we advocate that the various information dynamics measures should not be computed in isolation, but rather evaluated together as components of the total statistical dependence relevant to target process of a multivariate system.

The application to experimental data suggested the importance of adopting information decomposition methods to fully assess the cardiac, vascular and respiratory determinants of short-term heart rate variability. The analysis confirmed known findings about the variations in the complexity and causality of cardiovascular and cardiorespiratory variability, but also revealed novel interpretations related to how the overall predictability of the dynamics of a target system is modified due to possible interactions between the information sources. Since the proposed quantities are highly specific and take the form of indexes that can be computed very efficiently and robustly via traditional multivariate regression analysis of spontaneously varying variables, they appear to be suitable candidates for large scale applications to clinical databases recorded under uncontrolled conditions. Future studies should be directed to extend the decompositions to model-free frameworks that assess the role of nonlinear physiological dynamics in information storage, transfer and modification, to explore scenarios with more than two source processes, and to evaluate the importance of these measures for the assessment of cardiovascular and cardiorespiratory interactions in diseased conditions. Moreover, thanks to its generality, the approach might be applied not only to cardiovascular physiology, but also to any field of science in which interactions among realizations, representing the behaviour of interacting systems, are under scrutiny.

References

1. Malliani, A.: Principles of cardiovascular neural regulation in health and disease. Kluwer academic publishers, Norwell, MA, USA (2000)
2. Cohen, M.A., Taylor, J.A.: Short-term cardiovascular oscillations in man: measuring and modelling the physiologies. J. Physiol. **542**, 669–683 (2002)
3. Eckberg, D.L.: Point:counterpoint: respiratory sinus arrhythmia is due to a central mechanism vs. respiratory sinus arrhythmia is due to the baroreflex mechanism. J. Appl. Physiol. **106**, 1740–1742 (2009)
4. Koepchen, H.P.: History of studies and concepts of blood pressure waves. In: Miyakawa, K., Polosa, C., Koepchen, H.P. (eds.) Mechanisms of blood pressure waves. Springer, Berlin (1984)
5. Faes, L., Porta, A.: Conditional entropy-based evaluation of information dynamics in physiological systems. In: Vicente, R., Wibral, M., Lizier, J.T. (eds.) Directed Information Measures in Neuroscience. Springer-Verlag, Berlin (2014)
6. Wibral, M., Lizier, J.T., Priesemann, V.: Bits from biology for biologically-inspired computing. Front. Robot. AI 2. 5 (2015). doi:10.3389/frobt.2015.00005
7. Faes, L., Kugiumtzis, D., Nollo, G., Jurysta, F., Marinazzo, D.: Estimating the decomposition of predictive information in multivariate systems. Phys. Rev. E. **91**, 032904 (2015)
8. Faes, L., Porta, A., Nollo, G.: Information decomposition in bivariate systems: theory and application to cardiorespiratory dynamics. Entropy. **17**, 277–303 (2015)

9. Porta, A., Faes, L., Nollo, G., Bari, V., Marchi, A., De Maria, B., Takahashi, A.C.M., Catai, A.M.: Conditional self-entropy and conditional joint transfer entropy in heart period variability during graded postural challenge. PLoS One. **10**, e0132851 (2015)
10. Porta, A., Faes, L., Marchi, A., Bari, V., De Maria, B., Guzzetti, S., Colombo, R., Raimondi, F.: Disentangling cardiovascular control mechanisms during head-down tilt via joint transfer entropy and self-entropy decompositions. Front. Physiol. **6**, 301 (2015)
11. Triedman, J.K., Perrott, M.H., Cohen, R.J., Saul, J.P.: Respiratory sinus arrhythmia-time-domain characterization using autoregressive moving average analysis. Am. J. Physiol. Heart Circ. Physiol. **268**, H2232–H2238 (1995)
12. Baselli, G., Cerutti, S., Civardi, S., Malliani, A., Pagani, M.: Cardiovascular variability signals: towards the identification of a closed-loop model of the neural control mechanisms. I.E.E.E. Trans. Biomed. Eng. **35**, 1033–1046 (1988)
13. Porta, A., De Maria, B., Bari, V., Marchi, A., Faes, L.: Are nonlinear model-free approaches for the assessment of the entropy-based complexity of the cardiac control superior to a linear model-based one? IEEE Trans. Biomed. Eng. **64**(6), 1287–1296 (2017). doi:10.1109/TBME.2016.2600160
14. Faes, L., Porta, A., Nollo, G., Javorka, M.: Information decomposition in multivariate systems: definitions, implementation and application to cardiovascular networks, Entropy, special issue on Multivariate entropy measures and their applications, **19**(1), 5 (2017). doi: 10.3390/e19010005.
15. Cover, T.M., Thomas, J.A.: Elements of information theory. Wiley, New York (2006)
16. Barnett, L., Barrett, A.B., Seth, A.K.: Granger causality and transfer entropy are equivalent for Gaussian variables. Phys. Rev. Lett. **103**, 238701 (2009)
17. Montano, N., Gnecchi Ruscone, T., Porta, A., Lombardi, F., Pagani, M., Malliani, A.: Power spectrum analysis of heart rate variability to assess the change in sympathovagal balance during graded orthostatic tilt. Circulation. **90**, 1826–1831 (1994)
18. Nagaya, K., Wada, F., Nakamitsu, S., Sagawa, S., Shiraki, K.: Responses of the circulatory-system and muscle sympathetic-nerve activity to head-down tilt in humans. Am. J. Phys. Regul. Integr. Comp. Phys. **268**, R1289–R1294 (1995)
19. Standards of measurement, physiological interpretation, and clinical use. Task force of the European Society of Cardiology and the North American Society of Pacing and Electrophysiology: Heart rate variability. Eur. Heart J. **17**, 354–381 (1996)
20. Magagnin, V., Bassani, T., Bari, V., Turiel, M., Maestri, R., Pinna, G.D., Porta, A.: Non-stationarities significantly distort short-term spectral, symbolic and entropy heart rate variability indices. Physiol. Meas. **32**, 1775–1786 (2011)
21. Akaike, H.: A new look at the statistical model identification. IEEE Trans. Autom. Contr. **19**, 716–723 (1974)
22. Porta, A., Gnecchi-Ruscone, T., Tobaldini, E., Guzzetti, S., Furlan, R., Montano, N.: Progressive decrease of heart period variability entropy-based complexity during graded head-up tilt. J. Appl. Physiol. **103**, 1143–1149 (2007)
23. Taylor, J.A., Eckberg, D.L.: Fundamental relations between short-term RR interval and arterial pressure oscillations in humans. Circulation. **93**, 1527–1532 (1996)
24. Cooke, W.H., Hoag, J.B., Crossman, A.A., Kuusela, T.A., Tahvanainen, K.U.O., Eckberg, D.L.: Human response to upright tilt: a window on central autonomic integration. J. Physiol. **517**, 617–628 (1999)
25. London, G.M., Levenson, J.A., Safar, M.E., Simon, A.C., Guerin, A.P., Payen, D.: Hemodynamic-effects of head-down tilt in normal subjects and sustained hypertensive patients. Am. J. Phys. **245**, H194–H202 (1983)
26. Dick, T.E., Baekey, D.M., Paton, J.F.R., Lindsey, B.G., Morris, K.F.: Cardio-respiratory coupling depends on the pons. Respir. Physiol. Neurobiol. **168**, 76–85 (2009)
27. Porta, A., Bassani, T., Bari, V., Tobaldini, E., Takahashi, A.C.M., Catai, A.M., Montano, N.: Model-based assessment of baroreflex and cardiopulmonary couplings during graded head-up tilt. Comput. Biol. Med. **42**, 298–305 (2012)

28. Faes, L., Nollo, G., Porta, A.: Information domain approach to the investigation of cardio-vascular, cardio-pulmonary, and vasculo-pulmonary causal couplings. Front. Physiol. **2**, 1–13 (2011)
29. De Boer, R.W., Karemaker, J.M., Strackee, J.: Hemodynamic fluctuations and baroreflex sensitivity in humans: a beat-to-beat model. Am. J. Phys. **253**, H680–H689 (1987)
30. Barrett, A.B.: Exploration of synergistic and redundant information sharing in static and dynamical Gaussian systems. Phys. Rev. E. **91**, 052802 (2015)
31. Voss, A., Schroeder, R., Heitmann, A., Peters, A., Perz, S.: Short-term heart rate variability-influence of gender and age in healthy subjects. PLoS One. **10**, e0118308 (2015)

Chapter 4
Multiscale Entropy: Recent Advances

Meng Hu and Hualou Liang

Abstract Multiscale entropy is a widely used metric for characterizing the complexity of physiological time series. The fundamental difference to classical entropy measures is it enables quantification of nonlinear dynamics underlying physiological processes over multiple time scales. The basic idea of multiscale entropy was initially developed in 2002 and has since witnessed considerable progress in methodological expansions along with growing applications. Here, we provide an overview of some recent developments in the theory, identify some methodological constraints of the originally introduced multiscale entropy analysis, and discuss some improvements that we, and others, have made regarding the definition of the time scales, its multivariate extension and improved methods for estimating the basic technique. Finally, the application of multiscale entropy to the analysis of cardiovascular data is summarized.

4.1 Introduction

The last decade has witnessed considerable progress in the development of multiscale entropy method [1], especially for the analysis of physiological time series in revealing underlying complex dynamics of the system. The concept of entropy has been widely used to measure complexity of a system. The idea was initially developed from the classic Shannon entropy [2]. In mathematics, the Shannon entropy of a discrete random variable X is defined as: $H(X) = -\sum_x P(x)log_2\,[P(x)]$, where $P(x)$ is the probability that X is in the event x, and $P(x)log_2[P(x)]$ is defined as 0 if $P(x) = 0$. From the equation, if one of the events occurs more frequent than others, observation of that event bears less informative. Conversely, less frequent events provide more information when observed. Entropy is calculated as zero when the outcome is certain. Thus, given a known probability distribution of the source, Shannon entropy quantifies all these considerations exactly, providing a way to estimate the average minimum number of bits needed to encode a string of symbols, based on the frequency of the symbols, which also provides a basis

M. Hu • H. Liang (✉)
School of Biomedical Engineering, Drexel University, Philadelphia, PA, 19104, USA
e-mail: Hualou.Liang@drexel.edu

R. Barbieri et al. (eds.), *Complexity and Nonlinearity in Cardiovascular Signals*,
DOI 10.1007/978-3-319-58709-7_4

for the ensuing development within the framework of information theory. Among many entropy measures developed such as Kolmogorov–Smirnov (K–S) entropy [3], E–R entropy [4] and compression entropy [58], approximate entropy (ApEn) [5] has been highlighted as one of most effective methods to provide a relatively robust measure of entropy, particularly for short and noisy time series. Sample Entropy [6], as a further development, was used to correct the bias of the ApEn algorithm. All these entropy methods did not consider the multiscale nature of the underlying signals, thus could yield misleading results for complex multiscale system [1]. In reality the signals derived from physiological and complex systems usually have multiple spatiotemporal scales. Therefore, the multiscale entropy [1] was proposed as a new measure of complexity in order to scrutinize complex time series by taking into account different scales. Since its first application to analysis of human heartbeat time series signal [1, 7], the multiscale entropy has been increasingly applied to analyze not only cardiovascular signal [8–10], but also many different physiological time series signals for various applications. For instance, the multiscale entropy has been employed to perform the analysis of (1) electroencephalogram [11, 12], functional magnetic resonance imaging [13] and magnetoencephalogram [14] signals to investigate the nonlinear dynamics in brain; (2 the gait time series to explore the gait dynamic [15]; (3) the laser speckle contrast images to examine the aging effect on microcirculation [16]; (4) the lung sound signal to classify the different lung status [17] and (5) the alternative medicine study [18]. In addition to the physiological signal, the multiscale entropy recently has also been utilized to the mechanic fault detection [19–21], the characterization of the physical structure of complex system [22–24], nonlinear dynamic analysis of financial market [25–29], complexity examination of traffic system [30, 31] and the nonlinear analysis in geophysics [32–34].

In terms of methodological development, the multiscale entropy can be implemented in two steps: (1) extraction of different scales of time series, and (2) calculation of entropy over those extracted scales. In the original development of the multiscale entropy [1], the scales of data are determined by the so-called coarse-grained procedure, whereby the raw time series at individual scale is first divided into the nonoverlapping windows, and each window is then replaced by its average [3], upon which the sample entropy is applied. For instance, to generate the consecutive coarse-grained time series at the scale factor Γ', the original time series was first divided into nonoverlapping windows of length Γ' and the data points within each window were averaged. For scale one (i.e., $\Gamma' = 1$), the coarse-grained time series is simply the original time series. The length of each coarse-grained time series is equal to the length of the original time series divided by the scale factor. This 'coarse-graining' procedure essentially represents a linear smoothing and decimation to progressively eliminate the fast temporal component from the original signal, which nonetheless might be suboptimal due to the way the scale is extracted [35]. On the other hand, the sample entropy has been criticized for the inaccurate estimation or too sensitive to the choices of the parameters [36]. For example, in sample entropy, the Heaviside function is employed to assess the similarity between the embedded data vectors, which is like a two-state classifier: the contributions of all the data points inside the boundary are treated equally, while the data points outside the boundary are abandoned. As a result, sample entropy may vary dramatically when the tolerance parameter r is slightly changed. As such, different methods have been developed to improve the multiscale entropy estimation. In this chapter, we aim to provide a systematic review for these algorithms and offer some insight for future directions. In what follows, we specifically focus on two aspects of development: (1) extraction of the scales, and (2) estimation of the entropy. From each step of development of the multiscale entropy method, it has also been seen that the multiscale entropy method has been being utilized in the much expanded areas. Specifically, we provide a short overview on the application of multiscale entropy to the analysis of cardiovascular time series. We finally conclude the chapter with discussions.

4.2 Overview of Multiscale Entropy

The multiscale entropy [1] was developed to measure the complexity/irregularity over the different scales of time series. The scales are determined by the 'coarse-grained' procedure, where the length of the coarse-grained time series is equal to the length of original time series divided by the scale factor. At each time scale, sample entropy is used to determine the amount of irregularity to provide the entropy estimation. The sample entropy is thought of as a robust measure of entropy at a single scale due to its insensitivity to the data length and immunity to the noise in the data [6]. It is equal to the negative natural logarithm of the conditional probability that m consecutive points that repeat themselves, within some tolerance, r, will again repeat with the addition of the next $(m + 1)$ point.

To compute sample entropy, a time series $I = \{i(1), i(2), \ldots, i(N)\}$ is first embedded in a m-dimensional space, in which the m-dimensional vectors of the time series are constructed as $x_m(k) = (i(k), i(k+1), \ldots, i(k+m-1)), k = 1 \sim N - m + 1$. In the embedded space, the match of any two vectors is defined as their distance lower than the tolerance r. The distance between two vectors refers to as the maximum difference between their corresponding scalar components. $B^m(r)$ is defined as the probability that two vectors match within a tolerance r in the m-dimensional space, where self-matches are excluded. Similarly, $A^m(r)$ is defined in thc $m + 1$ dimensional space. Sample entropy is then defined as the negative natural logarithm of the conditional probability that two sequences similar for m points remain similar at the next $m + 1$ point in the data set within a tolerance r, which is calculated as:

$$SampEn\,(I, m, r) = -\ln\left(\frac{A^m(r)}{B^m(r)}\right)$$

As a result, regular and/or periodic signals have theoretical sample entropy of 0, whereas uncorrelated random signals have maximum entropy depending on the signal length.

The multiscale entropy can then be obtained by applying the sample entropy across multiple time scales. This is achieved through a coarse-graining procedure, whereby, i.e., at the scale n, the raw time series first is divided into the nonoverlaping windows with the length of n, and each window is then replaced by its average. For instance, at the first scale, the multiscale entropy algorithm evaluates sample entropy for the original time-series. For the second scale (i.e., $n = 2$), the original time-series (length L) is first divided into non-overlapping windows of length 2. Within each window the average is taken, resulting in a new time-series of length $L/2$, over which the sample entropy is computed. The procedure is repeated until the last time scale is accomplished. Therefore the coarse-grained procedure essentially represents a linear smoothing and decimation (down sampling) of the original time series. Despite the wide applications, the multiscale entropy approach however has several limitations:

1. The coarse-graining procedure has the effect of down sampling on original time series, which reduces the original sampling rate of a time series to a lower rate, losing the high-frequency component of the signal. Thus, the multiscale entropy only captures the low-frequency components and thus does not entail the high-frequency components as the scales were extracted. Yet there is little reason to ignore the high-frequency components in signal, which may retain significant information in the system [35, 37].
2. Due to the linear operations, the algorithm for extracting the different scales is not well adapted to nonlinear/nonstationary signals, particularly in physiological system. Considering the coarse-grained process as a filter, the features of its frequency response are poor since it shows side lobes in the stopband [35, 38].
3. With 'Coarse-graining', the larger the scale factor, the shorter the coarse-grained time series. As the scale increases, the number of data points is reduced [39, 40]. The variance of the estimated entropy gets quickly increased; hence the statistical reliability of entropy measure is greatly diminished.

As a result, several methods have been proposed to overcome these shortcomings. We will review those improvements by focusing on the extraction of the scales and the enhancements of entropy estimation. Table 4.1 below briefly summarizes the key information for these methods.

Table 4.1 Summary of developed multiscale entropy methods

Method	Scale extraction	Entropy estimation	Features	Application example
Multiscale entropy [1]	Coarse-graining procedure; (Linear)	Sample entropy	The first multiscale entropy method	Heartbeat interval data
Composite multiscale entropy [39]	Modified coarse graining procedure (i.e., generating multiple coarse-grained series by using the different starting points of time series); (Linear)	Sample entropy	1. More accurate multiscale entropy estimation, even when the time series is short; 2. Increased probability of inducing undefined entropy values	Vibration data (bearing fault data)

(continued)

Table 4.1 (continued)

Method	Scale extraction	Entropy estimation	Features	Application example
Refined composite multiscale entropy [40]	Modified coarse graining procedure (i.e., generating multiple coarse-grained series by using the different starting points of time series); (Linear)	Sample entropy	1. More accurate multiscale entropy estimation, even when the time series is short; 2. Reducing the probability of undefined entropy values	NA
Short-time multiscale entropy [42]	Modified coarse graining procedure (i.e., generating multiple coarse-grained series by using the different starting points of time series); (Linear)	Sample entropy	Same with Composite multiscale entropy	Pulse wave velocity signals
Modified multiscale entropy [43]	Moving average processing; (Linear)	Sample entropy	1. Better accuracy of entropy estimation and fewer undefined entropy, even when the time series is short; 2. Heavy computational cost and the limited improvement for longer time series	Bearing vibration data
Refined multiscale entropy [38]	Low-pass Butterworth filter; (Linear)	Sample entropy	1. More accurate elimination of the components and reducing aliasing for the coarse-graining procedure; 2. Calculation of Sample entropy on each filtered component	Heart rate variability
Generalized multiscale entropy [44]	Coarse-graining procedure with the higher moment calculation (e.g., variance) to replace average calculation for the nonoverlapped segments; (Linear)	Sample entropy	New way to explain the calculated entropy results	Heartbeat volatility time series
Intrinsic mode entropy [37]	Empirical mode decomposition; (Non-linear/nonstationary)	Sample entropy	Adaptive scale extraction, suitable for nonlinear/ nonstationary time series signal	Stabilogram signals

(continued)

Table 4.1 (continued)

Method	Scale extraction	Entropy estimation	Features	Application example
Adaptive multiscale entropy [35]	Multivariate empirical mode decomposition; (Nonlinear/ nonstationary)	Sample entropy	Adaptive scale extraction, suitable for nonlinear/ nonstationary time series signal; Two scale extraction directions: 1. fine-to-coarse and 2. coarse-to-fine	Electro-neurophysiological data from monkey
Multiscale permutation entropy [54, 55]	Coarse-graining procedure; (Linear)	Permutation entropy	1. Robust for the time series with nonlinear distortion; 2. Computationally efficient	Bearing fault data
Multiscale symbolic entropy [56]	Coarse-graining procedure with the median calculation to replace average calculation for the nonoverlapped segments; (Linear)	Symbolic entropy	1. Highly resilient to outliers; 2. Robust to noisy data	Human heartbeat recordings
Multiscale compression entropy [57]	Coarse-graining procedure; (Linear)	Compression entropy	Based on data compression technique	Microvascular blood flow data
Multiscale fuzzy entropy [36, 63]	Coarse-graining procedure; (Linear)	Fuzzy entropy	Improved entropy estimation with the stronger relative consistency and less dependence on the data length	Bearing vibration signal
Multivariate multiscale entropy [65]	Coarse-graining procedure; (Linear)	Multivariate sample entropy	Be able to analyze the multivariate time series data	Human stride interval fluctuations data; cardiac interbeat interval and respiratory interbreath interval data

4.3 Methods for Scale Extraction

4.3.1 Composite Multiscale Entropy

The composite multiscale entropy [39] was developed to address the statistical reliability issue of the multiscale entropy. As described above, in original development of multiscale entropy, the length of coarse-grained time series is shorter when the scale

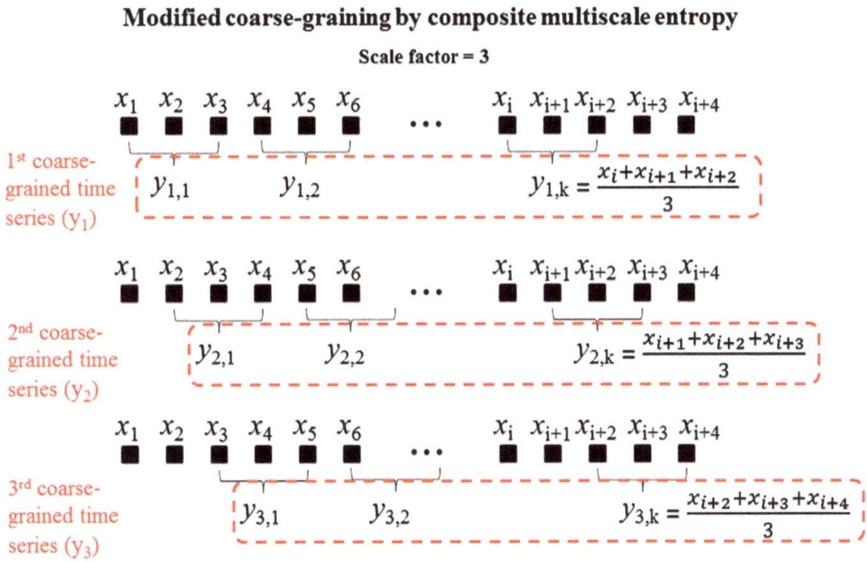

Fig. 4.1 The modified coarse-graining procedure used by composite multiscale entropy

factor becomes larger, leading to the larger variance of sample entropy estimation at greater scales. Thus large variance of estimated entropy values results in the reduction of reliability of multiscale entropy estimation.

To address this issue, the composite multiscale entropy was developed to first generate multiple unique coarse-grained time series at a certain scale factor (e.g., 3) by changing the starting point (e.g., 1st, 2nd, 3rd) of time series in the coarse-graining procedure, and then calculate sample entropy on each coarse-grained time series, and finally take average of all the calculated sample entropy values. Such a modified coarse-graining procedure is illustrated in Fig. 4.1.

As shown in [39], both the simulated (e.g., white noise and 1/f noise) and real vibration data (e.g., Figs. 8 and 10 in [39]) show that the composite multiscale entropy yielded the almost same mean estimation with the multiscale entropy, but with much more accurate individual estimation, which is manifested as the smaller standard deviation values of entropy estimation on scales by the composite multiscale entropy than those by the multiscale entropy.

Subsequently, the composite multiscale entropy [40] was refined to further improve the performance in terms of entropy estimation, as it was observed that in composite multiscale entropy, though the multiple coarse-grained time series at individual scales facilitate the accurate estimation of entropy, the method also increases the probability of inducing undefined entropy. Instead of calculating average of sample entropy values over all the coarse-grained time series at single scale, the refined composite multiscale entropy first collected the number of matched patterns over all the coarse-grained time series for the selected parameter m (i.e., collecting the numbers of the matched patterns for both m and $m + 1$, according to the sample entropy algorithm), and then applied the sample entropy algorithm to calculate the negative natural logarithm of the conditional probability, hence reducing the probability of undefined entropy values. In the Reference [40], the simulated data (white noise and 1/f noise) showed that the refined composite multiscale entropy is superior to both the multiscale entropy and composite multiscale entropy by increasing the accuracy of entropy estimation and reducing the probability of inducing undefined entropy, particularly for short time series data. This can be clearly seen from the Table 2 in [40] that the refined composite multiscale entropy can still work on the shorter time series data and provide the smaller standard deviation of entropy estimations on scales than multiscale entropy and composite multiscale entropy. Furthermore, the refined composite multiscale entropy can be further improved by incorporating with empirical mode decomposition (EMD) algorithm for the purpose of removing the data baseline before calculating the entropy [41].

It should be mentioned that by the same idea of composite multiscale entropy in generating multiple coarse-grained time series, the so-called short-time multiscale entropy [42] was independently developed to address the issue of multiscale entropy when facing with short time series data.

4.3.2 Modified Multiscale Entropy for Short Time Series

As indicated by its name, the modified multiscale entropy was proposed to address the reliability of multiscale entropy when time series is short [43]. With the same rationale as the above-mentioned composite multiscale entropy, this method tended to offer the better accuracy of entropy estimation or fewer undefined entropy when the time series is short. In essence, the modified multiscale entropy replaced the coarse-graining procedure with a moving-average process, in which a moving window with length of scale factor was slid through the whole time series point-by-point to generate the new time series at the scale. By doing so, data length was largely reserved, compared to the coarse-graining procedure. For instance, at the scale of 2, a time series with 1000 data points is changed into a new time series with 500 data points by the coarse graining procedure, while by the moving-average procedure, the new time series becomes 999 data points.

In [43], both simulation (white noise and 1/f noise) and real vibration data were performed to validate the effectiveness of proposed method for short time series data. For both white noise and 1/f noise with 1 k data points, the modified multiscale entropy provides the sample entropy curve with the less fluctuation than the conventional sample entropy in a typical example. By using 200 independent noise samples for both white noise and 1/f noise, each noise sample contained 500 data points, the results show that the modified multiscale entropy offers the nearly equal mean of entropy estimations with the sample entropy over independent samples, but with the less the standard deviation of entropy estimations than the sample entropy, indicating that the modified multiscale entropy and multiscale entropy methods are nearly equivalent statistically; however, the former method can provide a more accurate estimate than the conventional multiscale entropy. The applications of methods to the real bearing vibration data also show the consistent conclusions. However, it should be mentioned that considering the computational cost and the limited improvement for longer time series, the modified multiscale entropy is not suitable for direct application in analyzing longer time series. For instance, from the Table 2 in [43], it is seen that for the 1/f noise with data length of 30 k, the modified multiscale entropy provides the almost same entropy estimations (including mean and standard deviation over independent samples) with the sample entropy, but demanding significantly heavier computational cost.

4.3.3 Refined Multiscale Entropy

The refined multiscale entropy [38] is a method to address two drawbacks of the multiscale entropy: (1) the 'coarse-graining' procedure can be considered as applying a finite-impulse response (FIR) filter to the original series x and then downsampling the filtered time series with the scale factor, in which frequency response of the FIR filter can be characterized as a very slow roll-off of the main lobe, large transition band, and important side lobes in the stopband, thus introducing aliasing when the subsequent downsampling procedure is applied; and (2) the parameter r for determining the similarity between two patterns remains constant for all the scales as a percentage of standard deviation of the original time series, thus leading to artificially decreasing entropy rate, as the patterns more likely becomes indistinguishable at higher scale in such a setting.

In order to overcome both shortcomings, the refined multiscale entropy (1) replaced the FIR filter with a low-pass Butterworth filter, whose frequency response was flat in the passband, side lobes in the stopband were not present, and the roll-off was fast, thus ensuring a more accurate elimination of the components and reducing aliasing when the filtered series were downsampled; and (2) let parameter r be continuously updated as a percentage of the standard deviation (SD) of the filtered series, thus compensating the decrease of variance with the elimination of the fast temporal scales.

The performance of the refined multiscale entropy was verified and examined by Gaussian white noise, 1/f noise and autoregressive processes, as well as real 24-h Holter recordings of heart rate variability (HRV) obtained from healthy and aortic stenosis (AS) groups. It is worth noting that the results of Gaussian white noise and 1/f noise by the refined multiscale entropy are opposite to those obtained via the multiscale entropy. The main reason leading to this discrepancy is that two methods apply the different calculations of the similarity parameter, r for the coarse-graining procedure. In the original multiscale entropy, the parameter r remains constant for all scale factors as a percentage of standard deviation of the original time series [1], whereas the parameter r in the refined multiscale entropy is continuously updated as a percentage of the standard deviation of the filtered series [38]. It has been pointed out [38] that the coarse-graining procedure in original multiscale entropy eliminates the fast temporal scales as the scale factor increases, acting as a low-pass filter. As such, the coarse-grained series are usually characterized by a lower standard deviation compared to the original time series. Thus, if the parameter r is kept constant for all the scales, more and more patterns will be considered similar with the increasing scale factor, hence increasing estimated regularity and reporting artificially decreasing entropy with the scale factor.

4.3.4 Generalized Multiscale Entropy

The generalized multiscale entropy [44] was developed to generalize the multiscale entropy. In the coarse-graining procedure, a time series was first divided into non-overlapping segments with the length of scale factor, and then the average was taken within each segment, which can been thought as the first moment. Instead of the first moment, the generalized multiscale entropy was proposed to use an arbitrary moment (e.g., volatility when the second moment) to generate the coarse-grained time series for entropy analysis. This approach may offer a new way to explain the calculated entropy results. For instance, when using the second moment, the multiscale entropy represents the multiscale complexity of the volatility of data. This approach has been applied to study cardiac interval time series from three groups, comprising health young and older subjects, and patients with chronic heart failure syndrome [44]. The results show that the complexity indices of healthy young subjects were significantly higher than those of healthy older subjects, whereas the complexity indices of the healthy older subjects were significantly higher than those of the heart failure patients. These results support their hypothesis that the heartbeat volatility time series from healthy young subjects are more complex than those of healthy older subjects, which are more complex than those from patients with heart failure.

4.3.5 Intrinsic Mode Entropy

The intrinsic mode entropy [37] identified two shortcomings of the multiscale entropy: (1) the high-frequency components were eliminated by the coarse-graining procedure, which may contain relevant information for some physiological data, and (2) the coarse-graining procedure, while extracting the scales, may be not adapted to nonstationary/nonlinear signal, which is quite common in the physiological systems.

The intrinsic mode entropy was proposed to directly address these issues by exploiting a fully adaptive, data-driven time series decomposition method, namely empirical mode decomposition (EMD) [45], to extract the scales intrinsic to the data. The EMD adaptively decomposes a time series signal, by means of the so-called the sifting method, into a finite set of amplitude- and/or frequency-modulated (AM/FM) modulated components, referred to as intrinsic mode functions (IMFs) [45]. IMFs satisfy the requirements that the mean of the upper and lower envelops is locally zero and the number of extrema and the number of zero-crossing differ by at most one, and thus represent the intrinsic oscillation modes of data on the different frequency scales. By virtue of the EMD, the time series $x(t)$ can be decomposed as $x(t) = \sum_{j=1}^{k} C_j(t) + r(t)$, where $C_j(t), j = 1, \ldots, k$ are the IMFs and $r(t)$ is the monotonic residue. The EMD algorithm is briefly described as follows.

1. Let $\tilde{x}(t) = x(t)$;
2. Find all local maxima and minima of $\tilde{x}(t)$;
3. Interpolate through all the minima (maxima) to obtain the lower (upper) signal envelop $e_{\min}(t)$ $(e_{\max}(t))$;
4. Compute the local mean $m(t) = \frac{e_{\min}(t)+e_{\max}(t)}{2}$;
5. Obtain the detail part of signal $c(t) = \tilde{x}(t) - m(t)$;
6. Let $\tilde{x}(t) = c(t)$ and repeat the process from step 2 until $c(t)$ becomes an IMF.

Compute the residue $r(t) = x(t) - c(t)$ and go back to step 2 with $\tilde{x}(t) = r(t)$, until the monotonic residue signal is left.

As such, in the intrinsic mode entropy, a time series is first adaptively decomposed into several IMFs with distinct frequency bands by EMD, and then the multiple scales of the original signal can be obtained by the cumulative sums of the IMFs, starting from the finest scales and ending with the whole signal, over which the sample entropy is applied. The usefulness of intrinsic mode entropy was demonstrated by its application to real stabilogram signals for discrimination between elderly and control subjects.

It bears noting that relying on the EMD method, the intrinsic mode entropy inevitably suffers from both the mode-misalignment and mode-mixing problems, particularly in the analysis of multivariate time series data [46]. The mode-misalignment refers to a problem where the common frequency modes across a multivariate data appear in the different-index IMFs, thus resulting that the IMFs are not matched either in the number or in scale, whereas the mode-mixing is manifested by a single IMF containing multiple oscillatory modes and/or a single mode residing in multiple IMFs which may in some cases compromise the physical meaning of IMFs and practical applications [47]. Both problems can be simply illustrated by a toy example shown in Fig. 4.2. Because the intrinsic mode entropy is applied to each time series separately, these two problems may contribute to inappropriate operation (e.g., comparison) of entropy values at the 'same' scale factor from different time series, for the 'same' scale from different time series could be in the largely different frequency ranges.

Therefore, an improved intrinsic mode entropy method [46] was introduced to address the potential mode-misalignment and mode-mixing problems based on the multivariate empirical mode decomposition (MEMD), which directly works with multivariate data to greatly mitigate above two problems, generating the aligned IMFs. In this method, the MEMD [48] plays an important role for the extraction of the scales. The MEMD is the multivariate extension of EMD, thus mitigating

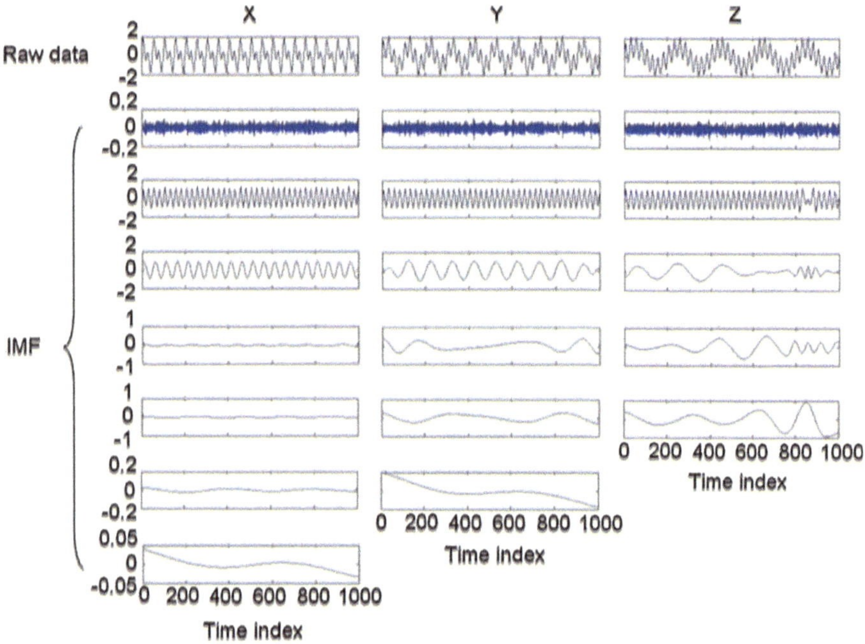

Fig. 4.2 Example of mode-mixing and mode-misalignment by original empirical mode decomposition

both mode-misalignment and mode-mixing problems. An important step in the MEMD method, distinct from the EMD, is the calculation of local mean, as the concept of local extrema is not well defined for multivariate signals. To deal with this problem, MEMD projects the multivariate signal along different directions to generate the multiple multidimensional envelops; these envelops are then averaged to obtain the local mean. The details of MEMD can be found in [48]. Figure 4.3 demonstrates the decomposition of same data as in Fig. 4.2 by the MEMD to show that MEMD is effective to produce the aligned IMFs. The application of the improved intrinsic mode entropy approach to real local field potential data from visual cortex of monkeys illustrates that this approach is able to capture more discriminant information than the other methods [49].

Thus far, the EMD family has been extensively used in multiscale entropy analysis [41, 50]. It should be noted that one important extension of EMD, ensemble EMD [47], has also been applied to multiscale entropy method to solve the mode-mixing problem [51, 52]. However, the aligned IMF numbers from different time series are not guaranteed by this method. In addition, to preserve the high frequency component in entropy analysis, the hierarchical entropy was used essentially based on wavelet decomposition [53], which is effective for the analysis of nonstationary time series data.

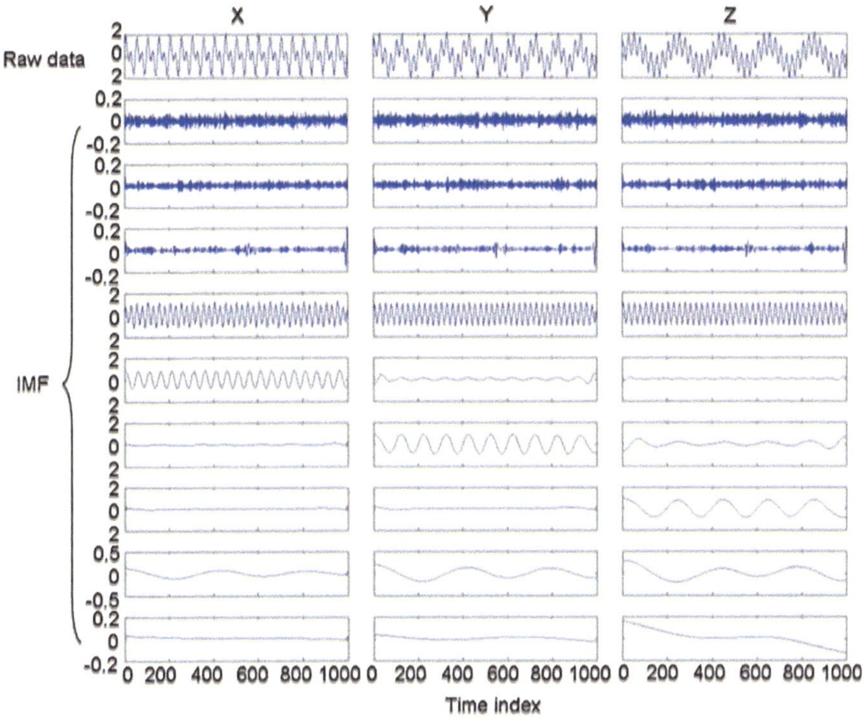

Fig. 4.3 Decomposition of same data in Fig. 4.2 by multivariate empirical mode decomposition (MEMD), showing the aligned decomposed components

4.3.6 Adaptive Multiscale Entropy

The adaptive multiscale entropy (AME) [35] was developed to provide a comprehensive means for the scales extracted from time series including both the 'coarse-to-fine' and 'fine-to-coarse' scales. The method was specifically to address the issue of multiscale entropy that the high frequency component is eliminated. With the AME, the scale extraction is adaptive, completely driven by the data via EMD, thus suitable for nonstationary/nonlinear time series data. Moreover, by employing the multivariate extension of empirical mode decomposition (i.e., MEMD), the proposed AME can fully address the mode-misalignment and mode-mixing problems induced by univariate empirical mode decomposition for the analysis of individual time series, thus producing the aligned IMF components and ensuring proper operations (e.g., comparison) of entropy estimation at one scale over multiple time series data.

To implement the AME, the MEMD method is first applied to decompose data into the aligned IMFs. The scales are then selected by consecutively removing either the high-frequency or low-frequency IMFs from the original data. The way to select the scales results in two algorithms, namely the fine-to-coarse AME and the coarse-to-fine AME, which essentially represent the multiscale low-pass and high-pass filtering of the original signal, respectively. The sample entropy is applied to the selective scales to estimate the entropy measure. When applying this approach to physiological time series signal (e.g., neural data), two issues [49] had to be solved: (1) the physiological data are often collected over certain time period from multiple channels across many trials, which can be represented as a three-dimensional matrix, i.e. TimePoints $\times$ Channels $\times$ Trials, thus not directly solvable by the MEMD; and (2) the physiological recordings are usually collected over many trials spanning from days to months, or even years, so that the dynamic ranges of multiple signals are likely to be of high degree of variability, which can have detrimental effect upon the final decomposition of MEMD. Therefore, the AME adopts two important preprocessing steps to accommodating the multivariate data. First, the high-dimensional physiological data (e.g. TimePoints $\times$ Channels $\times$ Trials) is first reshaped into such a two-dimensional time series as TimePoints $\times$ [Channels $\times$ Trials] before subject to the MEMD analysis. It is an important step to make sure that all the IMFs be aligned not only across channels, but also across trials. Second, in order to reduce the variability among neural recordings, individual time series in the reshaped matrix is normalized against their temporal standard deviation before the MEMD is applied. After the MEMD decomposition, those extracted standard deviations are then restored to the corresponding IMFs.

Simulations demonstrate that the AME is able to adaptively extract the scales inherent in the nonstationary signal and that the AME works with both the coarse scales and fine scales in the data [35]. The application to real local field potential data in the visual cortex of monkeys suggests that the AME is suitable for entropy analysis of nonlinear/nonstationary physiological data, and outperforms the multiscale entropy in revealing the underlying entropy information retained in the intrinsic scales. The AME method has been further extended as adaptive multiscale cross-entropy (AMCE) [49] to assess the nonlinear dependency between time series signals. Both the AME and AMCE are employed to neural data from the visual cortex of monkeys to explore how the perceptual suppression is reflect by neural activity within individual brain areas and functional connectivity between areas.

4.4 Methods for Entropy Estimation

4.4.1 Multiscale Permutation Entropy

The permutation entropy [54], instead of the sample entropy, has been applied to conduct the multiscale entropy analysis [55]. By using the rank order value of embedded vector, the permutation entropy is robust for the time series with nonlinear distortion, and is also computationally efficient. Specifically, with the m-dimensional delay, the original time series is transformed into a set of embedded vectors. Each vector is then represented by its rank order. For instance, a vector,

[12, 56, 0.0003, 100,000, 50] can be represented by its rank order, [1–5]. We can see that the rank order is insensitive to the magnitude of the data, though the values may be at largely different scales. Subsequently, for a given embedded dimension (e.g., m dimension), the rate of each possible rank order, out of m! possible rank orders, is calculated over all the embedded vectors, forming a probability distribution over all the rank orders. The Shannon entropy is finally applied to the probability distribution to obtain the entropy estimation.

In the similar fashion, the multiscale symbolic entropy [56] was developed to reliably assess the complexity in noisy data while being highly resilient to outliers. In the time series data, the outliers are often illustrated as those data points with the significantly higher or lower magnitudes than most of data points due to some internal/external uncontrolled influence. In this approach, the variation at a time point consists of both the magnitude (absolute value) and the direction (sign). It is hypothesized that dynamics in the sign time series can adequately reflect the complexity in raw data and that the complexity estimation based on the sign time series is more resilient to outliers as compared to raw data. To implement this method, the original time series at a given scale is first divided into non-overlapped segments, within each of which the median is calculated to generate a coarse-grained time series. The sign time series of the coarse grained signal is then generated by considering the direction of change at each point (i.e., 1 for increasing and 0 otherwise). The discrete probability count and the Shannon entropy are applied to the sign data to obtain the entropy estimation. The multiscale measure is obtained when the above procedure goes over all the predefined scales. This method has been successfully applied to the analysis of human heartbeat recordings, showing the robustness to noisy data with outliers.

4.4.2 *Multiscale Compression Entropy*

The multiscale compression entropy [57] has been reported, which replaced the sample entropy with the compression entropy [58] to conduct the multiscale entropy analysis. The basic idea of compression entropy is the smallest algorithm that produces a string is the entropy of that string from algorithmic information theory, which can be approximated by the data compression techniques. For instance, the compression entropy per symbol can be represented by the ratio of compressed text to the original text length, if the length of the text to be compressed is sufficiently large and if the source is an ergodic process. The compression entropy provides an indication to which degree the time series under study can be compressed using the detection of recurring sequences. The more frequent the occurrences of sequences (and thus, the more regular the time series), the higher the compression rate. The ratio of the lengths of the compressed to the uncompressed time series is used as a complexity measure and identified as compression entropy. The multiscale compression entropy exploits the coarse-graining procedure to generate the scales of data, over which the compression entropy is applied. This approach has been applied to the entropy analysis for the microvascular blood flow signals [57]. In this application, microvascular blood flow was continuously monitored with laser speckle contrast imaging (LSCI) and with laser Doppler flowmetry (LDF) simultaneously from healthy subjects. The results show that, for both LSCI and LDF time series, the compression entropy values are less than 1 for all of the scales analyzed, suggesting that there are repetitive structures within the data fluctuations at all scales.

4.4.3 *Fuzzy Entropy*

In the family of approximate entropy [5], for example, sample entropy [6], the similarity of vectors (or patterns) from a time series is a key component for accurate estimation of entropy. The sample entropy (and approximate entropy) assesses the similarity between vectors based on the Heaviside function, as shown in Fig. 4.4. It is evident that the boundary of Heaviside function is rigid, which leads to that all the data points inside boundary are treated equally, whereas the data points outside

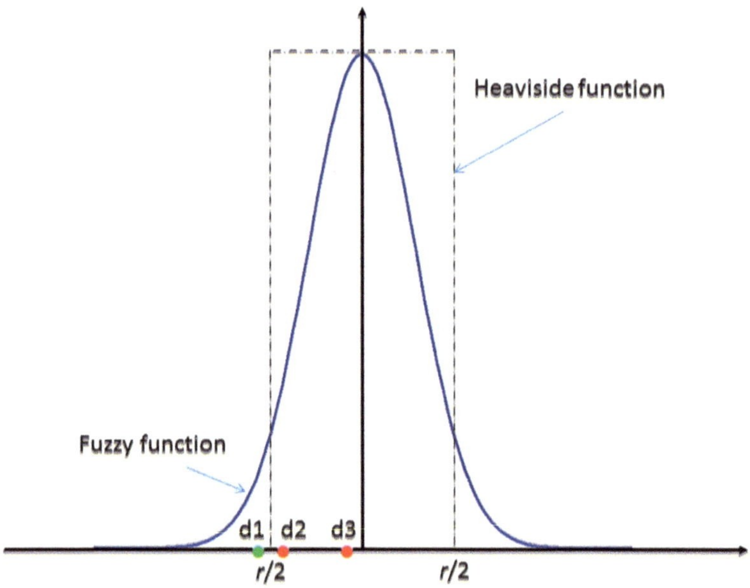

Fig. 4.4 Heaviside function (*dash dotted black*) and fuzzy function (*solid blue*) for similarity in entropy estimation (modified based on [21]. As figure shown, by the Heaviside function, both points d2 (*red*) and d3 (*red*) are considered within the boundary and both similar to the original point (i.e., '0'); whereas another point d1 (*green*), very close to d2 though, is considered dissimilar because it falls just outside the boundary. Thus a slight increase in *r* will make the boundary encloses the point d1, and then the conclusion totally changes. However, the fuzzy function will largely relieve this issue

the boundary are excluded no matter how close this point locates to boundary. Thus, the estimation of sample entropy is highly sensitive to the change of the tolerance, r or data point location.

The fuzzy entropy [36] aimed to improve the entropy estimation at just this point, based on the concept of fuzzy sets [59], which adopted the "membership degree" with a fuzzy function that associates each point with a real number in a certain range (e.g., [0, 1]). The fuzzy entropy employed the family of exponential function as the fuzzy function to obtain the fuzzy measurement of similarity, which bears two desired properties: (1) being continuous so that the similarity does not change abruptly; (2) being convex so that the self-similarity is the maximum. In addition, the fuzzy entropy removes the baseline of the vector sequences, so that the similarity measure more relies on vectors' shapes rather than their absolute coordinates, which makes the similarity definition fuzzier.

Extensive simulations and application [60] to experimental electromyography (EMG) suggest that the fuzzy entropy is a more accurate entropy measure than the approximate entropy and sample entropy, and exhibits the stronger relative consistency and less dependence on the data length, thus providing an improved evaluation of signal complexity, especially for short time series contaminated by noise. For instance, as shown from the Fig. 4.2 in [60], with only 50 data samples, the fuzzy entropy can successfully discriminate three periodical sinusoidal time series with different frequencies, outperforming the approximate entropy and sample entropy. And from the Fig. 5 in [60], it is observed that the fuzzy entropy can consistently separate the different logistic datasets contaminated by noises with different noise levels, while it becomes difficult for the approximate entropy and sample entropy to distinguish those data. In addition to the exponential function, the nonlinear sigmoid and Gaussian functions have also been used to replace the Heaviside function for similarity measure [61]. The fuzzy entropy was also extended to cross-fuzzy entropy to test nonlinear pattern synchrony of bivariate time series [62]. It has been shown that the fuzzy entropy can be also used in the multiscale entropy analysis by replacing the sample entropy [63, 64].

4.4.4 Multivariate Multiscale Entropy

The multivariate multiscale entropy [65] was introduced to perform entropy analysis for multivariate time series, which has been increasingly common for physiological data due to the advance in recording techniques. The multivariate multiscale entropy adopts the coarse-graining procedure to extract the scales of data, and then extends the sample entropy algorithm to multivariate sample entropy for entropy estimation of coarse-grained multivariate data. The detailed multivariate sample entropy algorithm can be found in [65]. The simulations and a large array of real-world applications [66] including the human stride interval fluctuations data, cardiac interbeat interval and respiratory interbreath interval data, 3D ultrasonic anemometer data taken in the north-south, east-west, and vertical directions etc., have demonstrated the effectiveness of this approach in assessing the underlying dynamics of multivariate time series.

It is worth to mention that copula can be used to implement multivariate multiscale entropy by incorporating the multivariate empirical mode decomposition and the Renyi entropy [67]. The Renyi entropy (or information) is a probability-based unified entropy measure, which is considered to generalize the Hartley entropy, the Shannon entropy, the collision entropy and the min-entropy. The Renyi entropy has been shown as a robust (multivariate) entropy estimation when its implementation algorithm is based on the copula of multivariate distribution [67]. The copula [68, 69] can be considered as the transformation which standardizes the marginal of multivariate data to be uniform on [0, 1] while preserving many of the distribution's dependence properties including its concordance measures and its information, which has been widely recognized in many fields, e.g., finance [70, 71], physiology [72–74], etc. Thus, the Renyi entropy estimation is based entirely on the ranks of multivariate data, therefore robust to outliers. As such, the application of Renyi entropy to the adaptively extracted scales of multivariate data by the multivariate empirical mode decomposition may provide a potential option to implement multivariate multiscale entropy.

The above multivariate entropy methods all assess the simultaneous dependency for multivariate data, while the transfer entropy [75] has been developed to measure the directed information transfer between time series, revealing a causal relationship between signal rather than correlation. The multiscale transfer entropy [76] makes use of the wavelet-based method to extract multiple scales of data, and then measure directional transfer of information between coupled systems at the multiple scales. This approach has demonstrated its effectiveness by extensive simulations and application to real physiological (heart beat and breathe) and robotic (composed of one sensor and one actuator) data.

4.5 Multiscale Entropy Analysis of Cardiovascular Time Series

Nonlinear analysis of cardiovascular data has been widely recognized to provide relevant information on psychophysiological and pathological states. Among others, entropy measure has been serving as a powerful tool to quantify the cardiovascular dynamics of a time series over multiple time scales [1] through approximate entropy [5], sample entropy [6] and multiscale entropy [1]. When the multiscale entropy was initially developed, it has been applied to analysis of heart beat signal for diagnostics, risk stratification, detection of toxicity of cardiac drugs and study of intermittency in energy and information flows of cardiac system [1, 44, 77]. In the original publication for the sample entropy [1], the multiscale entropy was applied to analyze the heartbeat intervals time series from (1) healthy subjects, (2) subjects with congestive heart failure, and (3) subjects with the atrial fibrillation. The analysis results of multiscale entropy show that at scale of 20 (note not the original scale) the entropy value for the coarse-grained time series derived from healthy subjects is significantly higher than those for atrial fibrillation and congestive heart failure, facilitating to address the longstanding paradox for the applications of traditional single-scale entropy methods to physiological time series, that is, the traditional entropy methods may yield the higher complexity for certain pathologic processes associated with random outputs than that for healthy dynamics exhibiting long-range correlations, but it is believed that disease states or aging may be defined by a sustained breakdown of long-range correlations and thus loss of information, i.e., less complexity. This work suggests that the paradox may be due to the fact that conventional algorithms fail to account for the multiple time scales inherent in physiologic dynamics, which can be discovered by the multiscale entropy. In addition, the multiscale entropy has been applied, but not limited, to analysis of heart rate variability for the objective quantification of psychological traits through autonomic nervous system biomarkers [78], detection of cardiac autonomic neuropathy [79] and assessing the severity of sleep disordered breathing [80], to analysis of microvascular blood flow signal for better understanding of the peripheral cardiovascular system [57], to analysis of interval variability from Q-waveonset to T-wave end (QT) derived from 24-hour Holter recordings for improving identification of condition of the long QT syndrome type 1 [81] and to analysis of pulse wave velocity signal for differentiating among healthy, aged, and diabetic populations [42]. A typical example of application of multiscale entropy to the heart rate variability analysis [80] is to investigate the relationship between the obstructive sleep apnea (OSA) and the complexity of heart rate variability to identify the predictive value of the heart rate variability analysis in assessing the severity of OSA. In the study, the R-R intervals from 10 segments of 10-min electrocardiogram recordings during non-rapid eye movement sleep at stage N2 were collected from four groups of subjects: (1) the normal snoring subjects without OSA, (2) mild OSA, (3) moderate OSA and (4) severe OSA. The multiscale entropy was applied to perform the heart rate variability analysis, in which the multiple scales were divided into the small scale (scale 1–5) and the large scale (scale 6–10). The analysis results show that the entropy at the small scale could successfully distinguish the normal snoring group and the mild OSA group from the moderate and severe groups, and a good correlation between the entropy at the small scale and the apnea hypopnea index was displayed, suggesting that the multiscale entropy analysis at the small scale may serve as a simple preliminary screening tool for assessing the severity of OSA. Except the heart rate variability signal, the multiscale entropy has been proved as a powerful analysis tool for many other physiological signals, e.g., for analysis of pulse wave velocity signal [42]. In the study, the pulse wave velocity series were recorded from 4 groups of subjects: (1) the healthy young group, (2) the middle-aged group without known cardiovascular disease, (3) the middle-aged group with well-controlled diabetes mellitus type 2 and (4) the middle-aged group with poorly-controlled diabetes mellitus type 2. By applying the multiscale entropy analysis, the results show that the multiscale entropy can produce significant differences in entropies between the different groups of subjects, demonstrating a promising biomarker for differentiating among healthy, aged, and diabetic populations. It is worth to mention that an interesting frontier in nonlinear analysis on heart rate variability (HRV) data was represented by the assessment of psychiatric disorders. Specifically, multiscale entropy was performed on the R-R interval series to assess the heartbeat complexity as an objective clinical biomarker for mental disorders [82]. In the study, the R-R interval data were acquired from the bipolar patients who exhibited mood states among depression, hypomania, and euthymia. Multiscale entropy analysis was applied to the heart rate variability to discriminate the three pathological mood states. The results show that the euthymic state is associated to the significantly higher complexity at all scales than the depressive and hypomanic states, suggesting a potential utilization of the heart rate variability complexity indices for a viable support to the clinical decision. Recently, an instantaneous entropy measure based on the inhomogeneous point-process theory is an important methodological advance [83]. This novel measure has been successfully used for analyzing heartbeat dynamics of healthy subjects and patients with cardiac heart failure together with gait recordings from short walks of young and elderly subjects. It therefore offers a promising mathematical tool for the dynamic analysis of a wide range of applications and to potentially study any physical and natural stochastic discrete processes [84]. With the further advance of multiscale entropy method, it is expected that this method will make much more contributions in discovering the nonlinear structure properties in cardiac systems.

4.6 Conclusion

This systematic review summarizes the multiscale entropy and its many variations mainly from two perspectives: (1) the extraction of multiple scales, and (2) the entropy estimation methods. These methods are designed to improve the accuracy and precision of the multiscale entropy method, especially when the time series is short and noisy. Given a large cohort of multiscale entropy methods rapidly developed within the past decade, it becomes clear that the multiscale entropy is an emerging technique that can be used to evaluate the relationship between complexity and health in a number of physiological systems. This can be seen, for instance, in the analysis of cardiovascular time series with the multiscale entropy measure. It is our hope that this review may serve as a reference to the development and application of multiscale entropy methods, to better understand the pros and cons of this methodology, and to inspire new ideas for further development for years to come.

References

1. Costa, M.D., Goldberger, A.L., Peng, C.K.: Multiscale entropy analysis of physiologic time series. Phys. Rev. Lett. **89**, 0621021–4 (2002)
2. Shannon, C.E.: A Mathematical Theory of Communication. Bell Syst. Tech. J. **27**(3), 379–423 (1948)
3. Grassberger, P., Procaccia, I.: Estimation of the Kolmogorov entropy from a chaotic signal. Phys. Rev. A. **28**, 2591–2593 (1983)
4. Eckmann, J.P., Ruelle, D.: Ergodic theory of chaos and strange attractors. Rev. Mod. Phys. **57**, 617–656 (1985)
5. Pincus, S.M.: Approximate entropy as a measure of system complexity. Proc. Natl. Acad. Sci. **88**, 2297–2301 (1991)
6. Richman, J.S., Moorman, J.R.: Physiological time-series analysis using approximate entropy and sample entropy. Amerian Journal of Physiology-Heart and Circulatory. Physiology. **278**, H2039–H2049 (2000)
7. Costa, M.D., Goldberger, A.L., Peng, C.K.: Multiscale entropy analysis of biological signals. Phys. Rev. E. **71**, 021906 (2005)
8. Humeau-Heurtier, A., Wu, C.W., Wu, S.D., Mahe, G., Abraham, P.: Refined multiscale Hilbert–Huang spectral entropy and its application to central and peripheral cardiovascular data. IEEE Trans. Biomed. Eng. **63**(11), 2405–2415 (2016)
9. Silva, L.E., Lataro, R.M., Castania, J.A., da Silva, C.A., Valencia, J.F., Murta Jr., L.O., Salgado, H.C., Fazan Jr., R., Porta, A.: Multiscale entropy analysis of heart rate variability in heart failure, hypertensive, and sinoaortic-denervated rats: classical and refined approaches. Am. J. Phys. Regul. Integr. Comp. Phys. **311**(1), R150–R156 (2016)
10. Liu, T., Yao, W., Wu, M., Shi, Z., Wang, J., Ning, X.: Multiscale permutation entropy analysis of electrocardiogram. Physica A. **471**, 492–498 (2017)
11. Liu, Q., Chen, Y.F., Fan, S.Z., Abbod, M.F., Shieh, J.S.: EEG artifacts reduction by multivariate empirical mode decomposition and multiscale entropy for monitoring depth of anaesthesia during surgery. Med. Biol. Eng. Comput. (2016). Online, doi: 10.1007/s11517-016-1598-2.
12. Shi, W., Shang, P., Ma, Y., Sun, S., Yeh, C.H.: A comparison study on stages of sleep: Quantifying multiscale complexity using higher moments on coarse-graining. Commun. Nonlinear Sci. Numer. Simul. **44**, 292–303 (2017)
13. Grandy, T.H., Garrett, D.D., Schmiedek, F., Werkle-Bergner, M.: On the estimation of brain signal entropy from sparse neuroimaging data. Sci. Rep. **6**, 23073 (2016)
14. Kuo, P.C., Chen, Y.T., Chen, Y.S., Chen, L.F.: Decoding the perception of endogenous pain from resting-state MEG. NeuroImage. **144**, 1–11 (2017)
15. Costa, M., Peng, C.K., Goldberger, A.L., Hausdorff, J.M.: Multiscale entropy analysis of human gait dynamics. Physica A. **330**, 53–60 (2003)
16. Khalil, A., Humeau-Heurtier, A., Gascoin, L., Abraham, P., Mahe, G.: Aging effect on microcirculation: a multiscale entropy approach on laser speckle contrast images. Med. Phys. **43**(7), 4008–4015 (2016)
17. Rizal, A., Hidayat, R., Nugroho, H.A.: Multiscale Hjorth descriptor for lung sound classification. International Conference on Science and Technology, 160008–1 (2015)
18. Ma, Y., Zhou, K., Fan, J., Sun, S.: Traditional Chinese medicine: potential approaches from modern dynamical complexity theories. Front. Med. **10**(1), 28–32 (2016)
19. Li, Y., Yang, Y., Li, G., Xu, M., Huang, W.: A fault diagnosis scheme for planetary gearboxes using modified multi-scale symbolic dynamic entropy and mRMR feature selection. Mech. Syst. Signal Process. **91**, 295–312 (2017)
20. Aouabdi, S., Taibi, M., Bouras, S., Boutasseta, N.: Using multi-scale entropy and principal component analysis to monitor gears degradation via the motor current signature analysis. Mech. Syst. Signal Process. **90**, 298–316 (2017)

21. Zheng, J., Pan, H., Cheng, J.: Rolling bearing fault detection and diagnosis based on composite multiscale fuzzy entropy and ensemble support vector machines. Mech. Syst. Signal Process. **85**, 746–759 (2017)
22. Zhuang, L.X., Jin, N.D., Zhao, A., Gao, Z.K., Zhai, L.S., Tang, Y.: Nonlinear multi-scale dynamic stability of oil–gas–water three-phase flow in vertical upward pipe. Chem. Eng. J. **302**, 595–608 (2016)
23. Tang, Y., Zhao, A., Ren, Y.Y., Dou, F.X., Jin, N.D.: Gas–liquid two-phase flow structure in the multi-scale weighted complexity entropy causality plane. Physica A. **449**, 324–335 (2016)
24. Gao, Z.K., Yang, Y.X., Zhai, L.S., Ding, M.S., Jin, N.D.: Characterizing slug to churn flow transition by using multivariate pseudo Wigner distribution and multivariate multiscale entropy. Chem. Eng. J. **291**, 74–81 (2016)
25. Xia, J., Shang, P., Wang, J., Shi, W.: Classifying of financial time series based on multiscale entropy and multiscale time irreversibility. Physica A. **400**(15), 151–158 (2014)
26. Xu, K., Wang, J.: Nonlinear multiscale coupling analysis of financial time series based on composite complexity synchronization. Nonlinear Dyn. **86**, 441–458 (2016)
27. Lu, Y., Wang, J.: Nonlinear dynamical complexity of agent-based stochastic financial interacting epidemic system. Nonlinear Dyn. **86**, 1823–1840 (2016)
28. Hemakom, A., Chanwimalueang, T., Carrion, A., Aufegger, L., Constantinides, A.G., Mandic, D.P.: Financial stress through complexity science. IEEE J. Sel. Topics Signal Process. **10**(6), 1112–1126 (2016)
29. Fan, X., Li, S., Tian, L.: Complexity of carbon market from multiscale entropy analysis. Physica A. **452**, 79–85 (2016)
30. Wang, J., Shang, P., Zhao, X., Xia, J.: Multiscale entropy analysis of traffic time series. Int. J. Mod. Phys. C. **24**, 1350006 (2013)
31. Yin, Y., Shang, P.: Multivariate multiscale sample entropy of traffic time series. Nonlinear Dyn. **86**, 479–488 (2016)
32. Guzman-Vargas, L., Ramirez-Rojas, A., Angulo-Brown, F.: Multiscale entropy analysis of electroseismic time series. Nat. Hazards Earth Syst. Sci. **8**, 855–860 (2008)
33. Zeng, M., Zhang, S., Wang, E., Meng, Q.: Multiscale entropy analysis of the 3D near-surface wind field. World Congress on Intelligent Control and Automation, pp. 2797–2801, IEEE, Piscataway, NJ (2016)
34. Gopinath, S., Prince, P.R.: Multiscale and cross entropy analysis of auroral and polar cap indices during geomagnetic storms. Adv. Space Res. **57**, 289–301 (2016)
35. Hu, M., Liang, H.: Adaptive multiscale entropy analysis of multivariate neural data. IEEE Trans. Biomed. Eng. **59**(1), 12–15 (2012)
36. Chen, W., Wang, Z., Xie, H., Yu, W.: Characterization of surface EMG signal based on fuzzy entropy. IEEE Trans. Neural Syst. Rehabil Eng. **15**(2), 266–272 (2007)
37. Amoud, H., Snoussi, H., Hewson, D., Doussot, M., Duchece, J.: Intrinsic mode entropy for nonlinear discriminant analysis. IEEE Signal Process.Lett. **14**(5), 297–300 (2007)
38. Valencia, J.F., Porta, A., Vallverdu, M., Claria, F., Baranowski, R., Orlowska-Baranowska, E., Caminal, P.: Refined multiscale entropy: application to 24-h Holter recordings of heart period variability in healthy and aortic stenosis subjects. IEEE Trans. Biomed. Eng. **56**, 2202–2213 (2009)
39. Wu, S.D., Wu, C.W., Lin, S.G., Wang, C.C., Lee, K.Y.: Time series analysis using composite multiscale entropy. Entropy. **15**, 1069–1084 (2013)
40. Wu, S.D., Wu, C.W., Lin, S.G., Lee, K.Y., Peng, C.K.: Analysis of complex time series using refined composite multiscale entropy. Phys. Rev. A. **378**, 1369–1374 (2014)
41. Wang, J., Shang, P., Xia, J., Shi, W.: EMD based refined composite multiscale entropy analysis of complex signals. Physica A. **421**, 583–593 (2015)
42. Chang, Y.C., Wu, H.T., Chen, H.R., Liu, A.B., Yeh, J.J., Lo, M.T., Tsao, J.H., Tang, C.J., Tsai, I.T., Sun, C.K.: Application of a modified entropy computational method in assessing the complexity of pulse wave velocity signals in healthy and diabetic subjects. Entropy. **16**, 4032–4043 (2014)

43. Wu, S.D., Wu, C.W., Lee, K.Y., Lin, S.G.: Modified multiscale entropy for short-term time series analysis. Physica A. **392**, 5865–5873 (2013)
44. Costa, M.D., Goldberger, A.L.: Generalized multiscale entropy analysis: Application to quantifying the complex volatility of human heartbeat time series. Entropy. **17**, 1197–1203 (2015)
45. Huang, N.E., Wu, M.L., Long, S.R., Shen, S.S., Qu, W.D., Gloersen, P., Fan, K.L.: A Confidence Limit for the Empirical Mode Decomposition and Hilbert Spectral Analysis. Proc. R. Soc. A. **459**(2037), 2317–2345 (2003)
46. Hu, M., Liang, H.: Intrinsic mode entropy based on multivariate empirical mode decomposition and its application to neural data analysis. Cogn. Neurodyn. **5**(3), 277–284 (2011)
47. Wu, Z., Huang, N.E.: Ensemble empirical mode decomposition: a noise-assisted data analysis method. Adv. Adapt. Data Anal. **1**(1), 1–41 (2009)
48. Rehman, N., Mandic, D.P.: Multivariate Empirical Mode Decomposition. Proc. R. Soc. A. **466**, 1291–1302 (2010)
49. Hu, M., Liang, H.: Perceptual suppression revealed by adaptive multi-scale entropy analysis of local field potential in monkey visual cortex. Int. J. Neural Syst. **23**(2), 1350005 (2013)
50. Manor, B., Lipsitz, L.A., Wayne, P.M., Peng, C.K., Li, L.: Complexity-based measures inform tai chi's impact on standing postural control in older adults with peripheral neuropathy. BMC Complement Altern. Med. **13**, 87 (2013)
51. Wayne, P.M., Gow, B.J., Costa, M.D., Peng, C.K., Lipsitz, L.A., Hausdorff, J.M., Davis, R.B., Walsh, J.N., Lough, M., Novak, V., Yeh, G.Y., Ahn, A.C., Macklin, E.A., Manor, B.: Complexity-based measures inform effects of tai chi training on standing postural control: cross-sectional and randomized trial studies. PLoS One. **9**(12), e114731 (2014)
52. Zhou, D., Zhou, J., Chen, H., Manor, B., Lin, J., Zhang, J.: Effects of transcranial direct current stimulation (tDCS) on multiscale complexity of dual-task postural control in older adults. Exp. Brain Res. **233**(8), 2401–2409 (2015)
53. Jiang, Y., Peng, C.K., Xu, Y.: Hierarchical entropy analysis for biological signals. J. Comput. Appl. Math. **236**, 728–742 (2011)
54. Bandt, C., Pompe, B.: Permutation entropy—a natural complexity measure for time series. Phys. Rev. Lett. **88**(17), 174102 (2002)
55. Wu, S.D., Wu, P.H., Wu, C.W., Ding, J.J., Wang, C.C.: Bearing fault diagnosis based on multiscale permutation entropy and support vector machine. Entropy. **14**, 1343–1356 (2012)
56. Lo, M.T., Chang, Y.C., Lin, C., Young, H.W., Lin, Y.H., Ho, Y.L., Peng, C.K., Hu, K.: Outlier-resilient complexity analysis of heartbeat dynamics. Sci. Rep. **6**(5), 8836 (2015)
57. Humeau-Heurtier, A., Baumert, M., Mahé, G., Abraham, P.: Multiscale compression entropy of microvascular blood flow signals: comparison of results from laser speckle contrast and laser Doppler flowmetry data in healthy subjects. Entropy. **16**, 5777–5795 (2014)
58. Baumert, M., Baier, V., Haueisen, J., Wessel, N., Meyerfeldt, U., Schirdewan, A., Voss, A.: Forecasting of life threatening arrhythmias using the compression entropy of heart rate. Methods Inf. Med. **43**(2), 202–206 (2004)
59. Zadeh, L.A.: Fuzzy sets. Inf. Control. **8**, 338–353 (1965)
60. Chen, W., Zhuang, J., Yu, W., Wang, Z.: Measuring complexity using FuzzyEn, ApEn, and SampEn. Med. Eng. Phys. **31**, 61–68 (2009)
61. Xie, H., He, W., Liu, H.: Measuring time series regularity using nonlinear similarity-based sample entropy. Phys. Lett. A. **372**, 7140–7146 (2008)
62. Xie, H., Zheng, Y., Guo, J., Chen, X.: Cross-fuzzy entropy: A new method to test pattern synchrony of bivariate time series. Inf. Sci. **180**, 1715–1724 (2010)
63. Zhang, L., Xiong, G., Liu, H., Zou, H., Guo, W.: Applying improved multi-scale entropy and support vector machines for bearing health condition identification. Proc. Inst. Mech. Eng. Part C. **224**, 1315–1325 (2010)
64. Xiong, G.L., Zhang, L., Liu, H.S., Zou, H.J., Guo, W.Z.: A comparative study on ApEn, SampEn and their fuzzy counterparts in a multiscale framework for feature extraction. J. Zhejiang Univ. Sci. A. **11**, 270–279 (2010)

65. Ahmed, M.U., Mandic, D.P.: Multivariate multiscale entropy: a tool for complexity analysis of multichannel data. Phys. Rev. E. **84**, 061918 (2011)
66. Ahmed, M.U., Mandic, D.P.: Multivariate multiscale entropy analysis. IEEE Signal Processing Letters. **19**, 91–94 (2012)
67. Poczos, B., Kirshner, S., Szepesvari, C.: REGO: rank-based Estimation of Renyi Information Using Euclidean Graph Optimization. Proceedings of the 13th International Conference on AI and Statistics, JMLR Workshop and Conference Proceedings, vol. 9, pp. 605–612, MIT Press, Cambridge, MA (2010)
68. Sklar, A.: Random variables, joint distributions, and copulas. Kybernetica. **9**, 449–460 (1973)
69. Nelsen, R.B.: An introduction to copulas. Springer, Berlin (2006)
70. Asai, M., McAleer, M., Yu, J.: Multivariate stochastic volatility: a review. Econ. Rev. **25**, 145–175 (2006)
71. Aas, K., Czado, C., Frigessi, A., Bakken, H.: Pair-copula constructions of multiple dependence. Insur. Math. Econ. **44**, 182–198 (2009)
72. Hu, M., Liang, H.: A copula approach to assessing Granger causality. Neuro. Image. **100**, 125–124 (2014)
73. Elidan, G.: Copula Bayesian networks. Adv. Neural Inf. Proces. Syst. **23**, 559–567 (2010)
74. Hu, M., Clark, K., Gong, X., Noudoost, B., Li, M., Moore, T., Liang, H.: Copula regression analysis of simultaneously recorded frontal eye field and inferotemporal spiking activity during object-based working memory. J. Neurosci. **35**, 8745–8757 (2015)
75. Schreiber, T.: Measuring Information Transfer. Phys. Rev. Lett. **85**, 461–464 (2000)
76. Lungarella, M., Pitti, A., Kuniyoshi, Y.: Information transfer at multiple scales. Phys. Rev. E. **76**, 056117 (2007)
77. Costa, M.D., Peng, C.K., Goldberger, A.L.: Multiscale analysis of heart rate dynamics: entropy and time irreversibility measures. Cardiovasc. Eng. **8**(2), 88–93 (2008)
78. Nardelli, M., Valenza, G., Cristea, I.A., Gentili, C., Cotet, C., David, D., Lanata, A., Scilingo, E.P.: Characterization of behavioral activation in non-pathological subjects through heart rate variability monovariate and multivariate multiscale entropy analysis. The 8th conference of the European study group on cardiovascular oscillations, pp. 135–136, IEEE, Piscataway, NJ (2014)
79. Cornforth, D., Herbert, F.J., Tarvainen, M.: A comparison of nonlinear measures for the detection of cardiac autonomic neuropathy from heart rate variability. Entropy. **17**, 1425–1440 (2015)
80. Pan, W.Y., Su, M.C., Wu, H.T., Lin, M.C., Tsai, I.T., Sun, C.K.: Multiscale entropy analysis of heart rate variability for assessing the severity of sleep disordered breathing. Entropy. **17**, 231–243 (2015)
81. Bari, V., Marchi, A., Maria, B.D., Girardengo, G., George Jr., A.L., Brink, P.A., Cerutti, S., Crotti, L., Schwartz, P.J., Porta, A.: Low-pass filtering approach via empirical mode decomposition improves short-scale entropy-based complexity estimation of QT interval variability in long QT syndrome type 1 patients. Entropy. **16**, 4839–4854 (2014)
82. Valenza, G., Nardelli, M., Bertschy, G., Lanata, A., Scilingo, E.P.: Mood states modulate complexity in heartbeat dynamics: a multiscale entropy analysis. Europhys. Lett. **107**(1), 18003 (2014)
83. Valenza, G., Citi, L., Scilingo, E., Barbieri, R.: Inhomogeneous point-process entropy: An instantaneous measure of complexity in discrete systems. Phys. Rev. E. **89**, 052803 (2014)
84. Valenza, G., Citi, L., Scilingo, E., Barbieri, R.: Point-process nonlinear models with Laguerre and Volterra expansions: instantaneous assessment of heartbeat dynamics. IEEE Trans. Signal Process. **61**, 2914 (2013)

Chapter 5
Introduction to Complex Systems Analysis with Wavelets

Nicola Vanello and Luigi Landini

Abstract In this chapter the authors deal with a few methods of nonlinear wavelet analysis for the characterization of nonstationary signals. The methods herein described can be used in a wide variety of biological signals including ECG, HRV, pressure waves and heart sounds. The reader can find both conventional methods of wavelet analysis such as linear and nonlinear denoising, as well as more sophisticated methods based on fractal analysis and entropy. Applications of such algorithms to the analysis of the heartbeat dynamics are also described.

5.1 Introduction

Wavelet based signal analysis provides a powerful means for the analysis of nonstationary signals. It is a prerequisite to analyse biological time series, such as the ones derived from the cardiovascular system, that are intrinsically nonstationary signals. Typical examples include heart rate variability (HRV) [1], pressure waves [2] and heart sound signals [3]. In this chapter we will introduce both classical and advanced methods used in the application of the wavelet analysis to biological signals. In particular, wavelet analysis results in a time-frequency description of biosignals that proved to be very efficient in handling nonstationary phenomena. A relevant aspect associated to the use of wavelet analysis is the possibility to optimize the temporal and the frequency resolution at any scale in the respect of the Nyquist theorem. It sacrifices the temporal resolution at high frequencies with the benefit of an increase in the frequency resolution at low frequencies, with important implications in signal interpretation. A crucial application of wavelet analysis is the linear and nonlinear wavelet denoising able to reduce the noise overimposed to the signal bandwidth, causing very low distortion. Moreover, in the wavelet domain it is also possible to localize and to filter out unwanted components of the signal at any scale. An important aspect of wavelets decomposition is that it behaves like a decorrelation transformation, thus optimizing the information associated to its

N. Vanello (✉) • L. Landini
Dipartimento di Ingegneria dell'Informazione, University of Pisa, Pisa, Italy
e-mail: nicola.vanello@unipi.it

R. Barbieri et al. (eds.), *Complexity and Nonlinearity in Cardiovascular Signals*,
DOI 10.1007/978-3-319-58709-7_5

coefficients across the different scales. Thus, several measures related to information theoretic criteria, as entropy [4], can be efficiently estimated and applied to biosignal interpretation. Moreover, the same property can be exploited to estimate nonlinear parameters as those describing the fractal [5, 6], or auto-similarity behaviour of different processes of interest, with relevant implications in the field of signal processing and modelling.

5.2 Continuous Wavelet Transform

The continuous wavelet transform (CWT) provides the optimal time and frequency resolution for a given signal [7]. This is achieved by imposing the Nyquist condition at any sub-band in which the signal is split. In the analysis of biomedical signals, a time signal can be made up of slowly and quickly (nonstationary signals) varying components. In these cases, it is necessary to dispose of an analysis technique that is able to provide the best compromise between time and frequency resolution at the different frequencies constituting the signal. For signals with low and high frequencies components the best compromise is:

- high temporal resolution (high degree of detail on the time signal) and adequate frequency resolution (on the basis of the Nyquist theorem) at high frequencies;
- high frequency resolution and adequate time resolution at low frequencies

These requirements satisfy the condition of maintaining constant the $\frac{\Delta f}{f}$ at each sub-band of spectrum of the signal, where f is the centre frequency of the sub-band. The CWT is therefore a signal analysis technique that provides:

- the time-frequency representation (such as the Short Time Fourier Transform STFT) of a signal, because it temporally locates the spectral components; this peculiarity is essential for the analysis of non-stationary signals;
- the signal is split into frequency sub-bands and at each sub-band the resolution dictated by the Nyquist theorem is attributed.

Consequently, the time resolution becomes non redundant.

The CWT is defined as it follows:

$$TWC\,(\tau, s) = \int_{-\infty}^{\infty} \left[x(t) \cdot \Psi^* \left(\frac{t-\tau}{s} \right) \right] \cdot e^{-j\frac{2\pi}{s}t} \cdot dt \tag{5.1}$$

where a wavelet function is included in the classical definition of the (Short-Time-Fourier-Transform (STFT) with the following relationship:

$$\Psi_{\tau,s}(t) = \frac{1}{\sqrt{s}} \Psi \left(\frac{t-\tau}{s} \right) \tag{5.2}$$

being s the scale factor (inverse of frequency) and τ the translation term. The symbol "*" indicates the complex conjugate The term 'wavelet' is synonymous of damped oscillating wave, typical of a pass-band signal.

An example of a wavelet function is described by the following formula, that defines a Morlet wavelet:

$$\Psi(t) = \frac{1}{\pi^{1/4}} e^{j\omega_0 t} e^{-t^2/2} \tag{5.3}$$

It is recalled that, given the scaling properties of the Fourier transform, the compression in the time domain produces an expansion in the frequency domain and vice versa, according to the following relationship:

$$F\left\{x\left(\frac{t}{a}\right)\right\} = |a|\, X(af) \tag{5.4}$$

Because the translation parameter τ introduces a delay in the frequency domain, the Fourier transform of the wavelet function can be written as follows:

$$F\left\{\frac{1}{\sqrt{s}}\Psi\left(\frac{t-\tau}{s}\right)\right\} = \sqrt{s}\widehat{\Psi}(sf)e^{-j\omega\tau} \tag{5.5}$$

being $F\{\Psi(t)\} = \widehat{\Psi}(f)$.

Then the wavelet compression (or expansion) corresponds to dilate (or compress) the corresponding Fourier transform. The time shift, necessary to analyse the entire signal introduces a phase term in the previous formula, not affecting the modulus. From the scaling properties of the Fourier transform it is also evident that the contraction of a factor s of a wavelet in the time domain is equivalent to a dilation of the spectrum at high frequencies by the same factor (remember that a wavelet is a bandpass signal). Therefore, the process of scaling achieved by varying s ensures the coverage of the signal spectrum; the time domain signal coverage is ensured by the operation of temporal translation.

So, if $\Psi_{\tau,s}(t)$ has a band Δf centred at f, then $\Psi\left(\frac{t}{s}\right)$ has a band centred at $s\Delta f$ and, also, the ratio is constant. The sub-bands are the frequency intervals in which the signal is divided due to the application of the wavelet transform.

The scale parameter s takes on a meaning similar to the scales used in the maps: large scales (low frequencies) provide a global characterization of the signal, while low scales (high frequency) provide detailed information on local trends.

Expansion in quadratic powers and time translations of an entire wavelet function constitute an orthonormal basis.

The map of Fig. 5.1 (bottom) is obtained by applying the previously procedure and it represents the CWT of the signal at the top. On the vertical axis the scale is shown (inverse of frequency).

It is observed that the lines (bottom) at lower scale values are characterized by a sufficient frequency resolution, while they show a high temporal resolution needed for localizing the transition zone on the time signal. At higher scales the opposite phenomenon is observed. The top rows correspond to high scales (low frequencies)

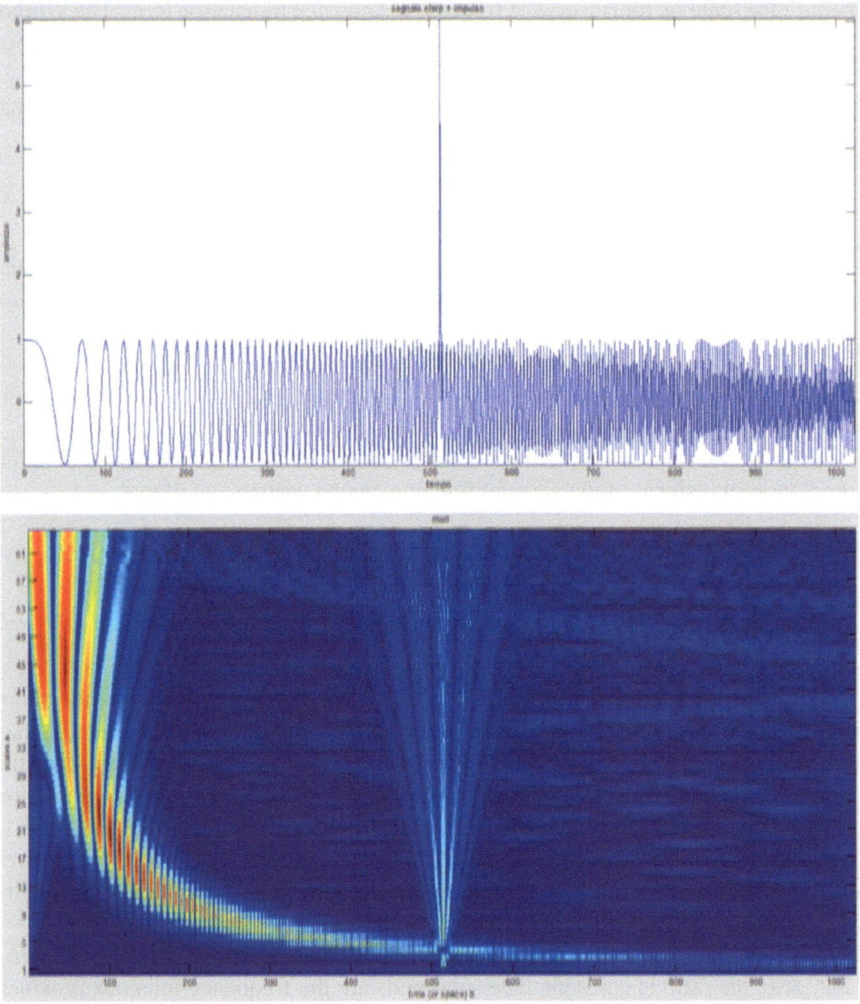

Fig. 5.1 An example of the application of the CWT to a chirp signal

and they show the frequency details of the chirp signal (which is predominantly localized in the low frequencies interval).

Therefore, exploiting the representation on the time-frequency plane, the CWT is capable of representing simultaneously temporal and frequency details of the same signal. The inverse CWT is calculated by the following formula:

$$x(t) = \frac{1}{c_\Psi} \iint_{-\infty}^{\infty} \left[X(\tau, s) \cdot \frac{1}{s^2} \Psi^* \left(\frac{t-\tau}{s} \right) \right] \cdot e^{j\frac{2\pi}{s}t} \cdot d\tau ds \tag{5.6}$$

where c_Ψ is the admissibility constant and X(.) is the Fourier transform of x(t).

5.3 Discrete Wavelet Transform

Since the wavelet functions are shifted and scaled versions of themselves, it is evident that a redundant information on the signal in the time-scale representation is produced. It is known that, in general, the redundancies can make easy the interpretation. On the other hand, a signal reconstruction based on the sampling theorem, and thus with a mathematically sufficient number of samples (two samples per period), can make difficult the visual interpretation of the signal as the signal morphology is not adequately represented.

It can be shown that a suitable discretization of delay and scale parameters (τ, s), while reducing redundancy in the CWT, allows to reduce its computational cost. A common approach to the discretization for s and τ is the following:

$$s = s_0^j; \tau = k\tau_0 s_0^j \tag{5.7}$$

with j, k integers and greater than zero. By setting the s_0 and τ_0 values, the wavelet function $\Psi_{\tau,s}(t) = \frac{1}{\sqrt{s}}\Psi\left(\frac{t-\tau}{s}\right)$ becomes a function of (j, k) as it follows:

$$\Psi_{j,k}(t) = s_0^{-j/2}\Psi\left[s_0^{-j}\left(t - k\tau_0 s_0^j\right)\right] \tag{5.8}$$

When $s_0 = 2$ and $\tau_0 = 1$, we obtain the dyadic sampling of the time-scale space, that is: $s = 2^{-j}; \tau = k2^{-j}$.

For example, for $j = 2$ then $s = ¼$, and the delay assumes multiple values of 1/4 as k increases; for $j = 3$ (third level of decomposition) then $s = 1/8$ and the delay assumes multiple values of 1/8 as k increases, and so on. So at low scales (high frequencies) the temporal sampling is high, while at high scales the temporal sampling is lower. In the frequency domain, the frequency resolution increases with decreasing the frequency, in agreement with the previous assumptions. The discrete wavelet family with dyadic sampling becomes:

$$\Psi_{j,k}(t) = \frac{1}{\sqrt{2^j}}\Psi\left(\frac{t}{2^j} - k\right) j, k > 0 \tag{5.9}$$

The wavelet basis functions are chosen so as to respect the orthonormality condition.

It is reasonable to ask whether the dyadic sampling allows to reconstruct the signal without loss of information. Daubechies shows that it is possible, provided some type of energy conditions are met [8], conditions which will not discussed here.

Among the properties of the wavelet transform it is worth to recall the linearity and the time and scale invariances. The admissibility condition implies that $\widehat{\Psi}(\omega)$ must be zero at zero frequency, that is, the wavelet function has a bandpass behaviour. In addition, the mean value in the time domain must be null.

A subsequent discretization of the time domain t using the variable n, allows to introduce the fully discretized version of the CWT known as DWT (Discrete Wavelet Transform). The transfer from the continuous time domain to the discrete-time domain in wavelet analysis allows a drastic reduction of the algorithm complexity and computation times.

5.3.1 Mallat Algorithm

Starting from the discrete dyadic wavelet family and their frequency behaviour as the scale index j changes, Mallat suggests to decompose the discrete signal using two families of wavelet functions [9]. A family composed of low-pass filters, i.e. $h_{j,k}[n] = 2^{j/2}h[2^j n - k]$, also called 'approximation', and a family composed of high pass functions, i.e. $g_{j,k}[n] = 2^{j/2}g[2^j n - k]$ also called 'detail'.

The discrete index k determines the time position of the wavelet function with respect to the signal, while the index j determines the extent of the temporal sub-sampling of the signal. The aforesaid pair of filters is known as 'quadrature mirror filters', since such filters have the following mirror symmetry:

$$g\,[L - 1 - n] = (-1)^n \cdot h\,[n] \tag{5.10}$$

where L is the number of filter samples.

Starting from the value $j = 1$, the Mallat algorithm decomposes the signal into two equal sub-bands, each one equal to half of the signal spectrum to be decomposed. A further signal subdivision into sub-bands may be obtained by maintaining fixed the two filters g[n] and h[n] and by performing a signal compression (also called decimation or subsampling) at the output of the g[n] and h[n] filters so as to produce an expansion of the signal bandwidth. The time decimation is allowed since the signal bandwidth after the application of the band filters has halved. It allows the subsampling of the signal according to the sampling theorem valid for low-pass signals. In the case of band-pass signals, available at the out of the $g_{j,k}(t)$ filters, the sampling theorem for bandpass signals is used. It is governed by the following relationship:

$$f_c = \frac{2f_{\max}}{m} \tag{5.11}$$

where f_c is the sampling frequency and m is the maximum integer not exceeding $\frac{f_{\max}}{B}$, where B is the frequency bandwidth of the signal. The above formula becomes the classic Nyquist theorem for low-pass signals, when f_{max} becomes equal to B.

The signal at the first level is under-sampled and hence its bandwidth doubles; subsequently it enters in the next level of filters, which have the same impulse response of the previous level. The filters output is still under-sampled and such operation is carried out on a number of levels determined by the experimenter based on the level of detail to be achieved. According to Mallat algorithm, the decomposed signal is represented by the last 'approximation' and all the sequences of 'detail'. At each level of decomposition the temporal resolution is halved due to the under-sampling, confirming that the signals at the lowest frequencies have a lower temporal resolution (but always respecting the Nyquist theorem) compared to the signal at higher frequencies. Consequently, the frequency resolution at the lower frequencies turns out to be higher than that for the highest frequency signals.

The signal reconstruction process at any step, involves: (a) an interpolation by a factor of 2; (b) the convolution with the reconstruction filters (the same as those used in the decomposition); (c) the sum of the resulting signals. At any level, the reconstruction algorithm is described by the following relationship:

$$x[n] = \sum_{j=1}^{J}\sum_{k=1}^{M}\left\{y_{high}[j,k]\cdot g\left[n-2^{j}k\right] + y_{low}\left[j,k\right]\cdot h\left[n-2^{j}k\right]\right\} \tag{5.12}$$

where J is the number of decomposition levels, M the number of delay step and y_{high} and y_{low} are respectively the high and low frequency components of the decomposed signal.

5.4 Non Linear Denoising

Let as assume the following additive degradation model:

$$y = f(x) + n \tag{5.13}$$

where *f(.)* models the degradation process, due for instance to a transmission channel or a measurement system, x is the unknown input and n is the noise. In the case of a linear degradation model, a convolutive relationship between the input and the noiseless output can hold. For a discrete system the convolution can be expressed using a Toeplitz matrix **H**, such that the previous equation can be written as

$$\mathbf{y} = \mathbf{Hx} + n \tag{5.14}$$

We recall that in the absence of knowledge about the noise model n, to solve the inverse problem in order to derive x from the knowledge of $\boldsymbol{H}$ and $\boldsymbol{y}$, the least squares method is used. The least squares method is the function that best approximates x minimizing the norm of the noise. However, in general, the $\boldsymbol{H}$ system can be ill-conditioned and the solution, if it exists, may not be unique. To find a solution to such problem, a regularization term is introduced in the minimization procedure.

Introducing the regularization term, we obtained:

$$\arg\min \frac{1}{2}\|\mathbf{H}\cdot\mathbf{x}-\mathbf{y}\|_2^2+\lambda\|\boldsymbol{\Psi}\cdot\mathbf{x}\|_1 \tag{5.15}$$

where $\boldsymbol{\Psi}$ is the regularization matrix or a matrix which determines which of the signal components we would penalize, λ is a scalar and acts as a regularization parameter, $\| \|_1$ is the norm one.

The regularization term privileges the solutions for which the product $\boldsymbol{\Psi}\cdot\mathbf{x}$ is 'sparse', i.e. characterized by a few significant coefficients, while the remaining coefficients may be assimilate to noise.

Using an orthonormal operator (that is: $\boldsymbol{\Psi}^{\mathrm{T}}\boldsymbol{\Psi}=\mathbf{I}$) the $\boldsymbol{\Psi}$ transformation can be applied to the data as follows:

$$\arg\min \frac{1}{2}\|\mathbf{y}_{\Psi}-\mathbf{z}\|_2^2+\lambda\|\mathbf{z}\|_1 \tag{5.16}$$

where: $\mathbf{y}_{\Psi}=\boldsymbol{\Psi}\cdot\mathbf{y}$, $\mathbf{z}=\boldsymbol{\Psi}\cdot\mathbf{x}$, $\widehat{\mathbf{x}}=\boldsymbol{\Psi}^{\mathrm{T}}\widehat{\mathbf{z}}$. If the matrix $\boldsymbol{\Psi}$ is the wavelet transform operator, then the product $\boldsymbol{\Psi}\cdot\mathbf{x}$ is typically 'sparse' or compressible. We name such operation as "Wavelet Regularization" [10]. It brings to the 'wavelet denoising', that is an example of nonlinear filtering in the wavelet domain [11]. In this case the solution has an explicit form called Soft Thresholding function, and is defined for each frequency as:

$$\widehat{z}_i=S_{\lambda}\left(y_{\Psi_i}\right)=\begin{cases} y_{\Psi_i}-\lambda \ \ if\ y_{\Psi_i}>\lambda \\ 0 \qquad\quad if\ |y_{\Psi_i}|\le\lambda \\ y_{\Psi_i}+\lambda\ \ if\ y_{\Psi_i}<-\lambda \end{cases} = \left(1-\frac{\lambda}{|y_{\Psi_i}|}\right)_+ y_{\Psi_i} \tag{5.17}$$

By a properly choice of the wavelet function, an efficient denoising can be obtained which preserves the high frequencies. In Fig. 5.2a an example of the denoising filter in the wavelet domain using the 'db1' windows is shown; in Fig. 5.3a the error due to the filtering operation is shown. The efficiency of the denoising process is related to the degree of compression of the signal. The more the signal is compressible, the better is the denoising.

A non-linear filter widely used for the analysis of biomedical signals is the 'Total variation denoising' [12]. The name means that the goal is to minimize the Total Variation function of a signal, which corresponds to the sum of all the derivatives of the signal.

$$TV(x)=\sum_{i=1}^{N_p}|x_{i-1}-x_i|=\|\mathbf{D}x\|_1 \tag{5.18}$$

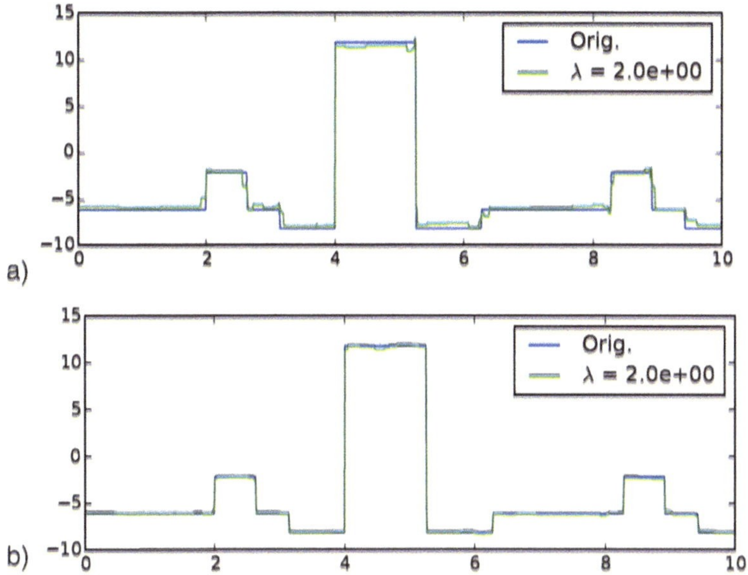

Fig. 5.2 (**a**) An example of denoising in the wavelet domain using the 'soft thresholding' filter; (**b**) the same as in (**a**) after the application of the 'Total variation' filter

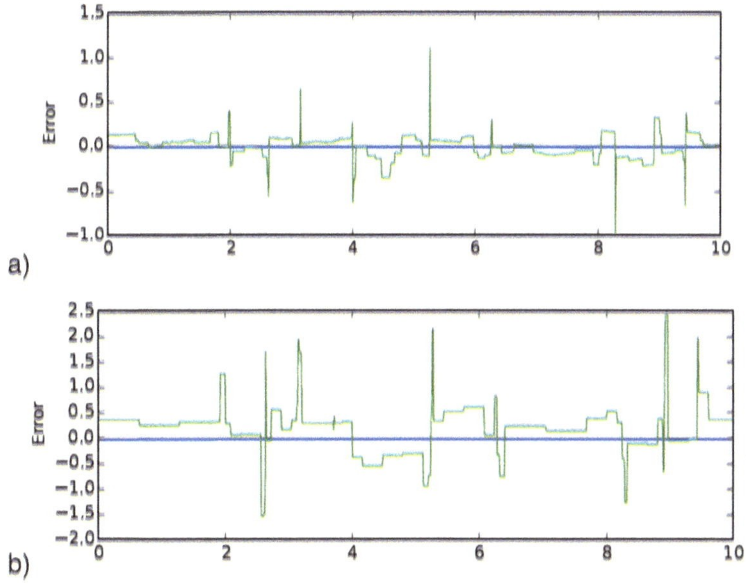

Fig. 5.3 (**a**) Error profile for the filtering in Fig. 5.2a; (**b**) error profile for the filtering in Fig. 5.2b

$$\mathbf{D} = \begin{bmatrix} -1 & 1 & 0 & 0 \;\ldots. & 0 \\ 0 & -1 & 1 & 0 \;\ldots. & 0 \\ 0 & 0 & -1 & 1 \;\ldots & 0 \\ \ldots & \ldots & \ldots & \ldots\ldots & \ldots \\ 1 & 0 & 0 & 0 \;\ldots. & -1 \end{bmatrix} \tag{5.19}$$

where N_p is the length of the signal. The Total Variation denoising is performed by solving the following problem:

$$\widehat{\mathbf{x}} = \arg\min_x \frac{1}{2} \|\mathbf{y} - \mathbf{x}\|_2^2 + \lambda TV(x) \tag{5.20}$$

To implement the TV denoising it is necessary to use iterative algorithms [13]. In Fig. 5.2b the effect of the Total variation filter is shown, with the relevant error shown in Fig. 5.3b. As expected the edges are well preserved, while the noise has been removed.

5.5 Wavelets and Nonlinear Signal Processing

Wavelet analysis has been shown to perform an efficient time-frequency description of biomedical signals. In fact, as we already discussed, it shows a superior time resolution at higher frequencies along with a good spectral resolution at lower frequencies. This property allows for instance to detect rapid and small phenomena superimposed on large fluctuations in the analysis of heart rate variability [14]. Moreover, it allows a fast description of the signal at different scales allowing to efficiently detect the phenomena at the bands of interest [15]. The application of wavelet to the analysis of electrocardiogram, has shown how this approach is efficient for the description of QRS abnormalities, ECG timing, HRV and arrhythmias [16]. Here, we will focus on the use of the wavelets in the field nonlinear signal analysis since wavelets have been found to be an efficient tool to estimate nonlinear measure of a process. Specifically, they are proposed as a preprocessing step to detect nonstationary phenomena and thus classify different regimes within the phenomenon under study. Moreover, they offer a time frequency representation that is suitable for the successive estimation of nonlinear statistics.

In [17] a double step approach to study nonlinear interactions in a process is introduced. In a first step, some coefficients $\{\beta_i\}$ related to a time-frequency representation of the signal under investigation are estimated using the DWT. The orthonormal signal basis is indicated as $\{b_i\}$. In a second step, the coefficients are combined using a nth-order polynomial operations and results in nonlinear coefficients $\{\theta_i\}$ (see Fig. 5.4). The second stage performs several operations on

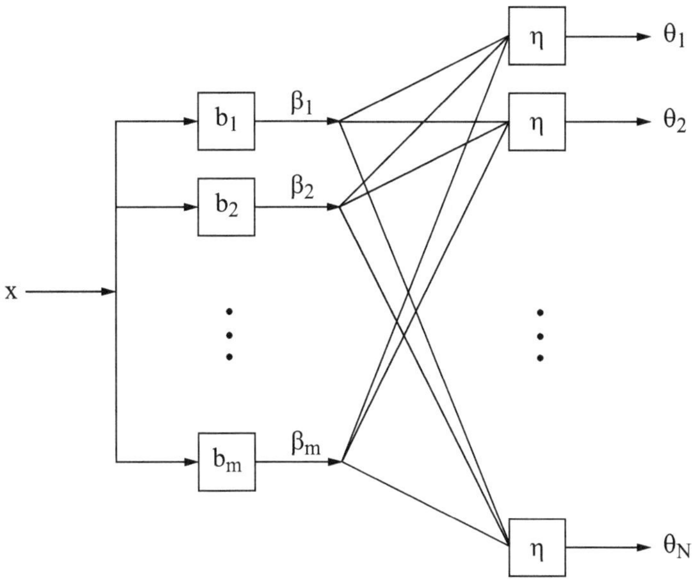

Fig. 5.4 Scheme of the two stages process to perform non linear transformation of the input signal x. The first stage performs linear transformation, while the second calculates non linear combinations of the first stage coefficients (from [17])

the first level coefficients, at different output nodes. Specifically, some nodes are product nodes such that their output can be written as:

$$\eta\left(\beta_1, \beta_2, \ldots, \beta_m\right) = \beta_{i_1}\beta_{i_2}\beta_{i_3}\ldots\beta_{i_n} \tag{5.21}$$

In this case the product is an n-fold product and takes into account all the possible n-combinations of the m coefficients, so that $(1 \leq i_1 \leq \cdots \leq i_n \leq m)$.

The other nodes are summing nodes, such that

$$\eta\left(\beta_1, \beta_2, \ldots, \beta_m\right) = \left(\sum_{i=1}^{m} \alpha_i \beta_i\right)^n \tag{5.22}$$

This approach was applied to the analysis of correlation and higher order statistics, Volterra filtering and nonlinear system identification. In all these applications Wavelet-based representation of the signal was found to outperform the Fourier transform-based representation [17].

These results were discussed in view of the fact that Wavelets represent an unconditional basis with good properties. When a signal is represented using an unconditional basis for a given space, it can be shown that its norm, is limited after the attenuation of any set of the coefficients describing the signal. Given a basis $\{z_i\}$ the signal in the space χ can be approximated as $x = \sum_{i=1}^{k} c_i z_i$, where k is the length of the time series The attenuation coefficients m_i can thus be applied to each element of the basis function, resulting in

$$\tilde{x} = \sum_{i=1}^{k} m_i c_i z_i \tag{5.23}$$

If the basis is unconditional in the given space we are sure that the norm of the attenuated reconstruction, is limited by the norm of the original signal, multiplied by some finite constant C

$$\left\|\tilde{x}\right\|_\chi \leq C \left\|x\right\|_\chi \tag{5.24}$$

This property guarantees that the signal approximation will not increase when some of the terms in the reconstruction are removed. Moreover, in the case of wavelets this property holds for many signal spaces [18], thus leading to algorithms with a good behavior. Specifically, Nowak and Baraniuk [17], showed that the higher order statistics and correlations were more robustly estimated using Wavelet than using Fourier Transform, moreover the Volterra filter output were more localized in time and frequency. Finally, regarding the identification of a nonlinear system, the Fourier transform resulted in a smoother estimate and in a higher mean squared error with respect to the one obtained using wavelets.

The wavelet transform was used also a preprocessing step for the study of the irregularities of the QRS complex [19]. To study such irregularities an analysis in the phase space was performed estimating the phase portrait, the Lyapunov exponent and the correlation dimension. The wavelet transform was thought to be efficient in detecting irregularities: in fact, it was found that when the scale of the wavelet is chosen properly, the absolute value of the wavelet transform at that scale can detect signal sharp changes [20]. The wavelet description of the signal was exploited to estimate a measure of regularity of the QRS complex, based on a Lipschitz exponent, that can be used to model signal singularities. The nonlinear measures were then obtained from a time series, with one regularity value estimated for each QRS complex.

In [21] a recurrence quantification analysis (RQA) was applied to a three lead vector cardiogram (VCG) signal. The goal was to classify healthy controls and subject that had a myocardial infarction. The discrete wavelet transform was used to obtain multiple filter banks. The Daubechies wavelets (db4) were chosen since they were thought to match the ECG signal pattern, thus resulting in a more compact description of the signal [22]. Seven details scales and the approximation were employed. One RQA was then obtained at each scale, merging the information obtained from the three leads. A feature selection step, for the feature dimensionality reduction, and a successive classification allowed to achieve a classification rate of 92.7%. Interestingly, the most informative features where RQA-based measures obtained from three different scales (D2 125-250 Hz, D3, 62.5–125 Hz, D5, 15.62-31.25 Hz).

5.5.1 Nonlinear Phase Coupling

Here, we report a finding by Ivanov et al. [23] who studied the heartbeat dynamics using wavelet functions. The goal was to reveal possible nonlinear phase coupling phenomena. Specifically, the R-R interval was analyzed using the derivatives of the Gaussian function as a mother wavelet, both in healthy and in apnoeas. A scale equal to eight beats was adopted, and it allowed to remove constant and linear trends in the time series, while smoothing high frequency changes. The choice of the scale was performed to optimize the detection of patterns lasting from 30 s to 1 min. The scale of the wavelet is in fact proportional to the interval over which the changes might occur. After this operation, different time intervals characterized by different dynamics can be observed in the transformed series. Using the Hilbert transform on this series, and studying the amplitude of the analytic signal it was possible to reveal the changes in the heartbeat dynamics. The results are a cumulative measure of inter-beat distance variations within a time interval proportional to the wavelet time scale. Ivanov et al. [23] observed that the distribution of the inter-beat intervals in healthy subject, after the proposed preprocessing steps, is described by a Gamma distribution family. Moreover, they found a normalization procedure that allowed to have a very good fit of the analyzed healthy subjects. This normalization procedure was not effective with patients suffering from apnea, since the different subjects' distributions did not collapse to the same values. After Fourier phase randomization of the inter-beat intervals time series in healthy subjects, the distribution was found to follow a Rayleigh distribution, while the original data showed a heavier tail. This distribution is found to describe the instantaneous amplitudes of a Gaussian process. Moreover, Fourier phase randomization destroys possible phase couplings due to non-linear dynamics. Taken together these observations lead the researchers to hypothesize a non-linear phase coupling in heartbeat dynamics in healthy subjects.

5.5.2 Wavelet Entropy

Spectral entropy measures how the information in the signal is distributed across different frequencies [24]. In a first application, this quantity was estimated through the Fourier Transform and STFT for nonstationary signals. However, the limitation of these procedure regarding the temporal and frequency resolution due to windowing operation are not optimal for nonstationary signals. An Entropy definition, based on the wavelet transform would allow to exploit the temporal and frequency characteristics of the wavelet analysis. For this reason, the entropy based entropy demonstrated to be effective in different applications within the field of biomedical signal processing [25].

5.5.2.1 Normalized Total Wavelet Entropy (NTWE)

Given the coefficient of the wavelet transform at level j, $C_j(k)$ where k is a time variable, the energy at level j is defined as

$$\mathcal{E}_j = \sum_{k=-N}^{-1} \left|C_j(k)\right|^2 \tag{5.25}$$

The relative energy at level j is defined as

$$p_j = \frac{\mathcal{E}_j}{\sum_{k=-N}^{-1} \mathcal{E}_k} \tag{5.26}$$

The values p_j can be considered as describing a probability distribution, with $\sum_j p_j = 1$ where the summation is performed across the scales ranging from −N to −1. According to these definitions the NTWE can be estimated as a function of the scales

$$S_W(N) = -\sum_{j=-N}^{-1} p_j log_2 p_j / S_{MAX} \tag{5.27}$$

where $S_{MAX} = log_2 N$. This quantity can be also estimated, for non-stationary signals, dividing the observed signal in N_T non overlapping temporal windows. In this case, it is possible to have a temporal average of *NTWE*

$$\langle S_W \rangle = \frac{1}{N_T} \sum_{i=1}^{N_T} S_W^{(i)} \tag{5.28}$$

while the mean *NTWE* can be estimated by evaluating the mean wavelet energy at level j as:

$$\langle \mathcal{E}_j \rangle = \frac{1}{N_T} \sum_{i=1}^{N_T} \mathcal{E}_j^{(i)} \tag{5.29}$$

and the total mean energy as:

$$\langle \mathcal{E}_{tot} \rangle = \sum_{j=-N}^{-1} \langle \mathcal{E}_j \rangle \tag{5.30}$$

A mean probability distribution can be defined as

$$q_j = \frac{\langle \mathcal{E}_j \rangle}{\sum_{k=-N}^{-1} \langle \mathcal{E}_k \rangle} \tag{5.31}$$

with the corresponding mean normalized wavelet entropy, defined as

$$\widetilde{S}_W(N) = -\sum_{j=-N}^{-1} q_j log_2 q_j / S_{MAX} \tag{5.32}$$

This approach was tested in [25], on fractional Brownian and fractional Gaussian processes. More details about these processes will be given below, but here we just highlight that these processes are characterized by a 1/f-like process and they have been proposed to model many biological phenomena, as brain and heart activity [26].

Both $\widetilde{S}_W$ and S_W were able to discern the processes families characterized by different scaling parameters. Specifically, S_W was more suitable than $\widetilde{S}_W$ to discern among *fBm* processes, since in the range of parameters describing such a process is monotonic function. However, the $\widetilde{S}_W$ was found to have its maximum for a gaussian white noise, and could distinguish more easily between white noise and 1/f-like process (see Fig. 5.5).

5.5.2.2 Relative Entropy

Given $\{p_j\}$ and $\{q_j\}$ two different probability distributions across scales ($\sum_j p_j = 1$ and $\sum_j q_j = 1$ with the summation ranging from −N to −1) of the wavelet coefficients that can be obtained from two different time intervals or from two different signals, it is possible to obtain the relative entropy, that describes the similarities between two distributions [24]

$$S_W\,(p|q) = \sum_{j=-N}^{-1} p_j log_2 \left[\frac{p_j}{q_j} \right] \tag{5.33}$$

5.5.2.3 Approximate Entropy

Entropy measures were also studied at each frequency scale, by using Approximate Entropy (ApEn) [27]. This measure is defined exploiting two phase space reconstructions with different embedding dimensions, m and $m + 1$. ApEn is related to the likelihood that vectors in a m-dimensional phase space will remain close, e.g. within a radius r, also in the m + 1 dimensional space. This measure thus

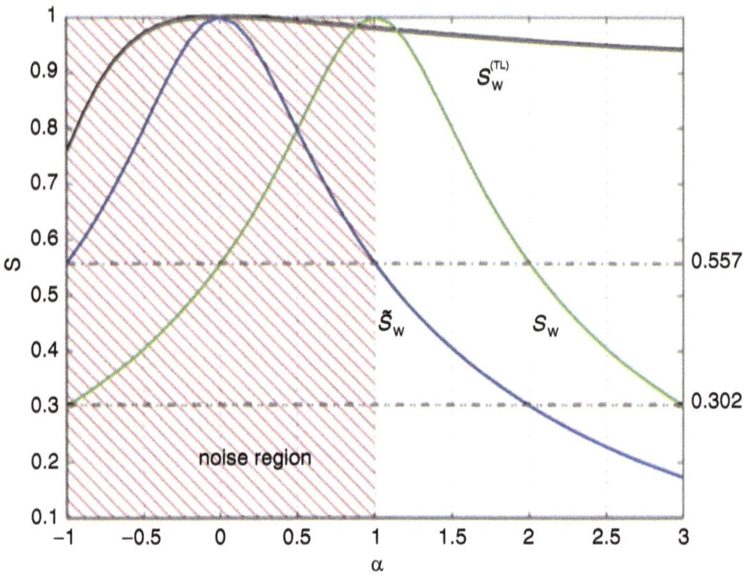

Fig. 5.5 Normalized wavelet entropy for fGn ($-1 < \alpha < 1$) and fBm process ($1 < \alpha < 3$). The performance of Tavares–Lucena wavelet entropy ($S_W^{(TL)}$ not discussed here) were found to fall behind those of NTWE. The value of $\alpha = 0$ corresponds to white gaussian noise. The figure is reprinted from [25]

reflects the regularity and the predictability of a time series. Given this definition, it's possible to estimate ApEn from different frequency bands of a signal, and within the context of wavelet analysis, from different scales. Specifically, in [28] the ApEn was estimated on the details and approximation coefficients obtained using DWT. One difficulty for the application of this approach, it the proper choice of the embedding parameters, and the radius r.

5.5.2.4 Application to the Analysis of Cardiac Function

In the field of study of cardiac function, the classification of subjects with congestive heart failure exploiting the analysis of HRV was found to be enhanced by concomitant use of wavelet entropy measures [4]. In this work, the wavelet entropy definition given in Rosso et al. [24] were used, adopting Daubechies (db4) and seven scales. In [29], the same approach was applied to the analysis of heart rate variability comparing resting and lower body negative pressure conditions. In this work, however the wavelet entropy measures were estimated on the original ECG signal within the low frequency band. ApEn, using the wavelet-based approach defined in [28] was also estimated. Both measures could detect significant differences between the two conditions.

5.5.3 Mono and Multi Fractal Processes

The analysis of processes with 1/f behaviour has a large interest within the field of biomedical signal analysis. As an example, this behaviour was found in resting state fluctuation of blood oxygenation level dependent signal in the brain, related to neuronal activity [30]. Also the inter-beat intervals dynamics in the hearth activity was found to be described by such a power law [23]. This power spectrum scaling law is related to long-range dependencies or temporal autocorrelation. While autoregressive models can describe serial correlations, they fail when long range dependencies occur. Fractional Gaussian noise (fGn) that is the increment process of a fractional Brownian Motion (fBm) [31] was introduced to have a parsimonious description of temporal autocorrelation, that could describe several behaviors. For a fGn the autocovariance at time lags τ can be seen as [30, 32]

$$c(\tau) = \frac{\sigma^2}{2}\left(|\tau + 1|^{2H} + 2|\tau|^{2H} + |\tau - 1|^{2H}\right) \tag{5.34}$$

where $c(\tau)$ is the process variance and τ is the discrete time lag From this equation it is possible to see that for $\tau \neq 0$ and H = 0.5 the autocovariance is zero. In this case the process reduces to a classical Brownian motion, and samples at different time lags are independent. For H smaller than 0.5 the dynamics of the underlying process is reversing along time, and an increase in the time series will be, on average, followed by a decrease and vice-versa. Such a process is said to be anti-persistent. A H exponent bigger than 0.5 is related to a persistent or positively correlated behaviour, and the process is said to have a long memory. When H is equal or close to 0.5 the time course is related to a process *without memory*. The frequency content of the time series reflects these behaviors. Since the power spectrum of the fGn process can be written as the Fourier Transform of the autocorrelation function so that

$$S(f) = \sum_{\tau=-\infty}^{+\infty} c(\tau)\, e^{-j2\pi f\tau} = 4\sigma^2 C_H sin^2(\pi f) \sum_{j=-\infty}^{+\infty} \frac{1}{|f + j|^{2H+1}} \tag{5.35}$$

with $C_H = \Gamma(2H + 1)\sin(\pi H)/(2\pi)^{2H+1}$ where $\Gamma(\bullet)$ is the gamma function. Using a Taylor expansion around the origin we have

$$S(f) \approx \sigma^2 C_H (\pi f)^2 |f|^{1-2H} \tag{5.36}$$

As a result the power spectrum of fGn can parameterized in terms of H as $S(f) \propto |f|^{-\gamma}$ with $\gamma = 1 - 2H$.

Thus, the power spectrum of a long memory, persistent process is thus 1/f-like. In this case the signal power is mainly located at low frequencies, and the signal is characterized by slowly varying components. A process with H < 0.5 is characterized by rapidly changing components and consequently its power spectrum is mainly located at higher frequencies.

fGn and fBm were introduced to model processes that show a fractal behavior [33], i.e. they are self-similar in that they show the same statistical properties at different scales of analysis. The Hurst exponent is also related to the fractal dimension by this relationship D = 2 – H. For a time series, D ranges from 1 to 2. For instance, 1 is the fractal dimension of a straight line, while a Gaussian random walk is characterized by D equal to 1.5. As D tends to 2, the time series becomes more rugged.

When the fractal behavior of a process is described by one parameter, the process is said to be monofractal. When the system is nonstationary, the scaling behavior can change along time and local Hurst exponent can be defined. In this case, the wavelet transform can be used to characterize the nonstationarity of a time series.

5.5.3.1 Estimation of 1/f Behavior Using Wavelet

This approach is based on the observation that the wavelet coefficients of a correlated 1/f process are uncorrelated both along and across the different scales [34]. This result is a valid approximation, assuming the wavelet filter as a band-pass filter. As a results, the wavelet transform can be seen as a decorrelating or whitening procedure.

The variance of the coefficients can be written as

$$var\left(d_j\right) = 2^{-j+1}\int_{2^{j-1}}^{2^j} S(f)df \text{ with } j = -j_0, \ldots, -1 \tag{5.37}$$

while for the approximation coefficients

$$var\left(a_{-j_0}\right) = 2^{j_0+1}\int_0^{2^{-j_0-1}} S(f)df \tag{5.38}$$

where *S(f)* is the power spectral density of the process under study. The variances can be approximated by the following expression, substituting the expression for the power spectral density

$$var\left(d_j\right) \approx \sigma^2 K_d(H) 2^{-j(2H-1)} \text{ with } j = -j_0, \ldots, -1$$

and

$$var\left(a_{-j_0}\right) \approx \sigma^2 K_a(H) 2^{j_0(2H-1)} \tag{5.39}$$

with $K_a(H)$ and $K_d(H)$ defined as a function of H. Calculating the base 2 logarithm of the variance a relationship showing an almost linear relationship between the log variance of the coefficient and the wavelet scale can be observed.

$$log_2\left(var\left(d_j\right)\right) \approx -j\,(2H-1) + \log_2\left(\sigma^2 K_d(H)\right) \tag{5.40}$$

with $j = -j_0, \ \ldots \ , -1$.

This observation can be used as the basis for the estimation of the Hurst exponent. In Wornell and Oppenheim [35] a maximum likelihood approach to the estimation of the Hurst exponent was proposed. The approach was developed under the hypothesis of 1/f process contaminated by stationary white Gaussian noise. Specifically, the wavelet coefficients of such a process are normal, independent and identically distributed [25], and show the dependence properties on H just seen The Maximum likelihood is expressed as a function of the $\tau = 2^{2H-1}$ and $\sigma^2 K_d(H)$. These parameters are estimated using an Expectation Maximization algorithm. This approach is prone to error when the number of observation is small.

5.5.3.2 Multifractal Processes

The time varying properties of biological processes, lead to the need of adopting a multifractal approach. Regarding this point, the time series can be divided in windows, under the hypothesis of stationary behavior within each time window. In Duverney et al. [36], a multifractal approach was used to classify periods of atrial fibrillation. The analysis of R-R using the discrete wavelet transform, allowed to estimate period of high variability. The Hurst exponent were thus estimated within each window. The Hurst exponent was estimated exploiting the relationship between the wavelet coefficient variance and scale, as shown above. Moreover, an analysis was made to check the frequency ranges where the differences among healthy sinus rhythm and atrial fibrillation were located. Specifically, the different information was spread among the higher frequencies (10^{-1}–10^{-2} Hz). For this reason, the Hurst exponent was estimated using the five higher frequency scales. Atrial fibrillation periods were detected using a threshold equal to 0.7 for the Hurst Exponent. The authors highlighted some limitations to their work as the need of having at least 64 valid consecutive heart beats. This represents a limit for the detection of short episodes. Moreover, false might occur in case of supraventricular extrasystoles and tachycardia, while atrial arrhythmias might induce false negatives. Ivanov et al. [23] studied the local Hurst exponent in interbeat interval time series obtained from healthy subject and from patients with congestive heart failure. The Hurst exponent was found to be in the range [0.07,0.17] for healthy subjects thus showing an anticorrelated behavior. Higher value, approximately equal to 0.22, was found for patients showing a less anti-correlated behavior with respect to healthy subjects. This result was interpreted as a signature of a more complex behavior of heartbeat dynamics with respect to the pathological conditions.

5.6 Concluding Remarks

In this chapter, the properties that have motivated the use of wavelet transform for the exploration of nonlinear measures for the characterization of heart activity were discussed. We have to stress that we did not introduce the use of wavelet packet decomposition and empirical mode decomposition. The former approach allows a finer description of the signal under investigation, using a more rich set of details and approximations [37]. This allows, for instance, to improve the frequency resolution at lower scales. The latter is often compared to wavelet transform, because it also allows to efficiently describe nonstationary phenomena and to reduce noise, using a data driven approach for the signal decomposition [38].

References

1. Porta, A., Guzzetti, S., Furlan, R., Gnecchi-Ruscone, T., Montano, N., Malliani, A.: IEEE Trans. Biomed. Eng. **54**(1), 94–106 (2007)
2. De Melis, M., Morbiducci, U., Rietzschel, E.R., et al.: Med. Biol. Eng. Comput. **47**, 165–173 (2009)
3. Babaei, S., Geranmayeh, A.: Comput. Biol. Med. **39**, 8–15 (2009)
4. Işler, Y., Kuntalp, M.: Comput. Biol. Med. **37**, 1502–1510 (2007)
5. Makowiec, D., Rynkiewicz, A., Wdowczyk-Szulc, J., et al.: Physiol. Meas. **32**, 1681–1699 (2011)
6. Makowieca, D., Dudkowska, A., Galaska, R.: Physica A. **388**, 3486–3502 (2009)
7. Unser, M., Aldroubi, A.: Proc. IEEE. **84**, 626–638 (1996)
8. Daubechies, I.: Ten Lectures on Wavelets. Society for Industrial and Applied Mathematics, Philadelphia (1992)
9. Mallat, S.: A Wavelet Tour of Signal Processing. Elsevier, Amsterdam (2009)
10. Beck, A., Teboulle, M.: SIAM J. Imaging Sci. **2**, 183–202 (2009)
11. Donoho, D.L.: IEEE Trans. Inf. Theory. **41**, 613–627 (1995)
12. Wang, Y., Yang, J., Yin, W., Zhang, Y.: SIAM J. Imaging Sci. **1**, 248–272 (2008)
13. Daubechies, I., Defrise, M., De Mol, C.: Commun. Pure Appl. Math. **57**, 1413–1541 (2004)
14. Gamero, L.G., Risk, M., Sobh, J.F., Ramirez, A.J., Saul, J.P.: Comput. Cardiol. **1996**, 177–180 (1996)
15. Gamero, L.G., Vila, J., Palacios, F.: Med. Biol. Eng. Comput. **40**, 72–78 (2002)
16. Addison, P.S.: Physiol. Meas. **26**, R155–R199 (2005)
17. Nowak, R.D., Baraniuk, R.G.: IEEE Trans. Signal Process. **47**, 1852–1865 (1999)
18. Meyer, Y., Salinger, D.H.: Wavelets and Operators. Cambridge University Press, Cambridge (1992)
19. Zhang, X.-S., Zhu, Y.-S., Zhang, X.-J.: Med. Biol. Eng. Comput. **35**, 467–473 (1997)
20. Mallat, S., Hwang, W.L.: IEEE Trans. Inf. Theory. **38**, 617–643 (1992)
21. Yang, H.: IEEE Trans. Biomed. Eng. **58**, 339–347 (2011)
22. Yang, H., Bukkapatnam, S.T., Komanduri, R.: Phys. Rev. E Stat. Nonlinear Soft Matter Phys. **76**, 026214 (2007)
23. Ivanov, P.C., Rosenblum, M.G., Peng, C.-K., Mietus, J., Havlin, S., Stanley, H.E., Goldberger, A.L.: Nature. **383**, 323–327 (1996)
24. Rosso, O.A., Blanco, S., Yordanova, J., Kolev, V., Figliola, A., Schürmann, M., Baar, E.: J. Neurosci. Methods. **105**, 65–75 (2001)

25. Zunino, L., Pérez, D.G., Garavaglia, M., Rosso, O.A.: Phys. A Stat. Mech. Appl. **379**, 503–512 (2007)
26. Eke, A., Herman, P., Kocsis, L., Kozak, L.R.: Physiol. Meas. **23**, R1–R38 (2002)
27. Pincus, S.M.: Proc. Natl. Acad. Sci. U. S. A. **88**, 2297–2301 (1991)
28. Ocak, H.: Expert Syst. Appl. **36**, 2027–2036 (2009)
29. Wachowiak, M.P., Hay, D.C., Johnson, M.J.: Comput. Biol. Med. **77**, 222–230 (2016)
30. Maxim, V., Şendur, L., Fadili, J., Suckling, J., Gould, R., Howard, R., Bullmore, E.: NeuroImage. **25**, 141–158 (2005)
31. Mandelbrot, B.B., Van Ness, J.W.: SIAM Rev. **10**, 422–437 (1968)
32. Beran, J.: Biometrika. **81**, 755–766 (1994)
33. Mandelbrot, B.B.: Fractals and Turbulence: Attractors and Dispersion. Springer-Verlag, Berlin (1977)
34. Wornell, G.: Signal Processing with Fractals: A Wavelet-Based Approach. Prentice Hall PTR, Upper Saddle River, NJ (1996)
35. Wornell, G.W., Oppenheim, A.V.: IEEE Trans. Signal Process. **40**, 611–623 (1992)
36. Duverney, D., Gaspoz, J.-M., Pichot, V., Roche, F., Brion, R., Antoniadis, A., Barthélémy, J.-C.: Pacing Clin. Electrophysiol. **25**, 457–462 (2002)
37. Tikkanen, P.E.: Biol. Cybern. **80**, 259–267 (1999)
38. Kabir, M.A., Shahnaz, C.: Biomed. Signal Process. Control. **7**, 481–489 (2012)

Chapter 6
Intermittency-Driven Complexity in Signal Processing

Paolo Paradisi and Paolo Allegrini

Abstract In this chapter, we first discuss the main motivations that are causing an increasing interest of many research fields and the interdisciplinary effort of many research groups towards the new paradigm of *complexity*. Then, without claiming to include all possible complex systems, which is much beyond the scope of this review, we introduce a possible definition of complexity. Along this line, we also introduce our particular approach to the analysis and modeling of complex systems. This is based on the ubiquitous observation of metastability of self-organization, which triggers the emergence of intermittent events with fractal statistics. This condition, named *fractal intermittency*, is the signature of a particular class of complexity here referred to as *Intermittency-Driven Complexity (IDC)*. Limiting to the IDC framework, we give a survey of some recently developed statistical tools for the analysis of complex behavior in multi-component systems and we review recent applications to real data, especially in the field of human physiology. Finally, we give a brief discussion about the role of complexity paradigm in human health and wellness.

P. Paradisi (✉)
Istituto di Scienza e Tecnologie dell'Informazione "A. Faedo" (ISTI-CNR), Via Moruzzi 1, 56124 Pisa, Italy

BCAM - Basque Center for Applied Mathematics, Alameda de Mazarredo, 14 E-48009, Bilbao, Basque Country, Spain
e-mail: paolo.paradisi@cnr.it

P. Allegrini
Center for Nonlinear Science, University of North Texas, 1155 Union Circle, #311427 Denton, TX 76203-5017, USA
e-mail: allegrip@gmail.com

R. Barbieri et al. (eds.), *Complexity and Nonlinearity in Cardiovascular Signals*,
DOI 10.1007/978-3-319-58709-7_6

6.1 What Is Complexity?

6.1.1 Complexity as Emergence of Self-Organization from Cooperation

In the past, monitoring the activity of systems with many degrees of freedom was typically limited to a very small portion of it, and often only a single basic component could be observed or, on the contrary, the overall activity of the system itself (bulk measurement) could be measured, but without any knowledge about the detailed evolution of the single components of the system. In the last two decades or so, many research fields have seen the development of new experimental techniques opening the way to much more accurate and complete observations of multi-component systems. More precisely, in many research fields it has become possible to get simultaneous observations, with high time/space resolution, of all the single unit activities in the network. As an example, the spiking activity of many neurons linked through a network can nowadays be simultaneously observed by means of large arrays of electrodes [1] or by recording calcium fluorescence through a high-speed high-resolution camera mounted on a microscope [2, 3]. In this example, the new experimental tools make it available a large set of neuron spiking data that are recorded in parallel from many units and with great accuracy (i.e., not bulk measurements).

These large datasets are nowadays available in many research fields: sociology and economy (e.g., social networks, internet data, GPS mobility data); biology and physiology with the "*-omics*" data (e.g., proteomics, genomics, metabolomics, connectomics) [4–9]. The details of a multi-component system are then known and typically represented as a complex network (graph) of interacting units, defined as a set of nodes and links among nodes, thus allowing for the analysis and modeling of different time and space scales of the system, ranging from the single unit to the global dynamics.

Many recent studies are unveiling some common features and behaviors (e.g., emergence of self-organization, multi-scaling, self-similarity) among data-sets collected from very different multi-component systems. A general opinion is taking momentum in the scientific community that these common features could be the signature of some "*universal*" behavior, which is generically denoted as *complexity*, while the systems displaying such behavior are denoted as *complex systems* [9–13]. This new paradigm of complexity is triggering many interdisciplinary research activities involving skill, expertise and ideas from different fields: statistical physics, probability theory and statistics, stochastic processes, nonlinear dynamical systems, network science, data mining, signal processing.

In spite of the extensive use of the term "*complexity*" in hundreds of papers regarding the study of multi-component systems, a definition of complexity that is universally accepted in the scientific community does not yet exist and a unified view of how Complexity should be defined is probably still far from being reached.

However, some aspects that, as said above, are common to many multi-component systems, are recognized to be signs or clues of complexity [7, 8, 10–14]:

- A complex system is a multi-component system with many degrees of freedom: individuals, particles, single units. Each unit is a node in a network with a complex topology of links among nodes, representing the (nonlinear) interactions among single units.
- Multi-component and nonlinearity are not sufficient to define complexity. **Complexity** is associated with the **emergence** of **self-organizing behavior**, i.e., the spontaneous formation of self-organized structures that are triggered by some **cooperative** mechanism in the nonlinear dynamics. This emergent behavior is not related to a master driving the system in a given self-organized state but, on the contrary, the self-organizing behavior emerges, in some sense, "*spontaneously*" from the overall cooperation on the single units.[1] A master can be some unit or hub in the network, or an external forcing, affecting directly all the units, or a great majority of the units, of the multi-component system, thus having a direct control over all the internal dynamics.
- **Non-reducibility**: self-organized states have features and space-time scales that are hardly obtained as a simple (linear) function of an external forcing or by means of linear coarse graining procedures (e.g., average or weighted sum over single components).
- The spontaneous emergence of self-organization by the cooperative behavior in a complex network and without a master is associated with **self-similar**, i.e., **mono-fractal** behavior, whose signature is seen in the power-law relationships among different physical quantities. A **multiscaling** (**multifractal**) behavior can also emerge in some physical observable quantities [15]. For example, self-organized states display **long-range** space and/or time correlation functions, given by slow **power-law** decays in space and/or time. Consequently, the correlation exponents are an important example of **emergent properties**. It is worth noting that the association between self-organization (without a master) and scaling is so ubiquitous that scaling exponents are often used as an indicator of complexity and self-organization itself.

The emergence of scaling exponents related to self-organizing behavior is a intriguing and crucial aspect of complexity. Critical phenomena are an example of complexity [16] where long-range correlations and mono- or multi-scaling are always observed in combination with the emergence of self-organized structures (e.g., clusters of different sizes in the Ising spin model). The origin of scaling behavior and long-range correlations is well understood in this case. However,

---

[1]Cooperative dynamics in multi-component systems always need an external energy source to sustain self-organization, i.e., the formation of coherent structures from the disordered background. However, this does not mean that the external forcing, even if pumping energy into the system, can control the inner mechanisms triggering the emergence of self-organizing behavior. Thus, the external forcing is not a master explicitly controlling the parameter of self-organized states, such as time and space scales, but only an external energy supply.

to our knowledge, for a generic complex cooperative system without a master the underlying mechanism determining the emergence of scaling behavior is not yet clear. Heuristically, it is reasonable to suppose that emergence of self-similar behavior is related to the absence of a master (unit, hub or external forcing). In fact, when a master determines the dynamics of all the units, then the self-organized structure should be driven by the same parameters, e.g., time and space scales, of the master.

As an example, in the laminar motion of a liquid pumped in a pipeline, a rotation is observed immediately downstream of the pump whose angular velocity is related to that of the pump itself, with possible differences can be related to the friction of pipeline wall. No scaling behavior can be observed in this case. On the contrary, in cooperative dynamics without a master, whatever the scale of motion, self-organized structures emerge spontaneously and, in this case, long-range correlations and, in general, mono- or multi-scaling behavior, are always observed. Then, we can heuristically argue that the "spontaneous" emergence of a macroscopic, ordered structure from microscopic units requires a set of intermediate levels of organization, from a few units interacting over local domains and short time intervals (small scales) to the global level (large scales).[2] The need for intermediate levels of organization is essentially the reason why self-organizing behavior is usually related to a scale-free condition and, thus, to the emergence of self-similarity, mono-/multi-scaling and fractality.

From the above discussion about the general features of a complex system, we here propose a definition where the scaling features are explicitly required, even if the simultaneous presence of cooperative dynamics and absence of a master should be sufficient to justify the emergence not only of self-organization, but also of the related scaling features.

Definition 6.1 (Complexity) A multi-component, nonlinear, system is defined to be **complex** if

(i) the dynamics are cooperative and trigger the emergence of self-organizing behavior;
(ii) there is not a master (unit or hub) whose features and parameters can be directly linked to the features and parameters of the emergent self-organized states.
(iii) The system's dynamics are monoscaling (self-similar) or multiscaling.

We do not claim to give here a general definition of complexity. However, in the following we refer to systems satisfying the definition of complexity given above, a definition which seems to include many real multi-component systems spanning from socio-economic systems to biological networks. The self-organizing

[2]Roughly speaking, the emergence of self-similarity is probably the most efficient way to carry information from the small to the large scales and this could be the reason for the emergence of this intermediate organizing levels filling the gap from the microscopic to the macroscopic scales. However, this intriguing problem is not well established and should deserve further investigations, which are beyond the scope of this chapter.

behavior associated with mono-/multi-scaling determines the ubiquitous emergence of power-law dependence of different physical quantities, such as: long-range time and/or space correlations; the scale-free distribution of the average degree of nodes in a complex network; the avalanche size distribution in self-organized critical systems; the cluster size distribution in percolation [16].

Another aspect that has to be mentioned regards the transport properties, which are usually characterized by anomalous scaling, i.e., non-linear time dependence in the growth of the variance. This condition is also known as *anomalous diffusion* [13, 17–20]. Many authors refer to the tools of *fractional calculus* to develop models that are able to reproduce the the power-law behavior of different observable quantities and, in particular, the anomalous transport properties of a complex system [17, 21–25].

6.1.2 Metastability of Self-Organizing States: Intermittency-Driven Complexity

Another property is often observed in multi-component complex systems. This property concerns the stability of self-organized states, which typically do not emerge as asymptotic equilibrium states, but are characterized by *metastability*. In our opinion, this property, which is often neglected or considered as a side effect, it is, on the contrary, a crucial aspect deserving a great attention when dealing with the emergence of self-organization. In more detail, we have the following general observations [11, 12, 18, 26–32]:

(a) Self-organized states are usually **metastable** states, i.e., relatively long, but not infinite, **life-times** characterize these states. The life-time is defined as the time interval between some **birth** time (**emergence** of self-organizing behavior) and a **death** time (**decay** of the self-organized structure).
(b) An ubiquitous observation in complex systems is that the transitions between a not-organized state to a self-organized one (birth) and *vice versa* (death) are very **rapid**, usually leaving a mark of their occurrence in some experimentally measured quantities (e.g., neuron spiking activity). The overall behavior is then given by an alternation of self-organized and not-organized conditions whose passages are marked by **fast transition events** among these two conditions. In some complex systems the rapid transition can also occur between two different self-organized states.
(c) The fast transitions events often determines a **fast memory drop** in the dynamics, so that self-organized states, and the transition events themselves, are statistically independent from each other. This is known as **renewal** condition [33].

The mathematical description of the metastability described above refers to tools of probability theory and stochastic processes [13, 18–20, 33–39]. In particular:

(i) The sequence of fast transition events among self-organized states is described by a **intermittent birth-death point process** of self-organization (i.e., coherence): $\{t_n\}$; $t_{n+1} > t_n; t_0 = 0$; $n = 0, 1, 2, \ldots$, being t_n the occurrence times of the nth transition event.
(ii) The life-times of coherent structures are defined by the time intervals between two successive transition events: $T_n = t_n - t_{n-1}$; $n = 1, 2, \ldots$. These times, hereafter denoted as *Waiting Times* (WTs), are mathematically treated as random variables and studied through the tools of probability and statistics. An ubiquitous feature of WTs is the emergence of a inverse **power-law tail** in the Probability Density Function (PDF): $\psi(\tau) \sim 1/\tau^{\mu}$ [26–32], being $\psi(\tau)d\tau = \text{Prob}\{\tau \leq T < \tau + d\tau\}d\tau$.
(iii) Due to the renewal condition, the sequence of transition events is mathematically described by a **renewal point process**, which is defined as a point process whose WTs T_n are mutually independent random variables [33]. Conversely, in the time interval (t_n, t_{n+1}) between two events, i.e., in correspondence of a self-organized state, the dynamics are strongly correlated.[3]

The inverse power-law tail in the WT distribution is the manifestation of a self-similar behavior in the cooperative dynamics of the complex system and is also a crucial emergent property, characterizing the capacity of the complex system to trigger self-organization. Below we will show how this emergent property can be exploited as a measure of complexity, at least for the class of complex systems displaying intermittency associated with metastable self-organized states.

Exploiting the above list, we can give the following

Definition 6.2 (Fractal Intermittency) Given a complex multi-component system, we define as **Fractal Intermittency (FI)** the condition emerging when the transition events between two metastable, self-organized states are described by the stochastic point process defined in the above Points (i)–(iii).

We chose here to include also the renewal condition (i.e., statistically independent events and WTs) in Definition 6.2, as this is, sometimes implicitly, the definition applied in many theoretical and experimental studies [26, 34–42]. However, the role of the renewal condition is not yet clear and would deserve further investigations.[4] FI is the signature of a particular class of complex systems [12, 13, 20, 30, 31]. This particular kind of complexity is defined by the following

---

[3]Surprisingly, even in the presence of the renewal condition, a complex system can display long-range correlation functions, and the slow power-law decay of the correlation is connected to the inverse power-law decay in the statistical distribution of the random life-times [26].

[4] It is rather intuitive that the fast transition events should always be associated with a memory drop (low predictability) in the system itself, so that the events should always satisfy the renewal condition. However, this is not experimentally verified in all complex signals. In spite of this, we are convinced that FI typically involves renewal events and that the renewal process driving the complexity could be sometimes hidden below a mixture of different contributions to the intermittency generated by the system, including also the presence of noisy, secondary events. However, it is possible that an extension of the renewal condition to a slightly non-renewal condition could be necessary in order to derive more robust models and algorithms for data analysis based on the FI and IDC paradigms.

Definition 6.3 (Intermittency Driven Complexity) Let us consider a complex system, i.e., a multi-component system satisfying Definition 6.1. We define the **Intermittency Driven Complexity (IDC)** as the particular class or subset of complex systems displaying Fractal Intermittency, that is, the kind of metastability described in the above points (a)–(c) and mathematically represented in the above points (i)–(iii).

In the following we will limit ourselves to the class of complexity given in Definition 6.3.[5]

6.1.3 *How Intermittent Event Are Generated: A Dynamical Explanation of Metastability*

We do not claim here that all complex systems belong to the IDC class. However, it is also true that this kind of complexity seems to emerge in a great majority of complex systems. In fact, fractal intermittency is observed in many multi-component systems where cooperative dynamics triggers the emergence of self-organized structures, which are typically metastable and self-similar. Examples of complex systems displaying a fractal intermittent behavior are: ecological systems [51], neural dynamics [52], blinking quantum dots [40, 41, 53], social dynamics [54], brain information processing [26, 28, 30, 31, 48, 49, 55–59], atmospheric turbulence [19, 20, 22, 39], earthquakes [60], single particle tracking in cell biology [61–64], molecular biology [18].

In order to give a possible explanation of the dynamical origin of metastability and of associated complex transition events, we refer to the models discussed in [65–67]. These authors propose a dynamical model for the brain information processing, but the paradigm of metastability introduced therein can be used also as a general paradigm for complex transition events. The brain dynamics are here modeled through a dynamical nonlinear system living most of the time on a *stable heteroclinic channel*. This is essentially a set of trajectories in the vicinity of a heteroclinic skeleton, consisting of saddle points and unstable separatrices. As known, this determines a slow motion towards the saddle points. This slow motion in the neighborhood of the saddle point can be interpreted as a metastable state that maintains its coherence (i.e., self-organizing behavior) for a long time interval. As the dynamical system approaches the saddle point a critical time occurs when the

[5]This complex behavior is also known as *Temporal Complexity* [43–47], a term underlining the difference of the intermittency-based complexity, focused on the study of the temporal structure of self-organization, with the more known approach focused on the topological and spatial features of complexity (e.g., the degree distribution in a complex network, the avalanche size distribution) [47–50].

motion suddenly changes from a slow to a fast one. In this passage, the system experiences a sudden acceleration and a consequent rapid motion towards another saddle point (i.e., another metastable state). This critical time corresponds to the fast transition event, also associated with a sudden decrease of predictability and, then, with the drop of self-organization and memory, thus corresponding to the renewal property.

This very simplified brain model also illustrate another important aspect of complexity, which is actually a crucial paradigm for living systems. From nonlinear dynamical systems theory it is known that the above situation is not really typical of so-called chaotic systems (existence of a *strange or fractal attractor*), neither of systems with well-defined stable points or structures, such as limit cycles giving rise to perfect periodic patterns. Complexity is neither totally disordered (completely random) nor totally ordered (completely deterministic) In fact, in the above model, complex behavior is given by an alternation of calm (laminar) and chaotic (turbulent) motions. Then, complexity is a particular condition emerging in a intermediate region between total disorder and total order, total randomness and total determinism.

6.2 Complexity in Biology and Human Physiology

The emergence of self-organization in biological systems is nowadays well established [12, 18, 43, 64, 68–70]. There is a lack of general leading principles, which is a very old problem of theoretical biophysics with respect to other field where the theoretical research can refer to guiding principles (e.g., the postulates of classical mechanics). Even though, the search of complexity in biology and physiological is increasing very rapidly, attracting the interest of many research groups, as it can be seen from the rapid increase in the rate of publications dedicated to these topics. For example, systems biology is a very fascinating research field where the paradigm of complexity could have deep implications. The meaning and the role of the complexity paradigm in systems biology is deeply discussed in [71], a interesting paper facing the recent epistemological questions arising in biology and reviewing the historical debate between the reductionist and the holistic view in systems biology.

The research activity in biology is very active since many years. Novel experiments are continuously carried out and new experimental data are often available, so that new findings are obtained and published very rapidly. Biology research is nowadays so active that it is not rare that some unexpected experimental findings determines the failure of existing paradigms and models, thus triggering the search of new paradigms, interpretations and models. An important example comes from cell biology and, in particular, the important finding of anomalous transport behavior in the cell environment, such as the motion of lipids or proteins in the cytoplasm and on the cell membrane [18, 61–64]. An interesting debate about the best modeling approach to describe anomalous transport in the cell is taking momentum in the field of statistical biophysics, as it is not yet clear which one of two modeling approaches is the best candidate: (i) a intermittency-based transport model (Continuous Time Random Walk, CTRW) or (ii) a long-range correlated model (Fractional Brownian Motion, FBM; Generalized Langevin Equation, GLE) [25, 64, 72].

6.2.1 The Challenge of Physiological Complexity

The idea that an integrated view of the different physiological functions is becoming more and more necessary to better characterize the healthy condition of a subject (positive or negative) is taking momentum in the scientific community. Making reference to the availability of *-omics* data, many research groups are focusing on the development of complex network models. The goal is not only linking the functions of different tissues and organs, but also trying to fill the gap between molecular biology and the physiology of human body by means of theoretical tools and instruments taken from different fields such as: network theory, statistical physics, data mining, information science, signal processing [73, 74].

In the following we give a brief survey of brain and heart physiology in the framework of the complexity paradigm.

6.2.1.1 Brain

The brain is an important example and prototype of complexity. The nodes of the brain network are the neurons, which are basic units of information transport by means of electro-chemical activity, and the astrocytes, mainly responsible for the nutrient supply [75], while the (anatomical) links are given by synapses (connecting axons and dendrites) and the metabolic pathway involving both astrocytes and neurons. In recent years the role of astrocytes is being reconsidered. Many studies are finding that astrocytes could play a more active role in brain dynamics, including the secretion or absorption of neural transmitters and the propagation of intercellular Ca^{2+} waves over long distances in response to stimulation [75–77]. In summary, the topology of brain network is very branched and inhomogeneous, while brain dynamics are very rich and span over many temporal and spatial scales. Many research groups are focusing their attention on understanding the basic self-organizing mechanisms of the neural information processing. This is done through the characterization of different signals measured in the brain, such as the functional magneto-resonance imaging (fMRI), measuring indirectly the neural activity through the oxygen supply by the blood flow, and electroencephalography (EEG), measuring the electrical neural activity.

In the last two decades, the Fingelkurts brothers deeply investigated the brain dynamics and developed a conceptual multi-scale model of the brain, the Operational Architectonics [58, 59, 78]. This is a model of the brain information processing that is based on the emergence of neural assemblies and operational moduli that are self-organized and metastable. Information processing in the brain is probably the most important example of IDC, as fast transition events and fractal intermittency were found to characterize the brain collective behavior [26, 28, 78, 79]. In Operational Architectonics neural assemblies are associated with transient information flow. The temporal evolution of an assembly is given by a fast self-organizing event (*birth*), a relatively long quasi-stationary period, and a critical time when the self-organizing behavior suddenly decays. These sudden changes can be detected through the EEG signal, thus qualitatively clarifying the relationship between Operational Architectonics, Fractal Intermittency and IDC.

6.2.1.2 Heart

The heart activity is usually recorded through the electrocardiogram (ECG), which, similarly to the EEG, is the measure of an electrical activity [80]. However, at variance with the EEG, the ECG has a well-defined counterpart in autorhythmic features of the myocardial heart activity, which is driven by a small set of cells having a *pacemaker* function, denoted as the sinus-atrial node. This corresponds to the master hub, controlling directly the heart rhythm, discussed in the FI and IDC Definitions 6.2 and 6.3. The ECG in a healthy subject is given by the normal sinus rhythm (NSR), given by a well-known sequence of waves: P, QRS complex and T. As known, the R wave is given by a sharp peak in the ECG, which is so easily identified that is used to define the heart beating as the sequence of RR time distances, which are exactly the time intervals between two successive R peaks. Thus, the heart rate is given as number of R waves (beats) per minute. The NSR is identified through a set of reference ranges for some given features of these waves, mainly time intervals such as, e.g., the time between the beginning of the P wave and that of the QRS complex (QR interval, about 0.12–0.20 s) Then, standard criteria commonly used in the clinical practice to evaluate the healthy/unhealthy cardiovascular status of a patient are given by the mean morphological parameters of the P-QRS-T sequence.

Another important diagnostic parameter is given by the heart rate variability (HRV), which involves variations in the RR sequence [81]. At variance with the mean morphological parameters, the HRV is strictly connected with the correlation features of the ECG. This does not only add further information to the evaluation of the subject's health condition, but it has also a potential predictive capability. At a first sight the R peaks, being rapid transitions in the ECG signal, could be considered as good candidates for a event-based description of heart dynamics and the RR distances as the WTs in the IDC framework. However, the RR time distances have slow variations from one beat to the next, so that there is a strong correlation among R peak events due to the quasi-periodicity of the RR time distances. This violates the renewal condition and the inverse power-law distribution of WTs in the FI and IDC Definitions 6.2 and 6.3.

It is also worth noting that the well-defined sequence of waves P-QRS-T, which is the manifestation of the controlling *pacemaker* function, is not compatible with power-law correlations in the original ECG signal. Thus, the RR peaks are not genuine complex events compatible with the IDC description. In fact, in HRV there are two kind of superposed dynamical systems. The first one is related to the sinus-atrial pacemaker, is strongly synchronized, generates the RR sequences, and is associated with the mechanics of the heart pump. The second, hidden, one is responsible for the HRV, operates through the modulation of heart beating by other physiological systems, mainly the autonomic nervous system (sympathetic and parasympathetic), and is associated with the adaptability of the heart rhythm to internal and environmental changes [81]. The non-complexity of the first dynamics, even if producing the very coherent and self-organized structure of heart pumping, is due to the presence of a master or control hub, i.e., the pacemaker. On the contrary, the second, hidden, dynamics is responsible for the complexity features of HRV. This concept is also denoted as *memory beyond memory* [82] and involves a proper definition of events marking the variations in the RR frequency. These genuine complex events are mutually independent and, at the same time, responsible for the emergence of long-range, power-law, correlations in the HRV.

6.3 Measuring IDC in Signal Processing: A Survey of Statistical Tools and Algorithms

In the complexity field, the main goal of many research groups working on complex systems is nowadays the development of algorithms, and associated models, for the extraction and interpretation of useful information from the big bunch of available data. Then, the main focus is on the development and testing of reliable and synthetic statistical indices (e.g., data mining, network analysis [7, 9, 14]). Following definition 6.1, we recall that the complexity paradigm is essentially based on the concept of emergence. In particular, we mean here emergence of self-organized structures from cooperative dynamics. The main idea is that self-organized structures are the main contributors to different features of the complex system: transport properties; relaxation curves; response to external stimuli, this last one involving the adaptability to environmental changes (e.g., homeostasis in biology and physiology).

Consequently, in the development of models and statistical tools for data analysis and signal processing the main focus is on the characterization and simulation of emerging self-organized structures. In this framework, the statistical indicators extracted from the data analysis usually refer to some global property associated with the dynamical evolution of coherent, self-organized structures. Along this line, a *complexity measure* should characterize the ability of the system to trigger self-organization from overall cooperation among units without a master.

Coming back to Definition 6.1, point (iii) suggests the main direction in the development of complexity indices, that is, the self-similarity (mono-scaling) or multi-scaling, multi-fractal behavior of several observable quantities. In practice, this leads us to exploit the power-law behavior of several functions, such as, for example, the correlation function or the power spectrum, as a measure of complexity. In the recent literature many complexity measures were proposed and applied in different contexts and, in particular, in the processing of physiological signals. The most involved research fields are probably network theory (network analysis) and data mining, involving information science and statistics [5, 7, 9, 14, 83]. These measures are usually focused on the topological or spatial structure of the complex network, involving concepts such as metrics and distances (e.g., connectivity measures).

We refer the reader to the cited literature for the above measures, while we focus here on the intermittency driven complexity (IDC). The IDC class can usually be described in terms of the above topological measures, but other measures can be introduced in this case. These measures essentially focus on the temporal structure of the complex system, are clearly inspired to the metastability of the self-organized states and, thus, exploit the intermittency measures and other indices related to intermittency.

It is worth noting that these IDC-specific measures, involving the temporal structure of the complex system, can be related with some topological measures, but the two kind of measures could be also independent from each other. It is quite obvious that the existence of these relationships (e.g., correlation) depend on the observed features and on their dynamical evolution. As an example, let us consider the motion of a random walker over a scale-free complex network [4], being the motion given by random jumps between connected nodes. As known, the complex network is characterized by a scale-free degree distribution with a given power-law decay. We expect that the temporal features of the walker, such as diffusivity or return times, can be related to some topological measure of the complex network. This was proven by the authors of [84], who found that the scale-free property of the network is inherited by the random walker, showing a power-law decay in the PDF of the return times and they also found an explicit relationship between the power exponents of the return time PDF and of the degree distribution. This is a simple example where temporal and topological measures can be related. However, we again underline that this depends on the particular system under consideration and on its dynamical behavior.

In the following we give a brief survey of some IDC-specific complexity measures, some of which were developed and/or applied by our research group.

6.3.1 The Search for Critical Events in Signal Processing

According to Definition 6.3, IDC system dynamics trigger fast transition events occurring at some critical time instant when it is almost probable that also a sudden drop of memory occurs, a mechanism that is related to the passage from two different self-organized states. Then, IDC-specific measures are based on the existence of these *critical events* and on the possibility of extracting these events, i.e., their intensities and, especially, their occurrence times, from the experimental time series. In many complex systems (see, e.g., the fluorescence intermittency in Blinking Quantum Dots, BQDs, [40, 41]) the emergence of self-organized states is quite evident and the definition of critical event is also clear and unambiguous. In these cases a well-defined event detection algorithm can be easily implemented.

This is not always the case, as it happens in brain dynamics or in turbulence, where more evident transitions are mixed with many ones that are much less evident. Even worst, the genuine complex events can be hidden by the presence of more evident fast transitions in the signal that, however, are not complex, which is the case of R peak events in HRV [82]. Up to our knowledge, there is no general event detection algorithm that can work for every system, but different methods can be applied depending on the particular kind of complex system considered. Different algorithms were in fact developed and applied to BQDs [29, 40–42], atmospheric turbulence [39], HRV [82, 85–87] and human EEGs [26–28, 57, 78, 79].

6.3.1.1 Brain Events

The search for critical brain events is based on the Operational Architectonics model by Fingelkurts and Fingelkurts [57, 78] and on the associated event detection algorithm [79]. As said above, the concept of metastable neural assemblies correspond to the existence of crucial birth/death events. Neural assemblies and crucial events have a direct manifestation in EEG records by means of an alternation of relatively long-time quasi-stationary periods (neural assemblies) and quasi-instantaneous fast transitions between a self-organized neural assembly and a not-organized condition (death event) or *vice versa* (birth event). In the not-organized condition the majority of neurons near the EEG electrode is not firing (hyper-polarization) or, in any case, the local neural activity is not coordinated or synchronized. Consequently, the fast transition can be detected in the signals recorded by the nearest EEG electrodes. In [79] these fast transition events are denoted as Rapid Transition Processes (RTPs). After the usual artifact removal, the event detection algorithm for RTPs is applied to the single EEG channels according to the following steps: (i) a Hilbert transform is applied to the EEG signal; (ii) a moving average is applied to the resulting transformed signal to obtain a sort of local mean signal. (iii) The crossings between the transformed signal and its local mean are computed and the associated crossing times stored. These crossing events are candidates to become the critical transition events that we are looking for. (iv) A local derivative is evaluated. This can be done by just taking the two nearest sampling times, let's say t_n and t_{n+1}, and evaluating the corresponding local derivative of the EEG signal: $(\mathrm{EEG}(n+1) - \mathrm{EEG}(n))/(t_{n+1} - t_n)$. To avoid noisy effects, a local mean of the derivative is evaluated. (v) The distribution of the local derivative (absolute values) at the crossing times is computed. (vi) Finally, the RTPs are selected from the totality of the crossing times considering only the extreme values of the distribution, thus retaining only the crossing times with the steepest derivative. In our applications, we chose to retain the 99th percentile of the derivative distribution.[6] In the case of brain data, RTPs represent the prototype of complex events that we are looking for.

6.3.1.2 Heart Events

The main goal of HRV analysis is characterizing the modulation of heart beating by the autonomic nervous system. This is the main focus of complexity studies as well, as HRV was found to display long-range, power-law, correlations and, thus, fractal or multifractal features [82, 88–91].

As discussed in Sect. 6.2.1, the R peaks are the wave of ECG that are most used to evaluate the heart beating. In order to detect the R peaks, and the associated QRS complexes, the algorithm are often divided into two main steps: (i) ECG signal preprocessing; (ii) decision rule for the QRS detection. The techniques used for Step (i) usually include a bandpass filtering to reduce noise coming by several sources, such as power line noise and muscle noise. The range 5–30 Hz usually covers most of the frequency content of the QRS complex [92]. Being Q, R and S sharp cusps, Step (i) can also include the evaluation of the signal derivative and/or even the squaring of the filtered signal, or of its derivative, in order to enhance the extreme values corresponding to local maximum/minimum time points. The decision rule in Step (ii) is usually given through an amplitude threshold, which is often determined with some adaptive procedure. The reference time point is generally selected to be the R-wave, and the sequence of R peak occurrence times $\{t'_n\}_{n=1}^{N}$ is given as the output of the detection algorithm.

The associated RR time distances are then easily computed from the sequence of R time points t'_n and represent the basic feature exploited for the estimation of HRV in terms of complexity measures [82, 88, 90]. We recall that, at the end of Sect. 6.2.1, we showed that the R peak events are not complex events according to the IDC Definition 6.3, as they do not satisfy neither the renewal condition nor the emergence of fractal WT statistics (inverse power-law WT distribution). We again underline that this aspect is strictly connected to the presence of a control or master hub (the pacemaker) and indicates the lack of a spontaneous emergence of self-organization from cooperation in the heart.

Even though, the sequence of R peak events remains the basic feature used to characterize the HRV, even in the IDC framework. Roughly speaking, we can say that the dynamics of the single heart beat is not complex, while the variability in the heart beating (HRV) is complex, at least in healthy subjects [74, 81]. For this reason, the authors of [82] focused on a different kind of event involving the variations of RR distances. The algorithm works as follows:

[6]For further details about the RTP detection algorithm, we refer the reader to [79] and to [26, 28] where our implementation is explained.

(i) A coarse graining procedure is applied by approximating the ideally continuous values of the RR distances by a set of discrete values with some given step ΔT. More precisely, let us consider the sequence of discrete values: $\tilde{T}_i = i \cdot \Delta T$. Then, the nth WT is approximated by the nearest $\tilde{T}_i$:

$$T_n \to \tilde{T}_i \text{ if } \tilde{T}_i - \frac{\Delta T}{2} \leq T_n < \tilde{T}_i + \frac{\Delta T}{2} \tag{6.1}$$

(ii) The selection rule is given by selecting the time t_n when a jump between two different $\tilde{T}_i$ occurs, usually: $\tilde{T}_i \to \tilde{T}_{i+1}$ or $\tilde{T}_i \to \tilde{T}_{i-1}$.

The event detection algorithm was shown to be robust in a neighborhood of $\Delta T = 1/30\,\text{s}$.

It is worth noting that the heart events extracted with the above algorithm are a prototype of the hidden complex events discussed at the end of Sect. 6.2.1 and driving the heart pacemaker. However, even this sequence of events contains both genuine complex events and pseudo-events and this is probably due to the unavoidable presence of false positives in the event detection algorithm [82].

6.3.2 Complexity Measures for the IDC Class

Let us recall that, according to definition 6.1, a reliable measure of complexity should be able to estimate the ability of the system to trigger self-organization. Always from definition 6.1 we guess that the best approach is to evaluate the scaling features, i.e., the power exponents (e.g., the critical exponents in critical phenomena). Complexity measures are often developed in the context of network science [5, 9] or in critical phenomena [8, 16, 93] and usually refer to the topology of the link structure or to concept associated with information transfer.

Before considering the IDC class, let us give two examples of topological measures of complexity. The most simple property that can be evaluated in a complex network is probably the covariance matrix. This can be used to give a first estimation of the connectivity (without causality relationships among nodes) and it is often used to define an adjacency matrix by a thresholding technique applied to the covariance matrix [50, 94]. An important topological measure is given by the degree distribution, which is defined as the distribution of the number of link is whatever node of the network. In other words, if we randomly choose a node, the probability distribution of the links of this same node is defined as the degree distribution. A crucial result is that, in scale-free complex networks, the degree distribution is a inverse power-law function, thus revealing some kind of self-organization characterized by a self-similar behavior among different scales in the network structure.

From now on we refer only to complex systems in the IDC class. We are interested in the temporal complexity generated by the sequence of intermittent transition events, which, as already said, is modeled through a birth-death, stochastic, point process of self-organization. After the application of the event detection algorithm, we get the experimental sequence of events:

$$\{t_n\}\ ; \quad n = 0, 1, 2, \dots\ ; \quad t_0 = 0\ , \tag{6.2}$$

being t_n the occurrence time of the nth event. The process is equivalently defined by the sequence of WTs $\{T_n\}$ with $\{T_n = t_n - t_{n-1};\ n = 1, 2, \dots\}$.

Then, the formal definition of the birth-death *stochastic point process* of self-organization, associated with the above event sequence, is given through the following counting process:

$$N(t) = \max\{n \in \mathbb{N} : t_n \leq t\}\ . \tag{6.3}$$

being $\mathbb{N}$ the set of positive integer numbers. The IDC indices must refer to the statistical features of the point process $N(t)$. However, in the general case, the complete characterization of the point process $N(t)$ from an experimental dataset is practically impossible. In fact, the process $N(t)$ is rigorously and completely defined when all the k-order statistical distribution of the sequences $\{t_n\}$ and/or $\{T_n\}$ are given[7]:

$$\begin{aligned}
P_1(\tau) &= \text{Prob}\{T_n < \tau; n = 1, 2, \dots\} \\
P_2(\tau_1, \tau_2) &= \text{Prob}\{T_{n_1} < \tau_1; T_{n_2} < \tau_2; n_1, n_2 = 1.2, \dots\} \\
P_3(\tau_1, \tau_2, \tau_3) &= \text{Prob}\{T_{n_1} < \tau_1; T_{n_2} < \tau_2; T_{n_3} < \tau_3\} \\
P_4(\tau_1, \tau_2, \tau_3, \tau_4) &= \dots\dots\dots
\end{aligned}$$

The numerical estimation of these distributions from the data is not only very demanding, but also clearly limited by the size of the statistical ensemble. For finite statistical samples it is also well known that the accuracy of the k-point distributions rapidly decreases as the order k increases. Even the development of theoretical models with general k-order statistics is a very difficult task.

However, in many models the WT-PDF $\psi(\tau) = P_1(\tau)$ and the $1|1$ conditional probability:

$$P_{1|1}(\tau_1|\tau_2) = \text{Prob}\{T_{n_1} < \tau_1 | T_{n_2} < \tau_2\}$$

are sufficient to characterize, at least approximately, the point process $N(t)$. This is exactly true for a Markovian sequence of WTs, and it is even more true for renewal point processes, whose conditional probabilities do not depend on the previous history [33]:

[7]It is also possible to characterize the point process $N(t)$ by using directly the k-point statistical distribution of $N(t)$ itself. The statistical features of $N(t)$ and $\{T_n\}$ are clearly linked to each other.

$$P_{1|n}\left(\tau_{n+1}|\tau_n, \tau_{n-1} \ldots, \tau_2, \tau_1\right) = P_1(\tau_{n+1}) = \psi(\tau_{n+1}) .$$

Being the WTs mutually independent variables, a renewal point process $N(t)$ is uniquely defined by the WT-PDF $\psi(\tau)$.

Let us now recall that, as given in definition 6.3, the IDC of a complex system is associated with the emergence of FI, which is defined in definition 6.2. By this definition, FI is described by a renewal process with self-similar WT statistics, i.e., FI is associated with the emergence of a power-law tail in the WT-PDF:

$$\psi(\tau) \sim \frac{1}{\tau^{\mu}} , \tag{6.4}$$

which is, as said above, the signature of a self-organization with self-similarity and, thus, without a master. As a consequence, the signature of IDC emergence is associated with:

(i) the occurrence of the renewal condition (at least approximately) and
(ii) a power-law tail in the WT-PDF.

The power exponent μ is then a fundamental feature of the self-organizing behavior of the system that has been denoted as *complexity index* in recent literature [12, 20, 29, 30]. With a more precise meaning, we here denote μ as **IDC index** and we claim that it can evaluate the capacity of the dynamics to trigger intermittent, complex events and, thus, self-organized, metastable structures whose self-similar behavior is not directly driven by a master.

The evaluation of the renewal condition and the computation of the IDC index requires some specific algorithms of statistical data analysis and signal processing. In the following we give a brief list of some algorithms developed and/or applied by our group to evaluate the renewal condition and the IDC index μ or scaling exponents related to μ itself. For further details about these algorithms, we refer the reader to the cited literature.

6.3.2.1 Analysis of Renewal Aging

Let us assume to observe a statistical ensemble of independent renewal processes with same fractal WT distribution. Suppose that the system's preparation (i.e., initial condition) is made at $t = 0$ and that the starting time of experimental observation is some t_a. Then, when the WT-PDF $\psi(\tau)$ of the single renewal process has a slow inverse power-law (i.e., $\mu < 3$), the ensemble averages depend on the *aging time* t_a. In particular, the WT-PDF derived from the ensemble $\psi_{t_a}(\tau)$ depends on t_a and it is different from the WT-PDF of the single renewal process, which corresponds to the WT-PDF for $t_a = 0$ and is denoted as brand new WT-PDF: $\psi(\tau) = \psi_0(\tau)$. The departure of $\psi_{t_a}(\tau)$ from $\psi_0(\tau)$ in a renewal process is a statistical feature that was named *renewal aging* [35, 36, 38–42, 95, 96]. In general, aging is a property of many complex systems associated with the very slow relaxation of the initial conditions, also determining a departure from the ergodic condition (e.g., weak ergodicity breaking [64]).

The above feature can be exploited to give an indication about the "*renewal content*" of a time series or, in other words, if a point process is also renewal. Usually only a single sequence of events is experimentally available, so that the method has to start with the building of a statistical ensemble. Then, given the series of event occurrence times $\{t_n\}$, the renewal aging algorithm works as follows:

(1) M sequences of *aged* events are built in the following way: the first sequence is the original one; the second sequence is given by the first one by removing the first WT; the third sequence is obtained from the second one by removing the first WT of the second sequence itself; and so on.
(2) A time window of duration t_a is superposed to all sequences and the first available event with $t_n > t_a$ is taken, i.e., $t_n < t_a < t_{n+1}$.
(3) The *aged* WTs are computed: $\mathrm{WT}_m(t_a) = t_n - t_a;\ m = 1, M$.
(4) Given the sequence $\{\mathrm{WT}_m(t_a)\}$, the experimental aged WT-PDF $\psi_{t_a}^{exp}(\tau)$ is evaluated. The brand new WT-PDF $\psi_0^{exp}(\tau)$ is also evaluated.
(5) In order to derive the aged WT-PDF in agreement with the renewal condition, we apply a random shuffling to the experimental WT sequence. The new sequence has exactly the same brand new WT-PDF $\psi_0^{exp}(\tau)$ of the original WT sequence, but the possible presence of inter-WT dependence has been destroyed, thus the shuffled WT sequence is renewal. Then, we repeat the steps from (1) to (4) to evaluate the renewal aged WT-PDF: $\psi_{t_a}^{ren}(\tau)$.

The comparison of the two WT-PDFs can be used to establish if the experimental WT sequence is renewal or, if not, how it departs from the renewal condition. The renewal aging analysis was implemented also by using the corresponding Survival Probability Functions (SPFs), which are defined by:

$$\Psi(\tau) = \mathrm{Prob}\{WT > \tau\} = \int_\tau^\infty \psi(\tau) d\tau = 1 - \int_0^\tau \psi(\tau) d\tau \,. \qquad (6.5)$$

6.3.2.2 Renewal, IDC Index and Diffusion Scaling: The EDDiS Algorithm

The analysis of Event-Driven Diffusion Scaling (EDDiS) is based on the building of event-driven random walks, also named Continuous Time Random Walks (CTRWs) [97–99], and on the estimation of the scaling exponents associated with the resulting diffusion process. In the case of fractal intermittency, the relationships among the different diffusion scaling exponents and the IDC index μ are known. Then, it is possible to derive independent estimations of the IDC index μ that can be compared with each other. When the differences between the values of μ so obtained are inside the statistical errors, we can reasonably argue that the process is renewal and we also get a robust estimation of the IDC index μ.

The EDDiS algorithm collects a number of different methods for scaling analysis and several theoretical results from the theory of CTRWs that are well known in literature by many years (for a review, see [13, 26] and references cited therein). However, the joined exploitation of these methods and results with the scope of evaluating both the renewal condition and estimating the IDC index was carried out by our group for the first time in [26–28] and the first application was on ElectroEncephaloGram (EEG) data.

We give here a brief explanation of the algorithm and refer the reader to [13, 20] and references cited therein for further details. The EDDiS algorithm works as follows:

(1) Given the experimental sequence of events, three different event-drive random walks are built by applying three walking rules for the instantaneous velocity $\xi(t)$:

  (i) Asymmetric Jump (AJ) walking rule: the walker makes a unitary step ahead at every event occurrence time: $\xi(t_n) = +1$; $\xi(t) = 0$ if $t_n < t < t_{n+1}$.
  (ii) Symmetric Jump (SJ) walking rule: similar to the AJ rule, but the walker can make positive or negative jumps: $\xi(t_n) = \pm 1$, being the sign $\pm$ chosen with a coin tossing prescription: $\text{Prob}\{+1\} = \text{Prob}\{-1\} = 1/2$.
  (iii) Symmetric Velocity (SV) walking rule: the walker moves with constant velocity in a given direction, until a new random direction is chosen in correspondence of an event by a coin tossing prescription: $\xi(t) = \pm 1$; $t_n \leq t < t_{n+1}$. This walking rule is also known as *telegraph signal.*

(2) For each walking rule, the associated CTRW is given by the diffusion variable:

$$X(t) = X_0 + \int_0^t \xi(t')dt' \tag{6.6}$$

(3) We estimate two different scaling exponent of the diffusion process $X(t)$:

  (a) the self-similarity index δ of the PDF of $X(t)$:

$$P(x,t) = \frac{1}{t^\delta} F\left(\frac{x}{t^\delta}\right) . \tag{6.7}$$

  The scaling δ is computed by applying the Diffusion Entropy (DE) analysis [100] (see Appendix).
  (b) the scaling exponent H of the second moment:

$$\sigma^2(t) = \langle \left(X(t) - \overline{X}\right)^2 \rangle \sim t^{2H} , \tag{6.8}$$

  where $\overline{X}$ denotes the mean value of $X(t)$. The scaling H is computed by applying the Detrended Fluctuation Analysis (DFA) [101] (see Appendix).

The scaling exponents δ and H are theoretically exactly known in the case of fractal intermittency with index μ (see [13] for and references therein) and can be used as reference values for the estimation of the system's IDC. In fact, as said above, these scaling exponents can be jointly used to estimate the renewal condition and, indirectly, the index μ [20]. However, δ and H can be also used directly as indicators of IDC, as they were built on the basis of the sequence of events extracted from the experimental times series.

The functional relationships $\delta = \delta(\mu)$ and $H = H(\mu)$ are summarized in Figs. 6.1 and 6.2. Normal scaling is defined by $\delta = H = 0.5$ and is the signature of absence of cooperation and memory, a condition compatible with the central limit theorem and the emergence of Gaussian PDF and markovian exponential correlations. On the contrary, $\delta \neq 0.5$ and/or $H \neq 0.5$ is associated with cooperation and spontaneous emergence of self-organization, and the distance from the reference value 0.5 is a measure of complexity. It is clear from Figs. 6.1, 6.2 that the interesting

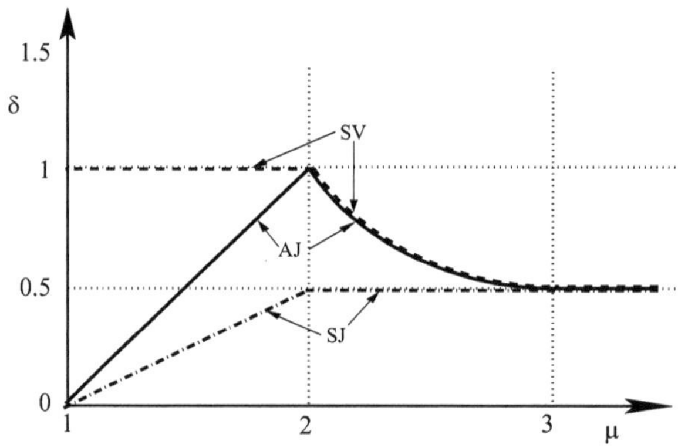

Fig. 6.1 Scaling δ vs. complexity index μ for the three walking rules: AJ (*continuous line*), SJ (*dotted–dashed line*) and SV (*dashed line*)

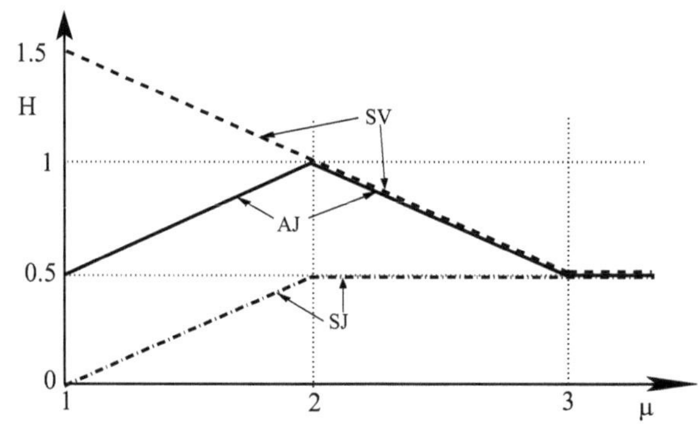

Fig. 6.2 Scaling H vs. complexity index μ for the three walking rules: AJ (*continuous line*), SJ (*dotted–dashed line*) and SV (*dashed line*)

range is $1 < \mu < 3$, where anomalous diffusion scalings are observed (for the SJ rule this occurs only in the range $1 < \mu < 2$), whereas normal diffusion ($H = \delta = 0.5$) emerges in the range $\mu \geq 3$.

6.3.2.3 The Problem of Noise and the PANDORA Algorithm

The application of the EDDiS method was originally thought (i) to obtain a second check about the renewal condition in combination with the renewal aging method and (ii) to get a reliable estimation of the IDC index μ [26]. However, we later understood that the EDDiS method allows also to take into account the possible presence of noisy secondary events in the experimental sequence in the estimation of the IDC index μ. In fact, during the application of the EDDiS method, a surprising normal diffusion regime was seen in the long-time behavior of the SV-CTRW, even if the AJ-CTRW showed superdiffusion ($H > 0.5$) [26, 27]. This observation was explained by assuming the presence of noisy non-complex events generated by a Poisson process with event rate r_p, thus generating normal diffusion, mixed with the genuine complex events, thus generating fractal intermittency. This model is called Time Mixed Model (TMM). From this assumption, we found that the long-time normal diffusion in the SV rule was related to the exponential cutoff emerging in the WT-PDF of the mixed events.

This first result allowed us to explain the divergence among our findings, based on the EDDiS method, and those of other authors, based on a best fit applied to the WT-PDF (see, e.g., [102]). In fact, the presence of noisy events does not only introduce an exponential cutoff in the tail of the WT-PDF, but can also affect its power-law decay, as an apparent power exponent, completely different by the genuine IDC index of the complex process, can also appear in the WT-PDF [13, 27]. This makes the application of the EDDiS method a very reliable statistical tool for the estimation of μ, a feature that is related to the capacity of the event-driven diffusion processes, especially the AJ rule, to separate the effect of noisy events from that of the genuine complex point process [19, 20].

Regarding TMM, we also found an interesting scaling law for the long-time diffusivity coefficient for the SV-CTRW:

$$D(\mu, r_p, T) \propto T^{\mu-2} \, r_p^{\mu-3} \, ; \quad \mu > 2 \, , \tag{6.9}$$

being r_p the Poisson rate of event occurrence and T the WT scale after which the complex behavior, marked by the passage to the power-law decay $1/\tau^{\mu}$, emerges in the WT-PDF. This scaling is valid for $\mu > 2$ and in the limit of small ratios between Poisson and complex events.

In [13] we suggested a possible algorithm of time series analysis that, exploiting Eq. (6.9), could be used to evaluate both the IDC index μ and the contribution of noisy events. This algorithm is here denoted as *Poisson Added Noise DiffusiOn*

Rescaling Analysis (PANDORA). Given the experimental sequence of event occurrence times $\{t_n\}$, the PANDORA algorithm works as follows:

(1) Let us generate N random sequences of Poisson events $\{t_j^k\}$, $j = 1, 2, 3, \ldots$ with the constrain that the occurrence time of the last Poisson event must be less than the total duration time of the experimental sequence. Each sequence has a different Poisson rate r_k, $k = 1, \ldots, N$.
(2) Then, for each k, let us build sequences that are given by the superposition of the experimental sequence $\{t_i\}$ and of the artificially generated Poisson sequence with rate r_k.
(3) For each k, we apply the SV rule on the generated sequence of events (real+artificial Poisson) and we evaluate the long-time diffusivity coefficient $D_k = D(r_k)$. In this way we obtain a set of couples (D_k, r_k) that can be plotted to get the graph of the numerical function $D_k vs. r_k$.
(4) It is known that the superposition of two Poisson processes with rates r_p and r_k results in a global Poisson process with total rate given by the sum of the single rates: $r_{\text{tot}} = r_p + r_k$. When the experimental sequence is affected by a Poisson noise with rate r_p, Eq. (6.9) can be rewritten substituting the rate r_p with the total rate r_{tot}:

$$D_k = \tilde{D}(r_k;\ \mu, r_p, T) \propto T^{\mu-2} \left(r_p + r_k\right)^{\mu-3} . \tag{6.10}$$

A best fit procedure can now be used to evaluate the three parameters of the function $\tilde{D}(r_k)$, i.e., the IDC index μ, the rate r_p of the noisy (Poisson) events and the *complexity emergence time* T. It is also possible to apply a best fit with respect to only two parameters: r_p and T. In this case, μ is evaluated from the EDDiS method.

In order to validate the PANDORA algorithm, we simulated a sequence of complex events according to a fractal renewal process with $\mu = 2.3$. Then, we generated several TMM sequences where different Poisson processes with rates r were superposed to the simulated sequence. The long-time diffusivity coefficients $D(r)$ of the SV-CTRWs were estimated for each r and the resulting function $D(r)$ has been reported in Fig. 6.3 (circles). The dashed line is a best fit with a power-law function $r^{3-\mu}$, whose power exponent is in agreement with that predicted by the theoretical result given in Eq. (6.10), thus confirming the validity of our approach.

6.4 Applications to Real Physiological Signals

The above algorithms (Renewal Aging, EDDiS and PANDORA), which are based on the FI and IDC paradigms given in Definitions 6.2 and 6.3, were applied to different experimental datasets, spanning from BQDs [40, 41] and turbulence [19, 20, 39] to ECGs and EEGs. In the following we give a brief survey of the

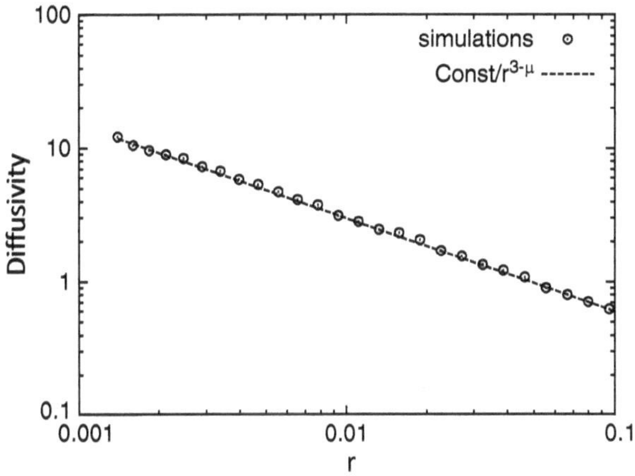

Fig. 6.3 Simulation study showing the application of the PANDORA algorithm to an artificially generated sequence of complex events with $\mu = 2.3$. The diffusivity $D(r)$ has been plotted with respect to the rate r of the added Poisson noise (*circles*)

applications carried out in the processing of two physiological signals, the EEG and the ECG, underlining the main results that were found within the IDC approach.

6.4.1 EEG Complexity

As explained above, the Operational Architectonics model [78] includes concepts such as neural assemblies and operational moduli that are in agreement with the paradigm of emerging metastable self-organized states given in the IDC paradigm (Definitions 6.2 and 6.3). Through the definition of RTP events that mark sudden changes in the EEG traces and are associated with the fast transition events among self-organized states, a RTP event detection algorithm was also developed [79]. A version of this algorithm was implemented and applied by our group to extract RTP events from observed human EEG records collected in two different experimental conditions: the resting state condition and the sleep condition. Below we give a brief survey of the main findings.

6.4.1.1 Resting State

As known, the *resting state* condition is obtained when the subject is relaxed with closed eyes. This condition is typically characterized by the simultaneous emergence of α-waves in several EEG channels. From the RTP events of the single channels it is possible to define global events as the simultaneous occurrence of a given minimum number of single EEG events. Both the single channel and the global RTP events were analyzed through the EDDiS method to characterize the complexity and, in particular, to estimate the IDC of the human brain during the resting state condition. The main experimental findings regarding IDC are [26–29]: (1) The sequences of global RTP events are compatible with a renewal fractal process for the global brain dynamics. The values of the global IDC index are in the range 2.05–2.4, while the values of the global diffusion scaling exponents span over the interval 0.8–1. (2) The probability of a EEG channel to be recruited into a global RTP event is higher for EEG channels belonging to the Default Mode Network (DMN) [103, 104]. (3) The EEG channels with high probabilities of being recruited into a global event are also those with the most complex behavior, with high values of the diffusion scaling δ that are compatible with superdiffusion. As said above, these channels belong to the DMN. On the contrary, the EEG channels with low δ (less complex or non-complex) have low probabilities to be recruited in a global complex brain event. (4) The variability among subjects of the δ scaling indices is much smaller for the EEG channels that are more complex (inside the DMN) with respect to the EEG channels that are less complex (outside the DMN). (5) The IDC indices of single EEG channels in the DMN have almost the same values of the global IDC index.

From these findings we can argue that the DMN drives the global complex behavior of the brain in the resting state condition. The small inter-subject variability suggests a sort of universal behavior of the DMN during the resting state, a result in agreement with previous findings, being the DMN the most active brain sub-network during the resting state condition [103, 104].

6.4.1.2 Sleep Condition: Conscious vs. Unconscious

The EDDiS analysis was also applied to EEG data collected during sleep. Twenty nine nights have been analyzed, focusing on the first cycle that is usually the longest one [30–32]. As known, sleep is divided into cycles, typically 4 or 5 per night, and each cycle is divided into four main stages, defined on the presence of different waves, or graphoelements, and specific rhythms: N_1, N_2, N_3 (also called Slow Wave Sleep, SWS) and Rapid Eye Movement (REM) [55]. We focused on the first cycle, and we considered also the pre-sleep WAKE condition, approximately collecting the 30 min before the beginning of stage N_1. Passing through stages N_1, N_2 and N_3 the subject goes from a shallow sleep in the N_1 stage to the deep sleep in the N_3 stage (SWS). The SWS stage is a unconscious state, as the brain dynamics are segregated during this stage. On the contrary, during the REM stage, when dreaming occurs, the bran is segregated and a global conscious state emerges similarly to the WAKE conscious condition [55].

The main results regarding the IDC paradigm in the sleep condition are reported in Fig. 6.4, where the diffusion scaling exponents H, associated with the global brain events, are compared for different sleep stages. The pre-sleep WAKE and REM stages show an anomalous diffusion scaling $H = 0.75$, which corresponds, in the renewal condition, to $\mu = 2.5$. On the contrary, the SWS (N_3) stage

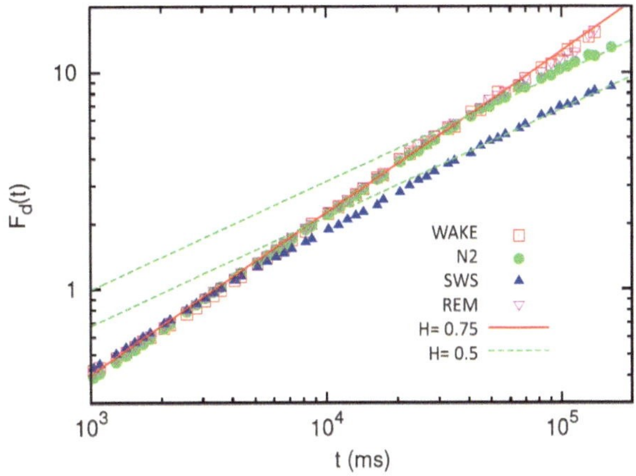

Fig. 6.4 Comparison of the scaling *H* for different sleep stages

reaches the asymptotic normal scaling $H = 0.5$ after a short-time transient. The N_2 stage also reaches an asymptotic normal scaling, but this occurs at times much longer than the SWS stage. In fact, it is evident an extended short-time range with an anomalous diffusion behavior. All these results are in agreement with the concept of segregation-integration in the brain network, associated with emergence of unconscious-conscious states. This confirms that the global IDC index and/or the related diffusion scaling exponents can be used to estimate the connectivity features of the brain neural network. Then, μ, H or δ can be used as indicators of consciousness.

6.4.2 *ECG Complexity*

HRV is recognized to be a useful clinical tool for the evaluation of cardiac autonomic changes. The relationship of HRV with the mortality risk following myocardial infarction has been established since many decades [105, 106]. More precisely, HRV is reduced in patients recovering from an acute myocardial infarction and, further, the mortality risk increases as the HRV decreases [107]. This important result was found by means of time and frequency methods, such as the estimation of signal variance, usually applied to the RR distance sequence. Clearly, a small variance indicates low HRV. However, it is well-known that HRV is affected the parasympathetic and the sympathetic autonomic systems. Further, even if these relationships are still not completely understood and quite controversial, it is well established that the heart-autonomic system interactions are strongly nonlinear, and the overall coupled heart-autonomous system is reminiscent of a globally cooperative, complex system. Consequently, it seems quite natural to characterize the HRV in terms of complexity features extracted from the sequence of RR distances in order to improve the prognosis of different clinical conditions and, in particular, the distinction between health and disease. These methods do not strictly estimate the amplitude of HRV, but its complexity by means of scaling exponents and long-range correlation features associated with self-similarity, i.e. fractality.

Following this idea, the authors of [86, 87] compared several complexity measures, also denoted as nonlinear indexes, to investigate the prognostic potentiality in patient with chronic heart failure (CHF), thus highlighting the importance of HRV complexity measures in the prognostic stratification of CHF patients. Among others, also the δ index, computed from the R sequence through DE (see Appendix), was evaluated. The DE and the δ scaling were also used in [82, 85] to evaluate the complexity and, thus, long-range correlations of RR distances. As said at the end of Sect. 6.2.1, these authors assumed that the hidden dynamics, related to the autonomic system and driving the heart pacemaker, are associated with the heart events extracted with the algorithm discussed at the end of Sect. 6.3.1. The main results of their analysis were:

(i) The comparison between δ, estimated with the DE, and μ, estimated by the WT-PDF, makes evident the existence of hidden dynamics, as the DE is able to detect the genuine scaling δ of the hidden dynamics themselves.
(ii) The scaling δ can be used as a measure of complexity.
(iii) Subjects with congestive Heart Failure (cHF) were compared with healthy subjects and it was found two well distinct regions in the plane (δ, ϵ), being ϵ the rate of pseudo-events.

Another powerful complexity measure extensively applied to HRV through the analysis of RR distances is given by the detrended fluctuation analysis (DFA), evaluating the second moment scaling H and whose algorithm and meaning are discussed in Appendix. DFA quantifies long-range correlations and, thus, the presence of (fractal) power-law memory in the signal. The first application to HRV proved the existence of universal fractal properties in healthy subjects and loss of fractal, complex, self-organization in heart failure patients [108]. In particular, it was found a general distinction between a short-time and a long-time behavior:

(i) In the range $L < 11$ beats, a short-term scaling exponent H_1, also denoted as α_1, can be seen. Typical values for healthy subjects are $H_1 \sim 1$.
(ii) In the range $L > 11$ beats, a long-term scaling exponent H_2, also denoted as α_2 and different from α_1, is evident.

Many authors found evidence that the short-term scaling H_Υ of heart beating is related to the subject condition (health status, aging) and can have a better prognostic capability than that obtained by more conventional measures of HRV (e.g., variance, power spectrum). Decrease in the H_1 scaling was shown to be associated with activation of both vagal and sympathetic outflow, thus resulting in a more random behavior, i.e., lower complexity, of heart beating [109]. Healthy elderly subjects display changes in the long-range correlation features of HRV [110]. The authors of [111] found that altered short-term DFA scaling in HRV precedes the onset of atrial fibrillation in patients without a structural heart disease. Similarly, changes in the short-term scaling are observed in ectopic tachycardia associated with disturbances in the autonomic system or in ectopic atrial pacemakers [112]. The application of DFA to patients with acute myocardial infarction (AMI) showed that lower values of short-term scaling H_1 are associated with a greater mortality [113]. DFA was observed to have a prognostic value even in the case of patients taking beta-blockers after acute myocardial infarction [114].

In general, all these studies indicate that, in subjects with several heart diseases, fractal or complexity measures change towards values indicating the presence of more random fluctuations and, thus, less complex behavior in HRV with respect to healthy subjects. Nonlinear, complexity, indexes have some advantages over the conventional measures of HRV in risk stratification purposes, such as (i) less dependency on heart rate, (ii) less inter-individual and intra-individual variations and (iii) smaller relative changes of individual values over time [86, 87, 115, 116] (see [117] for a review).

6.5 A Brief Discussion About Wellness, Health and Complexity

We conclude this chapter discussing an open issue in physiology, biomedicine and physiological signal processing that will probably attract the interest of the scientific community in the next years.

Diagnostic methods are based on chemical and physical analyses and on evaluation of symptoms. When there are no symptoms and the chemico-physical analyses give results within the normative values, the subject is considered to be in a healthy condition. However, it is very usual that the general health equilibrium of a subject with a good general diagnosis could be affected by some subtle, hidden stress working silently, i.e., much under the threshold of both subject self-perception and medical analyses. This silent external stress can work for years without any evidence, especially if it is almost continuous in time (daily) and relatively weak in intensity. In fact, human physiology is able to respond to a silent and weak stress factor by means of homeostatic mechanisms determining a shift in the general physiological equilibrium. This subtle, prolonged condition is probably the most hard to be detected when looking at single physiological parameters, which are usually associated with the functioning of single organs or tissues or with some cell metabolic process.

The above discussion highlights the need for a integrated view of the human physiology [74, 118] and, thus, for a model coupling the single physiological systems defined in the standard biomedical analyses. This coupled model surely requires ideas and concepts taken from the complexity paradigm, as it is based on the emergence of self-organizing behavior with scaling and self-similarity. Following this holistic view ("the whole is more than the sum of its parts"), but without neglecting the complementary reductionist, analytic approach [71], future work and effort will be probably devoted to the development of a integrated model based on complexity paradigm. We expect that this is the best approach to derive a reliable *wellness* index, which should be the signature of a general healthy condition. The wellness index could be considered a sort of pre-nosological global parameter, able to reveal possible departures from a healthy physiological equilibrium. In this view, the overall self-organization of the human body is the central aspect to be understood, and we believe that wellness will probably refer to some kind of global *physiological complex self-organization* or *physiological complex integration*.

An example of a physiological coupling, which is attracting the increasing interest of many research groups [119–122], is given by the heart-brain coupling, where self-similarity and scaling indices seem to play a crucial role [91]. The findings in this new challenging issue indicate that the IDC class of complex systems could play a crucial role in the study of this integrated view in human physiology, both through the theoretical development of IDC-based physiological models and the applications of related signal processing tools.

Appendix: Diffusion Entropy and Detrended Fluctuation Analysis

Given a diffusive variable $X(t)$; $t = 1, 2, \ldots$ [e.g., the event-driven random walks of the EDDiS algorithm, Eq. (6.6)], we are interested in evaluating the self-similarity index δ of the PDF, defined in Eq. (6.7), and the second moment scaling H, defined in Eq. (6.8). δ and H are evaluated by means of the Diffusion Entropy (DE) analysis [13, 20, 100] and of the Detrended Fluctuation Analysis (DFA) [101], respectively.

Diffusion Entropy

Given the PDF $P(x,t)$, the DE is defined as the Shannon entropy of the diffusion process:

$$S(\Delta t) \equiv -\int_{-\infty}^{+\infty} p(\Delta x, \Delta t) \ln p(\Delta x, \Delta t) d\Delta x \,, \tag{6.11}$$

where Δt here denotes a time lag and not the absolute laboratory time. Using the self-similarity condition (6.7), it is easy to prove that

$$S(t) = \delta \ln \Delta t + S_* \,, \tag{6.12}$$

where $S_* = -\int_{-\infty}^{+\infty} F(x) \ln F(x) dx$. Notice that the scaling is in fact asymptotic, namely it is only exact for $t \to \infty$, and an effective time T_* can be introduced as an additional fitting parameter:

$$S(\Delta t) = \delta \ln(\Delta t + T_*) + S_*. \tag{6.13}$$

It is possible to estimate δ by considering the graph $(\Delta t, S(\Delta t))$ in a log-lin plot and then fitting Eq. (6.13) to the data.

The computation of $S(\Delta t)$ requires the evaluation of the PDF $P(x, \Delta t)$. This is done by considering a moving window of length Δt, so that the set of pseudo-trajectories $X_r(k) = X(r+k) - X(r)$, with $0 \le k \le \Delta t$, $r = 1, 2, \ldots$, is considered. The pseudo-trajectories all start from $X_r(0) = 0$, and, for each Δt, it is possible to evaluate the histogram $P(x, \Delta t)$ of the sequence $X_1(\Delta t), X_2(\Delta t), \ldots$ and, then, the DE $S(\Delta t)$.

Detrended Fluctuation Analysis

Given the diffusive variable $X(t)$; $t = 1, 2, \ldots$, the DFA essentially estimates the second moment of a proper detrended time series $X(t) - \overline{X}(t)$. The detrending can be done with a n-order polynomial function and the most simple algorithm uses a linear detrending by a least-squares straight line fit [101]. The DFA algorithm works as follows: (i) for each discrete time L, the time series $X(t)$ is split into not-overlapping time windows of length L: $[kL + 1, kL + L]$, $k = 0, 1, \ldots$; (iii) for each time window $[kL+1, kL+L]$ the local trend is evaluated with a least-squares straight line fit: $\overline{X}_{k,L}(t) = a_{k,L}t + b_{k,L}$; $kL < t \le (k+1)L$; (iii) the fluctuation is derived in the usual way: $\widetilde{X}_{k,L}(t) = X(t) - \overline{X}_{k,L}(t) = X(t) - a_{k,L}t - b_{k,L}$; $kL < t \le (k+1)L$; (iv) the mean-square deviation of the fluctuation is calculated over every window:

$$F^2(k, L) = \frac{1}{L} \sum_{t=kL+1}^{(k+1)L} \widetilde{X}^2_{k,L}(t) = \frac{1}{L} \sum_{t=kL+1}^{(k+1)L} \left(X(t) - \overline{X}_{k,L}(t)\right)^2 ; \tag{6.14}$$

and, finally, averaged over all the time windows, thus getting $F^2(L)$.

In the case of a self-similar process, it results: $F(L) \sim L^H$. The parameter H can be derived by a linear fitting applied to the function $z = Hy + C$, with $z = \log(F(L))$ and $y = \log(L)$. The DFA output is $H = 0.5$ for the case of uncorrelated (white) noise (e.g., Brownian motion), where the integrated proccss $X(t)$ display the typical Gaussian PDF $G(x, t)$ with so-called *normal* scaling of the variance: $\langle X^2 \rangle(t) \sim t$. $H \neq 0.5$ is denoted as anomalous scaling, is a signature of long-range (power-law) correlations and, thus, cooperation and complexity. In particular, $H < 0.5$, also denoted as *subdiffusion*, corresponds to a anti-correlated (anti-persistent) signal, while $H > 0.5$, also denoted as *superdiffusion*, corresponds to a positively correlated (persistent) signal.

References

1. Segev, R., Goodhouse, J., Puchalla, J., Berry, M.J.: Recording spikes from a large fraction of the ganglion cells in a retinal patch. Nat. Neurosci. **7**(10), 1154–1161 (2004)
2. Stetter, O., Battaglia, D., Soriano, J., Geisel, T.: Model-free reconstruction of excitatory neuronal connectivity from calcium imaging signals. PLoS Comput. Biol. **8**(8), e1002653 (2012)
3. Orlandi, J.G., Soriano, J., Alvarez-Lacalle, E., Teller, S., Casademunt, J.: Noise focusing and the emergence of coherent activity in neuronal cultures. Nat. Phys. **9**(9), 582–590 (2013)
4. Albert, R., Barabási, A.-L.: Statistical mechanics of complex networks. Rev. Mod. Phys. **74**, 47–97 (2002)
5. Boccaletti, S., Latora, V., Moreno, Y., Chavez, M., Hwang, D.-U.: Complex networks: structure and dynamics. Phys. Rep. **424**(4–5), 175–308 (2006)
6. Newman, M.E.J.: The structure and function of complex networks. SIAM Rev. **45**, 167–256 (2003)
7. Barrat, A., Barthélemy, M., Vespignani, A.: Dynamical Processes on Complex Networks. Cambridge University Press, Cambridge (2008)
8. Mora, T., Bialek, W.: Are biological systems poised at criticality? J. Stat. Phys. **144**, 268–302 (2011)
9. Boccaletti, S., Bianconi, G., Criado, R., Genio, C.I.D., Gómez-Gardenes, J., Romance, M., Sendina-Nadal, I., Wang, Z., Zanin, M.: The structure and dynamics of multilayer networks. Phys. Rep. **544**(1), 1–122 (2014)
10. Kenett, D.Y., Perc, M., Boccaletti, S.: Networks of networks - an introduction. Chaos Solitons Fractals **80**, 1–6 (2015)
11. Paradisi, P., Kaniadakis, G., Scarfone, A.M. (eds.): The Emergence of Self-Organization in Complex Systems. Elsevier, Amsterdam (2015). Special Issue: Chaos Solitons Fractals **81**(B), 407–588
12. Paradisi, P., Kaniadakis, G., Scarfone, A.M.: The emergence of self-organization in complex systems - preface. Chaos Solitons Fractals **81**(Part B), 407–411 (2015)
13. Paradisi, P., Allegrini, P.: Scaling law of diffusivity generated by a noisy telegraph signal with fractal intermittency. Chaos Solitons Fractals **81**(Part B), 451–462 (2015)
14. Zanin, M., Papo, D., Sousa, P.A., Menasalvas, E., Nicchi, A., Kubik, E., Boccaletti, S.: Combining complex networks and data mining: why and how. Phys. Rep. **635**, 1–44 (2016)
15. Ott, E.: Chaos in Dynamical Systems, 2nd edn. Cambridge University Press, Cambridge (2002)
16. Sornette, D.: Critical Phenomena in Natural Sciences, 2nd edn. Springer Series in Synergetics. Springer, Berlin (2006)
17. Metzler, R., Klafter, J.: The random walk's guide to anomalous diffusion: a fractional dynamics approach. Phys. Rep. **339**, 1–77 (2000)
18. Paradisi, P., Chiarugi, D., Allegrini, P.: A renewal model for the emergence of anomalous solute crowding in liposomes. BMC Syst. Biol. **9**(Suppl. 3), s7 (2015)
19. Paradisi, P., Cesari, R., Donateo, A., Contini, D., Allegrini, P.: Scaling laws of diffusion and time intermittency generated by coherent structures in atmospheric turbulence. Nonlinear Process. Geophys. **19**, 113–126 (2012). Corrigendum. Nonlinear Process. Geophys. **19**, 685 (2012)
20. Paradisi, P., Cesari, R., Donateo, A., Contini, D., Allegrini, P.: Diffusion scaling in event-driven random walks: an application to turbulence. Rep. Math. Phys. **70**, 205–220 (2012)
21. Mainardi, F.: Fractional Calculus and Waves in Linear Viscoelasticity: An Introduction to Mathematical Models. World Scientific, Singapore (2010)
22. Paradisi, P., Cesari, R., Mainardi, F., Maurizi, A., Tampieri, F.: A generalized Fick's law to describe non-local transport effects. Phys. Chem. Earth **26**(4), 275–279 (2001)
23. Pramukkul, P., Svenkeson, A., Grigolini, P., Bologna, M., West, B.: Complexity and the fractional calculus. Adv. Math. Phys. **2013**, 498789 (2013)

24. Paradisi, P.: Fractional calculus in statistical physics: the case of time fractional diffusion equation. Commun. Appl. Ind. Math. (2015). doi:10.1685/journal.caim.530
25. Pagnini, G., Paradisi, P.: A stochastic solution with gaussian stationary increments of the symmetric space-time fractional diffusion equation. Fract. Calc. Appl. Anal. **19**(2), 408–440 (2016)
26. Allegrini, P., Menicucci, D., Bedini, R., Fronzoni, L., Gemignani, A., Grigolini, P., West, B.J., Paradisi, P.: Spontaneous brain activity as a source of ideal 1/f noise. Phys. Rev. E **80**(6) Pt 1, 061914 (2009)
27. Allegrini, P., Menicucci, D., Bedini, R., Gemignani, A., Paradisi, P.: Complex intermittency blurred by noise: theory and application to neural dynamics. Phys. Rev. E **82**(1) Pt 2, 015103 (2010)
28. Allegrini, P., Paradisi, P., Menicucci, D., Gemignani, A.: Fractal complexity in spontaneous EEG metastable-state transitions: new vistas on integrated neural dynamics. Front. Physiol. **1**, 128 (2010). doi:10.3389/fphys.2010.00128
29. Allegrini, P., Paradisi, P., Menicucci, D., Bedini, R., Gemignani, A., Fronzoni, L.: Noisy cooperative intermittent processes: from blinking quantum dots to human consciousness. J. Phys. Conf. Ser. **306**(1), 012027 (2011)
30. Paradisi, P., Allegrini, P., Gemignani, A., Laurino, M., Menicucci, D., Piarulli, A.: Scaling and intermittency of brain events as a manifestation of consciousness. AIP Conf. Proc. **1510**, 151–161 (2013)
31. Allegrini, P., Paradisi, P., Menicucci, D., Laurino, M., Bedini, R., Piarulli, A., Gemignani, A.: Sleep unconsciousness and breakdown of serial critical intermittency: new vistas on the global workspace. Chaos Solitons Fractals **55**, 32–43 (2013)
32. Allegrini, P., Paradisi, P., Menicucci, D., Laurino, M., , Piarulli, A., Gemignani, A.: Self-organized dynamical complexity in human wakefulness and sleep: different critical brain-activity feedback for conscious and unconscious states. Phys. Rev. E **92**(3), 032808 (2015)
33. Cox, D.R.: Renewal Theory. Methuen, London (1962)
34. Akin, O.C., Paradisi, P., Grigolini, P.: Periodic trend and fluctuations: the case of strong correlation. Phys. A **371**, 157–170 (2006)
35. Paradisi, P., Grigolini, P., Bianco, S., Akin, O.C.: Renewal aging in non-homogeneous poisson processes with periodic rate modulation. Int. J. Bifurcation Chaos **18**(9), 2681–2691 (2008)
36. Paradisi, P., Cesari, R., Grigolini, P.: Superstatistics and renewal critical events. Cent. Eur. J. Phys. **7**, 421–431 (2009)
37. Akin, O.C., Paradisi, P., Grigolini, P.: Perturbation-induced emergence of poisson-like behavior in non-poisson systems. J. Stat. Mech. Theory Exp. **2009**, P01013 (2009)
38. Bianco, S., Grigolini, P., Paradisi, P.: A fluctuating environment as a source of periodic modulation. Chem. Phys. Lett. **438**(4–6), 336–340 (2007)
39. Paradisi, P., Cesari, R., Contini, D., Donateo, A., Palatella, L.: Characterizing memory in atmospheric time series: an alternative approach based on renewal theory. Eur. Phys. J. Spec. Top. **174**, 207–218 (2009)
40. Paradisi, P., Allegrini, P., Barbi, F., Bianco, S., Grigolini, P.: Renewal, modulation and blinking quantum dots. AIP Conf. Proc. **800**(1), 92–97 (2005)
41. Bianco, S., Grigolini, P., Paradisi, P.: Fluorescence intermittency in blinking quantum dots: renewal or slow modulation? J. Chem. Phys. **123**(17), 174704 (2005)
42. Allegrini, P., Barbi, F., Grigolini, P., Paradisi, P.: Renewal, modulation, and superstatistics in times series. Phys. Rev. E **73**(4), 046136 (2006)
43. Grigolini, P.: Emergence of biological complexity: criticality, renewal and memory. Chaos Solitons Fractals **81**(Part B), 575–88 (2015)
44. Beig, M.T., Svenkeson, A., Bologna, M., West, B.J., Grigolini, P.: Critical slowing down in networks generating temporal complexity. Phys. Rev. E **91**, 012907 (2015)
45. Turalska, M., West, B.J., Grigolini, P.: Temporal complexity of the order parameter at the phase transition. Phys. Rev. E **83**, 061142 (2011)
46. Grigolini, P., Chialvo, D.R.: Emergent critical brain dynamics - editorial. Chaos Solitons Fractals **55**(1–2), 1–120 (2013)

47. Grigolini, P., Chialvo, D.R. (eds.): Emergent Critical Brain Dynamics. Elsevier, Amsterdam (2013). Special Issue Chaos Solitons Fractals **55**, 1–120
48. Beggs, J.M., Plenz, D.: Neuronal avalanches in neocortical circuits. J. Neurosci. **23**, 11167–11177 (2003)
49. Plenz, D., Thiagarjan, T.C.: The organizing principles of neuronal avalanches: cell assemblies in the cortex? Trends Neurosci. **30**, 101–110 (2007)
50. Fraiman, D., Balenzuela, P., Foss, J., Chialvo, D.R.: Ising-like dynamics in large-scale functional brain networks. Phys. Rev. E **79**, 061922 (2009)
51. Solé, R.V., Bascompte, J.: Self-Organization in Complex Ecosystems. Princeton University Press, Princeton (2006)
52. Gerstner, W., Kistler, W.M.: Spiking Neuron Models. Cambridge University Press, Cambridge (2002)
53. Stefani, F.D., Hoogenboom, J.P., Barkai, E.: Beyond quantum jumps: blinking nanoscale light emitters. Phys. Today **62**, 34–39 (2009)
54. Maillart, T., Sornette, D., Frei, S., Duebendorfer, T., Saichev, A.: Quantification of deviations from rationality with heavy tails in human dynamics. Phys. Rev. E **83**, 056101 (2011)
55. Laureys, S., Tononi, G. (eds.): The Neurology of Consciousness. Elsevier, Amsterdam (2009)
56. Allegrini, P., Paradisi, P., Menicucci, D., Gemignani, A.: Is temporal scaling at the basis of allometry? comment on 'physiologic time: a hypothesis' by west and west. Phys. Life. Rev. **10**, 233–234 (2013)
57. Fingelkurts, A.A., Fingelkurts, A.A., Neves, C.F.H.: Brain-mind operational architectonics imaging: technical and methodological aspects. Open Neuroimaging J. **2**, 73–93 (2008)
58. Fingelkurts, A.A., Fingelkurts, A.A., Neves, C.F.H.: Natural world physical, brain operational, and mind phenomenal space–time. Phys. Life Rev. **7**, 195–249 (2010)
59. Fingelkurts, A.A., Fingelkurts, A.A., Neves, C.F.H.: Consciousness as a phenomenon in the operational architectonics of brain organization: criticality and self-organization considerations. Chaos Solitons Fractals **55**, 13–31 (2013)
60. Saichev, A., Sornette, D.: Distribution of inter-earthquake times explained. Phys. Rev. Lett. **97**, 078501 (2006)
61. Wong, I.Y., Gardel, M.L., Reichman, D.R., Weeks, E.R., Valentine, M.T., Bausch, A.R., Weitz, D.A.: Anomalous diffusion probes microstructure dynamics of entangled f-actin networks. Phys. Rev. Lett. **92**, 178101 (2004)
62. Weigel, A.V., Simon, B., Tamkun, M.M., Krapf, D.: Ergodic and nonergodic processes coexist in the plasma membrane as observed by single-molecule tracking. Proc. Natl. Acad. Sci. USA **108**, 6438–6443 (2011)
63. Jeon, J.H., Tejedor, V., Burov, S., Barkai, E., Selhuber-Unkel, C., Berg-Sørensen, K., Oddershede, L., Metzler, R.: In vivo anomalous diffusion and weak ergodicity breaking of lipid granules. Phys. Rev. Lett. **106**, 048103 (2011)
64. Metzler, R., Jeon, J.H., Cherstvy, A.G., Barkai, E.: Anomalous diffusion models and their properties: non-stationarity, non-ergodicity, and ageing at the centenary of single particle tracking. Phys. Chem. Chem. Phys. **16**(4), 24128–24164 (2014)
65. Rabinovich, M.I., Huerta, R., Afraimovich, V.S.: Dynamics of sequential decision making. Phys. Rev. Lett. **97**, 188103 (2006)
66. Rabinovich, M.I., Huerta, R., Varona, P., Afraimovich, V.S.: Transient cognitive dynamics, metastability, and decision making. PLoS Comput. Biol. **4**(5), e1000072 (2008)
67. Rabinovich, M.I., Afraimovich, V.S., Bick, C., Varona, P.: Information flow dynamics in the brain. Phys. Life Rev. **9**, 51–73 (2012)
68. Bhadola, P., Deo, N.: Study of RNA structures with a connection to random matrix theory. Chaos Solitons Fractals **81**(Part B), 542–550 (2015)
69. Furstenberg, F., Dolgushev, M., Blumen, A.: Exploring the applications of fractional calculus: hierarchically built semiflexible polymers. Chaos Solitons Fractals **81**(Part B), 527–533 (2015)
70. Basak, U.K., Datta, A.: Fractal dimension and complexity in the long-term dynamics of a monomolecular layer. Chaos Solitons Fractals **81**(Part B), 534–541 (2015)

71. Gatherer, D.: So what do we really mean when we say that systems biology is holistic? BMC Syst. Biol. **4**, 22 (2010)
72. Geneston, E., Tuladhar, R., Beig, M.T., Bologna, M., Grigolini, P.: Ergodicity breaking and localization. Phys. Rev. E **94**, 012136 (2016)
73. Bassingthwaighte, J.B., Liebovitch, L.S., West, B.J.: Fractal Physiology. Springer, Berlin (1994)
74. Goldberger, A., Amaral, L.A.N., Hausdorff, J.M., Ivanov, P.C., Peng, C.-K., Stanley, H.E.: Fractal dynamics in physiology: alterations with disease and aging. Proc. Natl. Acad. Sci. USA **99**(Suppl. 1), 2466–2472 (2002)
75. Kolb, B., Whishaw, I.Q.: Fundamentals of Human Neuropsychology, 6th edn. Worth Publishers, New York (2008)
76. Cakir, T., Alsan, S., Selma, S., Hale, A., Ata, U., Kutlu, O.: Reconstruction and flux analysis of coupling between metabolic pathways of astrocytes and neurons: application to cerebral hypoxia. Theor. Biol. Med. Model. **4**(1), 48 (2007)
77. Sofroniew, M.V., Vinters, H.V.: Astrocytes: biology and pathology, Acta Neuropathol. **119**(1), 7–35 (2010)
78. Fingelkurts, A.A., Fingelkurts, A.A.: Operational architectonics of the human brain biopotential field: towards solving the mind-brain problem. Brain Mind **2**, 261–296 (2001)
79. Kaplan, A.Y., Fingelkurts, A.A., Fingelkurts, A.A., Borisov, B.S., Darkhovsky, B.S.: Nonstationary nature of the brain activity as revealed by EEG/EMG: methodological, practical and conceptual challenges. Signal Process. **85**, 2190–2212 (2005)
80. Zipes, D., Jalife, J.: Cardiac Electrophysiology: From Cell to Bedside, 6th edn. Saunders, Elsevier, Philadelphia (2014)
81. Billman, G.: Heart rate variability - a historical perspective. Front. Physiol. **2**, 86 (2011). doi:10.3389/fphys.2011.00086
82. Allegrini, P., Grigolini, P., Hamilton, P., Palatella, L., Raffaelli, G.: Memory beyond memory in heart beating, a sign of a healthy physiological condition. Phys. Rev. E **65**(4), 041926 (2002)
83. Tang, L., Lv, H., Yang, F., Yu, L.: Complexity testing techniques for time series data: a comprehensive literature review. Chaos Solitons Fractals **81**, 117–135 (2015)
84. Allegrini, P., Grigolini, P., Palatella, L.: Intermittency and scale-free networks: a dynamical model for human language complexity. Chaos Solitons Fractals **20**, 95–105 (2004)
85. Allegrini, P., Balocchi, R., Chillemi, S., et al.: Long- and short-time analysis of heartbeat sequences: correlation with mortality risk in congestive heart failure patients. Phys. Rev. E **67**(6), 062901 (2003)
86. Maestri, R., Pinna, G.D., Allegrini, P., et al.: Linear and non-linear indices of heart rate variability in chronic heart failure: mutual interrelationships and prognostic value. Comput. Cardiol. **32**, 981–984 (2005)
87. Maestri, R., Pinna, G.D., Accardo, A., et al.: Nonlinear indices of heart rate variability in chronic heart failure patients: redundancy and comparative clinical value. J. Cardiovasc. Electrophysiol. **18**(4), 425–433 (2007)
88. Ashkenazy, Y., Ivanov, P., Havlin, S., Peng, C., Goldberger, A., Stanley, H.: Magnitude and sign correlations in heartbeat fluctuations. Phys. Rev. Lett. **86**(9), 1900–1903 (2001)
89. Ivanov, P., Amaral, L.N., Goldberger, A., Havlin, S., Rosenblum, M., Stanley, H., Struzik, Z.: From 1/f noise to multifractal cascades in heartbeat dynamics. Chaos **11**(3), 641–652 (2001)
90. Chang, M., Peng, C., Stanley, H.: Emergence of dynamical complexity related to human heart rate variability. Phys. Rev. E **90**(6), 062806 (2014)
91. Valenza, G., Allegrini, P., Lanatá, A., Scilingo, E.P.: Dominant Lyapunov exponent and approximate entropy in heart rate variability during emotional visual elicitation. Front. Neuroeng. **5**, 3 (2012)
92. Pahlm, O., Sornmo, L.: Software QRS detection in ambulatory monitoring - a review. Med. Biol. Eng. Comput. **22**, 289–297 (1984)
93. Stanley, H.E.: Introduction to Phase Transitions and Critical Phenomena. Oxford University Press, Oxford (1987)

94. Tagliazucchi, E., Balenzuela, P., Fraiman, D., Chialvo, D.R.: Criticality in large-scale brain FMRI dynamics unveiled by a novel point process analysis. Front. Physiol. **3**, 15 (2012)
95. Allegrini, P., Bellazzini, J., Bramanti, G., Ignaccolo, M., Grigolini, P., Yang, J.: Scaling breakdown: a signature of aging. Phys. Rev. E **66**, 015101 (2002)
96. Allegrini, P., Barbi, F., Grigolini, P., Paradisi, P.: Aging and renewal events in sporadically modulated systems. Chaos Solitons Fractals **34**, 11–18 (2007)
97. Montroll, E.W.: Random walks on lattices. Proc. Symp. Appl. Math. Am. Math. Soc. **16**, 193–220 (1964)
98. Montroll, E.W., Weiss, G.H.: Random walks on lattices. II. J. Math. Phys. **6**, 167–181 (1965)
99. Weiss, G.H., Rubin, R.J.: Random walks: theory and selected applications. Adv. Chem. Phys. **52**, 363–505 (1983)
100. Scafetta, N., Grigolini, P.: Scaling detection in time series: diffusion entropy analysis. Phys. Rev. E **66**, 036130 (2002)
101. Peng, C.-K., Buldyrev, S.V., Havlin, S., Simons, M., Stanley, H.E., Goldberger, A.L.: Mosaic organization of dna nucleotides. Phys. Rev. E **49**, 1685–1689 (1994)
102. Gong, P., Nikolaev, A.R., van Leeuwen, C.: Intermittent dynamics underlying the intrinsic fluctuations of the collective synchronization patterns in electrocortical activity. Phys. Rev. E **76**(1), 011904 (2007)
103. Raichle, M.E., Leod, A.M.M., Snyder, A.Z., Powers, W.J., Gusnard, D.A., Shulman, G.L.: A default mode of brain function. Proc. Natl. Acad. Sci. USA **98**, 676–682 (2001)
104. Beckmann, C.F., Luca, M.D., Devlin, J.T., Smith, S.N.: Investigation into resting-state connectivity using independent component analysis. Philos. Trans. R. Soc. Lond. B Biol. Sci. **360**, 1001–1013 (2005)
105. Wolf, M., Varigos, G., Hunt, D., Sloman, J.: Sinus arrhythmia in acute myocardial infarction. Med. J. Aust. **2**, 52–53 (1978)
106. T. F. of the European Society of Cardiology, the North American Society of Pacing, and Electrophysiology: Heart rate variability: standards of measurement, physiological interpretation, and clinical use. Circulation **93**, 1043–1065 (1996)
107. Kleiger, R., Miller, J., Bigger, J., Moss, A.: Decreased heart rate variability and its association with increased mortality after acute myocardial infarction. Am. J. Cardiol. **59**, 256–262 (1987)
108. Peng, C., Havlin, S., Stanley, H., Goldberger, A.: Quantification of scaling exponents and crossover phenomena in nonstationary heartbeat time series. Chaos **5**, 82–87 (1995)
109. Tulppo, M., Kiviniemi, A., Hautala, A., Kallio, M., Seppanen, T., Makikallio, T., Huikuri, H.: Physiological background of the loss of fractal heart rate dynamics. Circulation **112**, 314–319 (2005)
110. Iyengar, N., Peng, C., Morin, R., Goldberger, A., Lipsitz, L.: Age-related alterations in the fractal scaling of cardiac interbeat interval dynamics. Am. J. Physiol. **271**, R1078–R1084 (1996)
111. Vikman, S., Makikallio, T., Yli-Mayry, S., Pikkujamsa, S., Koivisto, A., Reinikainen, P., Airaksinen, K., Huikuri, H.: Altered complexity and correlation properties of R-R interval dynamics before the spontaneous onset of paroxysmal atrial fibrillation. Circulation **100**, 2079–2084 (1999)
112. Huikuri, H., Poutiainen, A., Makikallio, T., Koistinen, M., Airaksinen, K., Mitrani, R., Myerburg, R., Castellanos, A.: Dynamic behaviour and autonomic regulation of ectopic atrial pacemakers. Circulation **100**, 1416–1422 (1999)
113. Makikallio, T., Hoiber, S., Kober, L., Torp-Pedersen, C., Peng, C., Goldberger, A., Huikuri, H.: Fractal analysis of heart rate dynamics as a predictor of mortality in patients with depressed left ventricular function after acute myocardial infarction. Am. J. Cardiol. **83**, 836–839 (1999)
114. Jokinen, V., Tapanainen, J., Seppanen, T., Huikuri, H.: Temporal changes and prognostic significance of measures of heart rate dynamics after acute myocardial infarction in the beta-blocking era. Am. J. Cardiol. **92**, 907–912 (2003)

115. Perkiomaki, J., Zareba, W., Ruta, J., Dubner, S., Madoery, C., Deedwania, P., Karcz, M., de Luna Bayes, A., Investigators, I.: Fractal and complexity measures of heart rate dynamics after acute myocardial infarction. Am. J. Cardiol. **88**, 777–781 (2001)
116. Maestri, R., Pinna, G., Porta, A., Balocchi, R., Sassi, R., Signorini, M., Dudziak, M., Raczak, G.: Assessing nonlinear properties of heart rate variability from short-term recordings: are these measurements reliable? Physiol. Meas. **28**, 1067–1077 (2007)
117. Perkiomaki, J.: Heart rate variability and non-linear dynamics in risk stratification. Front. Physiol. **2**, 81 (2011). doi:10.3389/fphys.2011.00081
118. Federoff, H.J., Gostin, L.O.: Evolving from reductionism to holism: is there a future for systems medicine? J. Am. Med. Assoc. **302**(9), 994–996 (2009)
119. Valenza, G., Toschi, N., Barbieri, R. (eds.): Uncovering Brain–Heart Information Through Advanced Signal and Image Processing. The Royal Society Publishers, London (2016). Theme Issue - Philos. Trans. R. Soc. A **374**(2067)
120. Valenza, G., Toschi, N., Barbieri, R.: Introduction: uncovering brain–heart information through advanced signal and image processing. Philos. Trans. R. Soc. A **374**(2067), 20160020 (2016)
121. Valenza, G., Greco, A., Gentili, C., Lanata, A., Sebastiani, L., Menicucci, D., Gemignani, A., Scilingo, E.P.: Combining electroencephalographic activity and instantaneous heart rate for assessing brain–heart dynamics during visual emotional elicitation in healthy subjects. Philos. Trans. R. Soc. A **374**(2067), 20150176 (2016)
122. Schulz, S., Bolz, M., Bar, K.-J., Voss, A.: Central and autonomic nervous system coupling in schizophrenia. Philos. Trans. R. Soc. A **374**(2067), 20150178 (2016)

Chapter 7
Self-Similarity and Detrended Fluctuation Analysis of Cardiovascular Signals

Paolo Castiglioni, Marco Di Rienzo, and Andrea Faini

Abstract The beat-by-beat dynamics of several cardiovascular signals, including heart rate and arterial blood pressure, shows statistically self-similar properties. This means that shorter segments extracted from these signals may have the same structure of the whole time series, whenever the vertical axis is rescaled by a proper scale coefficient. A "self-similar" signal is thus analogous to a geometric fractal that can be split into small fragments, each containing the whole complexity of the original object. The self-similar dynamics reflects the nature of the complex system generating the signal. Therefore, in recent years a large body of research investigated the self-similar characteristics of cardiovascular signals by estimating their "self-similarity" scale coefficients for better understanding the mechanisms of cardiovascular regulation in health and disease, and for risk stratification. This chapter illustrates the main methods used in literature for assessing the self-similarity of cardiovascular time series, especially focusing on methods based on the popular Detrended Fluctuation Analysis (DFA) algorithm. In particular, it reviews applications of DFA that describe the cardiovascular time series in terms of fractal models, with one or more scale coefficients. Furthermore, it illustrates the more recent advancements of the DFA method for describing self-similarity as a continuous temporal spectrum of scale coefficients and for deriving multifractal spectra.

7.1 Introduction

The cardiovascular time series show an intrinsic variability spanning over several time scales, even when recorded in healthy individuals and in steady state conditions. This is likely due to the activity of different control mechanisms that try to maintain the cardiovascular homeostasis in response to a multitude of external

P. Castiglioni (✉) • M. Di Rienzo
IRCCS Fondazione Don C. Gnocchi, Milan, Italy
e-mail: pcastiglioni@dongnocchi.it

A. Faini
Department of Cardiology, Istituto Auxologico Italiano, Milan, Italy

R. Barbieri et al. (eds.), *Complexity and Nonlinearity in Cardiovascular Signals*,
DOI 10.1007/978-3-319-58709-7_7

perturbations. When relatively long cardiovascular time series are recorded, one may often recognize a particular feature in their spontaneous variability, named "self-similarity" [1, 2]. A time series is "self-similar" if any data segment extracted from the series has the same appearance, independently of the time scale at which it is observed, provided that its amplitude is properly rescaled. Scale invariance also characterizes geometric fractals: one may select a small fragment of a fractal, properly enlarge it and obtain an object with the same structure and the same details of the original fractal (Fig. 7.1). For this reason, self-similar signals are also called "fractal" signals.

Investigating the existence and characterizing the properties of self-similarity in cardiovascular signals appears a potentially useful tool for assessing alterations in the cardiovascular regulation or for risk stratification. Physics and biology provide several examples of complex dissipative systems that naturally evolve toward a self-organized state without an intrinsic time scale [3]. These systems are typically composed by fractal structures connecting together several components strongly interacting with each other. Therefore, variables describing the state of such self-organized complex systems are expected to show scale-invariant dynamics. The cardiovascular system may be considered a complex system, because composed by a fractal network of branching vessels [4] connecting different vascular beds strongly interacting with each other through local and global regulatory mechanisms, based on fast neural and slower humoral effectors. For this reason, the cardiovascular dynamics is expected to behave like a self-similar process operating over different time scales [2], and alterations in local regulations or in the overall integrative control may lead to changes in cardiovascular self-similarity.

In more formal terms, a time series $x(n)$ is called *self-affine* if

$$x(n) \sim \frac{x(a \times n)}{a^H} \quad \text{for any } a > 0 \tag{7.1}$$

where the symbol "$\sim$" indicates that the series at the left- and right-hand sides of Eq. (7.1) have the same statistical properties. The exponent H, named Hurst exponent, defines the vertical rescaling factor, a^{-H}, that should be applied to preserve the statistical properties of $x(n)$ when the time axis is stretched by the horizontal scaling factor a. The Hurst exponent may range between 0 and 1, and only when H is equal to 1 the series should be called "self-similar" [5]. However, in the field of cardiovascular signals analysis, the term "self-similarity" is commonly used for any self-affine series $x(n)$ that satisfies Eq. (7.1) even with $H < 1$ [2], and we will follow this terminology.

In this chapter, we will first describe the main estimators of H, from the rescaled range method introduced by Hurst to methods forerunners of detrended fluctuation analysis, highlighting their applications in the field of cardiovascular signal analysis and introducing the families of fractional Gaussian noises and fractional Brownian motions as interpretative models. In the following paragraph, we will introduce the

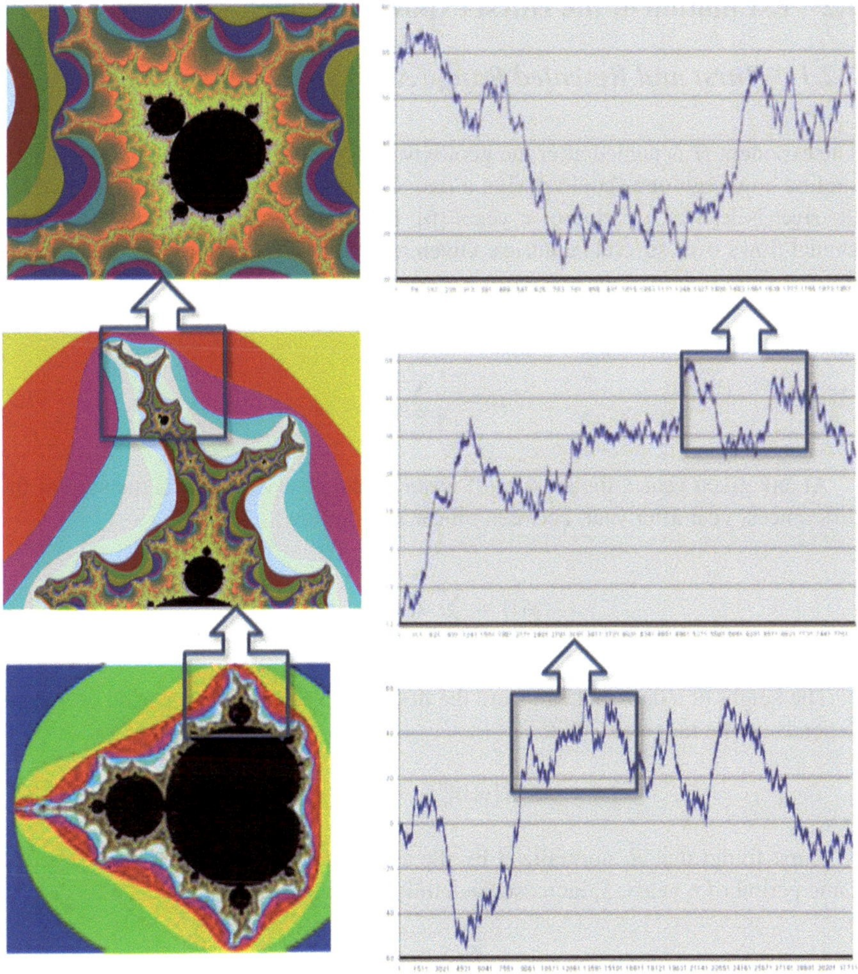

Fig. 7.1 Analogy between a fractal object and a self-similar signal. A small fragment of a fractal object (Mandelbrot's set, *left*) looks like the original, if properly enlarged; in the same way, a small segment of a self-similar time series (Brownian motion, *right*) resembles the original series, in statistical terms, if the horizontal and vertical axes are properly stretched

popular detrended fluctuation analysis algorithm, discuss its theoretical aspects and its applications in physiological and clinical studies of heart-rate self similarity. In the next and final paragraph, we will illustrate the more advanced methods for describing multifractality of heart rate and blood pressure based on detrended fluctuation analysis.

7.2 Estimation of the Hurst exponent

7.2.1 *Hurst and Rescaled Range Analysis*

The exponent H is named after the geologist H.E. Hurst who introduced it when he studied how high one should design a dam capable of storing the annual floods of the river Nile over a period of n years [6]. Hurst analyzed the records of the Nile annual flows over several centuries. Given an observation period of n years, he set the constant outflow of the reservoir, μ, equal to the average of the n annual inflows, $x(i)$:

$$\mu = \frac{1}{n}\sum_{i=1}^{n} x(i) \tag{7.2}$$

At any given year i, the amount of water in the reservoir, $y(i)$, is the sum of the differences, year after year, between inflow and outflow:

$$y(i) = \sum_{j=1}^{i} (x(j) - \mu) \tag{7.3}$$

The height of a dam that can store the floods of the river over a period of n years depends on the range of $y(i)$, R:

$$R = MAX[y(i)]_{i=1}^{n} - MIN[y(i)]_{i=1}^{n} \tag{7.4}$$

Hurst found that R, normalized by the standard deviation of the flow over the same period of n years, S, increases indefinitely as a power of n:

$$R/S \propto n^H \tag{7.5}$$

Equation (7.5) indicates that the dynamics of the annual flows of the river Nile is a scale invariant (or self-similar) phenomenon. In fact, the ratio between two rescaled ranges evaluated over periods of n_1 and n_2 years, does not depend on the absolute values n_1 and n_2, but only on the ratio $(n_1/n_2)^H$. Interestingly, H was greater than ½. In the same year, Feller demonstrated that the range of the sum of independent random variables increases as the square root of n, i.e., $H = ½$ [7]. Therefore Hurst found that the annual flows $x(i)$ are not independent each other, implying the existence of long-term correlations among periods of rain and drought.

With his work, Hurst introduced the first method for estimating H of a stationary time series $x(i)$, with $i = 1, \ldots, N$. The method is called Rescaled Range Analysis. First, (a) one splits $x(i)$ into M consecutive blocks of size $n=int(N/M)$. Next steps are to calculate, in each block j: (b) the mean μ_j as in Eq. (7.2); (c) the cumulative sum of deviations from μ_j, $y(i)$, as in Eq. (7.3); (d) the range of $y(i)$, R_j, as in Eq.

(7.4); (e) the standard deviation of $x(i)$, S_j; (f) the rescaled range, R_j/S_j. Then, (g) the M rescaled ranges are averaged obtaining a single value $(R/S)_n$ associated with the block size n. The (a–g) steps are repeated several times, splitting the series into blocks of different size n. If $x(i)$ is self-similar, then $(R/S)_n = k \times n^H$ and H can be estimated as the slope of the regression line between $\log(n)$ and $\log (R/S)_n$.

This method, important from a historical point of view, suffers from significant bias errors [8], and in the field of cardiovascular signals analysis it lost its initial popularity in favor to other estimators, described in the next paragraphs.

7.2.2 *Fractional Gaussian Noises and Fractional Brownian Motions*

Inspired by Hurst's work, Mandelbrot and van Ness proposed two general families of self-similar processes with H between 0 and 1: the family of fractional Gaussian noises, fG_H, and the family of their integral, or fractional Brownian motions, fB_H [9]. These families represent a simple model for describing self-similar signals. They generalize the classic white Gaussian noise, $G(i)$, and its cumulative sum, the Brownian motion $B(i)$. For a Gaussian noise, the value at any given time i has a Gaussian distribution and does not depend on the values that occurred before i; by contrast, for a Brownian motion, at any given time i the increment to the next value, $B(i+1) - B(i)$, does not depend on the values before i. According to [9], the fractional Brownian motion associated with the parameter H, $fB_H(i)$, is defined as:

$$fB_H(i) = fB_H(0) + \frac{1}{\Gamma\left(H+\frac{1}{2}\right)}\left\{\int_{-\infty}^{0}\left[(i-s)^{H-\frac{1}{2}} - (0-s)^{H-\frac{1}{2}}\right]dB(s) + \int_{0}^{i}(i-s)^{H-\frac{1}{2}}dB(s)\right\} \tag{7.6}$$

It can be easily verified that when $H = ½$, Eq. (7.6) gives $fB_{½}(i) = fB_{½}(0) + B(i)$. This means that the family of fractional Brownian motions contains the classic Brownian motion as special case for $H = ½$. For $H \neq ½$, the second integral makes $fB_H(i)$ a moving average of past $B(i)$ increments weighted by the kernel $(i - s)^{H - ½}$. Mandelbrot and van Ness demonstrated that the increments of $fB_H(i)$ are stationary and self-similar with parameter H, which means that $fG_{½}(i)$ is the classic white Gaussian noise.

An interesting property of the Fourier spectrum of fB_H is that it is proportional to a power of the frequency f:

$$PSD_{fBH}(f) \propto 1/f^{(2H+1)} \tag{7.7}$$

When a time series is differenced, its spectrum is multiplied by f^2, and thus the spectrum of fG_H is [9]:

$$\mathrm{PSD_{fGH}}(f) \propto 1/f^{(2H-1)} \tag{7.8}$$

When $H = ½$, Eqs. (7.7) and (7.8) give the well-known $1/f^{\beta}$ spectral shape with $\beta = 0$ for white noise and $\beta = 2$ for Brownian motion.

Therefore, each H exponent is associated with two processes. One is a fractal Gaussian noise with $1/f^{\beta}$ spectral shape and $\beta < 1$, condition that assures stationarity because the integral of the spectrum is finite. The other one is a fractal Brownian motion, also with $1/f^{\beta}$ spectral shape but exponent $\beta > 1$. In this case the process is nonstationary, because the integral of the spectrum diverges at $f = 0$ and the variance of the process increases indefinitely with time [10]. More precisely, Mandelbrot and van Ness demonstrated that the standard deviation of a sequence of n samples of fractional Brownian motion increases proportionally with the power H of n:

$$\sigma_{\mathrm{fBH}} \propto n^H \tag{7.9}$$

7.2.3 Spectral Estimators of H

The introduction of these two classes of fractal processes suggested a way for estimating the Hurst exponent directly from the power spectrum. First, the Fourier spectrum is estimated over a broad range of frequencies, *PSD(f)*. Second, the spectrum is represented on a *log-log* scale, to identify the frequency band where the plot looks like a straight line, i.e., where *PSD(f)* is proportional to $1/f^{\beta}$. Third, the slope of the straight line, $-\beta$, is calculated fitting a regression line over the identified frequency band. Finally, if $\beta > 1$, the time series is classified as a fractional Brownian motion and H is estimated as $½(\beta - 1)$. On the other hand, if $\beta < 1$ the time series is classified as a fractional Gaussian noise and H is estimated as $½(\beta + 1)$. Since the first discovery of a "1/f" spectral trend in heart rate recordings [11], a series of physiological [12–17] and clinical studies [18–28] aimed at deriving information on long-term cardiovascular regulation estimated β of heart rate or blood pressure spectra. Interestingly, some of these studies reported systematic deviations of the spectra from the "monofractal" model [12, 13, 16, 29, 30], implicitly providing hints of a possible multifractal nature of cardiovascular time series.

However, harmonic components, like respiratory oscillations or Mayer waves, are superimposed to the $1/f^{\beta}$ spectrum of cardiovascular time series. The harmonic components generate spectral peaks that should be avoided in the evaluation of the regression slope, β. This is not an easy task when the time series has short duration. To mitigate this problem, Yamamoto and Hughson [31] proposed to remove the harmonic components with a coarse-graining procedure. The idea is to coarse grain the time series, $x(i)$, $i = 1,\ldots,N$, by taking only n samples, one every M with

$M = int(N/n)$, and by calculating the cross-power spectrum between the coarse grained series of n samples and any segment of n consecutive samples of the original series. Because of the scale invariant properties of self-similar signals, the cross-spectrum of the fractal component equals the original fractal spectrum, while the cross-spectrum of the harmonic components tends to zero. In this way, it is possible to remove the harmonic components superimposed to the fractal spectrum, and to evaluate the β slope more easily. Some authors used this method as a "preprocessing" technique for better estimating the low and high frequency spectral powers [32], removing fluctuations of fractal nature [33–35]. Other authors applied coarse graining to estimate the spectral slope in physiological [36–45] and clinical [46, 47] studies where relatively short recordings did not allow to estimate β with a regression fitting over ultra- and very-low frequencies.

Spectral estimators of H do not require knowing *a priori* whether the time series belongs to the class of fractional Gaussian noises or to the class of fractional Brownian motion. In fact, β can be estimated for both families of fractal processes, and values greater or lower than 1 discriminate between fractional noises or motions. For this reason, it has been suggested to use spectral methods to preliminary evaluate whether the time series is a fractional noise or a fractional motion, in conjunction with more efficient methods (see below) that estimate H for stationary fractional Gaussian noises only or for nonstationary fractional Brownian motions only [48].

7.2.4 *H Estimators for Fractional Gaussian Noise: Autocorrelation Analysis*

Mandelbrot and van Ness also derived the expression of the correlation between two samples at times i_1 and i_2 of a fractional Gaussian noise [9]. They found that the correlation is positive for all i_1 and i_2 if $½ < H < 1$, and negative if $0 < H < ½$. The positive correlation when $H > ½$ means that if we observe a positive deviation from the mean at i_1, then it is more likely to observe a positive than a negative deviation at time i_2, and similarly, if we observe a negative deviation at i_1 it is more likely to observe a negative deviation at i_2. This statistical trend occurs independently from the distance between i_1 and i_2, (even if the probability to observe deviations with the same sign tends to decrease when $|i_1 - i_2|$increases). For this reason, fractional Gaussian noises with $H > ½$ are called "long-memory processes" with "persistent behavior". On the other hand, if $H < ½$, it is more likely that the deviation from the mean at i_2 has the opposite sign of the deviation at i_1. Therefore, fractional Gaussian noises with $H < ½$ are said to have an "anti-persistent behavior".

The theoretical analysis in [9] provides the basis for estimating H of fractional Gaussian noises by calculating the autocorrelation function. Following the notation in [1], the autocorrelation function ρ at lag k of a fractional Gaussian noise with exponent H is:

$$\rho(k) = \frac{1}{2}|k+1|^{2H} - |k|^{2H} + \frac{1}{2}|k-1|^{2H} \tag{7.10}$$

The estimator is based on the calculation of the autocorrelation function of the time series *x(i)*, with $i = 1,..,N$. The autocorrelogram is estimated as:

$$\overline{\rho}(k) = \frac{\frac{1}{N-k-1}\sum\limits_{i=k+1}^{N}\left(x(i)-\overline{\mu}\right)\left(x(i-k)-\overline{\mu}\right)}{\frac{1}{N-1}\sum\limits_{i=1}^{N}\left(x(i)-\overline{\mu}\right)^2} \tag{7.11}$$

with

$$\overline{\mu} = \frac{1}{N}\sum_{i=1}^{N} x(i) \tag{7.12}$$

Bassingthwaighte and Beyer extended Eq. (7.11) to any non-integer, rational number k [49]. If *x(i)* belongs to the class of fractional Gaussian noises, $\overline{\rho}(k)$ is calculated for different k values according to Eq. (7.11), and H is estimated by fitting Eq. (7.10) to $\overline{\rho}(k)$ [50]. If *x(i)* is a fractional Brownian motion, it should be differenced before applying the autocorrelation method. In the field of the cardiovascular system studies, the autocorrelation method has been employed for evaluating the *spatial* self-similarity of vascular blood flows, *x(i),* observed in successive tissue units i [49, 51].

7.2.5 *H Estimators for Fractional Gaussian Noise: Dispersional Analysis*

A well know property of the estimator of the mean μ of n independent Gaussian random variables with variance σ^2 is that the standard deviation of the estimator, σ_μ, decreases when n increases as:

$$\sigma_\mu = \frac{\sigma}{\sqrt{n}} \tag{7.13}$$

If the data are not independent each other and μ is the mean of n consecutive values of a stationary ergodic process, $x(i)$, with autocorrelation function $\rho(k)$, then σ_μ is [52]:

$$\sigma_\mu = \frac{\sigma}{\sqrt{n}}\left(1 + 2\sum_{k=1}^{n-1}\left(1 - {}^{k}\!/_{n}\right)\rho(k)\right) \tag{7.14}$$

When *x(i)* is a white Gaussian noise, $\rho(k) = 0$ for $k \geq 1$ and Eq. (7.14) coincides with Eq. (7.13). In this case, σ_μ decreases proportionally to $n^{-1/2}$. More in general, if *x(i)* is a fractional Gaussian noise, $\rho(k)$ follows Eq. (7.10) and:

$$\sigma_\mu = \frac{\sigma}{n^{1-H}} \tag{7.15}$$

Equation (7.15) suggests how to estimate *H* of fractional Gaussian noises from the dispersion (i.e., the standard deviation) of the mean of *n* consecutive samples [53]. Briefly, (a) the series *x(i)*, with $i = 1,\ldots, N$, is aggregated into consecutive, non-overlapping blocks of size *n*; (b) the mean is calculated in each block; (c) the standard deviation of the means, $\sigma(n)$, is evaluated over all the blocks of size *n*. The steps (a–c) are repeated for block sizes between $n=1$ and $n = \text{int}(N/4)$. For $n = 1$, $\sigma(n)$ is the standard deviation of *x(i)*. For the largest *n*, equal to int(*N*/4), $\sigma(n)$ is evaluated on at least four blocks, this being considered the smallest number of blocks that allows a meaningful estimation of the standard deviation. If *x(i)* is a fractional Gaussian noise, then Eq. (7.15) holds and the log-log plot of $\sigma(n)$ and *n* depicts a straight line, at least on a range of *n* values. Its slope is calculated by linear regression over the range of *n* where the relation between log $\sigma(n)$ and log *n* is linear. Finally, *H* is estimated as slope+1 (Fig. 7.2). In *relative dispersional*

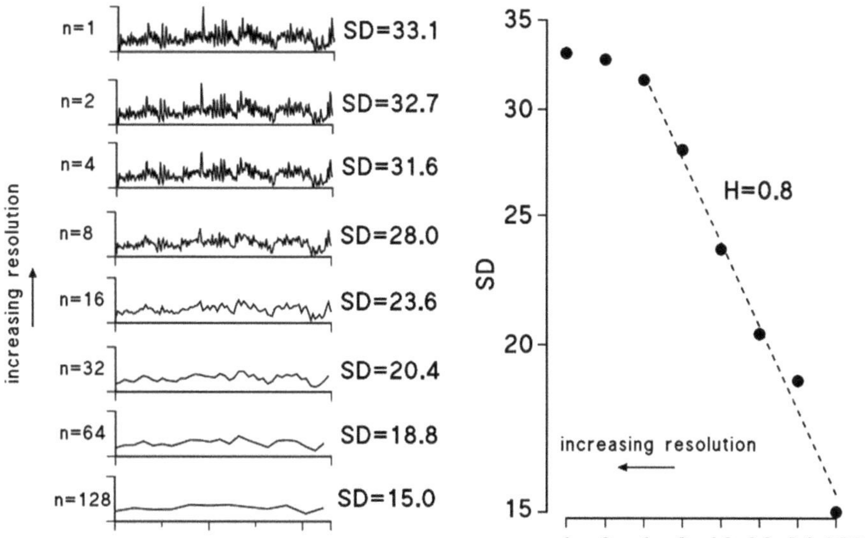

Fig. 7.2 Example of heart rate dispersional analysis. *Left*: beat-by-beat cardiac intervals plotted at decreasing resolutions, from $n = 1$ to $n = 128$ beats, with SD the standard deviation of each aggregated time series. *Right*: SD vs. *n* in a log–log scale; *H* is derived from the slope of the regression line evaluated where log SD decreases linearly with log *n* (from [54] with permission)

analysis, H is estimated as slope of the regression line between the coefficient of variation (=standard deviation/mean) and n, plotted in a log-log scale, for $n > 1$. If $x(i)$ belongs to the class of fractional Brownian motions, it should be differenced before applying dispersional or relative dispersional analysis.

In the field of cardiovascular signals analysis, dispersional and relative dispersional analysis characterized the fractal properties of heart rate, of sympathetic nerve discharge, of laser Doppler flowmetry and of membrane potentials of cardiac myocites [55–57].

7.2.6 H Estimators for Fractional Brownian Motion: Scaled Windowed Variance

This method exploits the property of the standard deviation of fractional Brownian motions to increase as the power H of the length of the time series, as indicated by Eq. (7.9). Given N consecutive samples of a fractional Brownian motion $x(i)$, (a) $x(i)$ is split into M consecutive, nonoverlapping windows of size n, with $M = \mathrm{int}(N/n)$; (b) mean and standard deviation are calculated in each window j, as:

$$\overline{\mu_j} = \frac{1}{n} \sum_{i=n(j-1)+1}^{nj} x(i) \tag{7.16}$$

$$\overline{\sigma_j} = \sqrt{\frac{1}{n-1} \sum_{i=n(j-1)+1}^{nj} \left(x(i) - \overline{\mu_j}\right)^2} \tag{7.17}$$

and (c) the M standard deviations are averaged:

$$SD_n = \frac{1}{M} \sum_{j=1}^{M} \overline{\sigma_j} \tag{7.18}$$

The steps (a–c) are repeated for window sizes increasing between $n = 2$ and $n = N$. Then, SD_n is plotted vs. n in a log–log scale, to identify the range where the plot follows a straight line. H is the slope of the regression line calculated over this range.

The *bridge-detrended* variant of this method subtracts from $x(i)$ the line connecting the first and last point of the window, before calculating Eqs. (7.16) and (7.17). The *line-detrended* variant subtracts from $x(i)$ the regression line in each window. The two detrended variants perform better than the standard method in terms of estimation bias [58]. The line-detrended variant deserves to be mentioned because of the close relation with the detrended fluctuation analysis method (see the next paragraph) in the way deviations from the regression line of each window are defined.

The scaled windowed variance method should be applied to fractional Brownian motions only, or on the cumulative sum of *x(i)*, if *x(i)* is a fractional Gaussian noise. In the field of studies of the cardiovascular system, the scaled-windowed variance method was used to estimate *H* of cerebral and skin blood flows and of arterial blood pressure [59–61].

7.3 Self-Similarity by Detrended Fluctuation Analysis

All previously described methods, with the exception of spectral methods, require knowing whether the time series belongs to the class of fractional Gaussian noises or to the class of fractional Brownian motions, before evaluating *H*. This is a major limitation because most cardiovascular signals show $1/f^{\beta}$ spectra with β slope close to 1, at the boundary between fractal noise and fractal motion. Therefore, when studying patients and controls, or when the same group is monitored under different conditions, it might be required to employ different *H* estimators or different data preprocessing procedures (like cumulative sum or differencing) within the same study, making critical the comparison of *H* estimates. The spectral estimator does not suffer from this limitation, but its sensitivity to harmonic components restricts its use to long-time recordings. The coarse-graining method may remove harmonics superimposed to a 1/f spectral trend. However, this method implicitly assumes that the scaling exponent is the same at all the scales while it is reported in literature that β, slope of the spectral trend, may not be the same over a broad range of frequencies.

The Detrended Fluctuation Analysis (DFA) method [62], originally proposed in 1994 for describing the mosaic structure of DNA sequences [63], overcomes the above limitations. In fact, one does not need to know *a priori* the fractional class to which the time series belongs, because DFA provides a scale exponent, α, equal to *H* for fractal Gaussian noises and to $H+1$ for fractional Brownian motions. Moreover, unlike the power spectrum, the detrended fluctuation function of DFA is relatively insensitive to harmonic components in *x(i)* (see Fig. 7.3). These factors, along with the availability of free code easily accessible to the community of cardiovascular signal analysis [64], rapidly made DFA the most popular method for the analysis of self-similarity in cardiovascular signals. As illustrated in the last part of this chapter, the DFA approach can also investigate specific aspects of multifractality.

7.3.1 DFA Theoretical Aspects

Preliminary step for DFA of a time series *x(i)* of *N* samples is to calculate its cumulative sum, *y(i)*, as in Eqs. (7.2) and (7.3). Then, (a) *y(i)* is split into *M* consecutive, nonoverlapping blocks of *n* samples, with $M=\text{int}(N/n)$; (b) a polynomial trend of order *p* is calculated in each block by least square fitting (let's call $y_n^p(i)$, with *i* from 1 to *N*, the piecewise function obtained by connecting

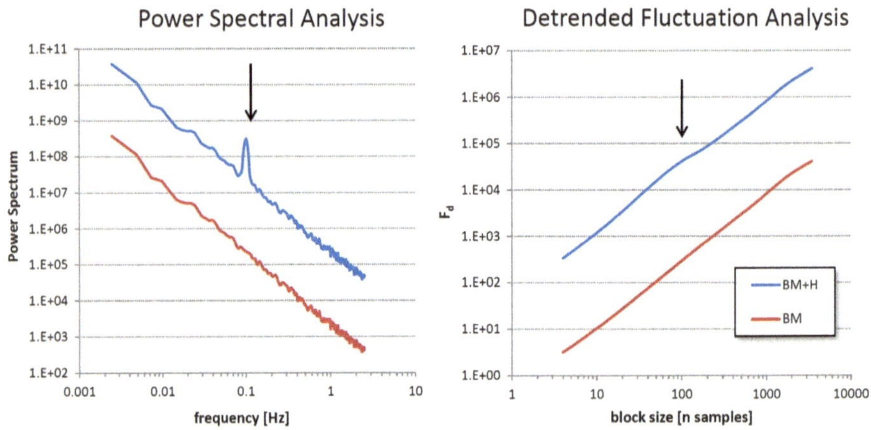

Fig. 7.3 Example of the different sensitivity to harmonic components of fractal methods based on power spectral analysis and based on detrended fluctuation analysis. A time series representing a Brownian motion process (BM) lasting 1683.5 s and sampled every 0.1 s was synthesized summing 2^{14} uncorrelated increments uniformly distributed between -1000 and 1000; a second time series was synthesized adding the harmonic component $3000 \times \sin(2\pi \times 0.1t)$ to the same Brownian motion process (BM+H). The power spectrum of BM+H (*left panel*, *blue line*) shows a clear peak at the frequency of the superimposed harmonic: this peak may substantially deviate the slope of the regression line characterizing the power spectrum of the pure Brownian motion (*red line*) in a log-log scale. By contrast, the detrended fluctuation function of BM+H (F_d, *right panel*) shows an almost imperceptible deviation from the BM slope at the scale corresponding to the period of the harmonic component (n = 100 samples). For clarity, BM curves were shifted down with respect to BM+H curves by a factor of 100

the trends of the *M* blocks); (c) $F_d(n)$, root mean square of the deviations of the integrated series $y(i)$ from the local trend $y_n^p(i)$, is evaluated as:

$$F_d(n) = \sqrt{\frac{1}{N}\sum_{i=1}^{N}\left(y(i) - y_n^p(i)\right)^2} \tag{7.19}$$

This function measures the variability of detrended fluctuations within blocks of size n. The steps (a–c) are repeated for different block sizes n. In a variant of this method, called "detrended moving average", DMA, the fitting stepwise polynomial of order p, $y_n^p(i)$, is substituted by a moving average [65], in order to not assume trends of polynomial nature.

The rationale for evaluating root mean square deviations from the polynomial trend in Eq. (7.19) derives from original observations on the correlation properties of DNA sequences. Segments of DNA can be represented as fractal random walks, $y(i)$, by summing random steps $x(i)$ equal to +1, if a "C" or "T" (=pyrimidine bases) occurs in position i of the DNA sequence, or equal to -1, if a "A" or "G" (=purine bases) occurs in i. Some works reported that the root mean square deviation of DNA random walks $y(i)$ of length n, $F(n)$, is not always characterized by the n^α

power law with α>½, as expected for sequences with long-range correlations. It was hypothesized that the lack of the power law relationship was due to the effect of nonstationarities in $x(i)$ associated with large DNA coding regions [66]. It was thus proposed to assess the power law relation n^{α} not for $F(n)$, but for $F_d(n)$, the root mean square deviation of $y(i)$ after the removal of nonstationary trends in each block (the demonstration that $F_d(n)$ of long-range correlated data follows the power law relation n^{α} for large n was provided in [67]).

Given the dependency of $F_d(n)$ on n^{α}, the exponent α is estimated by plotting log $F_d(n)$ vs. log n, and by calculating the slope of the regression line where the two log-quantities appear linearly related. If $x(i)$ is a fractional Gaussian noise, then $y(i)$ is a fractional Brownian motion and $\alpha=H$, being $F_d(n)$ similar to the standard deviation of fractional Brownian motions in the line-detrended scaled-windowed variance method of Eq. (7.18). However, the *DFA* can be applied also to nonstationary time series. If $x(i)$ is a nonstationary fractional Brownian motion with Hurst exponent H, then $y(i)$ is the cumulative sum of a fractional Brownian motion, and $F_d(n)$ increases as n^{α} with $\alpha = H + 1$.

Readers may find detailed descriptions of the effects of different types of noise, or of pre-elaboration procedure applied to real time series (filters, data editing, etc) on self-similarity coefficients estimated by DFA in [68–70].

The DFA method is often indicated as DFA_p if a detrending polynomial of order p is used. The shortest block size n to be considered in Eq. (7.19) depends on the order p. This is illustrated in Fig. 7.4, which shows $y_n^p(i)$ functions for $p = 1$, 2 and 3, and block sizes n between 16 and 4. If the order p is too high for a given n, the detrending function may overfit the data, removing not only the trend but also the fractal variability components of interest (in Fig. 7.4, this is the case for DFA_2, $n = 4$, and $9 \leq i \leq 12$). The extreme case occurs when $p = n - 1$, as for DFA_3 and $n = 4$ in Fig. 7.4. In this case, the detrending polynomial passes through all the $y(i)$ points, and the corresponding variability function $F_d(n)$ is null (see Fig. 7.5 for the deviations from the trends of Fig. 7.4). Therefore, the shortest block size n should be greater than $p + 1$ to avoid this extreme case, and not too close to $p + 1$ to also avoid data overfitting and underestimation of $F_d(n)$.

Kantelhardt *et al.* suggested a procedure for correcting the distortions induced in $F_d(n)$ by a too high fitting order p [71]. They empirically calculated $F_d(n)$ with different fitting orders p for synthesized series with α between ½ and 0.9, and compared the results with the theoretical n^{α} power law increase. They found that the distortion in $F_d(n)$, quantified by the ratio between empirical and theoretical trends, $F_d(n)/n^{\alpha}$, regards small n and depends only weakly on α. Therefore, they proposed a correction function for all α, $K(n)$, based on results obtained for α = ½, because for this value of α the correction function can be easily calculated by shuffling the original input data $x(i)$.

The correction function is:

$$K(n) = \frac{E\left[\left(F_{shuff}(n)\right)^2\right]^{1/2}}{n^{1/2}} \frac{1}{Q} \tag{7.20}$$

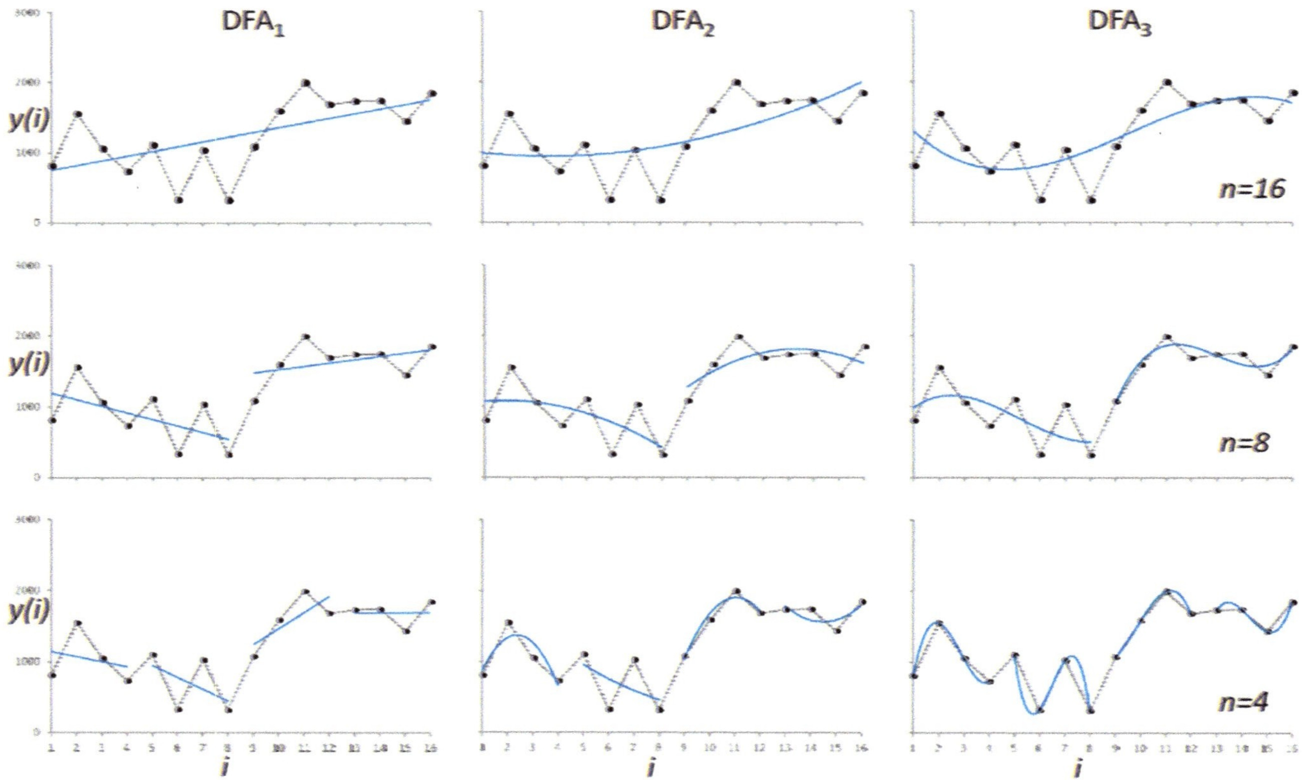

Fig. 7.4 Examples of piecewise detrending functions for DFA. Regression functions (continuous *blue line*) that fit the data in each block are straight lines (DFA$_1$, *left*), parabolas (DFA$_2$, *centre*) or cubic polynomials (DFA$_3$, *right*); block sizes decrease from $n = 16$ (*top panels*) to $n = 4$ (*bottom panels*)

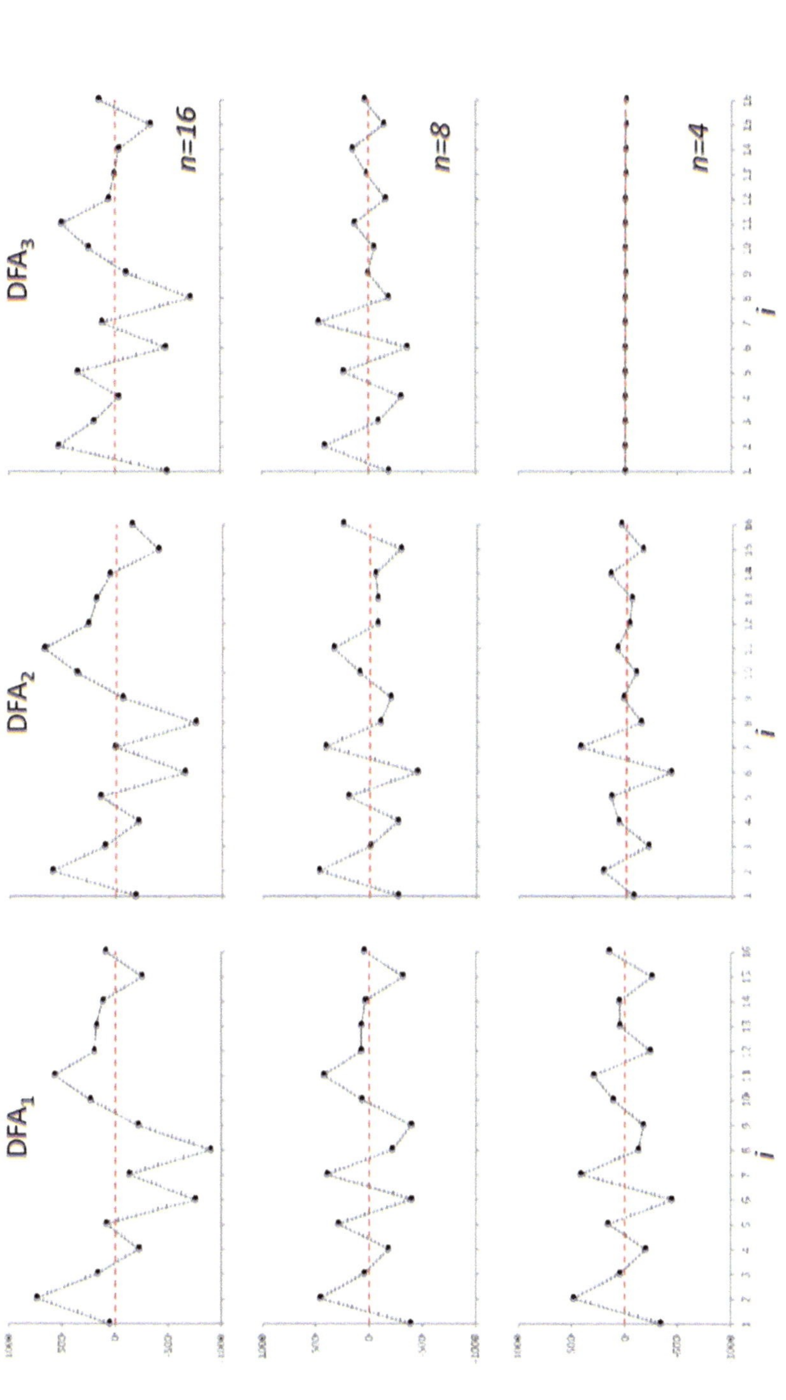

Fig. 7.5 Residuals after least-square polynomial fitting of the data of Fig. 7.4. The $F_d(n)$ function of DFA quantifies the root mean square amplitude of these deviations

where the expected value E[. . .] is the average over several shuffled configurations. The constant Q is a normalizing factor. Q is obtained by calculating $F_d(n)$ of shuffled data for an intermediate block size n, neither to small nor to large, in order not to introduce the distortions that may occur at extreme n values. Kantelhard *et al.* suggested to calculate Q for the size ν equal to $N/20$:

$$Q = \frac{E\left[\left(F_{shuff}(\nu)\right)^2\right]^{1/2}}{\nu^{1/2}} \tag{7.21}$$

In this way, a corrected detrended fluctuation function, $F_c(n)$, is obtained from $F_d(n)$ as:

$$F_c(n) = F_d(n)/K(n) \tag{7.22}$$

and α can be estimated as slope of log $F_c(n)$ vs. log n, also including the blocks of shorter size. Figure 7.6 illustrates the efficacy of this correction procedure on synthetized data.

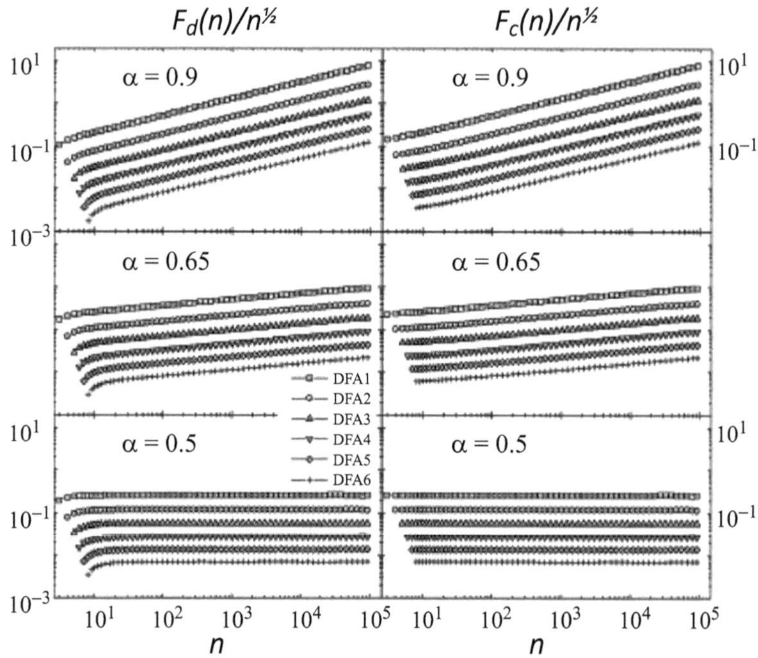

Fig. 7.6 Efficacy of $F_d(n)$ correction. DFA$_i$ fluctuation functions (with i from 1 to 6) for synthesized series, with α between 0.5 and 0.9, are normalized by the white noise case ($n^{1/2}$) and plotted before (*left*) and after (*right*) the Kantelhard's correction. This correction removes distortions at the shorter scales (redrawn from [71], with permission)

It is worth to mention that Kandelhardt *et al* [71] used a different notation for Eq. (7.19), successively extended for multifractal analysis (see the next paragraph). These authors observed that when *y(i)* is split into M consecutive, nonoverlapping blocks of n samples, with $M = \mathrm{int}(N/n)$, a short segment at the end of the series is not covered by any block if N is not a multiple of n. To consider also this last segment, other M blocks are generated starting from the end of the record. Therefore, for each block size n, $2 \times M$ blocks are considered. In each block j (with j between 1 and $2M$), the variance of the detrended fluctuations, $\sigma_n^2(j)$, is calculated. Finally, the detrended fluctuation function, *F(n)*, is the square root of the average of the variances in the $2M$ segments:

$$F(n) = \left[\frac{1}{2M} \sum_{j=1}^{2M} \sigma_n^2(j) \right]^{1/2} \tag{7.23}$$

7.3.2 DFA and Bi-Fractal Models of Cardiovascular Signals

The first use of DFA on cardiovascular times series dates back to 1995, when scale exponents from 24-h heart rate recordings were described in heart failure patients and healthy controls [62, 72]. Results highlighted that the heart rate has two scaling regions, defined by block sizes n shorter or longer than 10 beats, with different slopes both in patients and controls (Fig. 7.7). Therefore two scale coefficients were introduced, called α_1 and α_2, respectively associated with short-term (4–10 beats) [73] and long-term (>11 beats) [25] fractal dynamics. Successively other authors set different n ranges for estimating α_1 ($4 \le n \le 16$) and α_2 ($16 \le n \le 64$) [74], which are often used and implemented in commercial software for heart rate variability analysis [75].

Following studies investigated the autonomic determinants of α_1 and α_2 exponents of heart rate. In healthy volunteers, cardiac vagal blockade by atropine increases α_1 [76–78] and α_2 [76]. Moreover, α_1 increases when the sympatho/vagal balance increases following passive head-up tilt or low-intensity exercise [79]. The α_1 exponent also increases during cold hand immersion while it decreases during cold face immersion, both these maneuvers rising the sympathetic activity but only the latter one also rising the vagal activity [80]. The α_2 long-term exponent decreases when clonidine resets the overall sympathetic activity to a lower set-point [76], and is influenced by sleep stages, being lower in light and deep sleep compared to wake and REM sleep [81]. Some studies showed that α_1 tends to increase with ageing [82, 83] but this trend was not confirmed in another study [84], and a more recent work based on DFA_2 highlighted that α_1 initially increases with age, from 20 to 50–60 years, where it reaches a maximum, and then progressively decreases with age [81].

The α_1 and α_2 coefficients can be estimated also for systolic and diastolic blood pressure [85, 86]. Although vagal blockade by atropine does not affect the faster

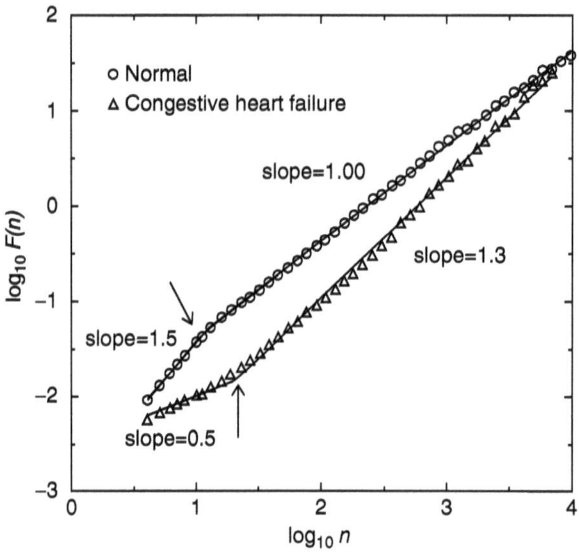

Fig. 7.7 DFA of heart rate recorded over the 24 hours in a patient and in a healthy subject. This figure showed for the first time the fluctuation functions of heart rate and documented changes induced by congestive heart failure in short-term and long-term slopes. Redrawn from [72], with permission

spectral components of blood pressure variability, it increases the short-term scale exponent of blood pressure [76]. The blockade of cardiac sympathetic activity by propranolol does not affect α_1 and α_2 of blood pressure, while the lowering of overall sympathetic activity by clonidine reduces α_2 of both systolic and diastolic blood pressure [76].

The vast literature on clinical applications of heart-rate DFA is summarized by three recent reviews [74, 87, 88]. The more interesting results regard the short-term exponent. A reduced α_1 predicts sudden cardiac death in elderly subjects [89] and mortality in heart-failure patients [90]. Low α_1 values are also associated with increased risk for heart failure in patients after acute myocardial infarction [91]. Evaluating α_1 after acute myocardial infarction is important also because it predicts mortality [25], coronary events [92], ventricular tachycardia [93] and fibrillation [73].

7.4 Multifractality by Detrended Fluctuation Analysis

As illustrated in the previous paragraph, cardiovascular self-similarity is rarely described by a single DFA scale exponent, and bifractal models are often applied. However, the cardiovascular signals may show more complicated fractal character-

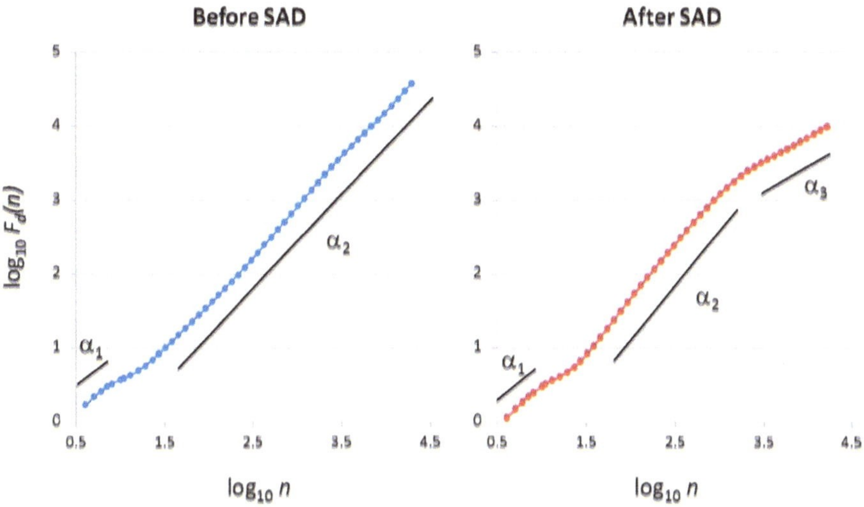

Fig. 7.8 DFA fluctuations function of systolic blood pressure before and few days after the surgical opening of the baroreflex loop by baroreceptors sino-aortic denervation (SAD). After SAD, the bifractal model cannot fully describe the self-similarity structure and a three-coefficient model appear more adequate

istics that require more than one or two scale coefficients. The DFA approach can be employed also in these cases.

Common situations where the bifractal model is inadequate occur when long-term physiological adaptations make the cardiovascular regulation change over time. For instance, changes in activity level and posture over the 24 h may influence α_1 and α_2 of heart rate [83, 94]; moreover, α_1 and α_2 may change during sleep following autonomic modulations associated with light, deep or paradoxical sleep [81]. In these cases, the bifractal DFA model has been sequentially applied over a running window, obtaining profiles of short-term and long-term exponents over time, $\alpha_1(t)$ and $\alpha_2(t)$.

In other cases, the relation between *log* $F_d(n)$ and *log(n)* shows more than two regions with a linear slope. For example, Fig. 7.8 shows that opening the baroreflex loop in an animal model breaks the regression line modeling α_2 of systolic blood pressure at block sizes around $n = 1000$ heart beats. In this case, the bifractal model cannot describe fully the blood pressure fractal dynamics and three rather than two scale coefficients should be used (coherently with this result, the opening of the baroreflex loop was previously reported to induce systematic deviations in the slope of the 1/f spectrum of blood pressure [13]). Examples like this suggest that if one rigidly defines "a priori" the number of self-similarity coefficients, α_j, and their range of scales, important details of the fractal structure may go undetected. Therefore, in some cases it is preferable to use a more flexible structure, which describes α as a continuous function of the time scale τ. This $\alpha(\tau)$ function represents self-similarity as a *temporal spectrum* of scale coefficients.

However, the dynamics of cardiovascular time series may also result from the superposition of interlaced fractal components. For instance, this is the case when two fractal components, one with a small and one with a large amplitude, overlap at the same scales. Also in these cases, methods based on DFA may describe features of the multifractal structure of cardiovascular series, in particular the *multifractal spectrum*. Details on the DFA approach for estimating the temporal spectrum of scale coefficients and the multifractal spectrum follow.

7.4.1 DFA and the Temporal Spectrum of Scale Coefficients

The group of Echevarria *et al.* was the first to highlight that DFA of heart rate variability should not simply evaluate a short-term and a long-term scale exponent, claiming that local deviations from the bifractal power-law model may be helpful for better evaluating pathological conditions [95]. They evaluated the slope of log $F_d(n)$ vs. log n as a continuous function of the scale n by applying a simplified Kalman filter, called αβ-filter. The αβ-filter assumes that the system is described by two internal states: the first state, $G(k)$, is the integral over the time k of the second state, $m(k)$. The αβ-filter also assumes that the available measures, $G_m(k)$, are observations corrupted by noise of the first internal state.

The filter is applied on $G_m(k)$, that is the logarithm of $F_d(n)$ after interpolation at points k equispaced over the log-transformed n axis. Therefore $G_m(k)$ are the measured observations of the internal variable $G(k)$, while $G_e(k)$ and $m_e(k)$ are estimates of the internal states $G(k)$ and $m(k)$. The prediction of $G(k)$ from the estimates at the step k-1, with δ the increment between lags, is:

$$G_p(k) = G_e\,(k-1) + m_e\,(k-1) \times \delta \tag{7.24}$$

The prediction error of the filter at lag k, $r(k) = G_m(k)$ - $G_p(k)$, is used to correct $G_p(k)$ and to estimate the gradient at lag k, $m_e(k)$:

$$\begin{cases} G_e(k) = G_p(k) + r(k)\alpha(k) \\ m_e(k) = m_e\,(k-1) + r(k)\beta(k)/\delta \end{cases} \tag{7.25}$$

The α and β coefficients in Eq. (7.25) give the name of the filter. Since α and β critically weight the prediction error, they should be selected carefully. Echevarria *et al.* proposed a dependency on k according to the following empirical relation:

$$\begin{cases} \alpha(k) = \frac{2(2k-1)}{k(k+1)} \\ \beta(k) = \frac{6}{k(k+1)} \end{cases} \text{for } k \le 500; \quad \begin{cases} \alpha(k) = \alpha(500) \\ \beta(k) = \beta(500) \end{cases} \text{for } k > 500 \tag{7.26}$$

and showed that scaling patterns derived from $m_e(k)$ distinguish congestive heart failure patients from controls better than α_1 and α_2 [96].

A more direct approach to the estimation of the DFA slope is the calculation of the derivative of log $F_d(n)$ vs. log n [97]. This approach avoids the empirical and somehow arbitrary definitions of α and β coefficients in Eq. (7.26). Let's consider the DFA estimate of $F_d(n)$ over a set of scales, n_k, with k between 1 and M, and let's assume that the n_k values are equispaced over the logarithmic scale. Then, the local scale exponents can be estimated at scales n_k, with k between 2 and M-1, as:

$$\alpha_B(n_k) = \frac{\log F_d(n_{k+1}) - \log F_d(n_{k-1})}{\log n_{k+1} - \log n_{k-1}} \tag{7.27}$$

(a variant of this method estimates $\alpha_B(n)$ by fitting regression lines within windows moving along n in order to provide estimates of local coefficients less sensitive to changes of $F_d(n)$ with n [98]).

Figure 7.9 shows an example of local scale exponents for R-R intervals derived from a Holter ECG recording of several hours in a healthy volunteer. Two periods of 7 h were selected: one during nighttime sleep and one during afternoon activities. The profiles of $\alpha_B(n)$ functions highlight that the bifractal approach may oversimplify the fractal dynamics. In fact, only at nighttime $\alpha_B(n)$ shows relatively constant values in the α_1 range ($n \leq 16$ beats). During daytime $\alpha_B(n)$ decreases substantially at scales larger than 7 beats from values typical of Brownian motions ($\alpha_B = 1.4$ at $n = 7$) to values typical of 1/f noise ($\alpha_B = 1.0$ at $n = 16$), and a single α_1 coefficient cannot describe such abrupt change of dynamics. Similarly, also a single α_2 coefficient does not adequately describe the long-term fractal dynamics, because $\alpha_B(n)$ appears relatively constant only for scales larger than 2000 beats during daytime (with α_B around 1.1) and between 1000 and 2500 beats during nighttime (with α_B around 0.85).

The $\alpha_B(n)$ functions are defined on the *beat-domain n*. Therefore, a comparison between conditions with markedly different heart rates may be critical, because block sizes expressed by the same number of beats, n, correspond to substantially different time scales, in seconds. To avoid this problem, $\alpha_B(n)$ can be mapped from the beat domain n to the time domain τ, by setting

$$a(\tau_k) = a_B(n_k) \text{ for } \tau_k = n_k \times \mu_{CI} \tag{7.28}$$

with μ_{CI} the mean cardiac interval, in seconds, evaluated over the whole series. Finally, the $\alpha(\tau_k)$ estimates can be linearly interpolated to obtain a continuous function of τ, $\alpha(\tau)$, named "temporal spectrum of scale coefficients".

For a given cardiovascular time series, the longest τ at which $\alpha(\tau)$ can be estimated depends on the duration of the recording. Empirical estimates are provided in [99] for heart rate recordings obtained during daily activities and during night sleep. The estimates suggest that τ can reach up to 17% the entire duration of the recording without introducing substantial estimation bias. The shortest τ value, for a given block size n, depends on the order p of the detrending function (see Fig. 7.5), and the lower is p, the shorter can be τ. For DFA_1 of heart rate and blood pressure, it has been suggested that $\alpha(\tau)$ can be estimated without bias for τ values

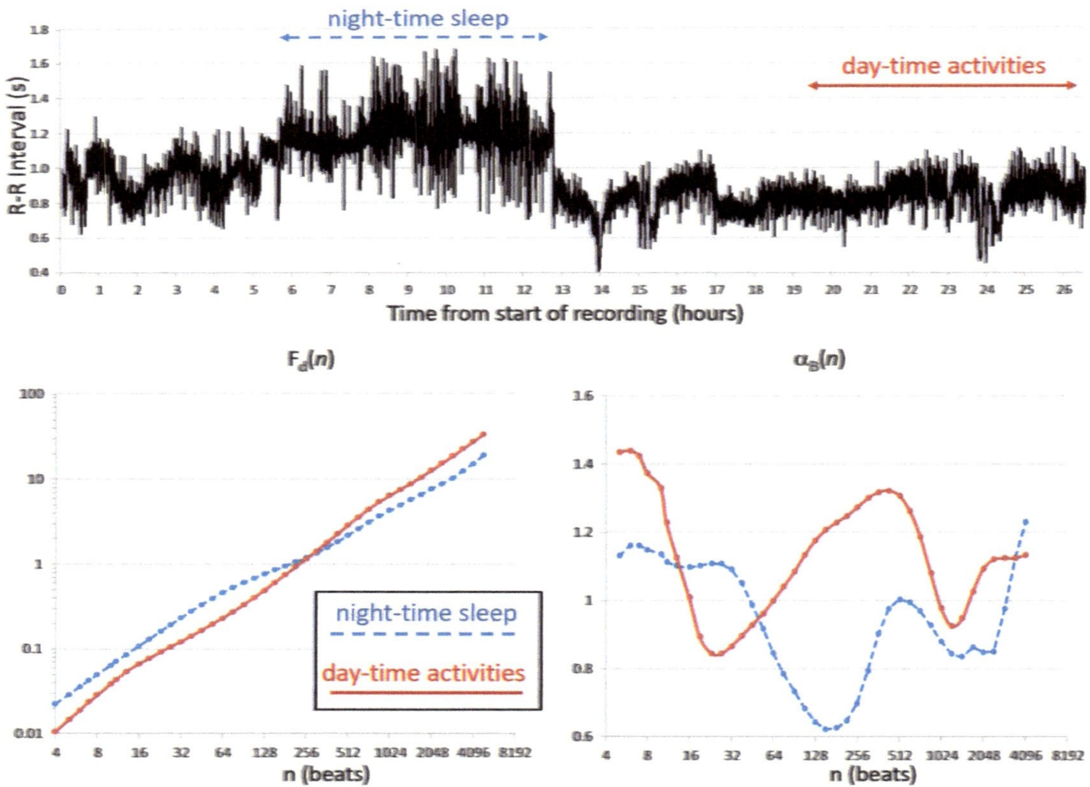

Fig. 7.9 DFA spectrum of scale coefficients, $\alpha_B(n)$, for R-R intervals recorded during daytime and nighttime. *Upper panel*: beat-by-beat R-R intervals in a healthy volunteer; the arrows indicate two segments of 7-h duration selected for DFA, one during nighttime sleep (*blue dashed line*) and one during daytime activities in the afternoon (*red continuous line*). *Lower panels:* the corresponding $F_d(n)$ functions (*left*) and $\alpha_B(n)$ scale exponents (*right*)

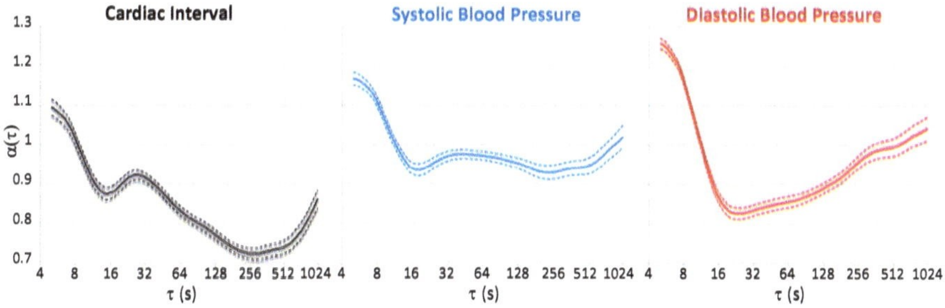

Fig. 7.10 Temporal spectrum of DFA scale coefficients. *Upper panel*: Mean ± standard error of the mean in 68 volunteers sitting at rest for 2 hours (drawn from data presented in [101])

as short as 5 s. Estimations up to $\tau = 4$ s were obtained by applying a variant of the Kantelhard correction method of Eq. (7.10) [100].

The shape of $\alpha(\tau)$ of blood pressure and heart rate was described in a population of healthy subjects sitting at rest, by analyzing beat-by-beat blood pressure data recorded noninvasively at the finger artery for 2 h [101]. The $\alpha(\tau)$ spectrum of cardiac intervals (left panel in Fig. 7.10) is slightly greater than 1 at $\tau = 5$ s, i.e., the α exponent that characterizes "1/f" fractal processes. It decreases when τ increases reaching the minimum value for τ around 256 s, where $\alpha(\tau)$ is closer to the value of white noise than to the value of 1/f noise. The decreasing trend between $\tau = 5$ s and $\tau = 256$ s is interrupted by a relative maximum at τ around 30 s. The temporal spectra of blood pressure scale exponents differ substantially from those of heart rate (Fig. 7.10, central and right panels). In particular, $\alpha(\tau)$ of systolic blood pressure is greater than 0.9 at all τ, $\alpha(\tau)$ of diastolic blood pressure decreases from $\alpha = 1.26$ at $\tau = 5$ s to a minimum $\alpha = 0.83$ at $\tau = 24$ s, and the relative maximum characterizing $\alpha(\tau)$ of cardiac intervals around $\tau = 30$ s is completely missing.

The influence of the autonomic regulation on $\alpha(\tau)$ of heart rate and blood pressure were studied making use of selective autonomic blockade in humans [76] and in an animal model [102]. Cardiac vagal, cardiac sympathetic and vascular sympathetic outflows contribute differently to heart rate and blood pressure fractal structures. The vagal outflow adds white-noise components to $\alpha(\tau)$ of heart rate, amplifying the differences with $\alpha(\tau)$ of systolic and diastolic blood pressure. By contrast, the cardiac sympathetic outflow adds Brownian motion components to $\alpha(\tau)$ of heart rate at the shorter scales. An overall vascular and cardiac sympathetic inhibition decreases short- and long-term exponents of heart rate, and short-term exponents of blood pressure, so that the $\alpha(\tau)$ spectra of heart rate and blood pressure have different mean values but a similar profile.

When only the cardiac vagal outflow modulates the heart rate dynamics (cardiac sympathetic outflow blockade with propranolol), the $\alpha(\tau)$ profile is relatively flat, on average below 0.7 and closer to a white noise dynamics than to a "1/f" dynamics (Fig. 7.11). By contrast, when the heart rate is modulated by the cardiac sympathetic outflow only (vagal blockade by atropine), the $\alpha(\tau)$ profile decreases from values typical of a Brownian motion at the shorter scales, to values closer to a "1/f" dynamics at scales larger than 60 s (Fig. 7.11).

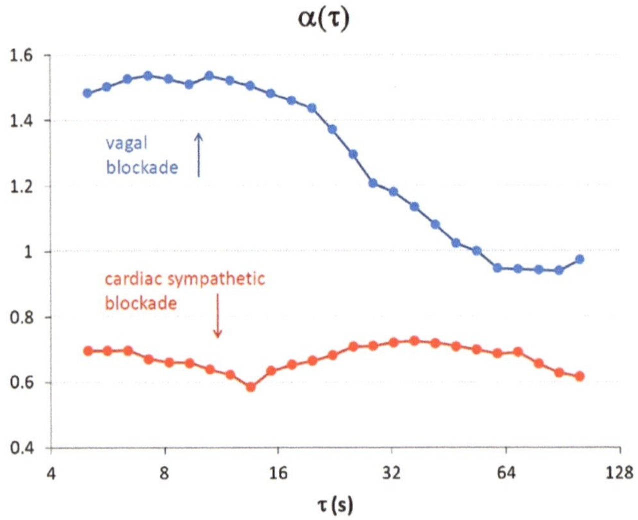

Fig. 7.11 Temporal spectra of DFA scale coefficients for heart rate after selective autonomic blockade. Median values over 9 volunteers (redrawn from data originally presented in [76])

The different patterns associated with vagal and sympathetic modulations of heart rate could be explained by differences in the transfer functions linking the vagal outflow with heart rate and linking the cardiac sympathetic outflow with heart rate (Fig. 7.12). It has been shown that the transfer function from vagal outflow to heart rate is almost flat [104]. This means that if the vagal outflow is white noise, then the corresponding heart-rate modulations are also white. By contrast, the transfer function from cardiac sympathetic outflow to heart rate was described as a one-pole low-pass filter with time constant equal to 66 s [105]. The time constant at 66 s reflects the slow removal from cardiac tissues of noradrenaline released by the sympathetic nerve endings (the removal of acetylcholine released by the vagal terminals is much faster) [106]. Therefore, if the sympathetic outflow is white, the corresponding heart rate modulation is the integral of white noise, i.e., a Brownian motion, at scales much shorter than 66 s, and is white noise at scales much larger than 66 s. This would explain the different $\alpha(\tau)$ profile between vagal and sympathetic heart-rate modulations (Fig. 7.11), in particular the Brownian motion dynamics of sympathetic modulations of heart rate at scales much shorter than 66 s.

Interestingly, an altered autonomic integrative control may be better identified by the $\alpha(\tau)$ fractal analysis of heart rate variability rather than by power spectral analysis. This was demonstrated in spinal cord injured individuals with lesion levels sufficiently high to guarantee an intact autonomic regulation of the heart and of the splanchnic district [107]. While heart-rate power spectra of paraplegic individuals were identical to power spectra of able-bodied controls, the temporal spectra of scale coefficients differed markedly, revealing the expected alterations in the integrative autonomic control.

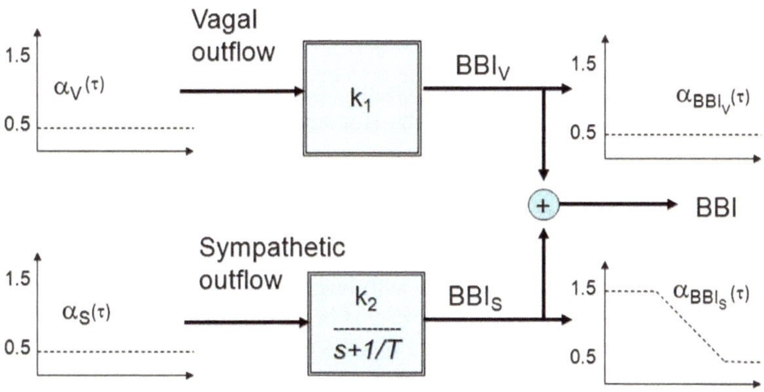

Fig. 7.12 Interpretative model of autonomic influences on $\alpha(\tau)$ of heart rate. The blocks are transfer functions, in terms of Laplace transforms, between cardiac vagal or cardiac sympathetic outflows and beat-to-beat intervals (BBI); BBI_V and BBI_S are vagal and sympathetic modulations of BBI. The sympathetic outflow passes through a one-pole low-pass filter with time constant $T = 66$ s (see text); this block can be approximated by a constant (k_2) for oscillations much slower than T, and by an integration block (k_2/s) for oscillations much faster than T. If sympathetic and vagal outflows are white noises ($\alpha_V(\tau) = \alpha_S(\tau) = 0.5$), then BBI_V would be white, and BBI_S would be white at scales much greater than T, and would be a Brownian motion at scales much shorter than T. Redrawn from [103], with permission

7.4.2 *DFA and the Multifractal Spectrum*

The approaches for assessing multifractality described in the previous paragraph assume that different self-similarity exponents can be identified in the same time series at different scales. To detect the presence of more fractal components interwoven at the same scale, Kantelhardt et al. [108] proposed to calculate the qth-order detrended fluctuation function by extending Eq. (7.23) as follows:

$$F_q(n) = \left(\frac{1}{2M} \sum_{j=1}^{2M} \left(\sigma_n^2(j) \right)^{q/2} \right)^{1/q} \quad \text{for } q \neq 0 \tag{7.29}$$

When $q = 2$, Eq. (7.29) coincides with the detrended fluctuation function introduced for the monofractal analysis, being $F_2(n) = F(n)$. Equation (7.29) cannot be calculated for $q = 0$ and the limit for $q \to 0$ is used:

$$F_0(n) = e^{\frac{1}{4M} \sum_{j=1}^{2M} \ln\left(\sigma_n^2(j)\right)} \tag{7.30}$$

If the time series $x(i)$ has a power-law, long-range, correlation, then $F_q(n)$ increases as a power of n:

$$F_q(n) \approx n^{h(q)} \tag{7.31}$$

Clearly, *h(2)* in Eq. (7.31) coincides with the α exponent of monofractal DFA. In particular, if *x(i)* is stationary, *h(2)* is the Hurst exponent *H*. For this reason, *h(q)* in Eq. (7.31) is called *generalized Hurst exponent.*

If *x(i)* is monofractal, then *h(q)* does not depend on *q*, and regression lines fitting log-log plots of $F_q(n)$ vs. *n* should have the same slope for different *q*. By contrast, *h(q)* would depend on *q* if *x(i)* is composed by at least two fractal components, one with small and one with large fluctuations, that have a different scaling behavior. In fact, $F_q(n)$ in Eq. (7.29) is the *q*-root-mean of σ_n^2 variances: if $q > 0$ then it mainly reflects the components with larger variances (i.e., those with the larger deviations from the fitting polynomial) while, if $q < 0$, it is dominated by components producing the smaller variances. In this way, the multifractal nature of time series in which small and large fluctuations scale differently, can be detected by estimating *h(q)*. The generalized Hurst exponent is calculated as the slope of the regression line fitting log $F_q(n)$ and log *n* (see an example in Fig. 7.13). If *h(q)* changes substantially when *q* ranges from negative to positive values (Fig. 7.14, *left*), then the time series is multifractal.

It should be considered, however, that multifractality is often described by evaluating exponents of partition functions, *τ(q)* (e.g., see [109–111]). The DFA approach allows obtaining also this type of representation. In fact, the exponents *τ(q)* can be derived from the generalized Hurst exponent as [108]:

$$\tau(q) = h(q) - 1 \tag{7.32}$$

Once that $\tau(q)$ is estimated by Eq. (7.32), the singularity strength, α_h, defined as:

$$\alpha_h = d\tau/dq \tag{7.33}$$

can be easily obtained. Therefore, DFA allows describing multifractality also in terms of the singularity spectrum, $D(\alpha_h)$, defined by the Legendre transform:

$$D(\alpha_h) = q\alpha_h - \tau(q) \tag{7.34}$$

The singularity spectrum represents the fractal dimension *D* for each fractal component included in the time series, ordered for increasing values of the singularity coefficient α_h (see Fig. 7.14, *right*). A free matlab library for evaluating *h(q)*, *τ(q)* and $D(\alpha_h)$ is described in [112].

Parameters typically used for concisely describing the singularity spectrum are $\alpha_h^{\max}$, the singularity value at which the spectrum reaches its maximum; α_h^{left} and α_h^{right}, the lowest and highest singularity values; $\alpha_h^{width} = \alpha_h^{right} - \alpha_h^{left}$, the range of singularity values [113, 114].

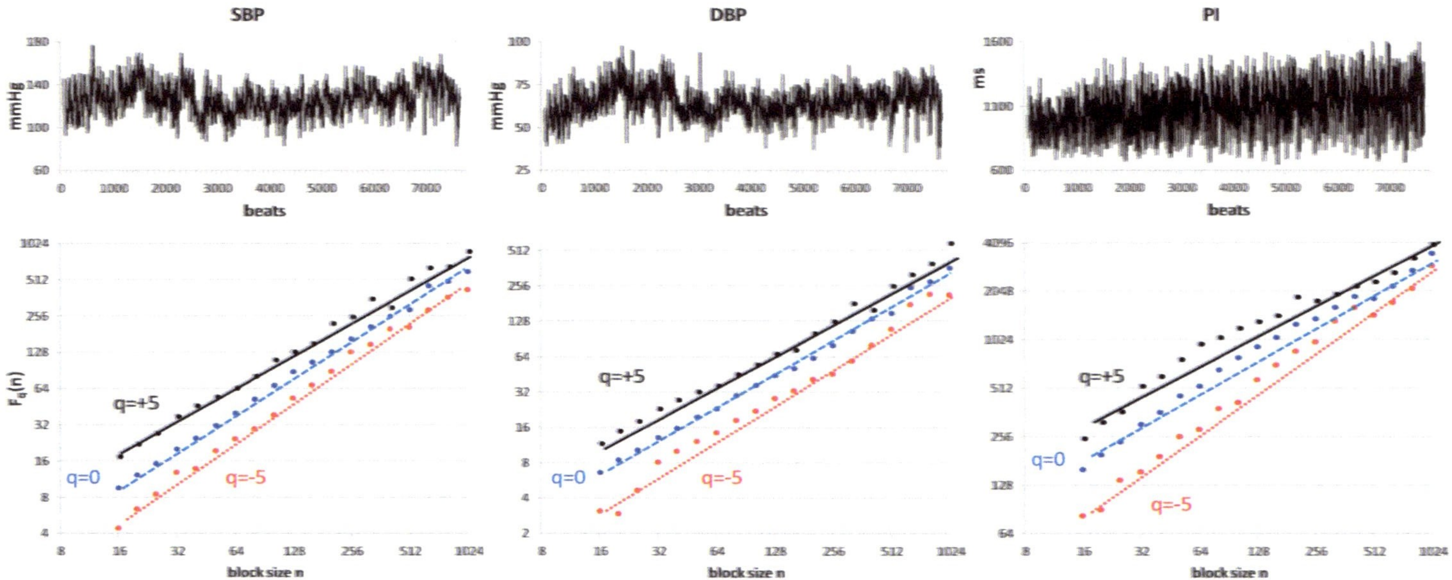

Fig. 7.13 Example of qth-order detrended fluctuation functions for systolic and diastolic blood pressure (SBP, DBP) and pulse interval (PI). *Upper panels:* 2-h long beat-by-beat series recorded in a healthy male volunteer sitting at rest; *lower panels*: corresponding $F_q(n)$ functions for $q = -5$, $q = 0$ and $q = 5$ plotted in a log–log scale, with corresponding regression lines. The slope of the regression lines provides an estimate of the generalized Hurst exponent, $h(q)$

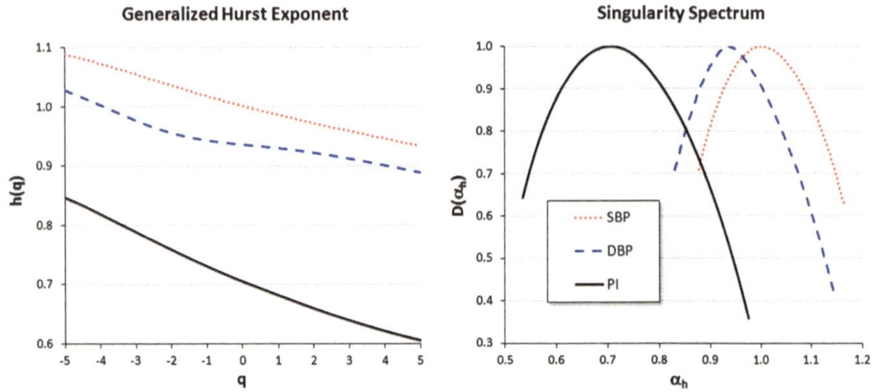

Fig. 7.14 Example of generalized Hurst exponent and singularity spectrum for systolic and diastolic blood pressure (SBP, DBP) and pulse interval (PI). Functions were estimated for the same time series employed in Fig. 7.13, with q between −5 and +5

7.4.3 Multiscale Multifractal DFA

The multifractal approach described in the previous paragraph assumes that the generalized Hurst exponent does not change over the scales considered for its estimation. In this regard, the multifractal DFA is complementary to the temporal spectrum of scale coefficients, $\alpha(\tau)$ in Eq. (7.28), that, by contrast, quantifies changes in the Hurst exponent among scales. Therefore, the two methods provide information of different nature. Obviously, it may be expected that the cardiovascular time series show a multifractal spectrum that in addition changes its characteristics at different scales. To take into account both these aspects, singularity spectra were estimated by separately considering time scales corresponding to some frequency bands usually employed for spectral analysis: the ultra-low frequency band (ULF, frequency $f < 0.003$ Hz), the very-low frequency band (VLF, $0.003 < f < 0.04$ Hz) and the low frequency band (LF, $0.04 < f < 0.15$ Hz) [115]. In particular, heart-rate singularity spectra evaluated for scales corresponding to the VLF band seem to describe changes in the cardiac autonomic control associated with healthy ageing concisely and reliably [113].

The singularity spectra calculated in this way describe only variability components that fall into specific frequency bands extensively studied in the past by frequency domain methods of heart rate variability. This is a clear advantage for interpreting the results. However, it is possible that a scaling region where multifractal DFA provides useful information does not correspond to any traditional frequency band of the power spectrum. Therefore, fixing "a priori" the scaling regions to be evaluated by multifractal DFA might prevent detecting important details of multifractal dynamics. This limit is overcome by calculating the generalized Hurst exponent as a continuous function of the block size n, as proposed by Gierałtowski *et al.* [116]. These authors extended the temporal spectrum of

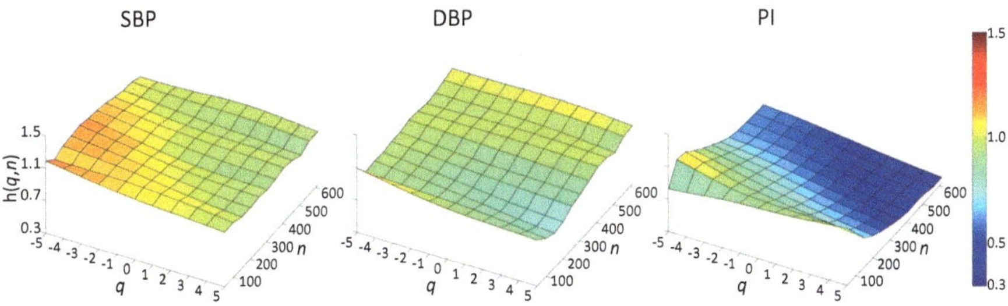

Fig. 7.15 Example of multiscale multifractal DFA for systolic and diastolic blood pressure (SBP, DBP) and pulse interval (PI). Surfaces of generalized Hurst exponents, h(q,n), were calculated for the same time series of Fig. 7.13, with code freely available on the Physionet web site [64]

scale coefficients evaluated for $q = 2$, i.e., $\alpha(\tau)$ in Eq. (7.28), to q values ranging from -5 to $+5$. Other differences from the $\alpha(\tau)$ analysis are the use of a second-order fitting polynomial in Eq. (7.19) and the estimation of h as the slope of the log-log regression line fitted in a moving window running over the n axis. In this way, the multifractality of cardiovascular time series can be described by an almost continuous surface of generalized Hurst exponents, $h(q,n)$, as in Fig. 7.15.

Examples of this approach include the assessment of the autonomic nervous system from fetal heart rate recordings [117, 118], the evaluation of risk in aortic valve replacement surgery [119] and the characterization of heart rate variability during sleep [120].

7.5 Summary and Conclusions

In the last decades, the analysis of the spontaneous variability of cardiovascular signals has been largely based on the estimation of powers in frequency bands. However, this chapter showed that specific aspects of variability, possibly quantifying fractal features of heart rate or blood pressure (like the slope of log-log spectra), actually accompanied the traditional spectral methods since the beginning of the ‘80s.

The development of increasingly more robust and effective estimators in the ‘90s provided the scientific community of cardiovascular signals analysis with tools, like DFA estimators of short-term and long-term scale coefficients, that are simple to calculate and relatively easy to interpret. Therefore, as this chapter reported, around the turn of the new millennium several studies based on DFA of heart rate variability were conducted on large populations of cardiac patients, demonstrating the capability of DFA bifractal models to provide information of clinical value and to assess the cardiovascular risk.

Self-similarity analysis is expected to provide complementary information, with respect to spectral methods, on the mechanisms responsible for cardiovascular regulation. Nevertheless, as reported in a recent review [88], clinical information derived from long-term correlations through the estimation of power law exponents, or from bifractal DFA models, appear sometimes to correlate with information from spectral methods. Thus, it is still unclear how much the use of such tools of fractal analysis, in addition to traditional methods, may actually increase the results of risk stratification. However, it should be considered that the more recent lines of research mentioned in this chapter are aimed at investigating multifractal aspects of cardiovascular signals by extending DFA-based methods. Some of these advanced tools were originally developed in the field of complex systems and statistical physics, while other were specifically designed for describing the cardiovascular signals from new multifractal and multiscale perspectives. The number of groups actively involved in this field of research is rapidly increasing. Therefore, it may be anticipated that in the next years multifractal DFA approaches will be more and more used in physiological studies and clinical settings. These trends of research will likely provide new analysis tools that, when used in conjunction with spectral methods, might be more effective in assessing the cardiovascular system in conditions of health and disease, and able to stratify more precisely the risk of cardiovascular events.

References

1. Eke, A., Herman, P., Kocsis, L., Kozak, L.R.: Fractal characterization of complexity in temporal physiological signals. Physiol. Meas. **23**(1), R1–38 (2002)
2. Goldberger, A.L.: Non-linear dynamics for clinicians: chaos theory, fractals, and complexity at the bedside. Lancet. **347**(9011), 1312–1314 (1996)
3. Bak, P., Tang, C., Wiesenfeld, K.: Self-organized criticality: An explanation of the 1/f noise. Phys. Rev. Lett. **59**(4), 381–384 (1987)
4. West, G.B., Brown, J.H., Enquist, B.J.: A general model for the origin of allometric scaling laws in biology. Science. **276**(5309), 122–126 (1997)
5. Mandelbrot, B.B.: Self-Affine Fractals and Fractal Dimension. Phys. Scr. **32**, 257–260 (1985)
6. Hurst, H.E.: Long Term Storage Capacity of Reservoirs. Trans. Am. Soc. Civ. Eng. **116**, 770–799 (1951)
7. Feller, W.: The Asymptotic Distribution of the Range of Sums of Independent Random Variables. Ann. Math. Stat. **22**(3), 427–432 (1951)
8. Caccia, D.C., Percival, D., Cannon, M.J., Raymond, G., Bassingthwaighte, J.B.: Analyzing exact fractal time series: evaluating dispersional analysis and rescaled range methods. Physica A. **246**(3–4), 609–632 (1997)
9. Mandelbrot, B.B., Van Nees, J.W.: Fractional brownian motions, fractional noises and applications. SIAM Rev. **10**(4), 422–437 (1968)
10. Keshner, M.S.: 1/f noise. Proc. IEEE. **70**(3), 212–218 (1982)
11. Kobayashi, M., Musha, T.: 1/f fluctuation of heartbeat period. IEEE Trans. Biomed. Eng. **29**(6), 456–457 (1982)
12. Brown, D.R., Cassis, L.A., Silcox, D.L., Brown, L.V., Randall, D.C.: Empirical and theoretical analysis of the extremely low frequency arterial blood pressure power spectrum in unanesthetized rat. Am. J. Physiol. Heart Circ. Physiol. **291**(6), H2816–H2824 (2006)
13. Di Rienzo, M., Castiglioni, P., Parati, G., Mancia, G., Pedotti, A.: Effects of sino-aortic denervation on spectral characteristics of blood pressure and pulse interval variability: a wide-band approach. Med. Biol. Eng. Comput. **34**(2), 133–141 (1996)

14. Hagerman, I., Berglund, M., Lorin, M., Nowak, J., Sylven, C.: Chaos-related deterministic regulation of heart rate variability in time- and frequency domains: effects of autonomic blockade and exercise. Cardiovasc. Res. **31**(3), 410–418 (1996)
15. Marsh, D.J., Osborn, J.L., Cowley Jr., A.W.: 1/f fluctuations in arterial pressure and regulation of renal blood flow in dogs. Am. J. Phys. **258**(5 Pt 2), F1394–F1400 (1990)
16. Wagner, C.D., Persson, P.B.: Two ranges in blood pressure power spectrum with different 1/f characteristics. Am. J. Phys. **267**(2 Pt 2), H449–H454 (1994)
17. Wagner, C.D., Nafz, B., Persson, P.B.: Chaos in blood pressure control. Cardiovasc. Res. **31**(3), 380–387 (1996)
18. Takalo, R., Korhonen, I., Sorsa, H., Majahalme, S., Tuomisto, M., Turjanmaa, V.: Wide-band spectral analysis of blood pressure and RR interval variability in borderline and mild hypertension. Clin. Physiol. **19**(6), 490–496 (1999)
19. Saul, J.P., Albrecht, P., Berger, R.D., Cohen, R.J.: Analysis of long term heart rate variability: methods, 1/f scaling and implications. Comput. Cardiol. **14**, 419–422 (1988)
20. Sassi, R., Cerutti, S., Hnatkova, K., Malik, M., Signorini, M.G.: HRV scaling exponent identifies postinfarction patients who might benefit from prophylactic treatment with amiodarone. IEEE Trans. Biomed. Eng. **53**(1), 103–110 (2006)
21. Omboni, S., Parati, G., Castiglioni, P., Di Rienzo, M., Imholz, B.P., Langewouters, G.J., Wesseling, K.H., Mancia, G.: Estimation of blood pressure variability from 24-hour ambulatory finger blood pressure. Hypertension. **32**(1), 52–58 (1998)
22. Makikallio, T.H., Hoiber, S., Kober, L., Torp-Pedersen, C., Peng, C.K., Goldberger, A.L., Huikuri, H.V.: Fractal analysis of heart rate dynamics as a predictor of mortality in patients with depressed left ventricular function after acute myocardial infarction. TRACE Investigators. TRAndolapril Cardiac Evaluation. Am. J. Cardiol. **83**(6), 836–839 (1999)
23. Jokinen, V., Tapanainen, J.M., Seppanen, T., Huikuri, H.V.: Temporal changes and prognostic significance of measures of heart rate dynamics after acute myocardial infarction in the beta-blocking era. Am. J. Cardiol. **92**(8), 907–912 (2003)
24. Huikuri, H.V., Makikallio, T.H., Airaksinen, K.E., Seppanen, T., Puukka, P., Raiha, I.J., Sourander, L.B.: Power-law relationship of heart rate variability as a predictor of mortality in the elderly. Circulation. **97**(20), 2031–2036 (1998)
25. Huikuri, H.V., Makikallio, T.H., Peng, C.K., Goldberger, A.L., Hintze, U., Moller, M.: Fractal correlation properties of R-R interval dynamics and mortality in patients with depressed left ventricular function after an acute myocardial infarction. Circulation. **101**(1), 47–53 (2000)
26. Guzzetti, S., La Rovere, M.T., Pinna, G.D., Maestri, R., Borroni, E., Porta, A., Mortara, A., Malliani, A.: Different spectral components of 24 h heart rate variability are related to different modes of death in chronic heart failure. Eur. Heart J. **26**(4), 357–362 (2005)
27. Gang, U.J., Jons, C., Jorgensen, R.M., Abildstrom, S.Z., Messier, M.D., Haarbo, J., Huikuri, H.V., Thomsen, P.E.: Risk markers of late high-degree atrioventricular block in patients with left ventricular dysfunction after an acute myocardial infarction: a CARISMA substudy. Europace. **13**(10), 1471–1477 (2011)
28. Bigger Jr., J.T., Steinman, R.C., Rolnitzky, L.M., Fleiss, J.L., Albrecht, P., Cohen, R.J.: Power law behavior of RR-interval variability in healthy middle-aged persons, patients with recent acute myocardial infarction, and patients with heart transplants. Circulation. **93**(12), 2142–2151 (1996)
29. Di Rienzo, M., Parati, G., Pedotti, A., Castiglioni, P.: 1/f modeling of blood pressure and heart rate spectra. In: Di Rienzo, M., Parati, G., Pedotti, A., Zanchetti, A. (eds.) Studies in Health Technology and Informatics, pp. 45–53. IOS Press, Amsterdam (1997)
30. Castiglioni, P., Frattola, A., Parati, G., Di Rienzo, M.: 1/f-Modelling of blood pressure and heart rate spectra: relations to ageing. In: Engineering in Medicine and Biology Society, 1992 14th Annual International Conference of the IEEE, vol. 2, pp. 465–466. (1992).
31. Yamamoto, Y., Hughson, R.L.: Coarse-graining spectral analysis: new method for studying heart rate variability. J. Appl. Physiol. **71**(3), 1143–1150 (1991)

32. Task Force of the European Society of Cardiology and the North American Society of Pacing and Electrophysiology: Heart rate variability. Standards of measurement, physiological interpretation, and clinical use. Eur. Heart J. **17**(3), 354–381 (1996)
33. Butler, G.C., Senn, B.L., Floras, J.S.: Influence of atrial natriuretic factor on heart rate variability in normal men. Am. J. Phys. **267**(2 Pt 2), H500–H505 (1994)
34. Butler, G.C., Naughton, M.T., Rahman, M.A., Bradley, T.D., Floras, J.S.: Continuous positive airway pressure increases heart rate variability in congestive heart failure. J. Am. Coll. Cardiol. **25**(3), 672–679 (1995)
35. Taranto, M.L., Floras, J.S., Picton, P., Kasai, T., Alshaer, H., Gabriel, J.M., Bradley, T.D.: Relationship of heart rate variability to sleepiness in patients with obstructive sleep apnea with and without heart failure. J. Clin. Sleep Med. **10**(3), 271–276 (2014)
36. Aoyagi, N., Ohashi, K., Tomono, S., Yamamoto, Y.: Temporal contribution of body movement to very long-term heart rate variability in humans. Am. J. Physiol. Heart Circ. Physiol. **278**(4), H1035–H1041 (2000)
37. Butler, G.C., Yamamoto, Y., Xing, H.C., Northey, D.R., Hughson, R.L.: Heart rate variability and fractal dimension during orthostatic challenges. J. Appl. Physiol. **75**(6), 2602–2612 (1993)
38. Fortrat, J.O., Sigaudo, D., Hughson, R.L., Maillet, A., Yamamoto, Y., Gharib, C.: Effect of prolonged head-down bed rest on complex cardiovascular dynamics. Auton. Neurosci. **86**(3), 192–201 (2001)
39. Koh, J., Hidaka, I., Miyata, M.: Effects of nicardipine and diltiazem on fractal features of short-term heart rate variability–application of coarse graining spectral analysis. J. Anesth. **16**(2), 108–113 (2002)
40. Nakamura, Y., Yamamoto, Y., Muraoka, I.: Autonomic control of heart rate during physical exercise and fractal dimension of heart rate variability. J. Appl. Physiol. **74**(2), 875–881 (1993)
41. Satoh, K., Koh, J., Kosaka, Y.: Effects of nitroglycerin on fractal features of short-term heart rate and blood pressure variability. J. Anesth. **13**(2), 71–76 (1999)
42. Togo, F., Yamamoto, Y.: Decreased fractal component of human heart rate variability during non-REM sleep. Am. J. Physiol. Heart Circ. Physiol. **280**(1), H17–H21 (2001)
43. Yamamoto, Y., Fortrat, J.O., Hughson, R.L.: On the fractal nature of heart rate variability in humans: effects of respiratory sinus arrhythmia. Am. J. Phys. **269**(2 Pt 2), H480–H486 (1995)
44. Yamamoto, Y., Nakamura, Y., Sato, H., Yamamoto, M., Kato, K., Hughson, R.L.: On the fractal nature of heart rate variability in humans: effects of vagal blockade. Am. J. Phys. **269**(4 Pt 2), R830–R837 (1995)
45. Yamamoto, Y., Hoshikawa, Y., Miyashita, M.: Effects of acute exposure to simulated altitude on heart rate variability during exercise. J. Appl. Physiol. **81**(3), 1223–1229 (1996)
46. Blaber, A.P., Bondar, R.L., Freeman, R.: Coarse graining spectral analysis of HR and BP variability in patients with autonomic failure. Am. J. Phys. **271**(4 Pt 2), H1555–H1564 (1996)
47. Fukusaki, C., Kawakubo, K., Yamamoto, Y.: Assessment of the primary effect of aging on heart rate variability in humans. Clin. Auton. Res. **10**(3), 123–130 (2000)
48. Eke, A., Herman, P., Bassingthwaighte, J.B., Raymond, G.M., Percival, D.B., Cannon, M., Balla, I., Ikrenyi, C.: Physiological time series: distinguishing fractal noises from motions. Pflugers Arch. **439**(4), 403–415 (2000)
49. Bassingthwaighte, J.B., Bever, R.P.: Fractal correlation in heterogeneous systems. Physica D. **53**(1), 71–84 (1991)
50. Schepers, H.E., van Beek, J.H.G.M., Bassingthwaighte, J.B.: Four methods to estimate the fractal dimension from self-affine signals. IEEE Eng. Med. Biol. Mag. **11**(2), 57–64 (1992)
51. Bassingthwaighte, J.B., Beard, D.A., Li, Z.: The mechanical and metabolic basis of myocardial blood flow heterogeneity. Basic Res. Cardiol. **96**(6), 582–594 (2001)
52. Flyvbjerg, H., Petersen, H.G.: Error estimates on averages of correlated data. J. Chem. Phys. **91**, 461–466 (1989)
53. Bassingthwaighte, J.B., Raymond, G.M.: Evaluation of the dispersional analysis method for fractal time series. Ann.Biomed. Eng. **23**(4), 491–505 (1995)

54. Castiglioni, P.: Scale structure of heart-rate dynamics: multifractality or artefacts? Auton. Neurosci. **137**(1–2), 92–93 (2007)
55. Aon, M.A., Cortassa, S., O'Rourke, B.: The fundamental organization of cardiac mitochondria as a network of coupled oscillators. Biophys. J. **91**(11), 4317–4327 (2006)
56. Carolan-Rees, G., Tweddel, A.C., Naka, K.K., Griffith, T.M.: Fractal dimensions of laser doppler flowmetry time series. Med. Eng. Phys. **24**(1), 71–76 (2002)
57. Das, M., Gebber, G.L., Barman, S.M., Lewis, C.D.: Fractal properties of sympathetic nerve discharge. J. Neurophysiol. **89**(2), 833–840 (2003)
58. Cannon, M.J., Percival, D.B., Caccia, D.C., Raymond, G.M., Bassingthwaighte, J.B.: Evaluating scaled windowed variance methods for estimating the Hurst coefficient of time series. Physica A. **241**(3-4), 606–626 (1997)
59. Lal, C., Unni, S.N.: Correlation analysis of laser Doppler flowmetry signals: a potential non-invasive tool to assess microcirculatory changes in diabetes mellitus. Med. Biol. Eng. Comput. **53**(6), 557–566 (2015)
60. Soehle, M., Czosnyka, M., Chatfield, D.A., Hoeft, A., Pena, A.: Variability and fractal analysis of middle cerebral artery blood flow velocity and arterial blood pressure in subarachnoid hemorrhage. J. Cereb. Blood Flow Metab. **28**(1), 64–73 (2008)
61. Herman, P., Eke, A.: Nonlinear analysis of blood cell flux fluctuations in the rat brain cortex during stepwise hypotension challenge. J. Cereb. Blood Flow Metab. **26**(9), 1189–1197 (2006)
62. Peng, C.K., Havlin, S., Stanley, H.E., Goldberger, A.L.: Quantification of scaling exponents and crossover phenomena in nonstationary heartbeat time series. Chaos. **5**(1), 82–87 (1995)
63. Peng, C.K., Buldyrev, S.V., Havlin, S., Simons, M., Stanley, H.E., Goldberger, A.L.: Mosaic organization of DNA nucleotides. Phys. Rev. E Stat. Phys. Plasmas Fluids Relat. Interdiscip. Topics. **49**(2), 1685–1689 (1994)
64. Goldberger, A.L., Amaral, L.A., Glass, L., Hausdorff, J.M., Ivanov, P.C., Mark, R.G., Mietus, J.E., Moody, G.B., Peng, C.K., Stanley, H.E.: PhysioBank, PhysioToolkit, and PhysioNet: components of a new research resource for complex physiologic signals. Circulation. **101**(23), E215–E220 (2000)
65. Xu, L., Ivanov, P.C., Hu, K., Chen, Z., Carbone, A., Stanley, H.E.: Quantifying signals with power-law correlations: A comparative study of detrended fluctuation analysis and detrended moving average techniques. Phys. Rev. E. **71**(5), 051101 (2005)
66. Buldyrev, S.V., Dokholyan, N.V., Goldberger, A.L., Havlin, S., Peng, C.K., Stanley, H.E., Viswanathan, G.M.: Analysis of DNA sequences using methods of statistical physics. Physica A. **249**(1), 430–438 (1998)
67. Taqqu, M.S., Teverovsky, V., Willinger, W.: Estimators for long-range dependence: an empirical study. Fractals. **3**(04), 785–798 (1995)
68. Chen, Z., Ivanov, P.C., Hu, K., Stanley, H.E.: Effect of nonstationarities on detrended fluctuation analysis. Phys. Rev. E. **65**(4), 041107 (2002)
69. Chen, Z., Hu, K., Carpena, P., Bernaola-Galvan, P., Stanley, H.E., Ivanov, P.C.: Effect of nonlinear filters on detrended fluctuation analysis. Phys. Rev. E. **71**(1), 011104 (2005)
70. Hu, K., Ivanov, P.C., Chen, Z., Carpena, P., Stanley, H.E.: Effect of trends on detrended fluctuation analysis. Phys. Rev. E. Stat. Nonlin. Soft. Matter Phys. **64**(1 Pt 1), 011114 (2001)
71. Kantelhardt, J.W., Koscielny-Bunde, E., Rego, H.H.A., Havlin, S., Bunde, A.: Detecting long-range correlations with detrended fluctuation analysis. Physica A. **295**(3–4), 441–454 (2001)
72. Peng, C.K., Havlin, S., Hausdorff, J.M., Mietus, J.E., Stanley, H.E., Goldberger, A.L.: Fractal mechanisms and heart rate dynamics. Long-range correlations and their breakdown with disease. J. Electrocardiol. **28**(Suppl), 59–65 (1995)
73. Makikallio, T.H., Koistinen, J., Jordaens, L., Tulppo, M.P., Wood, N., Golosarsky, B., Peng, C.K., Goldberger, A.L., Huikuri, H.V.: Heart rate dynamics before spontaneous onset of ventricular fibrillation in patients with healed myocardial infarcts. Am. J. Cardiol. **83**(6), 880–884 (1999)

74. Huikuri, H.V., Perkiomaki, J.S., Maestri, R., Pinna, G.D.: Clinical impact of evaluation of cardiovascular control by novel methods of heart rate dynamics. Philos. Transact. A Math. Phys. Eng. Sci. **367**(1892), 1223–1238 (2009)
75. Tarvainen, M.P., Niskanen, J.P., Lipponen, J.A., Ranta-Aho, P.O., Karjalainen, P.A.: Kubios HRV–heart rate variability analysis software. Comput. Methods Prog. Biomed. **113**(1), 210–220 (2014)
76. Castiglioni, P., Parati, G., Di Rienzo, M., Carabalona, R., Cividjian, A., Quintin, L.: Scale exponents of blood pressure and heart rate during autonomic blockade as assessed by detrended fluctuation analysis. J. Physiol. **589**(Pt 2), 355–369 (2011)
77. Tulppo, M.P., Makikallio, T.H., Seppanen, T., Shoemaker, K., Tutungi, E., Hughson, R.L., Huikuri, H.V.: Effects of pharmacological adrenergic and vagal modulation on fractal heart rate dynamics. Clin. Physiol. **21**(5), 515–523 (2001)
78. Perkiomaki, J.S., Zareba, W., Badilini, F., Moss, A.J.: Influence of atropine on fractal and complexity measures of heart rate variability. Ann. Noninvasive Electrocardiol. **7**(4), 326–331 (2002)
79. Tulppo, M.P., Hughson, R.L., Makikallio, T.H., Airaksinen, K.E., Seppanen, T., Huikuri, H.V.: Effects of exercise and passive head-up tilt on fractal and complexity properties of heart rate dynamics. Am. J. Physiol. Heart Circ. Physiol. **280**(3), H1081–H1087 (2001)
80. Tulppo, M.P., Kiviniemi, A.M., Hautala, A.J., Kallio, M., Seppanen, T., Makikallio, T.H., Huikuri, H.V.: Physiological background of the loss of fractal heart rate dynamics. Circulation. **112**(3), 314–319 (2005)
81. Schumann, A.Y., Bartsch, R.P., Penzel, T., Ivanov, P.C., Kantelhardt, J.W.: Aging effects on cardiac and respiratory dynamics in healthy subjects across sleep stages. Sleep. **33**(7), 943–955 (2010)
82. Iyengar, N., Peng, C.K., Morin, R., Goldberger, A.L., Lipsitz, L.A.: Age-related alterations in the fractal scaling of cardiac interbeat interval dynamics. Am. J. Phys. **271**(4 Pt 2), R1078–R1084 (1996)
83. Pikkujamsa, S.M., Makikallio, T.H., Sourander, L.B., Raiha, I.J., Puukka, P., Skytta, J., Peng, C.K., Goldberger, A.L., Huikuri, H.V.: Cardiac interbeat interval dynamics from childhood to senescence: comparison of conventional and new measures based on fractals and chaos theory. Circulation. **100**(4), 393–399 (1999)
84. Beckers, F., Verheyden, B., Aubert, A.E.: Aging and nonlinear heart rate control in a healthy population. Am. J. Physiol. Heart Circ. Physiol. **290**(6), H2560–H2570 (2006)
85. Beckers, F., Verheyden, B., Ramaekers, D., Swynghedauw, B., Aubert, A.E.: Effects of autonomic blockade on non-linear cardiovascular variability indices in rats. Clin. Exp. Pharmacol. Physiol. **33**(5–6), 431–439 (2006)
86. Voss, A., Schroeder, R.F., Truebner S FAU - Goernig, M., Goernig, M.F., Figulla HR FAU - Schirdewan, A., Schirdewan, A.: Comparison of nonlinear methods symbolic dynamics, detrended fluctuation, and Poincare plot analysis in risk stratification in patients with dilated cardiomyopathy. Chaos. **17**, 015120. 1054-1500 (Print) (2007)
87. Perkiomaki, J.S.: Heart rate variability and non-linear dynamics in risk stratification. Front. Physiol. **2**, 81 (2011)
88. Sassi, R., Cerutti, S., Lombardi, F., Malik, M., Huikuri, H.V., Peng, C.K., Schmidt, G., Yamamoto, Y.: Advances in heart rate variability signal analysis: joint position statement by the e-Cardiology ESC Working Group and the European Heart Rhythm Association co-endorsed by the Asia Pacific Heart Rhythm Society. Europace. **17**(9), 1341–1353 (2015)
89. Makikallio, T.H., Huikuri, H.V., Makikallio, A., Sourander, L.B., Mitrani, R.D., Castellanos, A., Myerburg, R.J.: Prediction of sudden cardiac death by fractal analysis of heart rate variability in elderly subjects. J. Am. Coll. Cardiol. **37**(5), 1395–1402 (2001)
90. Makikallio, T.H., Huikuri, H.V., Hintze, U., Videbaek, J., Mitrani, R.D., Castellanos, A., Myerburg, R.J., Moller, M.: Fractal analysis and time- and frequency-domain measures of heart rate variability as predictors of mortality in patients with heart failure. Am. J. Cardiol. **87**(2), 178–182 (2001)

91. Perkiomaki, J.S., Hamekoski, S., Junttila, M.J., Jokinen, V., Tapanainen, J., Huikuri, H.V.: Predictors of long-term risk for heart failure hospitalization after acute myocardial infarction. Ann. Noninvasive Electrocardiol. **15**(3), 250–258 (2010)
92. Perkiomaki, J.S., Jokinen, V., Tapanainen, J., Airaksinen, K.E., Huikuri, H.V.: Autonomic markers as predictors of nonfatal acute coronary events after myocardial infarction. Ann. Noninvasive Electrocardiol. **13**(2), 120–129 (2008)
93. Makikallio, T.H., Seppanen, T., Airaksinen, K.E., Koistinen, J., Tulppo, M.P., Peng, C.K., Goldberger, A.L., Huikuri, H.V.: Dynamic analysis of heart rate may predict subsequent ventricular tachycardia after myocardial infarction. Am. J. Cardiol. **80**(6), 779–783 (1997)
94. Ivanov, P.C., Bunde, A., Amaral, L.A., Havlin, S., Fritsch-Yelle, J., Baevsky, R.M., Stanley, H.E., Goldberger, A.L.: Sleep-wake differences in scaling behavior of the human heartbeat: analysis of terrestrial and long-term space flight data. Europhys. Lett. **48**(5), 594–600 (1999)
95. Echeverria, J.C., Woolfson, M.S., Crowe, J.A., Hayes-Gill, B.R., Croaker, G.D., Vyas, H.: Interpretation of heart rate variability via detrended fluctuation analysis and alphabeta filter. Chaos. **13**(2), 467–475 (2003)
96. Bojorges-Valdez, E.R., Echeverria, J.C., Valdes-Cristerna, R., Pena, M.A.: Scaling patterns of heart rate variability data. Physiol. Meas. **28**(6), 721–730 (2007)
97. Castiglioni, P., Parati, G., Civijian, A., Quintin, L., Di Rienzo, M.: Local scale exponents of blood pressure and heart rate variability by detrended fluctuation analysis: effects of posture, exercise, and aging. IEEE Trans. Biomed. Eng. **56**(3), 675–684 (2009)
98. Xia, J., Shang, P., Wang, J.: Estimation of local scale exponents for heartbeat time series based on DFA. Nonlinear Dynamics. **74**(4), 1183–1190 (2013)
99. Castiglioni, P., Parati, G., Lombardi, C., Quintin, L., Di Rienzo, M.: Assessing the fractal structure of heart rate by the temporal spectrum of scale exponents: a new approach for detrended fluctuation analysis of heart rate variability. Biomed. Tech. **56**(4), 175–183 (2011)
100. Liao, F., Liau, B.Y., Rice, I.M., Elliott, J., Brooks, I., Jan, Y.K.: Using local scale exponent to characterize heart rate variability in response to postural changes in people with spinal cord injury. Front. Physiol. **6**, 142 (2015)
101. Castiglioni, P., Brambilla, V., Brambilla, L., Gualerzi, M., Lazzeroni, D., Coruzzi, P.: The fractal structure of cardiovascular beat-to-beat series described over a broad range of scales: Differences between blood pressure and heart rate, and the effect of gender. Conf. Proc. IEEE Eng Med .Biol. Soc. **2005**, 290–293 (2015)
102. Castiglioni, P., Di Rienzo, M., Radaelli, A.: Effects of autonomic ganglion blockade on fractal and spectral components of blood pressure and heart rate variability in free-moving rats. Auton. Neurosci. **178**(1-2), 44–49 (2013)
103. Castiglioni, P.: Self-similarity in physiological time series: new perspectives from the temporal spectrum of scale exponents. In: Biganzoli, E., Vellido, A., Ambrogi, F., Tagliaferri, R. (eds.) Computational Intelligence Methods for Bioinformatics and Biostatistics: 8th International Meeting, CIBB 2011, Gargnano del Garda, Italy, June 30–July 2 2011, Revised Selected Papers. Berlin, Heidelberg: Springer Berlin Heidelberg, pp. 164–175 (2012)
104. Berger, R.D., Saul, J.P., Cohen, R.J.: Transfer function analysis of autonomic regulation. I. Canine atrial rate response. Am. J. Phys. **256**(1 Pt 2), H142–H152 (1989)
105. Malpas, S.C.: Neural influences on cardiovascular variability: possibilities and pitfalls. Am. J. Physiol. Heart Circ. Physiol. **282**(1), H6–20 (2002)
106. Levy, M.N., Yang, T., Wallick, D.W.: Assessment of beat-by-beat control of heart rate by the autonomic nervous system: molecular biology technique are necessary, but not sufficient. J. Cardiovasc. Electrophysiol. **4**(2), 183–193 (1993)
107. Castiglioni, P., Merati, G.: Fractal analysis of heart rate variability reveals alterations of the integrative autonomic control of circulation in paraplegic individuals. Physiol. Meas. **38**(5), 774–786 (2017)
108. Kantelhardt, J.W., Zschiegner, S.A., Koscielny-Bunde, E., Havlin, S., Bunde, A., Stanley, H.E.: Multifractal detrended fluctuation analysis of nonstationary time series. Physica A. **316**(1–4), 87–114 (2002)

109. Ivanov, P.C., Amaral, L.A., Goldberger, A.L., Havlin, S., Rosenblum, M.G., Struzik, Z.R., Stanley, H.E.: Multifractality in human heartbeat dynamics. Nature. **399**(6735), 461–465 (1999)
110. Makowiec, D., Galaska, R., Dudkowska, A., Rynkiewicz, A., Zwierz, M.: Long-range dependencies in heart rate signals-revisited. Physica A. **369**(2), 632–644 (2006)
111. Sassi, R., Signorini, M.G., Cerutti, S.: Multifractality and heart rate variability. Chaos. **19**(2), 028507 (2009)
112. Ihlen, E.A.F.: Introduction to multifractal wavelet and detrended fluctuation analyses. Front. physiol. **3**, 141 (2012)
113. Makowiec, D., Rynkiewicz, A., Wdowczyk-Szulc, J., Zarczynska-Buchowiecka, M., Galaska, R., Kryszewski, S.: Aging in autonomic control by multifractal studies of cardiac interbeat intervals in the VLF band. Physiol. Meas. **32**(10), 1681–1699 (2011)
114. Eke, A., Herman, P., Sanganahalli, B.G., Hyder, F., Mukli, P., Nagy, Z.: Pitfalls in fractal time series analysis: fMRI BOLD as an exemplary case. Front. Physiol. **3**, 417 (2012)
115. Makowiec, D., Dudkowska, A., Galaska, R., Rynkiewicz, A.: Multifractal estimates of monofractality in RR-heart series in power spectrum ranges. Physica A. **388**(17), 3486–3502 (2009)
116. Gieraltowski, J., Zebrowski, J.J., Baranowski, R.: Multiscale multifractal analysis of heart rate variability recordings with a large number of occurrences of arrhythmia. Phys. Rev. E Stat. Nonlin. Soft. Matter Phys. **85**(2 Pt 1), 021915 (2012)
117. Gieraltowski, J., Hoyer, D., Tetschke, F., Nowack, S., Schneider, U., Zebrowski, J.: Development of multiscale complexity and multifractality of fetal heart rate variability. Auton. Neurosci. **178**(1-2), 29–36 (2013)
118. Gieraltowski, J., Hoyer, D., Schneider, U., Zebrowski, J.J.: Formation of functional associations across time scales in the fetal autonomic control system–a multifractal analysis. Auton. Neurosci. **190**, 33–39 (2015)
119. Zebrowski, J.J., Kowalik, I., Orlowska-Baranowska, E., Andrzejewska, M., Baranowski, R., Gieraltowski, J.: On the risk of aortic valve replacement surgery assessed by heart rate variability parameters. Physiol. Meas. **36**(1), 163–175 (2015)
120. Solinski, M., Gieraltowski, J., Zebrowski, J.: Modeling heart rate variability including the effect of sleep stages. Chaos. **26**(2), 023101 (2016)

Chapter 8
Time-Varying Cardiovascular Complexity with Focus on Entropy and Lyapunov Exponents

Gaetano Valenza, Luca Citi, Enzo Pasquale Scilingo, and Riccardo Barbieri

Abstract Measures of nonlinearity and complexity, such as entropy measures and Lyapunov exponents, have been increasingly employed to characterize dynamical properties in a wide range of biological nonlinear systems, including cardiovascular control. In this chapter, we present recent methodological advances allowing to effectively estimate instantaneous approximate and sample entropy measures, as well as the instantaneous Lyapunov spectrum of a series of stochastic events, i.e., the heartbeats. Because the proposed measures are instantaneously defined by means of probability functions defined within a point-process framework, these indices are able to provide instantaneous tracking of the degree of complexity associated with the physiological system in question. Long-term information is taken into account by expanding the linear, quadratic, and cubic Wiener-Volterra kernels with the orthonormal Laguerre basis functions. Applications on experimental heartbeat interval datasets (i.e., healthy subjects undergoing postural changes and patients with severe cardiac heart failure) are reported.

8.1 Introduction

Heart contractions are regarded by many scientists as the foremost example of a physiological system showing predominantly nonlinear behavior, mainly generated through integration of multiple neural signaling at the level of the sinoatrial node [1]. Accordingly, the fluctuations in the interval between consecutive heartbeats, mediated by the autonomic nervous system (ANS) [2, 3], have been widely investigated as output of a nonlinear system revealing and quantifying the complexity

G. Valenza (✉) • E.P. Scilingo
Department of Information Engineering & Bioengineering and Robotics Research Center "E. Piaggio", University of Pisa, School of Engineering, Pisa, Italy
e-mail: g.valenza@ing.unipi.it

L. Citi
School of Computer Science and Electronic Engineering, University of Essex, Colchester, UK

R. Barbieri
Department of Electronics, Informatics and Bioengineering, Politecnico di Milano, Milan, Italy

R. Barbieri et al. (eds.), *Complexity and Nonlinearity in Cardiovascular Signals*,
DOI 10.1007/978-3-319-58709-7_8

of cardiovascular control [4–29]. As a result of the evidence brought up by this important body of work, it has been widely accepted that physiological systems are indeed "complex", and that quantification of complexity is able to provide relevant information on psychophysiological and pathological states [30] being modulated by external stimuli, aging, or disease [30–34].

Among all nonlinearity and complexity measures, measures of entropy are primarily devoted to address the randomness and regularity of a dynamical system given the analysis of time series originated by the observed system [35–37]. As a primary definition, the entropy $H(X)$ of a mono-dimensional discrete random variable X is $H(X) = -\sum_{x_i \in \phi} p(x_i) \log p(x_i)$, where ϕ is the set of values and $p(x_i)$ is the i-th probability function. Other important definitions include the Kolmogorov-Sinai entropy [38], the K_2 entropy [39], and the marginal redundancy algorithm given by Fraser [40]. To compute these theoretical entropy indices, a large number of data points are needed in order to achieve convergence. This is true even in the presence of low-dimensional space-state systems [41]. To this extent, Pincus proposed a family of formulas and statistics referred to as *approximate entropy* (A_E) [41], which is able to discern dynamical systems given finite, noisy data. A further modification has been recently proposed to overcome the dependency of A_E on the time series length, referred to as *sample entropy* (S_E) [42]. A measure of multiscale entropy has also been proposed [43] in order to take into account the inherent multiple time scale properties of dynamical systems.

Besides entropy measures, Lyapunov exponents (LEs) have been proven to provide an important mathematical tool in characterizing dynamical properties of a physiological nonlinear system [44–46]. Specifically, they refer to the average exponential rates of divergence or convergence of neighboring trajectories in the system phase space. Several methods for a reliable data-driven LEs estimation, even in short time data records, have been also proposed [47, 48]. In a deterministic nonlinear system with no stochastic inputs, a positive LE reflects sensitive dependence to initial conditions and can be interpreted as a definition of a chaotic system [49]. Although it is straightforward to consider chaotic mathematical systems where stochastic inputs are suppressed, in actual applications (particularly those related to physiological systems) it is not possible to eliminate such inputs, thus making the chaos assessment simply unreliable [5]. Stationary aperiodic behavior, in fact, can also arise in linear or nonlinear stochastic systems. Relying on the approach suggested by Chon et al. [44] and, later, by Armoundas et al. [45], we consider the cardiovascular system as both chaotic and stochastic. This concept is in agreement with current physiological knowledge, since healthy HRV dynamics can be considered/modeled as the output of a nonlinear deterministic system (the pacemaker cells of sinus node) being forced by a high-dimensional input (the activity in the nerves innervating the sinus node).

Despite the considerable achievements obtained by studying changes in complexity through entropy measures and LEs, three major methodological and applicative issues have not been satisfactorily addressed.

1. *Unevenly Sampled Observations:* The intrinsic discrete nature of experimental observations can lead to estimation errors, especially in studying heartbeat dynamics. R-R intervals, in fact, consist of unevenly spaced samples, thus often requiring the application of preliminary interpolation procedures that could affect complexity measures, whereas considering series as inter-events does not account for their time occurrences and may miss intrinsic generative properties as reflected in complex dynamics.
2. *Estimation Window:* Even in the case of reliable quantification of entropy and, more in general, of complexity, traditional algorithms provide a single value (or a set of values) within a predetermined time window. Therefore, given the experimental time series, these values represent averaged measures of the entire dynamics observed in that specific time window. However, a single estimation could not be sufficient to completely characterize system complexity in the face of non-stationary behavior. It is well known, in fact, that dynamical systems (particularly those associated to physiological processes) evolve and change at each moment in time.
3. *Noise Properties:* It has been shown that measures of complexity are usually higher in the presence of uncorrelated (e.g. white noise) rather than correlated (e.g. $1/f$) underlying dynamics. This issue may lead to overestimation of complexity in systems dynamics associated with uncorrelated noise. Exemplary cases have been reported in the presence of certain pathologies such as cardiac arrhythmias and atrial fibrillation [43].

In this chapter we highlight recent definitions of *instantaneous approximate and sample entropy*, A_I and S_I, respectively, as well as *instantaneous Dominant Lyapunov exponents*, *IDLE*, overcoming these limitations. As suggested by the names, these instantaneous measures are inspired by the already established respective algorithms, and they have the additional property to estimate time-varying complexity assessments of discrete physiological system complexity.

The originality of the new definitions relies in the fact that they are fully embedded in the probabilistic framework of the inhomogeneous point-process theory, and introduce important differences to the mathematical formulation of the phase-space vectors and to the definition of the distance between phase-space vectors. It has been demonstrated that, by means of a point process approach, it is possible to characterize the events' probabilistic generative mechanism and to obtain continuous estimates, even considering short recordings under nonstationary conditions. Recently [21, 50], we demonstrated the application of these concepts in estimating instantaneous linear and nonlinear heartbeat dynamics. The unevenly spaced heartbeat intervals are represented as observations of a state-space point process model defined at each moment in time, thus allowing to estimate instantaneous HR and HRV measures [50] without using any interpolation method. As related to complexity, we particularly focused our efforts on defining an ad-hoc framework to account for long-term memory and high-order nonlinearities using a reduced set of model parameters [21, 51].

8.2 Materials and Methods

The instantaneous complexity estimation is performed through a parametrized nonlinear combination of the time series using the discrete Wiener-Volterra series. Such a nonlinear model is embedded into the point-process framework in order to allow for instantaneous estimates of the A_I and S_I, as well as IDLE measures.

Mathematical and algorithmic details follow below.

8.2.1 Nonlinear System Identification and Nonlinear Autoregressive Models

A Nonlinear Autoregressive Model (NAR) Model can be expressed, in a general form, as follows:

$$y(k) = \mathbf{F}(y(k-1), y(k-2), \ldots, y(k-M)) + \epsilon(k). \tag{8.1}$$

Considering $\epsilon(k)$ as independent, identically distributed Gaussian random variables, the model can be can be written as a Taylor expansion:

$$\begin{aligned} y(k) = \gamma_0 \, + \, & \sum_{i=1}^{M} \gamma_1(i)\, y(k-i) \\ + & \sum_{n=2}^{\infty} \sum_{i_1=1}^{M} \cdots \sum_{i_n=1}^{M} \gamma_n(i_1, \ldots, i_n) \prod_{j=1}^{n} y(k-i_j) + \epsilon(k). \end{aligned} \tag{8.2}$$

The autoregressive structure of (8.2) allows for system identification with only exact knowledge on output data and with only a few assumptions on input data. We represent here the nonlinear physiological system by using nonlinear kernels up to the second order, i.e. γ_0, $\gamma_1(i)$, and $\gamma_2(i,j)$, and take into account the series of the derivatives in order to improve stationarity [52, 53]. Hence, the general quadratic form of a Nonlinear Autoregressive Integrative (NARI) model becomes:

$$\begin{aligned} y(k) = y(k-1) + \gamma_0 \, + \, & \sum_{i=1}^{M} \gamma_1(i)\, \Delta y(k-i) \\ + & \sum_{i=1}^{M} \sum_{j=1}^{M} \gamma_2(i,j)\, \Delta y(k-i)\, \Delta y(k-j) + \epsilon(k) \end{aligned} \tag{8.3}$$

where $\Delta y(k-i) = y(k-i) - y(k-i-1)$ and $\Delta y(k-j) = y(k-j) - y(k-j-1)$. The quadratic kernel $\gamma_2(i,j)$ is assumed to be symmetric. We also define the extended kernels $\gamma_1'(i)$ and $\gamma_2'(i,j)$ as:

$$\gamma_1'(i) = \begin{cases} 1, & \text{if } i = 0 \\ -\gamma_1(i) & \text{if } 1 \leq i \leq M \end{cases} \tag{8.4}$$

$$\gamma_2'(i,j) = \begin{cases} 0, & \text{if } ij = 0 \wedge i+j \leq M \\ -\gamma_2(i,j) & \text{if } 1 \leq i \leq M \wedge 1 \leq j \leq M \end{cases} . \tag{8.5}$$

and link the NARI model to a general input-output form, here defined by using the well-known Wiener-Volterra [54] series:

$$y(k) = h_0 + \sum_{i=1}^{M} h_1(i)\, \Delta\epsilon(k-i)$$
$$+ \sum_{n=2}^{\infty} \sum_{i_1=1}^{M} \cdots \sum_{i_n=1}^{M} h_n(i_1, \ldots, i_n) \prod_{j=1}^{n} \Delta\epsilon(k-i_j) . \tag{8.6}$$

where the functions $h_n(\tau_1, \ldots, \tau_n)$ are the Volterra kernels. In the next paragraph, the definition of the generic variable $y(k)$ is used to represent the first-moment statistic (mean) of the probabilistic generative mechanism of the heart rate. In other words, the generic formulation of the NARI model parametrizes the autonomic control on the cardiovascular system using linear and nonlinear terms according to the Wiener-Volterra representation and point-process theory.

8.2.2 *Point-Process Nonlinear Model of the Heartbeat*

The point process framework primarily defines the probability of having a heartbeat event at each moment in time. This probability accounts for the instantaneous estimation of features reflecting short-time emotional variations. Formally, defining $t \in (0, T]$, the observation interval, and $0 \leq u_1 < \cdots < u_k < u_{k+1} < \cdots < u_K \leq T$ the times of the events, we can define $N(t) = \max\{k : u_k \leq t\}$ as the sample path of the associated counting process. Its differential, $dN(t)$, denotes a continuous-time indicator function, where $dN(t) = 1$ when there is an event (the ventricular contraction), or $dN(t) = 0$ otherwise. The left continuous sample path is defined as $\widetilde{N}(t) = \lim_{\tau \to t^-} N(\tau) = \max\{k : u_k < t\}$. Given the R-wave events $\{u_j\}_{j=1}^{J}$ detected from the ECG, $RR_j = u_j - u_{j-1} > 0$ denotes the jth RR interval. Assuming history dependence, the inverse Gaussian probability distribution of the waiting time $t - u_j$ until the next R-wave event is [50]:

$$f(t|H_t, \xi(t)) = \left[\frac{\xi_0(t)}{2\pi(t-u_j)^3}\right]^{\frac{1}{2}} \times \exp\left\{-\frac{1}{2}\frac{\xi_0(t)[t-u_j-\mu_{\mathrm{RR}}(t,H_t,\xi(t))]^2}{\mu_{\mathrm{RR}}(t,H_t,\xi(t))^2(t-u_j)}\right\} \quad (8.7)$$

with $j = \widetilde{N}(t)$ the index of the previous R-wave event before time t, $H_t = (u_j, \mathrm{RR}_j, \mathrm{RR}_{j-1}, ..., \mathrm{RR}_{j-M+1})$ is the history of events, $\xi(t)$ the vector of the time-varying parameters, $\mu_{\mathrm{RR}}(t, H_t, \xi(t))$ the first-moment statistic (mean) of the distribution, and $\xi_0(t) > 0$ the shape parameter of the inverse Gaussian distribution. Since $f(t|H_t, \xi(t))$ indicates the probability of having a beat at time t given that a previous beat has occurred at u_j, $\mu_{\mathrm{RR}}(t, H_t, \xi(t))$ can be interpreted as the expected waiting time until the next event could occur. The use of an inverse Gaussian distribution $f(t|H_t, \xi(t))$, characterized at each moment in time, is motivated both physiologically (the integrate-and-fire mechanism initiating the cardiac contraction [50]) and by goodness-of-fit comparisons [55]. In previous work [55, 56], the instantaneous mean $\mu_{\mathrm{RR}}(t, H_t, \xi(t))$ was expressed as a linear and quadratic combination of present and past R-R intervals, based on a nonlinear Volterra-Wiener expansion [57]. Here we report on the NARI formulation in which the instantaneous RR mean is defined as:

$$\mu_{\mathrm{RR}}(t, H_t, \xi(t)) = \mathrm{RR}_{\widetilde{N}(t)} + \gamma_0(t) + \sum_{i=1}^{p} \gamma_1(i,t)\,(\mathrm{RR}_{\widetilde{N}(t)-i} - \mathrm{RR}_{\widetilde{N}(t)-i-1}) + \sum_{n=2}^{\infty}\sum_{i_1=1}^{M}\cdot\cdot\sum_{i_n=1}^{M} \gamma_n(i_1, ..., i_n, t) \prod_{j=1}^{n}(\mathrm{RR}_{\widetilde{N}(t)-i_j} - \mathrm{RR}_{\widetilde{N}(t)-i_j-1}) \quad (8.8)$$

In this chapter, we consider nonlinearities associated to Eq. (8.8) up to the third-order. Cubic terms, in fact, allow for the estimation of the dominant Lyapunov exponent, whereas quadratic terms account for entropy estimation. Considering the derivative RR interval series improves the achievement of stationarity within the sliding time window W (usually chosen $W = 70\,\mathrm{s}$) [52]. Since $\mu_{\mathrm{RR}}(t, H_t, \xi(t))$ is defined in continuous time, we can obtain an instantaneous RR mean estimate at a very fine timescale (with an arbitrarily small bin size Δ), which requires no interpolation between the arrival times of two beats. Given the proposed parametric model, all linear and nonlinear indices are defined as a time-varying function of the parameters $\xi(t) = [\xi_0(t), \gamma_0(t), \gamma_n(i_1, \ldots, i_n, t)]$ with $n = \{1, 2, 3\}$. The unknown time-varying parameter vector $\xi(t)$ is estimated by means of a local maximum likelihood method [50, 58, 59]. The model goodness-of-fit is based on the Kolmogorov-Smirnov (KS) test and associated KS statistics (see details in [50, 60]). Autocorrelation plots are considered to test the independence of the model-transformed intervals [50]. Optimal order of linear and nonlinear regressions is determined by prefitting the point process model to a subset of the data [50, 58] and comparing scores from KS tests, and autocorrelation plots. Once the order of linear and nonlinear regressions is determined, the initial NARI coefficients are estimated by the method of least squares [61].

8.2.3 Definition of the Inhomogeneous Point-Process Entropy Measures

The A_{I} algorithm has its foundation in correlation dimension analysis [62] and in the A_{E} computation [41]. Given a distance measure $d[.]$, let us define $C_k^m(r(t), t)$ as the number of points $x(j)$ such that

$$d[x(k), x(j)] \leq r(t)/(N - m + 1), \forall j \tag{8.9}$$

where $x(k)|_{k=1,2,\ldots,N-m+1}$ are vectors of the phase space defined as $x(k) = [\mu_{\mathrm{RR}}(t_k), \mu_{\mathrm{RR}}(t_{k+1}), \ldots, \mu_{\mathrm{RR}}(t_{k+m-1})]$ in $\mathbb{R}^m$ of the time series $\mu_{\mathrm{RR}}(t_1), \mu_{\mathrm{RR}}(t_2), \ldots, \mu_{\mathrm{RR}}(t_N)$. In this formulation, m and $r(t)$ are the embedding dimension and time delay of the phase space, respectively. The time-varying quantity $r(t)$ is instantaneously expressed as $r(t) = 0.2\sigma_{\mu_{\mathrm{e}(t)}}$, as suggested by the current literature [43]. According to point-process theory, it is possible to define the distance $d[x(k), x(j)]$ as the Kolmogorov-Smirnov (KS) distance (i.e. the maximum value of the absolute difference between two cumulative distribution functions) between the IG_k and IG_j probability distributions of $\mu_{\mathrm{e}}(t_{k+k_n})$ and $\mu_{\mathrm{e}}(t_{j+k_n})$ for $k_n = 0, 1, \ldots, m-1$. Then, from the $C_k^m(r, t)$ it is possible to define:

$$\Phi^m(r, t) = (N - m + 1)^{-1} \sum_{i=1}^{N-m+1} \ln C_k^m(r, t) \tag{8.10}$$

and obtain:

$$A_{\mathrm{I}}(m, r, N, t) = \Phi^m(r, t) - \Phi^{m+1}(r, t). \tag{8.11}$$

As the definition of the proposed entropy measure is fully embedded into the inhomogeneous point-process nonlinear framework (IGs are defined at each moment in time), it is possible to obtain instantaneous tracking of the system complexity as $A_{\mathrm{I}}(m, r, N, t)$, thus overcoming the need of defining a single estimation window (*issue (ii)*).

To compute the $S_{\mathrm{I}}(m, r, N, t)$ algorithm, $C_k^m(r(t), t)$ is defined as the number of j such that

$$d[x(k), x(j)] \leq r(t)/(N - m), \forall j \neq k. \tag{8.12}$$

Then, from $C_k^m(r,t)$ it is possible to define

$$\Phi^m(r,t) = (N-m+1)^{-1} \sum_{i=1}^{N-m+1} C_k^m(r,t) \tag{8.13}$$

and obtain

$$S_I(m,r,N,t) = \ln \Phi^m(r,t) - \ln \Phi^{m+1}(r,t). \tag{8.14}$$

Of note, original definition of A_I and S_I employed model optimization through Laguerre expansion of the Wiener-Volterra terms [27].

8.2.4 Definition of Instantaneous Dominant Lyapunov Exponent

Cubic nonlinearities, in the framework of the proposed NARI point-process model, account for the novel estimation of the instantaneous Lyapunov spectrum. By definition, the generic Lyapunov exponent (LE) of a real valued function $f(t)$ defined for $t > 0$ is defined as:

$$\lambda = \limsup_{t\to\infty} \frac{1}{t} \log(|f(t)|) \tag{8.15}$$

Let us consider a generic n-dimensional linear system in the form $y_i = Y(t)\,p_i$, where $Y(t)$ is a time-varying fundamental solution matrix with $Y(0)$ orthogonal, and $\{p_i\}$ is an orthonormal basis of $\mathbb{R}^n$. The key theoretical tools for determining the IDLE and the whole spectrum of LEs is the continuous QR factorization of $Y(t)$ [33, 51]:

$$Y(t) = Q(t)R(t) \tag{8.16}$$

where $Q(t)$ is orthogonal and $R(t)$ is upper triangular with positive diagonal elements R_{ii}, $1 \leq i \leq n$. Then, LEs are formulated as:

$$\begin{aligned}\lambda_i &= \lim_{t\to\infty} \frac{1}{t} \log \|Y(t)p_i\| \\ &= \lim_{t\to\infty} \frac{1}{t} \log \|R(t)p_i\| = \lim_{t\to\infty} \frac{1}{t} \log \|R_{ii}(t)\| \,.\end{aligned} \tag{8.17}$$

The cubic NAR model [Eq. (8.8)] can be rewritten in an M-dimensional state space canonical representation:

$$r_n^{(k)} = \begin{cases} r_{n-1}^{(k+1)} & \text{if} k < M \\ F(r_{n-1}^{(M)}, r_{n-1}^{(M-1)}, \cdots, r_{n-1}^{(2)}, r_{n-1}^{(1)}) & \text{if} k = M \end{cases} \tag{8.18}$$

By evaluating the Jacobian $J(n)$ over the time series, which corresponds to the matrix $Y(t)$, the LE can be determined using the QR decomposition:

$$J(n)Q_{(n-1)} = Q_{(n)}R_{(n)} \tag{8.19}$$

This decomposition is unique except in the case of zero diagonal elements. Then the LEs λ_i are given by

$$\lambda_i = \frac{1}{\tau H}\sum_{j=0}^{H-1} \ln R_{(j)ii} \tag{8.20}$$

where H is the available number of matrices within the local likelihood window of duration W, and τ the sampling time step. The estimation of the LEs is performed at each time t from the corresponding time-varying vector of parameters, $\xi(t)$ [51]. Here, the first LE, $\lambda_1(t)$ is considered as the instantaneous dominant Lyapunov exponent (IDLE).

8.2.5 *The* Complexity Variability *Framework*

Once instantaneous complexity series are available, basic time-domain features can be used to summarize the cardiovascular complex dynamics. In particular, a measure of central tendency, e.g., the median value, and a measure of variability, e.g., the median absolute deviation can be calculated. The former (central tendency) can be considered equivalent to standard complexity estimates which collapse data across time by design. The latter (median absolute deviation) represents an innovation in the field of complexity analysis [27, 46], by defining a measure of *complexity variability*.

8.2.6 *Definition of Standard Nonlinear Measures of Heartbeat Dynamics*

In order to perform a comparison analysis with standard and nonlinear estimates of heartbeat dynamics, let us define the standard mean of the RR intervals (Mean RR), the root mean square of successive differences of intervals (RMSSD) and the number of successive differences of intervals which differ by more than 50 ms (pNN50 % expressed as a percentage of the total number of heartbeats analyzed) [63]. Referring to morphological patterns of HRV, the triangular index is obtained as the integral of the histogram (i.e. total number of RR intervals) divided by the height of the histogram which depends on the selected bin width [63]. Moreover, we performed the estimation of the dominant Lyapunov exponent according to the algorithm described by Wolf et al. [47] (L_{D_1}) and Rosenstein et al. [64] (L_{D_2}). Both algorithms are suitably applied to experimental noisy data. Finally, other nonlinear measures such as the approximate entropy (ApEn) [41], Sample Entropy (SampEn) [42], and the Detrended Fluctuation Analysis (DFA) [65] were evaluated.

8.3 Experimental Data and Results

In this section, results on two paradigmatic experimental heartbeat datasets, i.e., healthy postural changes and cardiac heart failure, are reported. Given a generic index variable X that can be associated to a specific measure, all results in this chapter are referred to inter-subject analyses and expressed as $\text{Median}(X) \pm \text{Median}(|X - \text{Median}(X)|)$.

8.3.1 *Postural Changes*

We performed the instantaneous analysis in RR-interval time series recorded from 10 healthy subjects undergoing a tilt-table protocol: each subject, initially lying horizontally in a supine position, is then passively tilted to the vertical position. The study, fully described in [27, 46, 50, 66], was conducted at the Massachusetts Institute of Technology (MIT) General Clinical Research Center (GCRC) and was approved by the MIT Institutional Review Board and the GCRC Scientific Advisory Committee.

8.3.1.1 Instantaneous Entropy Measures

The first order moment $\mu_{RR}(t)$ and A_I instantaneous dynamics are shown for one representative subject in Fig. 8.1, whereas the averaged A_I for all ten subjects is shown in Fig. 8.2, providing a clear portrayal of how the postural stimulus elicits the expected changes in the dynamic signatures of complexity. On average (see Table 8.1), a significant statistical difference was found between median A_I and S_I values of resting and tilt phases ($p < 10^{-6}$ and $p < 0.01$, respectively, given by non-parametric Wilcoxon test for paired data with null hypothesis of equal

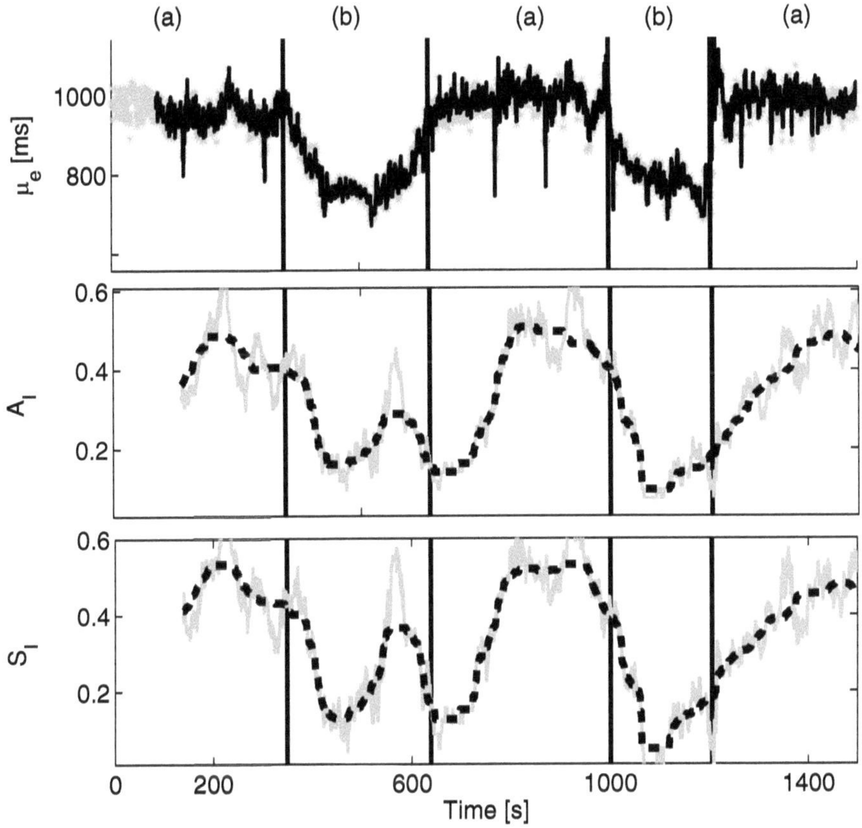

Fig. 8.1 Instantaneous heartbeat statistics computed from a representative subject of the Tilt-Table protocol using a NARL point-process model. In the first panel, the estimated $\mu_{RR}(t)$ (*top, continuous line*) is superimposed on the recorded R-R series (*gray asterisks*). Below, the instantaneous A_I and S_I complexity tracking are shown in *gray continuous lines* along with their low-pass (2nd-order Butterworth FIR Filter with cut-off 0.05 Hz) derived signal (*dotted line*). Resting phases (*a*) alternate with the gravitational changes (*b*)

medians). Of note, also the traditional ApEn and SampEn measures are able to discern between rest and tilted phases ($p < 10^{-3}$ given by non-parametric Wilcoxon test for paired data with null hypothesis of equal medians). However, it is important to note that these traditional measures are not able to follow changes in complexity. Remarkably, the averaged instantaneous measures provided by A_I resulted in a more discriminant statistical significance (A_I: $p < 10^{-6}$; *ApEn*: $p < 10^{-3}$). These results are in agreement with the current literature [67, 68], providing additional evidence for the observed progressive decrease in complexity as a function of tilt table inclination, thus indicating that the degree of complexity is highly correlated with sympathovagal response.

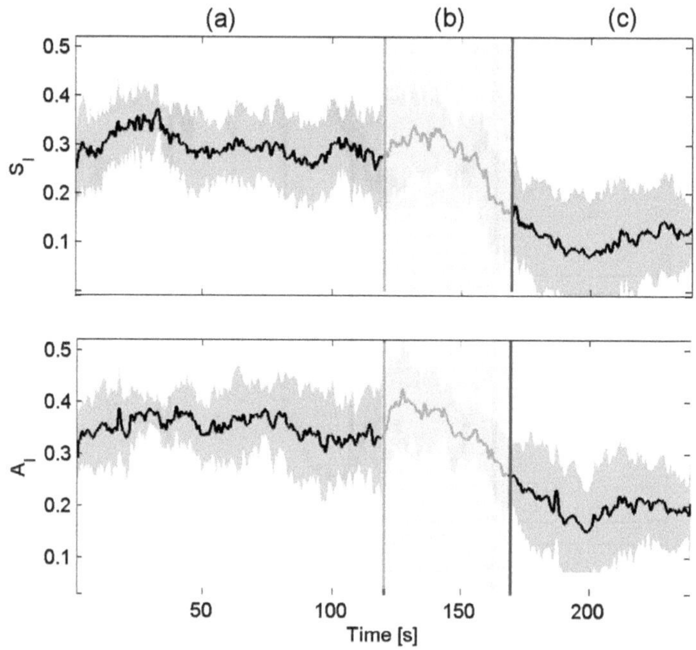

Fig. 8.2 Averaged S_I and A_I trends during resting (*a*) and tilting (*c*) phases, through slow-transitioning tilt (*b*). Considering data from all subjects, the plot shows the Median(X) $\pm$ Median($|X - \text{Median}(X)|$))

Table 8.1 Results from the experimental dataset related to postural changes

	Rest	Tilt	p-val.
ApEn	1.1671 ± 0.0912	0.9274 ± 0.1255	$p < 10^{-3}$
A_I	0.3062 ± 0.0422	0.2545 ± 0.0348	$p < 10^{-6}$
σ_{A_I}	0.0709 ± 0.0145	0.0722 ± 0.0115	n.s.
SampEn	1.4949 ± 0.1731	0.8998 ± 0.2471	$p < 10^{-3}$
S_I	0.3062 ± 0.0483	0.2514 ± 0.0543	$p < 10^{-3}$
σ_{S_I}	0.081 ± 0.0196	0.0883 ± 0.0141	n.s.

Comparison between standard and novel indices. Of note, the new instantaneous formulation allows for definition of the variance of A_I (σ_{A_I}) and S_I (σ_{S_I}) as indices of *complexity variability*
p-values from non-parametric Wilcoxon test for paired data with null hypothesis of equal medians
n.s. = not significant

8.3.1.2 Instantaneous Dominant Lyapunov Exponents

Group statistics of standard and proposed instantaneous measures are shown in Table 8.2, whose inter-subject analysis was performed using a non-parametric rank-sum test. Results on the proposed IDLE show a non-significant statistical difference between the stand-up epochs and their relative rest epochs ($p > 0.05$) and a significant difference for the slow-tilt epochs ($p < 0.05$). The highest significance

Table 8.2 Group statistics of standard and instantaneous heartbeat dynamics measures from the tilt-table experimental dataset

Feature	Rest	Stand-up	*p*-value	Rest	Slow tilt	*p*-value	Rest	Fast tilt *p*-value	
Standard and instantaneous time domain measures of HRV									
Mean RR (ms)	910.94 ± 123.08	781.92 ± 55.96	<0.003	871.86 ±74.27	772.82 ±46.10	<0.0002	860.50 ± 80.47	774.66 ±44.34	<0.0004
RMSSD	0.0279 ± 0.0123	0.0202 ± 0.0055	>0.05	0.0325 ±0.0127	0.0202 ±0.0048	<0.03	0.0296 ± 0.0127	0.0200 ±0.0042	<0.03
pNN50%	6.4335 ±6.4335	2.3472 ± 2.3472	>0.05	10.0744 ±9.4524	1.6860 ±1.5708	<0.05	7.6677 ± 7.2816	2.0339 ±2.0339	<0.05
HRV triangular index	4.1644 ± 0.6832	3.9065 ± 0.4754	>0.05	3.3012 ±0.5142	3.5347 ±0.3497	>0.05	3.6429 ± 0.9743	4.3333 ±0.6352	>0.05
$\boldsymbol{\mu}_{\mathbf{RR}}$(ms)	915.10 ± 122.16	769.77 ± 78.22	<0.02	879.60 ±74.80	773.23 ± 62.29	<0.001	890.17 ± 95.18	776.92 ± 57.40	<0.005
$\boldsymbol{\sigma}_{\mathbf{RR}^2}$(ms^2)	394.15 ± 319.26	233.09 ± 139.09	>0.05	435.36± 237.42	208.07 ± 112.84	<0.03	440.76 ± 302.71	220.71 ±108.79	>0.05
Standard and instantaneous nonlinear measures of HRV									
ApEn	1.122±0.055	0.944±0.079	$<10^{-3}$	1.167 ± 0.091	0.927 ± 0.125	$<10^{-3}$	1.087±0.116	0.964±0.072	<0.004
SampEn	1.501±0.192	1.243±0.245	<0.025	1.495±0.173	0.900±0.247	$<10^{-3}$	1.320±0.247	1.197±0.233	>0.05
DFA-α_1	0.9806±0.1039	0.9968±0.1624	>0.05	0.9892±0.0952	1.2128±0.1126	>0.05	0.9788±0.0990	1.0137±0.1425	>0.05
L_{D_1}	0.0128±0.0014	0.0125±0.0013	>0.05	0.0133±0.0017	0.0135±0.0015	>0.05	0.0137±0.0013	0.0128±0.0008	>0.05
L_{D_2}	0.0029±0.0005	0.0028±0.0005	>0.05	0.0037±0.0006	0.0024±0.0004	<0.005	0.0033±0.0006	0.0020±0.0007	<0.001
$\overline{\boldsymbol{\lambda}_{\mathbf{RR}}}$	−0.0128±0.0480	−0.0390±0.0330	>0.05	0.0205±0.0372	−0.0491±0.0236	<0.0005	−0.0022±0.0332	−0.0404 ±0.0232	<0.001
$\boldsymbol{\sigma}_{\boldsymbol{\lambda}_{\mathbf{RR}}}$	0.0585 ±0.0213	0.0652±0.0178	>0.05	0.0693±0.0098	0.0582±0.0113	>0.05	0.0617±0.0131	0.0500±0.0144	>0.05

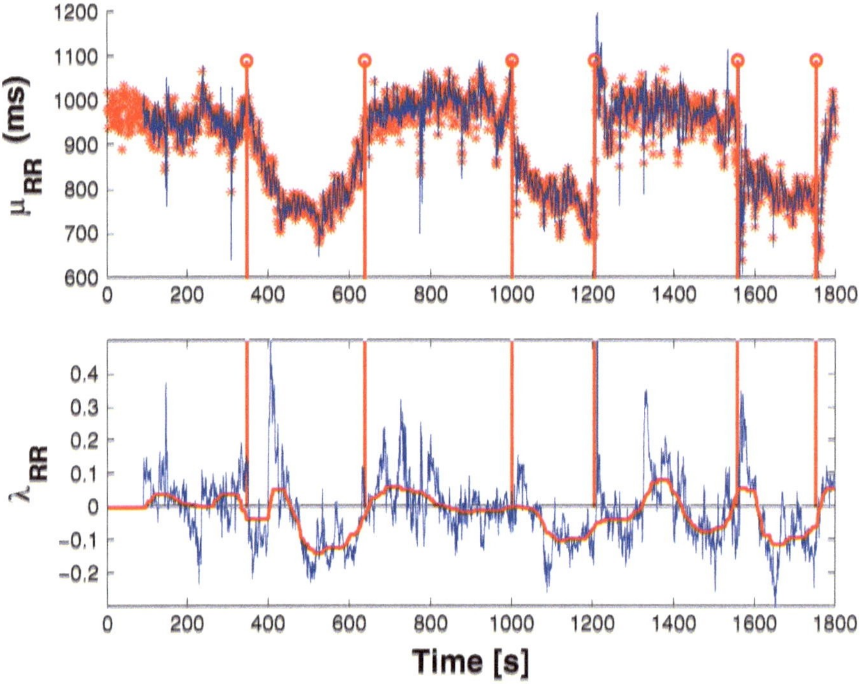

Fig. 8.3 Instantaneous heartbeat statistics computed from a representative subject (subject 1) from the tilt-table protocol. In the *top panel*, the estimated $\mu_{RR}(t)$ is superimposed on the recorded RR series. In the *bottom panel*, the instantaneous averaged IDLE is superimposed on the original IDLE

was found comparing the fast-tilt epochs with their relative rest ($p < 0.001$). These trends are confirmed by the standard DLE estimation according to the Rosenstein et al. [64] technique, whereas the one suggested by Wolf et al. [47] did not show such significant differences. IDLE dynamics for one representative subject are shown in Fig. 8.3, whereas the averaged IDLEs for all 10 subjects are shown in Fig. 8.4, providing a clear portrayal of how different postural stimuli elicit different changes in the dynamic signatures of complexity. Concerning other standard and instantaneous indices, we report significant differences on all three sessions for μ_{RR}, ApEn, and SampEn, whereas RMSSD and pNN50% showed significant differences during the slow and fast tilt sessions.

8.3.2 *Cardiac Heart Failure*

We further tested our instantaneous entropy, as well as IDLE indices to study the differences between healthy subjects and patients with severe congestive heart failure [69]. The dataset consists of RR interval time series recorded from 14 CHF

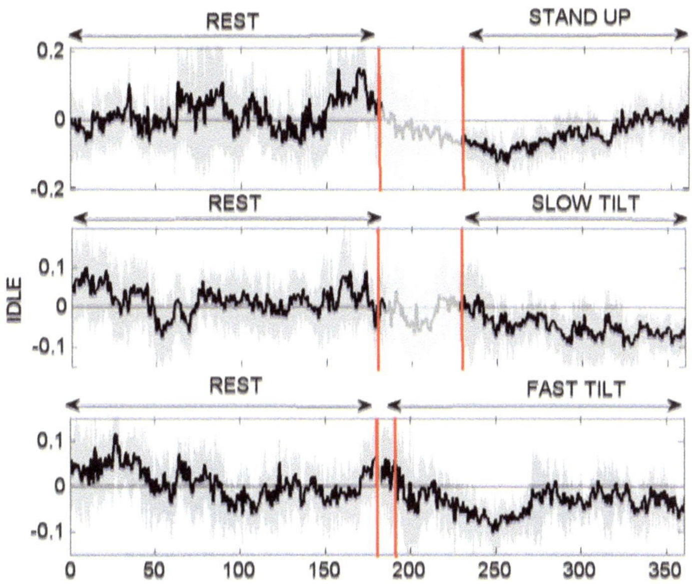

Fig. 8.4 IDLE dynamics averaged for all ten subjects. The *vertical red lines* indicate the transition from the supine to the upright position after stand up (*top panel*), after slow tilt (*middle panel*), and after fast tilt (*bottom panel*)

patients (from *BIDMC – CHF* Database) and 16 healthy subjects (from *MIT – BIH* Normal Sinus Rhythm Database). Each time series was artifact-free and lasted about 50 min. Further details can be found in [27, 46].

8.3.2.1 Instantaneous Entropy Measures

Table 8.3 shows lower values of both A_{I} and S_{I} in averaging the respective indices computed from time series gathered from CHF patients rather than healthy subjects (both $p < 10^{-3}$ given by non-parametric Mann-Whitney test with null hypothesis of equal medians). This result is in agreement with the significant difference found in multiscale entropy higher scales [43].

In this case the traditional *ApEn* and *SampEn* measures are not able to discern between healthy and pathological subjects ($p > 0.05$ given by non-parametric Mann-Whitney test with null hypothesis of equal medians), in agreement with the current literature [43] also reporting no complexity changes between healthy and CHF heartbeat dynamics (at least referring to traditional A_{E} which corresponds to multiscale entropy of scale 1). Moreover, the proposed complexity variability measures, $\sigma_{A_{\mathrm{I}}}$ and $\sigma_{S_{\mathrm{I}}}$, are able to provide discrimination power between the two mentioned populations ($p < 0.02$ and $p < 0.05$, respectively). Summary box plots of A_{I} and $\sigma_{A_{\mathrm{I}}}$ are shown in Fig. 8.5, whereas instantaneous entropy statistics computed from representative subjects are shown in Fig. 8.6.

Table 8.3 Results from the experimental healthy subjects and cardiac heart failure datasets

	Healthy	CHF	p-val.
ApEn	1.2177 ± 0.1066	1.2130 ± 0.1032	n.s.
A_{I}	0.3476 ± 0.032	0.1674 ± 0.0483	$p < 10^{-4}$
$\sigma_{A_{\mathrm{I}}}$	0.0655 ± 0.007	$0.0462 \pm 0.0148.$	$p < 0.02$
SampEn	1.4092 ± 0.1522	1.5670 ± 0.2690	n.s.
S_{I}	0.2876 ± 0.0353	0.1092 ± 0.0675	$p < 5 * 10^{-4}$
$\sigma_{S_{\mathrm{I}}}$	0.08 ± 0.0064	$0.0679 \pm 0.0082.$	$p < 0.05$

Comparison between standard *ApEn* and novel A_{I}, and standard *SampEn* and novel S_{I}
p-values from non-parametric Mann-Whitney tests with null hypothesis of equal medians
n.s. = not significant

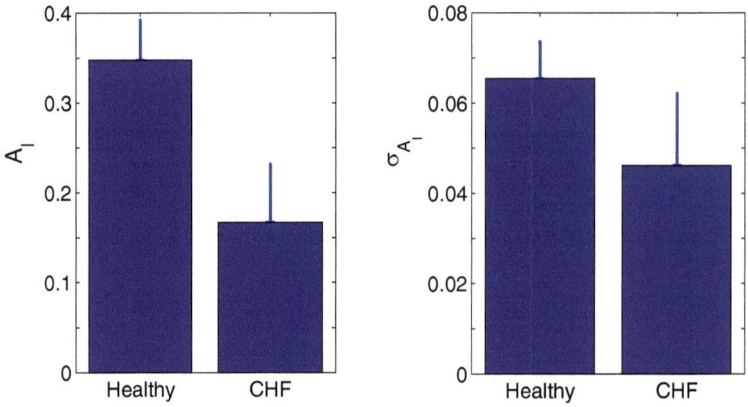

Fig. 8.5 Box plots of the averaged variable A_{I} and $\sigma_{A_{\mathrm{I}}}$ showing differences between healthy subjects and CHF patients. Each plot shows the $\mathrm{Median}(A_{\mathrm{I}}) \pm \mathrm{Median}(|A_{\mathrm{I}} - \mathrm{Median}(A_{\mathrm{I}})|))$

8.3.2.2 Instantaneous Dominant Lyapunov Exponents

For all the considered subjects, nearly all of the KS plots, as well as more than 97% of the autocorrelation samples, fell within the 95% confidence bounds. IDLE values averaged among groups are shown in Table 8.4. Overall, standard and instantaneous time domain measures were able to discern the two groups with high statistical significance ($p < 5e^{-4}$). In particular, on average, CHF patients show significantly lower μ_{RR} and σ_{RR}. On the other hand, comparing the standard and proposed instantaneous complexity measures, only the DFA-α_2 and the complexity variability $\sigma_{\lambda_{RR}}$ were able to provide significant discrimination capability between the two populations with $p < 0.05$. Concerning *IDLE* results, we observed lower complexity on time series from the pathological population with respect to the healthy subjects, although only $IDLE_{MAD}$ (i.e., the median absolute deviation over time of IDLE) shows results indicating significant higher complexity of cardiovascular dynamics in healthy patients.

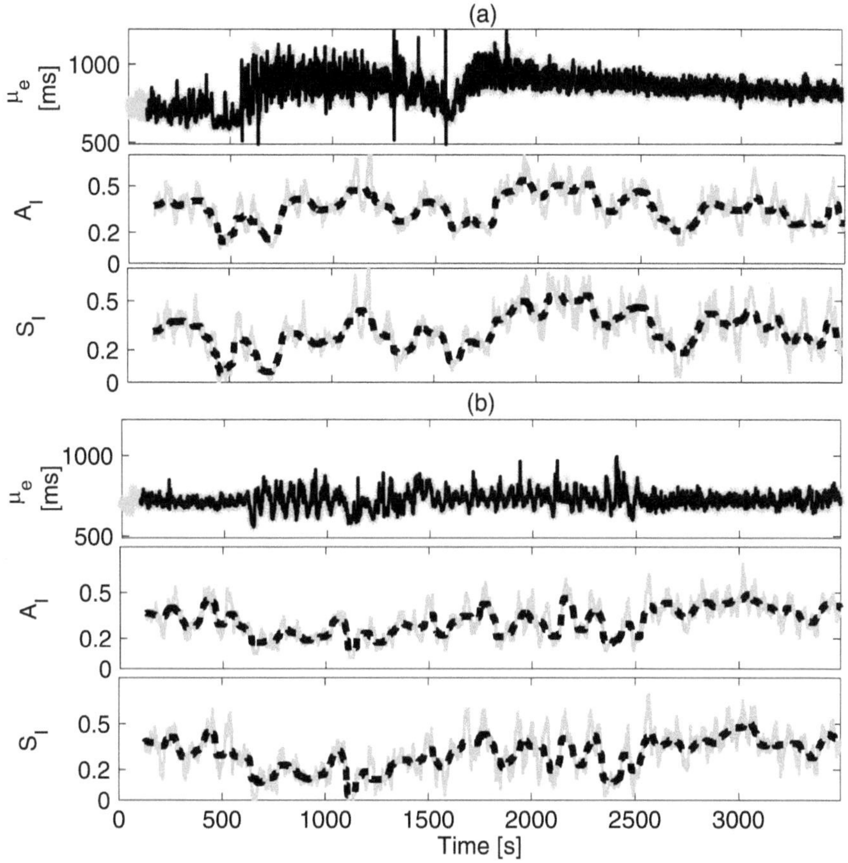

Fig. 8.6 Instantaneous statistics computed from a representative healthy subject (**a**) and a CHF patient (**b**). From the *top*, given the heartbeat interval *RR* (*gray asterisks*), the estimated $\mu_{\mathrm{RR}}(t)$ (*continuous line*) superimposed on the recorded RR series. Below, the instantaneous A_{I} and S_{I} complexity tracking are shown in *gray continuous lines* along with their low-pass derived signal (*dotted line*)

In Fig. 8.7, the RR interval series from one representative CHF and healthy subject is shown along with the related μ_{RR}, and *IDLE*.

8.4 Conclusion and Discussion

In this chapter, we have described the use of recent definitions of instantaneous approximate and sample entropy, as well as instantaneous dominant Lyapunov exponents, based on the inhomogeneous point-process theory [27, 46].

Table 8.4 Group statistics of standard and instantaneous heartbeat dynamics measures from CHF-healthy experimental dataset

	CHF (n=14)	Healthy (n=16)	p-value
Standard and instantaneous time domain measures of HRV			
Mean RR (ms)	669.73± 68.66	855.74± 56.14	< 0.002
RMSSD	0.0121±0.0036	0.0432±0.0145	$<4e^{-4}$
pNN50%	0.2357±0.2246	21.5406±15.4908	$<1e^{-4}$
HRV Triangular Index	2.9551±0.5769	2.5628±0.3593	> 0.05
μ_{RR}(ms)	671.55±69.6	864.7±53.3	$<4e^{-3}$
σ_{RR}(ms)	8.31±2.2	24.7±7.0	$<5e^{-4}$
Standard and instantaneous nonlinear measures of HRV			
ApEn	1.2130±0.1032	1.2177±0.1066	>0.05
SampEn	1.5670±0.2690	1.4092±0.1522	>0.05
DFA-α_1	0.8498±0.2191	1.0820±0.1467	>0.05
DFA-α_2	1.1552±0.1335	0.9286±0.0544	<0.05
L_{D_1}	0.0167±0.0025	0.0165±0.0012	> 0.05
L_{D_2}	0.0029±0.0008	0.0033±0.0005	> 0.05
IDLE	0.0014 ±0.0649	0.0135±0.0368	> 0.05
$IDLE_{MAD}$	0.0595±0.0120	0.0476±0.0066	< 0.05

p-values are obtained from the Mann-Whitney test with null hypothesis of equal medians between the CHF and healthy subject groups. Values are expressed as X = Median(X) ± MAD(X)

These measures were tested on synthetic data as well as on real data gathered from heartbeat dynamics of healthy subjects and patients with, e.g., cardiac heart failure [27, 46]. Results demonstrate that instantaneous complexity is able to effectively track the system dynamics and is not affected by statistical noise properties. The mathematical definition of inhomogeneous entropy relies on a continuous identification based on stochastic point processes, thus ensuring high resolution estimates in time without the use of any interpolation procedure. In this sense, the proposed indices could provide a more meaningful quantification than traditional entropy measures. Goodness-of-fit measures such as KS distance and autocorrelation plots quantitatively allow to verify the model fit and to choose the proper model order, thus addressing another open issue of current parametric approaches.

Unlike other paradigms for estimating nonlinearity indices developed in the literature [9, 41, 43, 70], our method is formulated within a probabilistic framework specifically developed for point-process observations (e.g. RR intervals), which has already produced important nonlinear quantifiers for autonomic assessment, based on second- and third-order statistics (instantaneous spectrum and bispectrum) [21]. Other most common nonlinearity indices are derived from non-parametric models, whereas our model is purely parametric and the analytically derived indices can be evaluated in a dynamic and instantaneous fashion. We believe these strengths enable our method as a useful tool for assessing nonlinear dynamics of heartbeat intervals in a non-stationary environment.

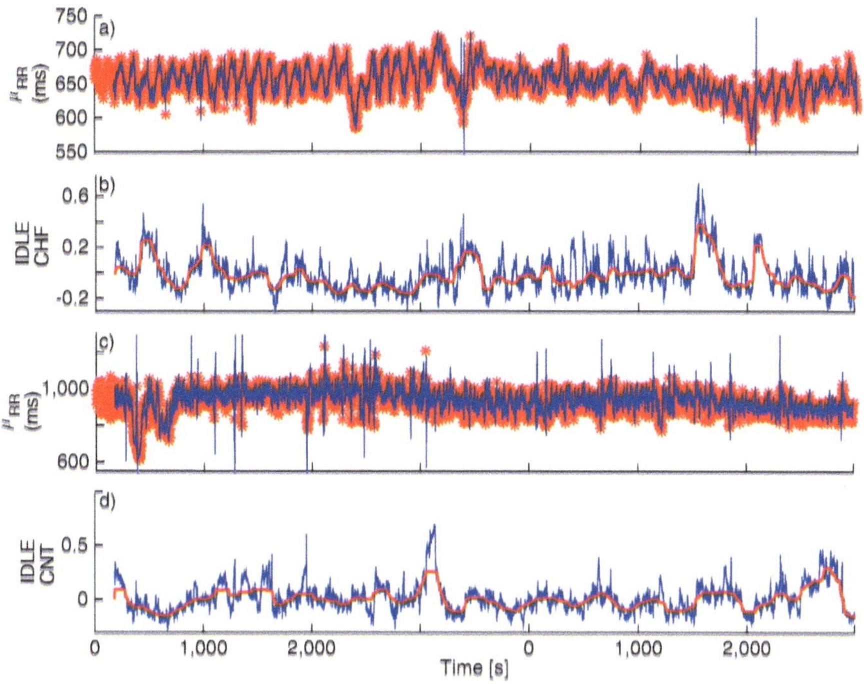

Fig. 8.7 Instantaneous heartbeat statistics computed from a representative CHF patient and healthy subject. In the panels (**a**) and (**c**), the estimated $\mu_{RR}(t)$ are superimposed on the recorded CHF and CNT series, respectively. The panels (**b**) and (**d**) show the IDLE on the recorded CHF and CNT series, respectively

To this extent, it is worth mentioning a pragmatic case in which traditional complexity measures do not allow a reliable assessment. Let us consider a generic experimental protocol comprising a 5-min resting state and, then, 5 s of stimuli. Standard, not time-varying measures such as *ApEn* and *SampEn* are not able to provide reliable quantifiers distinguishing between the two conditions, as they cannot be estimated on the latter experimental session (i.e. the 5 s of stimuli). However, the proposed A_l, S_l, and IDLE indices can be reliably used to perform instantaneous estimates starting from the length W (of 90 s for instance) on, thus proving an instantaneous complexity assessment also during the last 5 s.

The limitations of our methodology can be related to the need of a preliminary calibration phase before it can be effectively used to estimate the instantaneous entropy measures. Moreover, concerning noise dependency, it is important to remark that the proposed measures do not follow the traditional and intuitive interpretation of entropy (which implies that such a measure has to vary with the underlying system noise). Nevertheless, referring to a generalization of the definition of entropy, i.e., measures addressing the randomness and regularity of a dynamical system, it is proper to say that our indices simply have different properties than traditional entropy measures. In other words, if the traditional noise-dependent complexity indices are preferred in some instances where additive noise properties are relevant, there are many other cases where the use of a complexity measure that follows only the dynamics of the autoregressive system regardless of the observation time and the input modulating noise are recommended.

In validating the instantaneous entropy measures, we have shown that A_I and S_I promisingly provide helpful time-varying and adaptive indices for real-time monitoring of sympathovagal dynamics, which have also been proven in agreement with the current literature [67, 68]. In fact, our studies on healthy subjects undergoing postural changes confirm previous results [67, 68] and demonstrate that the instantaneous complexity measures appropriately reflect changes of ANS control on cardiovascular dynamics, thus improving the sympathovagal assessment in a non-stationary environment. We also found that pathological heartbeat dynamics are associated with decreased instantaneous complexity, confirming the common experimental experiences of reduced regularity, predictability, and sensitivity to initial conditions in pathological states.

In validating the instantaneous dominant Lyapunov exponents, we have demonstrated that our IDLE-based approach is useful in characterizing the inherent nonlinearity of the cardiovascular system. In resting condition, the dynamics driving the autonomic influence on the heart are more sensitive to the initial conditions (positive IDLE), whereas more regular dynamics (negative IDLE values) appear during tilt, i.e. under sympathetic control. These results are in agreement with previous findings that complex vagally-driven dynamics are blunted under sympathetic drive [71] and with more recent reports on loss of complexity during states of arousal [22]). Although other HRV-based measures are able to discern CHF from healthy subjects, no other measures have been proposed to characterize the time-varying complexity behavior occurring in pathological vs. a healthy cardiovascular system.

The proposed instantaneous measures also allow for the study of *complexity variability*, i.e., the analysis of complex systems referring to the fluctuations in complexity instead of analysis of central tendency, which has been recently explored in disease assessment of patients with severe congestive heart failure [72]. This concept can be linked to the current literature whereby cardiovascular disorders affect complexity and variability, and may lead to serious pathological events such as heart failure [73].

Of note, within the proposed framework, it is always possible to incorporate physiological covariates (such as respiration or blood pressure measures) and produce further instantaneous indices from their dynamic cross spectrum and cross bispectrum [74].

To conclude, the proposed methodology for time-varying complexity assessment offers a promising mathematical tool for the dynamic analysis of a wide range of applications and to potentially study any physical and natural stochastic discrete process (e.g. [21]).

Future works are related to pursue this direction in further investigating the potential of these high-order nonlinear models in producing new real-time measures for the underlying complexity of physiological systems, and to the investigation of the instantaneous complexity, along with baroreflex sensitivity and respiratory sinus arrhythmia, during postural changes in CHF subjects, thus yielding more insights into the complex physiological behavior of the cardiovascular system in CHF subjects.

References

1. Sunagawa, K., Kawada, T., Nakahara, T.: Dynamic nonlinear vago-sympathetic interaction in regulating heart rate. Heart Vessels **13**(4), 157–174 (1998)
2. E.T.F. of the European Society of Cardiology the North American Society of Pacing: Heart rate variability: standards of measurement, physiological interpretation, and clinical use. Circulation **93**(5), 1043–1065 (1996)
3. Acharya, U.R., Joseph, K.P., Kannathal, N., Lim, C.M., Suri, J.S.: Heart rate variability: a review. Med. Biol. Eng. Comput. **44**(12), 1031–1051 (2006)
4. Atyabi, F., Livari, M., Kaviani, K., Tabar, M.: Two statistical methods for resolving healthy individuals and those with congestive heart failure based on extended self-similarity and a recursive method. J. Biol. Phys. **32**(6), 489–495 (2006)
5. Glass, L.: Introduction to controversial topics in nonlinear science: is the normal heart rate chaotic? Chaos **19**(2), 028501 (2009). doi:10.1063/1.3156832
6. Valenza, G., Citi, L., Lanatá, A., Scilingo, E.P., Barbieri, R.: Revealing real-time emotional responses: a personalized assessment based on heartbeat dynamics. Nat. Sci. Rep. **4**, 4998 (2014)
7. Valenza, G., Citi, L., Lanata, A., Scilingo, E.P., Barbieri, R.: A nonlinear heartbeat dynamics model approach for personalized emotion recognition. In: 2013 35th Annual International Conference of the IEEE Engineering in Medicine and Biology Society (EMBC), pp. 2579–2582. IEEE, New York (2013)
8. Glass, L.: Synchronization and rhythmic processes in physiology. Nature **410**(6825), 277–284 (2001)
9. Goldberger, A., Peng, C., Lipsitz, L.: What is physiologic complexity and how does it change with aging and disease? Neurobiol. Aging **23**(1), 23–26 (2002)
10. Poon, C., Merrill, C.: Decrease of cardiac chaos in congestive heart failure. Nature **389**(6650), 492–495 (1997)
11. Lanata, A., Valenza, G., Mancuso, C., Scilingo, E.P.: Robust multiple cardiac arrhythmia detection through bispectrum analysis. Expert Syst. Appl. **38**(6), 6798–6804 (2011)
12. Wu, G., Arzeno, N., Shen, L., Tang, D., Zheng, D., Zhao, N., Eckberg, D., Poon, C.: Chaotic signatures of heart rate variability and its power spectrum in health, aging and heart failure. PloS one **4**(2), e4323 (2009)
13. Valenza, G., Lanata, A., Scilingo, E.P.: Oscillations of heart rate and respiration synchronize during affective visual stimulation. IEEE Trans. Inf. Technol. Biomed. **16**(4), 683–690 (2012)
14. Winfree, A.: Electrical turbulence in three-dimensional heart muscle. Science **266**(5187), 1003–1006 (1994)
15. Valenza, G., Nardelli, M., Lanata, A., Gentili, C., Bertschy, G., Paradiso, R., Scilingo, E.P.: Wearable monitoring for mood recognition in bipolar disorder based on history-dependent long-term heart rate variability analysis. IEEE J. Biomed. Health Inf. **8**(5), 1625–1635 (2014)
16. Chialvo, D.R., Gilmour Jr, R.F., Jalife, J.: Low dimensional chaos in cardiac tissue. Nature **343**(6259), 653–657 (1990)
17. Valenza, G., Lanatá, A., Scilingo, E.P.: Improving emotion recognition systems by embedding cardiorespiratory coupling. Physiol. Meas. **34**(4), 449 (2013)
18. Kanters, J.K., Højgaard, M.V., Agner, E., Holstein-Rathlou, N.-H.: Short-and long-term variations in non-linear dynamics of heart rate variability. Cardiovasc. Res. **31**(3), 400–409 (1996)

19. Schmidt, G., Monfill, G.: Nonlinear methods for heart rate variability assessment. In: Heart Rate Variability, pp. 87–98. Futura, Armonk (1995)
20. Jelinek, H.F., Khandoker, A.H., Quintana, D., Imam, M.H., Kemp, A.: Complex correlation measure as a sensitive indicator of risk for sudden cardiac death in patients with depression. In: Computing in Cardiology, 2011, pp. 809–812. IEEE, New York (2011)
21. Valenza, G., Citi, L., Scilingo, E., Barbieri, R.: Point-process nonlinear models with Laguerre and Volterra expansions: instantaneous assessment of heartbeat dynamics. IEEE Trans. Signal Process. **61**(11), 2914–2926 (2013)
22. Valenza, G., Allegrini, P., Lanatà, A., Scilingo, E.P.: Dominant Lyapunov exponent and approximate entropy in heart rate variability during emotional visual elicitation. Front. Neuroeng. **5**, 3 (2012)
23. Schmitt, D.T., Stein, P.K., Ivanov, P.C.: Stratification pattern of static and scale-invariant dynamic measures of heartbeat fluctuations across sleep stages in young and elderly. IEEE Trans. Biomed. Eng. **56**(5), 1564–1573 (2009)
24. Valenza, G., Citi, L., Gentili, C., Lanata, A., Scilingo, E., Barbieri, R.: Characterization of depressive states in bipolar patients using wearable textile technology and instantaneous heart rate variability assessment. IEEE J. Biomed. Health Inf. **19**, 263–274 (2014)
25. Bär, K.-J., Boettger, M.K., Koschke, M., Schulz, S., Chokka, P., Yeragani, V.K., Voss, A.: Nonlinear complexity measures of heart rate variability in acute schizophrenia. Clin. Neurophysiol. **118**(9), 2009–2015 (2007)
26. Tarvainen, M.P., Georgiadis, S., Laitio, T., Lipponen, J.A., Karjalainen, P.A., Kaskinoro, K., Scheinin, H.: Heart rate variability dynamics during low-dose propofol and dexmedetomidine anesthesia. Ann. Biomed. Eng. **40**(8), 1802–1813 (2012)
27. Valenza, G., Citi, L., Scilingo, E.P., Barbieri, R.: Inhomogeneous point-process entropy: an instantaneous measure of complexity in discrete systems. Phys. Rev. E **89**(5), 052803 (2014)
28. Orini, M., Laguna, P., Mainardi, L., Bail6n, R.: Assessment of the dynamic interactions between heart rate and arterial pressure by the cross time-frequency analysis. Physiol. Meas. **33**(3), 315 (2012)
29. Valenza, G., Nardelli, M., Bertschy, G., Lanata, A., Scilingo, E.: Mood states modulate complexity in heartbeat dynamics: a multiscale entropy analysis. Europhys. Lett. **107**(1), 18003 (2014)
30. Marmarelis, V.: Nonlinear Dynamic Modeling of Physiological Systems. Wiley/IEEE Press, New York (2004)
31. Ivanov, P.C., Amaral, L.A.N., Goldberger, A.L., Havlin, S., Rosenblum, M.G., Struzik, Z.R., Stanley, H.E.: Multifractality in human heartbeat dynamics. Nature **399**(6735), 461–465 (1999)
32. Kirsch, M.R., Monahan, K., Weng, J., Redline, S., Loparo, K.A.: Entropy-based measures for quantifying sleep-stage transition dynamics: relationship to sleep fragmentation and daytime sleepiness. IEEE Trans. Biomed. Eng. **59**(3), 787–796 (2012)
33. Valenza, G., Citi, L., Barbieri, R.: Instantaneous nonlinear assessment of complex cardiovascular dynamics by Laguerre-Volterra point process models. In: 2013 35th Annual International Conference of the IEEE Engineering in Medicine and Biology Society (EMBC), pp. 6131–6134. IEEE, New York (2013)
34. Valenza, G., Citi, L., Lanata, A., Scilingo, E.P., Barbieri, R.: A nonlinear heartbeat dynamics model approach for personalized emotion recognition. In: 2013 35th Annual International Conference of the IEEE Engineering in Medicine and Biology Society (EMBC), pp. 2579–2582. IEEE, New York (2013)
35. Joe, H.: Relative entropy measures of multivariate dependence. J. Am. Stat. Assoc. **84**(405), 157–164 (1989)
36. Karmeshu: Entropy Measures, Maximum Entropy Principle and Emerging Applications, vol. 119. Springer, Berlin (2003)
37. Renyi, A., On measures of entropy and information. In: Fourth Berkeley Symposium on Mathematical Statistics and Probability, pp. 547–561 (1961)
38. Eckmann, J.-P., Ruelle, D.: Ergodic theory of chaos and strange attractors. Rev. Mod Phys. **57**(3), 617 (1985)

39. Grassberger, P., Procaccia, I.: Estimation of the Kolmogorov entropy from a chaotic signal. Phys. Rev. A **28**(4), 2591–2593 (1983)
40. Fraser, A.M.: Information and entropy in strange attractors. IEEE Trans. Inf. Theory **35**(2), 245–262 (1989)
41. Pincus, S.: Approximate entropy as a measure of system complexity. Proc. Natl. Acad. Sci. **88**(6), 2297 (1991)
42. Richman, J., Moorman, J.: Physiological time-series analysis using approximate entropy and sample entropy. Am. J. Physiol. Heart Circ. Physiol. **278**(6), H2039 (2000)
43. Costa, M., Goldberger, A., Peng, C.: Multiscale entropy analysis of complex physiologic time series. Phys. Rev. Lett. **89**(6), 68102 (2002)
44. Chon, K., Yip, K., Camino, B., Marsh, D., Holstein-Rathlou, N.-H.: Modeling nonlinear determinism in short time series from noise driven discrete and continuous systems. Int. J. Bifurcation Chaos **10**(12), 2745–2766 (2000)
45. Armoundas, A., Ju, K., Iyengar, N., Kanters, J., Saul, P., et al.: A stochastic nonlinear autoregressive algorithm reflects nonlinear dynamics of heart-rate fluctuations. Ann. Biomed. Eng. **30**(2), 192–201 (2002)
46. Valenza, G., Citi, L., Barbieri, R.: Estimation of instantaneous complex dynamics through Lyapunov exponents: a study on heartbeat dynamics. PloS one **9**(8), e105622 (2014)
47. Wolf, A., Swift, J.B., Swinney, H.L., Vastano, J.A.: Determining Lyapunov exponents from a time series. Phys. D **16**(3), 285–317 (1985)
48. Abarbanel, H.D., Gollub, J.P.: Analysis of observed chaotic data. Phys. Today **49**, 86 (1996)
49. Ruelle, D.: Where can one hope to profitably apply the ideas of chaos? Phys. Today **47**, 24 (1994)
50. Barbieri, R., Matten, E., Alabi, A., Brown, E.: A point-process model of human heartbeat intervals: new definitions of heart rate and heart rate variability. Am. J. Phys. Heart Circ. Phys. **288**(1), H424 (2005)
51. Citi, L., Valenza, G., Barbieri, R.: Instantaneous estimation of high-order nonlinear heartbeat dynamics by Lyapunov exponents. In: Proceeding of the IEEE-EMBC, pp. 13–16 (2012)
52. Granger, C., Joyeux, R.: An introduction to long-memory time series models and fractional differencing. J. Time Ser. Anal. **1**(1), 15–29 (1980)
53. Chen, Z., Purdon, P., Brown, E., Barbieri, R.: A differential autoregressive modeling approach within a point process framework for non-stationary heartbeat intervals analysis. In: 2010 Annual International Conference of the IEEE Engineering in Medicine and Biology Society (EMBC), pp. 3567–3570. IEEE, New York (2010)
54. Billings, S.: Identification of nonlinear system - a survey. Proc. IEEE **127**, 272–285 (1980)
55. Chen, Z., Brown, E., Barbieri, R.: Assessment of autonomic control and respiratory sinus arrhythmia using point process models of human heart beat dynamics. IEEE Trans. Biomed. Eng. **56**(7), 1791–1802 (2009)
56. Barbieri, R., Brown, E.: Analysis of heartbeat dynamics by point process adaptive filtering. IEEE Trans. Biomed. Eng. **53**(1), 4–12 (2006)
57. Chen, Z., Brown, E.N., Barbieri, R.: Characterizing nonlinear heartbeat dynamics within a point process framework. IEEE Trans. Biomed. Eng. **57**(6), 1335–1347 (2010)
58. Loader, C.: Local Regression and Likelihood. Springer, Berlin (1999)
59. Tibshirani, R., Hastie, T.: Local likelihood estimation. J. Am. Stat. Assoc. **82**, 559–567 (1987)
60. Brown, E., Barbieri, R., Eden, U., Frank, L.: Likelihood methods for neural spike train data analysis. In: Computational Neuroscience: A Comprehensive Approach, pp. 253–286. CRC Press, London (2003). doi:10.1201/9780203494462.ch9
61. Westwick, D., Kearney, R.: Explicit least-squares methods. In: Identification of Nonlinear Physiological Systems, vol. 7. Wiley/IEEE Press, New York (2003)
62. Grassberger, P., Procaccia, I.: Measuring the strangeness of strange attractors. Phys. D **9**(1), 189–208 (1983)
63. Acharya, U.R., Joseph, K.P., Kannathal, N., Lim, C., Suri, J.: Heart rate variability: a review. Med. Biol. Eng. Comput. **44**(12), 1031–1051 (2006)

64. Rosenstein, M.T., Collins, J.J., De Luca, C.J.: A practical method for calculating largest Lyapunov exponents from small data sets. Phys. D **65**(1), 117–134 (1993)
65. Peng, C., Buldyrev, S., Havlin, S., Simons, M., Stanley, H., Goldberger, A.: Mosaic organization of DNA nucleotides. Phys. Rev. E **49**(2), 1685 (1994)
66. Heldt, T., Shim, E.B., Kamm, R.D., Mark, R.G.: Computational modeling of cardiovascular response to orthostatic stress. J. Appl. Physiol. **92**(3), 1239–1254 (2002)
67. Porta, A., Gnecchi-Ruscone, T., Tobaldini, E., Guzzetti, S., Furlan, R., Montano, N.: Progressive decrease of heart period variability entropy-based complexity during graded head-up tilt. J. Appl. Physiol. **103**(4), 1143–1149 (2007)
68. Tulppo, M.P., Mäkikallio, T.H., T. Seppänen, Shoemaker, K., Tutungi, E., Hughson, R.L., Huikuri, H.V.: Effects of pharmacological adrenergic and vagal modulation on fractal heart rate dynamics. Clin. Physiol. **21**(5), 515–523 (2001)
69. MIT-BIH normal sinus rhythm database and BIDMC congestive heart failure database. Available at http://www.physionet.org/physiobank/database/
70. Thuraisingham, R., Gottwald, G.: On multiscale entropy analysis for physiological data. Phys. A **366**, 323–332 (2006)
71. Hagerman, I., Berglund, M., Lorin, M., Nowak, J., Sylvén, C.: Chaos-related deterministic regulation of heart rate variability in time-and frequency domains: effects of autonomic blockade and exercise. Cardiovasc. Res. **31**(3), 410–418 (1996)
72. Valenza, G., Citi, L., Barbieri, R.: Instantaneous nonlinear assessment of complex cardiovascular dynamics by Laguerre-Volterra point process models. In: 2013 Annual International Conference of the IEEE Engineering in Medicine and Biology Society (EMBC). IEEE, New York (2013)
73. Schwartz, P.J., De Ferrari, G.M.: Sympathetic–parasympathetic interaction in health and disease: abnormalities and relevance in heart failure. Heart Fail. Rev. **16**(2), 101–107 (2011)
74. Chen, Z., Purdon, P.L., Harrell, G., Pierce, E.T., Walsh, J., Brown, E.N., Barbieri, R.: Dynamic assessment of baroreflex control of heart rate during induction of propofol anesthesia using a point process method. Ann. Biomed. Eng. **39**(1), 260–276 (2011)

Chapter 9
Time-Frequency Analysis of Cardiovascular Signals and Their Dynamic Interactions

Michele Orini, Pablo Laguna, Luca T. Mainardi, and Raquel Bailón

Abstract Cardiovascular signals are intrinsically non-stationary and interact through dynamic mechanisms to maintain blood pressure homeostasis in response to internal and external perturbations. The assessment of changes in cardiovascular signals and in their interactions provides valuable information regarding the cardiovascular function. Time-frequency analysis is a useful tool to study the time-varying nature of the cardiovascular system because it provides a joint representation of a signal in the temporal and spectral domain that allows to track the instantaneous frequency, amplitude and phase of non-stationary processes. The time-frequency distributions described in this chapter belong to the Cohen's class, and can be derived from the Wigner-Ville distribution, which represents the fundamental basis of this unified framework. Time-frequency analysis can be extended to the study of the dynamic interactions between two or more non-stationary processes. Time-frequency coherence, phase-delay, phase-locking and partial-spectra are estimators that assess changes in the coupling and phase shift of signals generated by a complex system.

This chapter introduces the reader to multivariate time-frequency analysis and covers both theoretical and practical aspects. The application of these methodologies in the study of the dynamic interactions between the most important variables of the cardiovascular function is discussed. In the introduction, classical spectral analysis of cardiovascular signals is reviewed along with its physiological interpretation. The limitations of this framework provides a motivation for implementing non-stationary tools. In the first section, time-frequency representations based on the Wigner-Ville distribution are introduced and important aspects, such as the interference cross-terms and their elimination, the time and frequency resolution and the estimation

M. Orini (✉)
Institute of Cardiovascular Science, University College of London, London, UK
e-mail: m.orini@ucl.ac.uk

P. Laguna • R. Bailón
GTC, Aragon Institute of Engineering Research, Universidad de Zaragoza, Zaragoza, Spain
CIBER-BBN, Zaragoza, Spain

L.T. Mainardi
Department of Electronics, Informatics and Bioengineering, Politecnico di Milano, 20133 Milan, Italy

R. Barbieri et al. (eds.), *Complexity and Nonlinearity in Cardiovascular Signals*,
DOI 10.1007/978-3-319-58709-7_9

of time-frequency spectra, are described. The second section describes algorithms to assess the dynamic interactions between non-stationary signals, including time-frequency coherence, phase delay and partial spectra, while the third section provides examples of multivariate time-frequency analysis of cardiovascular data recorded during a standard test to induce an autonomic response.

9.1 Introduction

A process is stationary if its probability distribution and therefore its main statistical properties do not change over time. There are different degrees and types of stationarity and a process can be stationary or non-stationary depending on the temporal scale of observation [1]. In the analysis of cardiovascular signals, stationarity is considered an exception rather than the rule. For example, the quantification of the degree of non-stationarity [1] of cardiovascular signals has demonstrated that heart rate variability is non-stationary even when recorded at rest [2]. Cardiovascular signals are intrinsically non-stationary and their interactions are time-varying because the system has to dynamically adjust to an ever-changing environment in order to maintain its function (blood pressure homeostasis). The quantification of changes and dynamic interactions within the cardiovascular system provides valuable information for understanding its function and recognizing possible dysfunctions.

The ultimate aim of most of signal processing methodologies is to provide a simple description of complex systems that captures the most relevant features for a given application. Over the years, two approaches have emerged to transform raw data into more useful representations. One includes model-based methodologies, where some a-priory knowledge is used to propose a model whose parameters fully characterize the system under investigation. The reliability of this parametric analysis depends on how well the model fits the data, which in cardiovascular applications are often assumed to be autoregressive processes [3, 4] or point-processes [5]. The second approach is not based on any assumption regarding the structure of the signal, and it transforms the data into a new representation that facilitates the analysis. This last approach is non-parametric.

The time-frequency analysis representations described in this chapter are non-parametric. They are used to transform non-stationary signals into joint time-frequency representations that reveal the temporal evolution of their spectral components.

Analysis of raw data through time-frequency representations can be achieved in three steps:

1. Representation of the signal or the interactions among signals in the joint time-frequency domain.
2. Localization of relevant patterns and regions of interest in this new domain.

3. Extraction of useful information, usually by transforming the relevant patterns observed in the region of interest into a time-varying physiological index or a single value.

9.1.1 Spectral Analysis and Its Limitations

The spectral analysis is based on the Fourier transform, which provides a decomposition of a signal $x(t)$ into a linear combination of complex exponential functions:

$$x(t) = \int X(f)e^{j2\pi tf}\,df \tag{9.1}$$

In this expression, the complex exponential functions $e^{j2\pi tf}$ are weighted by the complex function $X(f)$, that quantifies the spectral content of $x(t)$, i.e. the contribution of each oscillation to the signal $x(t)$. The function $X(f)$ is the Fourier transform of signal $x(t)$, and it is given by:

$$X(f) = \int x(t)e^{-j2\pi ft}\,dt \tag{9.2}$$

Expressions (9.1) and (9.2) operate transformations between the temporal and the spectral domains. The Fourier transform has a set of properties that makes it a widespread and versatile tool with many applications in signal processing. The most relevant are listed in Table 9.1.

The power spectral density is a representation of the distribution of the signal power over its spectral components, i.e. as a function of frequency. According to the Wiener-Khinchin theorem, the power spectral density of a stationary stochastic processes $x(t)$ is the Fourier transform of its autocorrelation function:

Table 9.1 Useful Fourier transform properties

Time		Frequency	
$h(t) = a \cdot x(t) + b \cdot y(t)$	$\leftrightarrow$	$H(f) = a \cdot X(f) + b \cdot Y(f)$	Linearity
$h(t) = x(t - t_0)$	$\leftrightarrow$	$H(f) = X(f)e^{-j2\pi t_0 f}$	t-Shifting
$h(t) = e^{j2\pi t f_0}x(t)$	$\leftrightarrow$	$H(f) = X(f - f_0)$	f-Shifting
$h(t) = x(a \cdot t)$	$\leftrightarrow$	$H(f) = \frac{1}{\lvert a\rvert}X\left(\frac{f}{a}\right)$	t-Scaling
$x(t) \in \mathbb{R}$	$\leftrightarrow$	$X(-f) = X^*(f)$	Conjugation
$h(t) = x^{(n)}(t)$	$\leftrightarrow$	$H(f) = (j2\pi f)^n X(f)$	Derivative
$h(x) = (x * y)(t)$	$\leftrightarrow$	$H(f) = X(f) \cdot Y(f)$	Product/Convolution
$\int \lvert x(t)\rvert^2 dt$	$\leftrightarrow$	$\int \lvert X(f)\rvert^2 df$	Energy conservation

$$S_{xx}(f) = \int r_{xx}(\tau)e^{-j2\pi\tau f}\,d\tau = \int \mathrm{E}\left[x(t)x^*(t-\tau)\right]e^{-j2\pi\tau f}\,d\tau \tag{9.3}$$

where $\mathrm{E}\,[\,]$ is the expectation value operator and $r_{xx}(\tau)$ the autocorrelation function of $x(t)$, a function that assesses the correlation between a signal and a copy of itself as a function of the delay between them. A simple estimator of the power spectral density that can be used when only one realization of finite length is available is the periodogram, which is proportional to the squared magnitude of the Fourier transform. More advanced estimators include the Welch [6] and the multitaper [7] spectra.

The power spectral density has been widely applied to the study of cardiovascular signals, in particular to the analysis of heart rate variability [8–10]. The spectral analysis of heart rate variability allows to non-invasively assess the autonomic nervous system modulation of the supra-ventricular cardiac structures responsible of heart rate regulation [11–13]. Figure 9.1 shows the power spectral density of cardiovascular signals simultaneously recorded at rest in supine position [14]. From left to right, this figure shows the spectrum of heart rate variability, respiratory signal measured as chest movement with a thoracic belt, and systolic arterial blood pressure variability. A respiratory component is present in all the three signals in the so called high-frequency spectral band (HF), 0.15–0.40 Hz [10], while a low-frequency component is present in the heart rate and blood pressure variability in the low-frequency spectral band (LF), 0.04–0.15 Hz. The analysis of these two spectral components provides information regarding the parasympathetic and sympathetic

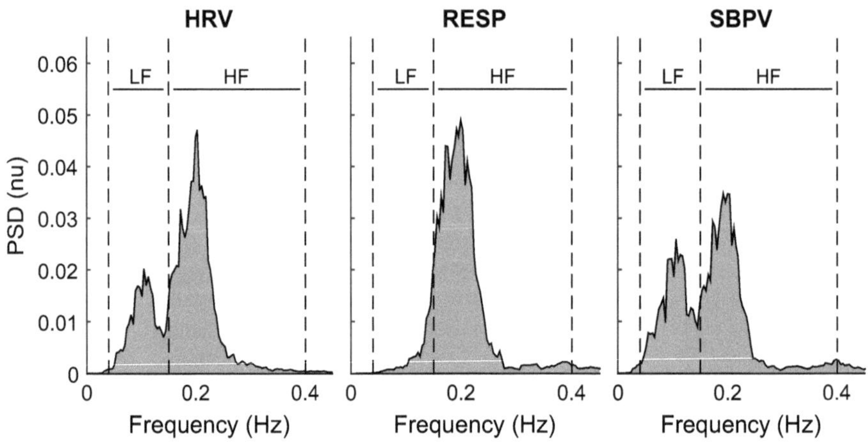

Fig. 9.1 Normalized power spectral density (PSD) of: Heart rate variability (HRV), respiratory signal (RESP) measured as chest movement with a thoracic belt, and systolic arterial blood pressure variability (SBPV). A high-frequency (HF) component reflecting respiration is present in all the three spectra, while a low-frequency component (LF) is present in the HRV and SBPV. These signals have been derived from simultaneous recordings in a subject at rest in supine position. A multitaper approach has been used to estimate the PSDs

activity. The HF component of the heart rate variability is considered a marker of parasympathetic activity [10, 15] reflecting respiratory sinus arrhythmia. The precise nature of its underlying mechanisms are still matter of debate [16]. The LF component of the heart rate variability is related to both the sympathetic and parasympathetic activity and its origin and interpretation are still not completely understood [17]. The proportion between the LF and HF component has been proposed and is often used as a marker of the sympathovagal balance [11], but its use remain controversial [18–20].

The main limitation of spectral analysis is that it requires a signal to be stationary and does not provide any temporal localization of events. Mathematically, the lack of localization is due to the decomposition of a signal into sinusoids and cosinusoids, i.e. $e^{j2\pi tf}$, that are equally spread along the entire temporal axis. Figure 9.2 shows two schematic examples of this limitation. In the first one, the frequency of a signal (on the left) is increasing during time, but this dynamic feature is lost in its spectral representation (on the right), where the effect of a time-varying instantaneous frequency results in a single widespread spectral component. In the second example, the amplitude of a sinusoid (on the left) is increasing over time, but this dynamic characteristic is again lost in its spectral representation (on the right), which only shows a single static spectral peak. Another classical example of spectral analysis limitations is that the spectrum of a signal and the spectrum of the same signal recorded backward are exactly the same. The temporal organization of a signal is coded in the phase of the Fourier transform, which is not used to construct a power spectrum.

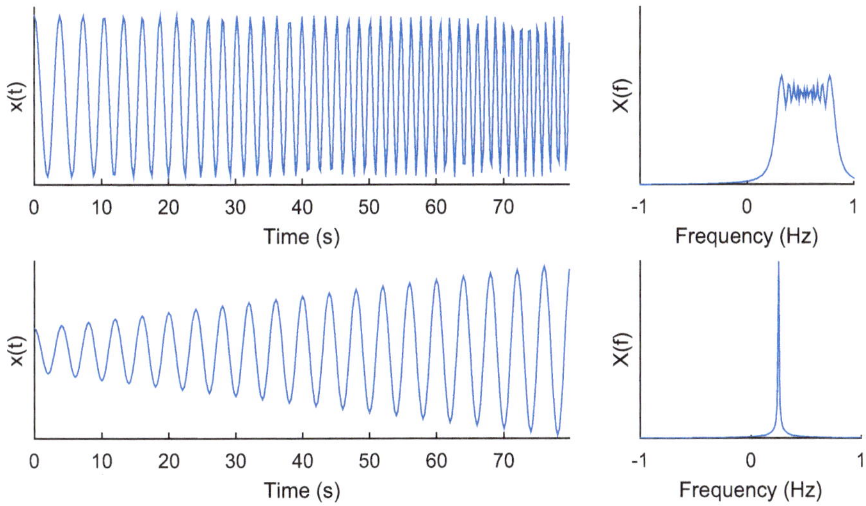

Fig. 9.2 Simple examples of spectral analysis limitations. *Top*: a signal with increasing instantaneous frequency (*on the left*) and its spectrum (*on the right*). *Bottom*: a sinusoid with increasing instantaneous amplitude (*on the left*), and its spectrum (*on the right*). The dynamic properties of these two signals are not reflected in their spectral representation

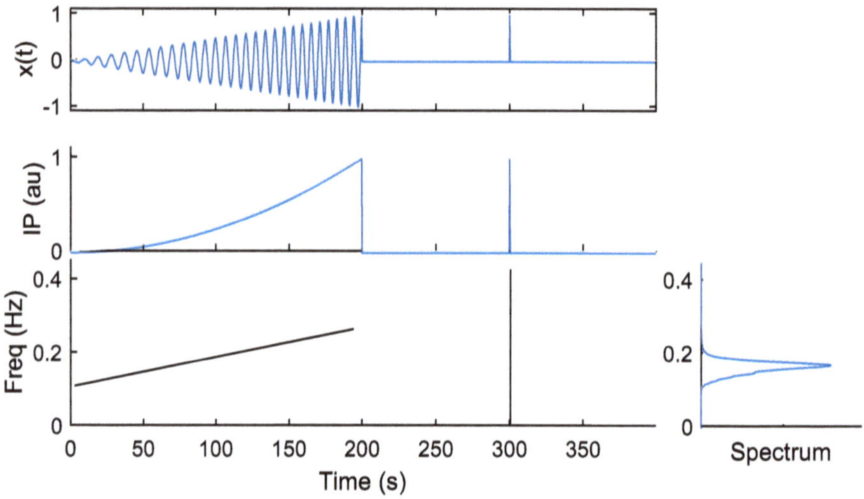

Fig. 9.3 Ideal joint time-frequency representation. A signal composed of sinusoid of finite length and time-varying amplitude and frequency, plus temporal impulse is shown on the *top*. On the *bottom*, its ideal time-frequency representation reveals its time-frequency structure and provides a perfect localization of the events. The marginals obtained by integrating the time-frequency representation over time (*right*) and frequency (*middle*) represents the instantaneous power and the spectrum or the signal, respectively

Ideally, a joint time-frequency representation should be able to track changes of the instantaneous frequencies and amplitudes of the signal's components with high time and frequency resolution. An example of an ideal time-frequency representation is reported in Fig. 9.3, where a sinusoid of finite length, time-varying amplitude and frequency, and a temporal impulse are all perfectly localized in time and frequency. The instantaneous power and the power spectral density can be obtained by integrating the time-frequency representation over time (right panel) and over frequency (middle panel), respectively. These representations are called marginals. However, as it will be shown in the next section, this ideal representation is not achievable because of fundamental compromise between localization in time and frequency.

9.2 Studying Signals: Time-Frequency Representations

This section provides a detailed description of time-frequency representations belonging to the Cohen's class. A more exhaustive presentation of this subject can be found in excellent books [21–23], and reviews [24, 25]. A useful toolbox with fundamental theory, examples and Matlab functions toolbox can be found in [26].

9.2.1 Introducing the Cohen's Class

The time-frequency (TF) representations that belong to the Cohen's class [21, 24] are energy distributions, in the sense that they provide a decomposition of the signal energy (for finite energy signals), E_x, in time and frequency. A fundamental property of an energy distribution, $C_x(t,f)$, is that it should preserve the energy of a signal $x(t)$, such that:

$$\iint C_x(t,f)\,dt\,df = E_x = \int |x(t)|^2\,dt = \int |X(f)|^2\,dt \tag{9.4}$$

Furthermore, Cohen's class distributions are energy distribution covariant by translation in time and frequency, meaning that they preserve temporal and spectral shifts, such that if a signal $x(t)$ is shifted in time and modulated in frequency, its time-frequency representation should be shifted by the corresponding amount both in time and frequency:

$$\text{if} \quad y(t) = x(t - t_0)e^{j2\pi f_0} \quad \Rightarrow \quad C_y(t,f) = C_x(t - t_0, f - f_0) \tag{9.5}$$

In this expression $C_x(t,f)$ and $C_y(t,f)$ are time-frequency representations of $x(t)$ and $y(t)$, respectively.

Ideally, an energy distribution should also satisfy the marginal properties, meaning that its integral over time should provide the energy spectral density of the signal, while its integral over frequency should give the instantaneous power of the signal:

$$\int C_x(t,f)\,dt = |X(f)|^2 \tag{9.6}$$

$$\int C_x(t,f)\,df = |x(t)|^2 \tag{9.7}$$

9.2.2 Equivalent Definitions of Cohen's Class Distributions

In its general form, each time-frequency representation belonging to the Cohen's class admits at least three equivalent definitions [21]:

1. *Time-delay* parametrization, via $\phi_{\text{t-d}}(t,\tau)$:

$$\begin{aligned} C_x(t,f) &= \iint \phi_{\text{t-d}}(t - v, \tau)x\left(v + \frac{\tau}{2}\right)x^*\left(v - \frac{\tau}{2}\right)e^{-j2\pi f\tau}dv\,d\tau \\ &= \int r_x(t,\tau)e^{-j2\pi f\tau}d\tau \end{aligned} \tag{9.8}$$

In this expression, the time-frequency distribution is seen as the Fourier transform of a non-stationary, deterministic, smoothed autocorrelation function $r_x(t, \tau)$:

$$r_x(t, \tau) = \int \phi_{\text{t-d}}(t - v, \tau) x\left(v + \frac{\tau}{2}\right) x^*\left(v - \frac{\tau}{2}\right) dv \quad (9.9)$$

where $\phi_{\text{t-d}}(t, \tau)$ is a kernel (subscripts t-d stand for time-delay) whose function is explained in the following. Note that (9.8) is similar to the Wiener-Khinchin theorem, which demonstrates that the autocorrelation function and the power spectral density of a stochastic stationary process are Fourier transform pairs.

2. *Delay-Doppler* parametrization, via $\phi_{\text{d-D}}(\tau, \nu)$:

$$C_x(t,f) = \iint \phi_{\text{d-D}}(\tau, \nu) A_x(\tau, \nu) e^{j2\pi(t\nu - f\tau)} d\nu\, d\tau \quad (9.10)$$

In this expression, the time-frequency distribution is seen as the two-dimensional Fourier transform of the narrow-band symmetric Ambiguity function, $A_x(\tau, \nu)$, modulated by the kernel function $\phi_{\text{d-D}}(\tau, \nu)$ (subscripts d-D stand for delay-Doppler, the Fourier pairs of frequency and time). The narrow-band symmetric Ambiguity function is defined as [27]:

$$A_x(\tau, \nu) = \int x\left(t + \frac{\tau}{2}\right) x^*\left(t - \frac{\tau}{2}\right) e^{-j2\pi\nu t} dt \quad (9.11)$$

and it can be seen as a two variable generalization of a correlation function, with the interesting property that its magnitude is invariant to temporal and spectral shifts:

$$\text{if} \quad y(t) = x(t - t_0) e^{j2\pi f_0} \quad \Rightarrow \quad |A_y(t,f)| = |A_x(t,f)| \quad (9.12)$$

3. *Time-frequency* parametrization, via $\phi_{\text{t-f}}(t,f)$:

$$C_x(t,f) = \iint \phi_{\text{t-f}}(t - v, f - u) W_x(v, u) dv\, du = \phi_{\text{t-f}}(t,f) \otimes W_x(t,f) \quad (9.13)$$

In this expression, the time-frequency distribution is obtained by low-pass filtering (symbol $\otimes$ stands for 2D convolution) in time and frequency the Wigner-Ville distribution (WVD), defined as the Fourier transform of the non-stationary, deterministic, autocorrelation function:

$$W_x(t,f) = \int x\left(t + \frac{\tau}{2}\right) x^*\left(t - \frac{\tau}{2}\right) e^{-j2\pi f\tau} d\tau \quad (9.14)$$

Expression (9.13) shows that the WVD can be considered as the central element of the Cohen's class, i.e. the distribution from which all the other ones can be derived through a simple convolution with a time-frequency kernel function $\phi_{\text{t-f}}(t,f)$. The

WVD (9.14) and the narrowband symmetric ambiguity function (9.11) are related by two dimensional Fourier transform [28]:

$$A_x(\tau,\nu) = \iint W_x(t,f)e^{-j2\pi(\nu t - f\tau)}dtdf \tag{9.15}$$

This implies that the WVD is itself a member of the Cohen's class, characterized by $\phi_{\text{d-D}}(\tau,\nu)=1$ and consequently $\phi_{\text{t-f}}(t,f) = \delta(t,f)$, i.e. a Dirac delta distribution centered at $t = 0$ and $f = 0$.

Every representation belonging to the Cohen's class is perfectly characterized by one of the three kernels previously introduced, i.e. $\phi_{\text{t-d}}(t,\tau)$, $\phi_{\text{d-D}}(\tau,\nu)$ or $\phi_{\text{t-f}}(t,f)$. These kernels completely define the properties of the TF distributions, their time-frequency resolution, and are mutually related by the Fourier transform [28]:

$$\phi_{\text{d-D}}(\tau,\nu) = \int \phi_{\text{t-d}}(t,\tau)e^{-j2\pi\nu t}dt = \iint \phi_{\text{t-f}}(t,f)e^{j2\pi(\tau f - \nu t)}dtdf \tag{9.16}$$

$$\phi_{\text{t-f}}(t,f) = \int \phi_{\text{t-d}}(t,\tau)e^{-j2\pi f\tau}d\tau = \iint \phi_{\text{d-D}}(\tau,\nu)e^{j2\pi(t\nu - \tau f)}d\tau d\nu \tag{9.17}$$

The main role of these kernels is to reduce the interference terms that affect the WVD and they have been extensively studied since the early Nineties [29–36].

Of note, from the WVD, by convolution with an appropriate kernel it is possible to derive:

(a) Any time-frequency distribution: $C_x(t,f) = \phi_{\text{t-f}}(t,f) \otimes W_x(t,f)$. This implies that the geometrical properties of $\phi_{\text{t-f}}(t,f)$ have a major effect on the representation $C_x(t,f)$.
(b) The spectrogram: $C_x(t,f) = W_h(t,f) \otimes W_x(t,f) = |\int x(s)h^*(s-t)e^{-j2\pi fs}\,ds|^2$, where $W_h(t,f)$ is the WVD of the sliding window $h(t)$.
(c) The scalogram: $C_x(t,f) = W_\psi(t,f) \otimes W_x(t,f)$, where $W_\psi(t,f)$ is the WVD of the mother wavelet.

9.2.3 The Wigner-Ville Distribution and the Interference Terms

The WVD defined in (9.14) is the Fourier transform of the non-stationary autocorrelation function and is the central element of the any time-frequency representation belonging to the Cohen's class. It has many interesting properties [21, 26], among which the most relevant are:

- Energy conservation: see (9.4).
- Satisfaction of marginal properties, see (9.6), meaning that its integration over time and frequency provides the power spectral density and the instantaneous power of the signal, respectively.

- Translation covariance, see (9.5), meaning that if a signal is shifted in time and modulated in frequency, its WVD will be shifted in time and frequency accordingly.
- Real valued: $W_x(t,f) \in \mathbb{R} \; \forall \; (t,f)$.
- Wide-sense support conservation:
  if $x(t) = 0$ for $t < |T| \Rightarrow W_x(t,f) = 0$ for $t < |T|$;
  if $X(f) = 0$ for $f < |\Omega| \Rightarrow W_x(t,f) = 0$ for $f < |\Omega|$.
- Perfect localization of linear chirps, i.e. a complex exponential whose instantaneous frequency $\theta(t)$ is a linear function of time :
  if $x(t) = e^{j2\pi\theta(t)t}, \theta(t) = a + \beta/2t \Rightarrow W_x(t,f) = \delta(f - (a + \beta t))$

However, its use is seriously limited by the presence of interference terms. Interferences are oscillatory terms inherent to the quadratic structure of the WVD, which do not represent signal spectral components but are due to their mutual interactions [37, 38]. Therefore, they are usually considered as artifacts that should be eliminated, or at least reduced, in order to improve the readability of the distribution. Interference terms are a consequence of the WVD's bilinear (or quadratic) structure and they can be identified mathematically with quadratic cross-terms [37, 39]. They appear whenever a signal can be described as the sum of different spectral components or distinct spectral atoms. For example, the WVD of a signal $x(t) = x_1(t) + x_2(t)$, is:

$$W_x(t,f) = W_1(t,f) + W_2(t,f) + 2\Re[W_{12}(t,f)] \tag{9.18}$$

where $W_1(t,f)$ and $W_2(t,f)$ are the WVDs of $x_1(t)$ and $x_2(t)$, also called auto-terms because they represent signal components, whereas $W_{12}(t,f)$ is an interference term that oscillates with a frequency and in a direction that depend on the relative position of the auto-terms in the TF plane [25, 37, 38].

The geometry of the interference terms can be explained considering two TF atoms $x_1(t) = h(t - t_1)e^{j2\pi f_1 t}$ and $x_2(t) = h(t - t_2)e^{j2\pi f_2 t}$, where $h(t)$ is a Gaussian window, well localized around (t_1,f_1) and (t_2,f_2). The WVD of their sum $x(t) = x_1(t) + x_2(t)$ includes an interference term with the following characteristics:

- It is centered around $(t_{12},f_{12}) = \left(\frac{t_2+t_1}{2}, \frac{f_2+f_1}{2}\right)$.
- It oscillates in a direction perpendicular to the axis connecting (t_1,f_1) and (t_2,f_2).
- Its frequency in time direction is proportional to $|f_2 - f_1|$.
- Its frequency in frequency direction is proportional to $|t_2 - t_1|$.

Therefore, the spectral and temporal distances between the TF atoms, $|f_2 - f_1|$ and $|t_2 - t_1|$, determine the velocity of the oscillation of the interference term.

Interference terms are conceptually divided in two groups: outer and inner interferences [38]. Outer interferences appear in multi-component signals, as the result of the interactions between spectral components that are present at the same time. They are usually located at midway between two components and they mainly oscillate in time direction. Inner interferences are due to non-linearities in the instantaneous frequency of a single spectral component. Their structure obeys the same geometric laws of outer interferences. Figure 9.4 provides a graphical explanation of the geometry of the interference terms using two examples. Both panels show the WVD of a signal composed of two TF atoms $x_1(t) = h(t-t_1)e^{j2\pi f_1 t}$ and $x_2(t) = h(t-t_2)e^{j2\pi f_2 t}$, with $t_1 < t_2$ and $f_1 < f_2$ and where $h(t)$ is a Gaussian window. In both cases the WVD provides a good localization of atoms $x_1(t)$ and $x_2(t)$, but it is affected by interference terms. The geometrical center of the

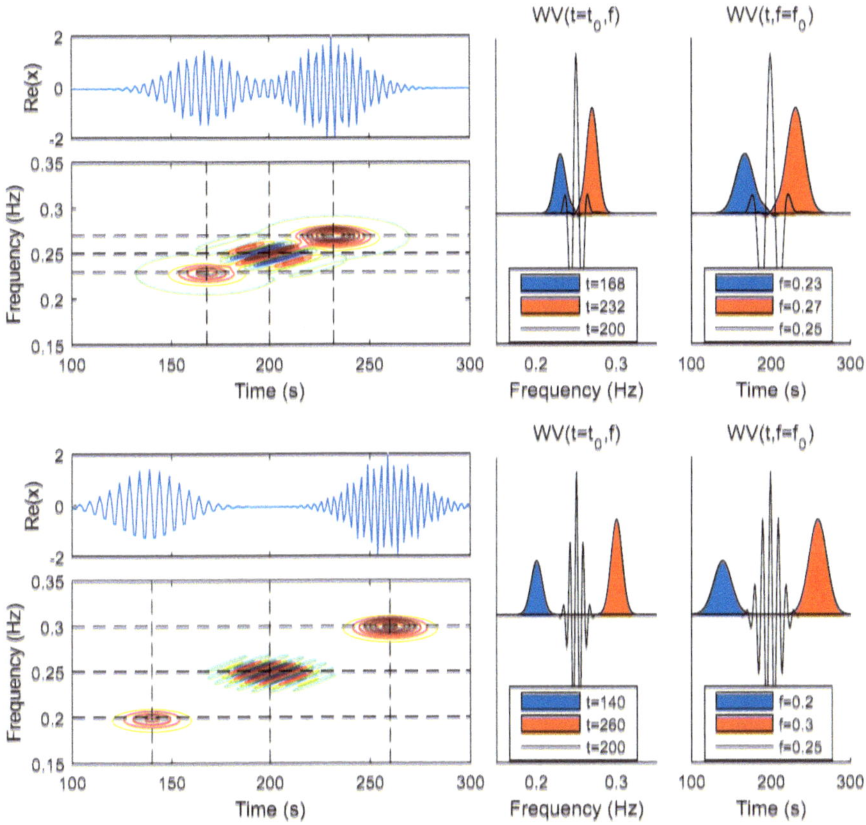

Fig. 9.4 Graphical explanation of the geometry of the interference terms. *On the left*: a signal composed of two time-frequency atoms $x_1(t)$ and $x_2(t)$, with $t_1 < t_2$ and $f_1 < f_2$, and where $h(t)$ is a Gaussian window, and its Wigner-Ville distribution. *On the right*: The Wigner-Ville distribution evaluated at time and frequency corresponding to the geometrical centers of auto- and interference terms. The WVD provides a good localization of atoms $x_1(t)$ and $x_2(t)$, but it is affected by interference terms. The geometrical center of the interference terms is in the middle of the auto-terms and the oscillation propagates along a direction perpendicular to the axis connecting the auto-terms. The interference oscillations in both time and frequency direction are faster in the example in the *upper* than in the *bottom panel* because the temporal and spectral distance between the atoms are larger in the latter. This is better represented in the panels on the *right-hand side*, which show slices of the WVD evaluated at time and frequency that correspond to the centers of the auto- and cross-terms. Here again it is clear that the velocity of the interference oscillation increases with the distance between the auto-terms

interference terms is in the middle of the auto-terms at $[(t_1 + t_2)/2, (f_1 + f_2)/2]$ and the oscillation propagates along a direction perpendicular to the axis connecting the auto-terms. The interference oscillations in both time and frequency directions are faster in the example shown in the upper panel than in the example shown in the lower one because both $|t_2 - t_1|$ and $|f_2 - f_1|$ are larger in the latter. This is better represented in the panels on the right-hand side, which show slices of the WVD evaluated at time and frequency that correspond to the centers of the auto- and cross-terms. Here again it is clear that the velocity of the interference oscillation increases with the distance between the auto-terms.

The oscillatory nature of the interference terms suggests that they can be selectively attenuated or eliminated by filtering the WVD with kernels that act as low-pass filters. The interpretation of the kernels in (9.16)–(9.17) as low-pass filters imposes geometrical constraints on their design, and algorithms that use specific signal properties to automatically determine an optimal kernel, so called signal-dependent kernels, have been proposed [30–33, 35, 40]. One strategy is to determine the properties of the low-pass filter based on the minimum time-frequency distance between the signal components, which is related to the lowest oscillation to be eliminated. For example, energy time-frequency distributions of cardiovascular signals are expected to present both outer and inner interferences. Outer interferences arise from the interactions between simultaneous LF and HF oscillations, while inner interferences may arise from non-linearities in the instantaneous frequencies of this components. Outer interferences may be eliminated by estimating the minimum distance between $|f_{\mathrm{HF}}(t) - f_{\mathrm{LF}}(t)|$, which is usually determined by the respiratory rate, and then by using it to determine the cut-off of the low-pass filter in the time direction [40].

Finally, it is worth mentioning that the complex analytic representation of the signal is often used instead of the original real signal to further reduce the interference terms [41]. This is because the power spectral density of the analytic representation of a signal vanishes for $f < 0$. Therefore, the use of the complex analytic signal assures that positive and negative spectral components do not produce interference terms and improves the readability of the TF distribution.

9.2.4 Time-Frequency Resolution

If we consider the general scheme for the analysis of non-stationary signals shown in Sect. 9.1, it is clear that the final result of the analysis is strictly related to the time-frequency resolution of the representation. For example, in the analysis of cardiovascular signals, the frequency resolution should be good enough to separate the LF and the HF components, even when low respiratory rate reduces their spectral distance. At the same time, time resolution should be good enough to follow quick variations, typically of the order of few seconds.

The time-frequency resolution of a distribution belonging to the Cohen's class depends on the degree of filtering provided by the kernel function, which, in turn, depends on its geometrical properties.

Although in theory the use of bespoke kernels offers the possibility to go beyond the Heisenberg principle, which states that time and frequency resolutions cannot be jointly arbitrarily low because they are inversely related, the violation of this principle introduces cross-terms. Thus, the classical trade-off between time and frequency resolution becomes a trade-off between joint time-frequency resolution and amount of cross-terms. As we will see in the following, this is particularly relevant when time-frequency distributions are used to estimate time-frequency spectra, mainly because positivity, a necessary condition for a distribution to be interpreted as an energy spectral density, requires an appropriate amount of filtering both in time and frequency [42]. Furthermore, the amount of smoothing also depends on the signals, being higher for those signals which present more interference terms [38, 43].

The quantification of the resolution is important because it allows to compare the performance of different distributions and to chose the most appropriate for a given application, and, importantly, it is necessary for a correct interpretation of the results. For example, to correctly assess the latency between a stimulus and a response in an index extracted from a time-frequency distribution it is necessary to know the effect of the smoothing kernel on the localization of the observed features.

Since a time-frequency distribution can be obtained by the convolution between the WVD of a signal and a kernel $\phi_{\text{t-f}}(t,f)$ (see (9.13)), and because the WVD is a non-filtered representation, the resolution of a time-frequency distribution can be calculated by studying $\phi_{\text{t-f}}(t,f)$. The full-width at half maximum or any other measurement of the spreading of the kernel in time and frequency provides an estimate of the time-frequency resolution [14]. This can be easily seen by considering the example of the time-frequency distribution of a temporal impulse $\delta(t-t_0)$ or of a complex exponential $e^{j2\pi f_0 t}$. The WVD of these two signals is a line along frequency and time, respectively. The convolution with $\phi_{\text{t-f}}(t,f)$ results in a time-frequency representation which is the same $\phi_{\text{t-f}}(t,f)$ centered around t_0 or f_0, respectively.

9.2.5 *The Smoothed-Pseudo Wigner-Ville Distribution (SPWVD)*

Several filtered versions of the WVD have been proposed to reduce the interference terms while maintaining a good TF resolution. Among those that have been applied to the study of the cardiovascular variability [44, 45], the smoothed pseudo Wigner-Ville distribution (SPWVD) is one of the most interesting, since it provides an independent control over time and frequency resolution. The SPWVD is the only member of the Cohen's class that makes use of separable kernels, i.e. $\phi_{\text{t-f}}(t,f) =$

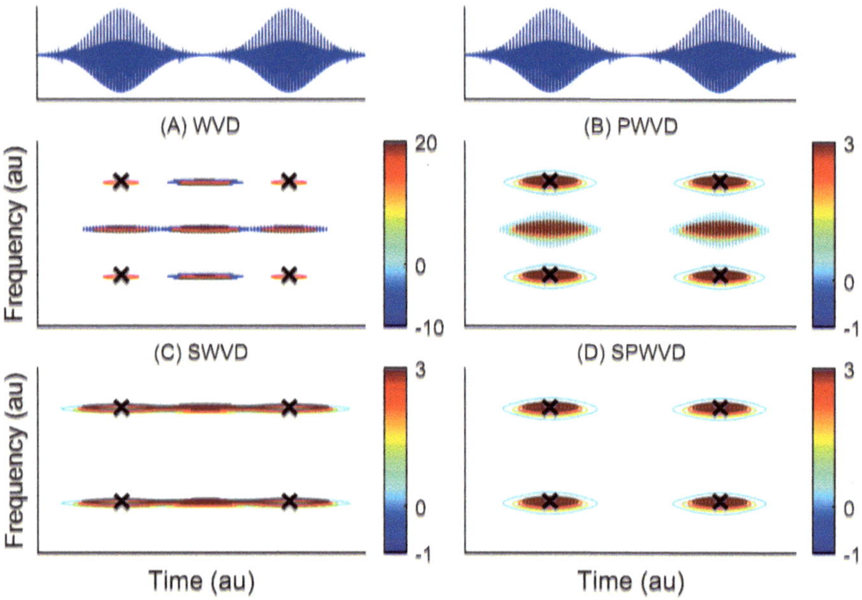

Fig. 9.5 Effect of time-frequency smoothing. *Upper line*: All panels represent a signal composed of four Gaussian atoms. (**a**)–(**d**): Iso-contours of time-frequency representation with different degree of filtering. *Black cross* localize the center of the Gaussian atoms in the time-frequency domain. Time-frequency distributions are normalized to the total energy of the signal. (**a**): Wigner-Ville distribution (no smoothing). (**b**): Pseudo Wigner-Ville distribution (frequency smoothing only). Note that interference terms oscillating in the frequency direction and centered between the *horizontal lines* connecting the atoms have been eliminated. (**c**): Smoothed Wigner-Ville distribution (time smoothing only). Note that interference terms oscillating in the time direction and centered between the *vertical line* connecting the atoms have been eliminated. (**d**): Smoothed pseudo Wigner-Ville distribution (both time and frequency smoothing). All interference terms have been removed at a cost of a increased smoothing

$\phi_t(t)\phi_f(f)$, and offers the possibility to independently control the amount of time and frequency filtering [21, 24, 41].

The SPWVD of a signal $x(t)$ can be estimated in the time-lag domain (t, τ) by means of (9.8) and by using the separable kernel $\phi_{t\text{-}d}(t, \tau) = \phi_t(t)\phi_d(\tau)$:

$$C_x(t,f) = \int \phi_d(\tau)\left[\int \phi_t(t-v)x\left(v+\frac{\tau}{2}\right)x^*\left(v-\frac{\tau}{2}\right)dv\right]e^{-j2\pi f\tau}d\tau \quad (9.19)$$

In this expression, the term in the brackets represents a smoothed version of the non-stationary deterministic autocorrelation function. The time and the frequency resolution of the SPWVD is separately determined by the shape of $\phi_t(t)$ and by the shape of the Fourier transform of $\phi_d(\tau)$, respectively.

Figure 9.5 provides an illustrative example of the effect of using different smoothing strategies. Panels on the upper line represent a signal composed of four

Gaussian atoms, i.e. four complex exponentials modulated by Gaussian functions. Graphics (a)–(d) show four different TF distributions of this signal. The centers of the TF atoms are localized by black crosses. Panel (a) shows the WVD. The four TF atoms are well localized both in time and frequency. However, several interference terms also appear in between the atoms. Importantly, some of these interference terms have an amplitude higher than the amplitude of the auto-terms related to the TF atoms. In panel (b) and (c) the WVD is only filtered in frequency and time, respectively. It is shown that the filtering in frequency cancels the interference terms arising from the interaction of TF atoms with the same central frequencies (see panel (b)), while the filtering in the time direction cancels the interferences terms arising from the interactions of TF atoms with the same central time (see panel (c)). If we compare the distributions in (b)–(c) with the WVD in (a), we observe that the filtering reduces the localization in frequency and time, respectively. Finally, in panel (d), the SPWVD obtained by combining the time and frequency filtering used in (b) and (c) is shown. In this representation, all the interference terms have been eliminated. Additionally, the smoothing applied to the distribution shown in (d) was sufficient to make the SPWVD positive, so that it can now be considered as an estimate of the non-stationary spectrum of signal $x(t)$.

9.2.6 From Time-Frequency Representations to Time-Frequency Spectra

The definition of the power spectral density of a stationary random process is given in (9.3). Contrary to the stationary case, a non-stationary spectrum for a time-varying random process is not uniquely defined. The Wigner-Ville spectrum (WVS), defined as the Fourier transform of a non-stationary auto-correlation function [21]:

$$S_{xx}(t,f) = \int \mathrm{E}\left[x\left(t+\frac{\tau}{2}\right)x^*\left(t-\frac{\tau}{2}\right)\right]e^{-j2\pi\tau f}\,d\tau \tag{9.20}$$

can be seen as a natural extension of the power spectral density in a non-stationary context. It can be shown that, under mild conditions, the Wigner-Ville spectrum (9.20) is equal to the ensemble average of the Wigner-Ville distributions (WVD), $W_{xx}(t,f)$ (see (9.14)), of the realizations of the processes [21]:

$$S_{xx}(t,f) = \mathrm{E}\left[W_{xx}(t,f)\right] \tag{9.21}$$

When only one observed realization is available, $S_{xx}(t,f)$ should be estimated by replacing the expectation $\mathrm{E}[W_{xx}(t,f)]$ with local averaging. This can be achieved by using Cohen's class distributions:

$$\hat{S}_{xx}(t,f;\phi) = W_{xx}(t,f) \otimes \phi_{\text{t-f}}(t,f) \tag{9.22}$$

where $\otimes$ represents the 2D convolution on t and f, and $\phi_{\text{t-f}}(t,f)$ is the smoothing function, which should guarantee the positiveness of the estimated spectra.

9.3 Studying Interactions: Multivariate Time-Frequency Representations

This section describes time-frequency methodologies utilized to study the dynamic interactions between non-stationary signals. These include:

1. Time-frequency coherence: For the analysis of the strength of the local coupling between two signals.
2. Time-frequency phase-difference: For the analysis of the degree of synchronization between two signals.
3. Time-frequency phase-locking: For assessing the degree of coordination between multiple responses.
4. Time-frequency partial spectrum: For assessing the variability of a given signal after removing the amount of variability in synchrony with another signal.

Most of the methodologies presented in this section have been specifically proposed to study dynamic interactions between cardiovascular signals [2, 14, 46].

In the following, when a time-frequency distribution represents a time-varying power spectrum, the symbol $S(t,f)$ will be used instead of $C(t,f)$.

9.3.1 *Time-Frequency Coherence*

Spectral coherence measures the degree of correlation between the spectral components of two signals [47]. This measure requires signals to be stationary and it is therefore inappropriate for studying non-stationary signals. In the analysis of cardiovascular signals, the estimation of spectral coherence in the joint time-frequency (TF) domain has many potential fields of application. For example, the localization of TF regions in which two signals are coupled can be applied in the time-varying characterization of baroreflex [48–50], or in the assessment of the degree of similarity between different signals to validate the use of one signal, and its derived measures, as surrogates of original ones [51].

The spectral coherence function between two stationary zero-mean stochastic processes $x(t)$ and $y(t)$ is a normalized version of the cross power spectral density, $S_{\text{xy}}(f)$, which measured the cross-correlation between two different signals as a function of frequency and is estimated as in (9.3). The spectral coherence is defined as:

$$\gamma(f) = \frac{S_{\text{xy}}(f)}{\sqrt{S_{\text{xx}}(f)S_{\text{yy}}(f)}}, \tag{9.23}$$

The magnitude of the spectral coherence is equal to one for those frequencies for which $x(t)$ and $y(t)$ are related by a linear time-invariant system and it is equal to zero for those frequencies for which $x(t)$ and $y(t)$ are uncorrelated, i.e. when $S_{xy}(f) = 0$ [47, 52].

By analogy with the stationary case, the time-frequency coherence (TFC) can be defined in non-stationary conditions as [52, 53]:

$$\gamma(t,f) = \frac{S_{xy}(t,f)}{\sqrt{S_{xx}(t,f)S_{yy}(t,f)}} \tag{9.24}$$

where $S_{xy}(t,f)$ is the non-stationary cross spectrum, and $S_{xx}(t,f)$ and $S_{yy}(t,f)$ are the non-stationary auto-spectra estimated as in (9.22). A critical aspect of the estimation of the TFC is the elimination of the interference terms, as these may cause the TFC to take values outside the range [0, 1], thus losing its physical meaning. This is shown by the Janssen's formula (also known as outer interference formula) [21, 38]:

$$|W_{xy}(t,f)|^2 = \iint W_{xx}\left(t + \frac{\tau}{2}, f + \frac{\nu}{2}\right) W_{yy}\left(t - \frac{\tau}{2}, f - \frac{\nu}{2}\right) d\tau d\nu \tag{9.25}$$

This expression shows that for given TF point (t_0,f_0), it is possible to have $|W_{xy}(t_0,f_0)| \neq 0$ while $W_{xx}(t_0,f_0) = W_{yy}(t_0,f_0) \approx 0$, and consequently $\gamma(t_0,f_0) > 1$.

To obtain TFC estimates bounded between zero and one, $\phi_{\text{t-f}}(t,f)$ should completely suppress the interference terms [52]. As long as the degree of TF filtering is strong enough, TF coherence can be obtained using (9.24).

In the analysis of the interactions between cardiovascular signals, a simplified version of the multiform–tiltable exponential kernel first proposed in [33], was successfully used [2, 14, 40, 46, 51]:

$$\phi_{\text{d-D}}(\tau, \nu) = \exp\left\{-\pi\left[\left(\frac{\nu}{\nu_0}\right)^2 + \left(\frac{\tau}{\tau_0}\right)^2\right]^{2\lambda}\right\} \tag{9.26}$$

In the ambiguity function domain, the iso-contours of this kernels are ellipses whose eccentricity depends on parameters ν_0 and τ_0. Parameters ν_0 and τ_0 are used to change the length of the ellipse axes aligned along ν (i.e. the degree of time filtering) and τ (i.e. the degree of frequency filtering), respectively. The parameter λ sets the roll off of the filter as well as the size of the tails of the kernel.

A signal-dependent kernel that provides a time-frequency coherence bounded between 0 and 1 can be empirically determined as follows. First, the target TF resolution corresponding to the minimum amount of TF smoothing is decided based on a-priori information about the signals and the experimental settings. The set of parameters $\{\tau_0, \nu_0, \lambda\}$ associated with this TF resolution are used as a starting point. If using this set of parameters $\gamma(t,f) \notin [0, 1]$, the degree of time (or frequency) smoothing is increased until reaching the minimum degree of smoothing associated with meaningful TFC estimates.

Despite the fact that the degree of time and frequency filtering can be tuned almost independently, the resulting time and frequency resolution can not be set arbitrarily, since the condition of having $\gamma(t,f) \in [0, 1]$ imposes a trade of on the minimum joint TF filtering.

A more sophisticated signal-dependent scheme to automatically find a kernel associated to TFC functions bounded between 0 and 1 was proposed in [40].

9.3.1.1 Statistical Assessment of Coherence Significance

As for the classical spectral coherence, time-frequency coherence estimates are not meaningful per se, but they strongly depend on the parameters used in its calculation [47, 54]. The local averaging provided by the smoothing function $\phi_{\text{t-f}}(t,f)$ introduces a bias in the TFC estimates, causing the TFC of two uncorrelated signals to be higher than zero. This dependence introduces an uncertainty in the interpretation of the coherence level. To reduce this uncertainty and to localize TF regions characterized by a significant coherence level, a hypothesis test can be used. The test is based on the point-by-point comparison of the time-frequency coherence estimates, $\gamma(t,f)$, with a threshold function, $\gamma_{\text{TH}}(t,f)$, obtained by estimating the time-frequency coherence between several realizations of locally uncoupled signals. The null hypothesis H_0 to be rejected is that two signals $x(t)$ and $y(t)$, with time-frequency coherence $\gamma(t,f)$, are uncorrelated around a point (t_0,f_0). The procedure is based on surrogate data analysis, where a surrogate is a signal that shares similar features as the original one but lack of the property that needs to be tested [54, 55]. In this case, the analysis consists of the following steps [14]:

1. Generate pairs of uncorrelated test signals $\{\tilde{x}_j(t),\tilde{y}_j(t)\}$.
2. Compute the distribution $\Gamma(t,f) = \{\gamma_1(t,f), \ldots, \gamma_j(t,f), \ldots\}$, where $\gamma_j(t,f)$ is the TFC between the jth realization of test signals $\tilde{x}_j(t)$ and $\tilde{y}_j(t)$;
3. Estimate the threshold $\gamma_{\text{TH}}(t,f;\alpha)$, corresponding to the desired significance level α. According to this framework, α is the probability of wrongly rejecting the null hypothesis, and $\gamma_{\text{TH}}(t,f;\alpha)$ is estimated point-by-point as the $(1-\alpha)$th percentile of $\Gamma(t,f)$;
4. Determine the TF region $\mathcal{R}(t,f;\alpha)$ in which the null hypothesis can be rejected as that where $\gamma(t,f) > \gamma_{\text{TH}}(t,f;\alpha)$.

When the test signals $\tilde{x}_j(t)$ and $\tilde{y}_j(t)$ are white noises the threshold function only depends on the kernel and it is expected to be constant over the entire time-frequency domain. If one of the test signals is replaced by one of the original signals, the threshold function will be signal-dependent and it should be estimated for each pair of signals even when the same kernel function is used for the analysis [14].

9.3.2 Time-Frequency Phase Differences

Phase differences represent the delay between two oscillations that share approximately the same frequency. This delay, which should be much smaller than the oscillation period, can vary during time as a response to a given stimulus and may provide valuable information about the system under investigation.

To introduce the time-frequency phase difference and its estimation, let's model two signals as AM-FM modulated complex exponentials with time-varying LF and HF components:

$$x(t) = A_{x,LF}(t)e^{j\theta_{x,LF}(t)} + A_{x,HF}(t)e^{j\theta_{x,HF}(t)} \tag{9.27}$$
$$y(t) = A_{y,LF}(t)e^{j\theta_{y,LF}(t)} + A_{y,HF}(t)e^{j\theta_{y,HF}(t)}$$

In these expressions, LF and HF indicate the low frequency component, traditionally inside 0.04–0.15 Hz, and the high frequency component, traditionally inside 0.15–0.4 Hz, respectively [10]; $\theta_{k,B}(t)$, with $B \in [LF, HF]$ and $k \in [x, y]$, is the instantaneous phase, related to the instantaneous frequency by $f_{k,B}(t) = (d\theta_{k,B}(t)/dt)/(2\pi)$.

The phase difference between these spectral components, $\theta_B(t) = \theta_{x,B}(t) - \theta_{y,B}(t)$, is expected to be time-varying. The time-course of the phase difference between two signals $x(t)$ and $y(t)$, evaluated for a specific spectral component B, can be estimated as:

$$\theta_B(t) = 2\pi \int_0^t [f_{x,B}(\tau) - f_{y,B}(\tau)]d\tau \tag{9.28}$$

However, the estimation of the phase difference from the instantaneous frequencies of the spectral components of two signals is very sensitive to estimation errors in $f_{k,B}(t)$, since an estimation error at t_0 affects all $\theta_B(t > t_0)$. The characterization of the phase differences in the joint time-frequency domain may therefore be preferable. The time-frequency representation of the phase difference between two signals is defined as:

$$\Theta(t,f) = \arg\,[S_{xy}(t,f)] = \arctan\left[\frac{\Im[S_{xy}(t,f)]}{\Re[S_{xy}(t,f)]}\right] \tag{9.29}$$

This expression provides a quantification of the phase difference between the spectral components of two signals in the time-frequency domain and is bounded $\Theta(t,f) \in [-\pi, \pi]$.

Other approaches to estimate the phase differences between two non-stationary signals have used the wavelet transform [56–59], the Rihaczek transform [60, 61] and the reduced interference distributions [29, 62].

A time-varying index describing the temporal evolution of the phase difference between a given spectral component B of two signals can be estimated by averaging the time-frequency representation of the phase difference in a specific time-frequency region $\Omega_{\mathrm{B}}^{(\theta)}$. Among a wide range of different possibilities, one that has been previously used is $\Omega_{\mathrm{B}}^{(\theta)}$ defined as a time-frequency region centered around the instantaneous frequency of a spectral component of the cross spectrum, $f_{\mathrm{B}}^{(\mathrm{xy})}(t)$ (estimated as the maximum of the instantaneous spectral peak) in which coherence is statistically significant [2]. These conditions are necessary to obtain robust and accurate estimates since the estimation of a phase difference is relevant only in those time intervals in which the two signals are locally coupled and therefore are sharing approximately the same instantaneous frequency. This region can be defined as [2]:

$$\Omega_{\mathrm{B}}^{(\theta)} \equiv \left\{(t,f)\middle| f = f_{\mathrm{B}}^{(\mathrm{xy})}(t) \pm \frac{\Delta_{\mathrm{f}}}{2} \cap \gamma(t,f) > \gamma_{\mathrm{TH}}(t,f)\right\}; \tag{9.30}$$

where Δ_{f} is the width of the spectral band centered around $f_{\mathrm{B}}^{(\mathrm{xy})}(t)$, which should be related to the frequency resolution of the TF distributions. The index that quantifies the phase difference, $\theta_{\mathrm{B}}(t)$, can be estimated (in radians) by averaging the time-frequency phase difference spectrum in $\Omega_{\mathrm{B}}^{(\theta)}$:

$$\theta_{\mathrm{B}}(t) = \left[\int_{\Omega_{\mathrm{B}}^{(\theta)}} \Theta(t,f)df\right] \bigg/ \left[\int_{\Omega_{\mathrm{B}}^{(\theta)}} df\right]; \tag{9.31}$$

The time delay associated to $\theta_{\mathrm{B}}(t)$ can be estimated (in seconds) by the index $\mathcal{D}_{\mathrm{B}}(t)$, defined as:

$$\mathcal{D}_{\mathrm{B}}(t) = \frac{\theta_{\mathrm{B}}(t)}{2\pi f_{\mathrm{B}}^{(\mathrm{xy})}(t)} \tag{9.32}$$

Assessing TF Coherence and Phase Differences In the following, the results of a simulation study designed to assess TF coherence and phase difference estimators is proposed. The purpose of the study is to demonstrate the potential of these methodologies to track fast changes in the dynamic interactions between cardiovascular signals.

The signals used in the simulation are modified versions of the RR variability signals recorded during the tilt table test [2]. They are obtained as:

$$x(t) = a_{\mathrm{RRV}}(t) + \xi_{\mathrm{x}}(t) \tag{9.33}$$

$$y(t) = [\gamma_0(t)\alpha_0(t)\exp(j\theta_0(t))]a_{\mathrm{RRV}}(t) + \xi_{\mathrm{y}}(t) \tag{9.34}$$

where $a_{\mathrm{RRV}}(t)$ is the complex analytic signal representation of the RRV signal, and $\xi_{\mathrm{x}}(t)$ and $\xi_{\mathrm{y}}(t)$ are complex white Gaussian noises whose amplitude is set to give a $SNR = 10\,\mathrm{dB}$. Indices $\gamma_0(t) \in \{0, 1\}$ and $\theta_0(t) \in [-2.2, -0.8]$ rad represent the reference time-course of the local coupling and phase difference, respectively. Index $\alpha_0(t)$ is a time-varying amplitude modulation. The estimation of these indices is challenging because: the TF structure of the signals reflects the complexity of real non-stationary biomedical signals; there are short intervals during which signals are not locally coupled and during which phase difference should not be estimated, the shortest lasting 30 s; phase difference varies non-linearly with time; the amplitude modulation first undergoes stepwise changes and then recovery toward higher values following a non-linear time-course; noise is added; finally, all these patterns imply simultaneous amplitude and frequency modulation of the signals. Results were obtained over 50 realizations and are shown in Fig. 9.6, where the time-course of $\gamma_0(t)$, $\theta_0(t)$ and $\alpha_0(t)$ is represented by a red-bold line, while estimates obtained through time-frequency analysis in represented in gray. Signals $x(t)$ and $y(t)$ are uncoupled during two intervals where $\gamma_0(t) = 0$, their phase difference changes as a sinusoids and the amplitude modulation presents slow and abrupt non-stationarities. Results demonstrate that the presented framework provides an accurate estimation of the changes in the signal properties and their dynamic interactions.

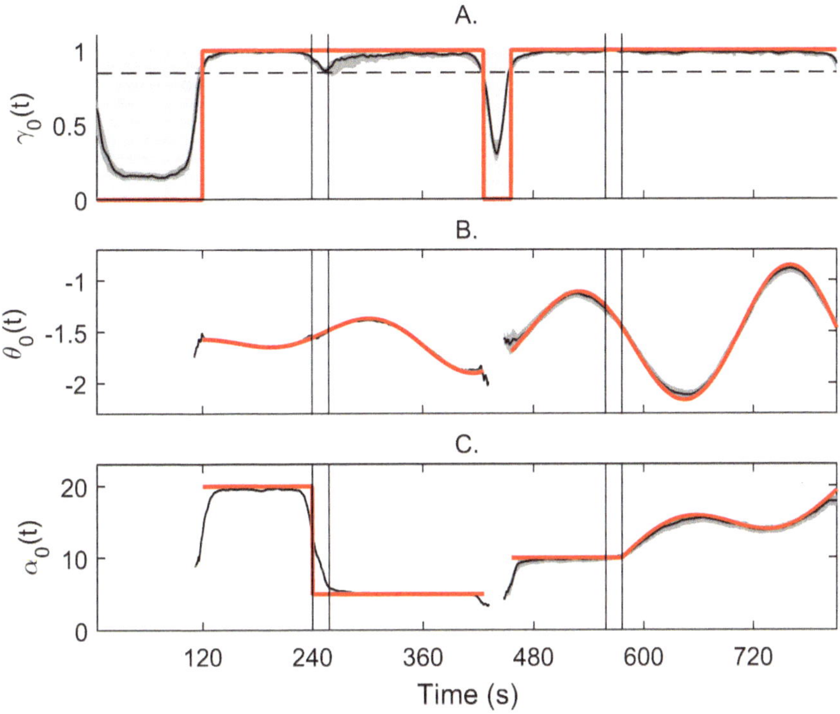

Fig. 9.6 Simulation: Time-course of (**a**) Coherence, (**b**) Phase differences, (**c**) Amplitude modulation. Reference values $\gamma_0(t)$, $\theta_0(t)$ and $\alpha_0(t)$ described in Eq. (9.33) are represented in *bold red lines*. Estimates are represented as median (*black line*) and interquartile range (*gray region*). *Vertical lines* mark supine positions and head-up tilt. *Light gray areas* represent the intervals during which signals are uncoupled. The *horizontal dashed line* in (**a**) represents the threshold for statistical significance

9.3.3 Time-Frequency Phase Locking

The time-frequency representation of phase differences offers the possibility of estimating the degree of phase-locking between different signals [56, 57, 60, 61, 63, 64]. The phase locking is used to assess whether a given stimulus provokes similar changes in the time-frequency phase difference spectrum. Its time-frequency representation is [65]:

$$\Psi(t,f) = \left| \frac{1}{L} \sum_{i=1}^{L} e^{j2\pi\Theta_i(t,f)} \right| = \left| \frac{1}{L} \sum_{i=1}^{L} \frac{S_{xy,i}(t,f)}{|S_{xy,i}(t,f)|} \right| \tag{9.35}$$

By construction, $\Psi(t,f)$ is bounded between 0 and 1, $\Psi(t,f) \in [0,1]$, being $\Psi(t,f) = 1$ if $\Theta_i(t,f)$ does not vary across different realizations i, and being $\Psi(t,f) = 0$ if $\Theta_i(t,f)$ is characterized by a zero mean random distribution across i. Indeed, if for a given point (t_0,f_0), the complex terms in the summation in (9.35) maintain similar phase for all i, $\Psi(t_0,f_0) \approx 1$.

Therefore, the phase locking measures the degree of similarity of the time-frequency structure of a group of signals, by pairwise comparing their time-frequency phase differences $\Theta_i(t,f)$.

9.3.4 Time-Frequency Partial Spectrum

The estimation of a time-frequency partial spectrum is interesting in the analysis of systems that can be modeled as the sum of two single-input/output models where inputs $u(t)$ and $v(t)$ are uncorrelated (see Fig. 9.7):

$$y(t) = x_1(t) + x_2(t) = h_1(t) * u(t) + h_2(t) * v(t) \tag{9.36}$$

where $*$ represents convolution with time-varying transfer functions $h_1(t)$ and $h_2(t)$.

In many applications, one needs to decompose the TF spectrum of $y(t)$ into a part related to $x_1(t)$ and another related to $x_2(t)$. This section describes how to obtain such a decomposition when only the output $y(t)$ and one of the inputs, i.e. either $u(t)$

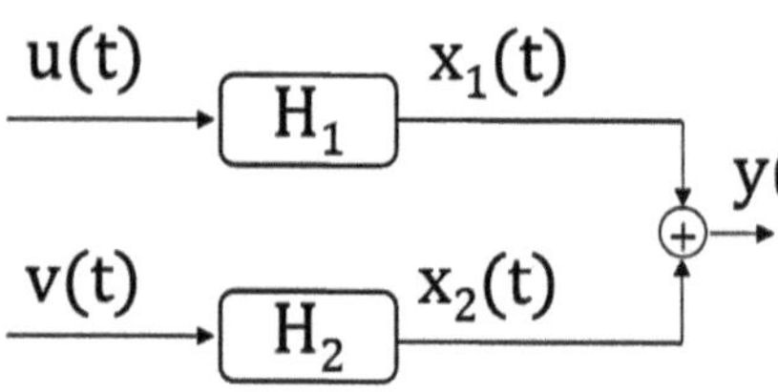

Fig. 9.7 A multiple-input single-output model. Inputs $u(t)$ and $v(t)$ are uncorrelated

or $v(t)$, are known. In the time-frequency domain, input/output auto spectrum and input/output cross spectrum relations are:

$$H_1(t,f) = \frac{S_{u,1}(t,f)}{S_{u,u}(t,f)} \tag{9.37}$$

$$|H_1(t,f)|^2 = \frac{S_{1,1}(t,f)}{S_{u,u}(t,f)} \tag{9.38}$$

Taking into account that the inputs are uncorrelated, the TF spectrum of the output is [66]:

$$S_{y,y}(t,f) = |H_1|^2 S_{1,1}(t,f) + |H_2|^2 S_{2,2}(t,f) \tag{9.39}$$

while TF coherence between $y(t)$ and input $u(t)$ is:

$$\gamma_{y,u}^2(t,f) = \frac{1}{1 + S_{2,2}(t,f)/S_{1,1}(t,f)} \tag{9.40}$$

where $S_{1,1}(t,f)$ and $S_{2,2}(t,f)$ are auto spectra of $x_1(t)$ and $x_2(t)$. In this formulation, $x_2(t)$ is considered a noise term uncorrelated with input $u(t)$. Similarly, the TF coherence between $y(t)$ and input $v(t)$ is

$$\gamma_{y,v}^2(t,f) = \frac{1}{1 + S_{1,1}(t,f)/S_{2,2}(t,f)} \tag{9.41}$$

where $x_1(t)$ is considered a noise term uncorrelated with input $v(t)$.

The TF coherence function can be used to separate the output spectrum into two partial spectra: the coherent output spectrum and the partial (or conditioned) spectrum [66]:

$$S_{y:1}(t,f) = S_{1,1}(t,f) = \gamma_{y,u}^2(t,f) S_{y,y}(t,f) = (1 - \gamma_{y,v}^2(t,f)) S_{y,y} \tag{9.42}$$

$$S_{y:2}(t,f) = S_{2,2}(t,f) = \gamma_{y,v}^2(t,f) S_{y,y}(t,f) = (1 - \gamma_{y,u}^2(t,f)) S_{y,y} \tag{9.43}$$

Note that $S_{1,1}(t,f)$ can be calculated from output $y(t)$ and input $u(t)$ even if $x_1(t)$ cannot be measured. Similarly, $S_{2,2}(t,f)$ can also be estimated from $y(t)$ and input $u(t)$ without measuring $x_2(t)$. Definitions (9.42)–(9.43) show that a partial TF spectrum is obtained from the original spectrum by using the TFC function as a TF filter that reduces the amplitude of the original spectrum by a factor proportional to the TFC. In regions where the TFC is equal to one, the TF partial spectrum is equal to zero. Whereas in regions where the TFC is zero, the original and the partial spectra will have the same amplitude. This approach has been used to estimate changes in the variability of the QT interval unrelated with heart rate variability, by using records of QT and RR intervals [67].

9.4 Example of Analysis of Cardiovascular Signals

The time-frequency framework described in this chapter has been applied to the study of cardiovascular signals and their dynamic interactions [2, 14, 46, 65], and references to specific studies have been given and briefly discussed in the previous sections. In this section, cardiovascular signals from one healthy volunteer undergoing a head-up tilt test are studied as a representative example. This test includes three phases: early supine position (4 min), head-up tilt at 70° (5 min) and late supine position (4 min) [2]. The RR interval was measured from the surface ECG, systolic arterial blood pressure was measured non-invasively from the finger using a Finometer device, and respiration was measured as chest movement using a thoracic belt. Results shown in Figs. 9.1 and 9.6 are from the same subject. Time-frequency analysis was performed using the kernel function described in (9.26). The time and frequency resolutions, measured as the full-width at half maximum of the kernel function evaluated at its center [14], were equal to 11 s and 0.039 Hz, respectively. A statistical test to estimate the threshold for significant coupling was conducted using white Gaussian noise as uncoupled surrogate data. The threshold associated with a 5% significance level was equal to 0.849.

Figure 9.8 shows the time-frequency spectra of RR variability (RRV), systolic arterial pressure variability (SAPV) and the respiratory signal (R). RRV and SAPV show spectral components in both the LF (0.04–0.15 Hz) and HF (0.15–0.40 Hz) bands, while the spectral power of the respiratory signal is in the HF band only. Interestingly, both amplitudes and frequencies of these signals clearly change during the test. The time-frequency resolution of these representations allows to track these changes. In particular, subtle changes in the instantaneous frequency of the respiratory oscillations can be followed accurately.

Time-frequency coherence (TFC) and time-frequency phase difference (TFPD) between each pair of signals is also reported. Regions of the time-frequency domain where the coherence between two signals is statistically significant are encircled by a black line. These time-frequency coherence representations demonstrate that the three signals are strongly locally coupled. Significant level of coupling between RRV and SAPV is found in both LF and HF, whereas respiration is strongly coupled with only HF components of RRV and SAPV. The coupling in HF is maintained during the entire test, demonstrating that respiratory oscillations in RRV and SAPV are persistent. Whereas the coupling in the LF band is intermittent.

The inspection of the TFPD representations reveals that the phase difference between coupled oscillations in these cardiovascular signals is affected by head-up tilt. A clear and sudden change in the phase difference after the onset of head-up tilt is registered between RRV and SAPV, RRV and R, and between SAPV and R. In the late supine position, baseline values are restored, suggesting that head-up tilt is provoking a shift in the degree of temporal synchronization of respiratory-related oscillations within the cardiovascular system.

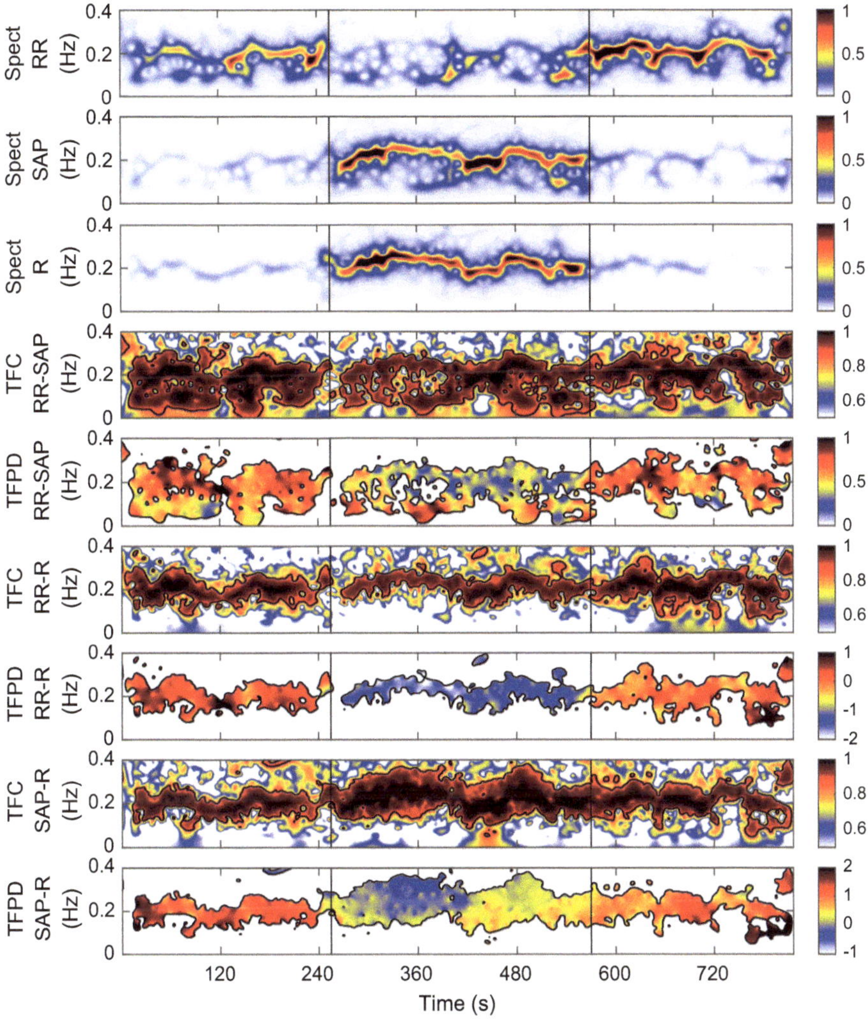

Fig. 9.8 Time-frequency analysis of RR interval variability (RRV), systolic arterial blood pressure variability (SAPV) and respiration (R) during head-up tilt table test. *Vertical lines* separate the early supine phase (4 min), from the head-up tilt phase (5 min), and the late supine phase (4 min). From *top to bottom*: The *first three panels* show the time-frequency spectra of RRV, SAPV and R, respectively. From the *forth panels onwards*, figures show the time-frequency coherence and phase difference of RRV and SAPV, RRV and R, and finally of SAPV and R. Black lines in the coherence and phase differences panels encircle the regions where coherence is statistically significant ($\alpha <$ 5%). Time-frequency phase is shown only in these regions. *Colors* represent arbitrary units for the TF spectra, normalized units for the TFC, and radians for the TFPD

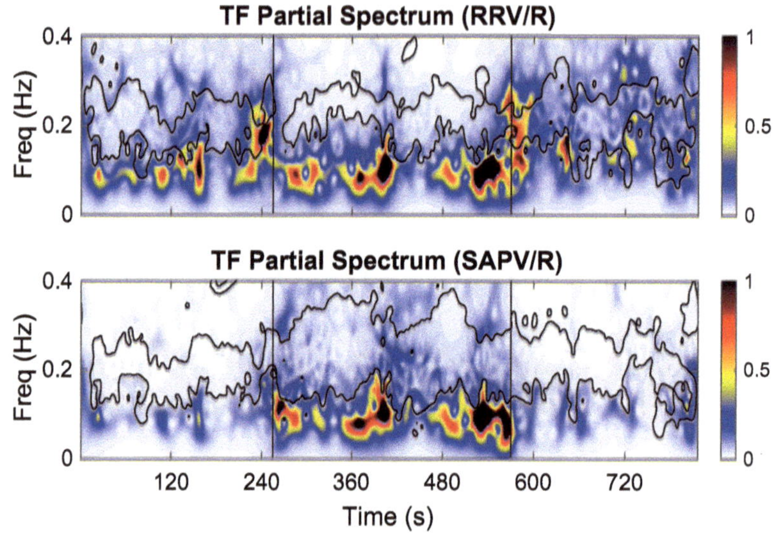

Fig. 9.9 Time-frequency partial spectra of RR interval variability (RRV) and systolic arterial pressure variability (SAPV) unrelated to respiration during head-up tilt table test. *Vertical lines* separate the early supine phase (4 min), from the head-up tilt phase (5 min), and the late supine phase (4 min). These representations greatly differ from the original TF spectra shown in Figure 9.8. In particular, HF components strongly coupled with respiration have been almost entirely removed. The black lines in these panels encircle the regions where coherence with respiration is statistically significant ($\alpha < 5\%$). Note that, as expected, inside these regions the time-frequency partial spectra show very little power. In the TF spectra, *colors* represent arbitrary units

Figure 9.9 shows the time-frequency partial spectra of RRV (upper panel) and SAPV (lower panel) after removing the components in synchrony with respiration. These representations have been obtained using the definition (9.42) and the kernel (9.26). The comparison between these representations and the original RRV and SAPV spectra shown in Fig. 9.8 suggests that the HF components of both signals are almost entirely due to respiration. The residual components still present in the TF spectra are mainly in the LF band. Black lines encircle the TF regions where TFC with respiration is statistically significant. As expected, in these regions the amplitude of the TF partial spectra is particularly low.

9.5 Final Considerations

In this chapter, we have described a framework for the assessment of the time-frequency structure of non-stationary signals and their dynamic interactions. We have used simulated data to demonstrate its reliability and we have provided examples based on cardiovascular signals recorded during a tilt-table test, a procedure that quickly modifies arterial blood pressure and activates a neural response (the baroreflex) to restore baseline values. The approach described in this chapter has been proven to be robust and accurate [2, 14, 46] and it is certainly a useful tool to track the cardiovascular response to different stimuli, such as emotions [68], orthostatic stress [2], exercise stress test [69] etc., and to detect subtle and transient dynamics such as repolarization alternans [70].

In recent years, different methods to estimate time-varying features of non-stationary signals have been proposed. Most of them are based on parametric autoregressive modeling [4, 71–73]. The performance of these methods is related to their capability of capturing signal dynamics and, in extremely non-stationary conditions, they have been observed to perform less accurately than non-parametric methods [45, 74].

Time-varying estimators based on non-parametric methods have the advantage of not requiring any assumption on the TF structure of the signals and are relatively easy to estimate. Beside those described in this chapter, others have been used in biomedical applications. These include the time-scale coherence [75] based on the continuous wavelet transform, which has been proposed to study cardiovascular dynamics [50, 76, 77] as well as neural connectivity [63, 78–80]. These methods are characterized by a non-homogeneous TF resolution and are suited for signals presenting high frequency components of short duration and low frequency components of long duration. In the late Nineties, estimators of time-frequency coherence based on multitaper spectrogram [81, 82] have also been proposed to study the coupling between neuronal signals [83] and atrial fibrillation [84].

The main limitation of the time-frequency framework described in this chapter is that it cannot be used to infer causality or to assess the direction of the coupling. In this case, other approaches, mainly parametric [85], should be used.

Acknowledgements The Matlab code to conduct the analysis and create the figures shown in this chapter is are available at http://www.micheleorini.com/matlab-code/.

References

1. Borgnat, P., Flandrin, P., Honeine, P., Richard, C., Xiao, J.: Testing stationarity with surrogates: A time-frequency approach. IEEE Trans. Signal Process. **58**(7), 3459–3470 (2010)
2. Orini, M., Laguna, P., Mainardi, L.T., Bail6n, R.: Assessment of the dynamic interactions between heart rate and arterial pressure by the cross time-frequency analysis. Physiol. Meas. **33**, 315–331 (2012)
3. Mainardi, L.T., Bianchi, A.M., Baselli, G., Cerutti, S.: Pole-tracking algorithms for the extraction of time-variant heart rate variability spectral parameters. IEEE Trans. Biomed. Eng. **42**(3), 250–259 (1995). ID: 1
4. Mainardi, L.T., Bianchi, A.M., Furlan, R., Piazza, S., Barbieri, R., di Virgilio, V., Malliani, A., Cerutti, S.: Multivariate time-variant identification of cardiovascular variability signals: a beat-to-beat spectral parameter estimation in vasovagal syncope. IEEE Trans. Biomed. Eng. **44**, 978–989 (1997)
5. Barbieri, R., Brown, E.: Analysis of heartbeat dynamics by point process adaptive filtering. IEEE Trans. Biomed. Eng. **53**(1), 4–12 (2006)

6. Welch, P.: The use of fast fourier transform for the estimation of power spectra: a method based on time averaging over short, modified periodograms. IEEE Trans. Audio Electroacoust. **15**, 70–73 (1967)
7. Thomson, D.J.: Spectrum estimation and harmonic analysis. Proc. IEEE **70**(9), 1055–1096 (1982)
8. Akselrod, S., Gordon, D., Ubel, F.A., Shannon, D.C., Barger, A.C., Cohen, R.J.: Power spectrum analysis of heart rate fluctuations: a quantitative probe of beat-to-beat cardiovascular control. Science **213**, 220–222 (1981)
9. Saul, J.P., Arai, Y., Berger, R.D., Lilly, L.S., Colucci, W.S., Cohen, R.J.: Assessment of autonomic regulation in chronic congestive heart failure by heart rate spectral analysis. Am. J. Cardiol. **61**(15), 1292–1299 (1988)
10. Task Force of the European Society of Cardiology the North American Society of Pacing: Heart rate variability: standards of measurement, physiological interpretation, and clinical use. Circulation **93**(5), 1043–1065 (1996)
11. Montano, N., Ruscone, T.G., Porta, A., Lombardi, F., Pagani, M., Malliani, A.: Power spectrum analysis of heart rate variability to assess the changes in sympathovagal balance during graded orthostatic tilt. Circulation **90**, 1826–1831 (1994)
12. Malliani, A., Pagani, M., Lombardi, F., Cerutti, S.: Cardiovascular neural regulation explored in the frequency domain. Circulation **84**, 482–492 (1991)
13. Piccirillo, G., Ogawa, M., Song, J., Chong, V., Joung, B., Han, S., Magra, D., Chen, L., Lin, S.-F., Chen, P.-S.: Power spectral analysis of heart rate variability and autonomic nervous system activity measured directly in healthy dogs and dogs with tachycardia-induced heart failure. Heart Rhythm. **6**(4), 546–552 (2009)
14. Orini, M., Bailon, R., Mainardi, L.T., Laguna, P., Flandrin, P.: Characterization of dynamic interactions between cardiovascular signals by time-frequency coherence. IEEE Trans. Biomed. Eng. **59**, 663–673 (2012)
15. Grossman, P., Taylor, E.W.: Toward understanding respiratory sinus arrhythmia: Relations to cardiac vagal tone, evolution and biobehavioral functions. Biol. Psychol. **74**, 263–285 (2007)
16. Eckberg, D.L.: Point: counterpoint: respiratory sinus arrhythmia is due to a central mechanism vs. respiratory sinus arrhythmia is due to the baroreflex mechanism. J. Appl. Physiol. **106**, 1740–1742 (2009); discussion 1744
17. Julien, C.: The enigma of Mayer waves: facts and models. Cardiovasc. Res. **70**, 12–21 (2006)
18. Eckberg, D.L.: Sympathovagal balance: a critical appraisal. Circulation **96**(9), 3224–3232 (1997)
19. Malliani, A., Pagani, M., Montano, N., Mela, G.S.: Sympathovagal balance: a reappraisal. Circulation **98**(23), 2640–2643 (1998)
20. Malliani, A., Julien, C., Billman, G.E., Cerutti, S., Piepoli, M.F., Bernardi, L., Sleight, P., Cohen, M.A., Tan, C.O., Laude, D., Elstad, M., Toska, K., Evans, J.M., Eckberg, D.L.: Cardiovascular variability is/is not an index of autonomic control of circulation. J. Appl. Physiol. **101**(2), 684–688 (2006)
21. Flandrin, P.: Time-Frequency/Time-Scale Analysis. Academic Press, New York (1999)
22. Cohen, L.: Time-Frequency Analysis, vol. 778. Prentice Hall, Englewood Cliffs (1995)
23. Hlawatsch, F., Auger, F., (eds.): Time-Frequency Analysis, Concepts and Methods. Wiley, London (2008)
24. Cohen, L.: Time-frequency distributions-a review. Proc. IEEE **77**, 941–981 (1989)
25. Hlawatsch, F., Boudreaux-Bartels, G.F.: Linear and quadratic time-frequency signal representations. IEEE Signal Process. Mag. **9**, 21–67 (1992)
26. Auger, F., Flandrin, P., Gonçalvès, P., Olivier, L.: Time-Frequency Toolbox. Technical Report http://www.nongnu.org/tftb/
27. Flandrin, P.: Ambiguity Function. Time-Frequency Signal Analysis and Processing, pp. 160–167. Elsevier, Amsterdam (2003)
28. Auger, F., Chassande-Mottin, É.: Quadratic Time-Frequency Analysis I: Cohen's Class, pp. 131–163. ISTE, London (2010)

29. Jeong, J., Williams, W.J.: Kernel design for reduced interference distributions. IEEE Trans. Signal Process. **40**(2), 402–412 (1992)
30. Baraniuk, R.G., Jones, D.L.: A signal-dependent time-frequency representation: optimal kernel design. IEEE Trans. Signal Process. **41**, 1589–1602 (1993)
31. Baraniuk, R.G., Jones, D.L.: A signal-dependent time-frequency representation: fast algorithm for optimal kernel design. IEEE Trans. Signal Process. **42**(1), 134–146 (1994)
32. Cunningham, G.S., Williams, W.J.: Kernel decomposition of time-frequency distributions. IEEE Trans. Signal Process. **42**(6), 1425–1442 (1994)
33. Costa, A., Boudreau-Bartels, G.: Design of time-frequency representations using a multiform, tiltable exponential kernel. IEEE Trans. Signal Process. **43**, 2283–2301 (1995)
34. Arce, G.R., Hasan, S.R.: Elimination of interference terms of the discrete wigner distribution using nonlinear filtering. IEEE Trans. Signal Process. **48**(8), 2321–2331 (2000)
35. Aviyente, S., Williams, W.J.: Multitaper marginal time-frequency distributions. Signal Process. **86**(2), 279–295 (2006)
36. Wahlberg, P., Hansson, M.: Kernels and multiple windows for estimation of the Wigner-Ville spectrum of gaussian locally stationary processes. IEEE Trans. Signal Process. **55**(1), 73–84 (2007)
37. Hlawatsch, F.: Interference terms in the Wigner distribution. In: Cappellini, V., Constantinides, A. (eds.) Digital Signal Processing, vol. 84, pp. 363–267. North-Holland, Amsterdam (1984)
38. Hlawatsch, F., Flandrin, P.: The interference structure of the Wigner distribution and related time-frequency signal representations. The Wigner Distribution - Theory and Applications in Signal Processing, pp. 59–113. Elsevier, Amsterdam (1997)
39. Flandrin, P.: Some features of time-frequency representations of multicomponent signals. In: Acoustics, Speech, and Signal Processing, IEEE International Conference on ICASSP'84, vol. 9, pp. 266–269 (1984)
40. Orini, M., Bail6n, R., Mainardi, L., Mincholé, A., Laguna, P.: Continuous quantification of spectral coherence using quadratic time-frequency distributions: error analysis and application. In: International Conference on Computers in Cardiology, pp. 681–684 (2009)
41. Martin, W., Flandrin, P.: Wigner-ville spectral analysis of nonstationary processes. IEEE Trans. Acoust. Speech Signal Process. **33**(6), 1461–1470 (1985), ID: 1
42. Janssen, A., Claasen, T.: On positivity of time-frequency distributions. IEEE Trans. Acoust. Speech Signal Process. **33**, 1029–1032 (1985)
43. Matz, G., Hlawatsch, F.: Nonstationary spectral analysis based on time-frequency operator symbols and underspread approximations. IEEE Trans. Inf. Theory **52**(3), 1067–1086 (2006). ID: 1
44. Mainardi, L.T.: On the quantification of heart rate variability spectral parameters using time-frequency and time-varying methods. Philos. Transact. A Math. Phys. Eng. Sci. **367**(1887), 255–275 (2009)
45. Pola, S., Macerata, A., Emdin, M., Marchesi, C.: Estimation of the power spectral density in nonstationary cardiovascular time series: assessing the role of the time-frequency representations (TFR). IEEE Trans. Biomed. Eng. **43**, 46–59 (1996)
46. Orini, M., Bailon, R., Laguna, P., Mainardi, L., Barbieri, R.: A multivariate time-frequency method to characterize the influence of respiration over heart period and arterial pressure. EURASIP J. Adv. Signal Process. **2012**(1), 214 (2012)
47. Carter, G.C.: Coherence and time delay estimation. Proc. IEEE **75**, 236–255 (1987)
48. Di Rienzo, M., Parati, G., Radaelli, A., Castiglioni, P.: Baroreflex contribution to blood pressure and heart rate oscillations: time scales, time-variant characteristics and nonlinearities. Philos. Trans. R. Soc. A Math. Phys. Eng. Sci. **367**, 1301–1318 (2009)
49. Orini, M., Mainardi, L.T., Gil, E., Laguna, P., Bailon, R.: Dynamic assessment of spontaneous baroreflex sensitivity by means of time-frequency analysis using either rr or pulse interval variability. Conf. Proc. IEEE Eng. Med. Biol. Soc. **1**, 1630–1633 (2010)
50. Keissar, K., Maestri, R., Pinna, G.D., Rovere, M.T.L., Gilad, O.: Non-invasive baroreflex sensitivity assessment using wavelet transfer function-based time-frequency analysis. Physiol. Meas. **31**, 1021–1036 (2010)

51. Gil, E., Orini, M., Bailon, R., Vergara, J.M., Mainardi, L., Laguna, P.: Photoplethysmography pulse rate variability as a surrogate measurement of heart rate variability during non-stationary conditions. Physiol. Meas. **31**(9), 1271–1290 (2010)
52. Matz, G., Hlawatsch, F.: Time-frequency coherence analysis of nonstationary random processes. In: Proceedings of the Tenth IEEE Workshop Statistical Signal and Array Processing, pp. 554–558 (2000)
53. White, L.B., Boashash, B.: Cross spectral analysis of nonstationary processes. IEEE Trans. Inf. Theory **36**, 830–835 (1990)
54. Faes, L., Pinna, G.D., Porta, A., Maestri, R., Nollo, G.: Surrogate data analysis for assessing the significance of the coherence function. IEEE Trans. Biomed. Eng. **51**, 1156–1166 (2004)
55. Schreiber, T., Schmitz, A.: Surrogate time series. Phys. D **142**(3–4), 346–382 (2000)
56. Lachaux, J.P., Rodriguez, E., Martinerie, J., Varela, F.J.: Measuring phase synchrony in brain signals. Hum. Brain Mapp. **8**(4), 194–208 (1999)
57. Lachaux, J., Rodriguez, E., Quyen, M.L.V., Lutzand, A., Martinerie, J., Varela, F.: Studying single-trials of phase-synchronous activity in the brain. Int. J. Bifurcation Chaos **10**, 2429–2439 (2000)
58. Quyen, M.L.V., Foucher, J., Lachaux, J., Rodriguez, E., Lutz, A., Martinerie, J., Varela, F.J.: Comparison of Hilbert transform and wavelet methods for the analysis of neuronal synchrony. J. Neurosci. Methods **111**, 83–98 (2001)
59. Rudrauf, D., Douiri, A., Kovach, C., Lachaux, J.-P., Cosmelli, D., Chavez, M., Adam, C., Renault, B., Martinerie, J., Quyen, M.L.V.: Frequency flows and the time-frequency dynamics of multivariate phase synchronization in brain signals. Neuroimage **31**, 209–227 (2006)
60. Aviyente, S., Bernat, E.M., Evans, W.S., Sponheim, S.R.: A phase synchrony measure for quantifying dynamic functional integration in the brain. Hum. Brain Mapp. **32**(1), 80–93 (2011)
61. Aviyente, S., Evans, W.S., Bernat, E.M., Sponheim, S.: A time-varying phase coherence measure for quantifying functional integration in the brain. In: Proceedings of IEEE International Conference on Acoustics, Speech and Signal Processing ICASSP 2007, vol. 4, pp. IV–1169–IV–1172, 15–20 (2007)
62. Shin, Y., Gobert, D., Sung, S.-H., Powers, E.J., Park, J.B.: Application of cross time-frequency analysis to postural sway behavior: the effects of aging and visual systems. IEEE Trans. Biomed. Eng. **52**, 859–868 (2005); Time-frequency phase cross time-frequency analysis
63. Lachaux, J.-P., Lutz, A., Rudrauf, D., Cosmelli, D., Quyen, M.L.V., Martinerie, J., Varela, F.: Estimating the time-course of coherence between single-trial brain signals: an introduction to wavelet coherence. Neurophysiol. Clin. **32**, 157–174 (2002)
64. Orini, M., Laguna, P., Mainardi, L., Bailón, R.: Characterization of the dynamic interactions between cardiovascular signals by cross time-frequency analysis: phase differences, time delay and phase locking. In: International Conference on Numerical Method in Engineering (2011)
65. Orini, M., Bailon, R., Mainardi, L.T., Laguna, P.: Time-frequency phase differences and phase locking to characterize dynamic interactions between cardiovascular signals. Conf. Proc. IEEE Eng. Med. Biol. Soc. **2011**, 4689–4692 (2011)
66. Bendat, J.S., Piersol, A.G.: Multiple-input/output relationships. In: Random Data, pp. 201–247. Wiley, New York (2012)
67. Orini, M., Taggart, P., Lambiase, P.D.: A multivariate time-frequency approach for tracking QT variability changes unrelated to heart rate variability. In: 2016 IEEE 38th Annual International Conference of the Engineering in Medicine and Biology Society (EMBC), pp. 924–927. IEEE, Orlando. doi:10.1109/EMBC.2016.7590852
68. Orini, M., Bailón, R., Enk, R., Koelsch, S., Mainardi, L., Laguna, P.: A method for continuously assessing the autonomic response to music-induced emotions through HRV analysis. Med. Biol. Eng. Comput. **48**, 423–433 (2010)
69. Bailon, R., Garatachea, N., De La Iglesia, I., Casajus, J.A., Laguna, P.: Influence of running stride frequency in heart rate variability analysis during treadmill exercise testing. IEEE Trans. Biomed. Eng. **60**, 1796–1805 (2013). doi:10.1109/TBME.2013.2242328

70. Orini, M., Hanson, B., Taggart, P., Lambiase, P.: Detection of transient, regional cardiac repolarization alternans by time-frequency analysis of synthetic electrograms. In: 2013 35th Annual International Conference of the IEEE Engineering in Medicine and Biology Society (EMBC), pp. 3773–3776 (2013)
71. Zhao, H., Lu, S., Zou, R., Ju, K., Chon, K.H.: Estimation of time-varying coherence function using time-varying transfer functions. Ann. Biomed. Eng. **33**, 1582–1594 (2005)
72. Zhao, H., Cupples, W.A., Ju, K.H., Chon, K.H.: Time-varying causal coherence function and its application to renal blood pressure and blood flow data. IEEE Trans. Biomed. Eng. **54**, 2142–2150 (2007)
73. Chen, Z., Purdon, P., Harrell, G., Pierce, E., Walsh, J., Brown, E., Barbieri, R.: Dynamic assessment of baroreflex control of heart rate during induction of propofol anesthesia using a point process method. Ann. Biomed. Eng. **39**, 260–276 (2011). doi:10.1007/s10439-010-0179-z
74. Orini, M., Bailon, R., Laguna, P., Mainardi, L.T.: Modeling and estimation of time-varying heart rate variability during stress test by parametric and non parametric analysis. In: Proceedings of Computers in Cardiology, pp. 29–32, Sept. 30 2007–Oct. 3 2007
75. Grinsted, A., Moore, J.C., Jevrejeva, S.: Application of the cross wavelet transform and wavelet coherence to geophysical time series. Nonlinear Process. Geophys. **11**, 561–566 (2004)
76. Gallet, C., Chapuis, B., Barrès, C., Julien, C.: Time-frequency analysis of the baroreflex control of renal sympathetic nerve activity in the rat. J. Neurosci. Methods **198**(2), 336–343 (2011)
77. Keissar, K., Davrath, L.R., Akselrod, S.: Coherence analysis between respiration and heart rate variability using continuous wavelet transform. Philos. Trans. R. Soc. A Math. Phys. Eng. Sci. **367**(1892), 1393–1406 (2009)
78. Zhan, Y., Halliday, D., Jiang, P., Liu, X., Feng, J.: Detecting time-dependent coherence between non-stationary electrophysiological signals–a combined statistical and time-frequency approach. J. Neurosci. Methods **156**, 322–332 (2006)
79. Bigot, J., Longcamp, M., Dal Maso, F., Amarantini, D.: A new statistical test based on the wavelet cross-spectrum to detect time-frequency dependence between non-stationary signals: application to the analysis of cortico-muscular interactions. Neuroimage **55**, 1504–1518 (2011)
80. Brittain, J.S., Halliday, D.M., Conway, B.A., Nielsen, J.B.: Single-trial multiwavelet coherence in application to neurophysiological time series. IEEE Trans. Biomed. Eng. **54**, 854–862 (2007)
81. Xiao, J., Flandrin, P.: Multitaper time-frequency reassignment for nonstationary spectrum estimation and chirp enhancement. IEEE Trans. Signal Process. **55**, 2851–2860 (2007)
82. Thomson, D.J.: Jackknifing multitaper spectrum estimates. IEEE Signal Process. Mag. **24**, 20–30 (2007)
83. Xu, Y., Haykin, S., Racine, R.J.: Multiple window time-frequency distribution and coherence of EEG using Slepian sequences and hermite functions. IEEE Trans. Biomed. Eng. **46**(7), 861–866 (1999)
84. Lovett, E.G., Ropella, K.M.: Time-frequency coherence analysis of atrial fibrillation termination during procainamide administration. Ann. Biomed. Eng. **25**(6), 975–984 (1997) multitaper, spectrogram,coherence.
85. Faes, L., Nollo, G.: Multivariate frequency domain analysis of causal interactions in physiological time series. Biomedical Engineering, Trends in Electronics, Communications and Software. InTech, Rijeka (2011)

Part III
Applications

Chapter 10
Measurements of Cardiovascular Signal Complexity for Advanced Clinical Applications

Roberto Sassi and Sergio Cerutti

Abstract Over the last decades, there has been an increasing interest in the analysis of Heart Rate Variability (HRV). Many parameters were proposed to properly describe the complex systems controlling the heart rate, while involving neural mechanisms (through the Autonomic Nervous System), as well as mechanical and humoral factors. After a first effort to rationalize all these parameters in 1996 (the "HRV Task Force"), a second paper in 2015, reported the consensus reached on the critical review of the new methods. The latter tried in particular to address the clinical impact of the nonlinear techniques, considering only studies with sufficiently sized populations. In this chapter, we relax the constrain on the number of patients and try to identify all those techniques which resonated in the scientific community and were applied and studied more than others. To guide our analysis, we considered a different set of objective criteria, based mainly on the number of citations received.

Our analysis show that all the parameters which were clinically relevant in the 2015 paper, proved also to have a significant impact in the methodological literature. However, other parameters received much more methodological interest than the clinical results they were capable to provide. Among these: several entropy measures and the metrics derived from nonlinear dynamical systems, multifractality and wavelets. The reasons of this lack of clinical results might be many, but the complexity of these techniques and the disconnection, which frequently happens between bioscientists/biomedical engineers and medical centers, are likely possible explanations.

R. Sassi (✉)
Dipartimento di Informatica, Università degli Studi di Milano, Crema, CR, Italy
e-mail: roberto.sassi@unimi.it

S. Cerutti
Department of Electronics, Information and Bioengineering, Politecnico di Milano, Milan, Italy

R. Barbieri et al. (eds.), *Complexity and Nonlinearity in Cardiovascular Signals*,
DOI 10.1007/978-3-319-58709-7_10

10.1 Introduction

About 20 years after the publication, in 1996, of the seminal Task Force paper on Heart Rate Variability (HRV) signal analysis [1], some of the major cardiological institutions (the European Society of Cardiology, the European Heart Rhythm Association and the Asia Pacific Heart Rhythm Society) took the initiative of publishing a paper on the subsequent advances in heart rate variability signal analysis [2]. The work was also endorsed as a "position paper" by the European Heart Rhythm Association (EHRA). It might be considered a kind of "update", aiming at identify those nonlinear/non traditional methods applied to HRV signal analysis in the last two decades, which have had a certain diffusion at a clinical level, according to some criteria which are well explained in [2]. In fact, in the original Task Force, besides the more traditional time-domain and frequency-domain metrics, only one parameter was suggested among the various nonlinear/non traditional approaches then available in literature, *i.e.*, α, the slope of linear interpolation of the entire 24 h HRV power spectrum plotted in a log–log scale (the so-called "exponent of the power-law spectrum") [3]. In 1996, this parameter was considered by the authors as the only one which had already manifested interesting real clinical applications, as well as promising performances [4, 5]. Nowadays, however, a series of new and advanced algorithms of HRV signal analysis have been introduced and many of these ones are classified as "nonlinear", aiming at a more precise quantification of the complexity behavior of the physiological underlying systems, mainly connected to the autonomic nervous system, but not only.

10.2 A Further Analysis

During the recent revision of the Task Force document [2], it soon appeared evident that some sort of "objective" criteria were necessary to guide the analysis of a huge body of literature, in order to avoid bias as much as possible. A simple list of the enormous number of methods proposed, and their relative fields of applications, was going to be both pointless and unfeasible. After a few discussions and with the approval of the board of the e-Cardiology group, the authors decided to start looking at those methods which had clinically proved as "relevant". The definition of a "clinically relevant" method is not straightforward and needed to be specified further. In practice, a method was considered "relevant" when it was employed in a large enough population (of at least 200 subjects) and which results were then published in a list of important clinical journals (high impact factor), which were historically connected with the topic of heart rate variability.

Clearly, any of these choices can be criticized: the threshold (set at 200) and the list of journals selected could have been different. However, to understand further the procedure it is important to state that the objective was not that of selecting the "best methods" (it was not a competition!) but more properly to help the scientific community in objectively appreciate what were the clinical results of a very large effort, put forward in many different research laboratories and hospitals worldwide.

In the discussions that followed the publications of the 2015 consensus document [2], one of the comments which the authors received often was more related to the list of methodologies themselves, than to their clinically efficacy. The comment was clearly coming from the engineering/computer science related communities: If we do not consider the clinical relevance, which techniques have "made an impact" in the scientific community and were applied and studied more than others? Were these techniques overlapping with the "clinically relevant" ones already listed in [2]?

To try to address this question, in this chapter we considered a different set of objective criteria. First, as a metric of "impact", we considered the number of citations received by each single paper (at least 200) and we selected Scopus, instead of Pubmed, as our reference search engine. We can frankly state that we have no conflict of interest here. Simply, Scopus keeps track also of citations received by books and conference papers. Second, mimicking what done before, we looked for papers which contained the words "HRV" or "Heart Rate Variability" in the title, the abstract or the keywords (keywords were not considered in [2]). Finally we considered those paper published since 1997, up to today.[1]

The number of papers located by Scopus under this query is huge (around 26,300 items) and about twice as large than what obtained from Pubmed using the same search criteria (around 13,900 items). This should not come as a surprise, as Pubmed covers only the biomedical literature, while Scopus indexes a broader sets of sources coming from diverse scientific fields. Instead it shows that concept related to HRV did not stay locked in the biomedical community, but percolated in many other applicative fields.

Of these 26,300 papers, only 178 received at least 200 citations and only 19 referred to methods of HRV analysis not listed in the original 1996 Task Force document ("non-standard methods"). They are listed in Table 10.1. For each paper, a short summary of the parameters considered is included. Two of them [6, 7] were also included among the papers considered in [2]. Overall, they can be subdivided into the following groups of methods and algorithms.

10.2.1 Long-Range Correlation and Fractal/Multifractal Scaling

Metrics considering the fractal behavior of HRV over long time scales (i.e., assessing their degree of long range dependence) were included in seven papers out of 19 [6, 8–13]. The slope of the power-law spectrum and the DFA α_2 parameters were the most represented, but also other fractal metrics were included (Higuchi

[1]The effective search query was: (TITLE-ABS-KEY ("HRV") OR TITLE-ABS-KEY ("Heart Rate Variability")) AND PUBYEAR > 1996 AND PUBYEAR <2017.

Table 10.1 Overview of the publications fulfilling the selection criteria

Reference	Methods [Patients]	Notes	Cited
Ho et al. (1997) [8] Circulation	α, DFA index, ApEn [CHF]	Nonstandard measurements intended to complement standard time- and frequency-domain statistics	328
Pikkujämsä et al. (1999) [9] Circulation	α, DFA $\alpha 1$, DFA $\alpha 2$, ApEn [Healthy/Aging]	Cardiac interbeat interval dynamics change markedly from childhood to old age in healthy subjects. Healthy aging is associated with R-R interval dynamics showing higher regularity and altered fractal scaling consistent with a loss of complex variability	269
Ivanov et al. (1999) [10] Nature	Multifractality spectrum, Degree of multifractality (difference between maximum and minimum local Hurst exponent) [CHF]	Evidence for multifractality and nonlinear properties in healthy HRV signals. Loss of multifractality for a life-threatening condition, such as congestive heart failure	870
Schmidt et al. (1999) [18] Lancet	HRT [AMI]	The absence of the heart rate turbulence after ventricular premature beats is a very potent postinfarction risk stratifier, independent of other known risk factors and stronger than other risk predictors	519
Huikuri et al. (2000) [6] Circulation	α, DFA $\alpha 1$, DFA $\alpha 2$, Poincaré plot [AMI]	Fractal analysis methods of HRV to predict death after acute myocardial infarction (MI) and depressed left ventricular function. Investigation for the prediction of either arrhythmic or nonarrhythmic cardiac death. Also considered in [2]	382
Pincus S.M. (2001) [20] Ann N Y Acad Sci	ApEn	ApEn intended as a parameter for measuring complexity in general time-series settings. ApEn is scale invariant and model independent and can be applied to discriminate both general classes of correlated stochastic processes, as well as noisy deterministic systems. ApEn is complementary to spectral and autocorrelation analyses, providing effective discriminatory capability in instances in which the aforementioned measures exhibit minimal distinctions	227
Brennan et al. (2001) [15] IEEE Trans Biomed Eng	Poincaré plot	Methodological paper showing that SD1 & SD2 measure linear aspects of HRV, despite Poincaré maps were developed in the context of nonlinear dynamical systems	344
Eke et al. (2002) [11] Physiol Meas	Various fractal measures and their relation	Review Paper providing an overview on the concept of fractal geometry with definitions and explanations of the most fundamental properties of fractal structures and processes like self-similarity, power law scaling relationship, scale invariance, scaling range and fractal dimensions	247

Marwan et al. (2002) [25] Phys Rev. E Stat Nonlin Soft Matter Phys	Recurrence plot derived complexity measures [Ventricular tachyarrhythmias]	Measures of complexity based on vertical structures in recurrence plots and applied to HRV. Laminar transition phases before a life-threatening cardiac arrhythmia can facilitate the prediction of malignant cardiac arrhythmias	382
Lake et al. (2002) [21] Am J Physiol Regul Integr Comp Physiol	SampEn, (ApEn for comparison) [new borns]	Optimal selection of SampEn and ApEn parameters in infant HRV. Explanation of why reduced variability and transient decelerations in fetal and neonatal HR lead to lower entropies	454
Niskanen et al. (2004) [16] Comput Methods Programs Biomed	Poincaré plot	HRV software, then evolved in Kubios	372
Addison P.S. (2005) [22] Physiol Meas	Wavelet	Review paper on the application of Wavelets to ECG signal and also to HRV	316
Bauer et al. (2006) [7] Lancet	AC/DC [AMI]	Also considered in [2]. Impaired heart rate deceleration capacity is a powerful predictor of mortality after MI and is more accurate than LVEF and conventional measures of HRV	249
Gamelin et al. (2006) [17] Med Sci Sports Exerc	Poincaré plot [healthy / tilt test]	Validation of a commercial heart rate monitor	235
Acharya et al. (2006) [12] Med Biol Eng Comput	Correlation dimension, DFA and other fractal metrics, ApEn, SampEn, Wavelet	Review paper on the various approaches to calculate HRV parameters	648
Friedman B.H. (2007) [23] Biol Psychol	Largest Lyapunov exponent [Anxiety]	Study of parameters which are correlated to lower level of vagal tone in anxiety: various non linear and chaotic approaches	272
Bauer et al. (2008) [19] J Am Coll Cardiol	HRT	Consensus document on Heart Rate Turbulence	217
Voss et al. (2009) [13] Philos Trans A Math Phys Eng Sci	α, DFA, multifractal spectrum, ApEn, SampEn, Multiscale entropy, Poincaré plot, symbolic dynamics	Review paper on nonlinear parameters applied to HRV	201
Kemp et al. (2010) [24] Biol Psychiatry	Largest Lyapunov exponent [Depression]	Meta-analysis for validating HRV parameters in depression also including nonlinear metrics	297

dimension, Hurst R/S coefficient, Nearest Neighbor correlation coefficient). As shown in [14], the scaling relation linking these different metrics (some of them are also listed in [11]) largely holds in the analysis of HRV signals.

The paper by Ivanov et al. [10] moved instead one step further and applied the multifractal formalism, which was originally developed in the context of fluid-dynamics (fully developed turbulence) to HRV, showing that the width of the multifractal spectrum might be influenced by pathological conditions.

10.2.2 Short-Term Complexity

Short-term fractal scaling (DFA α_1) showed to be one of the most effective non traditional clinical parameter in the analysis [2]. In particular in predicting post-AMI cardiac mortality (while long-term fractal scaling seemed to predict all-cause mortality). In the Table, it was considered in two studies: [6, 9].

While Poincaré plots were basically originated from Poincaré maps, and thus should be included in the (following) section "Metrics derived from the theory of nonlinear dynamical systems", in HRV analysis Poincaré plots were most often limited to a "tool for graphically representing summary statistics" [15]. Typical parameters derived from Poincaré plots, such as SD_1, SD_2, or SD_{12} are strictly related to standard temporal indices and do not express necessarily nonlinear properties of the series. Nevertheless, apart from the already cited critical methodological work of Brennan [15], over the years they still maintain quite an impact in the scientific community, as the four papers included in Table 10.1 seems to suggest [6, 13, 16, 17].

Two approaches which proved both effective in [2] and are also present in Table 10.1 are Acceleration and Deceleration Capacities AC/DC [7] and Heart Rate Turbulence (HRT) [18, 19]. In these nonlinear techniques, the HRV series is averaged synchronously with a trigger (i.e., an acceleration or a deceleration in AC/DC or a premature ventricular contraction—PVC—for HRT) evidencing behaviors otherwise hidden by noise and non-stationarities.

10.2.3 Entropy and Regularity

Starting with the work of Pincus, many entropy measures were employed in HRV analysis. While they were under represented in [2] (only Sample Entropy is used routinely in the clinical practice) their methodological impact has been much larger. Not only Approximate Entropy (ApEn) [8, 9, 12, 20], and Sample Entropy (SampEn) [12, 21] are listed in Table 10.1 but also Multiscale entropy [13] and Symbolic Dynamics [13].

10.2.4 Wavelet

The clinical usage of wavelet in [2] was not evident, but their methodological impact over the years was clearly more significant as shown by [12, 22] and implicitly in [10].

10.2.5 Metrics Derived from the Theory of Nonlinear Dynamical Systems

Since the beginning of the '80s, after the introduction of Takens' delay embedding theorem, a large research effort was finalized to seek for evidences of low order deterministic chaos involved in the cardiovascular system control and in particular in HRV series. Probably, due to the methodological complexity of the approach, the clinical applications of such metrics has been practically limited [2]. However, many are the parameters that had a strong impact in the technical literature as Correlation dimension [12], the largest Lyapunov exponent [23, 24] and chaotic metrics derived from recurrence plots analysis [25].

10.3 Discussion

A first interesting comment, is that all the parameters which proved to have clinical relevance in the 2015 paper [2] had also a significant impact in the methodological literature, as defined in the analysis we conducted in the present chapter. The opposite is not necessarily true. Some parameters received much more methodological interest than the clinical results they were capable to prove in the field. Multifractality, wavelets, the many entropic measures introduced (apart from ApEn and SampEn) and the metrics derived from the study of nonlinear dynamical systems fall in this category.

The reasons of this lack of clinical results might be many, but the complexity of these techniques and the disconnection, which frequently happens between bioscientists/biomedical engineers and medical centers, are likely possible explanations.

A second comment which can be derived from Table 10.1 is that linear parameters, like the SD_1 and SD_2 measures extracted from Poincaré plot were largely employed, often without realizing their relations with SDNN and SDSD [15] and misinterpreted as capable of offering a possible "nonlinear" characterization of the HRV series.

10.4 Conclusions

The objective of this chapter is to briefly summarize the most important methods which were actually applied to HRV signal analysis in order to enhance proper pathophysiological information via nonlinear/non traditional approaches, not considered by the Task Force document of 1996 and which have been instead the results of the experience of various groups of researchers in these recent years. Following a method of literature investigation, similar to the one employed in [2] for assessing those parameters which were considered successful from a clinical application standpoint, here the objective of the query was to find out those methods and parameters which exhibited the greater interest in the scientific community. In short time, it is reasonable to expect that some of these parameters might enter into the endowment of the most significant clinical centers of HRV signal analysis, not only relevant to the methodological and modelling aspects, but also for the more general clinical applications. Finally, a further source of enhanced information could be realized by fusing together the results obtained with these new parameters with those ones of the more traditional and well-established linear approaches, as also suggested in [2].

References

1. Heart rate variability standards of measurement, physiological interpretation, and clinical use. Task Force of the European Society of Cardiology and the North American Society of Pacing and Electrophysiology. Circulation. **93**(5), 1043–1065 (1996)
2. Sassi, R., et al.: Advances in heart rate variability signal analysis: joint position statement by the e-Cardiology ESC Working Group and the European Heart Rhythm Association co-endorsed by the Asia Pacific Heart Rhythm Society. Europace. **17**(9), 1341–1353 (2015)
3. Kobayashi, M., Musha, T.: 1/f fluctuation of heartbeat period. I.E.E.E. Trans. Biomed. Eng. **BME-29**(6), 456–457 (1982)
4. Saul, J.P., Albrecht, P., Berger, R.D., Cohen, R.J.: Analysis of long term heart rate variability: methods, 1/f scaling and implications. Comput. Cardiol. **14**, 419–422 (1988)
5. Lombardi, F., et al.: Linear and nonlinear dynamics of heart rate variability after acute myocardial infarction with normal and reduced left ventricular ejection fraction. Am. J. Cardiol. **77**(15), 1283–1288 (1996)
6. Huikuri, H.V., Mäkikallio, T.H., Peng, C.-K., Goldberger, A.L., Hintze, U., Møller, M.: Fractal correlation properties of R–R interval dynamics and mortality in patients with depressed left ventricular function after an acute myocardial infarction. Circulation. **101**(1), 47–53 (2000)
7. Bauer, A., et al.: Deceleration capacity of heart rate as a predictor of mortality after myocardial infarction: cohort study. Lancet. **367**(9523), 1674–1681 (2006)
8. Ho, K.K.L., et al.: Predicting survival in heart failure case and control subjects by use of fully automated methods for deriving nonlinear and conventional indices of heart rate dynamics. Circulation. **96**(3), 842–848 (1997)
9. Pikkujämsä, S.M., et al.: Cardiac interbeat interval dynamics from childhood to senescence comparison of conventional and new measures based on fractals and chaos theory. Circulation. **100**(4), 393–399 (1999)
10. Ivanov, P.C., et al.: Multifractality in human heartbeat dynamics. Nature. **399**(6735), 461–465 (1999)

11. Eke, A., Herman, P., Kocsis, L., Kozak, L.R.: Fractal characterization of complexity in temporal physiological signals. Physiol. Meas. **23**(1), R1 (2002)
12. Acharya, U.R., Joseph, K.P., Kannathal, N., Lim, C.M., Suri, J.S.: Heart rate variability: a review. Med. Biol. Eng. Comput. **44**(12), 1031–1051 (2006)
13. Voss, A., Schulz, S., Schroeder, R., Baumert, M., Caminal, P.: Methods derived from nonlinear dynamics for analysing heart rate variability. Philos. Trans. R. Soc. Lond. Math. Phys. Eng. Sci. **367**(1887), 277–296 (2009)
14. Cerutti, S., Esposti, F., Ferrario, M., Sassi, R., Signorini, M.G.: Long-term invariant parameters obtained from 24-h Holter recordings: a comparison between different analysis techniques. Chaos. **17**(1), 015108–015108-9 (2007)
15. Brennan, M., Palaniswami, M., Kamen, P.: Do existing measures of Poincaré plot geometry reflect nonlinear features of heart rate variability? I.E.E.E. Trans. Biomed. Eng. **48**(11), 1342–1347 (2001)
16. Niskanen, J.-P., Tarvainen, M.P., Ranta-aho, P.O., Karjalainen, P.A.: Software for advanced HRV analysis. Comput. Methods Prog. Biomed. **76**(1), 73–81 (Oct. 2004)
17. Gamelin, F.X., Berthoin, S., Bosquet, L.: Validity of the polar S810 Heart rate monitor to measure R-R intervals at rest. Med. Sci. Sports Exerc. **38**(5), 887–893 (2006)
18. Schmidt, G., et al.: Heart-rate turbulence after ventricular premature beats as a predictor of mortality after acute myocardial infarction. Lancet. **353**(9162), 1390–1396 (1999)
19. Bauer, A., et al.: Heart rate turbulence: standards of measurement, physiological interpretation, and clinical use: international society for holter and noninvasive electrophysiology consensus. J. Am. Coll. Cardiol. **52**(17), 1353–1365 (2008)
20. Pincus, S.M.: Assessing serial irregularity and its implications for health. Ann. N. Y. Acad. Sci. **954**, 245–267 (2001)
21. Lake, D.E., Richman, J.S., Griffin, M.P., Moorman, J.R.: Sample entropy analysis of neonatal heart rate variability. Am. J. Phys. **283**(3), R789–R797 (2002)
22. Addison, P.S.: Wavelet transforms and the ECG: a review. Physiol. Meas. **26**(5), R155 (2005)
23. Friedman, B.H.: An autonomic flexibility–neurovisceral integration model of anxiety and cardiac vagal tone. Biol. Psychol. **74**(2), 185–199 (2007)
24. Kemp, A.H., Quintana, D.S., Gray, M.A., Felmingham, K.L., Brown, K., Gatt, J.M.: Impact of depression and antidepressant treatment on heart rate variability: a review and meta-analysis. Biol. Psychiatry. **67**(11), 1067–1074 (Jun. 2010)
25. Marwan, N., Wessel, N., Meyerfeldt, U., Schirdewan, A., Kurths, J.: Recurrence-plot-based measures of complexity and their application to heart-rate-variability data. Phys. Rev. E. **66**(2), 026702 (2002)

Chapter 11
Applications of Complexity Analysis in Clinical Heart Failure

Chengyu Liu and Alan Murray

Abstract Heart failure is known to influence heart rhythm in patients. Complexity analysis techniques, including techniques associated with entropy, have great potential for providing a better understanding of cardiac rhythms, and for helping research in this area. We review the analysis principles of conventional time-domain analysis, frequency-domain analysis and of newer complexity analysis. We then illustrate the techniques using real clinical data, allowing a comparison of the techniques, and also of the differences between normal heart rate variability and that associated with heart failure.

11.1 Introduction

11.1.1 Introduction to Non-Linear Heart Rate Variability Methods for Cardiovascular Analysis

The application of non-linear dynamic methods to quantify the complexity of the cardiovascular system has opened up new ways to perform cardiac rhythm analysis, thereby enhancing our knowledge, and stimulating significant and innovative research into cardiovascular dynamics. Heart rate variability (HRV) has been conventionally analyzed with time- and frequency-domain methods, which allowed researcher to obtain information on the sinus node response to sympathetic and parasympathetic activities. However, heart rate (HR) regulation is one of the most complex systems in humans due to the variety of influence factors, e.g. parasympathetic and sympathetic ganglia, humoral effect, and respiration or mental load. Therefore, to extract the relevant properties of a non-linear cardiovascular

C. Liu (✉)
School of Instrument Science and Engineering, Southeast University, Nanjing, 210023, China
e-mail: bestlcy@sdu.edu.cn

A. Murray (✉)
Electrical and Electronic Engineering and Faculty of Medical Sciences, Newcastle University, Newcastle upon Tyne, UK
e-mail: alan.murray@newcastle.ac.uk

R. Barbieri et al. (eds.), *Complexity and Nonlinearity in Cardiovascular Signals*,
DOI 10.1007/978-3-319-58709-7_11

dynamic system, classical linear signal analysis methods (time- and frequency-domain) are often inadequate. HRV can exhibit very complex behaviour, which is far from a simple periodicity. Thus the application of non-linear HRV methods can provide essential information regarding the physiological and pathological states of cardiovascular time series [1–4].

Typical non-linear HRV analysis methods can be classified as four types:

(1) Poincare plot

The Poincare plot is a quantitative visual technique, whereby the shape of the plot is categorized into functional classes and provides detailed beat-to-beat information on the behaviour of the heart. Weiss et al. observed the respective Poincare plots of interbeat intervals from a variety of arrhythmia patients [5]. Brennan et al. proposed a physiological oscillator model of which the output could mimic the shape of the R–R interval Poincare plot [6]. By defining the quantitative indices for Poincare plots, different cardiovascular diseases can be identified from healthy control subjects, including: parasympathetic nervous activity [7], postoperative ischaemia [8], ventricular tachyarrhythmia [5], and heart failure [9].

(2) Fractal method

Fractal measures aim to assess self-affinity of heartbeat fluctuations over multiple time scales [4]. Kobayashi & Musha first reported the frequency dependence of the power spectrum of RR interval time series fluctuations [10]. The broadband spectrum, characterized by the slow HR fluctuations, indicates a fractal-like process with a long-term dependence [11]. Peng et al. developed this method to detrended fluctuation analysis (DFA) [12], which quantifies the presence or absence of fractal correlation properties in non-stationary time series data. Using DFA, healthy subjects revealed a scaling exponent of approximately 1, indicating fractal-like behaviour, and cardiovascular patients revealed reduced scaling exponents, suggesting a loss of fractal-like HR dynamics [13, 14]. The fractal method was also extended into multifractality applications [15].

(3) Symbolic dynamics measures

In 1898, Hadamard proposed a symbolic dynamics measure named as SDyn, which investigates short-term fluctuations caused by vagal and baroreflex activities and allows a simple description of a system's dynamics with a limited amount of symbols [16]. SDyn was applied in detecting the disease of sudden cardiac death [17, 18]. Later, Porta et al. developed this SDyn method into short-term 300 beat RR intervals [19] and 24 h Holter recording analysis [20].

Another popular symbolic dynamics measure, named LZ complexity, was proposed by Lempel and Ziv (LZ) [21]. LZ can evaluate the irregularity of RR interval time series. In most cases, LZ complexity algorithm is executed by transforming an original signal into a binary sequence. However, the binary coarse-graining process is associated with a risk of losing detailed information. Thus higher quantification levels (or symbols) are employed in the coarse-graining process. Abásolo et al. used three quantification levels for LZ calculation [22] and Sarlabous et al. developed a multistate LZ complexity algorithm [23]. Unlike the multistate coarse-graining, Zhang et al. recently proposed an encoding LZ algorithm aiming to explicitly discern between the irregularity and chaotic characteristics of RR interval time series [24].

(4) Entropy methods

Entropy can assess the regularity/irregularity or randomness of RR interval time series. Since Pincus proposed the approximate entropy (ApEn) method in 1991 [25], ApEn has achieved wide applications for analyzing physiological time series in clinic research. The popularity of ApEn stems from its capability in providing quantitative information about the complexity of both short- and long-term data recordings that are often corrupted with noise, and the calculation methods are relatively easy [26–29]. As an improved version of ApEn, Richman and Moorman proposed the sample entropy (SampEn) method in 2010 [29], which enhances the inherent bias estimation of ApEn method. SampEn quantifies the conditional probability that two sequences of m consecutive data points that are similar to each other (within a given tolerance r) will remain similar when one consecutive point is included. Self-matches are not included in calculating the probability. In a clinical application, reductions in SampEn of neonatal HR prior to the clinical diagnosis of sepsis and sepsis-like illness were reported in [30] and before the onset of atrial fibrillation [31]. The new improvements for entropy methods include: fuzzy-function based entropy methods [32, 33], multiscale entropy (MSE) methods [26] and multivariate multiscale entropy methods [34–36].

11.1.2 Common Heart Rate Variability Analysis in Congestive Heart Failure Patients

Congestive heart failure (CHF) is a typical degeneration of the heart function featured by the reduced ability for the heart to pump blood efficiently. CHF is also a difficult condition to manage in clinical practice, and the mortality from CHF remains high. Previous studies have proven that HRV indices relate to outcome for CHF patients [37–40].

For healthy subjects, it has been proven that the increased sympathetic and the decreased parasympathetic activity results in the decrease of mean RR interval, as well as the decrease of indices of the standard deviation of beat-to-beat intervals (SDNN), low frequency content (LF), and also non-linear indices VAI and VLI [7]. Moreover, the decreased parasympathetic activity has been proven to be a the major contributor to the increase in the index for high frequency (HF) content [41]. HRV analysis has also given an insight into understanding the abnormalities of CHF, and can also be used to identify the higher-risk CHF patients. Depressed HRV has been used as a risk predictor in CHF [42–44]. CHF patients usually have a higher sympathetic and a lower parasympathetic activity [42, 44].

Typical linear HRV analysis for CHF patients include the following publications: Nolan et al. performed a prospective study on recruited 433 CHF patients and found that SDNN was the most powerful predictor of the risk of death for CHF disease [38]. Binkley et al. studied 15 healthy subjects and 10 CHF patients, and reported that parasympathetic withdrawal, in addition to the augmentation of sympathetic drive, is an integral component of the autonomic imbalance characteristic for CHF patients and can be detected noninvasively by HRV spectral analysis [42]. Rovere et al. studied 202 CHF patients and reported that the LF component was a powerful predictor of sudden death in CHF patients [45]. Hadase et al. also confirmed that the very low frequency (VLF) content was a powerful predictor from a 54 CHF patient study [46]. All those studies have verified that decreased HRV was associated with the increased mortality in CHF patients.

Typical non-linear HRV analysis for CHF patients include the following publications: Woo et al. studied 21 patients with heart failure and demonstrated that Poincare plot analysis is associated with marked sympathetic activation for heart failure patients and may provide additional prognostic information and an insight into autonomic alterations and sudden cardiac death [44]. Guzzetti et al. tracked 20 normal subjects and 30 CHF patients for 2 years and found significantly lower normalized LF power and lower $1/f$ slope in CHF patients compared with controls. Moreover, the patients who died during the follow-up period presented further reduced LF power and steeper $1/f$ slope than the survivors [47]. Makikallio et al. studied 499 CHF patients and showed that a short-term fractal scaling exponent was the strongest predictor of mortality of CHF [48]. Poon and Merrill studied 8 healthy subjects and 11 CHF patients, and found that the short-term variations of beat-to-beat interval exhibited strongly and consistently chaotic behaviour in all healthy subjects but were frequently interrupted by periods of seemingly non-chaotic fluctuations in patients with CHF [43]. Peng et al. used DFA analysis and confirmed a reduction in HR complexity in CHF patients [12]. Liu et al. studied 60 CHF patients and 60 healthy control subjects, and reported decrease of ApEn values in CHF group [27]. Costa et al. [49] used the MSE method for classifying CHF patients and healthy subjects, and reported that the best discrimination between CHF and healthy HR signals with the scale 5 in MSE calculation [49].

11.1.3 Main Aims of This Review

We seek by the use of examples to illustrate the differences between normal cardiac physiology and those associated with CHF, determined from both time-domain, frequency-domain and typical short-term (5 min) complexity analysis, and also show the differences between the three types of analysis. These subjects can be seen as proof of the concept that, from the current literature on cardiovascular complexity and heart failure, the methods do have potential.

11.2 Review of Typical Methods and Indices for Heart Rate Variability Analysis

11.2.1 Time-Domain Indices

The typical and commonly used time-domain indices include: SDNN where SD is the standard deviation and NN was originally derived from the Normal-to-Normal interval, the square root of the mean squared successive differences in RR intervals (RMSSD), and the proportion of differences between successive RR intervals greater than 50 ms (PNN50) [50, 51].

11.2.2 Frequency-Domain Indices

Frequency-domain analysis is usually performed by a modern spectral estimation method, such as Burg's method, to produce the HRV spectrum [52]. The HRV spectrum is then integrated to derive a low-frequency power (in frequency range 0.04–0.15 Hz) and a high-frequency power (0.15–0.40 Hz). Indices are obtained by the calculation of the normalized low-frequency power (LFn), normalized high-frequency power (HFn), and the ratio of low-frequency power to high-frequency power (LF/HF) [51].

11.2.3 Non-Linear Methods

11.2.3.1 Poincare Plot

The Poincare plot analysis is a graphical non-linear method to assess the dynamics of HRV. This method provides summary information as well as detailed beat-to-beat information on the behaviour of the heart. It is a graphical representation of temporal correlations within RR intervals. The Poincare plot is known as a return map or scatter plot, where each RR interval from the RR time series is plotted against the next RR interval. Two paired parameters are commonly used as indices derived from the Poincare plot. One pair of parameters includes SD1 and SD2, another pair of parameters is vector length index (VLI) and vector angle index (VAI). SD2 is defined as the standard deviation of the projection of the Poincare plot on the line of identity ($y = x$), and SD1 is the standard deviation of projection of the Poincare plot on the line perpendicular to the line of identity. SD1 has been correlated with high frequency power, while SD2 has been correlated with both low and high frequency power [51]. VAI measures the mean departure of all Poincare points from the line of identity ($y = x$) and VLI measures the mean distance of all Poincare points from the centre poincare point. They are defined as:

$$\mathrm{VAI} = \sum_{i=1}^{N} |\theta_i - 45| / N \tag{11.1}$$

$$\mathrm{VLI} = \sqrt{\sum_{i=1}^{N} (l_i - L)^2 / N} \tag{11.2}$$

where, θ_i is the vector angle in degree, l_i is the vector length for each Poincare point, L is the mean vector length of all Poincare points, and N is the total number of Poincare points.

11.2.3.2 Histogram Plot

A histogram plot is another graphical method for the HRV analysis, which reflects directly the distribution of the RR sequence. There is no specific quantitative index in the traditional histogram plot. To resolve this, Liu et al. proposed a new histogram analysis for HRV, referred to as RR sequence normalized histogram [53]. The RR sequence normalized histogram divides all sequence elements into seven sections based on the element values. Then, three quantitative indices are defined from the normalized histogram, named as center-edge ratio (CER), cumulative energy (CE) and range information entropy (RIEn) respectively [53]. CER characterizes the element fluctuation apart from the sequence mean value. CE indicates the equilibrium of the percentage p_i in all seven sections and RIEn reflects the element distributions. If the element distribution in the RR sequence exhibits more uniformity in each section, RIEn is larger, the RR sequence is more complex, and the uncertainty of the sequence is higher. The general construction procedure for the RR sequence normalized histogram and the calculations for the three quantitative indices are summarized in the Appendix.

11.2.3.3 Entropy Measures

ApEn represents a simple index for the overall complexity and predictability of time series [25]. ApEn quantifies the likelihood that runs of patterns which are close, remain similar for subsequent incremental comparisons [54]. High values of ApEn indicate high irregularity and complexity in time-series data. However, inherent bias exists in ApEn due to the counting of self-matches, resulting in ApEn having a dependency on the record length and lacking relative consistency. SampEn, improving ApEn, quantifies the conditional probability that two sequences of m consecutive data points that are similar to each other (within a given tolerance r) will remain similar when one consecutive point is included. Self-matches are not included in SampEn when calculating the probability [29].

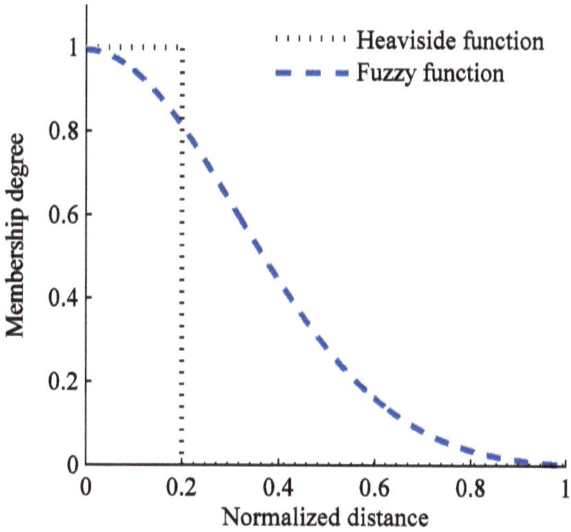

Fig. 11.1 The transient curve of the membership degree with the increase of normalized distance between two vectors. The black dotted line shows the determination criterion of the Heaviside function and the blue dashed line shows the determination criterion of the fuzzy function

Whether ApEn or SampEn is utilized, the decision rule for vector similarity is based on the Heaviside function and it is very rigid because two vectors are considered as similar vectors only when they are within the tolerance threshold r, whereas the vectors just outside this tolerance are ignored [32, 33, 55]. This rigid boundary may induce abrupt changes of entropy values when the tolerance threshold r changes slightly, and even fail to define the entropy if no vector-matching could be found for very small r. To enhance the statistical stability, a fuzzy measure entropy (FuzzyMEn) method was proposed in [33, 56], which used a fuzzy membership function to substitute the Heaviside function to make a gradually varied entropy value when r monotonously changes. Figure 11.1 illustrates the Heaviside and fuzzy functions that are mathematically given as

$$\text{For Heaviside function: Membership\_degree}\left(d_{i,j}, r\right) = \begin{cases} 1 & d_{i,j} \le r \\ 0 & d_{i,j} > r \end{cases}$$
$$\text{For Fuzzy function: Membership\_degree}\left(d_{i,j}, r\right) = \exp\left(-\frac{\left(d_{i,j}\right)^n}{r}\right) \tag{11.3}$$

where $d_{i,j}$ represents the distance of two vectors X_i and Y_j, r is the tolerate threshold and n is the vector similarity weight. The rigid membership degree determination in the Heaviside function could induce the weak consistency of SampEn, which means that the entropy value may have a sudden change when the parameter r changes slightly. This phenomenon has been reported in recent studies [32, 33, 57]. For fuzzy functions, this determination criterion exhibits a smooth boundary effect, while the traditional 0–1 judgment criterion of the Heaviside function is rigid in the boundary of the parameter r. Besides, FuzzyMEn also uses the information from both local and global vector sequences by removing both local baseline and global mean values, thus introducing the fact that FuzzyMEn has better consistency than SampEn.

The detailed calculation processes of SampEn and FuzzyMEn are given in the Appendix.

11.2.3.4 Lempel-Ziv (LZ) Complexity

Lempel-Ziv (LZ) complexity is a measure of signal complexity and has been applied to a variety of biomedical signals, including identification of ventricular tachycardia or atrial fibrillation [58]. In most cases, the LZ complexity algorithm is executed by transforming an original signal into a binary sequence by comparing it with a preset median [59] or mean value [58] as the threshold. So the HRV signal should be coarse-grained and be transformed into a symbol sequence before the LZ calculation. For generating the binary sequence, the signal x is converted into a 0–1 sequence R by comparing with the threshold T_h, producing the binary symbolic sequence $R = \{s_1, s_2, \ldots, s_n\}$ as follows:

$$s_i = \begin{cases} 0, if\ x(i) < T_h \\ 1, otherwise \end{cases}, \quad i = 1, 2, \ldots, n, \tag{11.4}$$

where n is the length of $x(n)$ and the mean value of the sequence is used as the threshold T_h in this chapter.

LZ complexity reflects the rate of new patterns arising within the sequence, and index $C(n)$ is usually denoted as the measure for the arising rate of new patterns in a normalized LZ complexity. The detailed calculation for index $C(n)$ is also attached in the Appendix to simple the text content.

11.3 Demonstration of Heart Rate Variability Analysis for Normal and Congestive Heart Failure RR time series

In this section, 5-min RR time series from a NSR subject and a CHF patient are employed as shown in Fig. 11.2, and are used to illustrate the various analysis techniques. Note that this is a “proof of the concept” about the current literature on cardiovascular complexity and heart failure, without statistical inference. The chapter focuses on demonstrating the differences of different techniques when dealing with the examples, using the same examples in each analysis to allow easy comparison. The graphical results for the various HRV methods are shown to facilitate the comparison between these example RR time series.

It can be seen in Fig. 11.2 that the CHF patient has a shorter mean RR, and hence, a faster heart rate. However, the more important difference can be seen in the variability.

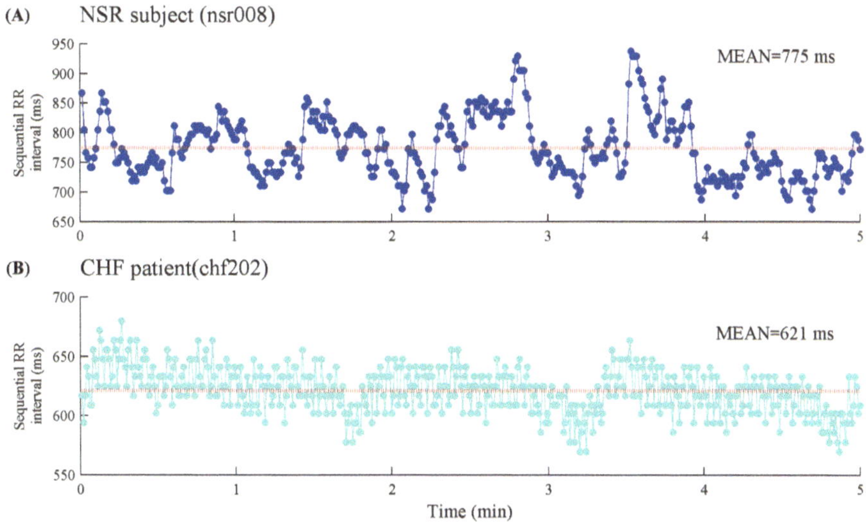

Fig. 11.2 Sequential 5-min RR interval time series with mean line (*red dotted line*) from (**a**) NSR subject and (**b**) CHF patient. The mean values of 5-min RR time series are 775 and 621 ms respectively

11.3.1 Time-Domain Indices

Figure 11.3 shows the calculations for the time-domain HRV indices. In each sub-figure, the upper panel shows the difference from the MEAN of the 5-min RR interval time series with ± SD lines (red dashed lines) and the lower panel shows the modulus of the sequential differences. RR intervals greater than 50 ms are marked as 'red squares'. The values of time-domain indices of SDNN, RMSSD and PNN50 are shown.

11.3.2 Frequency-Domain Indices

Figure 11.4 shows the analysis for the frequency-domain HRV indices. In each sub-figure, the left panel shows the classical spectrum estimation from the Fast Fourier Transform (FFT) method, and the right panel shows the modern spectrum estimation using the Burg method. Low frequency power (LF, 0.04–0.15 Hz) and high frequency power (HF, 0.15–0.40 Hz) areas are marked. The index values of LF/HF are also shown.

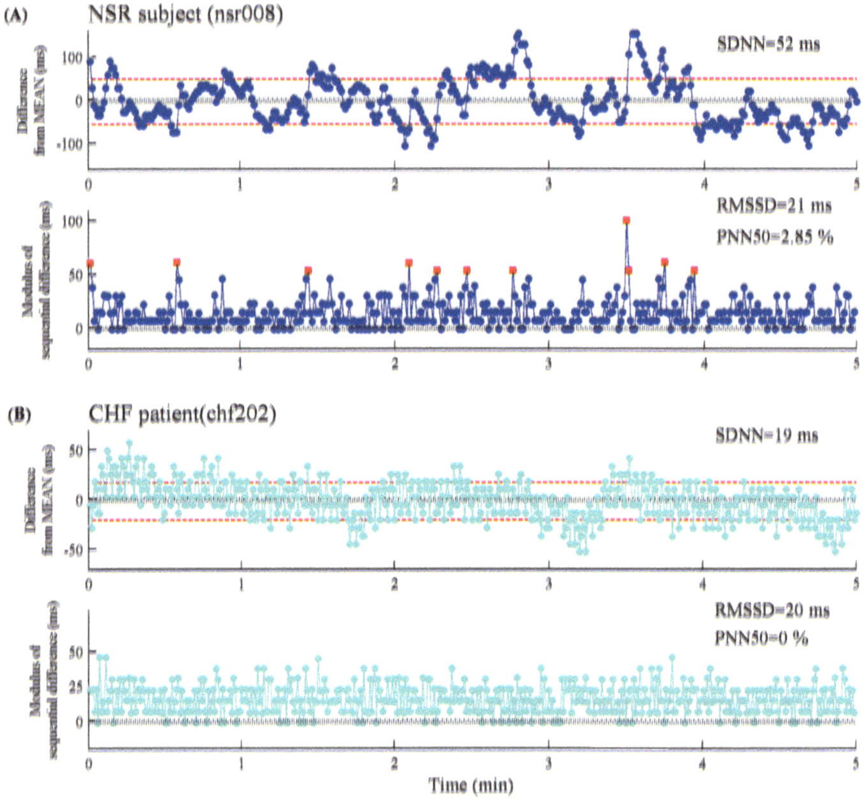

Fig. 11.3 Time-domain indices calculated for (**a**) NSR subject and (**b**) CHF patient

11.3.3 Non-Linear Methods

11.3.3.1 Poincare Plot

Figure 11.5 shows the analysis for the Poincare plot. In each sub-figure, the values of the four indices, i.e., SD1, SD2, VAI and VLI, are shown.

11.3.3.2 Histogram Plot

Figure 11.6 shows a traditional histogram plot. It is clear that the RR intervals have a more uniform distribution for the NSR subject, in comparison with a more concentrated distribution for the CHF patient.

Figure 11.7 demonstrates the RR sequence normalized histogram plot. The percentage p_i in NSR subject is fairly evenly distributed in all seven sections, while

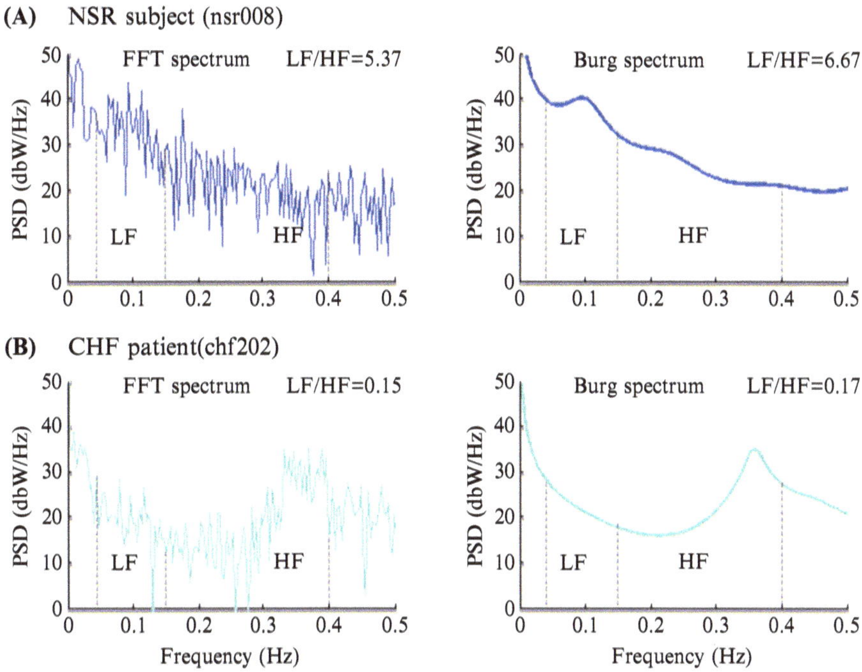

Fig. 11.4 Frequency-domain indices calculated for (**a**) NSR subject and (**b**) CHF patient

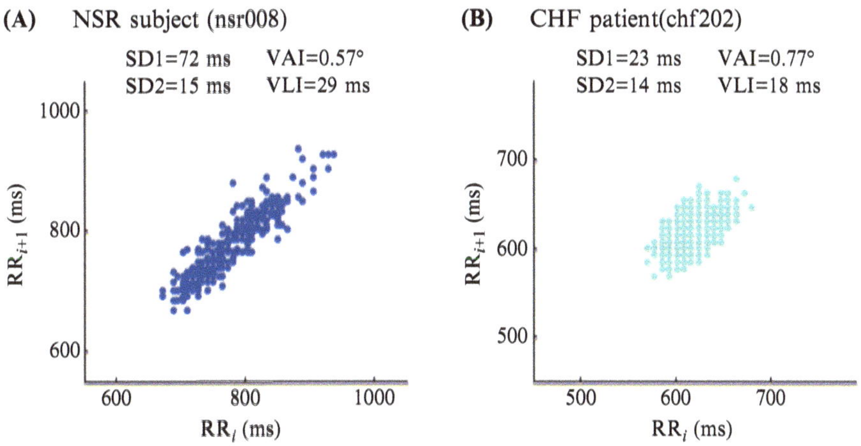

Fig. 11.5 Poincare plot for (**a**) NSR subject and (**b**) CHF patient

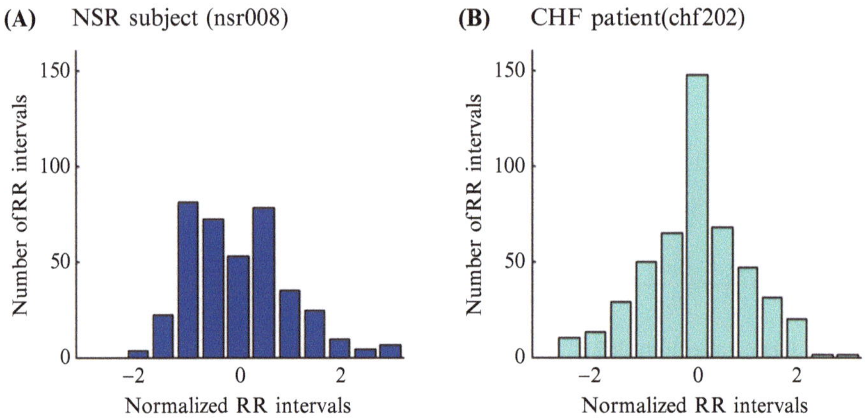

Fig. 11.6 Traditional histogram plot for (**a**) NSR subject and (**b**) CHF patient

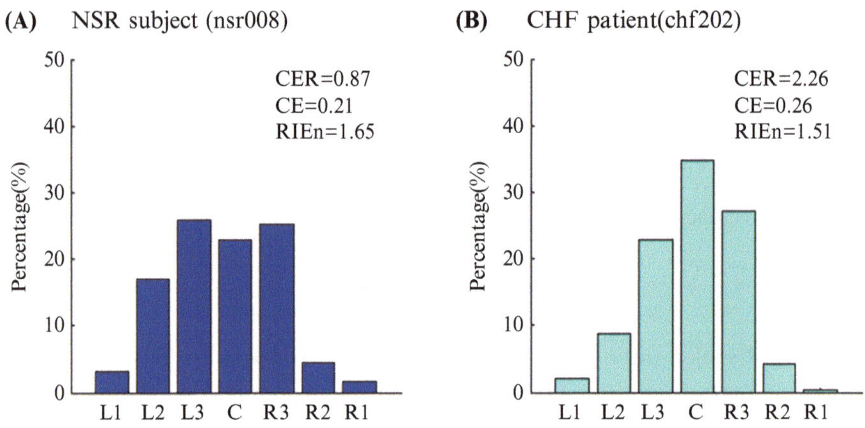

Fig. 11.7 RR sequence normalized histogram plot for (**a**) NSR subject and (**b**) CHF patient

the percentage p_i in the CHF patient has an uneven distribution. In each sub-figure, the values of the three indices CER, CE and RIEn are shown.

11.3.3.3 Entropy Measures

As an intermediate step, SampEn needs to determine the similarity degree between any two vectors X_i^m and Y_i^m at both embedding dimension m and $m + 1$ respectively, by calculating the distance between them. As shown in Appendix, the distance between X_i^m and Y_i^m is defined as $d_{i,j}^m = \max_{k=0}^{m-1} |x(i+k) - x(j+k)|$. In SampEn, if the distance is within the threshold parameter $r = 0.2$, the similarity degree between the two vectors is 1; if the distance is beyond the threshold parameter r, the

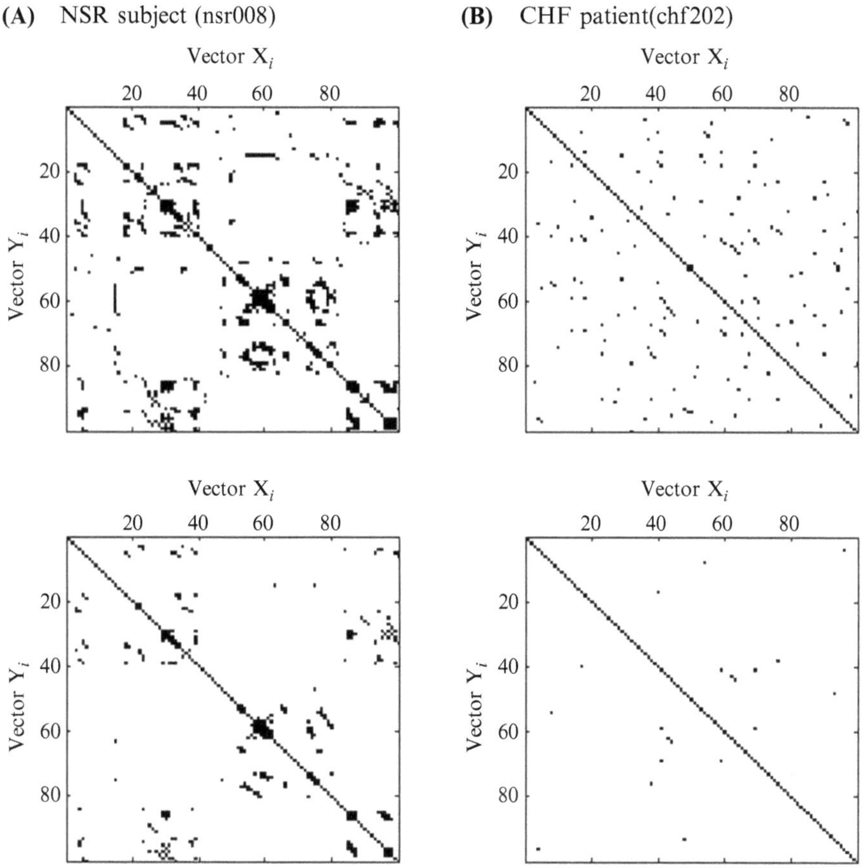

Fig. 11.8 Similarity degree matrices in the intermediate calculation procedure of SampEn for (**a**) NSR subject and (**b**) CHF patient

similarity degree is 0. There is absolutely a 0 or 1 determination. Figure 11.8 shows the demonstrations of the similarity degree matrices in this intermediate calculation procedure of SampEn. Only $1 \leq i \leq 100$ and $1 \leq j \leq 100$ are shown for illustrating the details. Black-coloured areas indicate the similarity degree = 1 and vice versa. In each sub-figure, the upper panel shows the results from the embedding dimension $m = 2$, and the lower panel shows the results from the embedding dimension $m + 1 = 3$. As shown in Fig. 11.8, it is clear that when the embedding dimension changes from m to $m + 1$, the number of similar vectors (i.e., matching vectors) decreases.

Unlike the 0 or 1 discrete determination for vector similarity degree in SampEn, FuzzyMEn permits the outputs of continuous real values between 0 and 1 for the vector similarity degree, by converting the absolute distance of $d_{i,j}^m =$

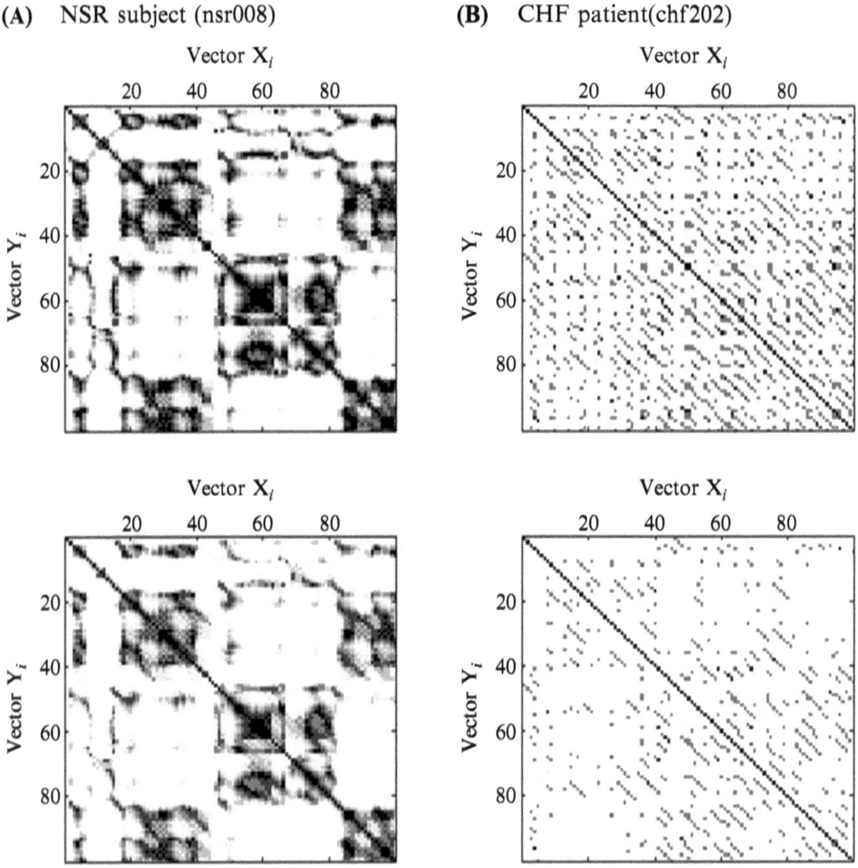

Fig. 11.9 The global similarity degree matrices in the intermediate calculation procedure of FuzzyMEn for (**a**) NSR subject and (**b**) CHF patient

$\max_{k=0}^{m-1} |x(i+k) - x(j+k)|$ using a fuzzy exponential function (see Appendix). Since FuzzyMEn not only measures the global vector similarity degree, but also refers to the local vector similarity degree, thus, Figs. 11.9 and 11.10 show the global and local similarity degree matrices in this intermediate calculation procedure of FuzzyMEn respectively. Furthermore, only $1 \leq i \leq 100$ and $1 \leq j \leq 100$ are shown to illustrate the details. Dark-colored areas indicate the higher similarity degree and vice versa. In each sub-figure, the upper panel shows the results from the embedding dimension $m = 2$, and the lower panel shows the results from the embedding dimension $m + 1 = 3$. As shown in Figs. 11.9 and 11.10, it is also evident that when the embedding dimension changes from m to $m + 1$, the similarity degrees in both Figures decrease.

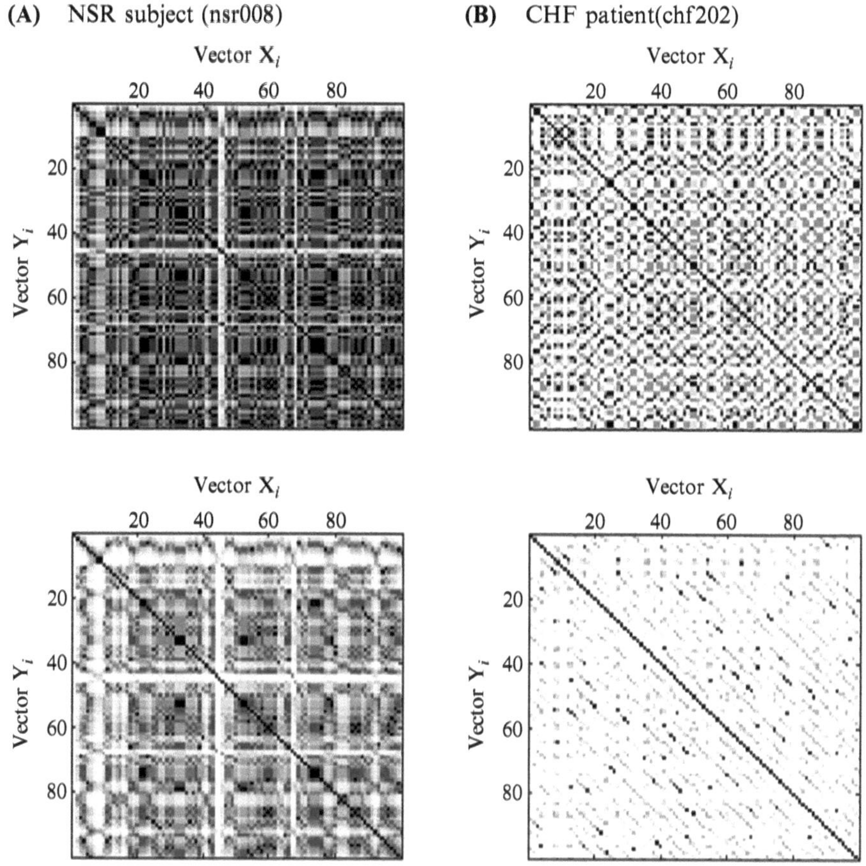

Fig. 11.10 The local similarity degree matrices in the intermediate calculation procedure of FuzzyMEn for (**a**) NSR subject and (**b**) CHF patient

Figure 11.11 demonstrates the results for the index of SampEn calculation. The SampEn was calculated by counting the mean percentages of matching vectors of the dimension $m = 2$ and $m + 1 = 3$, respectively. The percentages of matching vectors are illustrated for different threshold values r, and the corresponding SampEn values are shown. The NSR subject exhibits higher SampEn values than the CHF patient, indicating the more complicated component or irregular change in the RR time series.

Figure 11.12 demonstrates the results for the index of the FuzzyMEn calculation. Unlike SampEn identifying the vector similarity degree as binary determination (matching vector with similarity degree $= 1$ or not matching vector with similarity degree $= 0$), FuzzyMEn could output any value for similarity degree between 0 and 1. In each sub-figure, the upper panel shows the mean level of the global similarity degree, while the lower panel shows the mean level of the local similarity

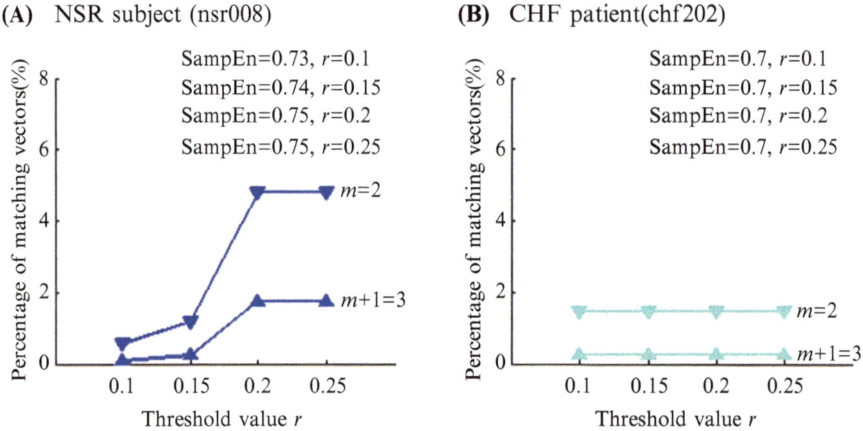

Fig. 11.11 SampEn calculation for (**a**) NSR subject and (**b**) CHF patient

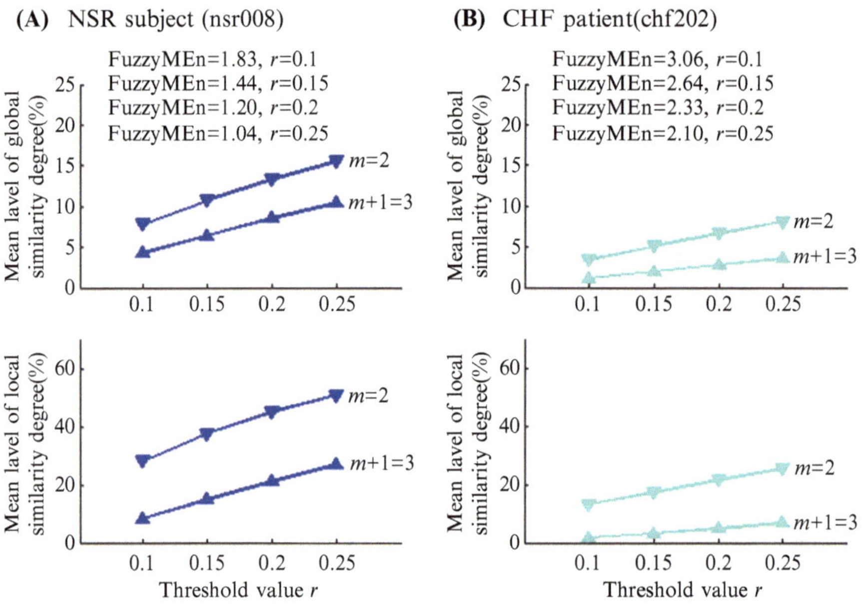

Fig. 11.12 FuzzyMEn calculation for (**a**) NSR subject and (**b**) CHF patient

degree, at the dimension $m = 2$ and dimension $m + 1 = 3$ respectively. The results corresponding to different threshold values r, and the corresponding FuzzyMEn values are shown. The NSR subject exhibits lower FuzzyMEn values than the CHF patient, although the mean levels of the vector similarity degree in the NSR subject are higher than those in the CHF patient, for both global and local aspects.

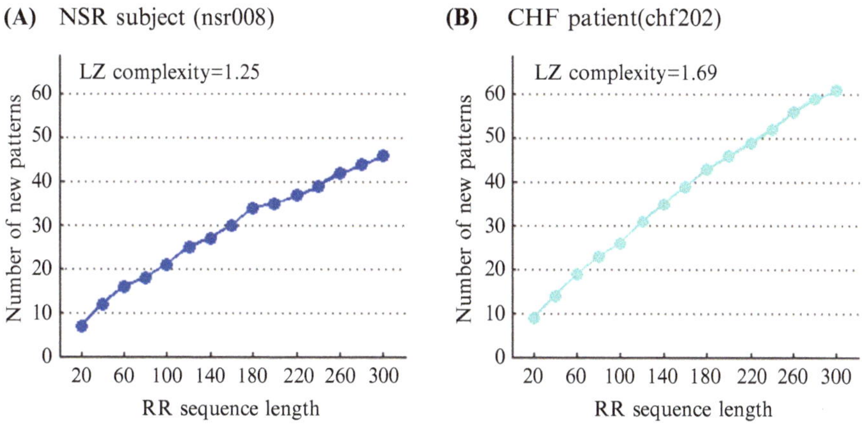

Fig. 11.13 LZ complexity calculation for (**a**) NSR subject and (**b**) CHF patient

11.3.3.4 LZ Complexity

Figure 11.13 shows the results for the LZ complexity index. With the increase of the RR sequence length, the number of the new patterns increases. The number of new patterns in the CHF patient increases more significantly than that in the NSR subject. The final LZ complexity values for the 5-min RR time series are depicted in this Figure.

11.4 Discussion

It is well accepted in the literature reviewed that congestive heart failure patients have a reduced heart rate variability in comparison with normal subjects, and also a different pattern in this variability. We have reviewed both the traditional linear analysis techniques (time- and frequency-domain) and the newer non-linear complexity techniques.

We have shown that this difference in heart rate variability in congestive heart failure has been identified by most of the techniques investigated. Those that were less successful in the examples analysed were the time-domain RMSSD index, but we have shown that other time-domain HRV indices were able to clearly separate heart failure from normal. For the newer non-linear complexity techniques, rather than only showing the values of the indices, we performed a detailed calculation progress analysis and revealed the inherent change of time series or patterns, providing further clear observations for these techniques. As expected, all newer non-linear complexity techniques were able to distinguish the differences between the illustrated examples.

We accept that this analysis is illustrative, and requires follow up analysis, but by illustrating the differences between normal and congestive heart failure rhythms we hope that this will encourage the use of the complete range of analysis techniques in research into both normal and abnormal cardiac rhythms.

Acknowledgement We acknowledge and thank Dr. Charalampos Tsimenidis of Newcastle University for his helpful review of the manuscript.

A. Appendix

A.1 RR Sequence Normalized Histogram

The general construction procedure for the RR sequence normalized histogram is summarized as follows [53]:

Given an RR sequence $\{RR_1, RR_2, \ldots, RR_N\}$, where N denotes the sequence length. The maximum ($RR_{\max}$) and minimum ($RR_{\min}$) values were firstly determined to calculate the range in the sequence:

$$RR_{\text{range}} = RR_{\max} - RR_{\min} \tag{11.5}$$

The threshold $a = 0.1 \times RR_{\text{range}}$ is set and then the left-step parameter H_l and right-step parameter H_r can be calculated as

$$H_l = \frac{RR_{\text{mean}} - (RR_{\min} + \alpha)}{5} \tag{11.6}$$

$$H_r = \frac{(RR_{\max} - \alpha) - RR_{\text{mean}}}{5} \tag{11.7}$$

where RR_{mean} denotes the mean value of the RR sequence. Based on these parameters, the RR_i is divided into seven sections. Table 11.1 details the element division rules.

The element percentage p_i in each of the sections is calculated as follows:

$$p_i = \frac{P_i}{N} \quad i = 1, 2, \cdots, 7 \tag{11.8}$$

In a rectangular coordinate system, the p_i corresponding to the seven sections (i.e. L1, L2, L3, C, R3, R2 and R1) is drawn to form the normalized histogram. Three quantitative indices can be obtained from the normalized histogram. They are respectively named as center-edge ratio (CER), cumulative energy (CE) and range information entropy (RIEn) and are defined as [53]:

Table 11.1 Element division rules for constructing the RR sequence normalized histogram

Section location	Section name	Element division rules	Element number	Element %
Leftmost section	L1	$RR_{\min} \leq RR_i < RR_{\min} + a$	P_1	p_1
Left second section	L2	$RR_{\min} + a \leq RR_i < RR_{\min} + a + 2 \times H_l$	P_2	p_2
Left third section	L3	$RR_{\min} + a + 2 \times H_l \leq RR_i < RR_{\min} + a + 4 \times H_l$	P_3	p_3
Middle section	C	$RR_{\text{mean}} - H_l \leq RR_i < RR_{\text{mean}} + H_r$	P_4	p_4
Right third section	R3	$RR_{\text{mean}} + H_r \leq RR_i < RR_{\text{mean}} + 3 \times H_r$	P_5	p_5
Right second section	R2	$RR_{\text{mean}} + 3 \times H_r \leq RR_i < RR_{\max} - a$	P_6	p_6
Rightmost section	R1	$RR_{\max} - a \leq RR_i \leq RR_{\max}$	P_7	p_7

$$\text{CER} = \frac{p_4}{p_1 + p_2 + p_6 + p_7} \tag{11.9}$$

$$\text{CE} = \sum_{i=1}^{7} p_i^2 \tag{11.10}$$

$$\text{RIEn} = -\sum\nolimits_{i=1}^{7} p_i * \ln p_i \tag{11.11}$$

A.2 Sample Entropy (SampEn)

The algorithm for SampEn is summarized as follows [29]: For the HRV series $x(i)$, $1 \leq i \leq N$, forms $N - m + 1$ vectors $X_i^m = \{x(i), x(i+1), \cdots, x(i+m-1)\}$, $1 \leq i \leq N - m + 1$. The distance between two vectors X_i^m and Y_i^m is defined as: $d_{i,j}^m = \max_{k=0}^{m-1} |x(i+k) - x(j+k)|$. Denote $B_i^m(r)$ the average number of j that meets $d_{i,j}^m \leq r$ for all $1 \leq j \leq N - m$, and similarly define $A_i^m(r)$ by $d_{i,j}^{m+1}$. SampEn is then defined by:

$$\text{SampEn}(m, r, N) = -\ln\left(\sum_{i=1}^{N-m} A_i^m(r) / \sum_{i=1}^{N-m} B_i^m(r)\right) \tag{11.12}$$

wherein the embedding dimension is usually set at $m = 2$ and the threshold at $r = 0.2 \times sd$ (sd indicates the standard deviation of the HRV series under-analyzed) [57, 60].

A.3 Fuzzy Measure Entropy (FuzzyMEn)

The calculation process of FuzzyMEn is summarized as follows [33, 56]:

For the RR or PTT segment $x(i)$ $(1 \leq i \leq N)$, firstly form the local vector sequences XL_i^m and global vector sequences XG_i^m respectively:

$$\begin{aligned} XL_i^m &= \{x(i), x(i+1), \cdots, x(i+m-1)\} - \overline{x}(i) \\ XG_i^m &= \{x(i), x(i+1), \cdots, x(i+m-1)\} - \overline{x} \end{aligned} \qquad 1 \leq i \leq N-m \quad (11.13)$$

The vector XL_i^m represents m consecutive $x(i)$ values but removing the local baseline $\overline{x}(i)$, which is defined as:

$$\overline{x}(i) = \frac{1}{m} \sum_{k=0}^{m-1} x(i+k) \quad 1 \leq i \leq N-m \quad (11.14)$$

The vector XG_i^m also represents m consecutive $x(i)$ values but removing the global mean value $\overline{x}$ of the segment $x(i)$, which is defined as:

$$\overline{x} = \frac{1}{N} \sum_{i=1}^{N} x(i) \quad (11.15)$$

Subsequently, the distance between the local vector sequences XL_i^m and XL_j^m, and the distance between the global vector sequences XG_i^m and XG_j^m are computed respectively as:

$$\begin{aligned} dL_{i,j}^m &= d\left[XL_i^m, XL_j^m\right] = \max_{k=0}^{m-1} \left|(x(i+k) - \overline{x}(i)) - (x(j+k) - \overline{x}(j))\right| \\ dG_{i,j}^m &= d\left[XG_i^m, XG_j^m\right] = \max_{k=0}^{m-1} \left|(x(i+k) - \overline{x}) - (x(j+k) - \overline{x})\right| \end{aligned} \quad (11.16)$$

Given the parameters n_L, r_L, n_G *and* r_G, we calculate the similarity degree $DL_{i,j}^m(n_L, r_L)$ between the local vectors XL_i^m and XL_j^m by the fuzzy function $\mu L\left(dL_{i,j}^m, n_L, r_L\right)$, as well as the similarity degree $DG_{i,j}^m(n_G, r_G)$ between the global vectors XG_i^m and XG_j^m by the fuzzy function $\mu G\left(dG_{i,j}^m, n_G, r_G\right)$:

$$\begin{aligned} DL_{i,j}^{m}\left(n_L, r_L\right) &= \mu L\left(dL_{i,j}^{m}, n_L, r_L\right) = \exp\left(-\frac{\left(dL_{i,j}^{m}\right)^{n_L}}{r_L}\right) \\ DG_{i,j}^{m}\left(n_G, r_G\right) &= \mu G\left(dG_{i,j}^{m}, n_G, r_G\right) = \exp\left(-\frac{\left(dG_{i,j}^{m}\right)^{n_G}}{r_G}\right) \end{aligned} \tag{11.17}$$

We define the functions $\phi L^m(n_L, r_L)$ and $\phi G^m(n_G, r_G)$ as:

$$\begin{aligned} \phi L^m\left(n_L, r_L\right) &= \tfrac{1}{N-m}\sum_{i=1}^{N-m}\left(\tfrac{1}{N-m}\sum_{j=1}^{N-m} DL_{i,j}^{m}\left(n_L, r_L\right)\right) \\ \phi G^m\left(n_G, r_G\right) &= \tfrac{1}{N-m}\sum_{i=1}^{N-m}\left(\tfrac{1}{N-m}\sum_{j=1}^{N-m} DG_{i,j}^{m}\left(n_G, r_G\right)\right) \end{aligned} \tag{11.18}$$

Similarly, we define the function $\phi L^{m+1}(n_L, r_L)$ for $m+1$ dimensional vectors XL_i^{m+1} and XL_j^{m+1} the function $\phi G^{m+1}(n_G, r_G)$ for $m+1$ dimensional vectors XG_i^{m+1} and YG_j^{m+1}:

$$\begin{aligned} \phi L^{m+1}\left(n_L, r_L\right) &= \tfrac{1}{N-m}\sum_{i=1}^{N-m}\left(\tfrac{1}{N-m}\sum_{j=1}^{N-m} DL_{i,j}^{m+1}\left(n_L, r_L\right)\right) \\ \phi G^{m+1}\left(n_L, r_L\right) &= \tfrac{1}{N-m}\sum_{i=1}^{N-m}\left(\tfrac{1}{N-m}\sum_{j=1}^{N-m} DG_{i,j}^{m+1}\left(n_G, r_G\right)\right) \end{aligned} \tag{11.19}$$

Then the fuzzy local measure entropy (FuzzyLMEn) and fuzzy global measure entropy (FuzzyGMEn) are computed as:

$$\begin{aligned} \text{FuzzyLMEn}\left(m, n_L, r_L, N\right) &= -\ln\left(\phi L^{m+1}\left(n_L, r_L\right)/\phi L^m\left(n_L, r_L\right)\right) \\ \text{FuzzyGMEn}\left(m, n_G, r_G, N\right) &= -\ln\left(\phi G^{m+1}\left(n_G, r_G\right)/\phi G^m\left(n_G, r_G\right)\right) \end{aligned} \tag{11.20}$$

Finally, the FuzzyMEn of RR segment $x(i)$ is calculated as follows:

$$\begin{aligned} \text{FuzzyMEn}\left(m, n_L, r_L, n_G, r_G, N\right) &= \text{FuzzyLMEn}\left(m, n_L, r_L, N\right) \\ &\quad + \text{FuzzyGMEn}\left(m, n_G, r_G, N\right) \end{aligned} \tag{11.21}$$

In this study, the local similarity weight was set to $n_L = 3$ and the global vector similarity weight was set to $n_G = 2$. The local tolerance threshold r_L was set equal to the global threshold r_G, i.e., $r_L = r_G = r$. Hence, the formula (11.21) becomes:

$$\text{FuzzyMEn}\left(m, r, N\right) = \text{FuzzyLMEn}\left(m, r, N\right) + \text{FuzzyGMEn}\left(m, r, N\right) \tag{11.22}$$

For both SampEn and FuzzyMEn, the entropy results were only based on the three parameters: the embedding dimension m, the tolerance threshold r and the RR segment length N.

A.4 Lempel-Ziv (LZ) Complexity

The calculation process of LZ complexity is summarized as follows [21, 59]:

1. For a binary symbolic sequence $R = \{s_1, s_2, \ldots, s_n\}$, let S and Q denote two strings, respectively, and SQ is the concatenation of S and Q, while the string $SQ\pi$ is derived from SQ after its last character is deleted (π means the operation to delete the last character in the string). $v(SQ\pi)$ denotes the vocabulary of all different substrings of $SQ\pi$. Initially, $c(n) = 1, S = s_1, Q = s_2$, and so $SQ\pi = s_1$;
2. In summary, $S = s_1\ s_2,\ \ldots,\ s_r$, $Q = s_{r+1}$, and so $SQ\pi = s_1\ s_2,\ \ldots,\ s_r$; if Q belongs to $v(SQ\pi)$, then s_{r+1}, that is, Q is a substring of $SQ\pi$, and so S does not change, and Q is updated to be $s_{r+1}s_{r+2}$, and then judge if Q belongs to $v(SQ\pi)$ or not. Repeat this process until Q does not belong to $v(SQ\pi)$;
3. Now, $Q = s_{r+1}s_{r+2},\ \ldots,\ s_{r+i}$, which is not a substring of $SQ\pi = s_1\ s_2,\ \ldots,\ s_rs_{r+1}, \ldots, s_{r+i\text{-}1}$, so increase $c(n)$ by one;
4. Thereafter, S is renewed to be $S = s_1\ s_2,\ \ldots, s_{r+i}$, and $Q = s_{r+i+1}$;
5. Then the procedures repeat until Q is the last character. At this time $c(n)$ is the number of different substrings contained in R. For practical application, $c(n)$ should be normalized. It has been proved that the upper bound of $c(n)$ is

$$c(n) < \frac{n}{(1-\varepsilon_n)\log_\alpha(n)}, \tag{11.23}$$

where ε_n is a small quantity and $\varepsilon_n \to 0$ ($n \to \infty$). In fact,

$$\lim_{n\to\infty} c(n) = b(n) = \frac{n}{\log_\alpha(n)}. \tag{11.24}$$

6. Finally, LZ complexity is defined as the normalized output of $c(n)$:

$$C(n) = \frac{c(n)}{b(n)}, \tag{11.25}$$

where $C(n)$ is the normalized LZ complexity, and denotes the arising rate of new patterns within the sequence.

References

1. Acharya, U.R., Joseph, K.P., Kannathal, N., Lim, C.M., Suri, J.S.: Heart rate variability: a review. Med. Biol. Eng. Comput. **44**, 1031–1051 (2006)
2. Huikuri, H.V., Perkiömaki, J.S., Maestri, R., Pinna, G.D.: Clinical impact of evaluation of cardiovascular control by novel methods of heart rate dynamics. Philos. Trans. A Math. Phys. Eng. Sci. **367**, 1223–1238 (2009)
3. Makikallio, T.H., Tapanainen, J.M., Tulppo, M.P., Huikuri, H.V.: Clinical applicability of heart rate variability analysis by methods based on nonlinear dynamics. Card. Electrophysiol. Rev. **6**, 250–255 (2002)

4. Voss, A., Schulz, S., Schroeder, R., Baumert, M., Caminal, P.: Methods derived from nonlinear dynamics for analysing heart rate variability. Philos. Trans. A Math. Phys. Eng. Sci. **367**, 277–296 (2009)
5. Weiss, J.N., Garfinkel, A., Spano, M.L., Ditto, W.L.: Chaos and chaos control in biology. J. Clin. Invest. **93**, 1355–1360 (1994)
6. Brennan, M., Palaniswami, M., Kamen, P.: Poincare plot interpretation using a physiological model of HRV based on a network of oscillators. Am. J. Physiol. Heart Circ. Physiol. **283**, H1873–H1886 (2002)
7. Kamen, P.W., Krum, H., Tonkin, A.M.: Poincare plot of heart rate variability allows quantitative display of parasympathetic nervous activity in humans. Clin. Sci. **91**, 201–208 (1996)
8. Laitio, T.T., Mäkikallio, T.H., Huikuri, H.V., Kentala, E.S., Uotila, P., Jalonen, J.R., Helenius, H., Hartiala, J., Yli-Mäyry, S., Scheinin, H.: Relation of heart rate dynamics to the occurrence of myocardial ischemia after coronary artery bypass grafting. Am. J. Cardiol. **89**, 1176–1181 (2002)
9. Kamen, P., Tonkin, A.: Application of the Poincare plot to heart rate variability: a new measure of functional status in heart failure. Aust. NZ J. Med. **25**, 18–26 (1995)
10. Kobayashi, M., Musha, T.: 1/f fluctuation of heartbeat period. I.E.E.E. Trans. Biomed. Eng. **29**, 456–457 (1982)
11. Lombardi, F.: Chaos theory, heart rate variability, and arrhythmic mortality. Circulation. **108**, 8–10 (2000)
12. Peng, C.K., Havlin, S., Stanley, H.E., Goldberger, A.L.: Quantification of scaling exponents and crossover phenomena in nonstationary heartbeat time series. Chaos. **5**, 82–87 (1995)
13. Huikuri, H.V., Mäkikallio, T.H., Peng, C.K., Goldberger, A.L., Hintze, U., Møller, M.: Fractal correlation properties of R-R interval dynamics and mortality in patients with depressed left ventricular function after an acute myocardial infarction. Circulation. **101**, 47–53 (2000)
14. Mäkikallio, T.H., Høiber, S., Køber, L., Torp-Pedersen, C., Peng, C.K., Goldberger, A.L., Huikuri, H.V.: Fractal analysis of heart rate dynamics as a predictor of mortality in patients with depressed left ventricular function after acute myocardial infarction. Am. J. Cardiol. **83**, 836–839 (1999)
15. Ivanov, P.C., Amaral, L.A.N., Goldberger, A.L., Havlin, S., Rosenblum, M.G., Struzik, Z.H.E.: Multifractality in human heartbeat dynamics. Nature. **399**, 461–465 (1999)
16. Hadamard, J.: Les surfaces a courbures opposees et leurs lignes geodesiques. J. Math. Pures. Appl. **4**, 27–73 (1898)
17. Kurths, J., Voss, A., Saparin, P., Witt, A., Kleiner, H.J., Wessel, N.: Quantitative analysis of heart rate variability. Chaos. **5**, 88–94 (1995)
18. Voss, A., Kurths, J., Kleiner, H.J., Witt, A., Wessel, N., Saparin, P., Osterziel, K.J., Schurath, R., Dietz, R.: The application of methods of non-linear dynamics for the improved and predictive recognition of patients threatened by sudden cardiac death. Cardiovasc. Res. **31**, 419–433 (1996)
19. Porta, A., Guzzetti, S., Montano, N., Furlan, R., Pagani, M., Malliani, A., Cerutti, S.: Entropy, entropy rate, and pattern classification as tools to typify complexity in short heart period variability series. I.E.E.E. Trans. Biomed. Eng. **48**, 1282–1291 (2001)
20. Porta, A., Faes, L., Masé, M., D'Addio, G., Pinna, G.D., Maestri, R., Montano, N., Furlan, R., Guzzetti, S., Nollo, G., Malliani, A.: An integrated approach based on uniform quantization for the evaluation of complexity of short-term heart period variability: application to 24 h Holter recordings in healthy and heart failure humans. Chaos. **17**, 015117 (2007)
21. Lempel, A., Ziv, J.: On the complexity of finite sequences. IEEE Trans. Inf. Theory. **22**, 75–81 (1976)
22. Abásolo, D., Hornero, R., Gómez, C., García, M., López, M.: Analysis of EEG background activity in Alzheimer's disease patients with Lempel-Ziv complexity and central tendency measure. Med. Eng. Phys. **28**, 315–322 (2006)
23. Sarlabous, L., Torres, A., Fiz, J.A., Morera, J., Jané, R.: Index for estimation of muscle force from mechanomyography based on the Lempel-Ziv algorithm. J. Electromyogr. Kinesiol. **23**, 548–557 (2013)

24. Zhang, Y.T., Wei, S.S., Liu, H., Zhao, L.N., Liu, C.Y.: A novel encoding Lempel-Ziv complexity algorithm for quantifying the irregularity of physiological time series. Comput. Methods Prog. Biomed. **133**, 7–15 (2016)
25. Pincus, S.M.: Approximate entropy as a measure of system complexity. Proc. Natl. Acad. Sci. U. S. A. **88**, 2297–2301 (1991)
26. Costa, M., Goldberger, A.L., Peng, C.K.: Multiscale entropy analysis of biological signals. Phys. Rev. E. **71**, 021906 (2005)
27. Liu, C.Y., Liu, C.C., Shao, P., Li, L.P., Sun, X., Wang, X.P., Liu, F.: Comparison of different threshold values r for approximate entropy: application to investigate the heart rate variability between heart failure and healthy control groups. Physiol. Meas. **32**, 167–180 (2011b)
28. Pincus, S.M., Goldberger, A.L.: Physiological time-series analysis: what does regularity quantify? Am. J. Physiol. Heart Circ. Physiol. **266**, H1643–H1656 (1994)
29. Richman, J.S., Moorman, J.R.: Physiological time-series analysis using approximate entropy and sample entropy. Am. J. Physiol. Heart Circ. Physiol. **278**, H2039–H2049 (2000)
30. Lake, D.E., Richman, J.S., Griffin, M.P., Moorman, J.R.: Sample entropy analysis of neonatal heart rate variability. Am. J. Phys. Regul. Integr. Comp. Phys. **283**, R789–R797 (2002)
31. Tuzcu, V., Nas, S., Börklü, T., Ugur, A.: Decrease in the heart rate complexity prior to the onset of atrial fibrillation. Europace. **8**, 398–402 (2006)
32. Chen, W.T., Zhuang, J., Yu, W.X., Wang, Z.Z.: Measuring complexity using FuzzyEn, ApEn, and SampEn. Med. Eng. Phys. **31**, 61–68 (2009)
33. Liu, C.Y., Li, K., Zhao, L.N., Liu, F., Zheng, D.C., Liu, C.C., Liu, S.T.: Analysis of heart rate variability using fuzzy measure entropy. Comput. Biol. Med. **43**, 100–108 (2013)
34. Ahmed, M.U., Mandic, D.P.: Multivariate multiscale entropy: a tool for complexity analysis of multichannel data. Phys. Rev. E. **84**, 061918 (2011)
35. Ahmed, M.U., Mandic, D.P.: Multivariate multiscale entropy analysis. IEEE Signal. Proc. Lett. **19**, 91–94 (2012)
36. Azami, H., Escudero, J.: Refined composite multivariate generalized multiscale fuzzy entropy: a tool for complexity analysis of multichannel signals. Phys. A. **465**, 261–276 (2017)
37. Arbolishvili, G.N., Mareev, V.I., Orlova, I.A., Belenkov, I.N.: Heart rate variability in chronic heart failure and its role in prognosis of the disease. Kardiologiia. **46**, 4–11 (2005)
38. Nolan, J., Batin, P.D., Andrews, R., Lindsay, S.J., Brooksby, P., Mullen, M., Baig, W., Flapan, A.D., Cowley, A., Prescott, R.J.: Prospective study of heart rate variability and mortality in chronic heart failure results of the United Kingdom heart failure evaluation and assessment of risk trial (UK-Heart). Circulation. **98**, 1510–1516 (1998)
39. Rector, T.S., Cohn, J.N.: Prognosis in congestive heart failure. Annu. Rev. Med. **45**, 341–350 (1994)
40. Smilde, T.D.J., van Veldhuisen, D.J., van den Berg, M.P.: Prognostic value of heart rate variability and ventricular arrhythmias during 13-year follow-up in patients with mild to moderate heart failure. Clin. Res. Cardiol. **98**, 233–239 (2009)
41. Malliani, A., Pagani, M., Lombardi, F., Cerutti, S.: Cardiovascular neural regulation explored in the frequency domain. Circulation. **84**, 482–492 (1991)
42. Binkley, P.F., Nunziata, E., Haas, G.J., Nelson, S.D., Cody, R.J.: Parasympathetic withdrawal is an integral component of autonomic imbalance in congestive heart failure: demonstration in human subjects and verification in a paced canine model of ventricular failure. J. Am. Coll. Cardiol. **18**, 464–472 (1991)
43. Poon, C.S., Merrill, C.K.: Decrease of cardiac chaos in congestive heart failure. Nature. **389**, 492–495 (1997)
44. Woo, M.A., Stevenson, W.G., Moser, D.K., Middlekauff, H.R.: Complex heart rate variability and serum norepinephrine levels in patients with advanced heart failure. J. Am. Coll. Cardiol. **23**, 565–569 (1994)
45. La Rovere, M.T., Pinna, G.D., Maestri, R., Mortara, A., Capomolla, S., Febo, O., Ferrari, R., Franchini, M., Gnemmi, M., Opasich, C., Riccardi, P.G., Traversi, E., Cobelli, F.: Short-term heart rate variability strongly predicts sudden cardiac death in chronic heart failure patients. Circulation. **107**, 565–570 (2003)

46. Hadase, M., Azuma, A., Zen, K., Asada, S., Kawasaki, T., Kamitani, T., Kawasaki, S., Sugihara, H., Matsubara, H.: Very low frequency power of heart rate variability is a powerful predictor of clinical prognosis in patients with congestive heart failure. Circ. J. **68**, 343–347 (2004)
47. Guzzetti, S., Mezzetti, S., Magatelli, R., Porta, A., De Angelis, G., Rovelli, G., Malliani, A.: Linear and non-linear 24 h heart rate variability in chronic heart failure. Auton. Neurosci. **86**, 114–119 (2000)
48. Mäkikallio, T.H., Huikuri, H.V., Hintze, U., Videbaek, J., Mitrani, R.D., Castellanos, A., Myerburg, R.J., Møller, M., DIAMOND Study Group: Fractal analysis and time- and frequency-domain measures of heart rate variability as predictors of mortality in patients with heart failure. Am. J. Cardiol. **87**, 178–182 (2001)
49. Costa, M., Goldberger, A.L., Peng, C.K.: Multiscale entropy analysis of complex physiologic time series. Phys. Rev. Lett. **89**, 068102 (2002)
50. Singh, J.P., Larson, M.G., Tsuji, H., Evans, J.C., O'Donnell, C.J., Levy, D.: Reduced heart rate variability and new-onset hypertension–insights into pathogenesis of hypertension: The Framingham Heart Study. Hypertension. **32**, 293–297 (1998)
51. Task Force of the European Society of Cardiology and the North American Society of Pacing and Electrophysiology: Heart rate variability: standards of measurement, physiological interpretation and clinical use. Circulation. **93**, 1043–1065 (1996)
52. Goldberger, J.J., Ahmed, M.W., Parker, M.A., Kadish, A.H.: Dissociation of heart rate variability from parasympathetic tone. Am. J. Phys. **266**, H2152–H2157 (1994)
53. Liu, C.Y., Li, P., Zhao, L.N., Yang, Y., Liu, C.C.: Evaluation method for heart failure using RR sequence normalized histogram. In: Murray, A. (ed.) Computing in Cardiology, pp. 305–308. IEEE, Hangzhou (2011a)
54. Ho, K.K., Moody, G.B., Peng, C.K., Mietus, J.E., Larson, M.G., Levy, D., Goldberger, A.L.: Predicting survival in heart failure case and control subjects by use of fully automated methods for deriving nonlinear and conventional indices of heart rate dynamics. Circulation. **96**, 842–848 (1997)
55. Liu, C.Y., Zhang, C.Q., Zhang, L., Zhao, L.N., L C, C., Wang, H.J.: Measuring synchronization in coupled simulation and coupled cardiovascular time series: a comparison of different cross entropy measures. Biomed. Signal. Process. Control. **21**, 49–57 (2015)
56. Liu, C.Y., Zhao, L.N.: Using Fuzzy Measure Entropy to improve the stability of traditional entropy measures. In: Murray, A. (ed.) Computing in cardiology, pp. 681–684. IEEE, Hangzhou (2011)
57. Zhao, L.N., Wei, S.S., Zhang, C.Q., Zhang, Y.T., Jiang, X.E., Liu, F., Liu, C.Y.: Determination of sample entropy and fuzzy measure entropy parameters for distinguishing congestive heart failure from normal sinus rhythm subjects. Entropy. **17**, 6270–6288 (2015)
58. Zhang, X.S., Zhu, Y.S., Thakor, N.V., Wang, Z.Z.: Detecting ventricular tachycardia and fibrillation by complexity measure. I.E.E.E. Trans. Biomed. Eng. **46**, 548–555 (1999)
59. Aboy, M., Hornero, R., Abasolo, D., Alvarez, D.: Interpretation of the Lempel-Ziv complexity measure in the context of biomedical signal analysis. I.E.E.E. Trans. Biomed. Eng. **53**, 2282–2288 (2006)
60. Yentes, J.M., Hunt, N., Schmid, K.K., Kaipust, J.P., McGrath, D., Stergiou, N.: The appropriate use of approximate entropy and sample entropy with short data sets. Ann. Biomed. Eng. **41**, 349–365 (2013)
61. Ruan, X.H., Liu, C.C., Liu, C.Y., Wang, X.P. and Li, P.: Automatic detection of atrial fibrillation using RR interval signal. In: 4th International Conference on Biomedical Engineering and Informatics: IEEE. pp. 644–647. (2011)

Chapter 12
Heart Rate Complexity Associated with Diabetic Cardiac Neuropathy

Herbert F. Jelinek and David J. Cornforth

Abstract Cardiac autonomic neuropathy (CAN) is a serious complication of diabetes mellitus that often leads to increased morbidity and mortality. An examination of heart rate provides an opportunity to investigate the functional attributes of the autonomic nervous system (ANS) and specifically cardiac rhythm. Measuring changes in heart rate is non-invasive and may indicate increased risk of arrhythmic events associated with cardiac autonomic neuropathy (CAN) leading to sudden cardiac death, especially in diabetes mellitus. However, it is also a surrogate marker for advancement of diabetic neuropathy affecting other organ systems, and pathologies that affect neural function, such as depression, schizophrenia, and Parkinson's disease. ANS modulation of the cardiac rhythm leads to short and long-term non-stationary and nonlinear changes in heart rate. In line with this inherent complexity, computational analytics are required that are sensitive yet robust enough to adequately describe this complexity. Entropy measures are a natural candidate for this application as they are able to estimate information content or complexity of the heart rate.

A number of different approaches have been used, as there are many variations of entropy measures. Empirical studies suggest that multiscale Rényi entropy allows a clear picture to emerge, that describes the advancement of diabetic neuropathy and specifically cardiac autonomic neuropathy from a preclinical, non-symptomatic stage to severe signs and symptoms of disease. This chapter describes the use of multiscale Rényi entropy to diagnose CAN progression. One hundred and forty nine ECGs were recorded at 400 samples/s of 71 controls, 67 participants with early CAN and 11 with definite CAN. All recordings were preprocessed and analysed using a number of nonlinear algorithms (Multiscale Entropy, multiscale DFA and multiscale Rényi Entropy). The results indicate that Rényi entropy is a better diagnostic tool to assess CAN progression with an area under the curve (AUC) of 0.723, 0.692 and 0.862 for differentiating no CAN from early CAN, early from definite CAN and no

H.F. Jelinek (✉)
School of Community Health, Charles Sturt University, Albury, NSW, Australia
e-mail: hjelinek@csu.edu.au

D.J. Cornforth
School of Electrical Engineering and Computing, University of Newcastle, Callaghan, NSW, Australia

R. Barbieri et al. (eds.), *Complexity and Nonlinearity in Cardiovascular Signals*,
DOI 10.1007/978-3-319-58709-7_12

CAN from definite CAN respectively $p < 0.05$. Using machine learning algorithms to identify the best subset of measures for CAN classification, an accuracy of 71% was obtained for differentiating no CAN from early CAN. Identification of early, asymptomatic disease is of great clinical importance as early intervention is known to have the best health outcomes. The current work suggests that Rényi entropy has the best discriminatory power for identifying early CAN.

12.1 Introduction

Normal cardiac rhythm is controlled by the sino-atrial node, neuro-hormonal and autonomic nervous system (ANS) modulation by affecting the output of the sympathetic and parasympathetic branches of the ANS [1, 2]. Diseases of the ANS will affect modulation of heart rhythm by affecting the output of the sympathetic and parasympathetic branches of the ANS, and therefore may reflect as changes in the heart rate. Cardiac Autonomic Neuropathy (CAN) is characterised by peripheral nerve damage and leads to abnormal regulation of the heartbeat by the ANS [3, 4]. CAN often presents as a complication of diabetes in up to 60% of individuals with type 2 diabetes at different stages during the disease progression [5]. The continuous dynamic changes in heart rate as it adjusts to internal and external stressors of the body are important in maintaining optimal cardiac function. The rapid frequency of vagal input tonically reduces heart rate from its intrinsic rate. The sympathetic influence has a lower frequency and the ensuing increase in heart rate not only depends on an increased sympathetic output but also parasympathetic withdrawal. Hence the sympathovagal balance is an important modulatory mechanism to retain appropriate heart rate variability and manifests in continuous changes in the beat-to-beat interval manifesting as nonlinear and nonstationary in the ECG or heart rate recordings [6, 7].

The clinical manifestations of CAN include orthostatic hypertension and abnormalities in heart rate variability (HRV). Poor diagnoses of CAN may result in increased incidence of silent myocardial infarction and ischemia, which can lead to sudden death. Silent ischaemia is significantly more frequent in patients with than in those without CAN (38% vs 5%) [8]. Early sub clinical detection of CAN for risk stratification and intervention for preventing the potentially serious consequences of CAN especially in people with diabetes are therefore of prime importance. New methods that are non-invasive and independent of patient cooperation are preferable in the diagnosis of CAN but still require further research to understand their sensitivity and specificity in risk stratification for CAN. The most common method used is heart rate variability analysis [9]. A change in HRV is regarded as one of the early signs of cardiac autonomic neuropathy [10]. However, conventionally used time and frequency domain parameters of HRV are not always suitable for analysis because of non-stationarity characteristic of the ECG recordings and the presence of nonlinear phenomena in the physiological signal's parameter variability. Only a few studies have applied new parameters based on nonlinear dynamics theory to HRV analysis in DM patients [11, 12].

The complexity of cardiac rhythm may be conducive to expression using entropy measures. Such changes in the beat-to-beat rhythm of the heart can be observed and measured from low fidelity data such as the interbeat (RR) interval only, and the full ECG recording is not required to identify phenomena that indicate physiological changes consistent with disease progression. CAN is a serious complication of DM and is caused by damage to autonomic nerve fibers that innervate the heart and blood vessels [2]. During progression of the disease, the parasympathetic nerve fibers are usually affected before the sympathetic nerve fibers, leading to reduced HRV [13]. HRV is often used in assessing the cardiac autonomic function [14, 15]. HRV is reduced in diabetes mellitus (DM) patients, suggesting dysfunction of cardiac autonomic regulation, which has been previously shown to be associated with increased risk for adverse cardiac events [16].

12.2 Measures of Heart Rate Variability

The raw RR interval is a time series, and HRV can be described by aggregate measures. These include time series measures such as mean and standard deviation of the RR interval, frequency domain measures such as the proportion of low to high frequencies present in the RR time series, and more complex non-linear measures, including fractal and entropy-based measures [17–19].

Previous studies have attempted to discern a relationship between blood glucose level (BGL) and HRV using frequency-domain measures. Some studies indicate a reduced HF component in subjects with DM [20], while others suggests a reduction of the LF component in diabetics and in subjects with impaired fasting BGL [21]. A negative correlation was shown between BGL and HRV (both LF and HF components [21]. In addition, an increased LF/HF power ratio has been shown during hyperglycaemia in controls and diabetics without CAN [22, 23].

Non-linear measures include Detrended Fluctuation Analysis (DFA), which is an estimate of the fractal correlation of the RR interval series and provides an exponent expressing short-term correlations (alpha1) and another expressing long-term correlations (alpha2). The correlation dimension (D2) of fractal analysis can also be considered. The following sections provide details about entropy measures. Entropy describes sub-states within a larger system or disorder of a system. Entropy may also be viewed as the amount of additional information required to characterise a system [24].

All of these nonlinear measures can be applied in an attempt to detect disease, yet from the large number of possible measures, it is difficult to know which to select in order to provide the best separation between disease and health. One promising approach, mentioned later in this chapter, uses machine learning to automatically choose measures that enhance the separation between classes.

12.3 Multiscale Sample Entropy

Multiscale Sample Entropy (MSE) is based on the Shannon entropy, and can applied to the RR interval time series:

$$H(X) = -\sum_{i=1}^{n} p(x_i)\log p(x_i) \tag{12.1}$$

where x_i is an RR interval and its probability is estimated by counting the number of RR intervals that occurs in a given recording. As the RR interval can assume many values, in practice a discretisation is required. This may be accomplished by defining a range of RR intervals, as in histogram estimation, or by using a threshold r and counting the number of RR intervals whose difference is less than r. The latter method provides an opportunity to consider sequences of m RR intervals rather than restricting analysis to a single RR interval. The number of similar sequences are counted having length m:

$$c_m = \sum_{i=1}^{n-m+1} c_m(i) \tag{12.2}$$

where n is the number of RR intervals in the recording and i is the position in the recording where a given sequence begins. Sample Entropy extends Shannon Entropy by estimating the probability that a sequence of length m that repeats itself (within a similarity threshold r) will also repeat itself if the sequence length is extended to $m + 1$ [25]. If $A = c_{m+1}$ and $B = c_m$ then Sample Entropy is given by:

$$SampEn = -\log\frac{A}{B} \tag{12.3}$$

Sample Entropy has been associated with changes to BGL and glycated haemoglobin (HbA1c), indicating that hyperglycaemia induces changes in RR time series complexity, which is not detectable by standard linear methods [23]. However this relationship is complex, as in this work Sample Entropy was negatively correlated with BGL, and positively correlated with HbA1c.

The multiscale extension of this measure uses coarse grained copies of the RR interval data by averaging over a number of RR intervals τ:

$$y_j^{(\tau)} = \frac{1}{\tau}\sum_{i=(j-i)\tau+1}^{j\tau} x_i \tag{12.4}$$

Multiscale Sample Entropy (MSE) is then calculated using Eq. (12.1) where the x is replaced by the y of Eq. (12.4) [26, 27].

12.4 Multi-Fractal Detrended Fluctuation Analysis

Detrended Fluctuation Analysis (DFA), like MSE, examines the self-similarity properties of a time series and is a fractal like measure. A transformation is used to subtract the mean and calculate the cumulative sum:

$$X_k = \sum_{i=1}^{k}(x_i - \bar{x}) \tag{12.5}$$

where x_1 is a single RR interval. Sequences of X of length m are extracted and a least square errors method used to fit a straight line to each sequence, where error is estimated as:

$$\epsilon_v = \sqrt{\sum_{i=mv}^{m(v+1)} (X_i - \acute{X}_i)^2} \tag{12.6}$$

where n/m is the number of segments in the complete series and $\acute{X}_i$ is the estimated value of X_i. The mean rms error is given by:

$$\epsilon_m = \sqrt{\frac{1}{n/m}\sum_{v=1}^{n/m} \epsilon_v^2} \tag{12.7}$$

DFA is calculated as the slope of the log ϵ_v against log m. The Multiscale DFA is a generalisation made by substituting an exponent:

$$\epsilon_{q,m} = \left(\frac{1}{n/m}\sum_{v=1}^{n/m} \epsilon_v^q\right)^{1/q} \tag{12.8}$$

12.5 Rényi Entropy

Rényi entropy is an extension of the Shannon entropy that preserves additivity of statistically independent systems and compatible with Kolmogorov's probability [28]. Rényi Entropy has appeared in the literature and examples are apparent for its efficacy in discriminating stages of ANS impairments, for example [29]. Rényi Entropy may be defined as:

$$H(\alpha) = \frac{1}{1-\alpha} log_2(\sum_{i=1}^{n} p_i^{\alpha}) \tag{12.9}$$

where p_i is the probability that a random variable takes a given value out of n values, and α is the exponent or order of the entropy measure. The choice of α determines the scale at which the entropy is evaluated, and so by choosing a selection of values for α, a multi-scale measure may be obtained. The special cases are $H(0)$, which is simply the logarithm of n, and $H(1)$ which is equivalent to the Shannon entropy [30]. As α increases, the measures become more sensitive to values of the random variable occurring at higher probability and less to values occurring at lower probability, which provides a picture of the RR length distribution within a signal.

However, the entropy requires an estimate of probabilities, and there are a number of ways in which this can be determined. Literature describes two main methods of estimating probabilities: the histogram method and the density method [31]. The cited work demonstrates the superiority of the density method in terms of providing a measure that can discriminate different classes of CAN. This method estimates the probability of a sequence of RR intervals of length m by considering each sequence as a point in m-dimensional space with its surrounding density estimated by comparing other RR intervals using a distance measure. A suitable kernel function can be centred on each point and the value of the kernel for all other points added to provide a probability for each point, or sequence of RR intervals.

Assuming an individual RR interval with index i, and using a Gaussian kernel, its probability is calculated as the sum of all contributions from other RR intervals with index j:

$$\rho_i = \frac{1}{\sigma\sqrt{2\pi}} \sum_{i=1}^{n} e^{\frac{dist_{ij}^2}{2\sigma^2}} \tag{12.10}$$

using a distance measure:

$$dist_{ij}^2 = \sum_{k=1}^{m} (x_{i+k} - x_{j+k})^2 \tag{12.11}$$

where σ is the dispersion of the kernel function.

12.6 Empirical Studies

Some results indicative of the efficacy of the Rényi Entropy have been obtained from studies using data from the Charles Sturt Diabetes Complications Screening Group (DiabHealth), Australia [32]. These studies were approved by the Charles Sturt University Human Ethics Committee and written informed consent was obtained from all participants. Participants were recruited from people attending the diabetes clinic and were comparable for age and gender. After initial screening, those with heart disease, presence of a pacemaker, kidney disease or polypharmacy (including multiple anti-arrhythmic medication) were excluded from the studies.

The status of CAN was determined using the Ewing battery criteria [19, 33, 34], and each participant was assigned as either Normal (without CAN, or control: 71 participants), Early (67 participants) or Definite Confirmed clinical CAN: 11 participants).

For each participant, a 20 min ECG recording was obtained in a temperature stable environment following a 5–10 min rest period in a supine position. The sampling rate was set to 400 samples/s and recordings pre-processed according to the method described in [35]. A 15 min segment was selected from the middle in order to remove start up artefacts and movement at the end of the recording. From this shorter recording, the RR intervals were extracted and detrended, and from this series a variety of measures was extracted [36].

12.6.1 Comparisons

Table 12.1 shows results from Mann-Whitney tests and from Receiver Operating Curve (ROC) analysis, comparing different participant groups when using time domain measures, as used in many studies. The heading row provides the abbreviations for time domain measures, which are explained as follows: SDNN (Standard Deviation of the RR intervals), Rmssd (Root Mean Square of the Successive Differences or RR intervals), pNN50 (the proportion of pairs of successive NN intervals that differ by more than 50 ms), Triang (integral of the density distribution of RR intervals divided by the maximum of the density), TINN (The triangular interpolation of NN interval histogram), SD1 and SD2 (short and long term variability from the Poincaré plot), REC (Recurrence Rate), DET (Determinism), and Lmean (mean length of diagonal lines o the recurrence plot exceeding a threshold). Results from Mann-Whitney are in the following three rows, and report the *p*-value. These rows show comparisons between Normal and Early, where the *p*-value is given in the row designated as pNE, Early vs. Definite (designated as pED), and Definite vs. Normal (designated as pDN).

Due to the number of comparisons in this work (144), we apply the Bonferroni correction, and set the significance level at 0.00035 (0.05 divided by 144). Some of the comparisons were significant at the $p \leq 0.00035$ level as indicated by bold face. These include SDNN, Triang, TINN, SD1 and SD2, but note that the apparent lesser significance of comparisons involving the Definite groups may be influenced by the fact that there are only 11 participants in that group. SD2 stands out as finding a significant difference for two comparisons, Normal vs Early and Definite vs Normal. Results from ROC analysis are in the last three rows of the table, and report the area under the ROC curve. Results showing area ≥ 0.7 are indicated by bold face. The largest values for area are provided by the measures pNN50 (0.891), TINN (0.891) and SD2 (0.872), suggesting these measures may be the most suited for separating participant groups. The smallest values for area correspond to the measures REC DET and Lmean, suggesting that these measures may not be the most suitable for separating participant groups.

Table 12.1 Results of Mann-Whitney and ROC analysis for comparison between different classes of participants using time domain measures calculated from RR interval series

Measure	SDNN	Rmssd	pNN50	Triang	TINN	SD1	SD2	REC	DET	Lmean
pNE	**0.000**	0.001	0.002	**0.000**	**0.000**	0.001	**0.000**	0.259	0.153	0.209
pED	0.011	0.007	0.002	0.022	0.008	0.007	0.017	0.713	0.640	0.343
pDN	0.011	0.007	0.002	0.022	0.008	**0.000**	**0.000**	0.925	0.788	0.757
aNE	0.698	0.657	0.648	0.691	0.693	0.657	**0.711**	0.555	0.570	0.561
aED	**0.739**	**0.755**	**0.763**	**0.717**	**0.737**	**0.755**	**0.726**	0.535	0.544	0.590
aDN	**0.869**	**0.850**	**0.891**	**0.838**	**0.891**	**0.850**	**0.872**	0.509	0.525	0.529

The p-value corresponding to a comparison of Normal and Early is provided in the row labelled pNE, while the p-value corresponding to a comparison of Early and Definite is shown in the row labelled pED, and so on. Significant results at the $p \leq 0.00035$ level are shown in bold face. The row labelled aNE provides the area under the ROC curve for the comparison between Normal and Early. Any area of 0.7 or more is indicated by bold face. Abbreviations for column headings are provided in the text

Table 12.2 Results of Mann-Whitney and ROC analysis for comparison between different classes of participants using frequency domain variables calculated from RR interval series

Measure	L pk	H pk	L pow	H pow	Lnorm	Hnorm	Ratio	Tot pow
pNE	0.012	0.375	**0.000**	0.001	0.108	0.108	0.108	**0.000**
pED	0.893	0.848	0.017	0.024	0.944	0.944	0.944	0.011
pDN	0.184	0.502	**0.000**	**0.000**	0.452	0.452	0.452	**0.000**
aNE	0.591	0.556	**0.718**	0.666	0.579	0.579	0.579	**0.701**
aED	0.533	0.533	**0.725**	**0.713**	0.507	0.507	0.507	**0.739**
aDN	0.658	0.548	**0.888**	**0.837**	0.571	0.571	0.571	**0.872**

Significant results at the $p \leq 0.00035$ level, and ROC area above 0.7 are indicated by bold face

Table 12.2 shows results from Mann-Whitney and Receiver Operating Curve (ROC) analysis, comparing different participant groups when using frequency domain measures. The top row gives abbreviations for measures, which are L pk (Low Frequency peak), HF pk (High Frequency peak), L pow (Low Frequency Power), H power (High Frequency Power), Lnorm (Low Frequency Power in normalised units), Hnorm (High Frequency Power in normalised units), Ratio (the LF/HF ratio) and Tot pow (Total power). The same comparisons were made as in the previous table. Very few of these comparisons were significant at the $p \leq 0.00035$ level. Of note are the results for Low Frequency Power and Total Power when comparing Normal participants to those in the Early or in the Definite CAN group ($p = 0.000$). Results from ROC analysis, as shown in the last three rows of the table, suggest that these same measures provide good separation between Normal and Definite groups, with Low Frequency Power achieving an ROC area of 0.888 and Total Power an area of 0.872.

Table 12.3 shows results from Mann-Whitney and Receiver Operating Curve (ROC) analysis, comparing different participant groups when using Multi-Scale Sample Entropy (MSE) measures. The Scale parameter is the m of Eq. (12.2). While

Table 12.3 Results of Mann-Whitney and ROC analysis for comparison between different classes of participants using MSE variables calculated from RR interval series

Scale	1	2	3	4	5	6	7	8	9	10
pNE	0.035	0.361	0.376	0.450	0.369	0.028	0.143	0.032	0.083	0.156
pED	0.746	0.509	0.872	0.782	0.936	0.819	0.647	0.652	0.882	0.661
pDN	0.661	0.812	0.665	0.976	0.740	0.420	0.749	0.455	0.307	0.270
aNE	0.602	0.545	0.543	0.537	0.544	0.606	0.571	0.604	0.584	0.569
aED	0.529	0.559	0.515	0.525	0.507	0.520	0.542	0.540	0.513	0.540
aDN	0.538	0.522	0.538	0.502	0.530	0.572	0.529	0.567	0.592	0.599

There are no significant results at the $p \leq 0.00035$ level

Table 12.4 Results of Mann-Whitney and ROC analysis for comparison between different classes of participants using MFDFA variables calculated from RR interval series

q	−5	−4	−3	−2	−1	0	1	2	3	4	5
pNE	0.065	0.018	0.004	0.011	0.108	0.282	0.141	0.068	0.156	0.375	0.500
pED	0.281	0.201	0.118	0.080	0.196	0.618	0.777	0.861	1.000	0.957	0.989
pDN	0.054	0.019	0.004	0.001	0.014	0.296	0.302	0.342	0.553	0.585	0.695
aNE	0.589	0.614	0.640	0.622	0.577	0.552	0.571	0.588	0.568	0.543	0.532
aED	0.598	0.616	0.642	0.659	0.618	0.545	0.526	0.516	0.500	0.505	0.501
aDN	0.673	**0.710**	**0.762**	**0.797**	**0.720**	0.594	0.593	0.585	0.553	0.549	0.535

Significant results at the $p \leq 0.00035$ level are shown in bold face

some of these results suggest a separation of class medians at the $p \leq 0.05$ level, none of them are significant at the corrected $p \leq 0.00035$ level. The final three rows show that area under the curve for these measures are little more than 0.5, indicating that these measures would not support discrimination of participant groups. This suggests that these measures are not the best entropy measures for detecting CAN in this dataset.

Table 12.4 shows results from Mann-Whitney and Receiver Operating Curve (ROC) analysis, comparing different participant groups when using MFDFA measures. None of these measures are significant at the $p \leq 0.00035$ level, but some have an area under the ROC curve greater than 0.7, and these relate to the difference between Definite and Normal participant groups, provided in the bottom row of the table. Comparing these results to those in Table 12.3, one could conclude that MFDFA provides measures that may be slightly more successful than MSE for discrimination of CAN in this cohort, but only for the comparison between Definite and Normal groups.

Table 12.5 shows results from Mann-Whitney and Receiver Operating Curve (ROC) analysis, comparing different participant groups when using Rényi Entropy, for values of α from -5 to $+5$. Some of the p-values resulting from comparisons between Normal and Early, and between Definite and Normal, were significant at the $p \leq 0.00035$ level for positive values of α. These results are supported by the area under the curve provided in the last three rows, however, these results more clearly show that the groups Normal and Definite are more well separated, and that positive

Table 12.5 Results of Mann-Whitney and ROC analysis for comparison between different classes of participants using RE variables calculated from RR interval series

α	−5	−4	−3	−2	−1	1	2	3	4	5
pNE	0.001	0.001	0.001	0.001	**0.000**	**0.000**	**0.000**	**0.000**	**0.000**	**0.000**
pED	0.063	0.059	0.061	0.051	0.036	0.056	0.038	0.041	0.037	0.034
pDN	0.001	0.001	0.001	0.001	0.000	**0.000**	**0.000**	**0.000**	**0.000**	**0.000**
aNE	0.657	0.658	0.660	0.667	0.685	**0.705**	**0.715**	**0.720**	**0.721**	**0.723**
aED	0.669	0.672	0.670	0.678	0.691	0.674	0.689	0.686	0.690	0.692
aDN	**0.792**	**0.794**	**0.798**	**0.801**	**0.821**	**0.828**	**0.847**	**0.854**	**0.859**	**0.862**

Significant results at the $p \leq 0.00035$ level are shown in bold face

values of α provide better separation. From these results one might conclude that Rényi Entropy provides the most promising measures for discrimination of CAN in this cohort, when compared to the other nonlinear analysis methods as considered above [36].

12.6.2 *Correlation*

The difference between Rényi entropy calculated using negative or positive values for the exponent α has been previously explored [31]. That work showed that although positive values of α provide higher discrimination of CAN classes, with corresponding lower p-values, these entropy values tend to be correlated with the standard deviation σ. Also σ has been shown to be less discriminatory for disease condition [37]. This suggests that Rényi entropy based on positive values of α is of less interest, and that measures based on negative α may prove more useful, as they discriminate between classes, and also provide information complimentary to that derived from time domain measures such as σ.

In order to clarify this it is desirable to investigate the correlation between the various measures discussed in this chapter. An analysis of correlation between measures was performed using Spearmans rank correlation. Table 12.6 shows correlation coefficients for the time domain measures considered. The correlation coefficients are shown in the upper right hand half of the table, while the lower left part of the table reports the p-values associated with the correlation. Due to the number of correlations examined, a reduced p-value has been chosen. On the basis that there are 250 correlations presented, the significance level has been set at $0.05/250 = 0.0002$. Most of the measures are highly correlated with each other, and have very low p-values. One example is the correlation between RMSSD and pNN50, which has a value of 0.97, and an associated p-value of 4e−93. However three measures stand out: REC, DET and Lmean appear to be poorly correlated with the others, but correlated with each other. For example, Lmean and DET has a value of 0.90 and an associated p-value of zero. This is important, as it is widely assumed that a robust method to select variables is to reject those that are mutually

Table 12.6 Spearmans rank correlation coefficients of time domain measures calculated from RR interval series

	SDNN	Rmssd	pNN50	Triang	TINN	SD1	SD2	REC	DET	Lmean
SDNN		**0.94**	**0.94**	**0.96**	**0.94**	**0.94**	**0.99**	−0.10	−0.09	−0.09
Rmssd	**0**		**0.97**	**0.91**	**0.91**	**1.00**	**0.89**	−0.21	−0.23	−0.16
pNN50	**1e−69**	**4e−93**		**0.89**	**0.91**	**0.97**	**0.89**	−0.15	−0.18	−0.13
Triang	**0**	**0**	**2e−51**		**0.88**	**0.91**	**0.95**	−0.25	−0.18	−0.19
TINN	**3e−72**	**1e−58**	**5e−59**	**2e−48**		**0.91**	**0.92**	0.06	0.01	0.04
SD1	**0**	**0**	**4e−93**	**0**	**1e−58**		**0.89**	−0.21	−0.23	−0.16
SD2	**0**	**0**	**5e−52**	**0**	**1e−62**	**0**		−0.07	−0.05	−0.08
REC	0.2	0.009	0.07	0.002	0.5	0.009	0.4		**0.88**	**0.85**
DET	0.3	0.004	0.03	0.03	0.9	0.004	0.6	**0**		**0.90**
Lmean	0.3	0.06	0.1	0.02	0.6	0.06	0.3	**0**	**0**	

Abbreviations are as in Table 12.1. For convenience all correlations greater than 0.8, and associated *p*-values less than 0.0002, are indicated by bold face

Table 12.7 Spearman's rank correlation coefficients of frequency domain measures calculated from RR interval series

	LFpk	HFpk	Lpow	Hpow	Lnorm	Hnorm	Ratio	TotP
LFpk		−0.01	0.36	0.24	0.11	−0.11	0.11	0.31
HFpk	0.9		−0.23	−0.02	−0.36	0.36	−0.36	−0.15
Lpow	**6e−06**	0.005		0.80	0.22	−0.22	0.22	**0.96**
Hpow	0.003	0.8	**0**		−0.35	0.36	−0.35	**0.92**
Lnorm	0.2	**8e−06**	0.008	**1e−05**		**−1.00**	**1.00**	−0.03
Hnorm	0.2	**8e−06**	0.008	**1e−05**	**0**		**−1.00**	0.03
Ratio	0.2	**8e−06**	0.008	**1e−05**	**0**	**0**		−0.03
TotP	**0.0001**	0.07	**0**	**0**	0.7	0.7	0.7	

Abbreviations are as in Table 12.2

correlated. In this case, one would conclude that discriminating these classes of disease would be affected by selecting one variable from the first group and one from the second group (REC, DET and Lmean). However, such an assumption is not always warranted and this approach has limitations in the presence of exceptions: sometime variables that appear to be highly correlated can still contribute additional information [38].

Table 12.7 shows correlation coefficients for frequency domain measures. Compared to the previous table, there are much fewer highly correlated variables. Notice the very high negative correlation (−1.00) between normalised LF and HF. This is a perfect negative correlation and is supported by a close examination of the data. This suggests that these particular measures have been designed is a way that means they are not independent. This suggests that a researcher should exercise caution when choosing variables that might be expected to be independent, but may turn out to be closely related. There are also moderately high correlations between Low Frequency, High Frequency, and Total Power.

Table 12.8 Spearman's rank correlation coefficients of MSE measures

	s01	s02	s03	s04	s05	s06	s07	s08	s09	s10
s01		0.65	0.37	0.25	0.32	0.32	0.19	0.24	0.14	0.15
s02	**2e−20**		0.72	0.54	0.43	0.37	0.30	0.32	0.26	0.24
s03	**2e−06**	**1e−26**		0.74	0.60	0.54	0.48	0.47	0.42	0.42
s04	0.001	**1e−13**	**3e−28**		0.75	0.68	0.69	0.64	0.60	0.55
s05	**4e−05**	**1e−08**	**4e−17**	**3e−30**		0.77	0.71	0.71	0.69	0.64
s06	**3e−05**	**2e−06**	**3e−13**	**3e−23**	**1e−31**		0.70	0.74	0.69	0.64
s07	0.02	**0.0001**	**2e−10**	**6e−24**	**9e−26**	**3e−24**		0.78	0.78	0.74
s08	0.002	**5e−05**	**3e−10**	**7e−20**	**7e−26**	**6e−29**	**1e−33**		0.79	0.78
s09	0.07	0.001	**5e−08**	**6e−17**	**3e−23**	**4e−24**	**4e−34**	**1e−35**		**0.80**
s10	0.06	0.003	**4e−08**	**5e−14**	**2e−19**	**1e−19**	**2e−28**	**2e−33**	**1e−36**	

Abbreviations are as in Table 12.3

Table 12.9 Spearman's rank correlation coefficients of MFDFA measures

	H−5	H−4	H−3	H−2	H−1	H−0	H+1	H+2	H+3	H+4	H+5
H−5		**1.00**	**0.99**	**0.93**	0.61	0.15	0.14	0.14	0.12	0.11	0.09
H−4	**0**		**0.99**	**0.94**	0.62	0.16	0.14	0.15	0.12	0.11	0.09
H−3	**0**	**0**		**0.97**	0.67	0.18	0.15	0.15	0.12	0.10	0.09
H−2	**0**	**0**	**0**		0.78	0.27	0.19	0.17	0.13	0.10	0.09
H−1	**0**	**0**	**0**	**0**		0.70	0.52	0.39	0.27	0.21	0.17
H−0	0.07	0.05	0.02	0.0006	**0**		**0.91**	0.74	0.56	0.45	0.39
H+1	0.09	0.07	0.06	0.02	**0**	**0**		**0.92**	0.78	0.67	0.61
H+2	0.08	0.07	0.06	0.03	**4e−07**	**0**	**0**		**0.95**	**0.88**	**0.83**
H+3	0.1	0.1	0.1	0.1	0.0005	**0**	**0**	**0**		**0.98**	**0.95**
H+4	0.2	0.2	0.2	0.2	0.01	**3e−09**	**0**	**0**	**0**		**0.99**
H+5	0.2	0.2	0.3	0.3	0.04	**4e−07**	**0**	**0**	**0**	**0**	

Abbreviations are as in Table 12.4

Table 12.8 shows correlation coefficients from the MSE analysis. Some of these measures are highly correlated, especially s09 with s10 (0.8, p-value = 1e−36), and the higher order measures tend to be well correlated with each other. Most of the p-values indicate that these correlations are significant.

Table 12.9 shows correlation coefficients from the MFDFA. As with the MSE measures, many of the measures are well correlated, and especially H(n) is well correlated with H(n+1).

Table 12.10 shows correlation coefficients from the Rényi Entropy. Correlation coefficients for MSE and MFDFA are not examined because the Rényi Exponents showed the best separation of classes. Rényi entropy measures calculated using negative α are strongly correlated, while those using positive α are also correlated with each other. But these two groups are not strongly correlated with each other.

As the correlation analysis shows, it is possible to identify several groups of measures in this dataset. In order to explore correlation between these groups, one measure was chosen from each group. From time domain measures, SDNN was

Table 12.10 Spearman's rank correlation coefficients of Rényi entropy measures

	H−5	H−4	H−3	H−2	H−1	H+1	H+2	H+3	H+4	H+5
H−5		1.00	1.00	0.99	0.97	−0.88	−0.85	−0.82	−0.81	−0.80
H−4	0		1.00	0.99	0.97	−0.89	−0.86	−0.83	−0.82	−0.81
H−3	0	0		1.00	0.98	−0.90	−0.87	−0.85	−0.83	−0.83
H−2	0	0	0		0.99	−0.92	−0.90	−0.88	−0.86	−0.85
H−1	3e−90	0	0	5e−133		−0.93	−0.91	−0.90	−0.89	−0.88
H+1	2e−48	1e−50	2e−54	9e−61	1e−65		0.99	0.99	0.98	0.98
H+2	3e−42	0	0	1e−53	0	6e−148		1.00	0.99	0.99
H+3	6e−38	0	0	3e−48	0	0	0		1.00	1.00
H+4	3e−35	0	0	6e−45	0	0	0	0		1.00
H+5	4e−34	7e−36	2e−38	1e−43	1e−49	0	0	0	0	

Abbreviations are as in Table 12.5

Table 12.11 Spearman's rank correlation coefficients of selected measures to compare time, frequency and entropy measures. The correlation coefficients are shown in the upper right hand half of the table, while the lower left part of the table reports the p-values associated with the correlation

	SDNN	Lmean	Ratio	TotP	mse06	DFA−5	DFA+5	RE−5	RE+5
SDNN		−0.09	−0.05	**0.99**	−0.23	−0.32	−0.19	0.16	−0.05
Lmean	0.3		0.22	−0.10	−0.11	0.03	0.04	−0.01	−0.01
Ratio	0.6	0.007		−0.03	0.36	−0.09	0.32	0.07	−0.06
TotP	**0**	0.2	0.7		−0.22	−0.33	−0.18	0.16	−0.06
mse06	0.004	0.2	**8e−06**	0.008		0.09	0.54	−0.04	−0.01
DFA−5	**7e−05**	0.7	0.3	**4e−05**	0.3		0.08	−0.17	0.14
DFA+5	0.02	0.6	**7e−05**	0.03	**1e−12**	0.3		0.02	−0.05
RE−5	0.05	0.9	0.4	0.05	0.6	0.04	0.8		−0.53
RE+5	0.6	0.9	0.5	0.5	0.9	0.09	0.5	**5e−12**	

chosen because it is highly correlated with most of the other time domain measures, and Lmean was chosen as it is highly correlated with REC and DET. From frequency domain measures, Total Power and Ratio were chosen. From MSE analysis s06 was chosen as it is well correlated with the other measures in this group. From MFDFA analysis, H−5 and H+5 were selected. From Rényi entropy, H−5 and H+5 were selected. Correlation coefficients between these selected measures are shown in Table 12.11.

These are mostly low values, with the exception of a high correlation between SDNN and Total Power. The low correlations in this table suggest that for the problem described in this chapter, discriminating between different classes of CAN, it is likely that a variety of measures from these three groups will provide complimentary information, and therefore such a selection should be considered when approaching this type of analysis.

It must be emphasised that if one were to attempt to identify a subset of measures to be used for discrimination of participant groups, the examination of correlation coefficients alone is fraught with difficulty. The foregoing material has not addressed in any way the interaction of multiple measures using multivariate analysis. In order to pursue such an identification, a variety of methods can be applied and software is available to assist in this endeavour. The next section reviews work on automated feature selection and the implementation of an automated discriminator using machine learning techniques.

12.7 Machine Learning

Earlier in this chapter, machine learning was mentioned as a potential method for selecting from the large number of measures. Several machine learning methods are relevant to this problem. First, a classifier is an algorithm that learns an association between a selection of measures and a discrete class label, in this case Definite CAN, Early CAN or Normal (control). When shown an unlabelled set of measures corresponding to a single participant, the classifier is able to suggest the most likely class or diagnosis based on the previous association [39]. A classifier is normally assessed by the proportion of participants assigned to the correct class, or from the specificity and sensitivity, or from the area under the Receiver Operating Characteristic (ROC) Curve. Second, a search algorithm is used to search through the list of candidate measures, generating a subset and evaluating it by determining the relative success of the classifier when operating on the subset only. This is known as a Wrapper method and is considerably more robust than a simple search that eliminates variables that are observed to be correlated [38]. In this case the dataset must be split into two, as the data used to train the classifier, or form the association, must be independent from that used to assess the accuracy of the classify, in order to eliminate training bias or over-fitting. An empirical study on the same dataset considered in this chapter shows that a variety of measures, including time-frequency and entropy measures, provide the most accurate classification [40]. In the work referred to, the classification task examined only a two class problem, but a difficult one: the differentiation of Early CAN from Normal participants. This work reported an accuracy of 71% correct using a classifier based on the Support Vector Machine. Table 12.12 provides the confusion matrix for this result. This was achieved after a Genetic Algorithm chose a subset of measures from those considered above.

Table 12.12 Results of a machine learning discrimination of Early CAN from Normal participants, after a multivariate search for a feature subset using independent data, as reported in [40]

		Predicted class	
		Normal	Early
Correct class	Normal	49	22
	Early	18	49

12.8 Recording Length

Identification of CAN from HRV is sensitive to the length of recording and the presence of artefacts including ectopic beats. This requires pre-processing and consideration of length of recording [41]. In this study, Early CAN could be identified using RR tachograms obtained from 10-s ECG recordings. However, a more effective screening is with RR intervals obtained from 5-min ECG recordings and RMSSD. Results suggested that more filtering provides a better outcome, but heavy filtering is impractical with a smaller sample, since too much information is lost, especially in the high frequency domain. That is because for short recordings, all the information is recorded in higher frequencies due to the limited recording length. A longer recording such as 5 min and longer contains information at lower frequencies also, so removal of more of the HF part of the spectrum (through heavier filtering) is not detrimental.

In other measures, LF and HF both separated Early CAN from Definite CAN. One of the first studies to report the effect of recording length on HRV compared RR tachograms from 2, 5, 10 and 15-min ECG recordings using frequency domain features and suggested no difference between results for shorter recordings compared to 24-h recordings (correlations > 0.75) for prediction of all-cause mortality, cardiac, and arrhythmic mortality and sudden death [42]. These authors and others note that a 2-min recording is not as reliable to determine VLF [43]. As mentioned above, the shorter the recording is, the less low frequency information can be contained in it. However, in that study [41] both 10-s and 5-min recordings provided significant classification results using VLF. de Bruyne reported the use of a standardized 10-s ECG recording as part of the Rotterdam study with a minimum of six beats recorded as a sensitive tool to identify risk of myocardial infarction using time domain measures. Thus, although very low frequencies are unlikely to be present in short RR tachograms, the possibility of classifying group based on VLF data has been corroborated by others and again shown in studies for CAN classification [44, 45]. Short time HRV analysis of less than 1 min also allows interpretation of dynamic changes associated with heart function such as recovery from exercise and effects of pharmacological intervention [46]. However, these previous studies do not report on pre-processing of the RR tachogram, which plays a role, if not in a 10-s recording then definitely in a 5-min recording. Results show that using the dynamic filtering algorithm proposed by Wessel, which is based on an infinite response filter [47], that the setting of the control variable needs to be taken into consideration when deciding the length of the RR tachogram to be used for HRV classification tasks [41]. This work, while using time and frequency based measures of HRV, does not make use of entropy or nonlinear measures, so it remains an open question as to how recording length affects those measures.

12.9 Conclusion

Analysis of HRV using time and frequency based measures makes assumptions about the linearity, stationary and equilibrium nature of heart rate signals. Nonlinear and multiscale studies using a variety of entropy measures suggest a complexity of HRV dynamics, with varying degree of success in discriminating different classes of neuropathology. Such work suggests that HRV analysis using time and frequency domain measures alone may not provide a full picture of the complex dynamics inherent in the heart rate time series [29, 37, 48]. Multiscale measures of HRV have generated a great deal of interest but a systematic approach is required in order to identify appropriate measures that may be of benefit in the context early identification of disease is of clinical importance to the practitioner.

The Rényi Entropy in particular has shown its efficacy as a complimentary measure that can extend and enhance the information that may be gleaned from sequences of RR intervals characterized by time and frequency analysis. In studies using Rényi Entropy, a very strong separation of disease classes has been demonstrated across all values of the parameter α, providing a suite of robust measures that can be used to identify CAN, especially early CAN, where treatment intervention is expected to be most effective. This conclusion is supported by low p-values obtained from the nonparametric Mann-Whitney test.

The physiological meaning of the measures such as multiscale exponents is not currently clear. In some of this work, Rényi Entropy was calculated for sequences of RR intervals, rather than single intervals, which complicates interpretation. For our dataset the choice of 16 beat sequences was optimal and suggests that this is an important time scale, which reflects the period of possible nonlinear oscillations involved in the pathophysiology of CAN progression [37]. Sequences of this length measured by Rényi Entropy can carry information of both sympathetic and parasympathetic functions of the autonomic system.

References

1. Valensi, P., Johnson, N., Maison-Blanche, P., Estramania, F., Motte, G., Coumel, P.: Influence of cardiac autonomic neuropathy on heart rate variability dependence of ventricular repolarization in diabetic patients. Diabetes Care **25**, 918–923 (2002)
2. Vinik, A., Maser, R., Mitchell, B., Freeman, R.: Diabetic autonomic neuropathy. Diabetes Care **26**, 1553–1579 (2003)
3. Tarvainen, M., Cornforth, D.J., Kuoppa, P., Lipponen, J., Jelinek, H.F.: Complexity of heart rate variability in type 2 diabetes-effect of hyperglycemia. In: Proceedings of the 35th Annual International Conference of the Engineering in Medicine and Biology Society (EMBS), Osaka, July 2013, pp. 5558–5561
4. Pop-Busui, R.: Cardiac autonomic neuropathy in diabetes. Diabetes Care **33**, 434–441 (2010)
5. Ziegler, D., Gries, F., Spuler, M., Lessmann, F.: Diabetic cardiovascular autonomic neuropathy multicenter study group. The epidemiology of diabetic neuropathy. J. Diabet. Complicat. **6**, 49–57 (1992)

6. Malliani, A., Pagani, M., Lombardi, F., Cerutti, S.: Cardiovascular neural regulation explored in the frequency domain. Circulation **84**(2), 482–492 (1991)
7. Goldberger, J.: Sympathovagal balance: how should we measure it? Am. J. Physiol. Heart Circ. Physiol. **276**(4), H1273–H1280 (1999)
8. Johnston, S., Easton, J.: Are patients with acutely recovered cerebral ischemia more unstable? Stroke **4**, 2446 (2003)
9. Rollins, M., Jenkins, J.G., Carson, D.,McGlure, B., Mitchell, R.H., Imam, S.: Power spectral analysis of the electrocardiogram in diabetic children. Diabetologia **35**, 452–455 (1992)
10. Spallone, V., Menzinger, G.: Diagnosis of cardiovascular autonomic neuropathy in diabetes. Diabetes **46**, S67–S76 (1997)
11. Ziegler, D.: Diabetic cardiovascular autonomic neuropathy: prognosis, diagnosis and treatment. Diabetes Metab. Rev. **10**, 339–383 (1994)
12. Jelinek, H., Pham, P., Struzik, Z., Spence, I.: Short term ECG recording for the identification of cardiac autonomic neuropathy in people with diabetes mellitus. In: Nineteenth International Conference on Noise and Fluctuations. IEEE Press, New York (2007), pp. 683–686
13. Ewing, D., Borsey, D., Bellavere, F., Clarke, B.: Cardiac autonomic neuropathy in diabetes: comparison of measures of R-R interval variation. Diabetologia **21**(1), 18–24 (1981)
14. Flynn, A., Jelinek, H.F., Smith, M.C.: Heart rate variability analysis: a useful assessment tool for diabetes associated cardiac dysfunction in rural and remote areas. Aust. J. Rural Health **13**, 77–82 (2005)
15. Vinik, A., Erbas, T., Casellini, C.: Diabetic cardiac autonomic neuropathy, inflammation and cardiovascular disease. J. Diab. Investig. **4**, 4–18 (2013)
16. Laitio, T., Jalonen, J., Kuusela, T., Scheinin, H.: The role of heart rate variability in risk stratification for adverse postoperative cardiac events. Anesth. Analg. **105**(6), 1548–1560 (2007)
17. Task Force of the European Society of Cardiology: Heart rate variability: standards of measurement, physiological interpretation, and clinical use. Circulation **93**, 1043–1065 (1996)
18. Sacre, J., Jellis, C., Marwick, T., Coombes, J.: Reliability of heart rate variability in patients with type 2 diabetes mellitus. Diabet. Med. **29**, e33–e40 (2012)
19. Khandoker, A., Jelinek, H., Palaniswami, M.: Identifying diabetic patients with cardiac autonomic neuropathy by heart rate complexity analysis. Biomed. Eng. OnLine **8**, 3 (2009)
20. Liao, D., Cai, J., Brancati, F., Folsom, A., Barnes, R., Tyroler, H., Heiss, G.: Association of vagal tone with serum insulin, glucose, and diabetes mellitus - the ARIC study. Diabetes Res. Clin. Pract. **30**(3), 211–221 (1995)
21. Singh, J., Larson, M., O'Donnell, C., Wilson, P., Tsuji, H., Lloyd-Jones, D., Levy, D.: Association of hyperglycemia with reduced heart rate variability (the Framingham heart study). Am. J. Cardiol. **86**(3), 309–312 (2000)
22. Santini, V., Ciampittiello, G., Gigli, F., Bracaglia, D., Baroni, A., Cicconetti, E., Verri, C., Gambardella, S., Frontoni, S.: QTc and autonomic neuropathy in diabetes: effects of acute hyperglycaemia and n-3 PUFA. Nutr. Metab. Cardiovasc. Dis. **17**(10), 712–718 (2007)
23. Tarvainen, M., Laitinen, T., Lipponen, J., Cornforth, D., Jelinek, H.: Cardiac autonomic dysfunction in type 2 diabetes - effect of hyperglycemia and disease duration. Front. Endocrinol. **5**, 1–9 (2014)
24. Balian, R.: Entropy, a protean concept. In: Dalibard, J. (ed.) Poincaré Seminar 2003: Bose-Einstein Condensation - Entropy, pp. 119–144. Birkhauser, Basel (2004)
25. Richman, J., Moorman, J.: Physiological time-series analysis using approximate entropy and sample entropy. Am. J. Physiol. Heart Circ. Physiol. **278**, H2039–H2049 (2000)
26. Costa, M., Goldberger, A., Peng, C.: Multiscale entropy analysis of complex physiologic time series. Phys. Rev. Lett. **89**, 068102 (2002)
27. Costa, M., Goldberger, A., Peng, C.K.: Multiscale entropy analysis of biological signals. Phys. Rev. E **71**, 021906 (2005)
28. Xu, D., Erdogmuns, D.: Renyi's Entropy, Divergence and Their Nonparametric Estimators, pp. 47–102. Springer, New York (2010)

29. Kurths, J., Voss, A., Saparin, P., Witt, A., Kleiner, H., Wessel, N.: Quantitative analysis of heart rate variability. Chaos **5**, 88–94 (1995)
30. Rényi, A.: On measures of information and entropy. In: Proceedings of the fourth Berkeley Symposium on Mathematics, Statistics and Probability, Berkely, CA (1960), pp. 547–561
31. Cornforth, D., Tarvainen, M., Jelinek, H.: How to calculate renyi entropy from heart rate variability, and why it matters for detecting cardiac autonomic neuropathy. Front. Bioeng. Biotechnol. **2**, 34 (2014)
32. Jelinek, H., Wilding, C., Tinley, P.: An innovative multi-disciplinary diabetes complications screening programme in a rural community: a description and preliminary results of the screening. Am. J. Public Health **12**, 14–20 (2006)
33. Javorka, M., Trunkvalterova, Z., Tonhajzerova, I., Javorkova, J., Javorka, K., Baumert, M.: Short-term heart rate complexity is reduced in patients with type 1 diabetes mellitus. Clin. Neurophysiol. **119**, 1071–1081 (2008)
34. Ewing, D., Martyn, C., Young, R., Clarke, B.: The value of cardiovascular autonomic functions tests: 10 years experience in diabetes. Diabetes Care **8**, 491–498 (1985)
35. Tarvainen, M., Ranta-Aho, P., Karjalainen, P.: An advanced detrending method with application to HRV analysis. IEEE Trans. Biomed. Eng. **49**, 172–175 (2002)
36. Cornforth, D, Jelinek, H., Tarvainen, M.: A comparison of nonlinear measures for the detection of cardiac autonomic neuropathy from heart rate variability. Entropy **17**, 1425–1440 (2015)
37. Goldberger, A., West, B.: Fractals in physiology and medicine. Yale J. Biol. Med. **60**, 421–435 (1987)
38. Guyon, I., Elisseeff, A.: An introduction to variable and feature selection. J. Mach. Learn. Res. **3**, 1157–1182 (2003) [Online]. Available: http://dl.acm.org/citation.cfm?id=944919.944968
39. Han, J., Kamber, M.: Data Mining: Concepts and Techniques. Morgan Kaufmann (2006). ISBN: 1-55860-901-6
40. Cornforth, D., Tarvainen, M., Jelinek, H.: Automated selection of measures of heart rate variability for detection of early cardiac autonomic neuropathy. In: Proceedings of Computers in Cardiology, Cambridge, MA, Sept 2014, pp. 93–96
41. Jelinek, H., Alothman, T., Cornforth, D., Khalaf, K., Khandoker, A.: Effect of biosignal preprocessing and recording length on clinical decision making for cardiac autonomic neuropathy. In: Eighth Conference of the European Study Group on Cardiovascular Oscillations (ESGCO), May 2014, pp. 3–4
42. Bigger, J., Fleiss, J., Rolnitzky, L., Steinman, R.: The ability of several short-term measures of RR variability to predict mortality after myocardial infarction. Circulation **88**, 927–934 (1993)
43. Sinnreich, R., Kark, J., Sapoznikov, D., Luria, M.: Five minute recordings of heart rate variability for population studies: repeatability and age-sex characteristics. Heart **80**, 156–163 (1998)
44. de Bruyne, M., Kors, J., Hoes, A., Klootwijk, P., Dekker, J., Hofman, A., van Bemmel, J., Grobbee, D.: Both decreased and increased heart rate variability on the standard 10-second electrocardiogram predict cardiac mortality in the elderly. Am. J. Epidemiol. **150**, 1282–1288 (1999)
45. Hodgart, E., Clark, E., Macfarlane, P., Latif, S.: Short term measures of heart rate variability. In: Proceedings of the 31st International Congress on Electrocardiology, pp. 174–181 (2005)
46. Smith, A., Owen, H., Reynolds, K.: Heart rate variability indices for very short-term (30 beat) analysis. Part 1: survey and toolbox. J. Clin. Monit. Comput. **27**, 569–576 (2013)
47. Wessel, N., Voss, A., Malberg, H., Ziehmann, C., Voss, H., Schirdewan, A., Meyerfeldt, U., Kurths, J.: Nonlinear analysis of complex phenomena in cardiological data. Herzschrittmacher Therapie Elektrophysiol **11**, 159–173 (2000)
48. Sassi, R., Signorini, M., Cerutti, S.: Multifractality and heart rate variabilit. Chaos **19**, 028507 (2009)

Chapter 13
Applications of Heartbeat Complexity Analysis to Depression and Bipolar Disorder

Gaetano Valenza, Luca Citi, Antonio Lanata, Claudio Gentili, Riccardo Barbieri, and Enzo Pasquale Scilingo

Abstract Nonlinear digital signal processing methods addressing system complexity have provided useful computational tools for helping in the diagnosis and treatment monitoring of a wide range of pathologies. In particular, heartbeat complexity measures have been successful in characterizing patients with mental disorders such as Major Depression and Bipolar Disorder. In this chapter, we describe the use of standard complexity measures such as sample entropy and multiscale entropy, as well as instantaneous measures of entropy to characterize pathological mood states when patients undergo affective elicitation or long-term monitoring. Results demonstrate that complexity measures of cardiovascular dynamics can be promising and viable tools to support clinical decision in mental health, improving on the diagnosis and management of psychiatric disorders.

13.1 Introduction

Nonlinear analysis of human physiological signals has been widely recognized to provide relevant information on psychophysiological and pathological states [1, 2]. Many physiological processes, in fact, involve nonlinear frequency modulation

G. Valenza (✉) • E.P. Scilingo
Department of Information Engineering & Bioengineering and Robotics Research Center "E. Piaggio", University of Pisa, School of Engineering, Pisa, Italy
e-mail: g.valenza@ing.unipi.it

L. Citi
School of Computer Science and Electronic Engineering, University of Essex, Colchester, UK

A. Lanata
Bioengineering and Robotics Research Centre "E. Piaggio" and Department of Information Engineering, University of Pisa, Pisa, Italy

C. Gentili
University of Padua, Padua, Italy

R. Barbieri
Department of Electronics, Informatics and Bioengineering, Politecnico di Milano, Milan, Italy

R. Barbieri et al. (eds.), *Complexity and Nonlinearity in Cardiovascular Signals*,
DOI 10.1007/978-3-319-58709-7_13

or multi-feedback interactions associated to long-range correlations that can be estimated with, e.g., Lyapunov exponents, 1/f slope, approximate entropy, and detrended fluctuation analysis [1–4]. Accordingly, many evidences in the literature on nonlinear dynamics of physiological signals show that complexity is a marker of health status of biological systems, and it is modulated by external stimuli, aging and presence of disease [1, 4–10]. It is also commonly thought that the complexity of healthy dynamics can be interpreted as an essential part of their ability to adapt to a varying environment.

A paradigmatic application of this methodological approach is given by the cardiovascular control dynamics, mediated by the Autonomic Nervous System (ANS). This system is very often investigated through the analysis of the series obtained by the time intervals between two consecutive R-waves detected from the Electrocardiogram, i.e. the R-R intervals, whose variability is defined as Heart Hate Variability (HRV). In the last decades, HRV studies using both linear and nonlinear modeling have been characterizing the influence of the ANS on the heartbeat, revealing the various nonlinear neural interactions and integrations occurring at the sinoatrial node and receptor levels [11]. Such phenomena include multiscale and fractal properties and much of nonlinear and non-stationary dynamics analysis provides significant cues for diagnostic and prognostic use [4, 8].

A recent frontier of nonlinear analysis on HRV series is represented by the assessment of psychiatric disorders. Most of the known mental disorders, in fact, are currently diagnosed relying on the clinicians' experience, who is supported only by verbal interviews and scores from specific questionnaires. Moreover, mental assessment through non-invasive, easy-to-record, and robust series such as HRV could open dramatic clinical perspectives in objectively managing the illness, helping patients, facilitating the interaction between patient and physician as well as to alert professionals in case of critical pathological episodes.

Studies reported in this chapter have been inspired by several works relating ANS markers to depression and mood swings in bipolar disorder [12–21]. In particular, it has been shown that linear-derived parameters are quite unreliable in effectively discerning healthy subjects and patients with major depressive disorder, as they have a high inter-subject variability [12–21]. On the contrary, nonlinear measures such as sample entropy, multiscale entropy, and instantaneous sample and approximate entropy by inhomogeneous point-process models allowed the characterization of patients with depression and patients with bipolar disorder, always showing a significant decrease of the complexity in such pathological groups [22, 23].

Brief methodological details of heartbeat complexity analysis, including standard Sample Entropy, Multiscale Entropy, and time-varying Entropy analysis through point-process modeling follow below. Next, experimental protocols and results on real data gathered from patients with major depression, and bipolar disorder are reported.

13.2 Methods

13.2.1 Sample Entropy

Calculation of Sample Entropy (SampEn) measures requires an estimation of the phase space of the cardiovascular system generating the HRV series. This can be performed through the Taken's theorem using two parameters: m, the embedding dimension; r, the margin of tolerance. Entropy calculation of short and noisy time series data was first studied by Pincus [24, 25] by defining the Approximate Entropy which has subsequently been followed by the sample entropy (SampEn) [26–29] .

Specifically, starting from the vectors $x(1), x(2), \ldots, x(N-m+1)$ in $\mathbb{R}^m$ defined by $x(i) = [u(i), u(i+1), \ldots, u(i+m-1)]$, the distance between two vectors x_1 and x_j is defined according to the Takens formulation in his studies on high dimensional deterministic system [30, 31] :

$$d[x(i), x(j)] = max_{k=1,2,\ldots,m}|u(i+k-1) - u(j+k-1)| \tag{13.1}$$

For each i, with $1 \leq i \leq N-m+1$, the variable $C_i^m(r)$ is defined as:

$$C_i^m(r) = \frac{\text{Number of j such that}(d[x(i), x(j)] \leq r)}{N-m+1} \tag{13.2}$$

end defined

$$C^m(r) = \frac{\sum_{i=1}^{N-m+1} \log C_i^m(r)}{N-m+1} \tag{13.3}$$

Then, *SampEn* is calculated by the expression[26].

$$\text{SampEn(m, r, N)} = -\ln \frac{C^{m+1}}{C^m} \tag{13.4}$$

Of note, standard values for r and m parameters have been suggested in previous studies dealing with RR interval series [26]: $m = 2$ and $r = 0.20 * std$ (std = standard deviation of time series).

13.2.2 MultiScale Entropy

MultiScale Entropy (MSE) is a powerful tool to quantify the nonlinear information of a time series over multiple time scales [32] through, e.g., sample entropy (*SampEn*) [26] algorithms. In the literature it can be found that MSE analysis has been applied to characterize human gait dynamics [33], Alzheimer's disease through EEG signals [34], drug-naive schizophrenia [35] and autism spectrum conditions [36]. Recently, MSE has been also applied to HRV data to study patients with diabetes mellitus type 1 [37] or with aortic stenosis subjects [38], effects of orthostatic stress [39]. Other studies involving healthy subjects point out some typical behaviors of MSE: entropy increases from small to large scales and then stabilizes its value at constant values; in patients with arrhythmia, the entropy decreases and then remains constant; in patients with congestive heart failure, the entropy decreases at the beginning and then gradually increases [32].

13.2.2.1 Multiscale Entropy (MSE) Methodology and Influence of the *r* Parameter

Here we briefly describe the theory behind the MSE methodology emphasizing the issue of choosing the optimal value of the *SampEn* radius. MSE is based on the calculation of the *SampEn* over several time series, which are constructed from the original discrete time series by averaging the data points within non-overlapping windows of increasing length, τ (see previous paragraph). Therefore, two parameters are mainly involved in the MSE estimation: the embedding dimension m of the series [30, 31], and r, a positive real number representing the margin of tolerance, i.e., the radius. As mentioned before, previous studies suggest a fixed straightforward choice of the parameters as $m = 2$, and $r = 0.15\,\sigma$ where σ is the standard deviation of the series [32]. While such a choice could be reasonable for the m values, it could be pretty hazardous for the r value. Inter-subject variability, in fact, can lead to non-effective results whether the parameters involved in the analysis are not adaptive and personalized. Moreover, some researchers recently pointed out that the recommended r does not fit all situations and may lead to wrong results [40, 41].

Therefore, in order to study the influence of the r-value on the MSE estimation and to improve the objectivity of the experimental results, here we show experimental results on real HRV data testing three r-choosing methodologies:

MSE-Method I consisting in the previously mentioned traditional choice for physiological data of $r = 0.15\,\sigma$ [23, 33] evaluated for each acquisition of each subject.

MSE-Method II considers the *grouped* standard deviation of the series belonging to all patients [42]. Such a series is obtained by concatenation of the series of each acquisition/mood state.

MSE-Method III considers different r values for each acquisition of each subject as to maximize the calculation of the Approximate Entropy (*ApEn*) [24] in the range $0.01\sigma \leq r \leq 1.2\sigma$ [40, 41, 43–46]. This popular method considers that the highest value $ApEn(r_k)$ is interpolated with the preceding and the following values, $ApEn(r_{k-1})$ and $ApEn(r_{k+1})$, with a parabola. The position of the vertex of the parabola gives r_{max}.

13.2.3 Inhomogeneous Point-Process Approximate and Sample Entropy

Despite the important achievements obtained studying the ANS-mediated cardiovascular control dynamics through *ApEn* and *SampEn* analysis of HRV series [5, 47], these algorithms are not able to characterize the dynamic nature of the system along time. *ApEn* and *SampEn*, in fact, compute a single value given a predetermined time window, thus providing only an averaged measured of system dynamics that is instead evolving at each moment in time.

To overcome these limitations and address the non-stationary dynamics commonly associated with real biological processes, recently we proposed two novel measures of entropy: the Inhomogeneous Point-process Approximate and Sample Entropy (*ipApEn* and *ipSampEn*, respectively). These measures are built on our previous probabilistic framework based on nonlinear inhomogeneous point processes, through which it is possible to effectively characterize the nonlinear probabilistic generative mechanism of heartbeat events, while obtaining instantaneous estimates of the underlying cardiovascular dynamics even considering short recordings under nonstationary conditions [8, 48, 49] and without using any interpolation method. Of note, this model accounts for long-term memory and high order nonlinearities using a reduced set of model parameters [8, 49] and it is used to perform instantaneous estimates of the phase-space vectors of the RR series. As detailed in the next section, the *ipApEn* and *ipSampEn* definitions are based on the distance measure between phase-space vectors as the Kolmogorov-Smirnov (KS) distance between two probability density functions (pdfs). As a result, the proposed *ipApEn* and *ipSampEn* indices, when estimated from RR interval series, are able to provide an instantaneous complexity assessment of the underlying ANS dynamics. Besides heartbeat data, we remark the general applicability of the proposed *ipApEn* and *ipSampEn* indices in further heterogeneous stochastic series of events such as milling inserts, neural activity, geyser geologic events, and gait from short walks [8].

The proposed measures of entropy are built on our previous Nonlinear AutoRegressive model with Laguerre expansion (NARL) of the Wiener-Volterra kernel, whose details are reported in [8, 49].

Briefly, assuming history dependence, the NARL point-process model addresses the pdf of the waiting time t until the next R-wave event occurs through an inverse Gaussian (IG) probability distribution [48]. Since this IG distribution is characterized at each moment in time, it is possible to obtain an instantaneous estimate of $\mu_{RR}(t)$ at a very fine timescale (with an arbitrarily small bin size Δ), which requires no interpolation between the arrival times of two beats, therefore addressing the problem of dealing with unevenly sampled observations.

The IG instantaneous mean $\mu_{RR}(t, \mathcal{H}_t, \xi(t))$ is defined as a combination of present and past R-R intervals based on the NARL model [8, 49]:

$$\mu_{\mathrm{RR}}(t, \mathcal{H}_{\mathrm{t}}, \xi(\mathrm{t})) = \mathrm{RR}_{\widetilde{N}(t)} + g_0(t) + \sum_{i=0}^{p} g_1(i,t)\, l_i(t^-) + \sum_{i=0}^{q}\sum_{j=0}^{q} g_2(i,j,t)\, l_i(t^-)\, l_j(t^-) \qquad (13.5)$$

where $\mathcal{H}_{\mathrm{t}}$ is the history given the past RR intervals,

$\xi(t) = [\xi_0(t), g_0(t), g_1(0,t), ..., g_1(p,t), g_2(0,0,t), ..., g_2(i,j,t)]$ with $\xi_0(t)$ as the shape parameter of the IG distribution, and

$l_i(t^-) = \sum_{n=1}^{\widetilde{N}(t)} \phi_i(n)(\mathrm{RR}_{\widetilde{N}(t)-n} - \mathrm{RR}_{\widetilde{N}(t)-n-1})$ is the output of the Laguerre filters just before time t, $\phi_i(n)$ is the ith Laguerre function, $\widetilde{N}(t)$ is the left continuous sample path of the associated counting process, and g_0,$\{g_1(i)\}$, and $\{g_2(i,j)\}$ correspond to the time-varying zero-, first-, second-order NARL coefficients, respectively [1, 8, 49].

The ith Laguerre function is defined as follows:

$$\phi_i(n) = \alpha^{\frac{n-i}{2}}(1-\alpha)^{\frac{1}{2}} \sum_{p=0}^{i} (-1)^p \binom{n}{p}\binom{i}{p} \alpha^{i-p}(1-\alpha)^p \qquad (13.6)$$

with $(n \geq 0)$, is the ith Laguerre function with $0 < \alpha < 1$, which determines the rate of exponential asymptotic decline of these functions, and g_0,$\{g_1(i)\}$, and $\{g_2(i,j)\}$ correspond to the time-varying zero-, first-, second-order NARL coefficients, respectively [1, 8, 49].

Therefore, given the original RR interval series, the output of the Laguerre filters is firstly evaluated through the convolution between the derivative RR series and the Laguerre functions. Then, the parameters of Eq. (13.5) are estimated to model the first order moment of the IG probability distribution. Moreover, Eq. (13.5) accounts for long-term memory and reduced number of parameters needed for the linear and quadratic functions [1, 8].

We estimate the parameter vector $\xi(t)$ using the Newton-Raphson procedure to compute the local maximum-likelihood estimate [8] within a sliding time-window of $W = 90$ s. Because there is significant overlap between adjacent local likelihood intervals, we start the Newton-Raphson procedure at t with the previous local maximum-likelihood estimate at time $t - \Delta$. Model goodness-of-fit is based on the KS tests and associated KS statistics [48], along with autocorrelation plots testing the independence of the model-transformed intervals [48].

Instantaneous estimates of *ipApEn* and *ipSampEn* measures are obtained by considering a novel definition of distance between phase-space vectors embedded within the standard ApEn and SampEn algorithms [24, 26].

In particular, let us define a distance measure $d[.]$ between two *IG* distributions of two heartbeat events according to the KS distance measures, i.e. maximum value of the absolute difference between two cumulative distribution functions. For each pair of phase space vectors, which are defined as $x(k) = [\mu_{\mathrm{RR}}(t_k), \mu_{\mathrm{RR}}(t_{k+1}), \ldots, \mu_{\mathrm{RR}}(t_{k+m-1})]$ in $\mathfrak{R}^m$ of the time series

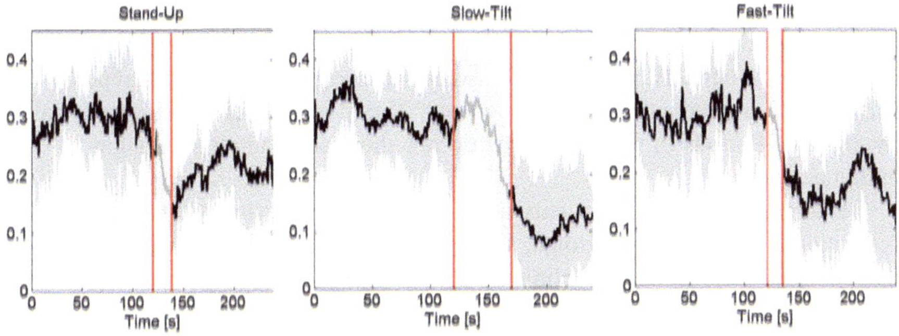

Fig. 13.1 Averaged *ipSampEn* trends during resting and tilting phases. Considering data from all subjects, the plot shows the Median(X) $\pm$ MAD(X)

$\mu_{RR}(t_1), \mu_{RR}(t_2), \ldots, \mu_{RR}(t_N)$ with embedding dimension $m = 2$, let us define $C_k^m(r(t), t)$ as the number of points $x(j)$ such that $d[x(k), x(j)] \leq r(t)/(N - m + 1)$ $\forall j$, with m and $r(t)$ as the embedding dimension and time delay of the phase-space, respectively. The time-varying quantity $r(t)$ is instantaneously expressed as $r(t) = 0.2\sigma_{\mu_{RR(t)}}$. Then, the $ipApEn(m, r, t)$ and $ipSampEn(m, r, t)$ are instantaneously derived following the standard ApEn and SampEn algorithms [24, 26], through the calculation of the normalized term $C^m(r, t)$.

Our instantaneous complexity assessment allows for the possibility of analyzing the proposed measures also in terms of variability of their evolution along time, which we refer to as *complexity variability*. Formally, if we consider $\overline{ipApEn}(m, r, N)$ and $\overline{ipSampEn}(m, r, N)$ as the average measures of $ipApEn(m, r, N, t)$ and $ipSampEn(m, r, N, t)$ within the N^* data points time window $T = [t_1, t_2, \ldots, t_{N^*}]$, then two novel *complexity variability* measures, σ_{ipApEn} and $\sigma_{ipSampEn}$, refer to the standard deviation of the $ipApEn(m, r, N, t)$ and $ipSampEn(m, r, N, t)$ series evaluated within T.

Exemplary tracking of the proposed time-varying complexity measures are shown in Fig. 13.1.

13.3 Application on Bipolar Disorder

In this section we show experimental results characterizing different pathological mood states in bipolar disorder through Sample Entropy and Multiscale Entropy analyses.

Bipolar Disorder (BD) is a chronic psychiatric condition [50] recognized to be one of the most common and dangerous disorder of the affectivity (Diagnostic and Statistical Manual of Mental Disorders (DSM-IV-TR) [51]). People affected by BD manifest drastically altered mood regulation, experiencing unbalanced mood shifts among depression, mania, and mixed states (where both symptoms of depression and mania are present at the same time). Specifically, depression in BD is characterized by sadness and hopelessness (including suicidal ideation), whereas mania leads to euphoria or irritability, excessive energy, hyperactivity, hypertrophic self-esteem, and reduction of sleeping need. The moderate form of mania is called hypomania. Periods in which patients do not show any pathological signs refer to euthymic states.

BD strongly affects the patients' quality of life, thus having a significant impact on patients' social, occupational, and general functioning and wellbeing, even during time periods free of clinical relevant symptomatology [52, 53]. Such a reduced quality of life can also be related to a significant loss of the cognitive performance [54]. In addition, BD patients often experience anxiety, which is associated to suicide attempts, lifetime alcohol abuse, and psychosis [55].

Despite the high managing costs and severity of the disease [56–59], in the current clinical practice BD diagnosis relies only on interviews and scores from psychological questionnaires, physician's own expertise as well as patient's subjective description of symptoms. Moreover, BD is always characterized by comorbidity, i.e., simultaneous presence of symptoms which are shared with other psychiatric disorders. All the mentioned issues associated with BD can likely lead to subjective interpretations, inconsistencies, and misdiagnoses [60].

To help overcoming these limitations, we recently proposed a personalized system comprised of a wearable monitoring system with embedded sensors and a smartphone to better manage patients affected by mental disorders such as BD: the PSYCHE system [2, 17, 18, 61, 62]. It is based on the novel paradigm related to the possibility to recognize mood status through information gathered form the Autonomic Nervous System (ANS). ANS dynamics, in fact, has been already proven to have a crucial role in discerning pathological mental states associated to BD [14, 20, 21, 63, 64]. However, none of the previous studies reached a significant clinical accuracy to be taken into account in the current clinical practice. This can be also due to the fact that they were focussing on biomarkers of BD regardless of a pervasive, comfortable, and wearable assisting living system for BD management.

In previous publications [2, 17, 18, 61, 62], we extensively described the PSYCHE (Personalized monitoring SYstems for Care in mental HEalth) system showing how a complex computational approach can be profitably used to discern different pathological mood states associated to BD.

In this chapter, after a brief overview of the experimental setup, data acquisition, and recruitment of eligible subjects, we show how heartbeat complexity analysis can be used in the following scenarios:

- Study 1: this concerns the monitoring of severe episodes in hospitalized patients with bipolar disorder in order to investigate cardiovascular complex dynamics associated with the response to treatment. Here we propose a simplified approach focusing on only a complexity measure, i.e. sample entropy, estimated on long-term HRV series.
- Study 2: this aims to exploit the use of MSE analysis applied to HRV series gathered from bipolar patients to gathered objective information on the clinical

mood state. As methodological contribution, we investigated three different approaches by choosing the most fundamental and sensible parameter of MSE analysis: the radius of the sample entropy.

In both studies, all patients were asked to wear the PSYCHE system for up to 24 h. HRV data were collected from each acquisition of each patient using the core system of the PSYCHE wearable monitoring platform. It consists in a comfortable, textile-based sensorized T-shirt developed by Smartex s.r.l., embedded with electrodes which are able to acquire electrocardiogram (ECG), from which the HRV signals are extracted after ad-hoc pre-processing and R-peak extraction steps [17, 18]. During all the acquisitions, the ECG sampling frequency was set to 250 Hz.

13.3.1 Experimental Setup, Data Acquisition, and Recruitment of Eligible Subject

Physiological data collected through the PSYCHE wearable system included long-term monitoring of ECG, Heart Rate Variability (HRV), respiration activity and, thanks to an internal tri-axial accelerometer, monitor movement activity including its classification (i.e. supine, sitting, sleep posture, etc.), see Fig. 13.2. It is worthwhile noting that the wearable system makes use of dry textile-based electrodes. These electrodes have a special multilayer structure that increases the amount of sweat and reduces the rate of evaporation reaching electrochemical equilibrium between the skin and electrodes after 10 s, thus showing a good and constant signal quality [65].

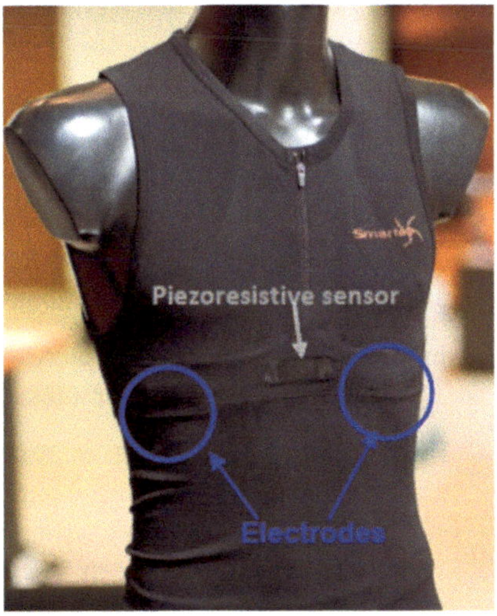

Fig. 13.2 Wearable system: sensorized T-shirt

The shirt was designed for both women and men and is made of elastic fibers allowing tight adhesion to the user's body, piezoresistive fibers to monitor fabric stretching (and consequently respiration activity), and metallic fibers knitted to create fabric electrodes to monitor the ECG. These materials are knitted together and are fully integrated in the garment without any mechanical and physical discontinuity, creating areas with different functionalities.

Shirt has been designed taking into account the thermal comfort, both in the selection of the yarns and in the stitch structure, several zones with an open net have been inserted, and a polyamide yarn with antibacterial properties and a natural feeling has been used as basic component. As a matter of fact, garments are made of commercial yarns, already tested (and certified) for contact with human skin, and can be easily washed and, in case, disinfected.

The electronic device of the wearable platforms sends the collected physiological data to a mobile device (i.e., a smartphone) through bluetooth communication channel. The mobile application, which is based on an object oriented java programming for Android 4.01 operating system, stores physiological data on a 128 GByte Secure Digital (SD) card and during inactive period of the psyche system forwarded them to the central server by means of a wi-fi internet TCP/IP protocol.

13.3.1.1 Study 1: Monitoring Severe Episodes in Hospitalized Patients with Bipolar Disorder

Ten patients were recruited and monitored during an acute episode and followed until clinical remission. Patients were recruited at the Pisa University Hospital within the Second Chair of Psychiatry and the chair of Clinical Psychology. Patients had to express their written informed consent to participate the study and fulfill the following exclusion/inclusion criteria:

- age between 18–65 years
- clinical diagnosis of bipolar disorder and of acute mood episode
- absence of high risk suicidal behaviors
- change of treatment (dosage and/or drugs) within the previous 2 weeks
- absence of delusions and hallucination
- absence of a conduct abuse disorders
- possibility to express informed consent to the study

In this study, we recruited acutely depressed and depressed/hypomanic, hospitalized patients. This choice was motivated by the following reasons:

- Hospitalized patients have, in general, full-bloomed pathological states. Thus, the changes in clinical states would have been more marked as, presumably, the effect of these changes on psychophysiological parameters.
- The PSYCHE system is in itself a naturalistic study that does not imply the control for medication intake. On the other hand, hospitalized patients can undergo a change in medication prescription, and therefore that can be

Table 13.1 Clinical mood states of the patients under study (*DP* depression; *MX* mixed state; *EU* euthymic state)

Patient	Acq.1	Acq.2	Acq.3	Acq.4	Acq.5
Pz01	DP	EU			
Pz02	MX	MX	DP	DP	EU
Pz03	MX	EU			
Pz04	DP	EU			
Pz05	DP	DP	EU		
Pz06	DP	DP	DP	EU	
Pz07	MX	MX	DP	EU	
Pz08	MX	MX	EU		
Pz09	DP	EU			
Pz10	DP	EU			

followed in terms of clinical and psycho-physiological changes due to response to treatment.

- The goal of this study was to assess if there are psycho-physiological changes that might predict remission. For this reason hospitalized patients are an ideal sample to this end. Giving the fact that they were under a constant supervision by health care professionals, they could be evaluated during remission.

The clinical history of the enrolled ten patients during this study is detailed in Table 13.1.

13.3.1.2 Study 2: Characterizing Different Pathological Mental States in Bipolar Patients

In this study, we show how to effectively characterize different pathological mental states using MSE analysis of HRV series gathered from bipolar patients. To this extent, we analyzed the nonlinear heartbeat dynamics of 8 bipolar patients selected according to the following features: age between 18–65, presence of a mood state between depressive and hypomanic at the moment of the recruitment, low risk of suicidality (as assessed as no thoughts of death and no previous attempts), no somatic or neurologic disorders that might be related to bipolar disorders (e.g. thyroid alterations), absence of cognitive impairment, absence of substance abuse disorders, necessity of a change in treatment (treatment change is defined as a augmentation of doses, introduction of or switch to other drugs, introduction of physical treatments), all the patients signed an informed consent already approved by the Ethical Committee of Strasbourg. The protocol planned a study entry visit when the patient was experiencing a depressive or hypomanic phase. Patients were studied with an average frequency of 2–3 times a month. Each patient was evaluated and monitored from the day of the hospital admission toward remission, i.e., until the reaching of an euthymic state as long as such a condition was presented within 3 months after the first visit. In any case, in this study no more than six evaluations per patient were performed. Accordingly, we analyzed a total amount of 16 night recordings, 6 out of which were associated to the label depression, 5 to label hypomania, and 5 to label euthymia. All clinical states have been evaluated according DSM-IV-TR criteria [51] and all patients were recruited in the out-patient University clinic of Strasbourg, France. The PSYCHE wearable system has been given to the patients in the afternoon and recollected the morning after.

13.3.2 Experimental Results using Sample Entropy Analysis and Discussion

Figures 13.3, 13.4, 13.5, 13.6, 13.7 and Table 13.2 show the SampEn trend for each of the considered BD. It is straightforward to notice that the SampEn values increase as the mood state goes from mixed state (the most severe pathological mental status) to euthymic state (the good affective balance), through depression. According to these considerations, the coherent trends highlighted using the intra-subject analysis are informative results.

In particular, we found that SampEn varies almost linearly when pathological states go towards euthymia including a clear distinction between psychopathological states with different levels of severity and in any case with different clinical characteristics (like mixed state and pure depressive states). Moreover following the trend of these parameters we were retrospectively able to state that the patients were going to remit. These results are in agreement with most of the current literature pointing out that complexity in physiological systems decreases in presence of pathologies [23, 66].

The achieved results show a coherent behavior of the Sample Entropy through all the patients. Although these results are very preliminary since obtained on a reduced, although consistent, population, they encourage us to state that it could be possible to hypothesize that this complexity measure can be used as an early and valid indicator of the treatment response of each patient toward remission.

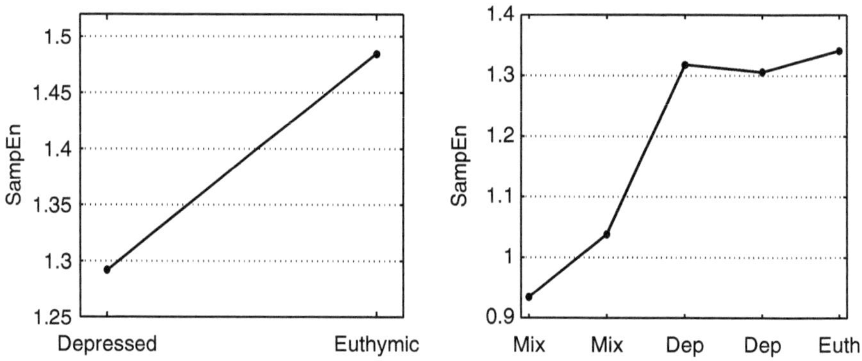

Fig. 13.3 Sample Entropy values computed on HRV series for patients Pz01 and Pz02

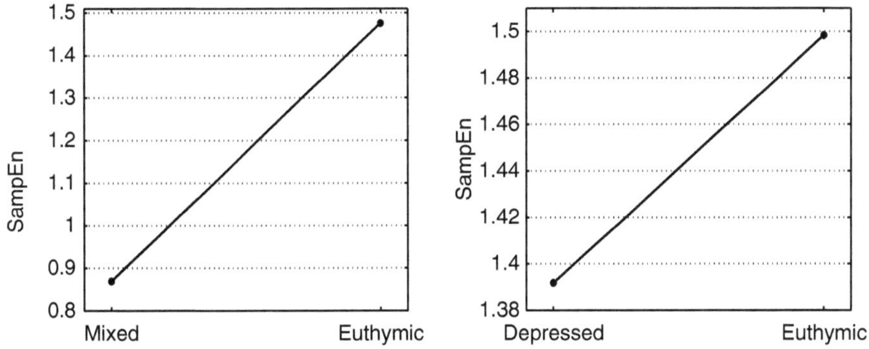

Fig. 13.4 Sample Entropy values computed on HRV series for patients Pz03 and Pz04

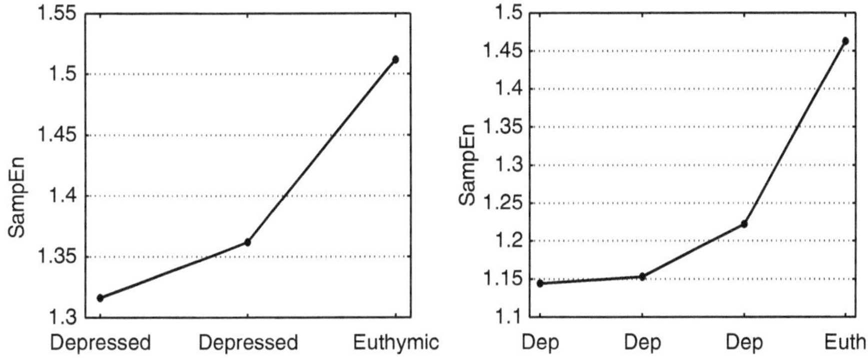

Fig. 13.5 Sample Entropy values computed on HRV series for patients Pz05 and Pz06

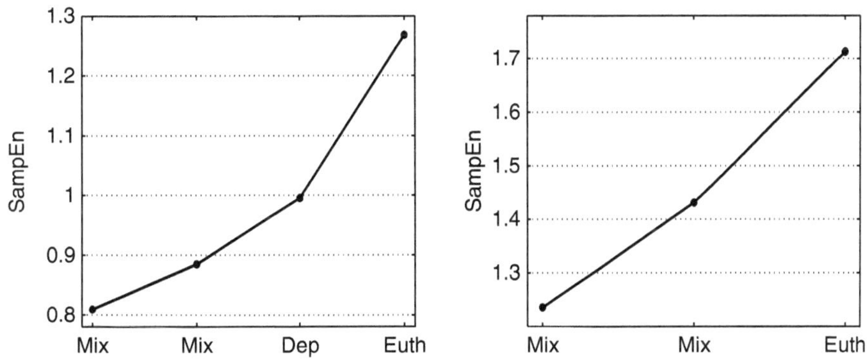

Fig. 13.6 Sample Entropy values computed on HRV series for patients Pz07 and Pz08

From a clinical perspective, such early indications would help greatly to improve treatment and care, since the adaptation of ineffective treatments would be fastened, and effective treatments could be optimized rapidly. Increasing the number of

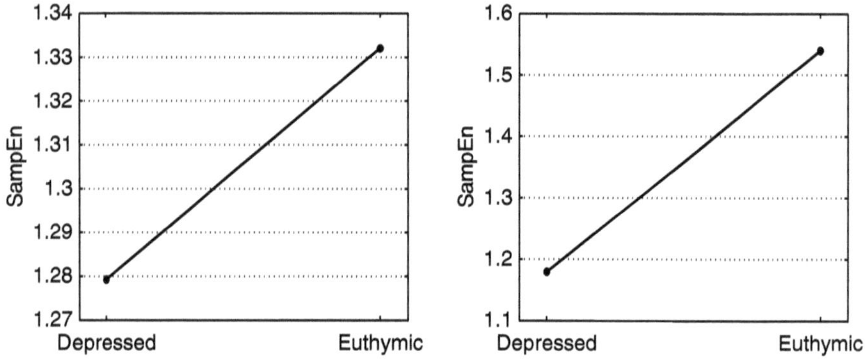

Fig. 13.7 Sample Entropy values computed on HRV series for patients Pz09 and Pz10

Table 13.2 SampEn values associated to BP patients under study

Patient	Acq.1	Acq.2	Acq.3	Acq.4	Acq.5
Pz01	1.29	1.49			
Pz02	0.93	1.04	1.32	1.31	1.34
Pz03	0.87	1.48			
Pz04	1.39	1.50			
Pz05	1.30	1.36	1.51		
Pz06	1.14	1.15	1.22	1.46	
Pz07	0.81	0.88	0.99	1.27	
Pz08	1.34	1.43	1.71		
Pz09	1.28	1.33			
Pz10	1.18	1.54			

Table 13.3 Experimental results on the inter-subject analysis among different mood states

Mood state	95% Confidence interval (t-Student)
Mixed-state	[0.8663, 1.1911]
Depression	[1.1890, 1.3045]
Euthymia	[1.3930, 1.5333]

acquisitions from each patient in a larger group of patients also might allow simple prediction approaches based on linear or nonlinear combinations of values of previous acquisitions. This approach would allow the determination of mood outcome not only retrospectively, as in the present paper, but also as a real forecast with tremendous impact on clinical course and quality of life of the patients.

In order to generalize the results, Table 13.3 shows the 95% Confidence Interval (CI) for each of the considered mood state. Of note, CI among the pathological mental states are not overlapped, suggesting the promising role of the cardiovascular complexity as biomarker for BD patients, also with severe symptoms.

13.3.3 Experimental Results Using Multiscale Entropy Analysis and Discussion

On all the analysis, Kruskal-Wallis non-parametric tests were used to test the null hypothesis of having no statistical difference in complexity indices among patients mood states (euthymic, depressed, hypomanic). Mann-Whitney non parametric U-tests were used to compare two samples belonging to two different groups on the post-hoc statistical analysis. The use of such non-parametric tests is justified by having non-gaussian distribution of the samples ($p < 0.05$ given by the Shapiro-Wilk test having the null hypothesis of Gaussian distributed samples).

MSE, estimating up to the twentieth scale factor, was calculated on the longest segment of consecutive artifact-free samples of each acquisition of each patient. Such a longest HRV series lasted for no less than 53 min and no more than 4 h and 24 min. All results are expressed as median and its respective absolute deviation [i.e. for a feature X, $X = \text{Median}(X) \pm \text{MAD}(X)$ where $\text{MAD}(X) = \text{Median}(|X - \text{Median}(X)|)$]. The m value is fixed for all cases to the standard value $m = 2$. Experimental results on using the MSE methods follow below. Figure 13.8 shows the HRV data for a representative patient while experiencing hypomania.

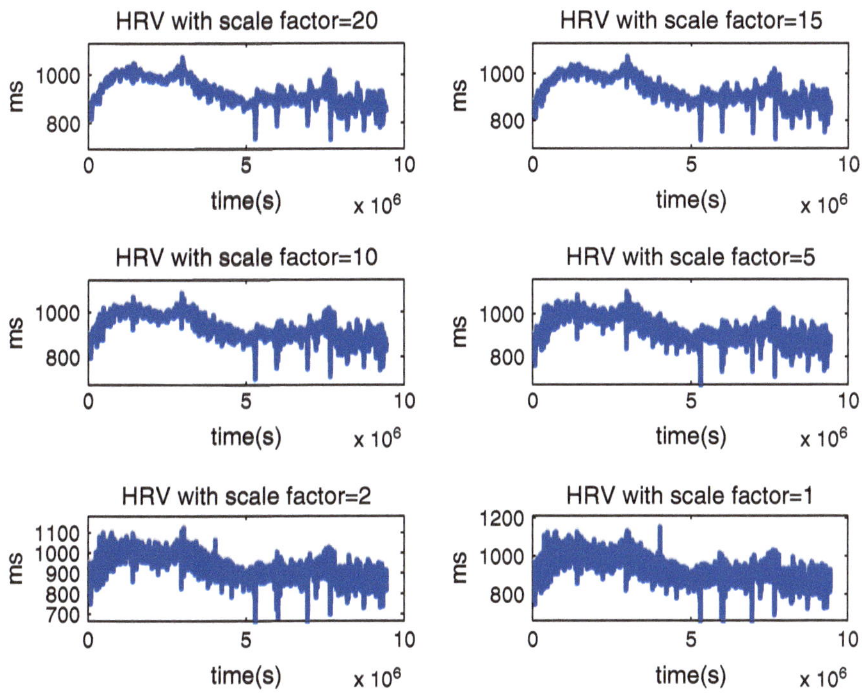

Fig. 13.8 Heartbeat dynamics gathered from a representative patient while experiencing the pathological mental state such as hypomania

MSE-Method I This method uses $r = 15\ \sigma$, where σ is the standard deviation of each HRV series, for the MSE calculation. The Kruskal-Wallis non-parametric test showed no statistical difference among the three pathological groups ($p > 0.05$). Figure 13.9 shows the MSE results over all the scale factors.

MSE-Method II This method chooses the radius as 15% of the standard deviation of all series. Likewise the previous method, the Kruskal-Wallis non-parametric test showed no statistical difference among the three pathological mental states ($p > 0.05$). Figure 13.10 shows the MSE results over all the scale factors.

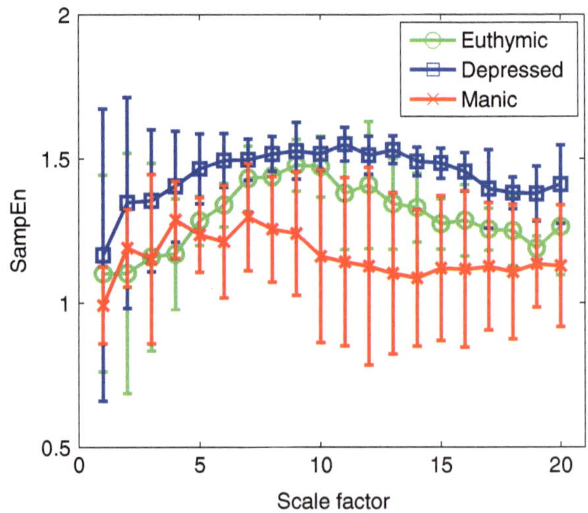

Fig. 13.9 MSE of heart rate dynamics in nocturnal period with method I, values are expressed as *median* ± *MAD*

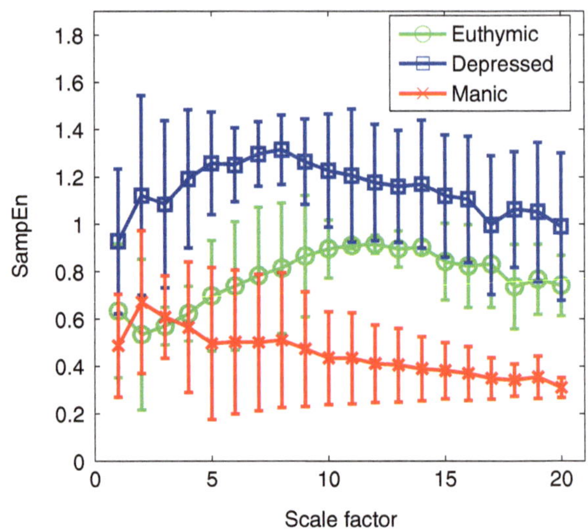

Fig. 13.10 MSE of heart rate dynamics in nocturnal period with method II, values are expressed as *median* ± *MAD*

MSE-Method III This method searches the objective r_{max} value which maximizes *ApEn* of each HRV series. We found that the maximum value of entropy was always within the range from 0.001σ to 0.30σ. Kruskal-Wallis test revealed statistical differences between the groups at all scales. In particular, when scale is equal to 1 and for scale values comprised between 7 and 19, the null hypothesis of having no difference was rejected with $p < 0.01$. At scales 2, 3, 4 and 6 the null hypothesis was rejected with $p < 0.05$, while at the remaining scale 5, the obtained p-value is less than 0.06. Moreover, the post-hoc analysis, performed using the Tukey procedure for the correction of the statistical significance, showed significant differences between the hypomanic and euthymic states with ($p < 0.05$) at scales 1,2 and from 4 to 20. At scales 1,9,10 hypomanic group data was also different from depressed. Figure 13.11 shows the MSE results over all the scale factors.

MSE: Complexity Index Analysis As a complementary feature, we evaluated the Complexity Index (CI) [23] of each series as the area under the curve of the MSE graph. CI is calculated on short time scales, from 1 to 8, and on higher time scales, from 1 to 20. Results of the CI index, referring to the best MSE estimations which are given by the MSE-Method III, are shown in Table 13.4 for the three considered pathological mental states (euthymic, depressed, hypomanic). The Kruskal-Wallis test revealed statistical differences between the three mood states on both short ($p < 0.03$) and higher time scales ($p < 0.001$). Concerning the post-hoc analysis,

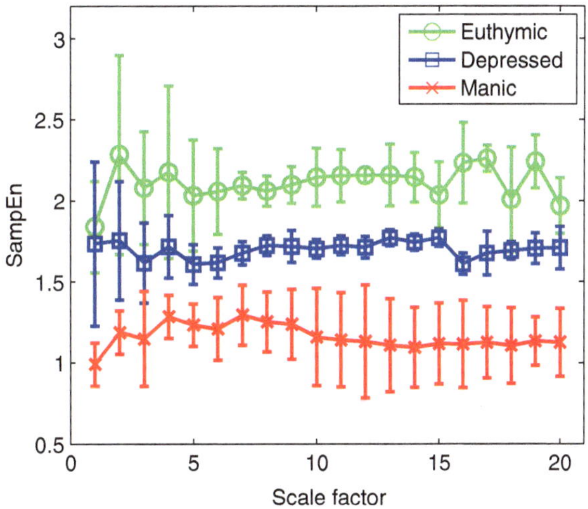

Fig. 13.11 MSE of heart rate dynamics in nocturnal period with method III, values are expressed as *median* $\pm$ *MAD*

Table 13.4 CI values estimated from MSE analysis of HRV series gathered during nocturnal recordings

	Short time	Higher time
Euthymic	15.18 ± 3.85	38.11 ± 2.40
Depressed	11.71 ± 1.50	31.39 ± 3.26
Hypomanic	9.14 ± 1.18	22.17 ± 4.17

Values are expressed as median and its respective absolute deviation also performed using the Tukey correction, the depressive group showed statistical difference with respect to the hypomanic group ($p < 0.02$) on short time scales, whereas significant variations were found between the hypomanic and euthymic states on both short and higher time scales ($p < 0.03$). It is worthwhile noting that the CI results are not biased by significant changes of the chosen r values of each considered recording. As a matter of fact, the following r_{max} statistics were found: 0.0024 ± 0.0012 s for the manic state, 0.0032 ± 0.0014 s for the euthymic state, 0.0023 ± 0.0006 s for the depressed. A probability of 0.4252 was associated to the null hypothesis of having no difference among the three mood states, according to the Kruskal-Wallis non-parametric test.

Complementary Analyses In order to further investigate how heartbeat dynamics is modulated during different pathological mental states, a multiscale Detrended fluctuation analysis (mDFA) was applied. According to the MSE processing, the mDFA, estimating up to the twentieth scale factor, was calculated on the longest segment of consecutive artifact-free samples of each acquisition of each patient. On both α_1 and α_2 indices, the Kruskal-Wallis non-parametric test showed no statistical difference among the three pathological groups ($p > 0.05$) for each of the twenty scales. These results suggest that heartbeat dynamics displays similar long- and short-range correlations between different pathological mental states.

As mood states could be associated to changes in the sympatho-vagal balance of ANS, we also tested the null hypothesis that standard HRV parameters defined in the frequency domain, as the power in the Low Frequency (LF) and High Frequency (HF) bands along with their ratio, are significantly altered among the three pathological mood states. Earlier studies, in fact, have shown a significant change in complexity measures and in the DFA scaling behavior of HRV in response to sympathetic or parasympathetic activation due to rest and exercise [67], sleep-wake and sleep-stage transitions [68, 69], circadian phases [70], beta-blockers and atropine drug administration [71], aging [72]. By calculating the HRV power spectra within non-overlapped moving time-windows of 5 min of length and averaging on each observation, we found that no statistical difference is associated to the LF, HF, and LF/HF indices different mood states, according to the Kruskal-Wallis non-parametric tests ($p > 0.05$). Concerning the aging effects, we report that a probability value of 0.435 (from the Kruskal-Wallis non-parametric test) is associated to the null hypothesis of having no significant difference in age among the pathological mood states.

Finally, in order to investigate whether the heartbeat complexity modulation is due to a loss of nonlinear properties of the cardiovascular system or changes in the parameters of the system (which would remain governed by nonlinear equations), we further applied an established time-domain method [73] to the RR interval series in order to test the presence of nonlinearity among the different mood states, as suggested by [74, 75]. The outcomes from the nonlinearity test demonstrate that all the considered long-term recordings are characterized by a relevant presence of non-linearity, over all the considered states (with $p < 0.05$).

In summary, using the objective estimation given by the so-called MSE-Method III, which searches maximum *ApEn* values in a parabola interpolating values $0.01\sigma \leq r \leq 1.2\sigma$ [40, 41, 43], we found an interesting complexity modulation coherent with both the current literature and different mood states. We found that higher complexity at all scales is associated to the euthymic state, whereas the depressive and hypomanic states show decreased complexity values ($p < 0.01$). Moreover, the complementary mDFA analysis suggests that pathological mental states modulate the pattern of signal complexity throughout different time scales without affecting patterns of signal amplitude with changing time scale. The nonlinearity tests also suggest that a loss of nonlinearity does not occur in case of mood disorders, as previously observed in the presence of, e.g., heart failure [76]. The HRV power spectral analysis also suggests that, although sympatho-vagal dynamics can be affected by pathological mental states, the inter-subject variability is too high to allow such changes to be revealed through a simple analysis in the frequency domain. From this point of view, the complexity analysis resulted to be a much more powerful tool. We also state that the differences in heartbeat complexity found among the three pathological mood states are not biased neither by the age of the patients enrolled in the study nor by the r_{max} values that are a function of the HRV standard deviation.

13.4 Application on Depression

Depression is a global public health problem of very high prevalence [77] characterized by persistent feelings of sadness and loss of interest or pleasure in daily activities for at least a 2 week period, with other symptoms including psychomotor retardation, fatigue, feelings of worthlessness and recurrent thoughts of death [78]. Clinical features related to autonomic dysfunction such as decreased appetite, gastrointestinal paresthesias, insomnia and increased sweating are also frequently reported [79]. All these alterations seem to be reinforced by a mood-congruent cognitive bias in which MD patients give preference to the processing of negative vs. positive emotional content material [23]. The exposure to negative emotional stimuli easily activates other negative thoughts and memories in MD patients, thus contributing to maintaining and enhancing depressive symptomatology [80].

Here, we present an attempt to characterize major depression (MD) by using a recently introduced computational method of nonlinear/complex digital signal processing. Accordingly, we hypothesize that instantaneous nonlinear analysis based on entropy measures can provide useful information about the clinical state of patients with MD while tracking the related complex cardiovascular dynamics. To this extent, we propose the application of point-process nonlinear models of heartbeat dynamics to derive instantaneous indices of complexity. In fact, we recently proposed a revised point-process framework, where the RR interval series is seen as a binary stochastic series characterized by inter-event probability functions [81], by embedding nonlinear autoregressive models with the Laguerre expansion of the Wiener-Volterra autoregressive terms. In doing so, we both achieved a more effective system identification [81], and derived novel instantaneous indices of complexity, i.e. the inhomogeneous point-process approximate entropy (ipApEn) and the inhomogeneous point-process sample entropy (ipSampEn) indices [81]. In this chapter, we describe the effectiveness of the instantaneous entropy measures in distinguishing patients with MD from healthy subjects.

13.4.1 Recruitment of Eligible Subjects

A number of 48 Hispanic outpatients with major depression (MD) were included in the study. All subjects were screened in local universities by applying the Zung-self-rating depression scale [82]. This scale is a 20-item questionnaire that measures the presence and severity of depressive symptomatology in the preceding 2 weeks. A score of 50/100 was considered compatible with a diagnosis of depression according to previous data in our population [83]. A Spanish Structured Clinical Interview for DSM-IV Axis 1 disorders, Clinical version, was applied by qualified psychiatrists to confirm the diagnosis of MD. All patients were experiencing their first MD episode and had not received psychotherapeutic or pharmacological treatment. A control group consisting of 48 age- and gender-matched healthy subjects was also included. Exclusion criteria for both MD and healthy subjects were: cardio-, cerebro-, or peripheral vascular diseases, the presence of neoplasm, diabetes mellitus, kidney or liver failure, infectious or systemic inflammatory disease and current neurological illnesses. All subjects received information about the study procedures and gave written informed consent approved by the local Institutional Review Board.

13.4.2 Experimental Protocol and Data Acquisition

The stimulus used in the study consisted of a set of eleven slides accompanied by an audio recording with two different narrative versions: an emotionally neutral recording (N) and an emotionally arousing one (E). These stories had been previously adapted and validated in a sample from the same Hispanic background and were kept as close as possible to the originals [84]. Both sets of slides showed a mother taking her young son to see his father at a nearby hospital where he works. The slides were identical, but the narrative differed in the N and E versions. In the N version of the story, the mother and son witness a minor car accident, which attracts the attention of the child, whereas in the E version, the child himself is critically injured and requires a surgical intervention at the hospital. The story content can be divided into three phases, with the second phase (slides 5–8) containing the emotionally arousing elements. Block scheme of the experimental protocol is shown in Fig. 13.12.

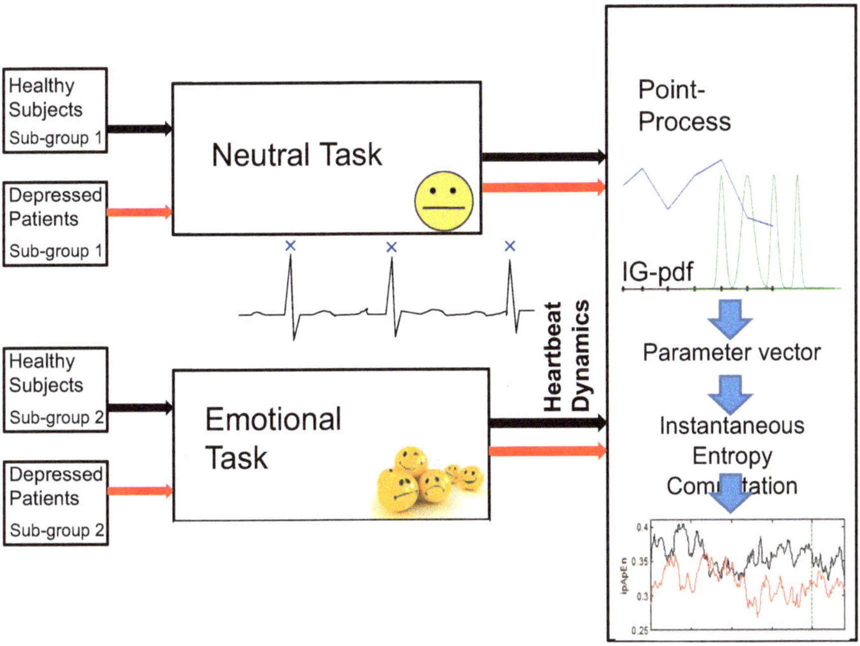

Fig. 13.12 Block scheme of the experimental protocol considered in this study. Healthy subjects and depressed patients are split into two sub-groups respectively. The first sub-group of healthy subjects and the first of depressed patients were randomly selected to undergo a neutral elicitation task, whereas the second respective sub-groups underwent an emotional elicitation task. Meanwhile, continuous ECG recording was performed in order to monitor the heartbeat dynamics of each subject/patient. Then, each RR interval series was modeled and processed through a point-process nonlinear model. Finally, starting from the time-varying parameter vector, the instantaneous entropy measures, ipApEn and ipSampEn, computation was performed

All recording sessions took place between 8 a.m. and 10 a.m. All tests were performed in a quiet, dimly lit room at a comfortable temperature (20–22 °C). Participants abstained from smoking or consuming beverages containing caffeine, xanthines or alcohol the day before evaluation. Subjects from each group (MD, HC) were randomly assigned to either undergo the N or E stimulus version, resulting in four different experimental groups (MD-N, MD-E, HC-N, HC-E). All subjects were told that the aim of the study was to evaluate how people pay attention to stories. It was explained that the slide presentation would be shown accompanied by a short narration. They were instructed to concentrate on each slide for the duration of its presentation and to watch the slide show as they would watch a movie. Continuous ECG monitoring (lead II) was performed with a Finometer device (Finapress Medical System, The Netherlands). Data was digitized and stored in a PC computer using a signal acquisition system DATAQ 720-WINDAQ PRO (DataQ

Instruments, Akron, OH, USA). Subjects were initially asked to rest for 10 min in a reclining position. Then, each subject underwent the stimulus elicitation previously described, followed by a 3-min recovery period, during which participants quietly rested.

13.4.3 Experimental Results Using Inhomogeneous Point Process Entropy

The primary goal of this study was to test the ability of instantaneous linear and complex nonlinear estimates of heartbeat dynamics in discriminating healthy subjects from the MD patients. We separated each task (neutral elicitation task or emotional elicitation task), and considered the median value over the whole acquisition for each subject. The averaged instantaneous tracking of the complex heartbeat dynamics, expressed as ipApEn and ipSampEn, during the emotional elicitation task are shown in Fig. 13.13. Values are averaged among healthy subjects and patients with MD.

Our working hypothesis was that the degree of complexity of cardiovascular dynamics differs between normal and MD subjects. Since we used two measures to verify our hypothesis, we appropriately adjusted the statistical significance level to account for multiple comparisons in testing each individual measure. In the case of emotional elicitation, we observed that the difference is statistically significant for ipApEn while not for ipSampEn. To confirm that this difference

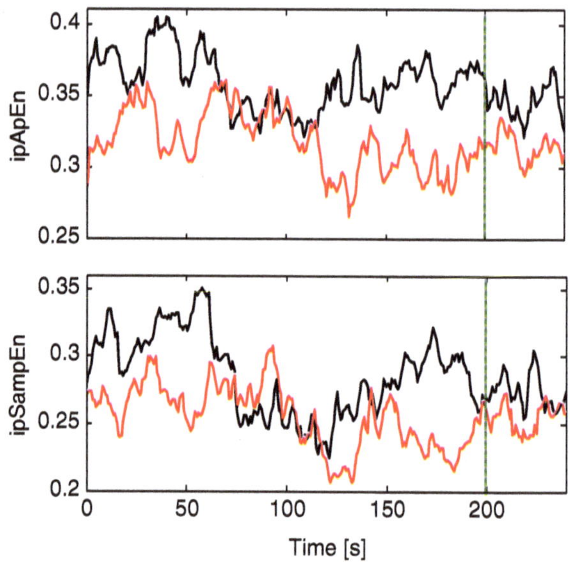

Fig. 13.13 Instantaneous complex heartbeat dynamics expressed as ipApEn (*top panel*) and ipSampEn (*bottom panel*) during the emotional elicitation task. Values are averaged among healthy subjects (*continuous black line*) and patients with MD (*continuous red line*). The *green dotter vertical line* indicates the end of the audio-visual stimulus

Table 13.5 Descriptive statistics of instantaneous heartbeat dynamics features during the neutral elicitation task

Feature	Statistics		
	Healthy	Depressed	p-Value
ipSampEn	0.286 ± 0.067	0.269 ± 0.023	0.970
ipApEn	0.345 ± 0.058	0.352 ± 0.046	0.447
μ_{RR}	761.62 ± 83.42	798.21 ± 61.57	0.447
σ^2_{RR}	319.11 ± 191.27	521.90 ± 187.03	0.442
LF	485.53 ± 253.45	763.60 ± 408.17	0.056
HF	321.12 ± 225.92	228.04 ± 109.43	0.799
LF/HF	1.818 ± 1.186	1.768 ± 0.862	0.817

p-Values from the Mann-Whitney non-parametric test with null hypothesis of equal medians

Table 13.6 Descriptive statistics of instantaneous heartbeat dynamics features during the emotional elicitation task

Feature	Statistics		
	Healthy	Depressed	p-Value
ipApEn	**0.361 ± 0.038**	**0.316 ± 0.041**	**0.016**
ipSampEn	0.298 ± 0.052	0.257 ± 0.041	0.212
μ_{RR}	809.17 ± 102.25	792.40 ± 84.05	0.328
σ^2_{RR}	596.71 ± 367.11	402.30 ± 209.61	0.093
LF	674.08 ± 417.99	616.29 ± 391.82	0.777
HF	518.34 ± 284.61	328.63 ± 214.38	0.054
LF/HF	1.068 ± 0.591	1.987 ± 0.751	0.095

Bold indicates statistically significant indices
p-Values from the Mann-Whitney non-parametric test with null hypothesis of equal medians

was due to nonlinear mechanisms rather than simply a reflection of linear changes on our nonlinear measures, we also applied the Mann-Whitney test on a batch of conventional HRV measures, such as the mean μ_{RR}, the standard deviation σ^2_{RR}, the low frequency power LF, the high frequency power HF, and the ratio LF/HF. None was statistically significant, even without accounting for multiple comparisons.

The statistics of each group and each instantaneous feature related to the neutral elicitation task are reported in Table 13.5, whereas those related to the emotional elicitation task are reported in Table 13.6.

Finally, we report that a $p = 0.15$ given by the Mann-Whitney non-parametric test was associated to the null hypothesis of having equal median age between the healthy and MD groups (age of healthy subjects: 21±2; age of MD: 23±2). This outcome confirms that our results are not affected by age differences between the groups.

This results demonstrate that instantaneous complex heartbeat dynamics is indeed modulated by emotionally-relevant elicitations in patients with MD. Of note, these results are in agreement with the current literature showing that pathological mental states modulate cardiovascular complexity [23, 85, 86]. The fact that this complexity modulation occurs only in case of emotionally-relevant stimuli is also in agreement with both clinical experience and the same research studies linking mood states, emotional regulation, and emotional response to ANS dynamics [6, 12, 14, 87]. Our statistical analysis performed on the linear instantaneous features defined in the time and frequency domains suggests that, although sympatho-vagal dynamics can be affected by pathological mental states [13, 88], the inter-subject variability is too high to allow such changes to be revealed through linear analysis. Conversely, the analysis on instantaneous complexity allowed to provide a much higher discriminating power. Our results further demonstrated that the differences in heartbeat complexity found among the healthy and MD groups are not biased by the age of the patients enrolled in the study nor by the r(t)values that are a function of the HRV standard deviation. Unlike other paradigms developed in the literature for characterizing human mental states, our approach is purely parametric, and the analytically-derived indices can be evaluated in a dynamic and instantaneous fashion. The presented point-process nonlinear analysis, in fact, represents a pioneering study in the field of mood assessment, as just being recently proposed for the characterization of heartbeat dynamics in bipolar patients [57].

13.5 Conclusion

In this chapter, we described effective applications of heartbeat complexity analysis for care in mental health. As in the current clinical practice the diagnosis of mental disorders does not rely on objective psycho-physiological markers, our results suggest the use of heartbeat complexity indices to give a viable support to the clinical decision. Our findings confirm the importance of nonlinear temporal patterns in mood recognition [22, 23]. Although the detailed physiology behind the complex dynamics of heartbeat variations has not been completely clarified, previous studies suggest that β-adrenoceptor system has little involvement in the generation of nonlinear HRV whereas α-adrenoceptors strongly influence the scaling properties of the time series [89]. Moreover, cholinergic iper-driving also induces changes in the complexity measures such as the MSE complexity index (seen in rats) [89].

We described the complexity of heartbeat dynamics in bipolar patients through sample entropy and MSE analysis of HRV series. We processed HRV series gathered from bipolar patients during night recordings by using a comfortable wearable monitoring system developed within the PSYCHE project [17, 18], aiming to differentiate among different mood states such as depression, hypomania, and euthymia, which is the good affective balance. The major methodological issue in estimating the *SampEn* of MSE over the scale factors, which involves the choice of the r value, has been deeply exploited comparing three different approaches.

Of note, our approach allowed to extract and analyze simple but effective complex measures such as the SampEn, which have been demonstrated to play an important role in several psychiatric and also cardiovascular disorders [23]. In particular, we demonstrated that the proposed HRV complexity-based approach can be profitably used in assessing the response to treatment in BD patients with severe symptoms. The uncertain positive effect of pharmacological treatments, in fact, is widely accepted to be one of the major clinical problems, even if appropriate drugs are administered. Clinicians often need to try and adjust their treatments if they realize that the expected clinical response is not coming.

Our results on patients with major depression suggested that a possible approach to investigate mood recognition is to explore emotional changes provoked by external stimuli. From a modeling point of view, such approach interprets the cardiovascular system as coupled with the central nervous system and governed by nonlinear dynamical equations which can be characterized by means of a "perturbation" analysis, i.e, analysis before and after short-time emotional elicitation. To this extent, we believe that the optimal mathematical framework for this analysis should be based on the point-process theory, which provides effective computational tools able to continuously estimate heartbeat dynamics without using any interpolation methods [8, 81, 90]. This powerful, fully-parametric statistical method accounts for the probabilistic complex generative mechanism of the heartbeat by considering a quadratic Wiener-Volterra representation of the first order moment of a physiological plausible inverse-Gaussian statistics. Moreover, goodness-of-fit measures such as KS distance and autocorrelation plots quantitatively allow to verify the model fit and to choose the proper model order, thus addressing another open issue of current parametric approaches. The main result was the validation of the utility of the presented computational framework by testing its ability in tracking nonlinear and non-stationary heartbeat dynamics of healthy subjects and MD patients undergoing affective elicitation.

From a physiological perspective, the inherent complexity of the cardiovascular system (e.g. the nonlinear neural signaling on the sinoatrial node) has been confirmed by our experimental results. Although the detailed physiology behind these complex dynamics has not been completely clarified, previous studies suggest that α-adrenoceptor system and cholinergic iper-driving might be involved (as seen in rats) [89]. Because mood disorders produce an altered emotional response, the achievements reported in this study could have a relevant impact on mood disorder psychopathology diagnosis and treatment. Monitoring fast emotional responses as the result of fast stimulation times through instantaneous heartbeat dynamics could make a continuous evaluation of disorder progression possible, thus representing an important scientific advancement. On a final note, as emotional state is presently determined in a clinical setting using questionnaires with limited accuracy and quantitative power, a more automated and objective assessment using a noninvasive physiological and easy-to-monitor signal such as the ECG would provide improved inpatient and outpatient care, thereby significantly reducing the time and cost of mental healthcare.

References

1. Marmarelis, V.Z.: Nonlinear Dynamic Modeling of Physiological Systems, vol. 10. Wiley, New York (2004)
2. Valenza, G., Scilingo, E.P.: Autonomic Nervous System Dynamics for Mood and Emotional-State Recognition: Significant Advances in Data Acquisition, Signal Processing and Classification. Springer, Cham (2014). http://www.springer.com/gp/book/9783319026381
3. Huikuri, H.V., Mäkikallio, T.H., Peng, C.-K., Goldberger, A.L., Hintze, U., Møller, M., Group, D.S., et al.: Fractal correlation properties of RR interval dynamics and mortality in patients with depressed left ventricular function after an acute myocardial infarction. Circulation **101**(1), 47–53 (2000)
4. Sassi, R., Signorini, M.G., Cerutti, S.: Multifractality and heart rate variability. Chaos: Interdiscip. J. Nonlinear Sci. **19**(2), 028507 (2009)
5. Valenza, G., Allegrini, P., Lanatà, A., Scilingo, E.P.: Dominant lyapunov exponent and approximate entropy in heart rate variability during emotional visual elicitation. Front. Neuroeng. **5**, 3 (2012)
6. Lanata, A., Valenza, G., Scilingo, E.P.: A novel EDA glove based on textile-integrated electrodes for affective computing. Med. Biol. Eng. Comput. **50**(11), 1163–1172 (2012)
7. Valenza, G., Lanata, A., Scilingo, E.P.: The role of nonlinear dynamics in affective valence and arousal recognition. IEEE Trans. Affect. Comput. **3**(2), 237–249 (2012)
8. Valenza, G., Citi, L., Scilingo, E.P., Barbieri, R.: Point-process nonlinear models with Laguerre and Volterra expansions: instantaneous assessment of heartbeat dynamics. IEEE Trans. Signal Process. **61**(11), 2914–2926 (2013)
9. Valenza, G., Lanatá, A., Scilingo, E.P.: Improving emotion recognition systems by embedding cardiorespiratory coupling. Physiol. Meas. **34**(4), 449 (2013)
10. Valenza, G., Lanata, A., Scilingo, E.P.: Oscillations of heart rate and respiration synchronize during affective visual stimulation. IEEE Trans. Inf. Technol. Biomed. **16**(4), 683–690 (2012)
11. Sunagawa, K., Kawada, T., Nakahara, T.: Dynamic nonlinear vago-sympathetic interaction in regulating heart rate. Heart Vessels **13**(4), 157–174 (1998)
12. Carney, R.M., Freedland, K.E., Miller, G.E., Jaffe, A.S.: Depression as a risk factor for cardiac mortality and morbidity: a review of potential mechanisms. J. Psychosom. Res. **53**(4), 897–902 (2002)
13. Agelink, M.W., Boz, C., Ullrich, H., Andrich, J.: Relationship between major depression and heart rate variability: clinical consequences and implications for antidepressive treatment. Psychiat. Res. **113**(1), 139–149 (2002)
14. Iverson, G., et al.: A new potential marker for abnormal cardiac physiology in depression. J. Behav. Med. **28**(6), 507–511 (2005)
15. Sponheim, S., Allen, J., Iacono, W.: Selected psychophysiological measures in depression: the significance of electrodermal activity, electroencephalographic asymmetries, and contingent negative variation to behavioral and neurobiological aspects of depression. In: Behavioral High Risk Paradigm in Psychopathology, pp. 222–249 (1995)
16. Taillard, J., et al.: Heart rate orcadian rhythm as a biological marker of desynchronization in major depression: a methodological and preliminary report. Chronobiol. Int. **7**(4), 305–316 (1990)
17. Valenza, G., Nardelli, M., Lanata, A., Gentili, C., Bertschy, G., Paradiso, R., Scilingo, E.P.: Wearable monitoring for mood recognition in bipolar disorder based on history-dependent long-term heart rate variability analysis. IEEE J. Biomed. Health Informatics **18**(5), 1625–1635 (2014)
18. Valenza, G., Gentili, C., Lanatà, A., Scilingo, E.P.: Mood recognition in bipolar patients through the psyche platform: preliminary evaluations and perspectives. Artif. Intell. Med. **57**(1), 49–58 (2013)
19. Voss, A., et al.: Linear and nonlinear methods for analyses of cardiovascular variability in bipolar disorders. Bipolar Disord. **8**(5p1), 441–452 (2006)

20. Levy, B.: Autonomic nervous system arousal and cognitive functioning in bipolar disorder. Bipolar Disord. **15**(1), 70–79 (2013)
21. Latalova, K., et al.: Autonomic nervous system in euthymic patients with bipolar affective disorder. Neuro Endocrinol. Lett. **31**(6), 829 (2010)
22. Schulz, S., Koschke, M., Bär, K.-J., Voss, A.: The altered complexity of cardiovascular regulation in depressed patients. Physiol. Meas. **31**(3), p. 303 (2010)
23. Leistedt, S.J., Linkowski, P., Lanquart, J., Mietus, J., Davis, R.B., Goldberger, A.L., Costa, M.D.: Decreased neuroautonomic complexity in men during an acute major depressive episode: analysis of heart rate dynamics. Transl. Psychiatry **1**(7), e27 (2011)
24. Pincus, S.: Approximate entropy as a measure of system complexity. Proc. Natl. Acad. Sci. **88**(6), 2297 (1991)
25. Pincus, S.: Approximate entropy (ApEn) as a complexity measure. Chaos: Interdiscip. J. Nonlinear Sci. **5**(1), 110–117 (1995)
26. Richman, J., Moorman, J.: Physiological time-series analysis using approximate entropy and sample entropy. Am. J. Physiol.- Heart Circul. Physiol. **278**(6), H2039 (2000)
27. Srinivasan, V., Eswaran, C., Sriraam, N.: Approximate entropy-based epileptic EEG detection using artificial neural networks. IEEE Trans. Inf. Technol. Biomed. **11**(3), 288–295 (2007)
28. Fusheng, Y., Bo, H., Qingyu, T.: Approximate entropy and its application to biosignal analysis. In: Nonlinear Biomedical Signal Processing: Dynamic Analysis and Modeling, vol. 2, pp. 72–91 (2001)
29. Chen, X., Solomon, I., Chon, K.: Comparison of the use of approximate entropy and sample entropy: applications to neural respiratory signal. In: Conference Proceedings: ... Annual International Conference of the IEEE Engineering in Medicine and Biology Society, vol. 4, pp. 4212–4215 (2004). IEEE Engineering in Medicine and Biology Society, Piscataway
30. Takens, F.: Detecting strange attractors in turbulence. In: Dynamical Systems and Turbulence, Warwick 1980, pp. 366–381. Springer, Berlin (1981)
31. Schouten, J.C., Takens, F., van den Bleek, C.M.: Estimation of the dimension of a noisy attractor. Phys. Rev. E **50**(3), 1851 (1994)
32. Costa, M., Goldberger, A.L., Peng, C.-K.: Multiscale entropy analysis of complex physiologic time series. Phys. Rev. Lett. **89**(6), 068102 (2002)
33. Costa, M., Peng, C.-K., Goldberger, A.L., Hausdorff, J.M.: Multiscale entropy analysis of human gait dynamics. Phys. A: Stat. Mech. Appl. **330**(1), 53–60 (2003)
34. Park, J.-H., Kim, S., Kim, C.-H., Cichocki, A., Kim, K.: Multiscale entropy analysis of EEG from patients under different pathological conditions. Fractals **15**(04), 399–404 (2007)
35. Takahashi, T., Cho, R.Y., Mizuno, T., Kikuchi, M., Murata, T., Takahashi, K., Wada, Y.: Antipsychotics reverse abnormal EEG complexity in drug-naive schizophrenia: a multiscale entropy analysis. Neuroimage **51**(1), 173–182 (2010)
36. Catarino, A., Churches, O., Baron-Cohen, S., Andrade, A., Ring, H.: Atypical EEG complexity in autism spectrum conditions: a multiscale entropy analysis. Clin. Neurophysiol. **122**(12), 2375–2383 (2011)
37. Trunkvalterova, Z., Javorka, M., Tonhajzerova, I., Javorkova, J., Lazarova, Z., Javorka, K., Baumert, M.: Reduced short-term complexity of heart rate and blood pressure dynamics in patients with diabetes mellitus type 1: multiscale entropy analysis. Physiol. Meas. **29**(7), 817 (2008)
38. Valencia, J.F., Porta, A., Vallverdú, M., Claria, F., Baranowski, R., Orlowska-Baranowska, E., Caminal, P.: Refined multiscale entropy: application to 24-h holter recordings of heart period variability in healthy and aortic stenosis subjects. IEEE Trans. Biomed. Eng. **56**(9), 2202–2213 (2009)
39. Turianikova, Z., Javorka, K., Baumert, M., Calkovska, A., Javorka, M.: The effect of orthostatic stress on multiscale entropy of heart rate and blood pressure. Physiol. Meas. **32**(9), 1425 (2011)
40. Castiglioni, P., Di Rienzo, M.: How the threshold "r" influences approximate entropy analysis of heart-rate variability. In: 2008 Computers in Cardiology, pp. 561–564. IEEE, New York (2008)
41. Chon, K.H., Scully, C.G., Lu, S.: Approximate entropy for all signals. IEEE Eng. Med. Biol. Mag. **28**(6), 18–23 (2009)

42. Magalhaes, F., de Sá, J.M., Bernardes, J., Ayres-de Campos, D.: Characterization of fetal heart rate irregularity using approximate entropy and wavelet filtering. In: 2006 Computers in Cardiology, pp. 933–936. IEEE, New York (2006)
43. Lu, S., Chen, X., Kanters, J.K., Solomon, I.C., Chon, K.H.: Automatic selection of the threshold value for approximate entropy. IEEE Trans. Biomed. Eng. **55**(8), 1966–1972 (2008)
44. Alcaraz, R., Abásolo, D., Hornero, R., Rieta, J.J.: Optimal parameters study for sample entropy-based atrial fibrillation organization analysis. Comput. Methods Progr. Biomed. **99**(1), 124–132 (2010)
45. Boskovic, A., Loncar-Turukalo, T., Sarenac, O., Japundzic-Zigon, N., Bajic, D.: Unbiased entropy estimates in stress: a parameter study. Comput. Biol. Med. **42**(6), 667–679 (2012)
46. Zurek, S., Guzik, P., Pawlak, S., Kosmider, M., Piskorski, J.: On the relation between correlation dimension, approximate entropy and sample entropy parameters, and a fast algorithm for their calculation. Phys. A: Stat. Mech. Appl. **391**(24), 6601–6610 (2012)
47. Acharya, U.R., Joseph, K.P., Kannathal, N., Lim, C.M., Suri, J.S.: Heart rate variability: a review. Med. Biol. Eng. Comput. **44**(12), 1031–1051 (2006)
48. Barbieri, R., Matten, E., Alabi, A., Brown, E.: A point-process model of human heartbeat intervals: new definitions of heart rate and heart rate variability. Am. J. Physiol.-Heart Circul. Physiol. **288**(1), H424 (2005)
49. Citi, L., Valenza, G., Barbieri, R.: Instantaneous estimation of high-order nonlinear heartbeat dynamics by lyapunov exponents. In: Proceedings of IEEE-EMBC, pp. 13–16. IEEE, New York (2012)
50. Merikangas, K., et al.: Prevalence and correlates of bipolar spectrum disorder in the world mental health survey initiative. Arch. Gen. Psychiatry **68**(3), 241 (2011)
51. A. P. Association: Diagnostic and statistical manual of mental disorders: DSM-IV-TR®. American Psychiatric Publishing, Arlington (2000)
52. Revicki, D.A., Matza, L.S., Flood, E., Lloyd, A.: Bipolar disorder and health-related quality of life. Pharmacoeconomics **23**(6), 583–594 (2005)
53. Michalak, E.E., Yatham, L.N., Lam, R.W.: Quality of life in bipolar disorder: a review of the literature. Health Qual. Life Outcomes **3**(1), 72 (2005)
54. Brissos, S., Dias, V.V., Kapczinski, F.: Cognitive performance and quality of life in bipolar disorder. Can. J. Psychiatry: Rev. Can. Psychiatr. **53**(8), 517–524 (2008)
55. Kauer-Sant'Anna, M., Frey, B.N., Andreazza, A.C., Cereser, K.M., Gazalle, F.K., Tramontina, J., da Costa, S.C., Santin, A., Kapczinski, F.: Anxiety comorbidity and quality of life in bipolar disorder patients. Can. J. Psychiatry: Rev. Can. Psychiatr. **52**(3), 175–181 (2007)
56. Kessler, R., McGonagle, K., Zhao, S., Nelson, C., Hughes, M., Eshleman, S., et al.: Lifetime and 12-month prevalence of DSM-III-R psychiatric disorders in the United States: results from the national comorbidity survey. Arch. Gen. Psychiatry **51**(1), 8 (1994)
57. Wittchen, H., Jacobi, F.: Size and burden of mental disorders in Europe–a critical review and appraisal of 27 studies. Eur. Neuropsychopharmacol. **15**(4), 357–376 (2005)
58. Pini, S., de Queiroz, V., et al.: Prevalence and burden of bipolar disorders in European countries. Eur. Neuropsychopharmacol. **15**(4), 425–434 (2005)
59. Chen, Y.-W., Dilsaver, S.C.: Lifetime rates of suicide attempts among subjects with bipolar and unipolar disorders relative to subjects with other axis I disorders. Biol. Psychiatry **39**(10), 896–899 (1996)
60. Vieta, E., Reinares, M., Rosa, A.: Staging bipolar disorder. Neurotox. Res. **19**(2), 279–285 (2011)
61. Greco, A., Valenza, G., Lanata, A., Rota, G., Scilingo, E.P.: Electrodermal activity in bipolar patients during affective elicitation. IEEE J. Biomed. Health Inform. **18**(6), 1865–1873 (2014)
62. Greco, A., Lanata, A., Valenza, G., Rota, G., Vanello, N., Scilingo, E.: On the deconvolution analysis of electrodermal activity in bipolar patients. In: Proceeding of IEEE-EMBC, pp. 6691–6694. IEEE, New York (2012)
63. Watkins, L., et al.: Association of anxiety with reduced baroreflex cardiac control in patients after acute myocardial infarction. Am. Heart J. **143**(3), 460–466 (2002)

64. Taillard, J., et al.: Sleep and heart rate circadian rhythm in depression: the necessity to separate. Chronobiol. Int. **10**(1), 63–72 (1993)
65. Scilingo, E., Gemignani, A., Paradiso, R., Taccini, N., Ghelarducci, B., De Rossi, D.: Performance evaluation of sensing fabrics for monitoring physiological and biomechanical variables. IEEE Trans. Inf. Technol. Biomed. **9**(3), 345–352 (2005)
66. Rajendra Acharya, U., Paul Joseph, K., Kannathal, N., Lim, C., Suri, J.: Heart rate variability: a review. Med. Biol. Eng. Comput. **44**(12), 1031–1051 (2006)
67. Karasik, R., Sapir, N., Ashkenazy, Y., Ivanov, P.C., Dvir, I., Lavie, P., Havlin, S.: Correlation differences in heartbeat fluctuations during rest and exercise. Phys. Rev. E **66**(6), 062902 (2002)
68. Ivanov, P.C., Bunde, A., Amaral, L.N., Havlin, S., Fritsch-Yelle, J., Baevsky, R.M., Stanley, H.E., Goldberger, A.L.: Sleep-wake differences in scaling behavior of the human heartbeat: analysis of terrestrial and long-term space flight data. Europhys. Lett. **48**(5), 594 (1999)
69. Kantelhardt, J.W., Ashkenazy, Y., Ivanov, P.C., Bunde, A., Havlin, S., Penzel, T., Peter, J.-H., Stanley, H.E.: Characterization of sleep stages by correlations in the magnitude and sign of heartbeat increments. Phys. Rev. E **65**(5), 051908 (2002)
70. Ivanov, P.C., Hu, K., Hilton, M.F., Shea, S.A., Stanley, H.E.: Endogenous circadian rhythm in human motor activity uncoupled from circadian influences on cardiac dynamics. Proc. Natl. Acad. Sci. **104**(52), 20702–20707 (2007)
71. Amaral, L.A.N., Ivanov, P.C., Aoyagi, N., Hidaka, I., Tomono, S., Goldberger, A.L., Stanley, H.E., Yamamoto, Y.: Behavioral-independent features of complex heartbeat dynamics. Phys. Rev. Lett. **86**(26), 6026 (2001)
72. Schmitt, D.T., Ivanov, P.C.: Fractal scale-invariant and nonlinear properties of cardiac dynamics remain stable with advanced age: a new mechanistic picture of cardiac control in healthy elderly. Am. J. Physiol.-Regul. Integr. Comp. Physiol. **293**(5), R1923–R1937 (2007)
73. Barnett, A., Wolff, R.: A time-domain test for some types of nonlinearity. IEEE Trans. Signal Process. **53**(1), 26–33 (2005)
74. Ivanov, P.C., Ma, Q.D., Bartsch, R.P., Hausdorff, J.M., Amaral, L.A.N., Schulte-Frohlinde, V., Stanley, H.E., Yoneyama, M.: Levels of complexity in scale-invariant neural signals. Phys. Rev. E **79**(4), 041920 (2009)
75. Ashkenazy, Y., Havlin, S., Ivanov, P.C., Peng, C.-K., Schulte-Frohlinde, V., Stanley, H.E.: Magnitude and sign scaling in power-law correlated time series. Phys. A: Stat. Mech. Appl. **323**, 19–41 (2003)
76. Ivanov, P.C., Amaral, L.A.N., Goldberger, A.L., Havlin, S., Rosenblum, M.G., Struzik, Z.R., Stanley, H.E.: Multifractality in human heartbeat dynamics. Nature **399**(6735), 461–465 (1999)
77. Nierenberg, A.A.: Major depressive disorder: epidemiology, course of illness, and treatment. CNS Spectr. **13**(S8), 4–7 (2008)
78. Soleimani, L., Lapidus, K.A., Iosifescu, D.V.: Diagnosis and treatment of major depressive disorder. Neurol. Clin. **29**(1), 177–193 (2011)
79. Rakofsky, J.J., Schettler, P.J., Kinkead, B.L., Frank, E., Judd, L.L., Kupfer, D.J., Rush, A.J., Thase, M.E., Yonkers, K.A., Rapaport, M.H.: The prevalence and severity of depressive symptoms along the spectrum of unipolar depressive disorders: a post hoc analysis. J. Clin. Psychiatry **74**(11), 1084–1091 (2013)
80. Mathews, A., MacLeod, C.: Cognitive vulnerability to emotional disorders. Annu. Rev. Clin. Psychol. **1**, 167–195 (2005)
81. Valenza, G., Citi, L., Scilingo, E.P., Barbieri, R.: Inhomogeneous point-process entropy: an instantaneous measure of complexity in discrete systems. Phys. Rev. E **89**(5), 052803 (2014)
82. Zung, W.W.: A self-rating depression scale. Arch. Gen. Psychiatry **12**(1), 63–70 (1965)
83. Campo-Arias, A., Díaz-Martínez, L.A., Eduardo, G., Jaimes, R., Cadena, L.D.P., Hernández, N.L.: Validation of Zung's self-rating depression scale among the colombian general population. Soc. Behav. Personal.: Int. J. **34**(1), 87–94 (2006)
84. Cahill, L., McGaugh, J.L.: A novel demonstration of enhanced memory associated with emotional arousal. Conscious. Cogn. **4**(4), 410–421 (1995)

85. Yang, A.C., Tsai, S.-J.: Is mental illness complex? From behavior to brain. Progr. Neuro-Psychopharmacol. Biol. Psychiatry **45**, 253–257 (2013)
86. Valenza, G., Nardelli, M., Bertschy, G., Lanata, A., Scilingo, E.: Mood states modulate complexity in heartbeat dynamics: a multiscale entropy analysis. Europhys. Lett. **107**(1), 18003 (2014)
87. Thayer, J.F., Friedman, B.H., Borkovec, T.D.: Autonomic characteristics of generalized anxiety disorder and worry. Biol. Psychiatry **39**(4), 255–266 (1996)
88. LeMoult, J., Yoon, K.L., Joormann, J.: Affective priming in major depressive disorder. Front. Integr. Neurosci. **6**, 76 (2012)
89. Beckers, F., Verheyden, B., Ramaekers, D., Swynghedauw, B., Aubert, A.E.: Effects of autonomic blockade on non-linear cardiovascular variability indices in rats. Clin. Exp. Pharmacol. Physiol. **33**(5–6), 431–439 (2006)
90. Valenza, G., Citi, L., Barbieri, R.: Estimation of instantaneous complex dynamics through lyapunov exponents: a study on heartbeat dynamics. PLoS One **9**(8), e105622 (2014)

Chapter 14
Cardiac Autonomic Changes in Epilepsy

Andrea Romigi and Nicola Toschi

Abstract The term "Epilepsy" encompasses a broad spectrum of medical and social disorders that affect about 65 million people worldwide and is commonly defined as a tendency to suffer recurrent seizures. In patients with epilepsy, ictal discharges that occur in (or propagate to) the anterior cingulate, insular, posterior orbito-frontal, and the pre-frontal cortices, along with the amygdala and hypothalamus play a key role in influencing the autonomic nervous system (ANS) at the cortical level. In turn, this can result in cardiac effects which are widespread and range from subtle changes in heart rate variability (HRV) to ictal sinus arrest, and from QT-interval shortening to atrial fibrillation. In addition, cardiac events are the main hypothesized mechanisms underlying sudden unexpected death in epilepsy (SUDEP), which occurs in absence of a known structural cause. Patients with epilepsy also experience long-lasting changes in the regulation of the ANS and target organs. Heart rate (HR) and HRV can be easily measured/estimated when compared to other biomarkers that are commonly associated with seizures (i.e., long-term EEG), and are therefore potentially valuable biomarkers when it comes to characterizing seizures. In this context, a number of linear and nonlinear analysis techniques have been applied in order to detect and characterize epilepsy-related ANS changes. While the physiological and clinical applicability of nonlinear analyses like fractal and complexity measures of HR dynamics are not yet completely understood, in view of recent experimental findings it is reasonable to assume that such indices highlight abnormal patterns of RR interval behaviour

A. Romigi
Stroke Unit Department Epilepsy Center, San Giovanni Addolorata General Hospital Neurology, Viale Amba Aradam 9, Rome, Italy

Neurophysiopathology Unit, Sleep Medicine Centre, Department of Systems Medicine, Tor Vergata University and Hospital, Rome, Italy

N. Toschi (✉)
Medical Physics Section, Department of Biomedicine and Prevention, University of Rome, Rome, Italy

Department of Radiology, Athinoula A. Martinos Center for Biomedical Imaging, Boston, MA, USA

Harvard Medical School, Boston, MA, USA
e-mail: toschi@med.uniroma2.it

R. Barbieri et al. (eds.), *Complexity and Nonlinearity in Cardiovascular Signals*, DOI 10.1007/978-3-319-58709-7_14 that are not easily detected by commonly used moment statistics of HR variation. These findings may provide new insight regarding physiological and seizure-induced states of the complex brain-heart network underlying epilepsy and related autonomic modifications. A better understanding of the autonomic manifestations of seizures would provide practical added value to clinical epileptologists dealing with differential diagnosis of epilepsy and related disorders, as well as aiding in designing more sensitive seizure detection and prediction algorithms.

14.1 Epilepsy and Its Cardiac Effects

The term "Epilepsy" encompasses a broad spectrum of medical and social disorders that affect about 65 million people worldwide. Epilepsy usually presents with a wide variety of features [1] and is commonly defined as a tendency to suffer recurrent epileptic seizures. The commission on classification and terminology belonging to the International League Against Epilepsy (ILAE) has recently revised the concepts, terminology, and approaches to the classification of epilepsies and has subdivided seizures into those "originating at some point within, and rapidly engaging bilaterally distributed networks" (generalized) and those "originating in networks limited to one hemisphere", either "discretely localized or more widely distributed" (focal) [2]. Epileptic seizures last no more than 2 min but, on rare occasions, may be organized in temporal clusters or may occur in rapid succession without a full recovery (i.e. *status epilepticus*).

In addition the definition of epilepsy has recently been critically revised and clarified, and the current, practical definition of this disorder implies a diagnosis of epilepsy after at least one unprovoked seizure in subjects affected by any other factors which may result in persistently lowered seizure threshold and therefore a high recurrence risk. The probability to present a further epileptic seizures is thought to be approximately equal to the risk of suffering a third unprovoked seizure (i.e. the previous definition of epilepsy) and is estimated at around 60%. The probability of seizure relapse increases in the presence of structural brain lesions (i.e. brain tumor, stroke, central nervous system (CNS) infection, traumatic brain injury), a diagnosis of a specific epilepsy syndrome, or in the presence of specific risk factors [3].

Cardiac effects of epilepsy are widespread and range from subtle changes in heart rate variability (HRV) to ictal sinus arrest, and from QT-interval shortening to atrial fibrillation [4]. In addition, cardiac events are the main hypothesized mechanisms underlying sudden unexpected death in epilepsy (SUDEP), a category of death which occurs in people with epilepsy in absence of a known structural cause [5]. SUDEP, whose etiology is still unclear [6], is the most prevalent epilepsy-related cause of death, accounting for up to 17% of all deaths in the general population with epilepsy and up to 50% of deaths in people suffering from refractory epilepsy [7, 8]. In this context, it has been hypothesized that repeated activation of the autonomic nervous system (ANS) may lead to sympathovagal imbalance, ultimately triggering fatal arrhythmias during or between ictal events [8, 9], especially in patients with higher risk (i.e. short QT interval) [10]. This hypothesis is supported by the seizure-induced sympathovagal imbalance demonstrated in drug-refractory and newly diagnosed temporal epilepsy [11, 12]. Also, while arrhythmias may represent the main cardiac-related risk factor for SUDEP, epilepsy-related ANS alterations are not only associated with vagal suppression (which is a risk factor for arrhythmias) but also with sympathetic activation, vagal activation, as well as with sympato-vagal suppression [4, 9]. In this context it is however important to note that most studies of ANS function in epilepsy are carried out in patients receiving antiepileptic drugs (AEDs), which are known to influence cardiac excitability and conduction, hence potentially interfering with autonomic cardiac activity. Additionally, hypoventilation of central origin is also considered a predominant factor in SUDEP, as confirmed by experimental models where central apnoea, sometimes with added peripheral respiratory obstruction, have been described [13] in SUDEP.

14.2 Pathophysiology of ANS Changes in Epilepsy

As is well know, the parasympathetic and sympathetic systems act in concert to maintain homeostasis and regulate key visceral functions such as heart rate. In particular, the anterior cingulate, insular, posterior orbito-frontal, and the pre-frontal cortices play key roles (along with the amygdala and hypothalamus) in influencing the ANS at the cortical and subcortical level [4]. In this context, the evolutionary aspects of ANS anatomy can be of aid in understanding its hierarchical organization as well as in interpreting its adaptive physiological responses. As described by Porges in what has been termed "polyvagal theory", the ANS passes through three stages. After the first stage (unmyelinated visceral vagus), the second stage is characterized by the sympathetic nervous system, whose function is to increase metabolic output and inhibit the visceral vagus. Finally, the third stage (which is unique to mammals), is characterized by a myelinated vagus that can effect quick cardiac output regulation and hence to foster engagement and/or disengagement with the environment. As the ANS changed through the process of evolution, so did the interplay between the ANS and the other physiological systems that respond to stress, including the previously reported cortical and subcortical structures, the neuropeptides of oxytocin and vasopressin, and the immune system. [14].

In patients with epilepsy, ictal discharges that occur in (or propagate to) these structures can lead to changes in ANS outflow, hence impacting autonomic function. In this respect, changes in cardiac signals represent potential biomarkers that may provide an extra-cerebral indicator of ictal onset in selected patients or patient populations. The first description of the occurrence of asystole during an epileptic seizure was performed 110 years ago by Russel (a medical doctor belonging to the Royal Society of Medicine) as follows: "he uttered a cry and was seen to be rubbing his hands together. His pulse was immediately examined for but was not palpable" [15]. Since then, cardiac arrhythmias have been described during and after epileptic seizures, hence acknowledging that epilepsy and seizures may directly modulate ANS activity [16]. Still, while this brain–heart connection has been explored by a number of studies, overall the results remained mixed and puzzling.

The interaction between seizures and the central ANS is extremely complex. A seizure can include autonomic symptoms at its initiation, during its propagation, as well as during the aftermath (i.e. post-ictal period). The probable paths of propagation of ictal electrical activity appear to involve mainly temporal and/or frontal areas. Also, the insular cortex is involved in both temporal and frontal seizures, and the hippocampus is involved in temporal seizures. Seizure activity can spread through the limbic system involving the amygdala, hypothalamus, and thalamus, hence stimulating the ANS nuclei in the medulla, including the nucleus tractus solitarius (NTS) and nucleus ambiguus. As a consequence, both sympathetic and parasympathetic efferent discharges can be generated [16–19], and epileptic seizures can lead to changes in autonomic function affecting the sympathetic, parasympathetic, and enteric nervous systems. In particular, research involving experimental stimulation of various neural structures suggests that the propagation of epileptic discharges to the right insular cortex may represent a driver of sympathetic-parasympathetic changes that influence heart rate [4]. Also, seizures with a right-sided epilepsy focus induce tachycardia and seizures with a left-sided epilepsy focus induce bradycardia, and this is confirmed by the observation that left insular cortex stimulation is more frequently associated with bradycardia and right insular cortex stimulation is more frequently associated with tachycardia [18]. However, studies regarding brainstem localization and lateralization of tachycardia and bradycardia did not confirm the lateralization theory [4], which therefore cannot fully explain why seizures are often associated with combined tachycardia/bradycardia and/or oscillatory HR patterns. As Leung [20] observed, "the pattern of seizure spread, hand dominance and the presence of a lesion" may modulate the ictal ANS response, which makes "the simplistic right-left dichotomy less likely". Finally, patients with epilepsy experience long-lasting changes in the regulation of the ANS and target organs (e.g., the heart) [4]. A better understanding of the autonomic manifestations of seizures would therefore provide practical added value to clinical epileptologists dealing with differential diagnosis of epilepsy and related disorders.

14.3 Heart Rate Dynamics in Epilepsy

Heart rate (HR) and HRV can be easily measured/estimated when compared to other biomarkers that are commonly associated with seizures (i.e., long-term EEG), and are therefore potentially valuable biomarkers when it comes to characterizing seizures. While HRV is swiftly measured with standard electrocardiography (ECG) devices and specific software, so far HRV changes in epilepsy have shown an unclear picture [4, 9], and have mostly been investigated in the context of SUDEP and of seizure detection/prediction [21, 22]. Understanding the prevalence and magnitude of HRV changes associated with seizures, as well as the timing of such changes relative to seizure onset, is a fundamental step towards developing possible heart rate-based algorithms for seizure detection.

14.3.1 Ictal and Post Ictal Arrhythmias

The most common cardiac changes associated with seizures are ictal tachycardia (IT) and ictal bradycardia (IB), along with resulting blood pressure changes [6]. As indicated by numerous studies, IT and IB are primarily sinus tachycardia and bradycardia, with heart rate exceeding 100 bpm or dropping below 50 bpm [23, 24], respectively. IT may be involved in up to 80% of seizures [4], while IB (<5%) and asystole (0.3–0.4%) are less frequent [4], although the occurrence of IB and asystole may be underestimated because a) and/or they do not necessarily occur in each seizure [16, 24] and b) patients with bradyarrhythmias are usually admitted to coronary care units lacking routine EEG monitoring [25]. In addition, in spite of being more rare, IB and asystole are more clinically relevant, since they may induce syncope, falls, fractures, head injuries and car accidents [26], and SUDEP [13]. Still, literature reports of this type of cardiac disorders are mostly confined to isolated case reports or case series with often unclear definitions of e.g. ictal asystole. For example, a recent review [16] attempted to identify clinical profiles associated with various ictal cardiac arrhythmias, uncovering great dishomogeneity in methodological criteria as well as clinical findings. Overall, ictal cardiac asystole, starting approximately 30s after seizure onset, was described in 103 patients affected by focal epilepsy. The threshold RR interval for defining asystole was >3 s in 2 case series, >4 s in another series and not explicitly defined in the remaining four series. In 90% of cases, seizures originated from temporal regions, irrespective of laterality. In turn, only few cases (13 patients) with post-ictal asystole after a focal seizure with convulsive generalization have been described, out of which 54% (7/13) died due to probable SUDEP, pointing towards a fundamental difference between the mechanisms underlying ictal and post-ictal asystole. Additionally, ictal bradycardia was reported in 25 well-documented cases with temporal onset and discognitive seizures (previously defined as complex partial seizures). Also, ictal atrioventricular conduction block was described in 11 cases, five of which evolved to asystole. These patients showed focal epilepsy, non-convulsive seizures and left sided ictal paroxysmal activity. Atrial fibrillation and atrial flutter were described in 13 and 1 subjects, respectively, and only 4 cases were well documented with videoEEG recordings.

In summary, the mechanisms underlying seizure-related asystole may involve a direct central ANS dysregulation induced through the temporal lobe and/or an indirect vagal reflex (which may evolve to asystole) due to catecholamine release during seizures. Promoting factors such as ion channel mutations, negative inotropic or proarrhythmogenic drugs (i.e. antiepileptic drugs, AEDs) as well as preexisting heart pathology (i.e. conduction disorders), may modulate this propensity and may explain why ictal asystole does not occur during every seizure. Focal stimulation of the limbic system (cingulate gyrus, amygdala, insular and orbitofrontal cortices) may also contribute to pathogenic ANS modulation which evolves to asystole.

14.3.2 Ictal Autonomic Changes

While there are a number of studies which performed ictal HRV analysis in polytreated, drug-resistant and surgery candidate populations [10, 11, 27–31], HRV data regarding untreated, drug-naive, newly diagnosed epilepsy is sparse [12]. In addition, most HRV studies in epilepsy do not report detailed clinical features (i.e. number and type of AEDs, disease duration, epilepsy syndrome). While HRV analysis can act as a promising and non-invasive biomarker for seizure detection and perhaps even prediction, to date most studies include patients selected for EEG evaluation due to the refractory nature of their seizures and hence affected by several confounds as far as ANS modulation is concerned [3]. Accordingly, the authors of [3] posit that high prevalence of tachycardia is likely to be associated with drug-resistant epilepsy. In classical HRV analysis, a number of indices derived from power spectrum analysis of RR interval time series are commonly employed, e.g. low frequency (LF, 0.04–0.15 Hz) HRV components, reflecting a mixture of sympathetic and parasympathetic activity, and high frequency (HF, 0.15–0.4 Hz) components, reflecting vagal (parasympathetic) tone (although often overlapping with respiratory frequencies and hence possibly reflecting vagal efferent control of respiratory sinus arrhythmia) [32]. Our recent study comparing HRV parameters in newly diagnosed and untreated TLE between the interictal, preictal, ictal, and postictal states showed higher preictal, ictal, and postictal LF/HF ratio compared to the interictal state. Similarly, the LF component increased progressively and was significantly higher during the ictal state compared to interictal and preictal states. Overall, our data showed an ictal sympathetic overdrive with partial recovery in the postictal state in patients affected by TLE without common confounding factors related to drug resistance, AEDs and long disease duration [12].

14.3.3 Interictal Autonomic Changes

The prevalence of interictal cardiac arrhythmias in patients with epilepsy is similar to what is observed in the general population [33], and is also confirmed in TLE and extratemporal epilepsy [30]. However, the higher interictal HR observed in patients with various types of epilepsy suggests a disease-related alteration of autonomic balance [34, 35]. In this context, interictal HRV analysis showed partially contrasting results, including combined inhibition of sympathetic and parasympathetic systems [30, 35–39], inhibition of either sympathetic or parasympathetic [37, 39], or low parasympathetic activation associated with sympathetic prevalence [34, 40]. The autonomic alterations showed by HRV studies were confirmed by autonomic tests (i.e. abnormal blood pressure variability in focal epilepsy [41]; decreased baroreflex sensitivity in TLE [40]; diminished BP response to isometric work, deep breathing and Valsalva maneuver in different types of epilepsy [36]).Autonomic alterations were both diurnal and nocturnal [42], and alterations in circadian dynamics of autonomic activity in the epilepsy population were also described [43, 44]. In addition, right-sided epilepsy appears to be more prone to involve ANS alterations as compared to left-sided epilepsy [30, 37], and the entity of these alterations may be of minor importance in early epilepsy stages and grow proportionally to disease duration [45]. This could potentially suggest a role of HRV as a biomarker of disease progression. Also, interictal autonomic alterations are independent of sex, age, and type of epilepsy [4]. Therefore, interictal studies regarding refractory or drug-naïve epileptic patients have provided controversial findings, ranging from no HRV interictal changes [36, 45] to subtle autonomic dysfunction in vagal tone [34, 46]- their clinical significance is therefore considered limited.

14.3.4 Influence of AEDs on Autonomic Changes in Epilepsy

Although in one study HRV changes in TLE were not influenced by drug-resistance [38], interictal baroreflex alterations may be modified by the degree of seizure control [39]. The modulatory potential of AEDs on ANS activity needs to be taken into account when investigating epilepsy-related autonomic dysfunction. While some AEDs have been associated with near-fatal and fatal arrhythmias [47, 48], this finding has been questioned in later studies [49, 50]. A recent comprehensive metanalysis [51] observed a trend towards higher LF values in association with AEDs, however potential confounds such as type of AED, dosage, and treatment duration were not taken into account. While previous studies indicated that polytherapy may represent an independent risk factor for SUDEP [52, 53], further studies investigating HRV changes due to therapeutic interventions are necessary to explain putative treatment effects on autonomic imbalance. Some studies suggested that carbamazepine, a widely utilised AED, involves depressant dromotropic and chronotropic effects [54]. In this context however, Kasarskis et al. [55] identified two types of cardiac changes associated with carbamazepine treatment, namely tachycardia in cases of carbamazepine overdose and bradycardia associated with therapeutic or moderate elevated doses in elderly women. Carbamazepine may therefore increase risk (in vulnerable patients) of developing ictal bradycardia during seizures [56]. Similar findings have been reported for phenytoin [25]. In addition to carbamazepine and phenytoin, other AEDs may also alter autonomic functions or induce proarrhythmic effects [6, 57], although and seizure control with AEDs may also be beneficial in the case of impaired cardiac autonomic function associated with epilepsy [58]. It remains unclear whether this is due to combined side effects of concomitant antiepileptic treatment or it simply reflects a more severe epileptic condition, which often treated with polytherapy [59, 60].

14.4 Nonlinear HRV Analysis in Epilepsy

A recent meta-analysis on HRV indices in epilepsy [51] did not find statistically significant changes in LF versus controls (g – 0.18; 95% CI −0.71 to 0.35). In addition, lower HRV and lower vagal activity were reported in patients with epilepsy (as measured by time domain variables, lower SDNN and lower RMSSD in epileptic patients when compared to controls). Also, drug-refractory patients showed lower SDNN values (g = −0.44; 95% CI = −0.76 to −0.13) [51]. In this context, indices derived from a Poincare plot [61] in which each RR interval is plotted against the subsequent RR interval may convey additional information about ANS function. Specifically, the length of the transverse axis (T) reflects beat-to-beat variability with deviations along this axis predominantly reflecting parasympathetic system influence, and the length of the longitudinal axis (L) reflects the overall range of RRs due to both sympathetic and parasympathetic influences. The measures derived from this plot are called cardiovagal index [CVI = Log10 (LT)] and cardiosympathetic index (CSI = L/T). It has been suggested that these two indices provide complementary information about parasympathetic and sympathetic contributions to HRV as compared to spectral analysis alone [61]. CSI was seen to be higher in temporal lobe seizures when compared with psychogenic non epileptic episodes [62, 63]. In addition, we found that ictal CSI was significantly higher when compared to preictal and interictal states, whereas preictal CVI was lower than in basal and ictal states, confirming a combined seizure-induced activation of the sympathetic nervous system, possibly preceded by a preictal reduced vagal tone. This could suggest that inputs from cortical and/or subcortical structures involved in and/or affected by the epileptic discharge may account for sympathetic imbalance [12]. In addition we also found that interictal CVI was lower in left TLE compared to right TLE. In turn, in left TLE, ictal CVI was higher than interictal CVI, whereas in right TLE, CVI was lower in the preictal state compared to all other states, confirming asymmetric representation of autonomic function [12, 14]. Although some authors have proposed that left-sided forebrain structures, such as the anterior cingulate cortex and anterior insula, are predominantly involved with maintaining parasympathetic tone whereas homotopic right (nondominant) forebrain regions control sympathetic tone and responses [18, 64], the lateralization model of autonomic control remains controversial. This is reflected in the challenge of detecting lateralized fMRI activations within a system characterized by opponent organization (coordinated and balanced coactivation of physiologically complementary modules) [65, 66].

While overtly nonlinear estimates/techniques (e.g. detrended fluctuation analysis, correlation function, Hurst exponents, fractal dimension, Lyapunov exponent, Poincarè plots, and various entropies [67]) applied to HRV analysis have proven to be useful predictors of morbidity and mortality in several clinical studies, very few studies have employed such techniques to the study of ANS changes in epilepsy. Approximate entropy (ApEn) and sample entropy (SampEn) are measures of regularity whose changes have been associated with disease-states such as fetal distress, neonatal acidosis, or postoperative ventricular dysfunction in adults [62, 68, 69]. Interictal ApEn was seen to be reduced in TLE patients, in particular in patients with refractory TLE when compared to patients with well-controlled TLE [38]. A recent study regarding TLE seizures and psychogenic non epileptic episodes found lower ictal ApEnas compared to interictal ApEN, thus indirectly confirming the activation of the sympathetic system and/or the inactivation of the parasympathetic system during epileptic seizures [62]. Similarly, we found in that in untreated TLE ictal ApEn was significantly reduced as compared to both interictal and preictal values, with a partial recovery towards interictal values after a seizure. When combined with data derived from time and frequency-domain HRV studies, the observation of a reduction of ApEn in RR series during seizures may suggest an ictal predominance of the sympathetic nervous system without a postictal recovery to the basal condition and may also aid in designing an initial, easy-to-use approach to develop future tools for seizure detection [12]. Detrended fluctuation analysis is a technique which quantifies fractal properties (i.e. statistical self-affinity) of a time series. One study [38] examined the short term ($\leq$11 beats, $\alpha 1$) and long term (>11 beats, $\alpha 2$) scaling exponents of RR interval data in TLE and found that $\alpha 2$ was lower in patients with well controlled TLE as compared to patients with refractory TLE ($p < 0.05$), indicating additional abnormalities in HR dynamics which may partly contribute to the occurrence of adverse cardiovascular events such as life threatening arrhythmias. Recently, non-linear interactions between EEG and HRV were investigated in children with TLE by means of a Convergent Cross Mapping (CCM) approach [70] (a technique which employs historical records of value of response Y to estimate values of the driver X [71]). This paper found statistical significant directed nonlinear interactions between delta activity (mainly) and alpha activity (less) and HRV in the preictal, ictal and postictal periods. Recently, we employed a probabilistic point-process model for heartbeat dynamics to investigate the alterations in instantaneous autonomic complexity in untreated TLE [72]. Instantaneous ApEn (ipApEn) and Instantaneous SampEn (ipSampEn) were lower during seizures as compared to interictal periods, and the variability (median absolute deviation) of ipApEn was also significantly lower during seizures as compared to basal values. In line with previous findings obtained with static complexity estimates, these findings suggest that ictal events in untreated TLE are associated with a decrease in heartbeat complexity and its variability, possibly pointing toward subtle autonomic changes which may accompany or precede seizures, and can only be detected using an instantaneous, time resolved approach to quantifying autonomic complexity.

In conclusion, while the physiological and clinical applicability of fractal and complexity measures of HR dynamics are not yet completely understood, it is reasonable to assume that such indices highlight abnormal patterns of RR interval behaviour that are not easily detected by commonly used moment statistics of HR variation [38]. These findings may provide new insight regarding physiological and seizure-induced states of the complex brain-heart network underlying epilepsy and related autonomic modifications [71].

References

1. Ngugi, A.K., et al.: Estimation of the burden of active and life-time epilepsy: a meta-analytic approach. Epilepsia. **51**(5), 883–890 (2010)
2. Berg, A.T., et al.: Revised terminology and concepts for organization of seizures and epilepsies: report of the ILAE Commission on Classification and Terminology, 2005–2009. Epilepsia. **51**(4), 676–685 (2010)
3. Eggleston, K.S., Olin, B.D., Fisher, R.S.: Ictal tachycardia: the head-heart connection. Seizure. **23**(7), 496–505 (2014)
4. Sevcencu, C., Struijk, J.J.: Autonomic alterations and cardiac changes in epilepsy. Epilepsia. **51**(5), 725–737 (2010)
5. Nashef, L., et al.: Unifying the definitions of sudden unexpected death in epilepsy. Epilepsia. **53**(2), 227–233 (2012)
6. Devinsky, O.: Effects of seizures on autonomic and cardiovascular function. Epilepsy Curr. **4**(2), 43–46 (2004)
7. Sperling, M.R.: Sudden unexplained death in epilepsy. Epilepsy Curr. **1**(1), 21–23 (2001)
8. Scorza, F.A., et al.: The brain-heart connection: implications for understanding sudden unexpected death in epilepsy. Cardiol. J. **16**(5), 394–399 (2009)
9. Jansen, K., Lagae, L.: Cardiac changes in epilepsy. Seizure. **19**(8), 455–460 (2010)
10. Jeppesen, J., et al.: Heart rate variability analysis indicates preictal parasympathetic overdrive preceding seizure-induced cardiac dysrhythmias leading to sudden unexpected death in a patient with epilepsy. Epilepsia. **55**(7), e67–e71 (2014)
11. Surges, R., Jordan, A., Elger, C.E.: Ictal modulation of cardiac repolarization, but not of heart rate, is lateralized in mesial temporal lobe epilepsy. PLoS One. **8**(5), e64765 (2013)
12. Romigi, A., et al.: Heart rate variability in untreated newly diagnosed temporal lobe epilepsy: evidence for ictal sympathetic dysregulation. Epilepsia. **57**(3), 418–426 (2016)
13. Shorvon, S., Tomson, T.: Sudden unexpected death in epilepsy. Lancet. **378**(9808), 2028–2038 (2011)
14. Porges, S.W.: The polyvagal theory: phylogenetic substrates of a social nervous system. Int. J. Psychophysiol. **42**(2), 123–146 (2001)
15. Russell, A.E.: The pathology of epilepsy. Proc. R. Soc. Med. **1**(Med Sect), 72–118 (1908)
16. van der Lende, M., et al.: Cardiac arrhythmias during or after epileptic seizures. J. Neurol. Neurosurg. Psychiatry. **87**(1), 69–74 (2016)
17. Pool, J.L., Ransohoff, J.: Autonomic effects on stimulating rostral portion of cingulate gyri in man. J. Neurophysiol. **12**(6), 385–392 (1949)
18. Oppenheimer, S.M., et al.: Cardiovascular effects of human insular cortex stimulation. Neurology. **42**(9), 1727–1732 (1992)
19. Altenmuller, D.M., Zehender, M., Schulze-Bonhage, A.: High-grade atrioventricular block triggered by spontaneous and stimulation-induced epileptic activity in the left temporal lobe. Epilepsia. **45**(12), 1640–1644 (2004)
20. Leung, H., Kwan, P., Elger, C.E.: Finding the missing link between ictal bradyarrhythmia, ictal asystole, and sudden unexpected death in epilepsy. Epilepsy Behav. **9**(1), 19–30 (2006)
21. Moseley, B., et al.: Autonomic epileptic seizures, autonomic effects of seizures, and SUDEP. Epilepsy Behav. **26**(3), 375–385 (2013)
22. Doyle, O.M., et al.: Heart rate based automatic seizure detection in the newborn. Med. Eng. Phys. **32**(8), 829–839 (2010)
23. Rocamora, R., et al.: Cardiac asystole in epilepsy: clinical and neurophysiologic features. Epilepsia. **44**(2), 179–185 (2003)
24. Schuele, S.U., et al.: Video-electrographic and clinical features in patients with ictal asystole. Neurology. **69**(5), 434–441 (2007)
25. Tinuper, P., et al.: Ictal bradycardia in partial epileptic seizures: autonomic investigation in three cases and literature review. Brain. **124**(Pt 12), 2361–2371 (2001)

26. Moseley, B.D., et al.: The treatment of ictal asystole with cardiac pacing. Epilepsia. **52**(4), e16–e19 (2011)
27. Adjei, P., et al.: Do subclinical electrographic seizure patterns affect heart rate and its variability? Epilepsy Res. **87**(2–3), 281–285 (2009)
28. Di Gennaro, G., et al.: Ictal heart rate increase precedes EEG discharge in drug-resistant mesial temporal lobe seizures. Clin. Neurophysiol. **115**(5), 1169–1177 (2004)
29. Jeppesen, J., et al.: Detection of epileptic-seizures by means of power spectrum analysis of heart rate variability: a pilot study. Technol. Health Care. **18**(6), 417–426 (2010)
30. Massetani, R., et al.: Alteration of cardiac function in patients with temporal lobe epilepsy: different roles of EEG-ECG monitoring and spectral analysis of RR variability. Epilepsia. **38**(3), 363–369 (1997)
31. Toth, V., et al.: Periictal heart rate variability analysis suggests long-term postictal autonomic disturbance in epilepsy. Eur. J. Neurol. **17**(6), 780–787 (2010)
32. Heart rate variability. Standards of measurement, physiological interpretation, and clinical use. Task Force of the European Society of Cardiology and the North American Society of Pacing and Electrophysiology. Eur. Heart J. **17**(3), 354–381 (1996)
33. Blumhardt, L.D., Smith, P.E., Owen, L.: Electrocardiographic accompaniments of temporal lobe epileptic seizures. Lancet. **1**(8489), 1051–1056 (1986)
34. Evrengul, H., et al.: Time and frequency domain analyses of heart rate variability in patients with epilepsy. Epilepsy Res. **63**(2–3), 131–139 (2005)
35. Harnod, T., et al.: Heart rate variability in children with refractory generalized epilepsy. Seizure. **17**(4), 297–301 (2008)
36. Isojarvi, J.I., et al.: Interictal cardiovascular autonomic responses in patients with epilepsy. Epilepsia. **39**(4), 420–426 (1998)
37. Tomson, T., et al.: Heart rate variability in patients with epilepsy. Epilepsy Res. **30**(1), 77–83 (1998)
38. Ansakorpi, H., et al.: Heart rate dynamics in refractory and well controlled temporal lobe epilepsy. J. Neurol. Neurosurg. Psychiatry. **72**(1), 26–30 (2002)
39. El-Sayed, H.L., et al.: Non-invasive assessment of cardioregulatory autonomic functions in children with epilepsy. Acta Neurol. Scand. **115**(6), 377–384 (2007)
40. Dutsch, M., Hilz, M.J., Devinsky, O.: Impaired baroreflex function in temporal lobe epilepsy. J. Neurol. **253**(10), 1300–1308 (2006)
41. Devinsky, O., Perrine, K., Theodore, W.H.: Interictal autonomic nervous system function in patients with epilepsy. Epilepsia. **35**(1), 199–204 (1994)
42. Ferri, R., et al.: Heart rate variability during sleep in children with partial epilepsy. J. Sleep Res. **11**(2), 153–160 (2002)
43. Ronkainen, E., et al.: Suppressed circadian heart rate dynamics in temporal lobe epilepsy. J. Neurol. Neurosurg. Psychiatry. **76**(10), 1382–1386 (2005)
44. Persson, H., et al.: Circadian variation in heart-rate variability in localization-related epilepsy. Epilepsia. **48**(5), 917–922 (2007)
45. Persson, H., Ericson, M., Tomson, T.: Heart rate variability in patients with untreated epilepsy. Seizure. **16**(6), 504–508 (2007)
46. Mativo, P., et al.: Study of cardiac autonomic function in drug-naive, newly diagnosed epilepsy patients. Epileptic Disord. **12**(3), 212–216 (2010)
47. Nilsson, L., et al.: Risk factors for sudden unexpected death in epilepsy: a case-control study. Lancet. **353**(9156), 888–893 (1999)
48. Langan, Y., Nashef, L., Sander, J.W.: Case-control study of SUDEP. Neurology. **64**(7), 1131–1133 (2005)
49. Sander, J.W.: The natural history of epilepsy in the era of new antiepileptic drugs and surgical treatment. Epilepsia. **44**(Suppl 1), 17–20 (2003)
50. Colugnati, D.B., et al.: Carbamazepine does not alter the intrinsic cardiac function in rats with epilepsy. Arq. Neuropsiquiatr. **68**(4), 573–578 (2010)
51. Lotufo, P.A., et al.: A systematic review and meta-analysis of heart rate variability in epilepsy and antiepileptic drugs. Epilepsia. **53**(2), 272–282 (2012)

52. Surges, R., et al.: Sudden unexpected death in epilepsy: risk factors and potential pathomechanisms. Nat. Rev. Neurol. **5**(9), 492–504 (2009)
53. Hesdorffer, D.C., Tomson, T.: Adjunctive antiepileptic drug therapy and prevention of SUDEP. Lancet Neurol. **10**(11), 948–949 (2011)
54. Kenneback, G., et al.: Electrophysiologic effects and clinical hazards of carbamazepine treatment for neurologic disorders in patients with abnormalities of the cardiac conduction system. Am. Heart J. **121**(5), 1421–1429 (1991)
55. Kasarskis, E.J., et al.: Carbamazepine-induced cardiac dysfunction. Characterization of two distinct clinical syndromes. Arch. Intern. Med. **152**(1), 186–191 (1992)
56. Persson, H., Ericson, M., Tomson, T.: Carbamazepine affects autonomic cardiac control in patients with newly diagnosed epilepsy. Epilepsy Res. **57**(1), 69–75 (2003)
57. Tomson, T., Kenneback, G.: Arrhythmia, heart rate variability, and antiepileptic drugs. Epilepsia. **38**(11 Suppl), S48–S51 (1997)
58. Hallioglu, O., et al.: Effects of antiepileptic drug therapy on heart rate variability in children with epilepsy. Epilepsy Res. **79**(1), 49–54 (2008)
59. Tomson, T., Nashef, L., Ryvlin, P.: Sudden unexpected death in epilepsy: current knowledge and future directions. Lancet Neurol. **7**(11), 1021–1031 (2008)
60. Ryvlin, P., Montavont, A., Kahane, P.: Sudden unexpected death in epilepsy: from mechanisms to prevention. Curr. Opin. Neurol. **19**(2), 194–199 (2006)
61. Toichi, M., et al.: A new method of assessing cardiac autonomic function and its comparison with spectral analysis and coefficient of variation of R-R interval. J. Auton. Nerv. Syst. **62**(1–2), 79–84 (1997)
62. Ponnusamy, A., Marques, J.L., Reuber, M.: Comparison of heart rate variability parameters during complex partial seizures and psychogenic nonepileptic seizures. Epilepsia. **53**(8), 1314–1321 (2012)
63. Jeppesen, J., et al.: Comparing maximum autonomic activity of psychogenic non-epileptic seizures and epileptic seizures using heart rate variability. Seizure. **37**, 13–19 (2016)
64. Craig, A.D.: Forebrain emotional asymmetry: a neuroanatomical basis? Trends Cogn. Sci. **9**(12), 566–571 (2005)
65. Thayer, J.F., Lane, R.D.: Claude Bernard and the heart-brain connection: further elaboration of a model of neurovisceral integration. Neurosci. Biobehav. Rev. **33**(2), 81–88 (2009)
66. Guo, C.C., et al.: Dominant hemisphere lateralization of cortical parasympathetic control as revealed by frontotemporal dementia. Proc. Natl. Acad. Sci. U. S. A. **113**(17), E2430–E2439 (2016)
67. Vanderlei, L.C., et al.: Basic notions of heart rate variability and its clinical applicability. Rev. Bras. Cir. Cardiovasc. **24**(2), 205–217 (2009)
68. Pincus, S.M., Viscarello, R.R.: Approximate entropy: a regularity measure for fetal heart rate analysis. Obstet. Gynecol. **79**(2), 249–255 (1992)
69. Fleisher, L.A., Pincus, S.M., Rosenbaum, S.H.: Approximate entropy of heart rate as a correlate of postoperative ventricular dysfunction. Anesthesiology. **78**(4), 683–692 (1993)
70. Sugihara, G., et al.: Detecting causality in complex ecosystems. Science. **338**(6106), 496–500 (2012)
71. Schiecke, K., et al.: Nonlinear directed interactions between heart rate variability and EEG activity in children with temporal lobe epilepsy. I.E.E.E. Trans. Biomed. Eng. **63**(12), 2497–2504 (2016)
72. Barbieri, R., et al: Lower instantaneous entropy of heartbeat dynamics during seizures in untreated temporal lobe epilepsy. In: 2015 Computing in Cardiology Conference (CinC). (2015).

Chapter 15
Applications of Nonlinear Methods to Atrial Fibrillation

Raúl Alcaraz and José J. Rieta

Abstract Atrial fibrillation (AF) is the most common arrhythmia in clinical practice, affecting roughly 1.5–2% of the general population and around 15% of the elderly. Despite its high prevalence, mechanisms triggering and maintaining the arrhythmia are not fully known, thus making current treatments often ineffective. Nonetheless, in last years intensive research in signal processing applied both to surface electrocardiogram (ECG) and intracardiac electrogram (EGM) recordings has resulted in an improved knowledge of this cardiac disorder. Indeed, by taking advantage of the fact that the heart during AF behaves in an inherently complex, heterogeneous and non-stationary way, relevant advances has been reached by nonlinear analysis of that recordings. The main goal of this chapter is hence to review the most significant achievements in some important AF facets, such as identification of events associated to its natural progression, accurate quantification of atrial substrate electrophysiological features and customization of current therapies.

15.1 Introduction

Atrial Fibrillation (AF) is nowadays a major cardiovascular challenge in the developed world [1]. It is the most common supra-ventricular arrhythmia in clinical practice and affects approximately 1.5–2% of the general population [2]. Over 33 million people worldwide suffer currently from this arrhythmia [3]. Moreover, it is expected that AF may affect over 6 million people in the USA and 18 million in Europe by the middle of this century [1]. Indeed, the lifetime risk of developing AF after 40 years of age is 26% for men and 23% for women [4]. Also, AF-related symptoms and morbidity are responsible for frequent visits to the physician and

R. Alcaraz (✉)
Research Group in Electronic, Biomedical and Telecommunication Engineering, University of Castilla-La Mancha, Cuenca, Spain
e-mail: raul.alcaraz@uclm.es

J.J. Rieta
BioMIT.org, Electronic Engineering Department, Universitat Politécnica de Valencia, Gandía, Spain
e-mail: jjrieta@upv.es

R. Barbieri et al. (eds.), *Complexity and Nonlinearity in Cardiovascular Signals*,
DOI 10.1007/978-3-319-58709-7_15

hospitalizations leading to substantial and rising costs [5]. Thus, this arrhythmia is associated with a fivefold risk of stroke and a threefold incidence of congestive heart failure, provoking that AF patients have twice the risk of death than a healthy person of the same age [6]. It is also a frequent complication of cardiothoracic surgery in 60% of the cases [7]. Hence, given the increasing incidence of this arrhythmia, at least 15% of the healthcare budget in cardiac diseases is spent on AF management [8].

AF has been traditionally described in terms of the duration of its episodes [7]. Firstly, paroxysmal AF (PAF) is often the first manifestation of the arrhythmia, with episodes that terminate spontaneously or with intervention in less than 1 week. In persistent AF, the episodes last for more than 7 days and intervention is required for its termination. If the arrhythmia persists for more than a year, then is named longstanding persistent AF. Finally, the term permanent AF is used when patient and clinician make a joint decision to stop more attempts of sinus rhythm (SR) restoration and/or maintenance [7]. About 20% of PAF evolves to permanent AF in 4 years. Therefore, early interventions aiming at mitigating AF progression are recommended to prevent cardiovascular complications [9].

Despite its high prevalence, the physiological mechanisms underlying AF are not fully explained yet [10]. Nowadays, it is known that AF produces electrophysiological atrial alterations, which provoke that this arrhythmia presents a progressive nature and perpetuates its-self [9]. In fact, several experimental studies have proven that AF promotes both electrical (changes in ion channel expression or activity) and structural (interstitial fibrosis, myocyte hypertrophy, apoptosis of the atria, etc.) remodeling [7], which provoke that normal electrical activation degenerates into a chaotic pattern depolarizing uncoordinatedly the atrial cells [11]. However, it is still unknown how this chaotic pattern starts, sustains and sometimes terminates spontaneously. There exist several theories which can be divided into three groups: focal activation (repetitive ectopic discharges), reentrant mechanisms (mother-wave, rotor, multiple wavelets) and longitudinal/transmural dissociated conduction [10]. Nonetheless, notable controversies among them make current AF therapeutic approaches poorly effective [11], moreover presenting significant adverse effect liability [12].

Intensive research in computational methods to characterize AF has been developed during past years with the aim of gaining new insights about AF mechanisms as well as reaching tailored therapeutical solutions. Thus, the application of nonlinear methods to AF recordings has recently received great attention. Indeed, mathematical models of electrophysiological dynamics in cardiac myocytes mainly consider that nonlinear dynamics occur on multiple spacial scales: sub-cellular (e.g., at the single-channel level), cellular and in the whole-heart [13]. For instance, the currents moving through the different ionic channels and transporters are almost always a nonlinear function of voltage [13]. Another example of a nonlinearity in action potential dynamics is the threshold-range effect seen in the generation of the action potential upstroke [13]. Furthermore, it has also been shown that the typical steep conduction velocity dispersion during AF represents one way of forming a spatially heterogeneous pattern in a completely homogeneous tissue [14], the theory of such pattern formation being a well-described part of nonlinear systems theory [15, 16]. Thus, given that a non-uniform and anisotropic atrial conduction occurs during AF [15, 16], it is clear that the heart's behavior is far from being linear during this arrhythmia [7]. The purpose of this chapter is therefore to review the most recent advances released through the use of nonlinear methods in the analysis of AF.

The document is structured in four parts. How nonlinear indices are applied to the most typical invasive and non-invasive recordings used to study AF will be firstly described. The ability of these metrics to detect useful events associated to this arrhythmia, such as its own spontaneous occurrence, onset and termination, will then be outlined. Thereafter, the most famous methodologies proposed to study dynamics and repetitiveness of AF patterns, which have been useful to quantify alterations in atrial substrate or arrhythmia's progression, will be presented. Finally, the main advances reached by nonlinear indices applied to improve the effectiveness of current AF treatments will be introduced.

15.2 Recording Methodologies in AF

AF has been commonly studied by two kinds of recordings, namely the surface electrocardiogram (ECG) and invasive intraatrial recordings or electrograms (EGM). Nonlinear methods have been applied on both types of recordings to gain new insights about the mechanisms of this arrhythmia. This section reviews the most widely used preprocessing steps required by both recording methodologies as well as how nonlinear metrics are computed on each one of them.

15.2.1 Surface Recording of AF

The ECG recording is a noninvasive and widely used tool to analyze AF, which records the heart's electrical activity through several leads placed on the body surface. In contrast to the EGM, interesting advantages of the ECG are its ability to record data for long periods of time as well as the reduced costs and risks involved for the patient [17]. However, this signal is easily corrupted by noise because it is acquired on the patient's thorax. Thus, to make its analysis easier, the ECG needs to be preprocessed. Filtering operations are typically used to reduce baseline wandering, high-frequency noise and power-line interference [18].

Another interesting feature of the ECG is its ability to simultaneously record atrial and ventricular activity during each cardiac cycle, thus allowing to study the isolated features of each one or, contrarily, their synchronization. Thus, atrial activity (AA) for a healthy subject is represented by the P-wave and ventricular activity by the QRS complex and the T-wave. More precisely, the P-wave and QRS complex represent atrial and ventricular depolarizations, respectively, whereas

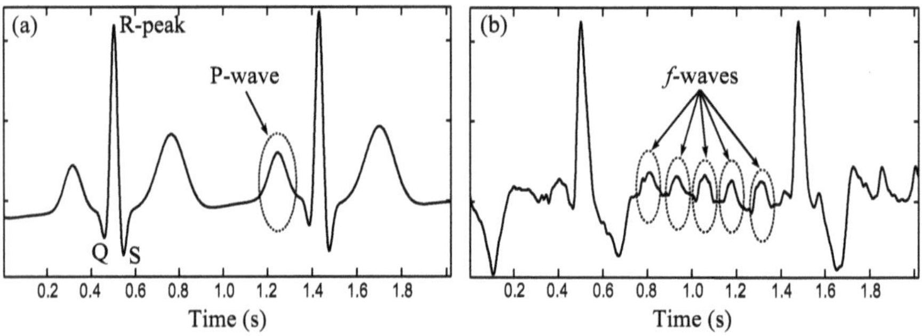

Fig. 15.1 Example of surface ECG recordings corresponding to (**a**) a subject in SR and (**b**) a patient in AF [19]

the T-wave indicates ventricular repolarization. Note that atrial repolarization is overlapped with ventricular depolarization and, therefore, masked by the QRS complex.

When AF occurs, the P-wave is replaced by a sawtooth-like pattern, which is variable in time, shape and timing, named fibrillatory (*f*) waves [17], such as Fig. 15.1 shows. This wave is overlapped in time and frequency with the ventricular activity and, therefore, their dissociation is required for further analysis of AF. Numerous methods for this purpose are available in the literature. On the one hand, some algorithms exploit multi-lead ECG spatial diversity through tools such as principal component analysis or blind source separation to obtain a global AA signal [20, 21]. On the other hand, for ECG recordings with a reduced number of leads (even only one), the most used methods are based on average beat subtraction from each individual QRS complex [21]. Although with these algorithms *f* waves are tipically extracted from the closest leads to the atria (i.e. leads V1 and II), to highlight their waveform and reduce noise as much as posible, the main atrial wave (MAW) has also been widely analyzed in previous works [19]. Thus, this signal is computed from the AA by applying a 3 Hz bandwidth filtering centered on its dominant frequency [19]. Nonetheless, it has to be noted that a variety of nonlinear metrics have been applied on both AA and MAW waveforms.

During AF a fast and irregular ventricular rhythm appears in most of the cases [22]. Indeed, heart rate variability (HRV) has been widely used to study AF patients, since this analysis is nowadays a clinically accepted tool to characterize the autonomic nervous system's performance [23]. HRV is based on the variability study of successive R-peaks, which can be automatically detected by means of a wide variety of different algorithms [24]. However, the obtained RR interval time series can sometimes contain several artifacts, including those of both physiological and technical origins [25]. On the one hand, technical artifacts may result from poorly fastened electrodes or be due to motion of the subject. On the other hand, ectopic beats are the most common physiological artifact. Many previous works have reported that the presence of these false beats mask the reliable HRV analysis, thus their removal or correction being strongly advised [25]. To this respect, numerous algorithms have been proposed to date [25, 26], such that the resulting edited RR interval series have been widely characterized from a nonlinear point of view by mainly entropy-based measures and detrended fluctuation analysis (DFA) [27].

15.2.2 Invasive Recording of AF

Several invasive ways of intra-atrial activity recording are currently used to study AF. They include bipolar and unipolar signals from endocardial and epicardial electrodes, optical as well as non-contact mapping techniques [28]. Whereas unipolar signals are commonly featured by a notable far-field contamination of ventricular activity, bipolar recordings mainly acquire local atrial activations from the region where electrodes are placed. However, bipolar signals from some areas near to the ventricles, such as the coronary sinus, can also be slightly contaminated by ventricular activity. Hence, a first step for the processing of both unipolar and bipolar recordings is to cancel out ventricular activity. To this respect, few methods based on average beat subtraction and blind signal separation have been proposed to date [29, 30]. The resulting signal has been commonly considered as a time series and nonlinear indices such as entropy-based measures or from recurrent pattern analysis have been directly applied to this signal in numerous clinical contexts, as will be described in the next sections.

On the other hand, the identification of local activation waves (LAWs) in EGMs has also proven to be useful in the study of AF. Indeed, the reconstruction of activation and frequency maps by using these waves has allowed to detect areas with delayed or blocked conduction, fibrosis and other electrophysiological alterations [31]. Hence, the most widely known algorithm for identification of LAWs is commonly used as a preproccessing step for EGMs [32]. This approach consists of three steps (EGM high-pass filtering between 40 and 250 Hz, rectification and low-pass filtering with a cut-off frequency of 20 Hz). Surprisingly, although this algorithm was only designed to identify LAWs [32], some authors have applied different indices on the resulting signal to quantify the presence of nonlinear dynamics, as will be described later.

Finally, given that current recording systems have the ability to acquire simultaneously multiple signals, some authors have also used nonlinear synchronization metrics to judge the relationship between the electrical activity registered at different atrial regions. Approaches derived in such a way emphasize the concepts of relative temporal behavior and spatial coordination between atrial activations occurring at different sites. Thus, the use of cross entropy-based measures as well as features obtained from cross recurrence plots and joint symbolic analysis has revealed complementary information to the yielded by nonlinear indices computed from single EGMs [33].

15.3 Identification of Events in AF

Automatic detection of events such as the spontaneous onset or termination of PAF can be useful to improve and optimize the management of PAF patients. To this respect, the anticipation of the arrhythmia's onset may allow the use of a proper therapy in order to avoid the recurrence of AF, thus reducing the risk of its perpetuation. Similarly, unnecessary treatments could be prevented for patients with a high likelihood of spontaneous termination of PAF and a low risk of its recurrence, thus reducing physical and emotional distress for patients as well as considerable costs for health care services. Within this context, in the last years some nonlinear indices have reported a promising ability to predict these events, as will be described next.

15.3.1 Automatic Detection of AF

The electrophysiological alterations associated with AF can sometimes provoke symptoms such as palpitations, weakness, dizziness, fatigue or chest pain [7]. However, some studies have reported that up to 40% of AF patients [34] and up to 90% of PAF episodes [35] may be asymptomatic. As a consequence, AF is often found during routine clinical examinations, thus leading to a delayed diagnosis and treatment. However, as previously mentioned, the early diagnosis of PAF is a priority to prevent more serious complications [9, 36]. Although the use of long-term ECG recordings can be useful for that purpose, the early identification of PAF is not an easy task. Indeed, the first PAF episodes often consist of only a few beats in length and, therefore, their detection by visual inspection is very time-consuming and, hence, not affordable [37].

Automatic PAF detection involves a significant advance, because arrhythmic episodes can be early detected and managed by using real-time continuous ECG monitoring systems. In fact, a wide variety of automatic AF detectors have been proposed during last years, such as Table 15.1 summarizes. Most of them are based on identifying the irregular ventricular rhythm during AF by means of entropy-based metrics, mainly Shannon entropy (ShEn) and sample entropy (SampEn). Thus, Dash et al. [37] constructed a histogram of RR interval series and, after discarding some outliers, they computed ShEn by considering 16 equally spaced bins. A simple combination of this entropy with other two parameters, such as the root mean square of successive RR differences and turning points ratio, achieved a diagnostic accuracy about 90 and 95% from two freely available databases, i.e., the MIT-BIH Arrhythmia and MIT-BIH AF databases. Lee et al. [38] also used the same entropy-based indices to identify AF episodes from the pulsatile photoplethysmogram signal recorded with an iPhone 4S. In this case, ShEn provided a slightly poorer result than SampEn for several datasets. Thus, whereas SampEn showed a discriminant ability about 95%, ShEn only reached an accuracy around 90%. More recently, Zhou et al. [39] have also used ShEn to discern between PAF and SR episodes. However,

Table 15.1 Comparison of the most relevant methods proposed to automatically detect AF

Works	Methodology	Diagnostic accuracy
Dash et al. [37]	Application of ShEn to RR interval series	90–95%
Lee et al. [38]	Application of ShEn and SampEn to RR interval series (computed from pulsatile photoplethysmogram signals)	85–100%
Zhou et al. [39]	Application of ShEn to symbolized RR interval series	≈98%
Kikillus et al. [40]	Characterization of Poincaré plots obtained from RR interval series	85–90%
Esperer et al. [41]	Characterization of Poincaré plots obtained from RR interval series	95–100%
Park et al. [42]	Characterization of Poincaré plots obtained from RR interval series	≈92%
Lian et al. [43]	Characterization of scatter plots of RR interval series	≈95%
Costa et al. [44]	Application of MSE to RR interval series	–
Lake and Moorman [45]	Computation of COSEn from RR interval series	≈90%
Petrenas et al. [46]	Computation of COSEn from RR interval series together with other metrics from P- and *f*-waves	≈90%
Ródenas et al. [47]	Application of WavEn to median TQ interval	≈93%

this index was not computed directly from the RR interval series, because it was firstly symbolized and transformed to a 3 symbol-length sequence. A discriminant ability around 97% was achieved for several freely datasets.

Another nonlinear tool widely used to analyze RR series is the Poincaré plot (PP). This tool characterizes the geometrical structure of the time series throughout its graphical representation, thus not being an entropy-based metric. More precisely, the time series is displayed in a phase space, where the embedding dimension plays a key role. Although several algorithms have been proposed for the optimal selection of this parameter, dimensions between 2 and 4 are often used to deal with physiological signals [48]. Regarding PP characterization, Kikillus et al. [40] estimated the density of points within every segment of the PP and derived an AF detector by using the standard deviation of the temporal differences of consecutive inter-beat intervals. The method's sensitivity depended on the AF burden, but it was always higher than 83%. Esperer et al. [41] also analyzed PP from 2700 patients with atrial and ventricular tachyarrhythmias together with 200 healthy controls in SR. Each PP was characterized according to its shape and other geometrical parameters, which were able to report different shapes both for AF and SR. Thus, a fan pattern indicated AF with a sensitivity of 100% and a specificity of 97%, and a comet shape was exclusively associated to SR. Additionally, Park et al. [42] also proposed the combination of two indices from PP, such as the mean stepping increment and dispersion of the points, by a support vector machine (SVM) classifier to identify around 92% of both AF and SR episodes. On the other hand, Lian et al. [43] analyzed a similar graph to the PP by mapping RR interval series versus its changes. Precisely, the authors divided the map by a grid with 25 ms resolution in two axes and nonempty cells were counted to discern between AF and SR episodes with an ability around 95%.

As described, RR-based algorithms have shown a high discriminant power. However, they also required notably long heart rate recordings, because their performance was significantly reduced under short data sets. To this respect, the same behavior was observed for multi-scale entropy indices [44]. Thus, under long RR time series and for frequent matches, regularity-based nonlinear metrics were able to distinguish quite efficiently AF from SR [44]. However, the challenge is to assure a sufficient number of matches under short data sets [45]. In fact, brief AF episodes are quite common in the early stages of PAF [46] and, furthermore, recent works have reported a significant relationship between risk of thrombus formation and the existence of brief AF episodes [49]. Hence, only by analyzing time series as short as the length of the shortest AF episodes, their proper detection would be possible. Within this context, Lake and Moorman [45] developed generalized standards for the proper selection of the template length m and the tolerance matching r to detect as quick as possible the transition of AF to SR and vice versa. The most relevant contribution was to propose the variation of r, so that enough matches could be found for a confident estimation of entropy, including the adaptation to a density of the final probability, by division of the matching region volume, $2r^m$. This optimized estimation of SampEn together with the estimation of the mean heart beat interval contributed to reach an improved detection of AF with only 12 heart beats. This methodology, named the coefficient of SampEn (COSEn), yielded an accuracy for the discrimination between AF and SR of about 90%. Furthermore, the most frequent mistakes of the method were due to atrial or ventricular ectopy, which increased SampEn despite the presence of SR, and atrial flutter, which can provoke low or high results of SampEn as a function of the dynamics of the atrio-ventricular (AV) conduction.

In order to overcome the need of long AF episodes to achieve a reliable detection of AF, other authors have also combined recently the information from the RR series variability with the P-wave absence analysis. Thus, Jiang et al. [50] introduced a method that first detected the suspected AF transitions and then classified them using a combination analysis of P-wave and RR interval. However, this algorithm still provided some problems to detect PAF episodes shorter than 30 s. Contrarily, in a more recent work an algorithm combining information from RR interval irregularity, P-wave absence, f-wave presence and noise level has shown to be able to identify PAF episodes as brief as 5 beats with an accuracy around 90% [46]. However, these methods still fail to work acceptably under regular ventricular rhythms. Unfortunately, this situation is frequent in AF patients with AV block as well as in AV junctional or ventricular tachycardia [50, 51]. In addition, the implantation of pacemakers and the administration of drugs to stabilize the heart rate during AF also reduce notably the ventricular irregularity normally associated to AF [50, 51]. Similarly, these algorithms are also unable to properly identify SR within an irregular ventricular response, which can occur in the presence of a high number of ectopic beats [52] or in sinus arrhythmia [50].

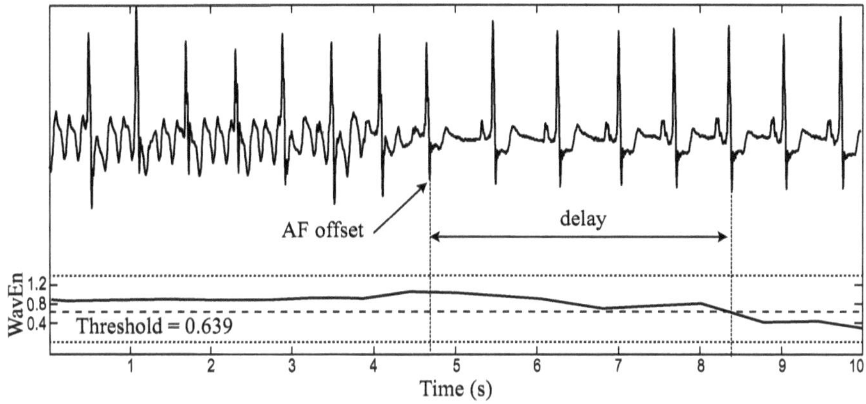

Fig. 15.2 Example where the established threshold of WavEn presents a delay of 5 beats to detect the offset of an AF episode [47]

In contrast to these aforesaid methods, those others based on exclusively detecting the P-wave absence do not suffer from these problems [51]. Indeed, the beat interval containing the AA, i.e., the TQ interval, is only analyzed and, therefore, the performance of these methods is independent on the heart rate and its variability. However, not many algorithms working under this strategy can be found in the literature [47, 51, 53], because the low signal-to-noise ratio obtained from the atrial activation waves in the surface ECG degrades notably their performance [54]. Interestingly, a recent work has overcome this problem by averaging ten noise-free TQ intervals and characterizing the resulting waveform with wavelet entropy (WavEn) [47]. In this way, the P- and f-waves are highlighted or attenuated, respectively, thus achieving a successful identification of the patient's rhythm for around 95% of PAF and SR episodes. Moreover, the algorithm was also able to identify the transition of AF to SR or vice versa with a mean delay of around 5 beats, such as Fig. 15.2 shows. Hence, this method also allowed to detect PAF or SR episodes as short as 5 beats in length.

On the other hand, AF identification from other arrhythmias is also an interesting challenge. To this respect, although AF and atrial flutter are associated with different electrophysiological mechanisms, they use to appear together in a high number of patients. Moreover, given that AF and atrial flutter are treated with different therapies, their discrimination is crucial. With that purpose, Kao et al. [55] computed the correlation dimension (CD), the largest Lyapunov exponent (LLE) and the Lempel-Ziv Complexity (LZC) directly from the AA signal. According to the organization levels of AF that were expected, their results proved that during AF, the analyzed nonlinear indices reported high values, which decreased for typical flutter and were in between for atypical flutter. Furthermore, these parameters were combined through a neural network classifier providing the ability to discriminate between these arrhythmias with a remarkable diagnostic accuracy of about 95%.

15.3.2 Prediction of PAF Termination

After the onset of a PAF episode, the early prediction of its maintenance or termination is crucial to manage the most appropriate patient's treatment [56]. This interesting challenge has been widely addressed in recent years through the application of nonlinear metrics both to atrial and ventricular activities recorded from the surface ECG, such as Table 15.2 overviews. Thus, RR interval series regularity during different times within PAF episodes has been quantified using mainly entropy-based metrics. For instance, Mainardi et al. [57] applied conditional entropy (ConEn) to 1 min-length RR interval series just before PAF and, at least, 1 h far from that event, thus finding higher regularity for non-terminating PAF episodes than for terminating ones. Similarly, Sun and Wang [58] characterized RR interval series by means of eleven PP-based features and used a sequential forward search algorithm to obtain a discriminant ability able to predict PAF termination. Another work in this line was also presented by Sun and Wang [59], who analyzed the sign sequence obtained by thresholding RR interval series. Indeed, PAF termination was successfully predicted by means of a SVM classifier that combined the complexity and ShEn obtained from that sequence.

Regarding the direct analysis of AA, numerous predictors of PAF termination based on quantifying its organization can be found in the literature. Although no standard definition of AA organization exists, nonlinear methods have been used to identify the amount of repetitive patterns which exist in the signal. Indeed, previous works have demonstrated that recordings acquired during organized atrial rhythms provide f waves presenting clearly defined and repetitive morphological patterns. In contrast, recordings acquired during highly disorganized AA with presence of

Table 15.2 Comparison of the most relevant methods proposed to predict PAF termination

Works	Methodology	Diagnostic accuracy (%)
Mainardi et al. [57]	Application of ConEn to RR interval series	≈87
Sun and Wang [58]	Characterization of Poincaré plots obtained from RR interval series	≈92
Sun and Wang [59]	Application of ShEn to the sign sequence computed from RR interval series	90
Alcaraz and Rieta [60]	Application of SampEn to the MAW	90
Alcaraz and Rieta [61]	Application of SampEn to the MAW and its wavelet decomposition	≈93
Alcaraz and Rieta [62]	Application of CTM to the AA wavelet decomposition	≈96
Sun and Wang [63]	Characterization of recurrent plots obtained from the AA signal	≈90
Alcaraz and Rieta [64]	Application of WavEn to the AA signal	≈94
Julián et al. [65]	Application of FuzzEn, SpEn, LZC and HEs to the MAW	≈95

wavefront fragmentation as well as chaotic activations provide surface *f* waves with very variable morphologies [66].

The first method introduced to non-invasively estimate the temporal organization of AF was based on the application of SampEn to the MAW [19]. Note that the computation of SampEn directly from the AA without extracting the MAW was also investigated. However, unsuccessful results for accurately estimate AF organization were reported by several authors [19]. The existence of ventricular residua as wells as other nuisance signals in the recording together with the high sensitivity of SampEn to noise were considered as the most important reasons of this poor result [19]. In contrast, the MAW-SampEn tandem demonstrated a significant ability to reflect reliably the fibrillatory activity dynamics [67]. In fact, this approach has been validated by several studies dealing with the prediction of a variety of AF organization-dependent events. In this regard, the methodology demonstrated a high diagnostic accuracy in the prediction of spontaneous PAF termination, reporting that *f* waves presented a more regular pattern for terminating than for non-terminating episodes [19]. This result agrees with the observed decrease in the amount of reentries preceding the restoration of SR that has been reported in previous invasive studies, where the termination of AF was achieved through the use of different strategies [68].

On the other hand, *f* waves regularity was also estimated by applying SampEn to the AA projected into the wavelet domain [19]. For this case, the proposed methodology reported a slightly lower discriminant ability than the MAW-SampEn tandem in the prediction of spontaneous PAF termination. Anyway, these two methodologies demonstrated to be able to yield complementary information, so that their combination improved the identification of AF events related to the organization time course of the arrhythmia [19]. To this respect, a similar result has been reported recently through the application of the central tendency measure (CTM) to study the wavelet coefficients variability associated to the AA [62]. This nonlinear metric is able to provide the percentage of points falling within a certain radius from the centre of the PP of the first difference of the original time series [69], such as Fig. 15.3 displays for a pair of examples. In view of its results, this new strategy has also been considered as an interesting non-invasive way of estimating temporal organization of AF.

Sun and Wang [63] also investigated PAF spontaneous termination by quantifying the AA structure trough the use of recurrent plots. In their study, a set of eleven features were defined from the recurrent plots which included, for example, the point recurrence rate, the patterns along the 135° diagonal, the patterns along the main diagonal as well as the square-like patterns. Afterwards, they applied a sequential forward search algorithm with the aim to select the subset of features that could better predict the termination of PAF. Finally, they applied a multilayer perceptron neural network to anticipate that event with an accuracy higher than 90%. A similar result was also obtained by applying WavEn to the AA signal [64] as well as Fuzzy entropy (FuzEn), spectral entropy (SpEn), LZC and several Hurst exponents (HEs) to the MAW signas [65]. All these methods reported results in line with the MAW-SampEn tandem, thus suggesting the presence of more organized

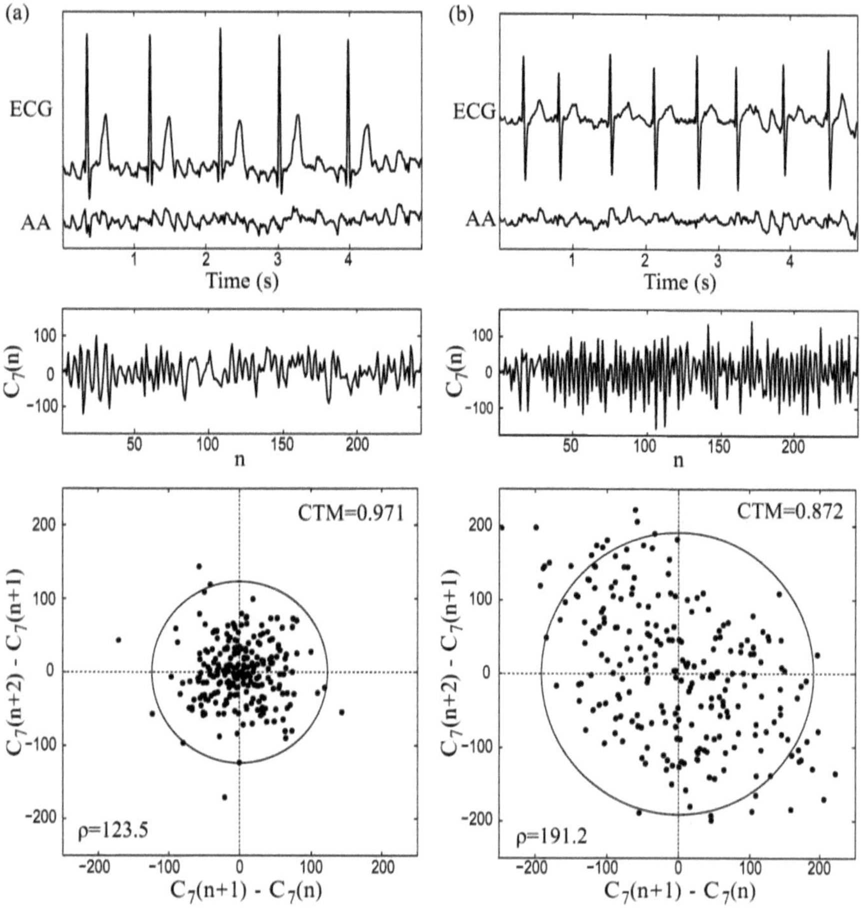

Fig. 15.3 Typical ECG interval together with its extracted AA signal and the wavelet coefficient vector corresponding with the seventh discrete scale ($\mathbf{C}_7$) together with its scatter plot first differences for (**a**) a non-terminating and (**b**) terminating PAF episode [70]

f-waves for terminating than for non-terminating recordings. However, HEs provided a considerably higher diagnostic accuracy than the MAW-SampEn approach with a discriminant ability of about 95%.

15.3.3 Prediction of PAF Onset

When a PAF episode terminates spontaneously, another very relevant clinical challenge starts, i.e., to predict the arrival of the next episode [71]. In this way, the possibility of avoiding PAF recurrence by using early pacing and drugs could hamper the progression of PAF to a more established stage [7]. Several works have reported that the autonomic nervous system plays a relevant role in the arrhythmia's onset, since increased both vagal or sympathetic tones can predispose to AF development [72, 73]. Therefore, several nonlinear metrics related to entropy, such as approximate entropy (ApEn) and SampEn, combined with DFA have been proposed with the aim to study the complexity evolution of HRV in the minutes before the onset of PAF. In this respect, Vikman et al. [74] studied the combination of DFA and ApEn along intervals of 20 min before PAF onset in 92 episodes of 22 patients with no structural heart disease. A progressive decrease of complexity was yielded by both indices before the onset of PAF episodes. Furthermore, they also reported that a lower complexity appeared before the onset of AF in comparison with the values obtained from age and gender-matched healthy control subjects.

Dealing with the same challenge, Tuzcu et al. [72] analyzed with SampEn the complexity of HRV in 30 min-length segments of the ECG immediately before a PAF episode as well as 30 min-length segments of a period, at least, 45 min away from any PAF episode. The HRV complexity was found to be notably reduced in those segments before PAF in comparison with those other distant from any PAF episode. In addition, the authors repeated the same study with the extra feature of removing premature atrial complexes before the analysis. In this case, lower differences were obtained. The authors reported that a decreased HRV complexity, in both cases, was able to reflect a change in cardiovascular autonomic regulation that predisposed to the onset of PAF. Also, the recorded ECG segments before AF onset were split into three successive 10 min-length intervals and were analyzed with SampEn to find a possible trend. Effectively, a decreasing trend in complexity as the onset of PAF approximated was reported independently of the presence or absence of premature atrial complexes. However, in this latter case, the trend was less pronounced. According to this work, a decrease in complexity of SampEn before the onset of PAF was mainly provoked by atrial ectopy. Furthermore, this complexity decrease was consistent with the observed significance of ectopic firing that, in fact, could serve as a trigger for PAF in those patients with no evidence of any other structural cardiac disease [75].

Given that PAF is often classified into sympathetically-mediated and vagally-mediated types, Shin et al. [76] also studied the complexity of HRV before these kinds of episodes. Precisely, they studied 44 episodes of PAF divided into three subgroups (vagal, sympathetic and non-related types of PAF). The segment of SR 60 min before the onset of PAF was divided into six intervals of 10 min in length. Through the application of DFA they reported a poor decreasing trend before the onset of PAF and, furthermore, the variation of this index was divergent as a function of the AF type. By contrast, both ApEn and SampEn offered a linear decreasing trend in complexity irrespective of the type of PAF. Moreover, this last result was independent on ectopic beats removal. The authors considered that this progressive reduction in entropy before the onset of PAF was due to the more regular heart rate behavior before PAF; that is, the normal "healthy" complexity of the heart is lost, thus leading to a vulnerable cardiac environment where the occurrence of PAF is more likely.

Despite the described decrease in HRV entropy before the onset of PAF, its ability to identify accurately this event was far from being optimal. In fact, some more recent woks have tried to combine entropy with other classical HRV indices in order to increase that ability. In this regard, Chesnokov [77] studied a combination of different spectral features together with SampEn and multi-scale SampEn (MSE) of the HRV through the use of different artificial intelligence classifiers. Similarly, Mohebbi and Ghassemian introduced two different combinations of parameters to detect the onset of PAF in 95% of episodes. Thus, using a SVM-based classifier they firstly combined five features computed from recurrent plots of the RR-interval series: the recurrence rate, the length of the longest diagonal segments, the average length of the diagonal lines, the entropy and the trapping time [78]. On the other hand, the second alternative also consisted of a SVM-based classifier used to combine spectrum and bispectrum characteristics together with SampEn and PP-extracted features from the HRV [79].

More recently, the P-wave variability during the 2 h before the onset of PAF has been proven as a promising predictor of that event [80]. Thus, Martínez et al. [80] have been able to quantify successfully the presence of non-linear dynamics in the P-wave progression before the onset of PAF by means of CTM. Precisely, they analyzed the variability of features like the duration, energy and area for all the P-waves from the 2 h just before the arrhythmia. According to the invasively observed inhomogeneous atrial conduction preceding the onset of PAF [81], CTM for all the considered P-wave features showed higher variability when the arrhythmia was closer to its onset (see Fig. 15.4). Moreover, a diagnostic accuracy around 80% was obtained when ECG segments 1 h far from PAF were discerned from those others close to PAF via the CTM values computed from the P-wave features. As a summary, Table 15.3 displays a comparison among all these described works.

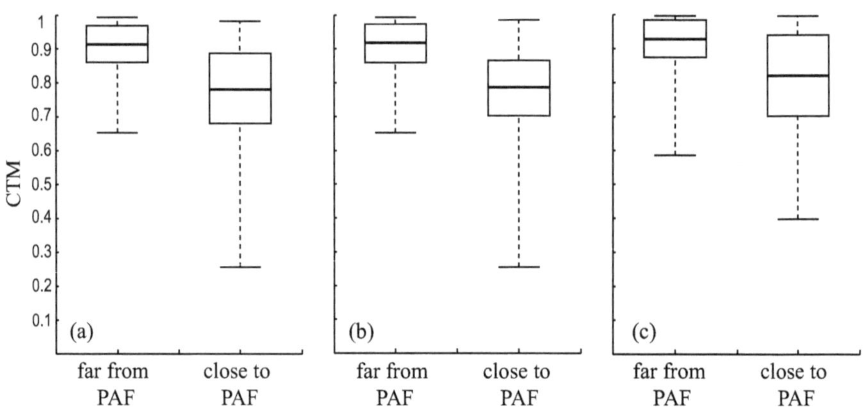

Fig. 15.4 Boxplots of the CTM values computed from some P-wave features such as its (**a**) length arc, (**b**) duration and (**c**) length of the initial part [80]

Table 15.3 Comparison of the most relevant methods proposed to predict the onset of PAF

Works	Methodology	Diagnostic accuracy
Vikman et al. [74]	Application of DFA and ApEn to RR interval series	–
Tuzcu et al. [72]	Application of DFA and SampEn to RR interval series	–
Shin et al. [76]	Application of DFA, ApEn and SampEn to RR interval series	–
Chesnokov [77]	Application of SampEn and MSE to RR interval series	≈90%
Mohebbi and Ghassemian [78]	Characterization of recurrent plots obtained from RR interval series	≈98%
Mohebbi and Ghassemian [79]	Computation of bispectrum features, SampEn, and PP-extracted metrics from RR interval series	≈95%
Martínez et al. [80]	Application of CTM to compute P-wave variability time course	90%

15.3.4 Prediction of Postoperative AF

Despite current improvements in anesthesia, surgical procedures and medical treatments, AF occurs with prevalence of up to 60% after open heart surgery [82]. Although the nature of this arrhythmia is benign and self-limiting, AF is a major contributing factor of increased postoperative morbidity [83]. Moreover, AF is related to the prolongation of the hospital stay, leading to higher costs of operative treatment [82]. Therefore, an early prediction of this arrhythmia is clinically interesting to reduce its incidence by applying a timely prophylactic treatment. Moreover, patients with a low risk of developing AF could be prevented from the side effects of that therapy.

In this context, AF prediction after coronary artery bypass grafting (CABG) has been addressed in a few works by characterizing HRV with nonlinear indices. Thus, Hogue et al. [84] reported that patients who developed postoperative AF presented a reduced HRV regularity (estimated via ApEn) just before the onset of the arrhythmia, while typical measures of HRV were unable to distinguish any kind of alteration. However, ApEn did not correlate with any classical HRV feature, so that the study provided reduced evidence of a direct relationship between ApEn and the level of autonomic modulation of the heart rate.

In contrast to ApEn, other nonlinear indices such as the short-term scaling exponent of DFA, symbolic entropy-based measures [85] or the peak point CD [86] have also been used to look for alterations in HRV before the surgery. Thus, all these metrics have been able to identify preoperatively most of the patients developing AF after CABG. However, they have only provided a poor ability to envisage the presence of AF after off-pump CABG [87]. In the same line, these metrics have also been unable to discern between patients with and without postoperative AF after isolated aortocoronary bypass surgery, isolated aortic valve surgery, and combined aortic valve replacement and bypass surgery [82].

15.4 Discovering Electrophysiological Insights

As previously described, no standard definition of AF organization can be found in the literature. Anyway, its most widely considered definition is related to the repetitiveness of the AF signal pattern [88]. Indeed, the level of organization, both for invasive and non-invasive recordings, has been directly associated with their morphology changes [88]. Therefore, considering that different morphologies reflect different activation patterns such as slow conduction, wave collision, and conduction blocks [89], AF organization analysis plays a key role to understand the mechanisms responsible for induction and maintenance of this arrhythmia. The main insights obtained in the last years by quantifying dynamics and repetitiveness of the AF pattern both from ECG and EGM signals are summarized in the next subsections.

15.4.1 Quantifying Atrial Substrate Alterations

Gaining information about the electrophysiological changes occurring in the atrial substrate during AF has received great interest with the aim to predict AF progression and decide the most appropriate therapy [3]. With this aim, Wells et al. [90] defined three types of AF (i.e., types I, II and III), which proved to reveal an increasing level of atrial alterations as well as different atrial propagation patterns [68]. In type I AF, the EGMs displayed a clear isoelectric baseline with discrete beats of variable morphology. In contrast, the EGMs for type II AF presented baseline continuous perturbations, although discrete atrial beat-to-beat complexes were still discerned. Finally, complex and fragmented atrial EGMs were observed with no presence of discrete beats or isoelectric intervals during type III AF [90].

This classification of AF patterns has been widely used in clinical and experimental studies to identify different levels of organization in paroxysmal and chronic states of the arrhythmia as well as to support its ablative treatment [91]. Thus, automatic identification of these AF types has been widely analyzed by making use a variety of different methodologies, including nonlinear metrics. To this respect, Hoekstra et al. [92] presented the first exhaustive nonlinear analysis of AF in human. Precisely, chaos degree of unipolar epicardial EGMs during induced AF was quantified by indices such as CD and correlation entropy (CorEn). Although nonlinear dynamics were observed in type I AF, no chaos was seen in type II and type III AF, thus making their proper discrimination impossible. On the contrary, more recently the three AF types have been successfully discerned through a regularity index based on ConEn [93, 94].

On the other hand, spatial estimates of AF organization have been reached by applying approaches based on nonlinear analysis to each one of the bipolar signals collected by basked catheters. Indeed, Pitschner et al. [95] computed CD from depolarization wavefronts registered during PAF, thus noticing the most significant nonlinear dynamics in the area anterior to the tricuspid valve. Similarly, regional organization differences between paroxysmal and persistent AF episodes have been recently analyzed through some entropy-based measures, such as ShEn, SampEn, MSE and causal entropy (CauEn) [96]. In this case, whereas no relevant differences were observed between EGMs from the same atrium (both for right and left atria), higher entropy values were seen in the left atrium compared to the right one, thus suggesting a more organized atrial activation in this latter chamber. Nonetheless, these differences were only statistically significant for PAF episodes, thus suggesting that an organization gradient from right to left atrium can be found in PAF but not in persistent AF.

In a similar way, some nonlinear measures have also been exploited to investigate synchronization among different atrial regions, thus gaining information about electrical waveforms propagation during AF as well as the spatial dispersion of electrophysiological features like conduction velocity or refractory period. To this respect, some entropy-based metrics have been readapted to be applied to several simultaneous EGMs. Thus, some previous authors have quantified nonlinear coupling between these signals by using multivariate embedding approaches. Indeed, Censi et al. [97] defined two novel indices based on CD and CorEn, named independence of complexity and independence of predictability, to estimate the organization of atrial activations during AF. Results showed a relevant level of nonlinear coupling in AF segments from both types I and II. In contrast, only a weak coupling was observed for type III EGMs. On the other hand, Mainardi et al. [93] proposed another nonlinear coupling approach based on ConEn. It was compared with other linear methods and reported a superior performance to discern between organized AF rhythms (type I) and non-organized ones (types II and III AF). Moreover, this metric was also able to identify subtle changes in atrial dynamics, thus providing new insights about how the sympathetic nervous system works during both PAF and persistent AF [98]. Finally, this index has also revealed a higher ability to quantify the coupling effect between different atrial regions by adrenergic stimulation than other linear and nonlinear measures [99].

Synchronization between atrial activations in different atrial regions has also been widely studied. Thus, Censi et al. [100] used recurrence plots to report a moderate degree of coupling between atrial depolarizations during AF. Other similar method was introduced by Masé et al. [101]. In this case, the authors used ShEn to study the properties of time delay distribution between atrial depolarization periods at two atrial regions. More precisely, the histogram of these propagation delays was divided into several bins and the entropy of this distribution was computed. The sampling frequency effect was corrected to avoid an entropy underestimation and, additionally, the index was validated with a computer model of AF. The metric was able to discriminate among different AF types and reveal notable spatial heterogeneities in the synchronization among atrial areas. Furthermore, simulation results were compared to real data and the observed time delay distributions were associated with different underlying electrophysiological propagation patterns. More recently, Cervigón et al. [96] quantified nonlinear coupling among different atrial regions by using CauEn. In this case, a higher level of synchronization in right atrium areas than in left atrium ones was observed in PAF patients, thus supporting the idea that high-frequency periodic sources in the left atrium can perpetuate the arrhythmia [102]. In contrast, similar levels of synchronization were observed in both atria during persistent AF.

Finally, some attempts to quantify alterations in the atrial substrate from the ECG can also be found in the literature. Thus, the MAW-SampEn tandem was also analyzed in this context, showing a notably high ability to discern very short ECG intervals of PAF and persistent AF [103]. Given the promising results of sensitivity, specificity and accuracy around 90%, the method was later validated on approximately one million of 10 s-length ECG intervals from PAF and persistent AF Holter recordings [104]. Similar classification results were again reported, thus demonstrating the method's ability to identify successfully different degrees of electrophysiological alterations in the atrial substrate from ambulatory recordings. More recently, other authors have shown that ECG phase information also seems to be useful to discern between both types of AF [105, 106]. However, this algorithm has been designed to work with digitized ECGs from PDF images and, therefore, its validation on directly acquired ECG recordings has to be developed yet. This last step is essential to really validate the method's ability to work under real-time monitoring systems and, thus, in real clinical environments.

15.4.2 Quantifying AF Progression

Some nonlinear metrics have also been used to analyze the progressive evolution of AF, providing useful insights to avoid the arrhythmia's perpetuation. To this respect, a nonlinear measure with ability to quantify the regularity of atrial activations has shown a deterioration of electrical organization in both atria during the first minutes of induced PAF [107]. Interestingly, a very similar finding has also been reported by the MAW-SampEn tandem from the ECG [108]. Indeed, in this case a progressive organization deterioration in the first 3 min after the onset of PAF was observed, such as Fig. 15.5a displays. According to these observations, to minimize the likelihood of AF perpetuation, a pacing treatment should be early delivery after the onset of the arrhythmia.

Contrarily, a progressive organization increase during the last minute before PAF termination was also observed by the MAW-SampEn tandem, such as Fig. 15.5b shows [108]. More recently, a similar observation has also been obtained by the application of a more robust AF organization estimator, such as the second-order generalized HEs [65], to the MAW. Thus, although more than 15 min before PAF

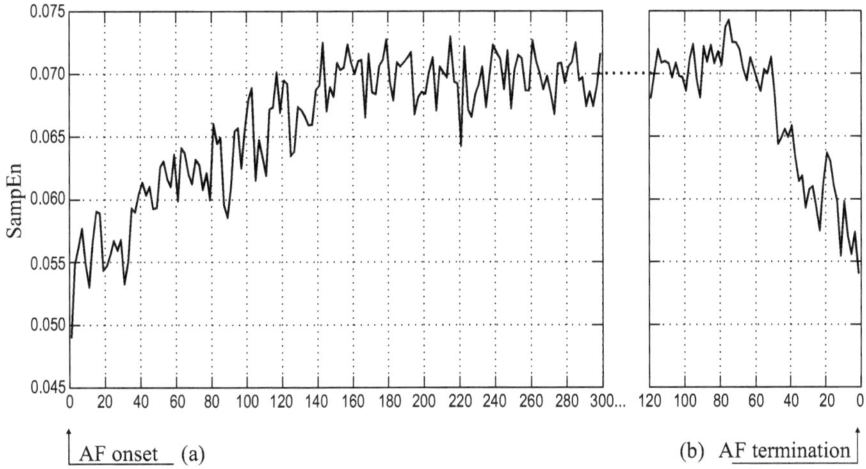

Fig. 15.5 Average progressive evolution of some PAF episodes during (**a**) their onset and (**b**) termination revealed by the MAW-SampEn tandem from the surface ECG [108]

termination were studied, HEs only reported a statistically significant organization variation within the last 2 min [109]. This results have also been supported by other authors, but making use of linear approaches. Thus, analyzing the AA signal from a frequency point of view, Bollmann et al. [110] suggested a higher AF organization in the last minute before PAF termination than 5 min after its onset for episodes lasting more than 15 min. The authors did not observe any disparity among organization values from different time instants far enough from PAF onset and termination. However, the time interval between 15 min after PAF onset and the last minute preceding its termination was not analyzed in that work. Additionally, several authors have also noticed that AF termination by anti-arrhythmic drugs is preceded by a gradual slowing of AF organization a few minutes before termination and, therefore, by a progressive increase in the atrial electrical organization [111, 112].

In another work [113], the time course of organization along onward episodes of non-induced PAF has also been analyzed using the MAW-SampEn tandem. In this way, relevant information about intra-patient evolution of AF disorganization, episode by episode, was obtained. Thus, results reported a progressive disorganization increase along successive episodes of PAF for 63% of the analyzed patients, whereas a stable AF organization degree was seen in the remaining 37%. Moreover, a statistically significant relationship between duration of episodes and their organization was also noticed [113], such as Fig. 15.6 shows. This finding is in line with previous invasive works, where it has been suggested that atrial remodeling is dependent on AF episode duration and may promote the arrhythmia perpetuation [114]. On the other hand, the study also reported that the SR interval preceding each PAF episode influenced its initial organization [113]. Indeed, higher values of organization during the first instants of the AF episode were observed for longer preceding SR intervals, thus supporting the well-known idea that reverse

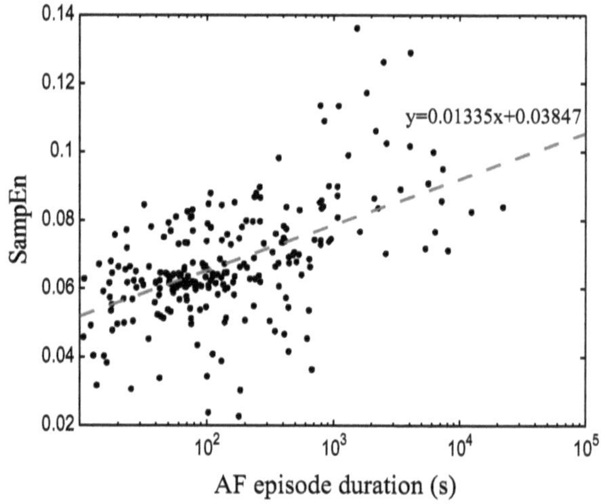

Fig. 15.6 Relationship between the organization obtained for each PAF episode by the MAW-SampEn tandem and its duration. Note that the PAF episode duration is expressed on a logarithmic scale for easy visualization. After logarithmic transformation of PAF episodes duration, the linear regression line was computed [113]

atrial remodeling occurs during SR [114]. Similarly, an interesting relationship between a specific interval of SR and the organization of its preceding AF episode was also noticed. In this case, higher values of AF disorganization were associated with shorter SR intervals. In contrast, no statistically association between the highest AF disorganization reached during a PAF episode with its preceding and subsequent SR intervals was revealed. Although some of these findings have not been corroborated by electrophysiological studies, they seem to be coherent with the progressive nature of PAF [113].

15.4.3 Atrio-Ventricular Node Characterization

The atrio-ventricular (AV) node is the part of the heart's electrical conduction system that coordinates the conduction of atrial impulses to the ventricles. Hence, fibrillatory impulses occurring in the atria during AF continuously bombard this atrial structure, which is unable to conduct all of them. Some atrial impulses are then blocked to keep the heart beating at a sustainable rate, thus provoking shorter and more irregular RR interval series than during SR [22]. Given that arrhythmia cardioversion is impossible or not recommended for some permanent AF patients, in this case it is vital to properly control rapid ventricular rates [7]. However, the AV conduction mechanisms are not completely known yet and their better knowledge would allow to improve current rate control strategies [115].

In the last years, a wide variety of signal processing tools have been used to explore characteristics of the AV conduction. However, the most interesting and useful information has been revealed by nonlinear indices, such as entropy-based measures and features obtained from PP. In fact, the lower envelope for the Poincaré graph of RR interval series has allowed to quantify non-invasively features of the

AV nodal conduction such as its functional refractory period as well as its rate dependence of conduction [116, 117]. Moreover, the dispersion of points in this plot, measured as the mean square distance between each RR interval and the lower envelope, has been considered as an accurate estimation of the AV concealed conduction [117]. It is interesting to remark that, making use of these parameters, Hayano et al. [117] reported circadian rhythms both in the AV node refractoriness and in the level of concealed AV conduction during AF.

Several authors have also observed two clearly differentiated sections in histograms of RR interval series computed from 24-h recordings of AF [118, 119]. Indeed, this is the case for about 40% of AF patients, thus suggesting the existence of two predominant conduction routes throughout the AV node [118]. They have also suggested that the functional refractoriness is less pronounced for the slow than the fast pathway. More recently, a novel method based on 3D PP to quantify the preferential AV conduction ways has also been proposed [119]. It removes AV node memory effects as well as dependencies between beats to quantify short-term variation of the preferential AV conduction ways and their link to the atrial conduction. Using this algorithm a statistically significant relationship between the predominant RR intervals and the dominant pattern of the AA was noticed for all AF patients showing bimodal histograms of RR interval series and for approximately 50% of patients presenting unimodal ones [120]. In view of these results, the authors suggested that the predominant RR intervals are provoked by different conduction ratios of the atrial rate.

On the other hand, given that it is well-kwon that the autonomic nervous system can modulate AV node refractoriness [22], some entropy-based measures have been applied to the RR series to gain information about the AV node behavior under several scenarios. Thus, Mainardi et al. [57] have used the previously described ConEn-based regularity index to quantify the predictability degree of ventricular response in persistent AF patients during exercise. Although a slight trend was only observed, an increased predictability was noticed, thus suggesting a lower refractory period for more intensive exercise [57]. Additionally, the same authors have recently assessed the effects of several rate-control and β-blocker drugs on some AV node features by estimating variability and irregularity of RR interval series [121, 122]. Whereas variability of this time series was computed from classical time HRV parameters, its irregularity was quantified via ApEn and SampEn. Interestingly, results showed that modification of the AV node conduction by rate-control drugs can increase variability of the RR interval series but not its irregularity [121]. Contrarily, β-blockers only provided an alteration of the RR interval series irregularity [122].

Finally, nonlinear interdependency between atrial and ventricular activation series has not been widely studied to date. Nonetheless, some pilot works have reported that the rate of atrial activations and their organization during AF harm the ventricular response [120]. Thus, a pioneer algorithm based on cross-SampEn has been recently introduced for the first time to quantify synchronization and coupling between atrial and ventricular activation series [123]. The algorithm considers instantaneous atrial and ventricular activations to compute the presence

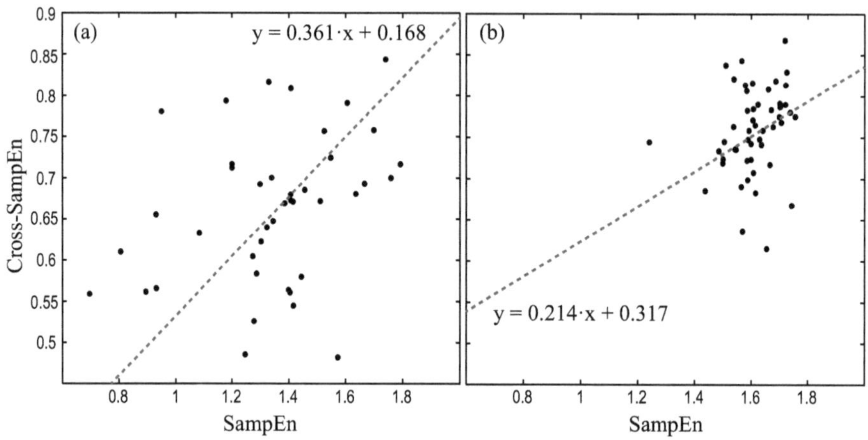

Fig. 15.7 Relationship between cross-SampEn computed from atrial and ventricular activation series and SampEn computed from the AA signal for (**a**) paroxysmal and (**b**) persistent episodes [123]

and repetitiveness of atrial patterns into the ventricular response. In this way, by analyzing atrial activation series with different organization properties from paroxysmal and persistent AF patients, a statistically significant positive correlation between atrial organization and the synchronization between atrial and ventricular activations has been reported, such as Fig. 15.7 displays. Moreover, a greater synchronization between atrial and ventricular activations was found in paroxysmal than in persistent ECG intervals. According to the authors, these findings could be useful to improve current rate-control therapies, which are completely focused on managing the ventricular response without considering the atrial fibrillatory process [22].

15.5 Improvement of AF Therapies

Management of AF patients is aimed at reducing symptoms and preventing severe complications associated with the arrhythmia [124]. Both therapeutic goals have to be pursued in parallel, especially for newly detected AF. Prevention of complications related to the arrhythmia relies on antithrombotic therapy, control of ventricular rate, and adequate therapy of concomitant cardiac diseases. Although these treatments can sometimes alleviate the symptoms, more aggressive therapies such as pharmacological or electrical cardioversion and ablation are also required many times. However, they are only partially effective and, moreover, are associated with a notably high morbidity and mortality [3]. With the aim to improve their efficacy, numerous studies based on nonlinear methods have been conducted in the last years. The next subsections summarize the main contributions reached in this context.

15.5.1 Drugs Therapy

Pharmacological cardioversion of AF is often considered as the first therapeutical option for patients who remain symptomatic despite an adequate rate control or for those others in whom rhythm control therapy is pursued [124]. Although its success rate is limited, this approach does not requiere any kind of sedation or anesthesia, such as the remaining AF therapies. Nonetheless, most of patients under this treatment require continuous medical supervision as well as ECG monitoring during the drug's infusion and a period afterwards to detect proarrhythmic events, such as ventricular proarrhythmia, sinus node arrest, or atrioventricular block. Within this context, several nonlinear indices have been used to monitor the effects caused by different antiarrhythmic drugs on the atrial substrate.

To this respect, Hoekstra et al. [125] introduced for the first time nonlinear analysis to study how dynamics of local atrial EGMs registered from conscious goats were modified by the administration of cibenzoline. Thus, coarse-grained CD and CorEn computed from the correlation integral reported no significant changes, compared with controls, during the drug's administration. Contrarily, slightly slower atrial activations were observed after the infusion of cibenzoline, thus suggesting a mild increase in the global atrial organization. Some years later, the same authors also studied the effect of cibenzoline on the coupling among several atrial regions through a nonlinear spatial redundancy function [126]. In this case, results showed that the drug was able to increase spatial organization, thus inducing more clear and repetitive activation patterns.

Berkowitsch et al. [127] also analyzed the complexity of the spatial distribution registered at different right atrial regions after the administration of another anti-arrhythmic drug, such as propafenone. In this case, an algorithm quantifying symbolic dynamics was used to notice that the low-complexity areas were enlarged with the drug's infusion. More recently, the effect on both atria of the most frequently used intravenous anesthetic agent, i.e. propofol, has also been analyzed using several entropy-based measures [128]. Thus, global organization of each atrium was quantified by averaging values of MSE and ShEn obtained from EGMs located at different regions. As expected, both indices reported different levels of organization in both atria before and after the drug's administration. However, whereas right atrium organization increased with the infusion of propofol in PAF patients, a decrease was noticed in the left atrium. Although this result was observed for all the considered scales for MSE as well as for ShEn, the authors did not suggest any explanation. On the contrary, a more coherent organization decrease in both atria after the propofol administration was noticed for persistent AF patients [128].

On the other hand, the adrenergic activation effect, induced by isoproterenol infusion (a sympathomimetic drug), on the repetitiveness of atrial activation patterns as well as on the synchronization among different atrial regions during SR and AF has also been analyzed from a nonlinear point of view. Thus, several indices based on ConEn and cross-ConEn were used to compute regularity and synchronization both from EGMs and LAW series [98, 129]. In this case, results evidenced the presence of a reduced regularity and coupling after the drug's infusion both in SR and AF, thus suggesting an important influence of the adrenergic activation to favor the arrhythmia's maintenance. More recently, a result in the same line has also been obtained during AF through a novel algorithm which computes atrial organization by considering the direction of propagation of the wavefronts [130]. However, this method was unable to quantify changes in the organization of EGMs registered at SR, thus revealing the need of a wider validation on longer databases.

On the other hand, in vivo studies with horses applied PP from the RR interval series to assess the effect of another anti-arrhythmic drug, such as the quinidine [131]. In this case, whereas a L-shaped figure was observed before the drug's administration, the RR intervals were mainly clustered around the identity line after that event. Thus, the drug was able to suppress the typical alternation of long and short RR intervals during AF. Moreover, the lower PP envelope also revealed a notable decrease in the functional refractory period of the AV node after the drug's infusion. Thus, these findings contributed to the authors' suggestion that ventricular response during AF could be predictable in a short-term fashion by considering the frequency-dependent conduction properties of the AV node [131].

15.5.2 Electrical Cardioversion

As previously described, pharmacological cardioversion is often ineffective and, moreover, is associated with notable side effects [124]. However, nowadays there are a number of non-pharmacological options for the management of AF, which can improve patient's health and wellbeing [124]. To this respect, electrical cardioversion (ECV) is a common alternative for patients under AF for more than 24 h [132]. The procedure has a high efficacy to restore SR and low complication rates [11]. However, the long-term risk of AF recurrence is high, especially for persistent AF patients [133]. Therefore, a priori selection of the patients with a high likelihood of maintaining SR would be desirable to improve this therapy [134].

For that purpose, numerous signal processing studies have been presented in the literature, many of them applying nonlinear indices both to atrial and ventricular activities. A summary of these works is presented in Table 15.4. Regarding the AA analysis, the MAW-SampEn tandem has shown an ability around 90% to identify early AF recurrence 1 month after ECV [135]. However, its short-term promising diagnostic accuracy has been drastically reduced for a longer post-cardioversion follow-up of around 1 year [134]. In this case, a discriminant model based on SampEn and other linear metrics was only able to discern about 65% of patients with and without AF recurrence during the follow-up. Nonetheless, entropy values in both studies suggest that patients presenting a more organized AA signal before ECV have a higher likelihood of AF maintenance after the procedure [134, 135]. A similar result has been also obtained by analyzing repetitive patterns from wavelet decomposition of the AA signal. Thus, WavEn provided more organized f-waves in patients with successful SR restoration during the first month post-cardioversion

Table 15.4 Comparison of the most relevant methods proposed to predict ECV outcome before the procedure is attempted

Works	Methodology	Diagnostic accuracy (%)
Alcaraz and Rieta [135]	Application of SampEn to the MAW for ECV outcome prediction after a 1 month-length follow-up	≈90
Lankveld et al. [134]	Application of SpEn and SampEn to the MAW for ECV outcome prediction after a 1 year-length follow-up	≈65
Alcaraz and Rieta [64]	Application of WavEn to the AA signal for ECV outcome prediction after a 1 month-length follow-up	≈84
Alcaraz and Rieta [70]	Application of CTM to the AA wavelet decomposition for ECV outcome prediction after a 1 mont-length follow-up	≈86
Zohar et al. [136]	Application of ApEn to RR interval series for ECV outcome prediction after a 3 month-length follow-up	≈86
Berg et al. [137]	Characterization of Poincaré plots obtained from RR interval series for immediate ECV outcome prediction	≈52
Guo et al. [138]	Characterization of histograms of RR interval series for ECV outcome prediction after 1 week-length and 1 month-length follow-ups	≈70

than those who relapsed to AF [64]. Also, CTM for the AA wavelet coefficients revealed the presence of more repetitive patterns for patients maintaining SR than for those who presented recurrent AF after a follow-up of 4 weeks [70]. In both cases, a discriminant ability about 85% was reported, however a wider validation of these methods with a longer follow-up is still a pending task.

On the other hand, the analysis of ventricular activity before ECV has also provided useful clinical information. Thus, Zohar et al. [136] analyzed regularity of the RR interval series using ApEn together with other linear and clinical variables. These indices were combined using an advanced classifier based on non-deterministic genetic programming, thus reaching a diagnostic accuracy around 85% for a post-cardioversion follow-up of 3 months. Similarly, Berg et al. [137] studied the ventricular rhythm during 24 h before ECV through 3D PP and observed that the procedure was more effective to immediately restore SR in AF patients with clustering of RR intervals. Moreover, they also noticed that SR persisted for a shorter period in patients without that clustering. Guo et al. [138] have also reported a significant relationship between the histogram shape of the RR interval series and the recurrence of AF during the first month post-cardioversion. Thus, patients with a bimodal histogram provided the greatest risk of developing post-cardioversion AF.

The ventricular response after ECV has also been analyzed from a nonlinear point of view, the most relevant works being summarized in Table 15.5. In this case, an optimal selection of patients for the procedure cannot be developed a priori, but useful information about who will relapse to AF during the follow-up can be

Table 15.5 Comparison of the most relevant methods proposed to predict ECV outcome after the procedure

Works	Methodology	Diagnostic accuracy
Vikman et al. [139]	Application of ApEn and DFA to RR interval series for ECV outcome prediction after a 1 month-length follow-up	–
Rademacher et al. [140]	Computation of the heart rate turbulence, deceleration capacity and some symbolic metrics from RR interval series to predict ECV outcome after a 1 year-length follow-up	≈85%
Brisinda et al. [141]	Application of ApEn, SampEn, CD, DFA and PP-based metrics to RR interval series for ECV outcome prediction after a 2 year-length follow-up	90–100%
Seeck et al. [142]	Coupling analysis between RR interval series and continuous blood pressure recording through symbolic and PP-based metrics to predict ECV outcome after a 1 year-length follow-up	≈90%

elucidated. This kind of analysis has a high clinical value, because it could be helpful to understand the mechanisms causing AF recurrence after ECV [141]. Additionally, it may help in the adjustment of a therapeutic alternative for patients with a high risk of AF recurrence [142]. Thus, Vikman et al. [139] analyzed for the first time regularity and complexity of the RR interval series by means of ApEn, short-term fractal scaling exponent of DFA and long-term power-law slope β. Whereas the two first indices did not provide statistically significant differences between patients who responded successfully to the procedure during the first post-cardioversion month, the third one was steeper among the patients who remained in SR. Another work in the same line was conduced by Rademacher et al. [140], who analyzed the RR interval series making use of uncommon nonlinear metrics such as the heart rate turbulence, deceleration capacity and measures based on quantifying symbolic dynamics. The combination of these metrics using a linear discriminant classifier reached a diagnostic accuracy around 85% for a long-term follow-up of around 1 year.

More recently, Brisinda et al. [141] have also reported the presence of different nonlinear dynamics in the RR interval series from patients with and without AF recurrence after ECV. In this case a longer follow-up of about 2 years was considered and a wider variety of nonlinear metrics were studied. Indeed, numerous indices from Poincaré and recurrence plots as well as other parameters such as CD, ApEn and SampEn were analyzed during two clear time intervals, i.e., daily activity and night sleep. Without paying attention to the administration or absence of anti-arrhythmic treatment, no significant differences between groups of patients were revealed at the first hour after ECV as well as during the time of daily activity. However, depressed vagal modulation and prevalence of sympathetic modulation were suggested by most of the studied metrics during the time of night sleep. On the contrary, considering only the patients without anti-arrhythmic therapy, a higher number of nonlinear metrics provided differences between groups of patients during both daily activity and night sleep. Thus, a discriminant analysis combining all the analyzed nonlinear metrics provided accuracies of 89 and 100% for both time intervals, respectively.

Interestingly, the interaction between ventricular activity and blood pressure after ECV has also revealed to be useful to predict the procedure outcome. Indeed, the coupling mechanisms between heart rate and blood pressure have been successfully characterized by two novel nonlinear approaches, such as the joint analysis of symbolic dynamics and the study of segmented PP [142]. Indeed, the combination of some parameters from these methodologies through a linear discriminant function has reported a predictive ability around 90% to discern between patients with stable SR and AF recurrence after a post-cardioversion follow-up of around 1 year.

A few works have also analyzed the effectiveness of every single electrical shock during internal [143] or external [144] cardioversion making use of the MAW-SampEn tandem. Nowadays, common ECV protocols do not pay attention to the effect caused by unsuccessful shocks. Indeed, the procedure starts with a limited energy shock and, in case of fail, later attempts (up to four or five) with increasing energy are applied to the patient after few minutes. However, these studies have provided that after each unsuccessful shock, a relative SampEn decrease was only observed for the patients who finally reverted back to SR, such as Fig. 15.8 displays. Moreover, the largest variation was noticed after the first attempt, thus indicating that this shock plays the most important role in the procedure [143, 144]. Therefore, attending to the effect caused by this shock, conventional ECV protocols could be improved and, consequently, the patient's suffering as well as the consequent risk associated to the delivery of excessive electrical energy to the heart could be drastically reduced [133].

15.5.3 Catheter Ablation

In contrast to pharmacological and electrical cardioversion, catheter ablation (CA) has offered encouraging results in the treatment of AF during the last decade, thus becoming the first-line therapy recommended in the current guidelines for the arrhythmia's management [145]. Given that ectopic foci arising from the pulmonary veins (PV) are a well-known source of initiation and maintenance of AF [75], nowadays its wide circumferential isolation using radiofrequency energy is the cornerstone of interventional therapy for AF [145]. This approach has shown ability to restore SR in most patients with PAF, although repeated procedures are required in some of them [146]. Moreover, CA extension beyond PV in patients with PAF has not proven to improve clinical results [145]. Contrarily, the arrhythmogenic substrate for persistent AF seems to shift away from the PV, thus their electrical isolation has only reported lower success rates than for PAF [147].

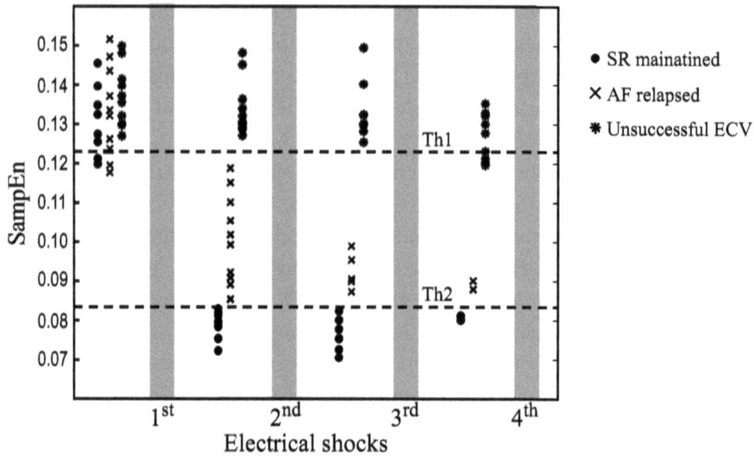

Fig. 15.8 SampEn values obtained after each unsuccessful shock during a common ECV protocol. After the first shock a notable decrease of entropy can only be observed for patients who will maintain SR at least 1 month after the procedure [144]

To improve CA outcome in persistent AF patients, a wide variety of additional ablation strategies to the PV isolation have been proposed from the multiple theories existing about the mechanisms of AF [145]. To this respect, complex and fractionated atrial EGMs (CFAEs) have been considered to represent areas of slowed conduction, wavefront collision or anchor points for re-entrant circuits and have therefore been proposed as targets for CA [148]. However, given the lack of objective criteria to define CFAEs, contradictory CA outcomes have been reported in different studies [149] and automatic quantification of these kinds of EGMs has been widely sought. Thus, numerous algorithms based on measurements of time intervals and voltage levels from the EGM can be found in the literature. Also, a few novel methods based on characterizing nonlinear dynamics from the EGM have revealed a great ability to discern among different levels of atrial fractionation, such as Table 15.6 overviews. Thus, Ng et al. [150] introduced for the first time the application of ShEn on the EGM amplitudes histogram to discriminate between CFAEs and non-CFAEs. Although results comparable to other typical time-based indices were only obtained, the method presented the additional advantage of not requiring amplitude thresholding [150]. Maintaining this benefit, SampEn has recently proven a notably better ability to identify CFAEs [151]. In this case, by computing regularity of EGMs with optimized key parameters for SampEn, values of sensitivity and specificity around 80% were reported.

In addition to these entropy-based measures, Navoret et al. [152] have also proposed the recurrence quantification analysis to detect CFAEs. Precisely, a set of five parameters were computed from CFAEs and non-CFAEs with very short sliding windows. Whereas every single metric was unable to identify successfully fractionated EGMs, the combination of the five features with advanced classifiers

Table 15.6 Comparison of the most relevant methods proposed to discern between CFAEs and non-CFAEs

Works	Methodology	Diagnostic accuracy (%)
Ng et al. [150]	Application of ShEn to the EGM amplitude histogram	≈85
Cirugeda-Roldán et al. [151]	Application of SampEn to the EGM	≈80
Navoret et al. [152]	Characterization of recurrent plots obtained from the EGM	≈75
Orozco-Duque et al. [153]	Application of generalized HE to the EGM	≈85

reported a diagnostic accuracy higher than 75%, with well-balanced values of sensitivity and specificity. However, the methodology was unable to discern between both active and passive CFAEs in the maintenance of AF [152]. Another recent work dealing with the nonlinear detection of CFAEs has been presented by Orozco-Duque et al. [153]. In this study, some multifractal properties of the EGM such as the generalized HEs were analyzed to obtain an index able to discern more than 85% of CFAEs and non-CFAEs. However, despite the described great results, these indices have not been still implemented into commercially available systems and, therefore, they have not been validated yet in daily clinical practice.

Recently, rotors have also been proposed as targets for CA [154]. Specifically, ablation at the rotor's pivot point (which is also called the phase singularity) has proven to lead to acute AF termination [154] as well as an improved clinical outcome [154, 155]. However, the accurate identification of rotors is still a challenging task. Indeed, rotors have often been detected manually from isochronic activation maps, but this approach presents a high computational cost and is unable to detect meandering rotors [156]. Moreover, the reconstruction of activation maps requires the establishment of EGM-dependent thresholds to identify properly the LAWs, thus making this approach extremely subjective [156]. To avoid these limitations, the ability of ShEn and ApEn to detect the rotor's pivot point has been validated on synthetic EGMs, generated by computational atrial models. To this respect, Ganesan et al. [157] have associated high values of ShEn with the rotor's tip in 2D arrays of bipolar EGMs. Similarly, Ugarte et al. [156] have shown that dynamic ApEn color maps may be useful to identify stable and meandering rotors in a 3D atrial model. Nonetheless, both methodologies need to be validated on experimental studies with humans.

On the other hand, some entropy-based metrics have also been applied on EGMs of both atria for the prediction of CA outcome before the procedure is attempted. The interest of this analysis relies on the possibility of selecting those patients with a high probability of responding positively to the treatment. In this way the number of unsuccessful CA procedures may be significantly reduced, thus saving costs to the health care service as well as avoiding the corresponding risk to the patient [158]. Within this context, Cervigón et al. [159] have recently reported that high SampEn, especially at the right atrium, is related to a great risk of AF recurrence after CA.

Indeed, higher entropy was found in both atria for patients who relapsed to AF than for those maintaining SR after a follow-up of about 3 months. Moreover, this index was able to identify properly all the patients with AF recurrence [159]. The same authors have also analyzed the application of ShEn on bipolar EGMs to predict preoperatively the outcome of CA [160]. According to their previous results, entropy computed from the right atrium again provided an interesting ability to discern between patients with and without AF recurrence after the procedure. Nonetheless, this methodology was less predictive than the previous one.

15.5.4 Cox-Maze Surgery

Despite the recent popularity of CA, Cox-MAZE surgery is still the most effective treatment for AF, with a reported long-term success rate higher than 80% [161]. Although initially this method was performed as an isolated procedure, nowadays it is concomitantly applied with other cardiac interventions [162]. Two main reasons support this approach. On the one hand, AF persists in patients suffering jointly from this arrhythmia and other heart disorder even after a successful resolution of the underlying disease [162]. On the other hand, most of patients undergoing the Cox-Maze procedure also need a cardiopulmonary bypass [11]. Additionally, several studies have reported the efficacy and safety of such concomitant procedures and, in fact, this combination of procedures has been incorporated into the guidelines for the prevention of postoperative AF [124].

Recently, the MAW-SampEn tandem has revealed an interesting ability to anticipate the result of this procedure at different postoperative moments. Precisely, the method has proven to be able to predict the surgery outcome with a diagnostic accuracy around 70% at discharge [163] and after a follow-up of 3 [164] and 6 [165] months. Moreover, the combination of this entropy with the f-wave amplitude has notably improved the result, thus reaching an predictive ability close to 85%. Although this kind of prediction is clinically relevant, it has been amazingly addressed for the first time from a signal processing point of view in the described works. Nonetheless, the promising reported results may be useful to reach a better management of AF patients under this procedure. In fact, they are normally treated with oral anticoagulants and antiarrhythmic drugs at discharge, regardless of the rhythm (SR or AF). Moreover, they are routinely assessed after 3, 6 and 12 months and, if AF reappears during this time, ECV is applied to restore SR. Contrarily, drugs are withdrawn if a stable SR is observed. Hence, gaining preoperative information about the risk of AF recurrence after Cox-Maze surgery could be helpful in planning an stricter follow-up as well as the ECV procedure for those patients with a low likelihood of SR maintenance and to avoid aggressive pharmacological treatments in the remaining patients.

15.6 Conclusions

A variety of pathophysiologic processes interact with the heart during AF in an inherently complex, nonlinear and non-stationary way, thus masking the underlaying mechanisms triggering and maintaining this arrhythmia. As a consequence, nonlinear analysis of AF recordings has recently arisen as a powerful tool to gain novel insights about the arrhythmia's understanding as well as to improve the management of patients and the effectiveness of current therapies. This chapter has reviewed the most recent advances in this context, thus highlighting that nonlinear analysis has been able to reveal much and very useful clinical information for multiple AF scenarios in a complementary or even improved way to the classical linear algorithms. Indeed, it has been described as entropy-based measures and parameters computed from Poincaré and recurrent plots are especially useful to detect and quantify interesting events and electrophysiological alterations in the arrhythmia's progression as well as to anticipate helpful information for optimal selection of patients under the current therapeutical alternatives. Nonetheless, the main pending challenge is the integration of this kind of analyses in monitoring, diagnostic and therapy systems to fully exploit their potential in everyday clinical practice.

Acknowledgements Research funded by grants TEC2014-52250-R from the Spanish Ministry of Economy and Competitiveness and PPII–2014–026–P from Junta de Comunidades de Castilla-La Mancha.

References

1. Potter, B.J., Le Lorier, J.: Taking the pulse of atrial fibrillation. Lancet **386**(9989), 113–115 (2015)
2. Zoni-Berisso, M., Lercari, F., Carazza, T., Domenicucci, S.: Epidemiology of atrial fibrillation: European perspective. Clin. Epidemiol. **6**, 213–220 (2014)
3. Van Wagoner, D.R., Piccini, J.P., Albert, C.M., Anderson, M.E., Benjamin, E.J., Brundel, B., Califf, R.M., Calkins, H., Chen, P.-S., Chiamvimonvat, N., Darbar, D., Eckhardt, L.L., Ellinor, P.T., Exner, D.V., Fogel, R.I., Gillis, A.M., Healey, J., Hohnloser, S.H., Kamel, H., Lathrop, D.A., Lip, G.Y.H., Mehra, R., Narayan, S.M., Olgin, J., Packer, D., Peters, N.S., Roden, D.M., Ross, H.M., Sheldon, R., Wehrens, X.H.T.: Progress toward the prevention and treatment of atrial fibrillation: a summary of the Heart Rhythm Society research forum on the treatment and prevention of atrial fibrillation, Washington, DC, December 9–10, 2013. Heart Rhythm**12**(1), e5–e29 (2015)
4. Woods, C.E., Olgin, J.: Atrial fibrillation therapy now and in the future: drugs, biologicals, and ablation. Circ. Res. **114**(9), 1532–1546 (2014)
5. Sheikh, A., Patel, N.J., Nalluri, N., Agnihotri, K., Spagnola, J., Patel, A., Asti, D., Kanotra, R., Khan, H., Savani, C., Arora, S., Patel, N., Thakkar, B., Patel, N., Pau, D., Badheka, A.O., Deshmukh, A., Kowalski, M., Viles-Gonzalez, J., Paydak, H.: Trends in hospitalization for atrial fibrillation: epidemiology, cost, and implications for the future. Prog. Cardiovasc. Dis. **58**(2), 105–116 (2015)

6. Yiin, G.S.C., Howard, D.P.J., Paul, N.L.M., Li, L., Luengo-Fernandez, R., Bull, L.M., Welch, S.J.V., Gutnikov, S.A., Mehta, Z., Rothwell, P.M., Oxford Vascular Study: Age-specific incidence, outcome, cost, and projected future burden of atrial fibrillation-related embolic vascular events: a population-based study. Circulation **130**(15), 1236–1244 (2014)
7. January, C.T., Wann, L.S., Alpert, J.S., Calkins, H., Cigarroa, J.E., Cleveland Jr., J.C., Conti, J.B., Ellinor, P.T., Ezekowitz, M.D., Field, M.E., Murray, K.T., Sacco, R.L., Stevenson, W.G., Tchou, P.J., Tracy, C.M., Yancy, C.W., American College of Cardiology/American Heart Association Task Force on Practice Guidelines: 2014 AHA/ACC/HRS guideline for the management of patients with atrial fibrillation: a report of the American College of Cardiology/American Heart Association Task Force on practice guidelines and the Heart Rhythm Society. J. Am. Coll. Cardiol. **64**(21), e1–76 (2014)
8. Wodchis, W.P., Bhatia, R.S., Leblanc, K., Meshkat, N., Morra, D.: A review of the cost of atrial fibrillation. Value Health **15**(2), 240–248 (2012)
9. Nattel, S., Guasch, E., Savelieva, I., Cosio, F.G., Valverde, I., Halperin, J.L., Conroy, J.M., Al-Khatib, S.M., Hess, P.L., Kirchhof, P., De Bono, J., Lip, G.Y.H., Banerjee, A., Ruskin, J., Blendea, D., Camm, A.J.: Early management of atrial fibrillation to prevent cardiovascular complications. Eur. Heart J. **35**(22), 1448–1456 (2014)
10. Schotten, U., Dobrev, D., Platonov, P.G., Kottkamp, H., Hindricks, G.: Current controversies in determining the main mechanisms of atrial fibrillation. J. Intern. Med. **279**, 428–438 (2016)
11. Fuster, V., Rydén, L.E., Cannom, D.S., Crijns, H.J., Curtis, A.B., Ellenbogen, K.A., Halperin, J.L., et al.: 2011 ACCF/AHA/HRS focused updates incorporated into the ACC/AHA/ESC 2006 guidelines for the management of patients with atrial fibrillation: a report of the American College of Cardiology Foundation/American Heart Association Task Force on practice guidelines. Circulation **123**(10), e269–367 (2011)
12. Dunning, J., Nagendran, M., Alfieri, O.R., Elia, S., Kappetein, A.P., Lockowandt, U., Sarris, G.E., Kolh, P.H., Committee, E.C.G.: Guideline for the surgical treatment of atrial fibrillation. Eur. J. Cardiothorac. Surg. **44**(5), 777–791 (2013)
13. Krogh-Madsen, T., Christini, D.J.: Nonlinear dynamics in cardiology. Annu. Rev. Biomed. Eng. **14**, 179–203 (2012)
14. Qu, Z., Garfinkel, A., Chen, P.S., Weiss, J.N.: Mechanisms of discordant alternans and induction of reentry in simulated cardiac tissue. Circulation **102**(14), 1664–1670 (2000)
15. Qu, Z.: Chaos in the genesis and maintenance of cardiac arrhythmias. Prog. Biophys. Mol. Biol. **105**(3), 247–257 (2011)
16. Qu, Z., Hu, G., Garfinkel, A., Weiss, J.N.: Nonlinear and stochastic dynamics in the heart. Phys. Rep. **543**(2), 61–162 (2014)
17. Petrutiu, S., Ng, J., Nijm, G.M., Al-Angari, H., Swiryn, S., Sahakian, A.V.: Atrial fibrillation and waveform characterization. A time domain perspective in the surface ECG. IEEE Eng. Med. Biol. Mag. **25**(6), 24–30 (2006)
18. Sörnmo, L., Laguna, P.: Bioelectrical Signal Processing in Cardiac and Neurological Applications. Elsevier, Academic, Amsterdam (2005)
19. Alcaraz, R., Rieta, J.J.: A review on sample entropy applications for the non-invasive analysis of atrial fibrillation electrocardiograms. Biomed. Signal Process. Control **5**, 1–14 (2010)
20. Rieta, J.J., Castells, F., Sánchez, C., Zarzoso, V., Millet, J.: Atrial activity extraction for atrial fibrillation analysis using blind source separation. IEEE Trans. Biomed. Eng. **51**, 1176–1186 (2004)
21. Sörnmo, L., Stridh, M., Husser, D., Bollmann, A., Olsson, S.B.: Analysis of atrial fibrillation: from electrocardiogram signal processing to clinical management. Philos. Trans. A Math. Phys. Eng. Sci. **367**(1887), 235–253 (2009)
22. Zhang, Y., Mazgalev, T.N.: Ventricular rate control during atrial fibrillation and AV node modifications: past, present, and future. Pacing Clin. Electrophysiol. **27**(3), 382–393 (2004)
23. Rajendra Acharya, U., Paul Joseph, K., Kannathal, N., Lim, C.M., Suri, J.S.: Heart rate variability: a review. Med. Biol. Eng. Comput. **44**(12), 1031–1051 (2006)
24. Merah, M., Abdelmalik, T.A., Larbi, B.H.: R-peaks detection based on stationary wavelet transform. Comput. Methods Prog. Biomed. **121**(3), 149–160 (2015)

25. Peltola, M.A.: Role of editing of R-R intervals in the analysis of heart rate variability. Front. Physiol. **3**, 148 (2012)
26. Nabil, D., Reguig, F.: Ectopic beats detection and correction methods: a review. Biomed. Signal Process. Control **18**, 228–244 (2015)
27. Sassi, R., Cerutti, S., Lombardi, F., Malik, M., Huikuri, H.V., Peng, C.-K., Schmidt, G., Yamamoto, Y.: Advances in heart rate variability signal analysis: joint position statement by the e-cardiology ESC working group and the European heart rhythm association co-endorsed by the Asia Pacific heart rhythm society. Europace **17**(9), 1341–1353 (2015)
28. Stevenson, W.G., Soejima, K.: Recording techniques for clinical electrophysiology. J. Cardiovasc. Electrophysiol. **16**(9), 1017–1022 (2005)
29. Rieta, J.J., Hornero, F.: Comparative study of methods for ventricular activity cancellation in atrial electrograms of atrial fibrillation. Physiol. Meas. **28**, 925 (2007)
30. Corino, V.D.A., Rivolta, M.W., Sassi, R., Lombardi, F., Mainardi, L.T.: Ventricular activity cancellation in electrograms during atrial fibrillation with constraints on residuals' power. Med. Eng. Phys. **35**(12), 1770–1777 (2013)
31. Verheule, S., Tuyls, E., van Hunnik, A., Kuiper, M., Schotten, U., Allessie, M.: Fibrillatory conduction in the atrial free walls of goats in persistent and permanent atrial fibrillation. Circ. Arrhythm. Electrophysiol. **3**(6), 590–599 (2010)
32. Castells, F., Cervigón, R., Millet, J.: On the preprocessing of atrial electrograms in atrial fibrillation: understanding Botteron's approach. Pacing Clin. Electrophysiol. **37**(2), 133–143 (2014)
33. Chen, Y., Wu, Z., Yang, C., Shao, J., Wong, K.K.L., Abbott, D.: Investigation of atrial vulnerability by analysis of the sinus node EG from atrial fibrillation models using a phase synchronization method. IEEE Trans. Biomed. Eng. **59**(9), 2668–2676 (2012)
34. Xiong, Q., Proietti, M., Senoo, K., Lip, G.Y.H.: Asymptomatic versus symptomatic atrial fibrillation: a systematic review of age/gender differences and cardiovascular outcomes. Int. J. Cardiol. **191**, 172–177 (2015)
35. Ahmad, N., Kamal, A.K.: Asymptomatic atrial fibrillation and stroke risk. J. Pak. Med. Assoc. **64**(3), 362 (2014)
36. Keach, J.W., Bradley, S.M., Turakhia, M.P., Maddox, T.M.: Early detection of occult atrial fibrillation and stroke prevention. Heart **101**(14), 1097–1102 (2015)
37. Dash, S., Chon, K.H., Lu, S., Raeder, E.A.: Automatic real time detection of atrial fibrillation. Ann. Biomed. Eng. **37**(9), 1701–1709 (2009)
38. Lee, J., Reyes, B.A., McManus, D.D., Maitas, O., Mathias, O., Chon, K.H.: Atrial fibrillation detection using an iPhone 4S. IEEE Trans. Biomed. Eng. **60**(1), 203–206 (2013)
39. Zhou, X., Ding, H., Ung, B., Pickwell-MacPherson, E., Zhang, Y.: Automatic online detection of atrial fibrillation based on symbolic dynamics and shannon entropy. Biomed. Eng. Online **13**(1), 18 (2014)
40. Kikillus, N., Hammer, G., Wieland, S., Bolz, A.: Algorithm for identifying patients with paroxysmal atrial fibrillation without appearance on the ECG. Conf. Proc. IEEE Eng. Med. Biol. Soc. **2007**, 275–278 (2007)
41. Esperer, H.D., Esperer, C., Cohen, R.J.: Cardiac arrhythmias imprint specific signatures on lorenz plots. Ann. Noninvasive Electrocardiol. **13**(1), 44–60 (2008)
42. Park, J., Lee, S., Jeon, M.: Atrial fibrillation detection by heart rate variability in poincare plot. Biomed. Eng. Online **8**, 38 (2009)
43. Lian, J., Wang, L., Muessig, D.: A simple method to detect atrial fibrillation using RR intervals. Am. J. Cardiol. **107**(10), 1494–1497 (2011)
44. Costa, M., Goldberger, A.L., Peng, C.-K.: Multiscale entropy analysis of biological signals. Phys. Rev. E Stat. Nonlinear Soft Matter Phys. **71**(2 Pt 1), 021906 (2005)
45. Lake, D.E., Moorman, J.R.: Accurate estimation of entropy in very short physiological time series: the problem of atrial fibrillation detection in implanted ventricular devices. Am. J. Physiol. Heart Circ. Physiol. **300**(1), H319–H325 (2011)
46. Petrėnas, A., Sörnmo, L., Lukoševičius, A., Marozas, V.: Detection of occult paroxysmal atrial fibrillation. Med. Biol. Eng. Comput. **53**(4), 287–297 (2015)

47. Ródenas, J., García, M., Alcaraz, R., Rieta, J.: Wavelet entropy automatically detects episodes of atrial fibrillation from single-lead electrocardiograms. Entropy **17**, 6179–6199 (2015)
48. Erem, B., Martinez Orellana, R., Hyde, D.E., Peters, J.M., Duffy, F.H., Stovicek, P., Warfield, S.K., MacLeod, R.S., Tadmor, G., Brooks, D.H.: Extensions to a manifold learning framework for time-series analysis on dynamic manifolds in bioelectric signals. Phys. Rev. E **93**(4), 042218 (2016)
49. Kishore, A., Vail, A., Majid, A., Dawson, J., Lees, K.R., Tyrrell, P.J., Smith, C.J.: Detection of atrial fibrillation after ischemic stroke or transient ischemic attack: a systematic review and meta-analysis. Stroke **45**(2), 520–526 (2014)
50. Jiang, K., Huang, C., Ye, S.-M., Chen, H.: High accuracy in automatic detection of atrial fibrillation for Holter monitoring. J. Zhejiang Univ. Sci. B **13**(9), 751–756 (2012)
51. Ladavich, S., Ghoraani, B.: Rate-independent detection of atrial fibrillation by statistical modeling of atrial activity. Biomed. Signal Process. Control **18**, 274–281 (2015)
52. Carrara, M., Carozzi, L., Moss, T.J., de Pasquale, M., Cerutti, S., Lake, D.E., Moorman, J.R., Ferrario, M.: Classification of cardiac rhythm using heart rate dynamical measures: validation in MIT-BIH databases. J. Electrocardiol. **48**(6), 943–946 (2015)
53. Slocum, J., Sahakian, A., Swiryn, S.: Diagnosis of atrial fibrillation from surface electrocardiograms based on computer-detected atrial activity. J. Electrocardiol. **25**(1), 1–8 (1992)
54. Larburu, N., Lopetegi, T., Romero, I.: Comparative study of algorithm for atrial fibrillation detection. Comput. Cardiol. **38**, 265–268 (2011)
55. Kao, T., Su, Y.-Y., Lu, C.-C., Tai, C.-T., Chen, S.-A., Lin, Y.-C., Tso, H.-W.: Differentiation of atrial flutter and atrial fibrillation from surface electrocardiogram using nonlinear analysis. J. Med. Biol. Eng. **25**(3), 117–122 (2005)
56. Alcaraz, R., Rieta, J.J., Hornero, F.: Non-invasive characterization of atrial activity immediately prior to termination of paroxysmal atrial fibrillation. Rev. Esp. Cardiol. **61**(2), 154–160 (2008)
57. Mainardi, L.T., Matteucci, M., Sassi, R.: On predicting the spontaneous termination of atrial fibrillation episodes using linear and non-linear parameters of ECG signal and RR series. Conf. Proc. IEEE Comput. Cardiol. **31**, 665–668 (2004)
58. Sun, R.R., Wang, Y.Y.: Predicting spontaneous termination of atrial fibrillation based on the RR interval. Proc. Inst. Mech. Eng. H **223**(6), 713–726 (2009)
59. dan Yuanyuan Wang, R.S.: Predicting termination of atrial fibrillation based on sign sequence of RR interval differences. Chin. J. Sci. Instrum. **7**, 021 (2009)
60. Alcaraz, R., Rieta, J.J.: Sample entropy of the main atrial wave predicts spontaneous termination of paroxysmal atrial fibrillation. Med. Eng. Phys. **31**(8), 917–922 (2009)
61. Alcaraz, R., Rieta, J.J.: Wavelet bidomain sample entropy analysis to predict spontaneous termination of atrial fibrillation. Physiol. Meas. **29**(1), 65–80 (2008)
62. Alcaraz, R., Rieta, J.J.: Central tendency measure and wavelet transform combined in the non-invasive analysis of atrial fibrillation recordings. Biomed. Eng. Online 11, 46 (2012)
63. Sun, R., Wang, Y.: Predicting termination of atrial fibrillation based on the structure and quantification of the recurrence plot. Med. Eng. Phys. **30**(9), 1105–1111 (2008)
64. Alcaraz, R., Rieta, J.J.: Application of wavelet entropy to predict atrial fibrillation progression from the surface ECG. Comput. Math. Methods Med. **2012**, 245213 (2012)
65. Julián, M., Alcaraz, R., Rieta, J.J.: Comparative assessment of nonlinear metrics to quantify organization-related events in surface electrocardiograms of atrial fibrillation. Comput. Biol. Med. **48**, 66–76 (2014)
66. Alcaraz, R., Hornero, F., Martínez, A., Rieta, J.J.: Short-time regularity assessment of fibrillatory waves from the surface ECG in atrial fibrillation. Physiol. Meas. **33**(6), 969–984 (2012)
67. Alcaraz, R., Hornero, F., Rieta, J.J.: Assessment of non-invasive time and frequency atrial fibrillation organization markers with unipolar atrial electrograms. Physiol. Meas. **32**(1), 99–114 (2011)

68. Konings, K.T., Kirchhof, C.J., Smeets, J.R., Wellens, H.J., Penn, O.C., Allessie, M.A.: High-density mapping of electrically induced atrial fibrillation in human. Circulation **89**(4), 1665–1680 (1994)
69. Kantz, H., Schreiber, T.: Nonlinear Time Series Analysis. Cambridge University Press, Cambrigde (2003)
70. Alcaraz, R., Rieta, J.J.: Central tendency measure and wavelet transform combined in the non-invasive analysis of atrial fibrillation recordings. Biomed. Eng. Online **11**, 46 (2012)
71. de Vos, C.B., Pisters, R., Nieuwlaat, R., Prins, M.H., Tieleman, R.G., Coelen, R.-J.S., van den Heijkant, A.C., Allessie, M.A., Crijns, H.J.G.M.: Progression from paroxysmal to persistent atrial fibrillation clinical correlates and prognosis. J. Am. Coll. Cardiol. **55**(8), 725–731 (2010)
72. Tuzcu, V., Nas, S., Börklü, T., Ugur, A.: Decrease in the heart rate complexity prior to the onset of atrial fibrillation. Europace **8**(6), 398–402 (2006)
73. Tan, A.Y., Verrier, R.L.: The role of the autonomic nervous system in cardiac arrhythmias. Handb. Clin. Neurol. **117**, 135–145 (2013)
74. Vikman, S., Mäkikallio, T.H., Yli-Mäyry, S., Pikkujämsä, S., Koivisto, A.M., Reinikainen, P., Airaksinen, K.E., Huikuri, H.V.: Altered complexity and correlation properties of R-R interval dynamics before the spontaneous onset of paroxysmal atrial fibrillation. Circulation **100**(20), 2079–2084 (1999)
75. Haïssaguerre, M., Jaïs, P., Shah, D.C., Takahashi, A., Hocini, M., Quiniou, G., Garrigue, S., Mouroux, A.L., Métayer, P.L., Clémenty, J.: Spontaneous initiation of atrial fibrillation by ectopic beats originating in the pulmonary veins. N. Engl. J. Med. **339**(10), 659–666 (1998)
76. Shin, D.-G., Yoo, C.-S., Yi, S.-H., Bae, J.-H., Kim, Y.-J., Park, J.-S., Hong, G.-R.: Prediction of paroxysmal atrial fibrillation using nonlinear analysis of the R-R interval dynamics before the spontaneous onset of atrial fibrillation. Circ. J. **70**(1), 94–99 (2006)
77. Chesnokov, Y.V.: Complexity and spectral analysis of the heart rate variability dynamics for distant prediction of paroxysmal atrial fibrillation with artificial intelligence methods. Artif. Intell. Med. **43**(2), 151–165 (2008)
78. Mohebbi, M., Ghassemian, H.: Prediction of paroxysmal atrial fibrillation using recurrence plot-based features of the RR-interval signal. Physiol. Meas. **32**(8), 1147–1162 (2011)
79. Mohebbi, M., Ghassemian, H.: Prediction of paroxysmal atrial fibrillation based on non-linear analysis and spectrum and bispectrum features of the heart rate variability signal. Comput. Methods Programs Biomed. **105**(1), 40–49 (2012)
80. Martínez, A., Abásolo, D., Alcaraz, R., Rieta, J.J.: Alteration of the P-wave non-linear dynamics near the onset of paroxysmal atrial fibrillation. Med. Eng. Phys. **37**(7), 692–697 (2015)
81. Dilaveris, P.E., Gialafos, J.E.: P-wave dispersion: a novel predictor of paroxysmal atrial fibrillation. Ann. Noninvasive Electrocardiol. **6**(2), 159–165 (2001)
82. Bauernschmitt, R., Malberg, H., Wessel, N., Brockmann, G., Wildhirt, S.M., Kopp, B., Kurths, J., Bretthauer, G., Lange, R.: Autonomic control in patients experiencing atrial fibrillation after cardiac surgery. Pacing Clin. Electrophysiol. **30**(1), 77–84 (2007)
83. Lahtinen, J., Biancari, F., Salmela, E., Mosorin, M., Satta, J., Rainio, P., Rimpiläinen, J., Lepojärvi, M., Juvonen, T.: Postoperative atrial fibrillation is a major cause of stroke after on-pump coronary artery bypass surgery. Ann. Thorac. Surg. **77**(4), 1241–1244 (2004)
84. Hogue Jr., C.W., Domitrovich, P.P., Stein, P.K., Despotis, G.D., Re, L., Schuessler, R.B., Kleiger, R.E., Rottman, J.N.: RR interval dynamics before atrial fibrillation in patients after coronary artery bypass graft surgery. Circulation **98**(5), 429–434 (1998)
85. Tarkiainen, T.H., Hakala, T., Hedman, A., Vanninen, E.: Preoperative alterations in correlation properties and complexity of R-R interval dynamics predict the risk of atrial fibrillation after coronary artery bypass grafting in patients with preserved left ventricular function. J. Cardiovasc. Electrophysiol. **19**(9), 907–912 (2008)
86. Chamchad, D., Djaiani, G., Jung, H.J., Nakhamchik, L., Carroll, J., Horrow, J.C.: Nonlinear heart rate variability analysis may predict atrial fibrillation after coronary artery bypass grafting. Anesth. Analg. **103**(5), 1109–1112 (2006)

87. Chamchad, D., Horrow, J.C., Samuels, L.E., Nakhamchik, L.: Heart rate variability measures poorly predict atrial fibrillation after off-pump coronary artery bypass grafting. J. Clin. Anesth. **23**(6), 451–455 (2011)
88. Faes, L., Nollo, G., Antolini, R., Gaita, F., Ravelli, F.: A method for quantifying atrial fibrillation organization based on wave-morphology similarity. IEEE Trans. Biomed. Eng. **49**(12 Pt 2), 1504–1513 (2002)
89. Konings, K.T., Smeets, J.L., Penn, O.C., Wellens, H.J., Allessie, M.A.: Configuration of unipolar atrial electrograms during electrically induced atrial fibrillation in humans. Circulation **95**(5), 1231–1241 (1997)
90. Wells Jr., J.L., Karp, R.B., Kouchoukos, N.T., MacLean, W.A., James, T.N., Waldo, A.L.: Characterization of atrial fibrillation in man: studies following open heart surgery. Pacing Clin. Electrophysiol. **1**(4), 426–438 (1978)
91. Gaita, F., Calò, L., Riccardi, R., Garberoglio, L., Scaglione, M., Licciardello, G., Coda, L., Di Donna, P., Bocchiardo, M., Caponi, D., Antolini, R., Orzan, F., Trevi, G.P.: Different patterns of atrial activation in idiopathic atrial fibrillation: simultaneous multisite atrial mapping in patients with paroxysmal and chronic atrial fibrillation. J. Am. Coll. Cardiol. **37**(2), 534–541 (2001)
92. Hoekstra, B.P., Diks, C.G., Allessie, M.A., DeGoede, J.: Nonlinear analysis of epicardial atrial electrograms of electrically induced atrial fibrillation in man. J. Cardiovasc. Electrophysiol. **6**(6), 419–440 (1995)
93. Mainardi, L.T., Porta, A., Calcagnini, G., Bartolini, P., Michelucci, A., Cerutti, S.: Linear and non-linear analysis of atrial signals and local activation period series during atrial-fibrillation episodes. Med. Biol. Eng. Comput. **39**(2), 249–254 (2001)
94. Mainardi, L.T., Porta, A., Calcagnini, G., Censi, F., Bartolini, P., Michelucci, A., Cerutti, S.: Discrimination of atrial rhythms by linear and non-linear methods. Ann. Ist. Super. Sanita **37**(3), 335–340 (2001)
95. Pitschner, H.F., Berkovic, A., Grumbrecht, S., Neuzner, J.: Multielectrode basket catheter mapping for human atrial fibrillation. J. Cardiovasc. Electrophysiol. **9**(Suppl.), S48–S56 (1998)
96. Cervigón, R., Moreno, J., Reilly, R.B., Millet, J., Pérez-Villacastín, J., Castells, F.: Entropy measurements in paroxysmal and persistent atrial fibrillation. Physiol. Meas. **31**(7), 1011–1020 (2010)
97. Censi, F., Barbaro, V., Bartolini, P., Calcagnini, G., Michelucci, A., Cerutti, S.: Non-linear coupling of atrial activation processes during atrial fibrillation in humans. Biol. Cybern. **85**(3), 195–201 (2001)
98. Mainardi, L.T., Corino, V.D.A., Lombardi, L., Tondo, C., Mantica, M., Lombardi, F., Cerutti, S.: Assessment of the dynamics of atrial signals and local atrial period series during atrial fibrillation: effects of isoproterenol administration. Biomed. Eng. Online **3**(1), 37 (2004)
99. Mainardi, L.T., Corino, V.D.A., Lombardi, L., Tondo, C., Mantica, M., Lombardi, F., Cerutti, S.: Linear and nonlinear coupling between atrial signals. Three methods for the analysis of the relationships among atrial electrical activities in different sites. IEEE Eng. Med. Biol. Mag. **25**(6), 63–70 (2006)
100. Censi, F., Barbaro, V., Bartolini, P., Calcagnini, G., Michelucci, A., Gensini, G.F., Cerutti, S.: Recurrent patterns of atrial depolarization during atrial fibrillation assessed by recurrence plot quantification. Ann. Biomed. Eng. **28**(1), 61–70 (2000)
101. Masè, M., Faes, L., Antolini, R., Scaglione, M., Ravelli, F.: Quantification of synchronization during atrial fibrillation by Shannon entropy: validation in patients and computer model of atrial arrhythmias. Physiol. Meas. **26**(6), 911–923 (2005)
102. Sanders, P., Berenfeld, O., Hocini, M., Jaïs, P., Vaidyanathan, R., Hsu, L.-F., Garrigue, S., Takahashi, Y., Rotter, M., Sacher, F., Scavée, C., Ploutz-Snyder, R., Jalife, J., Haïssaguerre, M.: Spectral analysis identifies sites of high-frequency activity maintaining atrial fibrillation in humans. Circulation **112**(6), 789–797 (2005)

103. Alcaraz, R., Rieta, J.J.: The application of nonlinear metrics to assess organization differences in short recordings of paroxysmal and persistent atrial fibrillation. Physiol. Meas. **31**(1), 115–130 (2010)
104. Alcaraz, R., Sandberg, F., Sörnmo, L., Rieta, J.J.: Classification of paroxysmal and persistent atrial fibrillation in ambulatory ECG recordings. IEEE Trans. Biomed. Eng. **58**(5), 1441–1449 (2011)
105. Ortigosa, N., Cano, Ó., Ayala, G., Galbis, A., Fernández, C.: Atrial fibrillation subtypes classification using the general fourier-family transform. Med. Eng. Phys. **36**(4), 554–560 (2014)
106. Ortigosa, N., Fernández, C., Galbis, A., Cano, Ó.: Phase information of time-frequency transforms as a key feature for classification of atrial fibrillation episodes. Physiol. Meas. **36**(3), 409–424 (2015)
107. Ravelli, F., Masè, M., Del Greco, M., Faes, L., Disertori, M.: Deterioration of organization in the first minutes of atrial fibrillation: a beat-to-beat analysis of cycle length and wave similarity. J. Cardiovasc. Electrophysiol. **18**(1), 60–65 (2007)
108. Alcaraz, R., Rieta, J.J.: Non-invasive organization variation assessment in the onset and termination of paroxysmal atrial fibrillation. Comput. Methods Programs Biomed. **93**(2), 148–154 (2009)
109. Julián, M., Alcaraz, R., Rieta, J.J.: Application of hurst exponents to assess atrial reverse remodeling in paroxysmal atrial fibrillation. Physiol. Meas. **36**(11), 2231–2246 (2015)
110. Bollmann, A., Sonne, K., Esperer, H.D., Toepffer, I., Langberg, J.J., Klein, H.U.: Non-invasive assessment of fibrillatory activity in patients with paroxysmal and persistent atrial fibrillation using the Holter ECG. Cardiovasc. Res. **44**(1), 60–66 (1999)
111. Ropella, K.M., Sahakian, A.V., Baerman, J.M., Swiryn, S.: Effects of procainamide on intra-atrial electrograms during atrial fibrillation: implications for detection algorithms. Circulation **77**(5), 1047–1054 (1988)
112. Bollmann, A., Kanuru, N.K., McTeague, K.K., Walter, P.F., DeLurgio, D.B., Langberg, J.J.: Frequency analysis of human atrial fibrillation using the surface electrocardiogram and its response to ibutilide. Am. J. Cardiol. **81**(12), 1439–1445 (1998)
113. Alcaraz, R., Hornero, F., Rieta, J.J.: Surface ECG organization time course analysis along onward episodes of paroxysmal atrial fibrillation. Med. Eng. Phys. **33**(5), 597–603 (2011)
114. Daoud, E.G., Bogun, F., Goyal, R., Harvey, M., Man, K.C., Strickberger, S.A., Morady, F.: Effect of atrial fibrillation on atrial refractoriness in humans. Circulation **94**(7), 1600–1606 (1996)
115. Corino, V.D.A., Sandberg, F., Mainardi, L.T., Sornmo, L.: An atrioventricular node model for analysis of the ventricular response during atrial fibrillation. IEEE Trans. Biomed. Eng. **58**(12), 3386–3395 (2011)
116. Hayano, J., Ishihara, S., Fukuta, H., Sakata, S., Mukai, S., Ohte, N., Kimura, G.: Circadian rhythm of atrioventricular conduction predicts long-term survival in patients with chronic atrial fibrillation. Chronobiol. Int. **19**(3), 633–648 (2002)
117. Hayano, J., Sakata, S., Okada, A., Mukai, S., Fujinami, T.: Circadian rhythms of atrioventricular conduction properties in chronic atrial fibrillation with and without heart failure. J. Am. Coll. Cardiol. **31**(1), 158–166 (1998)
118. Oka, T., Nakatsu, T., Kusachi, S., Tominaga, Y., Toyonaga, S., Ohnishi, H., Nakahama, M., Komatsubara, I., Murakami, M., Tsuji, T.: Double-sector lorenz plot scattering in an R-R interval analysis of patients with chronic atrial fibrillation: incidence and characteristics of vertices of the double-sector scattering. J. Electrocardiol. **31**(3), 227–235 (1998)
119. Climent, A.M., de la Salud Guillem, M., Husser, D., Castells, F., Millet, J., Bollmann, A.: Poincaré surface profiles of RR intervals: a novel noninvasive method for the evaluation of preferential AV nodal conduction during atrial fibrillation. IEEE Trans. Biomed. Eng. **56**(2), 433–442 (2009)
120. Climent, A.M., Guillem, M.S., Husser, D., Castells, F., Millet, J., Bollmann, A.: Role of the atrial rate as a factor modulating ventricular response during atrial fibrillation. Pacing Clin. Electrophysiol. **33**(12), 1510–1517 (2010)

121. Corino, V.D.A., Holmqvist, F., Mainardi, L.T., Platonov, P.G.: Beta-blockade and a1-adenosine receptor agonist effects on atrial fibrillatory rate and atrioventricular conduction in patients with atrial fibrillation. Europace **16**(4), 587–594 (2014)
122. Corino, V.D.A., Ulimoen, S.R., Enger, S., Mainardi, L.T., Tveit, A., Platonov, P.G.: Rate-control drugs affect variability and irregularity measures of RR intervals in patients with permanent atrial fibrillation. J. Cardiovasc. Electrophysiol. **26**(2), 137–141 (2015)
123. Alcaraz, R., Rieta, J.J.: Nonlinear synchronization assessment between atrial and ventricular activations series from the surface ECG in atrial fibrillation. Biomed. Signal Process. Control **8**(6), 1000–1007 (2013)
124. European Heart Rhythm Association, European Association for Cardio-Thoracic Surgery, Camm, A.J., Kirchhof, P., Lip, G.Y.H., Schotten, U., Savelieva, I., Ernst, S., Van Gelder, I.C., Al-Attar, N., Hindricks, G., Prendergast, B., Heidbuchel, H., Alfieri, O., Angelini, A., Atar, D., Colonna, P., De Caterina, R., De Sutter, J., Goette, A., Gorenek, B., Heldal, M., Hohloser, S.H., Kolh, P., Le Heuzey, J.-Y., Ponikowski, P., Rutten, F.H.: Guidelines for the management of atrial fibrillation: the task force for the management of atrial fibrillation of the European Society of Cardiology (ESC). Eur. Heart J. **31**(19), 2369–429 (2010)
125. Hoekstra, B.P.T., Diks, C.G.H., Allessie, M.A., DeGoede, J.: Nonlinear analysis of the pharmacological conversion of sustained atrial fibrillation in conscious goats by the class Ic drug cibenzoline. Chaos **7**(3), 430–446 (1997)
126. Hoekstra, B.P., Diks, C.G., Allessie, M.A., DeGoede, J.: Spatial correlation analysis of the pharmacological conversion of sustained atrial fibrillation in conscious goats by cibenzoline. Arch. Physiol. Biochem. **108**(4), 332–348 (2000)
127. Berkowitsch, A., Carlsson, J., Erdogan, A., Neuzner, J., Pitschner, H.F.: Electrophysiological heterogeneity of atrial fibrillation and local effect of propafenone in the human right atrium: analysis based on symbolic dynamics. J. Interv. Card. Electrophysiol. **4**(2), 383–394 (2000)
128. Cervigón, R., Moreno, J., Sánchez, C., Reilly, R.B., Villacastín, J., Millet, J., Castells, F.: Atrial fibrillation organization: quantification of propofol effects. Med. Biol. Eng. Comput. **47**(3), 333–341 (2009)
129. Corino, V.D.A., Mantica, M., Lombardi, F., Mainardi, L.T.: Assessment of spatial organization in the atria during paroxysmal atrial fibrillation and adrenergic stimulation. Biomed. Technol. (Berlin) **51**(4), 260–263 (2006)
130. Rivolta, M., Mainardi, L., Sassi, R.: A novel method of atrial fibrillation organization based on symbolic analysis. Comput. Cardiol. Conf. **39**, 813–816 (2012)
131. Gelzer, A.R., Moïse, N.S., Vaidya, D., Wagner, K.A., Jalife, J.: Temporal organization of atrial activity and irregular ventricular rhythm during spontaneous atrial fibrillation: an in vivo study in the horse. J. Cardiovasc. Electrophysiol. **11**(7), 773–784 (2000)
132. Ferrari, R., Bertini, M., Blomstrom-Lundqvist, C., Dobrev, D., Kirchhof, P., Pappone, C., Ravens, U., Tamargo, J., Tavazzi, L., Vicedomini, G.G.: An update on atrial fibrillation in 2014: from pathophysiology to treatment. Int. J. Cardiol. **203**, 22–29 (2016)
133. Gall, N.P., Murgatroyd, F.D.: Electrical cardioversion for AF-the state of the art. Pacing Clin. Electrophysiol. **30**(4), 554–567 (2007)
134. Lankveld, T., de Vos, C.B., Limantoro, I., Zeemering, S., Dudink, E., Crijns, H.J., Schotten, U.: Systematic analysis of ECG predictors of sinus rhythm maintenance after electrical cardioversion for persistent atrial fibrillation. Heart Rhythm **13**(5), 1020–1027 (2016)
135. Alcaraz, R., Rieta, J.J.: A non-invasive method to predict electrical cardioversion outcome of persistent atrial fibrillation. Med. Biol. Eng. Comput. **46**(7), 625–635 (2008)
136. Zohar, P., Kovacic, M., Brezocnik, M., Podbregar, M.: Prediction of maintenance of sinus rhythm after electrical cardioversion of atrial fibrillation by non-deterministic modelling. Europace **7**(5), 500–507 (2005)
137. Van Den Berg, M.P., Van Noord, T., Brouwer, J., Haaksma, J., Van Veldhuisen, D.J., Crijns, H.J.G.M., Van Gelder, I.C.: Clustering of RR intervals predicts effective electrical cardioversion for atrial fibrillation. J. Cardiovasc. Electrophysiol. **15**(9), 1027–1033 (2004)

138. Guo, X.H., Gallagher, M.M., Bland, J.M., Camm, A.J.: A distinctly bimodal distribution pattern in the RR interval histogram predicts early recurrence of atrial fibrillation after electrical cardioversion. Int. J. Cardiol. **145**(2), 244–245 (2010)
139. Vikman, S., Mäkikallio, T.H., Yli-Mäyry, S., Nurmi, M., Airaksinen, K.E.J., Huikuri, H.V.: Heart rate variability and recurrence of atrial fibrillation after electrical cardioversion. Ann. Med. **35**(1), 36–42 (2003)
140. Rademacher, W., Seeck, A., Surber, R., Lauten, A., Prochnau, D., Voss, A., Figulla, H.R.: Multidimensional ECG-based analysis of cardiac autonomic regulation predicts early AF recurrence after electrical cardioversion. J. Electrocardiol. **45**(2), 116–122 (2012)
141. Brisinda, D., Venuti, A., Iantorno, E., Efremov, K., Cataldi, C., Sorbo, A., Fioravanti, F., Fenici, R.: Discrimimant analysis of heart rate variability after electrical cardioversion predicts atrial fibrilation recurrence. Int. J. Clin. Cardiol. **1**, 10 (2014)
142. Seeck, A., Rademacher, W., Fischer, C., Haueisen, J., Surber, R., Voss, A.: Prediction of atrial fibrillation recurrence after cardioversion-interaction analysis of cardiac autonomic regulation. Med. Eng. Phys. **35**(3), 376–382 (2013)
143. Castro, N., Grimaldos, J., Kodoth, V., Lau, E., Manoharam, G., Escalona, O.: Assessment of atrial fibrillation organization during internal electrical cardioversion using sample entropy. In: Andean Region International Conference, pp. 191–194 (2012)
144. Alcaraz, R., Rieta, J.J., Hornero, F.: Non-invasive atrial fibrillation organization follow-up under successive attempts of electrical cardioversion. Med. Biol. Eng. Comput. **47**(12), 1247–1255 (2009)
145. Maurer, T., Lundqvist, C.B., Tilz, R., Mont, L., Chierchia, G.-B., Malmborg, H., Metzner, A., Kuck, K.-H.: What have we learned of ablation procedures for atrial fibrillation? J. Intern. Med. **279**(5), 439–448 (2016)
146. Lin, D., Santangeli, P., Zado, E.S., Bala, R., Hutchinson, M.D., Riley, M.P., Frankel, D.S., Garcia, F., Dixit, S., Callans, D.J., Marchlinski, F.E.: Electrophysiologic findings and long-term outcomes in patients undergoing third or more catheter ablation procedures for atrial fibrillation. J. Cardiovasc. Electrophysiol. **26**(4), 371–377 (2015)
147. Tilz, R.R., Chun, K.R.J., Schmidt, B., Fuernkranz, A., Wissner, E., Koester, I., Baensch, D., Boczor, S., Koektuerk, B., Metzner, A., Zerm, T., Ernst, S., Antz, M., Kuck, K.-H., Ouyang, F.: Catheter ablation of long-standing persistent atrial fibrillation: a lesson from circumferential pulmonary vein isolation. J. Cardiovasc. Electrophysiol. **21**(10), 1085–1093 (2010)
148. Nademanee, K., McKenzie, J., Kosar, E., Schwab, M., Sunsaneewitayakul, B., Vasavakul, T., Khunnawat, C., Ngarmukos, T.: A new approach for catheter ablation of atrial fibrillation: mapping of the electrophysiologic substrate. J. Am. Coll. Cardiol. **43**(11), 2044–2053 (2004)
149. Wynn, G.J., Das, M., Bonnett, L.J., Panikker, S., Wong, T., Gupta, D. : Efficacy of catheter ablation for persistent atrial fibrillation: a systematic review and meta-analysis of evidence from randomized and nonrandomized controlled trials. Circ. Arrhythm. Electrophysiol. **7**(5), 841–852 (2014)
150. Ng, J., Borodyanskiy, A.I., Chang, E.T., Villuendas, R., Dibs, S., Kadish, A.H., Goldberger, J.J.: Measuring the complexity of atrial fibrillation electrograms. J. Cardiovasc. Electrophysiol. **21**(6), 649–655 (2010)
151. Cirugeda-Roldán, E.M., Novák, D., Kremen, V., Cuesta-Frau, D., Keller, M.W., Luik, A., Srutová, M.: Characterization of complex fractionated atrial electrograms by sample entropy: an international multi-center study. Entropy **17**(11), 7493–7509 (2015)
152. Navoret, N., Jacquir, S., Laurent, G., Binczak, S.: Detection of complex fractionated atrial electrograms using recurrence quantification analysis. IEEE Trans. Biomed. Eng. **60**(7), 1975–1982 (2013)
153. Orozco-Duque, A., Novak, D., Kremen, V., Bustamante, J.: Multifractal analysis for grading complex fractionated electrograms in atrial fibrillation. Physiol. Meas. **36**(11), 2269–2284 (2015)

154. Narayan, S.M., Baykaner, T., Clopton, P., Schricker, A., Lalani, G.G., Krummen, D.E., Shivkumar, K., Miller, J.M.: Ablation of rotor and focal sources reduces late recurrence of atrial fibrillation compared with trigger ablation alone: extended follow-up of the confirm trial (conventional ablation for atrial fibrillation with or without focal impulse and rotor modulation). J. Am. Coll. Cardiol. **63**(17), 1761–1768 (2014)
155. Haissaguerre, M., Hocini, M., Shah, A.J., Derval, N., Sacher, F., Jais, P., Dubois, R.: Noninvasive panoramic mapping of human atrial fibrillation mechanisms: a feasibility report. J. Cardiovasc. Electrophysiol. **24**(6), 711–717 (2013)
156. Ugarte, J.P., Orozco-Duque, A., Tobón, C., Kremen, V., Novak, D., Saiz, J., Oesterlein, T., Schmitt, C., Luik, A., Bustamante, J.: Dynamic approximate entropy electroanatomic maps detect rotors in a simulated atrial fibrillation model. PLoS One **9**(12), e114577 (2014)
157. Ganesan, A.N., Kuklik, P., Gharaviri, A., Brooks, A., Chapman, D., Lau, D.H., Roberts-Thomson, K.C., Sanders, P.: Origin and characteristics of high Shannon entropy at the pivot of locally stable rotors: insights from computational simulation. PLoS One **9**(11), e110662 (2014)
158. Balk, E.M., Garlitski, A.C., Alsheikh-Ali, A.A., Terasawa, T., Chung, M., Ip, S.: Predictors of atrial fibrillation recurrence after radiofrequency catheter ablation: a systematic review. J. Cardiovasc. Electrophysiol. **21**(11), 1208–1216 (2010)
159. Cervigón, R., Moreno, J., García-Quintanilla, J., Pérez-Villacastín, J., Castells, F.: Entropy at the right atrium as a predictor of atrial fibrillation recurrence outcome after pulmonary vein ablation. Biomed. Technol. (Berlin) **61**(1), 29–36 (2016)
160. Cervigón, R., Moreno, J., Castells, F.: Entropy analysis of atrial activity morphology to study atrial fibrillation recurrences after ablation procedure. In: Bioinformatics and Biomedical Engineering - Third International Conference, IWBBIO 2015, Granada, 15–17 April 2015 Proceedings, Part I, pp. 146–154 (2015)
161. Albåge, A., Johansson, B., Kennebäck, G., Källner, G., Scherstén, H., Jidéus, L., Swedish Arrhythmia Surgery Group: Long-term follow-up of cardiac rhythm in 320 patients after the Cox-Maze III procedure for atrial fibrillation. Ann. Thorac. Surg. **101**(4), 1443–1449 (2016)
162. Lawrance, C.P., Henn, M.C., Damiano Jr., R.J.: Surgery for atrial fibrillation. Heart Fail. Clin. **12**(2), 235–243 (2016)
163. Hernández, A., Alcaraz, R., Hornero, F., Rieta, J.J.: Preoperative study of the surface ECG for the prognosis of atrial fibrillation maze surgery outcome at discharge. Physiol. Meas. **35**(7), 1409–1423 (2014)
164. Hernandez, A., Alcaraz, R., Hornero, F., Rieta, J.: Preoperative prognosis of atrial fibrillation concomitant surgery outcome after the blanking period. In: Mediterranean Conference on Medical and Biological Engineering and Computing, vol. 41, pp. 1879–1882 (2014)
165. Hernandez, A., Alcaraz, R., Hornero, F., Rieta, J.: Application of the preoperative ECG to predict Cox-Maze surgery mid-term outcome. In: Computing in Cardiology Conference, vol. 40, pp. 723–726 (2013)

Chapter 16
Complex and Nonlinear Analysis of Heart Rate Variability in the Assessment of Fetal and Neonatal Wellbeing

Maria G. Signorini, Giovanni Magenes, Manuela Ferrario, and Maristella Lucchini

Abstract Study of fetal and neonatal heart rate variability offers a powerful tool to monitor, even indirectly, the physiological control mechanisms playing a crucial role in the early development of human life. Moreover, sleep state analysis is a challenging topic as both quality of the sleep and its relationship with neural development and control can provide information potentially useful to the clinical environment. This contribution aims at presenting major advances in heart rate variability analysis in both fetuses and newborn population during sleep. The important role of quantitative multi-parameter analysis methods will be discussed with reference to new monitoring and diagnostic solutions. Results will be presented and discussed showing how sleep dynamics are related to cardiovascular control systems with the major focus of deeper investigating the development of human life in its early stages.

M.G. Signorini (✉) • M. Ferrario
Dipartimento di Elettronica, Informazione e Bioingegneria, DEIB, Politecnico di Milano, piazza Leonardo da Vinci 32, 20133, Milano, Italy
e-mail: mariagabriella.signorini@polimi.it

G. Magenes
Dipartimento di Ingegneria Industriale e dell'Informazione, Università degli Studi di Pavia, Pavia, Italy

M. Lucchini
Dipartimento di Elettronica, Informazione e Bioingegneria, DEIB, Politecnico di Milano, piazza Leonardo da Vinci 32, 20133, Milano, Italy

Department of Psychiatry, Columbia University Medical Center, New York, NY, USA

R. Barbieri et al. (eds.), *Complexity and Nonlinearity in Cardiovascular Signals*,
DOI 10.1007/978-3-319-58709-7_16

16.1 Introduction

16.1.1 Heart Rate Variability: Fetal and Neonatal Physiology and Control

Heart Rate Variability (HRV) signal is a well-known marker of heart functions. In addition to the ECG signal, from which it is derived, it contains information related both to heart functioning (regularity of beats morphology) and to physiological control systems that modulates it.

Activity of sympathetic and parasympathetic branches of the Autonomic Nervous System (ANS) are now clearly recognizable through processing of HRV time series and extraction of indices connected, even indirectly, to its activity [1].

Information extraction has been allowed by the evolution of analytic methods. At first, they were based on time domain methods only, mainly related to measurement of the signal variance. In the 1980s, spectral analysis in the frequency domain started to classify contribution of the ANS to HR regulation [2].

Moreover, these encouraging and important results did not conclude the information extraction process from the HR. In the late 1980s, new methods based on nonlinear analysis started to demonstrate that the HR content was more complex than expected. In time, it has been recognized that HR does not only contain information which can be explained by classical and linear methods. Nonlinear characteristics, such as unpredictability, complexity, similarity in time scales and pattern characterization were recognized in HR and measured with the new methods provided by the emerging nonlinear chaotic time series analysis [3].

The large use of these approaches characterized the HR analysis in the recent years, providing interesting features that demonstrated their usefulness in the classification and diagnostic process of several diseases [4].

In this chapter we want to summarize the main steps in the analysis of nonlinear and complex characteristics of HR signal in fetuses and newborn infants. Our main objective is to show how complexity and nonlinearity could contribute in understanding the dynamics of HR signal, thus providing additional tools in monitoring fetal and neonatal wellbeing.

16.1.1.1 Fetal: Development, Wellbeing, Disease, Risk, Prevention, Stress

Monitoring Fetal Heart Rate (FHR) during the pregnancy, starting from 24 weeks Gestational Age (GA) to delivery, is the main clinical and diagnostic tool employed in all pregnancies.

Pregnancy is not defined as a disease; nevertheless, it is undoubtedly a period of time when both mother and fetus can be at risk and for this reason great attention is dedicated to know if fetal development proceeds correctly or if possible pathological events are occurring. The main goal of monitoring FHR is thus to collect information about fetal development without interfering invasively.

The clinical relevance of HRV was first appreciated in 1965 when Hon and Lee [5] noted that fetal distress was preceded by alterations in inter-beat intervals before any appreciable change occurred in mean HR itself.

Around 24–25 weeks, the baby weighs about 600 g and measures about 30 cm. Patterns of sleeping and waking are becoming more defined, although still unpredictable in time and duration. Movements increase too, involving legs and arms. The brain is growing rapidly and even facial muscles start to have different expressions. All this events reveal that not only the baby body size is growing, but other nervous circuits at central and peripheral level are developing, influencing the physiological growth.

Unfortunately, stressful condition can affect the pregnancy, sometimes hidden even for long periods. Reduced fetal growth due to both maternal and fetal factors, diabetes and high levels of blood pressure in pregnancy are elements that can affect early fetal life. Some of them can generate risky conditions, till the fetal death or impose urgent hospitalization with cesarean section and premature delivery. Among them the most diffused fetal disease is the Intra Uterine Growth Restriction (IUGR), the second leading cause of perinatal morbidity and mortality. Incidence of IUGR is around 5% in the general obstetric population. Timely diagnosis and management of IUGR is one of the major achievements in contemporary obstetrics as its early identification leads to perinatal mortality reduction.

16.1.1.2 Neonatal: Cardiovascular Control, Prematurity, SIDS

Even in the population of newborn infants, HR monitoring and analysis represent one of the most used diagnostic tool.

Relative dominance of the sympathetic and parasympathetic system existent at birth, changes as the newborn infant develops. The parasympathetic system is fully formed shortly after birth, while the sympathetic innervation is incomplete [6].

While maturation of the ANS allows the full-term newborn to adapt its respiratory and hemodynamic responses to internal and external demands, a premature infant (born <37th weeks of gestation) may inappropriately adapt to environmental, nutritional or external conditions due to ANS immaturity.

Moreover, many external conditions can affect the development and the functioning of the ANS. Among them there are prenatal factors, such as alcohol and smoking exposure, but also postnatal environmental influences, such as room temperature and position during sleep.

Many studies have explored differences in cardiovascular control in the context of risk conditions, in particular during sleep in the first months of life, in order to monitor possible dangers in this helpless age.

As noticed for fetuses and adult subjects, even in newborns HRV changes can be seen as a first marker of distress. These issues are particularly relevant in Sudden Infant Death Syndrome (SIDS), as both cardiorespiratory and maturational factors have been implicated in SIDS etiology, suggesting that alterations in ANS function may be involved. Morphological anomalies have been shown in brainstem areas associated with autonomic control, and physiological evidence for autonomic dysfunction includes reduced HRV and prolonged QT interval in infants subsequently succumbing to SIDS [7, 8].

16.2 Collection of Fetal Heart Rate Signal: From Cardiotocographic Technique to Wearable Fetal ECG

Antenatal monitoring of FHR is commonly performed by means of Cardiotochography (CTG), which represents the most widespread technique for assessing the health status of the fetus during pregnancy. CTG consists of the simultaneous measurement of FHR and uterine contractions, by an external US probe and a pressure transducer both applied to the maternal abdomen. FHR is detected by exploiting the Doppler effect of an US beam directed towards the fetus's heart. The frequency of the echo signal is changed due to heart tissue movements, and its variations allow the computation of the heart cycle duration. As a matter of fact, beat-to-beat FHR was initially calculated through a peak detection algorithm of the US echo demodulated signal. However, since two heartbeats could have multiple peaks, different time intervals could be detected and consequently different FHR values could be provided. The introduction of autocorrelation techniques solved this problem, since they are based on the comparison of two successive pulse waveforms. This methodological improvement has led the US technology to closely approach the "gold standard" of the direct ECG in detecting the FHR signal. By means of a peak position interpolation algorithm, the effective resolution in detecting the heart period (the equivalent of RR period) can be better than 2 ms [9].

CTG monitors usually display only the FHR expressed in number of beats per minute (bpm) and the actual FHR value is updated depending on the model and the producer of the CTG monitor. All HP-like commercial monitors (HP series 13XX, Corometrics, Philips, General Electric) measure FHR values in bpm every 250 ms, store them in a buffer and finally compute the average on ten consecutive samples, thus producing a new FHR update each 2.5 s. This update rate is not enough for HRV analysis and we decided to read the FHR values each 0.5 s, which represents a reasonable compromise between bandwidth and accuracy of the FHR signal.

Although new devices have been recently proposed to the market of CTG systems, a CTG examination requires in all cases both a quite expensive equipment and a trained paramedical staff at the hospital, because a proper positioning of the ultrasound probe on the pregnant woman abdomen is compulsory. These constraints severely limit the possibility of recording frequent and long lasting sessions on patients at risk. Furthermore, despite the minimal invasiveness of ultrasound techniques in diagnostics, their use is not deprived of any risk, in particular when they should be applied frequently and for long durations on fetuses [10]. For those reasons, even disregarding the cost of the equipment, CTG is not a diagnostic technique which can be accomplished at home by the patient herself as a way to increase the frequency of its use.

To overcome the problem, less invasive approaches are being pursued, in order to balance the need of frequent home investigations, directly managed by the patient herself with safety. Those are aimed either at extracting FHR using alternative techniques such as phonocardiography [11], or even at assessing a fetus's wellbeing based on different acquisitions, such as measuring the oscillations of the maternal abdominal wall caused by fetus's movements through capacitive sensors [12]. Even abdominal ECG can be exploited for FHR monitoring, despite the difficulty of revealing and separating the fetal signal from the maternal one. With a multi electrode wearable belt, such as proposed in [13] (Telefetalcare), it will be possible to select from multiple derivations the best source for the fetal ECG signal and thus to obtain a reliable FHR time series. Moreover, the development and the diffusion of similar wearable devices would solve both the problems of the high costs of the CTG instrumentation and the necessity of performing the examination at the hospital.

16.3 Methods for Monitoring Fetal HRV Antepartum

Analysis of FHR signal as obtained from CTG recordings was characterized by several, incremental steps that followed the development of signal processing knowledges about time series structures. This section wants to briefly summarize this path connecting the different indices to physiological states and pathological conditions affecting fetal life.

The first step of the Fetal Herat rate traces recorded by CTG analysis was obtained by eye inspection of changes in the signal behavior. Even in this starting case, variability was still considered as containing information potentially useful for the diagnosis of fetal states. The development of computerized systems allowed to transfer these measurement in a numeric digital context. Nevertheless, variability analysis did not cover completely the information extraction from HR signals. Presence of several interacting mechanisms generate very different HR signal patterns that required analysis not only in time domain. HRV analysis received a strong improvement by the introduction of frequency domain methods. These tools allow to attribute variability contribution seen in the HR signal to physiological mechanisms controlling it.

Even more interesting, is the possibility of identifying by differences in the HRV the presence of respiration patterns during fetal life. The breathing-like activity of the fetus is a positive sign of reactivity. Moreover, fetal distress is often connected with oxygen reduction and asphyxia. These traditional analytic approaches did not become "per se" obsolete. They are still in use, accompanied by new parameters more recently developed.

16.3.1 Time Domain

Interpretation of the HR pattern is usually performed by the physician who analyses the deviations of the signal from the baseline. In the construction of an automated system for the evaluation of the CTG recordings, a reproducible determination of the baseline is a fundamental starting point. Several attempts in this direction have been made starting from the work of Dawes et al. [14] to the approach of Mantel et al. [15]. Accelerations and decelerations are deviations of FHR from the baseline lasting a sufficient amount of time. They are correlated with the normal activities of the fetus, who "trains," moves, and exercises to breath. Decelerations are usually correlated with uterine contraction. Unfortunately, different quantifications of the words "deviations" and "sufficient" led each medical school to develop its own method to evaluate, by means of a ruler, these quantities on the monitoring strip.

We applied a quantitative procedure not only fully consistent with the definition of Mantel et al. [8], but also holding the suggestions of Arduini et al. [16].

In the following, inter-beat sequences will be used instead of heart rate sequences in beats per minute, usually employed in cardiotocography: inter-beat sequences are computed as $T(i) = 60000/S(i)$ ms, where $S(i)$ is FHR in beats/s. Moreover, in order to be compatible with previous works [16] we also computed some indices on the basis of the undersampled time series $T_{24}(i) = 60000/S_{24}(i)$ ms, obtained by taking $S_{24}(i)$ as the average of five consecutive FHR values of S(i).

Long Term irregularity (LTI) was the first index ever introduced; it was proposed by De Haan et al. [17]. It is usually computed on a three-minute segment of inter-beat sequence in milliseconds. Large accelerations and decelerations are usually excluded from the calculation, as suggested by Arduini et al. [16]. LTI is defined as the interquartile range of the distribution $m_{24}(j)$ with

$$m_{24}(j) = \sqrt{T_{24}(j)^2 + T_{24}\,(j+1)^2}.$$

Short Term Variability (STV) quantifies FHR variability over a very short time scale, usually on a beat-to-beat basis [16, 18]. By considering one minute of inter-beat sequence, $T_{24}(j)$ in ms,

$$STV\ (j) = \text{mean}\left[\left|T_{24}\,(j+1) - T_{24}(j)\right|\right].$$

Interval Index (II) was introduced just after LTI and it is certainly one of the most used variability indices [19]. It is a long term variability statistic, with

$$II(j) = \frac{\text{std}\ [T_{24}\ (j+1) - T_{24}\ (j)]}{STV(j)}.$$

16.3.2 Frequency Domain

Estimation of the power spectral density (PSD), with both direct FFT and Autoregressive power spectrum estimation in the FHR signal, provides parameters related to the ANS activity.

In FHR analysis it is established to calculate power and central frequencies for three frequency bands, Low Frequency (LF), Movement Frequency (MF), and High Frequency (HF) power components as well as the ratio LF/(MF + HF) [20], differently from the bands usually adopted for standard adult HRV analysis [1].

LF contributions (0.03–0.15 Hz) can be associated with the sympathetic control and vasomotor activity. HF (0.5–1 Hz) is mostly driven by respiration mediated by vagal activity. MF (0.15–0.5 Hz) should quantify the movement activity of the fetus and the mechanical influences of the maternal breathing. LF/(HF + MF) ratio could represent a synthetic index of the balance between physiological control components and fetus activity level, representing the equivalent of the so called sympatho-vagal balance in standard HRV analysis.

Frequency domain analysis is suited for a short time scale (3–5 min) as the fetal HRV signal needs to maintain its stationarity [21].

16.3.3 Complexity and Nonlinear Parameters

The introduction of nonlinear approaches to signal processing led the consideration of a set of methods investigating geometric and dynamic properties of time series. Their statistical use can be of great importance, even in diagnostic field and in clinical knowledge related to different cardiovascular pathologies. Various techniques exist aimed at quantifying the degree of similarity and/or complexity in time series, which can be computed directly on the sequence of inter-beat intervals as Table 16.1 illustrates

- Approximate entropy (ApEn) quantifies regularity and complexity of a time series. The index was firstly proposed by Pincus [22] and further improvements and corrections were introduced with the Sample Entropy (SampEn) [23].

The rationale is to quantify the degree of regularity or loss of regularity in a time series without a priori information on its structure.

Given a discrete time series $x(n)$ of length N, a threshold r, and an embedded dimension m (usually $m = 1, 2, 3$, $r = 0.1 - 0.25$ std of the input data), $x(n)$ is split into a series of subsequences $X_m(i)$ of length m.

A distance is defined between two subsequences, X(i) and X(j), as the maximum absolute difference between their samples. The total number of subsequences of length m whose distance is less than r is denominated $B_i(r)$, whereas $A_i(r)$ accounts for subsequences of length $m + 1$. SampEn does not count self-matches.

Table 16.1 Analysis techniques in FHR. Each method is associated with details on application as well as hypothesis of physiological correlates to mechanisms performing control of the FHR signal

Method	Parameters	Sequence length	Hypothesis
Frequency domain analysis Autoregressive model estimation from data and measurement of spectral components in defined frequency bands.	% of spectral power (ms^2) in frequency bands: Low frequency 0.03–0.15 Hz Movement frequency 0.15–0.5 Hz High frequency 0.5–1 Hz LF/(MF + HF)	3 min 360 values	Quantification of the activity of the autonomic nervous system.
Time domain analysis: morphological HR modification and variability	STV (ms) II	1 min 120 values	Variability in the short period
	Delta FHR avg (ms) LTI (ms)	3 min 360 values	Variability in the long period
Approximate Entropy and Sample entropy	ApEn (m,r,N) Sampen(m,r,N)	3 min 360 values	Presence of recurrent patterns in a single scale
Multiscale entropy	MSE: entropy estimator as a function of a scale factor τ_{sf}	7000 values	Investigation of signal structure: repetitive patterns are present at different time scales.
Detrended fluctuation analysis (DFA)	DFA scaling exponent: linear relationship between the detrended integrated time series F(n) and the time scale on a log-log plot	1200 values	Quantification of long-range correlations in time series. Presence of power law (fractal) scaling Differences activity vs. quiet state
Lempel Ziv complexity (LZC)	LZC binary coding	7000 values	Rate of new patterns arising with the evolving of the signal
Phase rectified signal averaging (PRSA)	APRS, DPRS	3–30 min	Detect quasi-periodicities in noisy, non-stationary signals

ApEn(m, r, N) then averages the natural logarithms of A_i (r) and B_i(r) and calculates their difference. SampEn(m, r, N) instead is obtained as the natural logarithm of the ratio between the averages of A_i (r) and B_i (r).

SampEn does not count self-matches thus improving the estimation performed by ApEn removing the bias introduced by self-counts. SampEnconstitutes the basis for Multiscale Entropy (MSE) entropy estimation, where parameters are calculated at different scales in coarse-grained time series [24]

- Detrended Fluctuation Analysis (DFA) technique is used to quantify the fractal correlation properties of R–R interval data. This method is a modified root mean square analysis of a random walk [25].
- Lempel Ziv Complexity (LZC) [26] assesses the so-called algorithmic complexity, which is defined according to the Information Theory as the minimum quantity of information needed to define a binary string. In order to estimate the LZC for the FHR time series, the signal needs to be transformed into symbolic sequences. Each character forms a string and the LZC measures the number of characters, which contribute to form new substrings in the sequence.
- Phase Rectified Signal Average (PRSA) is a technique presented by Bauer et al. [27]. It aims at compressing a signal into a shorter sequence without losing any relevant quasi-periodicities, and eliminating at the same time non-stationarities, artifacts, and noise.

This technique consists of three simple steps. At first, anchor points (AP) are chosen based on a certain property of the signal, such as increases or decreases in the signal. Typically half of all points of the signal will be APs by these definitions. Thanks to this process we can extract the phase information of the oscillations from the signal itself. Afterwards, windows of length 2L, are defined around each AP. It is important to make a proper choice for the parameter L; it should be larger than the period of the slowest oscillation that one wants to detect. Lastly, the windows are aligned at the AP and the PRSA curve is obtained by averaging the aligned windows. Thanks to this average, components that are not phase synchronized with the AP, will have zero mean and thus they will cancel out; instead, the events that have a fixed phase relationship with the AP in all the 2L windows will have the same pattern and thus will be kept in the average.

Proposed parameters for these curves are the Acceleration Phase Rectified Slope (APRS) and the Deceleration Phase Rectified Slope (DPRS) [28].

What clearly happens, is that the complexity parameters increased classification and diagnostic powerfulness in the FHR analysis based on data processing. Their introduction produced new tools and generated a novel way to approach the heart rate classification. At present, complexity and nonlinear parameters are perceived even by clinicians as an essential part of a multi-parameter diagnostic process. This has also generated technological solutions embedding the estimation of some complexity indices.

16.4 Methods for Monitoring HRV in Newborns

As for the fetuses, newborn HRV analysis encompasses techniques from various domains.

There are still no standard norms developed for this particular population, thus parameters from adult and fetal HR analysis are usually adapted.

The Task Force of 1996 suggests the evaluation of parameters on five-minute windows for adults' HR [1]. However, since neonatal HR is higher, a similar number of beats occurs in about 3 min and this could be used as a reference length.

Moreover, in newborn infants the relationship between HR and breathing is particularly relevant and thus methods to investigate it will be presented.

16.4.1 Time Domain

For the time domain analysis, measures adapted from adult studies have been long used: SDNN, which is the standard deviation of normal to normal intervals (NN), defined as the RR distances excluding anomalous beats (e.g. ectopic beats), standard deviation of the average NN interval, SDANN, the HRV triangular index, HRVTI and root mean of successive differences, RMSSD.

SDNN and HRVTI estimate overall HRV, SDANN estimates long-term components of HRV and RMSSD estimates short-term components of HRV [1].

Fetal parameters listed in the previous paragraph can also be applied with success to this population.

Regarding breathing rate analysis, the Inter Breath Interval series (IBI) can be analyzed calculating mean, standard deviation and the coefficient of variation (CV) defined as the standard deviation of the respiration rate divided by the mean respiration rate.

16.4.2 Frequency Domain

Spectral analysis can be performed with parametric and nonparametric approaches on RR series and respiration, usually resampled at 2–5 Hz.

Cross-spectral analysis can also be performed, estimating cross-spectrum and coherence.

The literature does not provide precise values for frequency bands ranges, thus they are commonly chosen based on prior experience and observation of the spectra, with consideration of the general breathing characteristics of premature infants [6, 29, 30]. Commonly used frequency bands are Low Frequency (LF), 0.05–0.2 Hz, and High Frequency (HF), 0.2–1.5 Hz.

In the 1990s, surrogate series have been proposed to test for non-linearity [31]. Recently they have been applied to identify an adaptive significance threshold for the coupling between two time series expressed in the coherence function [32, 33].

By this approach, N pairs of surrogate time series, with features of the original series but completely uncoupled, are generated as realizations of independent stochastic processes, such as the Iterative Amplitude Adjusted Fourier Transform (IAAFT) [34]. The coherence is then estimated between each pair of surrogate series and its empirical sampling distribution (frequency histogram) computed accordingly at each frequency of the signal bandwidth. The threshold for zero coherence is then defined at the 95th percentile of the coherence sampling distribution.

16.4.3 Complexity and Nonlinear Parameters

Many of the non-linear measures described for fetal HR analysis have also been applied to newborns HR, such as ApEn, SampEn.

An improvement of SampEn was presented by Lake in 2006 [35]. He approached the problem from the standpoint of stochastic processes, applying concepts of probability density estimation. Thus, the measured conditional probability was converted to a density, by normalizing to the volume of the matching region, leading to the simple addition of ln(2r) to SampEn. The result was called Quadratic Sample Entropy (QSE) and allows to compare estimates obtained with different *r* values.

Furthermore, inaccuracy in the probability estimate may arise: as explained in the paragraph of fetal HR monitoring methods, SampEn is obtained from A/B ratio, thus the estimate is dependent on the magnitude of the numerator A and the denominator B. For this reason, it is important to maximize this value: min(A, B − A). Thanks to the flexibility to vary *r* introduced by QSE, inaccurate probability estimates can often be avoided, for instance with the method of the Minimum Numerator Count [35]. This approach calculate ratios with specified minimum values of the numerator and denominator, in the attempt to also minimize *r*.

Another useful method is the univariate PRSA technique, previously introduced among the methods for FHR analysis. The Bivariate PRSA (BPRSA) is an evolution of the PRSA, which highlights interrelationships between two signals supposed synchronous, defined as a trigger and a target signal [36].

The BPRSA algorithm consists in three major steps similar to the ones for PRSA, except for the fact that the anchor points are detected on the trigger signal and then transferred on the target signal, where the rest of the analysis is carried on.

Few of the advantages of PRSA are that it does not assume stationarity and it claim to be less affected by data quality than traditional methods. Moreover, BPRSA transforms the target signal in a compressed version, maintaining only (quasi)periodicities that are coupled to the trigger signal, eliminating uncoupled periodicities, artifacts, or noise. Comparing the BPRSA transformation of the target signal with the PRSA transformation of the trigger signal is possible to infer potential trigger and target signal interaction.

16.5 Classification of Fetal Wellbeing States

Monitoring fetal conditions during pregnancy to assure fetuses wellbeing state as the pregnancy develops is a challenging aspect for people working in the obstetric field.

New results were introduced when FHR time series, as it happened for adult HRV, started to be considered as complex signals in which different mechanisms even nonlinear can be recognized. [4].

This novel approach opened a new route to classify the FHR signal showing that nonlinear parameters related to mechanisms controlling FHR, reveal reliable information about fetal states. It has allowed the identification and analysis of different physiological conditions such as activity, quiet states, stress and sufferance signs, up to the distinction between normal and pathological subjects.

All these considerations led to adopt new nonlinear and complexity methods to perform a deeper investigation of FHR signals towards a novel monitoring approach.

CTG analysis with advanced signal processing techniques has shown that the FHR signal content is complex and needs a multi-parametric approach to be completely understood.

Moreover, parameters linked to physiology are more effective in FHR signal analysis than other without clear relationship with control systems of the heart. The problem is always the identification of potential risk conditions which are not associated to a clearly identified pathology, but which may lead complications for the fetus.

Studies developed in recent years, have shown that different wellbeing states (activity vs. quiet states) can be discriminated. The same happened when an external vibroacoustic stimulation was applied [20, 37].

In our CTG analysis, the first introduced nonlinear parameter was the ApEn. In association with other indices ApEn succeeded in separating physiological epochs such as active and quiet periods in FHR [20]. Other Entropy estimators were introduced furtherly such as SampEn, followed by ApEn and SampEn calculation in multiscale basis [38, 39]. They were employed to stratify suffering vs. non suffering fetuses even when no clear pathologies were identified.

Examples in Figs. 16.1 and 16.2 show how complexity indices allowed a different classification of fetal stress states.

16.6 Fetal Disease and Fetal Stress (Normal and IUGR Populations)

Complexity indices performance in FHR analysis has been widely recognized in recent years.

The challenge was to increase the knowledge about FHR dynamics. The results brought to a steep improvement toward the reduction of fetal morbidity and mortality through a better management of the pregnancy.

An example is provided by the results obtained by the methods shortly described in previous paragraphs, which were applied to a selected population of Normal and IUGR fetuses.

The upper part of Table 16.2 reports time domain indices related to short and long term HRV, the middle part summarizes results for Entropy and complexity parameters, the lower part shows PRSA derived slope parameters. A paired T-test was applied; * (* = 0.05%) marks the level of statistical significance.

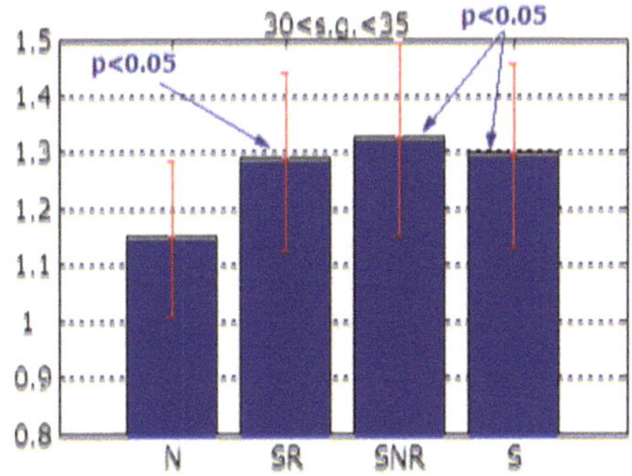

	Suffering			Not Recognized Suffering			Recognized Suffering		
	S.rank	KS	t_test	S.rank	KS	t_test	S.rank	KS	t_test
30<=g.w.<35	0,0092	1	0,0042	0,0153	1	0,0065	0,0073	1	0,0048
35<=g.w.<39	0,5148	0	0,3855	0,9543	0	0,9371	0,4573	0	0,3473
g.w.>=39	0,9108	0	0,9405	0,6755	0	0,7556	0,7252	0	0,7749

Fig. 16.1 Values of the ApEn parameter (avg ± std) are summarized in the *upper panel*. The table shows the results for the statistical tests: *t*-test, sum rank and Kolmogorov Smirnov test. In the *lower table* (*1*) and (*0*) indicate positive and negative results respectively. For the *t*-test the p value is indicated. N means normal, *SR* recognized as suffering, *SNR* suffering but not recognized, *S* all the suffering SR + SNR

Fig. 16.2 Results of MSE computed by SampEn in Normal and recognized as Suffering fetuses. Differences are noticed in population that will show pathological conditions with disease at delivery

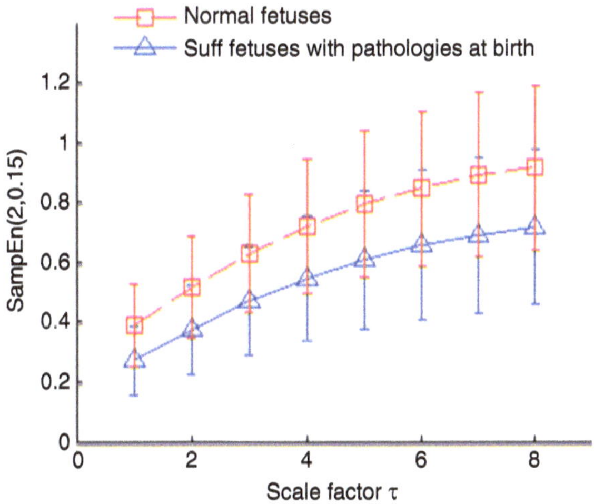

It is quite clear that all parameters succeed in separating Normal and IUGR fetal populations even with different level of performance, some of them reaching high levels of statistical significance.

Table 16.2 Results of the analysis performed in a population of 61 Normal and 61 IUGR fetuses. FHR signals were collected through Cardiotocographic recordings

Parameter	Healthy (mean ± std)	IUGR (mean ± std)	*t*-test	p-value
Time parameters				
Delta (ms)	41.5 ± 6.4	29.6 ± 9.3	**	6.3×10^{-7}
STV (ms)	6.7 ± 2.24	4.29 ± 1.62	***	$1{,}22 \times 10^{-9}$
LTI (ms)	21.46 ± 6.53	17.17 ± 5.37	***	1.5×10^{-11}
Nonlinear parameters				
ApEn(1,0.1)	1.33 ± 0.13	1.21 ± 0.11	**	5.14×10^{-7}
Lempel Ziv complexity (2,0)	1.00 ± 0.08	0.94 ± 0.09	*	0.00078
SampEn(1,0.1)	1.3 ± 0.19	1.13 ± 0.15	**	2.08×10^{-7}
PRSA parameters				
APRS	0.17 ± 0.041	0.12 ± 0.042	***	7.76×10^{-12}
DPRS	−0.18 ± 0.046	−0.12 ± 0.042	***	1.08×10^{-13}

Parameters values are reported as Mean ± Std and are grouped in Time domain (upper), Non linear (medium) and PRSA (low)
Last right column reports p-values. All parameters allow separating Normal from IUGR. It can be noticed that the best performance is obtained by PRSA derived parameters

In particular, the Short Time Variability together with the Long Term Irregularity measured on the FHR provide separation of the two fetal conditions. The novel indices from PRSA methods, proposed for FHR classification, succeeded in the classification of the two populations as well. They mark the best performance in the separation task. Moreover, Entropy estimators also demonstrate their reliability in FHR analysis thus suggesting a possible use in multiparameter analysis approaches.

These are only examples by our research activity: other research groups in the field are working on with similar approaches that provide promising results [40–42].

In order to complete the view, Fig. 16.3 underlines that by correlating indices from different groups (variability in time domain, LTI, and regularity/complexity estimation, ApEn) we can obtain a good separation between IUGR and Normal fetuses.

This result suggests that a combined use of different parameters can provide further discriminant capability to the CTG diagnosis process.

16.7 Newborn HRV Analysis: The SIDS Challenge

Sudden Infant Death Syndrome (SIDS) is the leading cause of death between 1 month and 1 year of age in developed countries. It is defined as an unexplained death even after autopsy and careful examination of the death scene. Its causes are still unknown, but researchers have suggested that SIDS could be the result of three factors coming together: firstly, a critical developmental period, secondly, a vulnerable infant, with underlying abnormalities in an area of their brainstem,

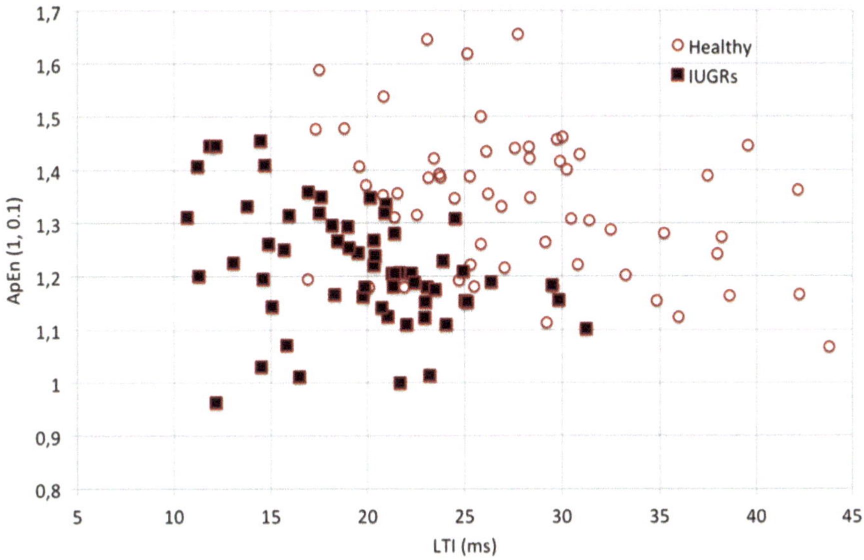

Fig. 16.3 Individual data of APEn(1,0.1) vs. LTI (time domain variability index). IUGRs and Healthy fetuses occupy different subspaces in the diagram. Separation is easily possible with few errors

leading to an impaired autonomic cardio-respiratory functioning. Lastly, a trigger presented by an environmental challenge, including prone sleeping [43].

For this reason, HR and ECG data have been the subject of considerable study. Most investigations are relative either to retrospective analysis of infants who died of SIDS, or to populations that are known to be at risk based on epidemiological findings (infants who had an Apparent Life Threatening Events, premature infants, infants exposed to smoking or alcohol during pregnancy).

Since the late 1970s, researchers have applied time and frequency domain techniques in order to characterize HR features that could provide insight about this syndrome.

- *Time domain analysis*: it showed increase in mean HR and decrease in HRV when comparing infants who subsequently died of SIDS with age-matched control infants [44, 45]. Moreover, time domain measures were also used to characterize populations at risk for SIDS, such as premature infants in prone position, highlighting higher HR and lower HRV in this position compared to supine [46].
- *Spectral analysis*: one study found that the area under the RR spectrum and the cross-spectrum in the range of the respiratory frequency showed a lower extent of respiratory sinus arrhythmia in SIDS victims across all sleep-waking states, persistent even after adjusting for HR [47].

  In another interesting work, spectral analysis revealed increase of low frequency power in the SIDS group compared to control and a dispersion of the respiratory frequency as determined by the respiratory bandwidth [48].

Still, researchers pointed out how none of the markers proved to be conclusive in distinguishing the SIDS cases from the controls in the population at large [49, 50].

In 1993, Pincus et al approached the problem with a non-linear method, ApEn. examining the capability of this parameter to detect abnormal HR dynamics in a group of healthy infants and age-matched aborted-SIDS infants, of ages 0–4 months [51].

Half of aborted-SIDS infants showed greater ApEn instability across quiet sleep than normal infant, suggesting that autonomic regulation of HR occasionally becomes abnormal in a high-risk subject. Moreover, there was an association between low ApEn values and aborted-SIDS event [51].

Following this interesting hint, a combination of time domain, frequency domain and non-linear parameters have been used to characterize populations in conditions that have been classified at risk for SIDS. In particular, term and preterm infants were studied to evaluate the effects of sleep state and position during sleep. From epidemiological studies it is known that infants sleeping in prone position are at higher risk for SIDS and physiological studies suggested that prone position and quiet sleep might increase the threshold for arousal [52–54].

The major findings showed that:

- *Sleep state*: in a study with 60 newborns (Gestational Age 39.6 ± 1 week), LTI, CV for respiration rate and SampEn for HR were capable of distinguishing active from quiet sleep. Results reported in Table 16.3 show that all the parameters obtained higher values in active sleep compared to quiet sleep, except for entropies which behaved oppositely [55].
- *Position during sleep*: a recent analysis of sleep position in premature newborns (Gestational Age 28.7 ± 2 weeks and Post Menstrual Age 37.7 ± 2.4 weeks) showed clear differences both in the long-term variability parameters (LTI and PRSA) and in the short-term variability parameters (SampEn and QSE) between prone and supine position. Our findings are reported in Table 16.4. In the follow-up study after 2 months, whose results are summarized in Table 16.5, the most

Table 16.3 Parameters' results from 60 newborn studies (mean ± SD). P-values obtained with Anova analysis

Newborn	Quiet	Active	P-val
LTI	23.6 ± 13.6	46.07 ± 22.0	<0.01
SampEn HR	0.15 ± 0.03	0.11 ± 0.03	<0.01
CV resp rate	0.13 ± 0.05	0.23 ± 0.06	<0.001

Table 16.4 Parameters' results for the newborn study with patients in active sleep (mean ± SD). P-values were obtained with Student's *t*-test on 20 subjects

Newborn	Prone	Supine	P-val
LTI	0.040 ± 0.018	0.049 ± 0.023	<0.001
SDNN	0.0212 ± 0.008	0.0237 ± 0.009	<0.05
PRSA ΔY	15.66 ± 10.18	18.71 ± 10.68	<0.01
SampEn	0.554 ± 0.183	0.443 ± 0.133	<0.001
QSE m = 1	−4.201 ± 0.415	−4.042 ± 0.304	<0.001

Table 16.5 Parameters' results for follow-up study with patients in AS (median ± IQR). P-values were obtained with Wilcoxon signed-rank test on ten subjects

Follow-up	Prone	Supine	P-val
Mean RR	0.414 ± 0.048	0.447 ± 0.072	<0.05
RMSSD	0.014 ± 0.009	0.015 ± 0.011	<0.05
STV	0.01 ± 0.006	0.011 ± 0.008	<0.05
QSE m = 2	−3.27 ± 0.848	−3.15 ± 7.783	<0.05

robust state differences were observed only for the parameters quantifying short-term variability, specifically, RMSSD, STV and [56], confirming and expanding previous findings [57].

These findings support the thesis of prone position being an intensive stressor factor in the autonomic regulation.

Furthermore, the parameters that characterized the autonomic control system differed at the two stages of development analyzed, with a predominance of long term variability right after birth and an increased short term variability influence later on. This is relevant due to the fact that SIDS has a peak precisely between the second and the fourth month of age.

Another interesting aspect of neonatal HR analysis is related to the bivariate approach considering together HR and respiratory signals. Traditional cross-spectral analysis has been applied in the past to evaluate the interactions between the cardiac and the respiratory system variability [47]. Recently, improvements for the bivariate analysis have been introduced with nonlinear approaches. For instance, with the use of surrogate series it is possible to define an adaptive threshold for evaluating significance in coherence analysis. This allows to dynamically follow frequency by frequency the strength of the coupling between HR and respiration.

Moreover, other methods such as Bivariate PRSA (BPRSA) were specifically designed to highlight interrelationship of quasi-periodic oscillations between different non stationary signals.

The PRSA technique applied to one signal only, characterizes quasi-periodic oscillations and nonlinear effects in non-stationary signals.

The BPRSA is a generalization of the previous method, with the goal of highlighting interrelationships between two signals supposed synchronous, defined as a trigger and a target signal [36, 58].

The BPRSA algorithm involves three main steps:

- Detection of anchor points (APs). APs are defined as increments (or decrements) in the trigger signal. Increments are estimated on series averaged every T samples. The choice of T can be optimized depending on the frequency of interest, according to this empirical relationship $T = 1/(2.5 * F)$. The APs are used to identify the corresponding samples on the target signals.
- The identification of all the time windows of length 2L around each AP by considering L samples before and L samples after the AP on the target series.
- Averaging of all 2L windows.

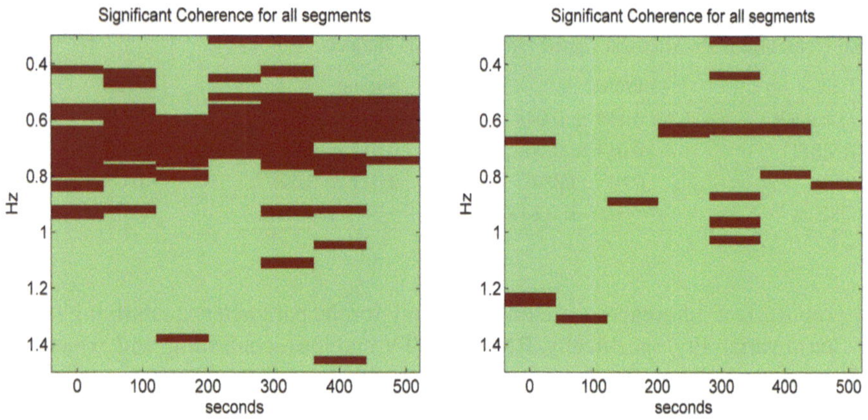

Fig. 16.4 Heat maps by assigning 1 when the coherence function exceeds the threshold estimated with the surrogate techniques (*purple boxes*) and 0 elsewhere (*green boxes*). The first heat map refers to an infant in quiet sleep while the second one to an infant in active

PRSA does not assume stationarity and it claim to be less affected by data quality than traditional methods. Furthermore, BPRSA transforms the target signal in a compressed version, maintaining only (quasi)periodicities that are coupled to the trigger signal, theoretically eliminating uncoupled periodicities, artifacts, or noise.

Comparing the BPRSA transformation of the target signal with the univariate PRSA transformation of the trigger signal, potential trigger and target signal interactions can be observed.

In a pilot study, cross-spectral analysis (coherence) and BPRSA were applied on 10 min recording from four infants.

Results show that both these methods can provide significant information about the occurrence and the strength of cardiorespiratory coupling.

Figure 16.4 summarizes these preliminary results regarding coherence. They showed a dynamic behavior appearing stable and centered around one single frequency in quiet sleep, while being more variable and less consistent in active sleep [33].

Figure 16.5 summarizes preliminary findings in the application of PRSA analysis in Bivariate form BPRSA.

An approach for SIDS screening and prevention is still far from being univocally identified. Certainly, HR analysis, in particular with insights introduced with recent nonlinear approaches, has provided useful information about possible mechanisms and pathways of this syndrome.

Given the complex picture, a trans-disciplinary approach combining data from HR analysis with information from animal models, autopsies, imaging and epidemiological studies might be a winning strategy to proceed further in the understanding and prevention against SIDS.

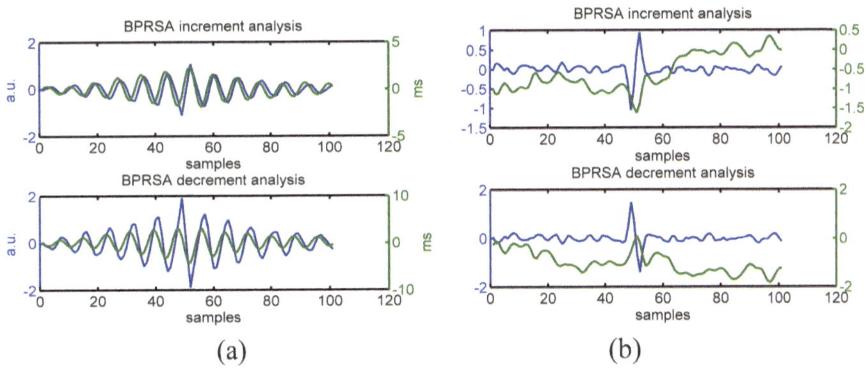

Fig. 16.5 The curve of PRSA of respiratory phase signal is in blue, the curve of BPRSA is in *green*, the *upper panels* refer to increment APs, the lower panels to decrement APs. (**a**) A subject in quiet sleep shows a clear and synchronous coupling (**b**) a subject in active sleep shows a weak coupling

16.8 Conclusions

Objective of this chapter was to propose an overview of what contribution the complexity analysis of fetal and neonatal HR can add to the understanding of signal dynamics and control.

Complexity has been identified as one of the interesting properties of biological signals and systems. Moreover many evidences were collected about its role in the generation of signal patterns both related to normal physiological behavior and pathological states.

This awareness pushed many researchers to investigate how to measure complexity properties in biological signals. Particular attention was dedicated to HRV analysis as the analysis traditionally employed in the linear approach always left something unexplained or at least open to further investigations.

This was true even for fetal and neonatal HR as in both cases subjects are not able to collaborate nor to provide explicit answers to stimuli. In the fetal case, however, the HR measure is indirect as mediated by the mother uterus and surrounding tissues. As a matter of fact, fetal HR might be the only way to investigate the wellbeing of the fetus. Similar condition applies to newborns in which the HR measurement is an extremely informative and minimally invasive way to evaluate the baby conditions.

The analysis we proposed suggests that possible indications on fetal sufferance leading to a pathology at birth and in newborns can be derived from advanced signal processing methods. The aim of the research work we described was to identify indices that can help the early identification of diseases arising during gestation and in the early life both in normal and premature newborns.

Several methods have been described and discussed together with example of applications. Several parameters with different and complementary capability to capture information were proposed and applied to an example population both in physiological state and in disease conditions.

This point deserves special remarks. Advanced analysis methods including complexity parameters enhance HR information content for recognizing and predicting pathological events.

In this context, a multiparametric approach shows its usefulness. Moreover, results are even more positive if parameter set considers simultaneously morphological signal characteristics, linear parameters and nonlinear and complexity estimators. For the fetal HR analysis, parameter estimation strongly depends on the gestational age following the physiological development of the fetus.

Most of the parameters examined in this study are highly significant in discriminating healthy fetuses from IUGR pathological subjects. In particular nonlinear indices have shown good performance in separating the two populations. Furthermore, these parameters allowed a thorough investigation on ANS response to environmental stimuli, such as sleep position, in newborn population at risk due to their prematurity.

However, if we consider each parameter as a single entity, its measurement will provide only information related to a single feature. The contribution of the complexity analysis allows to keep into account that many controlling mechanisms can affect HRV and they may act linearly and/or nonlinearly on the HR in pathological situations. One index at a time is unable to capture information connected with several features involved in pathophysiological processes driving the development of a fetus or a newborn towards her disease condition.

Multivariate approach, considering both linear and nonlinear complexity parameters, can really improve the discrimination of healthy and pathological fetuses and has contribute to understand newborn cardiovascular control mechanisms through the Autonomic Nervous System.

Results show that a careful analysis of HR signals in newborns during early stage life, and in fetuses, performed in standard CardioTocoGraphic tracings, by means of both linear and non-linear approaches, can provide helpful indications to describe pathophysiological mechanisms involved in the cardiovascular and neural system controlling the heart. However, only a multivariate approach has well performed improving the discrimination of healthy and pathological groups.

As a matter of fact, strong efforts are also necessary to reach the level of quality and robustness expected by mothers, children and clinical personnel as midwives, obstetrics and neonatologists.

Our work underlines that even some critical points are still open. Among them unavailability of annotated reference data for both populations complicates the road toward the use of new analysis tools in clinical environments. Moreover some parameters, proposed in the analysis, need a deeper investigation. Reference values in a selected population could be useful to introduce reference standards. For example, in fetal analysis, results are influenced by the different gestational ages both for physiological and pathological states. The same can be said for preterm newborns, whose development in utero is interrupted by a premature birth: as a matter of fact an infant born at 30 weeks and examined at 5 weeks of age is extremely different from an infant born and examined at 35 weeks. For this reason, it is challenging to obtain reliable reference values.

The results reported in the manuscript show that these objectives can be obtained mainly through a methodological improvement consisting of a detailed analysis of HRV with different computational approaches that can provide simple, fast and reliable tools to be used in clinical environment.

After a few years since we developed the tools implementing the illustrated FHR analysis procedure, our experience confirms the key role of analysis tools that implement the features extraction from FHR.

They are what a nurse needs in a crowded obstetric ward, when performing a CTG exam on a pregnant woman. They represent also a solution for the physician who wants to perform a complex real-time analysis on the fetal recordings, taking advantage from availability of mathematical parameters in the diagnostic process.

Research is still in progress and new features, in particular concerning nonlinear parameters, are still developing. The future for these analysis tools is to increase their capability in detecting fetal sufferance and illness states during the pre and post-natal period. Predicting models derived from data analysis could be the natural evolution toward novel diagnostic and predictive tools taking advantage from complexity methods.

References

1. Task force of the European Society of Cardiology, the North American Society of Pacing, and Electrophysiology: Heart rate variability standards of measurement, physiological interpretation, and clinical use. Eur. Heart J. **17**, 354–381 (1996)
2. Akselrod, S., Gordon, D., Ubel, F.A., Shannon, D.C., Barger, A.C., Cohen, R.R.: Power spectrum analysis of heart rate fluctuation: a quantitative probe of beat-to-beat cardiovascular control. Science. **213**(4504), 220–222 (1981)
3. Goldberger, A.L., Rigney, D.R., Mietus, J., Antman, E.M., Greenwald, S.: Nonlinear dynamics in sudden cardiac death syndrome: heartrate oscillations and bifurcations. Experientia. **44**(11–12), 983–987 (1988)
4. Sassi, R., Cerutti, S., Lombardi, F., Malik, M., Huikuri, H.V., Peng, C.K., Schmidt, G., Yamamoto, Y.: Advances in heart rate variability signal analysis: joint position statement by the e-Cardiology ESC Working Group and the European Heart Rhythm Association co-endorsed by the Asia Pacific Heart Rhythm Society. Europace. **17**(9), 1341–1353 (2015)
5. Hon, E.H., Lee, S.T.: The fetal electrocardiogram. Am. J. Obstet. Gynecol. **91**(1), 56–60 (1965)
6. Sahni, R., Schulze, K.F., Kashyap, S., Ohira-Kist, K., Fifer, W.P., Myers, M.M.: Maturational changes in heart rate and heart rate variability in low birth weight infants. Dev. Psychobiol. **37**(2), 73–81 (2000)
7. Fifer, W.P., Myers, M.M.: Sudden fetal and infant deaths: shared characteristics and distinctive features. Semin. Perinatol. **26**(1), 89–96 (2002)
8. Kahn, A., Groswasser, J., Franco, P., Scaillet, S., Sawaguchi, T., Kelmanson, I., Dan, B.: Sudden infant deaths: stress, arousal and SIDS. Pathophysiology. **10**(3–4), 241–252 (2004)

9. Hewlett-Packard, G.: Fetal Monitor Test – A Brief Summary, pp. 1–6. Hewlett-Packard, Boeblingen (1995)
10. Church, C.C., Miller, M.W.: Quantification of risk from fetal exposure to diagnostic ultrasound. Prog. Biophys. Mol. Biol. **93**(1–3), 331–353 (2007)
11. Yang, K., Jiang, H., Dong, J., Zhang, C., Wang, Z.: An adaptive real-time method for fetal heart rate extraction based on phonocardiography. In: Biomedical Circuits and Systems Conference (BioCAS), 2012 IEEE, pp. 356–359 (2012)
12. Ryo, E., Nishihara, K., Matsumoto, S., Kamata, H.: A new method for long-term home monitoring of fetal movement by pregnant women themselves. Med. Eng. Phys. **34**(5), 566–572 (2012)
13. Fanelli, A., Ferrario, M., Piccini, L., Andreoni, G., Matrone, G., Magenes, G., Signorini, M.G.: Prototype of a wearable system for remote fetal monitoring during pregnancy. In: Engineering in Medicine and Biology Society (EMBC), 2010 Annual International Conference of the IEEE, pp. 5815–5818 (2010)
14. Dawes, G., Houghton, C., Redman, C., Visser, G.: Pattern of the normal human fetal heart rate. BJOG. **89**(4), 276–284 (1982)
15. Mantel, R., van Geijn, H.P., Caron, F.J.M., Swartjes, J.M., van Woerden, E.E., Jongswa, H.W.: Computer analysis of antepartum fetal heart rate: 2. Detection of accelerations and decelerations. Int. J. Biomed. Comput. **25**(4), 273–286 (1990)
16. Arduini, D., Rizzo, G., Romanini, C.: Computerized analysis of fetal heart rate. J. Perinat. Med. **22**(1), 22–27 (1994)
17. de Haan, J., van Bemmel, J.H., Versteeg, B., Veth, A.F.L., Stolte, L.A.M., Janssens, J., Eskes, T.K.A.B.: Quantitative evaluation of fetal heart rate patterns: I. Processing methods. Eur. J. Obstet. Gynecol. **1**(3), 95–102 (1971)
18. Dalton, K., Dawes, G., Patrick, J.: Diurnal, respiratory, and other rhythms of fetal heart rate in lambs. Am. J. Obstet. Gynecol. **127**(4), 414–424 (1977)
19. Yeh, S.Y., Forsythe, A., Hon, E.H.: Quantification of fetal heart beat-to-beat interval differences. Obstet. Gynecol. **3**(41), 355–363 (1973)
20. Signorini, M.G., Magenes, G., Cerutti, S., Arduini, D.: Linear and nonlinear parameters for the analysis of fetal heart rate signal from cardiotocographic recordings. IEEE Trans. Biomed. Eng. **50**(3), 365–374 (2003)
21. Signorini, M.G., Fanelli, A., Magenes, G.: Monitoring fetal heart rate during pregnancy: contributions from advanced signal processing and wearable technology. Comput. Math. Methods Med. **2014**, 707581 (10p) (2014). doi:10.1155/2014/707581
22. Pincus, S.M.: Approximate entropy as a measure of system complexity. Proc. Natl. Acad. Sci. **88**(6), 2297–2301 (1991)
23. Richman, J.S., Moorman, J.R.: Physiological time-series analysis using approximate entropy and sample entropy. Am. J. Physiol. Heart Circ. Physiol. **278**(6), H2039–H2049 (2000)
24. Costa, M., Goldberger, A.L., Peng, C.-K.: Multiscale entropy analysis of complex physiologic time series. Phys. Rev. Lett. **89**(6), 068102 (2002)
25. Peng, C.-K., Havlin, S., Stanley, H.E., Goldberger, A.L.: Quantification of scaling exponents and crossover phenomena in nonstationary heartbeat time series. Chaos. **5**(1), 82–87 (1995)
26. Ferrario, M., Signorini, M.G., Magenes, G.: Complexity analysis of the fetal heart rate for the identification of pathology in fetuses. Comput. Cardiol. **32**(1), 989–992 (2005)
27. Bauer, A., Kantelhardt, J.W., Bunde, A., Barthel, P., Schneider, R., Malik, M., Schmidt, G.: Phase-rectified signal averaging detects quasi-periodicities in non-stationary data. Phys. A Stat. Mech. Appl. **364**, 423–434 (2006)
28. Fanelli, A., Magenes, G., Campanile, M., Signorini, M.G.: Quantitative assessment of fetal well-being through ctg recordings: A new parameter based on phase-rectified signal average. *IEEE J. Biomed. Health Inform.***17**(5), 959–966 (2013)
29. Sahni, R., Schulze, K.F., Kashyap, S., Ohira-Kist, K., Fifer, W.P., Myers, M.M.: Postural differences in cardiac dynamics during quiet and active sleep in low birthweight infants. Acta Paediatr. **88**(12), 1396–1401 (1999)

30. Rosenstock, E.G., Cassuto, Y., Zmora, E.: Heart rate variability in the neonate and infant: analytical methods, physiological and clinical observations. Acta Paediatr. **88**(5), 477–482 (1999)
31. Theiler, J., Eubank, S., Longtin, A., Galdrikian, B., Farmer, J.D.: Testing for nonlinarity in time series: the method of surrogate data. Physica D. **58**(1–4), 77–94 (1992)
32. Faes, L., Pinna, G.D., Porta, A., Maestri, R., Nollo, G.: Surrogate data analysis for assessing the significance of the coherence function. IEEE Trans. Biomed. Eng. **51**(7), 1156–1166 (2004)
33. Lucchini, M., Fifer, W.P., Ferrario, M., Signorini, M.G.: Feasibility study for the assessment of cardio-respiratory coupling in newborn infants. In: Engineering in Medicine and Biology Society (EMBC), 2016 38th Annual International Conference of the IEEE, pp. 5509–5512 (2016)
34. Schreiber, T., Schmitz, A.: Improved surrogate data for nonlinearity tests. Phys. Rev. Lett. **77**, 635–638 (1996)
35. Lake, D.E.: Renyi entropy measures of heart rate Gaussianity. IEEE Trans. Biomed. Eng. **53**(1), 21–27 (2006)
36. Bauer, A., Barthel, P., Müller, A., Kantelhardt, J., Schmidt, G.: Bivariate phase-rectified signal averaging-a novel technique for cross-correlation analysis in noisy nonstationary signals. J. Electrocardiol. **42**(6), 602–606 (2009)
37. Magenes, G., Signorini, M.G., Arduini, D., Cerutti, S.: Fetal heart rate variabilitydue to vibroacoustic stimulation: linear and nonlinear contribution. Methods Inf. Med. **43**(1), 47–51 (2004)
38. Ferrario, M., Signorini, M.G., Magenes, G.: Comparison between Fetal Heart Rate Standard Parameters and Complexity Indexes for the Identification of Severe Intrauterine Growth Restriction. Methods Inf. Med. **46**(2), 186–190 (2007)
39. Ferrario, M., Signorini, M.G., Magenes, G., Cerutti, S.: Comparison of entropy-based regularity estimators: application to the fetal heart rate signal for the identification of fetal distress. IEEE Trans. Biomed. Eng. **53**(1), 119–125 (2006)
40. Frank, B., Pompe, B., Schneider, U., Hoyer, D.: Permutation entropy improves fetal behavioural state classification based on heart rate analysis from biomagnetic recordings in near term fetuses. Med. Biol. Eng. Comput. **44**(3), 179–187 (2006)
41. Li, X., Zheng, D., Zhou, S., Tang, D., Wang, C., Wu, G.: Approximate entropy of fetal heart rate variability as a predictor of fetal distress in women at term pregnancy. Acta Obstet. Gynecol. Scand. **84**(9), 837–843 (2005)
42. Gonçalves, H., Rocha, A.P., Ayres-de-Campos, D., Bernardes, J.: Linear and nonlinear fetal heart rate analysis of normal and acidemic fetuses in the minutes preceding delivery. Med. Biol. Eng. Comput. **44**(10), 847–855 (2006)
43. Ronald, M.H., Hannah, C.K.: Potential mechanisms of failure in the sudden infant death syndrome. Curr. Pediatr. Rev. **6**(1), 39–47 (2010)
44. Schechtman, V., Harper, R., Kluge, K., Wilson, A., Hoffman, H., Southall, D.: Cardiac and respiratory patterns in normal infants and victims of the sudden infant death syndrome. Sleep. **11**(5), 413–424 (1988)
45. Schechtman, V.L., Raetz, S.L., Harper, R.K., Garfinkel, A., Wilson, A.J., Southall, D.P., Harper, R.M.: Dynamic analysis of cardiac R-R intervals in normal infants and in infants who subsequently succumbed to the sudden infant death syndrome. Pediatr. Res. **31**(6), 606–612 (1992)
46. Sahni, R., Schulze, K.F., Kashyap, S., Ohira-Kist, K., Myers, M.M., Fifer, W.P.: Body position, sleep states, and cardiorespiratory activity in developing low birth weight infants. Early Hum. Dev. **54**, 197–206 (1999)
47. Kluge, K.A., Harper, R.M., Schechtman, V.L., Wilson, A.J., Hoffman, H.J., Southall, D.P.: Spectral analysis assessment of respiratory sinus arrhythmia in normal infants and infants who subsequently died of sudden infant death syndrome. Pediatr. Res. **24**(6), 677–682 (1988)

48. Gordon, D., Cohen, R.J., Kelly, D., Akselrod, S., Shannon, D.C., Kenet, B., Ubel, F.A., Williams, O.: Sudden infant death syndrome: abnormalities in short term fluctuations in heart rate and respiratory activity. Pediatr. Res. **18**(10), 921 (1984)
49. Gordon, D., Southall, D.P., Kelly, D.H., Wilson, A., Akselrod, S., Richards, J., Kenet, B., Kenet, R., Cohen, R.J., Shannon, D.C.: Analysis of heart rate and respiratory patterns in sudden infant death syndrome victims and control infants. Pediatr. Res. **20**(7), 680–684 (1986)
50. Antila, K.J., Välimäki, I.A.T., Mäkelä, M., Tuominen, J., Wilson, A.J., Southall, D.P.: Heart rate variability in infants subsequently suffering sudden infant death syndrome (SIDS). Early Hum. Dev. **22**(2), 57–72 (1990)
51. Pincus, S.M., Cummins, T.R., Haddad, G.G.: Heart rate control in normal and aborted-SIDS infants. Am. J. Phys. **264**, R638–R646 (1993)
52. Horne, R.S.C., Ferens, D., Watts, A.M., Vitkovic, J., Lacey, B., Andrew, S., Cranage, S.M., Chau, B., Adamson, T.M.: The prone sleeping position impairs arousability in term infants. J. Pediatr. **138**(6), 811–816 (2001)
53. Phillipson, E.A., Sullivan, C.E.: Arousal: the forgotten response to respiratory stimuli. Am. Rev. Respir. Dis. **118**(5), 807–809 (1978)
54. Read, P.A., Horne, R.S.C., Cranage, S.M., Walker, A.M., Walker, D.W., Adamson, T.M.: Dynamic changes in arousal threshold during sleep in the human infant. Pediatr. Res. **43**(5), 697–703 (1998)
55. Lucchini, M., Fifer, W.P., Perez, A., Signorini, M.G.: Influence of sleep state and position on cardio-respiratory regulation in newborn babies. In: Proceedings of the Annual International Conference of the IEEE Engineering in Medicine and Biology Society EMBS, vol. 2015, pp. 302–305 (2015)
56. Lucchini, M., Fifer, W.P., Sahni, R., Signorini, M.G.: Novel heart rate parameters for the assessment of autonomic nervous system function in premature infants. Physiol. Meas. **37**(9), 1436–1446 (2016)
57. Galland, B.C., Hayman, R.M., Taylor, B.J., Bolton, D.P.G., Sayers, R.M., Williams, S.M.: Factors affecting heart rate variability and heart rate responses to tilting in infants aged 1 and 3 months. Pediatr. Res. **48**(3), 360–368 (2000)
58. Schumann, A.Y., Kantelhardt, J.W., Bauer, A., Schmidt, G.: Bivariate phase-rectified signal averaging. Physica A. **387**, 5091–5100 (2008)

Chapter 17
ARFIMA-GARCH Modeling of HRV: Clinical Application in Acute Brain Injury

Rute Almeida, Celeste Dias, Maria Eduarda Silva, and Ana Paula Rocha

Abstract In the last decade, several HRV based novel methodologies for describing and assessing heart rate dynamics have been proposed in the literature with the aim of risk assessment. Such methodologies attempt to describe the non-linear and complex characteristics of HRV, and hereby the focus is in two of these characteristics, namely long memory and heteroscedasticity with variance clustering. The ARFIMA-GARCH modeling considered here allows the quantification of long range correlations and time-varying volatility. ARFIMA-GARCH HRV analysis is integrated with multimodal brain monitoring in several acute cerebral phenomena such as intracranial hypertension, decompressive craniectomy and brain death. The results indicate that ARFIMA-GARCH modeling appears to reflect changes in Heart Rate Variability (HRV) dynamics related both with the Acute Brain Injury (ABI) and the medical treatments effects.

17.1 Introduction

The combination of sympathetic nervous system and parasympathetic nervous system activation defined as Autonomic Nervous System (ANS) tonus, is a function of both central mechanisms and feedback from peripheral systems that might be impaired in critically ill patients with Acute Brain Injury (ABI) [9, 15]. The incidence of ANS dysfunction after ABI is not exactly known, but changes are

R. Almeida
CMUP and BSICoS, Universidade do Porto, Porto, Portugal
e-mail: rbalmeid@fc.up.pt

C. Dias
CHSJ and FMUP, Universidade do Porto, Porto, Portugal
e-mail: mcdias@med.up.pt

M.E. Silva
CIDMA and FEP, Universidade do Porto, Porto, Portugal
e-mail: mesilva@fep.up.pt

A.P. Rocha (✉)
CMUP and FCUP, Universidade do Porto, Porto, Portugal
e-mail: aprocha@fc.up.pt

R. Barbieri et al. (eds.), *Complexity and Nonlinearity in Cardiovascular Signals*,
DOI 10.1007/978-3-319-58709-7_17

thought to be related to the severity of injury, to clinical management and, more importantly, seem to correlate independently with increased morbidity and mortality [6, 34, 55]. Heart rate variability (HRV) is currently defined by the fluctuating time between normal sinus beats (RR intervals) [51], and indicates modulation of the heart rate by the ANS [46], baroreceptor function, thermoregulation, vasomotor tone and circadian rhythms [38]. HRV measurements are easy to perform, non-invasive, and have good reproducibility, if used under standardized conditions such as those set by the Task Force of the European Society of Cardiology and the North American Society of Pacing and Electrophysiology in 1996 [49].

Traditional HRV techniques are mostly based in the analysis of RR intervals either in time or frequency domains. The resulting indexes include standard deviation measures, power spectral measures in clinically relevant frequency bands, and ratio between low and high frequency activity (ANS branches balance). Since 1965 when Vallbona et al. [50] linked HRV with the severity of brain injury and brain death risk, several other authors reported correlations between HRV, neuromonitoring variables and outcome after ABI [6, 13, 15, 21, 33, 34, 47, 55]. However, traditional approaches are unable to deal with high non stationarity of long HRV segments, specially marked in critically ill patients. Therefore early studies were restricted to short intervals limiting the development of dynamic prognostic tools in ABI [19].

In the last decades several new methodologies to analyse HRV have been proposed in the literature accompanied by their application in different physiological and clinical studies [21, 38, 44]. These methods comprehend, by a broad definition in [44], estimation of long-range correlation and fractal scaling, quantifying short-term complexity, entropy and regularity measures, analysis of chaotic behaviour with nonlinear dynamical systems. Alternatively, several authors proposed model based approaches to HRV analysis which cope with the nonstationarity of the data. One approach is time-variant AutoRegressive (AR) analysis using exponentially smoothed recursive least squares estimation, with fixed and varying forgetting factors [5, 31, 40]. Another approach is based on the adaptive segmentation of the non stationary record into approximately stationary records [36] which are usually modeled with short memory AR models. This procedure leads to linear parametric models for the conditional mean which, however, do not capture the long range correlations of HRV data. To overcome this drawback, Leite et al. [22] proposed the use of Fractionally Integrated AutoRegressive Moving Average (ARFIMA) models, an extension of the well-known AutoRegressive Moving Average (ARMA) models, to represent both short and long term behaviours of HRV series. The authors use adaptive segmentation of 24-h recordings of HRV to find that the long memory parameter changes with time and between day and night periods. These findings were later corroborated by Baillie et al. [3].

ARFIMA models describe the conditional mean of the data, however HRV exhibits also changes in variance over time, with periods of large variability followed by periods of stability, suggesting heteroscedasticity (varying variance). These volatility clustering phenomena may be well described by conditional volatility models such as the Generalized AutoRegressive Conditionally Heteroscedastic

(GARCH) [8]. Consequently, Leite et al. [24, 25] considered the joint modeling of long-memory and heteroscedasticity characteristics of HRV using fractionally integrated ARFIMA models with GARCH innovations. In these works ARFIMA-GARCH modeling is used to capture and remove long-range correlation and estimate conditional volatility in 24-h HRV recordings from Noltisalis database [45] and the database *Is the Normal Heart Rate Chaotic?*[1] from Physionet [14], respectively, allowing to discriminate between health and disease. A further empiric characteristic of HRV volatility is asymmetry in response to shocks. Leite et al. [27, 42] used exponential GARCH (ARFIMA-EGARCH) models to capture these effects and found that the parameters of the models are promising in differentiating health and disease. Note that GARCH models were first applied in the HRV context to develop an HRV based apnea screening tool [17], with the mean of the data described by a simple AR(1) process.

Rocha et al. [41] applied the ARFIMA-GARCH approach to the analysis of pediatric patients with acute brain injury. In this study, the parameters for long-memory and conditional variance estimated from cases with posterior confirmation of brain death (BD) differ significantly from those estimated from survivors and seem able to contribute to characterize disease severity in children with acute brain injury.

In spite of the dynamics of the ABI and the complexity of the Neurocritical Care Unit (NCCU) management, HRV analysis is a valuable tool to investigate the sympathetic and parasympathetic dysfunction of the ANS after ABI [47]. HRV has not been widely adopted by neurointensivists, nevertheless its enormous potential for application at bedside after ABI.

In this chapter are considered some clinical conditions that frequently occur in the NCCU where ARFIMA-GARCH HRV analysis integrated with multimodal brain monitoring may help clinicians to interpret or predict, in a more comprehensible way, acute cerebral phenomena such as intracranial hypertension with: plateau waves, decompressive craniectomy and brain death.

17.2 ARFIMA-GARCH Models

This section describes a class of models appropriate to characterize the persistence (long memory) and nonlinear characteristics of RR series.

The characteristic of persistence, long-memory or long range dependence appears in many natural phenomena and is characterized by slowly decaying serial correlations, or equivalently by a spectrum with an hyperbolic behaviour at the origin, $f(\lambda) \sim |\lambda|^{-2d}$. Long memory is thus related to self-similar processes also known as fractals and the so called $1/f$ noise. Heuristic methods, such as the Detrended Fluctuation Analysis (DFA), the Rescaled Range (R/S) and Rescaled

[1] https://www.physionet.org/challenge/chaos/.

Variance (R/V) are often used for assessing long memory in data but are mainly useful for descriptive purposes rather than concrete statistical inference or model building. The alternative taken herein is to consider a model based approach with ARFIMA models (sometimes also designated by FARIMA). These models extend the well known ARMA models for time series and are, thus, able to model the short as well as the long memory of the data.

GARCH models were introduced in the literature to account for the following nonlinear behaviour of many data sets: heteroscedasticity in the conditional variance with clusters of volatility. This same phenomena has been observed in RR series [24, 25]. To combine the persistence in mean with conditional time-varying variance in a model based approach, ARFIMA models with GARCH innovations are considered. The ARFIMA-GARCH models is a wide class of models which is able to cater for complex nonlinearities in dynamics of the data.

ARFIMA$(p, d, 0)$-GARCH$(1, 1)$ models are defined as follows

$$\phi(B)(1-B)^d x_t = \epsilon_t, \tag{17.1}$$

$$\epsilon_t = \sigma_t z_t, \tag{17.2}$$

$$\sigma_t^2 = \mathrm{Var}(\epsilon_t|\mathscr{F}_{t-1}) = u_0 + v_1\sigma_{t-1}^2 + u_1\epsilon_{t-1}^2 \tag{17.3}$$

where B is the backward-shift operator, $d \in \mathbb{R}$ is a real number, $(1-B)^d = \sum_{k=0}^{\infty}\binom{d}{k}(-1)^k B^k$ is the fractional difference operator, [4], $\phi(B) = 1 - \phi_1 B - \ldots - \phi_p B^p$ is a polynomial in B, $u_0 > 0$, $u_1, v_1 \geq 0$, $p \in \mathbb{N}_0$, z_t are independent and identically distributed random variables with zero mean and unit variance and $\mathscr{F}_t$ denotes the history of the process up to time t. Equation (17.1) describes the conditional mean of the process with serially uncorrelated residuals or shocks and is called an ARFIMA$(p, d, 0)$ model [4]. Equations (17.2) and (17.3) describe the conditional variance of the process and define a GARCH$(1, 1)$ model [8].

In Eq. (17.1) parameter d determines the long-term behaviour in mean,[2] whereas p and the coefficients in $\phi(B)$ allow for the modeling of short-range properties. The model is stationary for $-0.5 < d < 0.5$, nonstationary but mean reverting for $0.5 \leq d < 1$. Equation (17.3) describes the dynamics of the conditional variance (volatility) of the process: σ_t^2 is dependent on its own lagged value and on the past squared shocks or residuals from the mean equation. Parameter u_1 characterizes the short-range properties in volatility and parameter v_1 characterizes persistence in the volatility. The GARCH$(1, 1)$ model is second order stationary if $v_1 + u_1 < 1$ [8]. Therefore the ARFIMA$(p, d, 0)$-GARCH$(1, 1)$ is stationary if $-0.5 < d < 0.5$, all the roots of $\phi(B)$ lie outside the unit circle and $v_1 + u_1 < 1$.

---

[2]The long memory parameter is related to the Hurst coefficient, $H = d + 0.5$, to the fractal dimension, $D = 2 - H$ and to the slope of the (generalized) spectral density in the low frequency range by $\alpha = 2d$.

The spectral density function of a stationary ARFIMA$(p, d, 0)$ process is given by Hosking [16]

$$f_\omega = f_\omega^* |1 - e^{-i\omega}|^{-2d}, \quad -\pi \leq \omega \leq \pi, \tag{17.4}$$

$$f_\omega^* = \frac{\sigma_\epsilon^2}{|\phi(e^{-i\omega})|^2} \tag{17.5}$$

where

$$\sigma_\epsilon^2 = \mathrm{var}(\epsilon_t) = \frac{u_0}{1 - v_1 - u_1} \tag{17.6}$$

and f_ω^* is the spectral density of the AR(p) process. Since the autocorrelation function of a GARCH(1,1) process is the same as that of white noise [8], the spectral density function of a stationary ARFIMA$(p, d, 0)$-GARCH$(1, 1)$ process is given by Eqs. (17.4) and (17.5) with σ_ϵ^2 defined by (17.6). For $0.5 \leq d < 1$ Eq. (17.4) corresponds to a pseudo-spectral density [18, 52], since the process is nonstationary but mean reverting.

A final note regarding ARFIMA$(p, d, 0)$ models is due. The model may be extended to non-stationary settings in which $d = d_{LM} + D > 1$ with $-0.5 < d_{LM} < 0.5$ and $D \in \{1, 2, \ldots\}$. The most usual case occurs with $D = 1$, when the process is said to have a unit root: the ARFIMA$(p, d, 0)$ is, then, used to model the increments of the series, that is the differences between consecutive observations.

Given a time series $x_1, \ldots, x_n$, to estimate the parameters of an ARFIMA$(p, d, 0)$-GARCH$(1, 1)$, model proceed as follows [25]:

(i) estimate $-0.5 < d < 0.5$ using the semi-parametric local Whittle estimator [39];
(ii) define the filtered data $y_t = (1 - B)^d x_t$;
(iii) estimate the AR(p)-GARCH$(1, 1)$ parameters in the filtered data y_t by maximum likelihood.

The semi-parametric local Whittle estimator of the long memory, step (i), is an estimator in the Fourier domain, and acknowledged in the literature for its statistical properties, efficiency and asymptotic normality, and also easiness of implementation [4]. Other estimators, such as the log-wavelet regression estimator which is a semiparametric estimator in the wavelet domain, have the disadvantage of complicated asymptotic distributions. It is worth noting that the standard error of the estimator in (i) is quite large, $\sqrt{1/(4n^{0.8})}$: for example, the standard error associated with the estimate of d from a data set with $n = 2000$ observations is 0.024. Under the ARFIMA model based approach it is also necessary determine the order p of the AR component of the model. This is accomplished in step (iii) using Akaike Information Criterion, AIC, under maximum likelihood estimation. The conditional heteroscedasticity in the series, that is the need for the GARCH component, is assessed by the Ljung-Box test in the squared residuals [28]. If the series does not present heteroscedasticity then the final model is an ARFIMA$(p, d, 0)$. In any

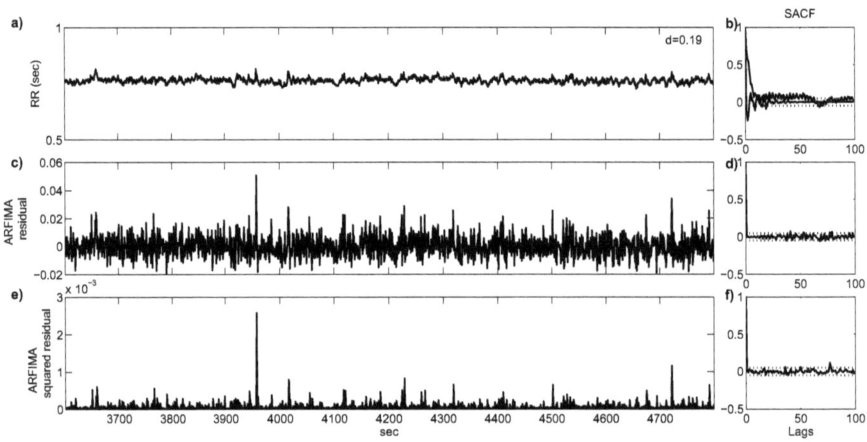

Fig. 17.1 (**a**) Tachogram (20 min segment with 1571 beats of RR series) with $d = 0.19$, (**b**) SACF (*black*) and AR ACF (*grey*), (**c**) and (**d**) ARFIMA residuals and corresponding SACF, (**e**) and (**f**) squared residuals and SACF. The *horizontal lines* show the 95% confidence limits for SACF

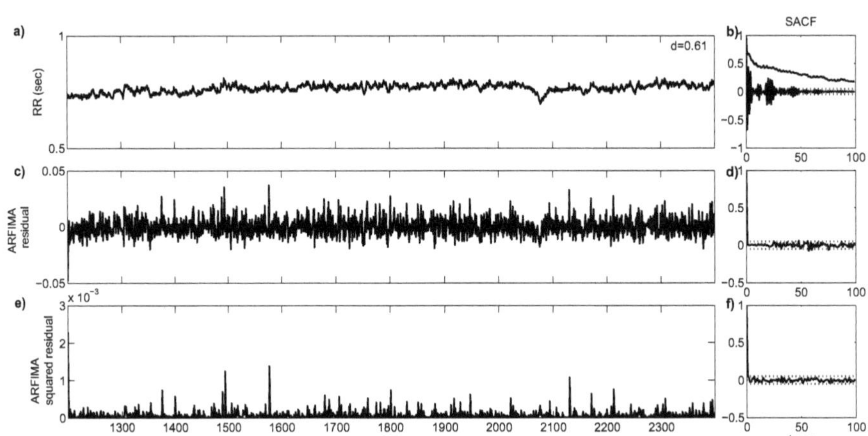

Fig. 17.2 (**a**) Tachogram (20 min segment with 1569 beats of RR series) with $d = 0.61$, (**b**) SACF (*black*) and AR ACF (*grey*), (**c**) and (**d**) ARFIMA residuals and corresponding SACF, (**e**) and (**f**) squared residuals and SACF. The *horizontal lines* show the 95% confidence limits for SACF

case, only models with uncorrelated residuals, checked by the Ljung-Box test, are considered valid models.

To illustrate the use of ARFIMA(p,d,0)-GARCH(1,1) models in HRV data consider Figs. 17.1, 17.2, and 17.3, each one representing the analysis of a segment of the tachogram of a patient in intensive care. Figure 17.1 represents in (a) the tachogram of a segment with 1571 beats and in (b) the corresponding sample autocorrelation function (SACF) in black and the ACF of the best AR model for the data in gray. The SACF shows a very slow decay indicating that the dependence between distant observations is not negligible and thus displaying long memory characteristics. An ARFIMA($p = 26, d = 0.19, 0$) model is fitted to this segment. The resulting residuals (ϵ_t), displayed in (c) exhibit little correlation, (d), indicating that the ARFIMA model is adequate to explain the dynamics and conditional mean of the data. Also, the squared residuals in (e) exhibit no significant autocorrelation in (f), indicating absence of time-varying conditional variance or heteroscedasticity. These results are confirmed by the p-value < 0.001 of Ljung-Box test applied to the squared residuals of ARFIMA. A similar analysis is depicted in Fig. 17.2 but it is worth noting that the SACF in (b) indicates a much stronger long-memory effect which is confirmed by the value $d = 0.61$ estimated from the data. The last illustration is depicted in Fig. 17.3. Once again the SACF in (b) of the tachogram segment in (a) indicates the existence of long memory. The residuals ϵ_t displayed in (c) and the corresponding SACF in (d), exhibit little correlation indicating that the estimated ARFIMA($p = 19, d = 0.82, 0$) model is adequate to explain the dynamics and conditional mean of the data. However, the squared residuals in (e) exhibit significant autocorrelation in (f), indicating time-varying conditional variance or heteroscedasticity. These results are confirmed by the p-value > 0.05

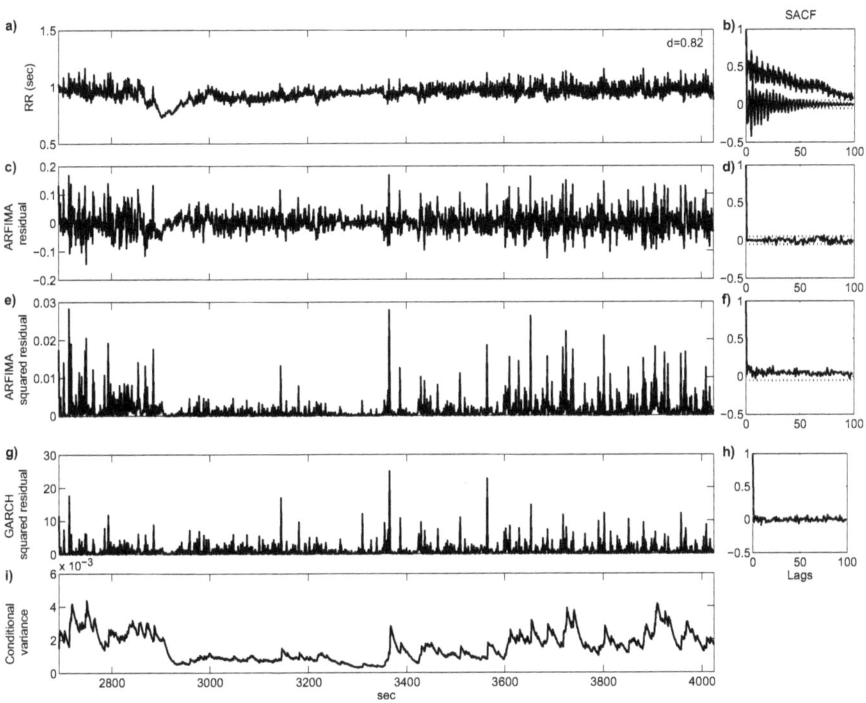

Fig. 17.3 (**a**) Tachogram (22 min segment with 1420 beats of RR series) with $d = 0.82$, (**b**) SACF (*black*) and AR ACF (*grey*), (**c**) and (**d**) ARFIMA residuals and corresponding SACF, (**e**) and (**f**) squared residuals and SACF, (**g**) and (**h**) ARFIMA-GARCH residuals and SACF, (**i**) conditional volatility. The *horizontal lines* show the 95% confidence limits for SACF

of Ljung-Box test applied to the squared residuals of ARFIMA. Now, to model this effect a GARCH(1,1) model is entertained for the ARFIMA residuals. The estimates for the parameters are $u_1 = 0.05$, $v_1 = 0.94$. The squared residuals from the ARFIMA-GARCH(1,1) model and corresponding SACF, (g) and (h), indicate absence of heteroscedasticity. Finally, the volatility of this HRV segment is depicted in (i).

ARFIMA-GARCH models provide a set of parameters $(\phi_1, \ldots, \phi_p, p, d, u, v)$ that can be used to describe linear and non-linear characteristics of HRV. The ARFIMA parameter d determines the long-term behaviour in mean, while GARCH parameters u and v describe the non-linear behaviour of conditional time-variant variance. The order p and the coefficients $\phi_1, \ldots, \phi_p$ of the AR component of the ARFIMA-GARCH model describe the short memory characteristics, as the usual AR approach to HRV modeling. Consequently as the ARFIMA-GARCH spectrum is the same as that of ARFIMA, the spectral indexes extracted from the ARFIMA-GARCH model provide the same information on HRV as that provided by AR models. The advantage is that the estimation of d allows an individualized data based removal of the very low frequency component and thus avoiding any additional filtering.

ARFIMA-GARCH models are applied in Sect. 17.4 to data from adult traumatic brain injury patients described in the next section.

17.3 Data, Pre-processing and HRV Modeling

The data analysed in the remainder of this chapter regard a retrospective study of five adult severe traumatic brain injury patients (TBI), admitted to a NCCU who developed intracranial hypertension. The exclusion criteria were pregnancy and age ($\leq$18 years). Standard management of patients included sedation and analgesia with propofol and/or midazolam and fentanyl and normoventilation. Intracranial Pressure (ICP) and Cerebral Perfusion Pressure (CPP) management was achieved according to the NCCU protocol previously published [11], and approved by the hospital Ethics Committee. The physiological variables continuously monitored included ECG at 250 Hz, heart rate (HR), arterial blood pressure (ABP), ICP, amplitude of ICP (AMP), CPP, End-tidal CO2 (ETCO2), brain tissue oxygenation (pbtO2), cerebral blood flow (CBF) and cerebral autoregulation (CAR) evaluated with cerebrovascular pressure reactivity index (PRx). Software ICM+® was used to collect primary data and continuously record raw signals that were analysed offline. Automatic ECG annotation was performed with a multi-scale wavelet-based ECG annotator previously developed and validated [32] using MATLAB® and BioSigBrowser [7] tools. The RR series were defined as the time interval, in seconds, between consecutive QRS automatic locations; RR values admitted have a maximum of 3 s and are within a 3/0.6745 tolerance on the difference to Median Absolute Deviation[54]. The series were divided into segments with minimum 1024 and maximum 4096 consecutive beats and assumed to be sampled at the local mean heart rate.

ARFIMA-GARCH modeling of each RR series segment, x_t, described in Sect. 17.2 by steps (i), (ii) and (iii) assumes that the segments x_t are approximately stationary. However, since this is not always the case, additional steps must be performed to ensure that the models are estimated over stationary segments. First, the segment is tested for stationarity using the Augmented-Dickey-Fuller test for unit roots [30]. If there is a unit root, then $D = 1$, c.f. Sect. 17.2 and the RR series is differenced. This means that step (i) is performed over the differenced RR or, in other words, the series of RR increments, $\Delta\text{RR} = x_t - x_{t-1}$. The resulting estimate $d_{LM}^{\Delta\text{RR}}$ is expected to belong to the interval $]-0.5, 0.5[$ otherwise over-differencing may have occurred. However, in view of the large standard error associated to the estimates for d, see Sect. 17.2, a shorter, more robust interval is considered for this purpose. Thus, if $d_{LM}^{\Delta\text{RR}} > -0.45$ then set $d = 1 + d_{LM}^{\Delta\text{RR}}$ in step (ii), and proceed to step (iii). Otherwise, if $d_{LM}^{\Delta\text{RR}} < -0.45$ steps (i), (ii), (iii) are performed over the RR series x_t. Also, if there is no unit root, steps (i), (ii), (iii) are performed over the RR series x_t. The order of the AR(p) model is step (iii) is selected in the range 6–36 according AIC. An RR segment is said to have a valid ARFIMA model if the residuals of the ARFIMA model are uncorrelated, tested by Ljung-Box at 5% level of significance. Similarly, an RR segment is said to have a valid ARFIMA-GARCH(1,1) model if the residuals of the ARFIMA model are uncorrelated but its squared residuals are correlated, all tested by Ljung-Box at 5% level of significance.

Finally and for comparison purposes, the RR segments are firstly detrended using smooth priors ($\lambda = 50$), [48] and then modeled by an AR model with adequate order chosen by AIC.

For each fitted model, parametric HRV measures were obtained using spectral AR decomposition [1] by assigning each pole contribution to the spectral band in which the pole is located according to the standard fixed bands defined in [49]: low frequency (LF: 0.04–0.15 Hz) and high frequency (0.15–1.4 Hz) bands, all frequencies in spectra (TP). LFn and HFn are measures, respectively at LF and HF bands, normalized by power in the band above 0.04 Hz. Autonomic balance B was obtained as LF/HF.

17.4 Clinical Applications in Acute Brain Injury

ANS dysfunction in ABI leads to changes in HRV which appear to be particularly marked in patients subsequently declared with Brain Death (BD). It is well known that reduced HRV, assessed in time, frequency and complexity, is a risk index after trauma for both morbidity and mortality and an early predictor of BD. Previous studies on cardiovascular series variability indicate that the derived indexes may provide an early complementary tool for time course prediction and prognostic in critical illness [2, 43]. Although some authors have proposed HRV as an auxiliary tool in trauma, it is not usually used in the clinical practice for ABI patients, nor considered as a decision support tool for the commencement of the proceeding for BD declaration [33, 35, 43]. This work aims at illustrating the potential of

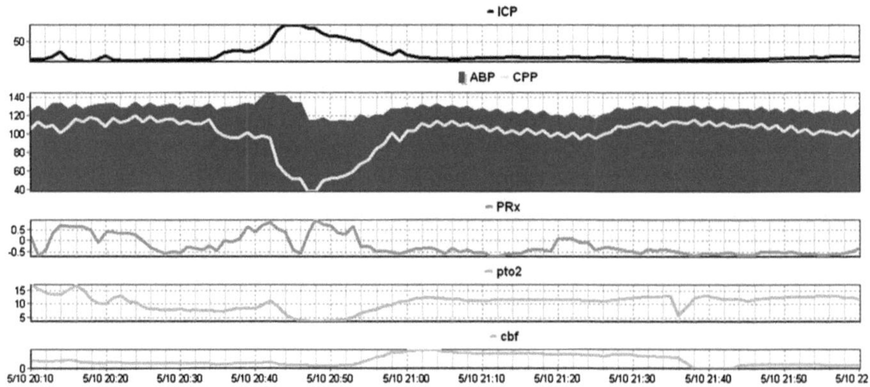

Fig. 17.4 Plateau wave (PW) documentation with multimodal brain monitoring using intracranial pressure (ICP), mean arterial blood pressure (ABP), cerebral perfusion pressure (CPP), brain tissue oxygenation pressure (pbtO2), cerebral blood flow (CBF) and cerebral autoregulation (CAR) evaluation with pressure reactivity index (PRx). During PW while ICP increases, CPP, pbtO2 and CBF decreases along with impairment of CAR (PRx >0.3). After PW, CPP and PRx recover but pbtO2 and CBF increase because of hyperaemic response after ischemia

ARFIMA-GARCH modeling of HRV, integrated with multimodal brain monitoring, for helping clinicians to interpret or predict in a more comprehensible way acute cerebral phenomena such as intracranial hypertension with: (a) plateau waves, (b) decompressive craniectomy, (c) brain death.

Plateau waves (PW) are sudden rises in ICP, up to 40–100 mmHg, lasting more than 5 min, terminating either spontaneously or in response to active treatment [10, 29]. Meanwhile arterial blood pressure remains stable or varies modestly, as illustrated in Fig. 17.4. Decompressive craniectomy (DC) is a surgical option for managing refractory intracranial hypertension that remains controversial. DC is efficacious in reducing ICP, as illustrated in Fig. 17.5, by the removal of a bone flap allowing brain volume expansion. The focus of current debate surrounds identification of the subset of patients for whom it is appropriate to offer DC [20]. Intracranial hypertension may also evolve to brain death (BD) due to the absence of blood flow, zero flow (ZF), in consequence of cerebral circulatory arrest leading to the complete and irreversible loss of brain stem function [53] as illustrated in Fig. 17.6.

The clinical examples presented in this section regard five adult patients with severe traumatic brain injury. All patients, as already mentioned, developed intracranial hypertension. Patient P1 is a noteworthy example of intracranial hypertension with plateau waves due to the huge number of occurrences, along with a good response to osmotherapy with hypertonic saline medical treatment thus avoiding DC. Four patients, hereafter designated by P2 to P5 were submitted to DC. Two of these, P2 and P3 survived with good outcome while the other two, P4 and P5, evolved to BD.

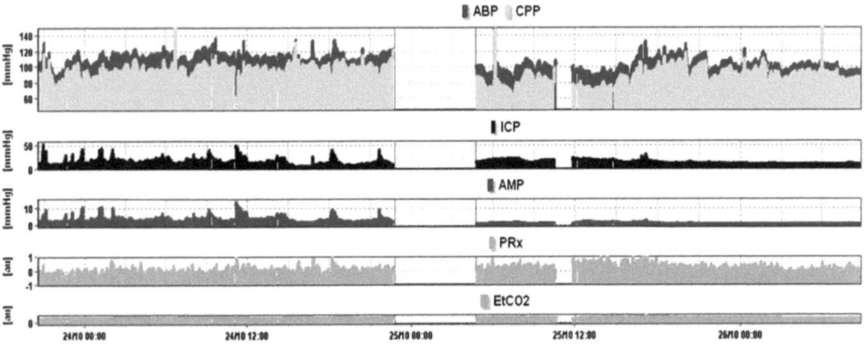

Fig. 17.5 Multimodal brain monitoring changes before and after decompressive craniectomy represented by mean arterial blood pressure (ABP), cerebral perfusion pressure (CPP), intracranial pressure (ICP), amplitude of ICP (AMP), cerebrovascular pressure reactivity (PRx) and endtidal carbon dioxide (ETCO2)

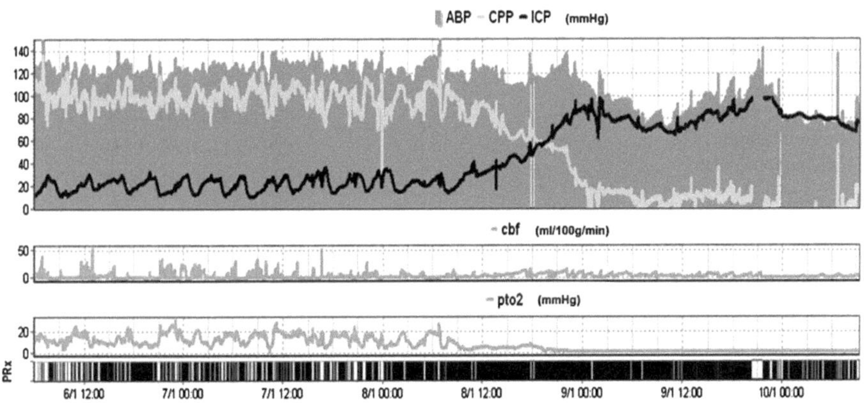

Fig. 17.6 Multimodal brain monitoring documentation of the evolution to irreversible zero flow and brain death. While intracranial pressure (ICP) increases and cerebral perfusion pressure (CPP) compromises, cerebral blood flow (CBF) and brain tissue oxygen pressure (pbtO2) decline to ischemic thresholds. Arterial blood pressure (ABP) remains almost constant. Cerebrovascular reactivity pressure (PRx) in the lower panel becomes a *solid line* clearly meaning severe impairment of cerebral autoregulation (CAR)

The results of analysing a total of 2123 segments across all available data from the five patients are summarized in Table 17.1. The low signal quality, with many corrupted excerpts, typically found in intensive care recordings, is the cause of the reduced number of segments per day. Firstly, note that not only the percentage of segments fitted with ARFIMA models is higher but also the median length of those segments is larger. Moreover, the order of the ARFIMA models is relevantly lower, with only approximately half of the segments needing an order higher than $p = 30$. These findings indicate that the ARFIMA model inherent detrending is superior to the current advanced HRV detrending techniques using smooth priors,

Table 17.1 Features resulting from the modeling of RR series segments, during the whole period of monitoring, of patients P1 to P5 with AR models and ARFIMA-GARCH models

		Patients					
		P1	P2	P3	P4	P5	Overall
Monitorization (approximate length)		21 days	7 days	16 days	12 days	10 days	66 days
Total number of segments		672	247	405	265	534	2123
% Segments with valid models	AR	60%	53%	56%	48%	49%	54%
	ARFIMA	86%	95%	88%	84%	87%	87%
Median length (beats)	AR	2162.5	1893	1776.5	1679	2870	2016
	ARFIMA	2926.5	3047	2672	2424.5	4096	3031
Median order	AR	36	36	35	36	36	
	ARFIMA	29	27	31	31	30	
% Models with maximum order chosen above 30	AR	90%	86%	82%	88%	81%	85%
	ARFIMA	46%	40%	51%	51%	47%	47%

Table 17.2 ARFIMA-GARCH model parameters estimates in RR segments *during PW* and *after PW* plateau waves with minimum duration of approximately 9 min (events)

	During PW			*After PW*		
Event	d	u	v	d	u	v
1	0.55	0.07	0.9	0.33	0	0
2	0.54	0.45	0.54	0.23	0.20	0.44
3	0.78	0.11	0.82	0.30	0	0
4	0.82	0.07	0.9	0.47	0.03	0.95

leading to a better adjustment to each patient. Thus ARFIMA-GARCH modeling is advantageous to describe HRV dynamics.

To analyse the effect of PW in the dynamics of HRV consider the occurrence of four plateau waves with minimum duration of approximately 9 min (events) for patient P1. For each event two segments are considered: the period of time with ICP >40 mmHg, designated by *during PW* and 1024 beats after ICP normalized below 20 mmHg, designated by *after PW*. The results are summarized in Table 17.2. In segment *during PW*, $d > 0.54$ and the GARCH parameters u, v satisfy $u+v > 0.9$, while in segment *after PW*, d decreases to values below 0.33 in the first three events and $d = 0.47$ in the 4th event. The volatility, measured by $u + v$ also decreases to low values, $u + v < 0.65$, in the first three events, with two of them presenting absence of volatility, $u+v = 0$. The 4th event, presenting the higher values for long memory $d = 0.47$ and volatility $u+v = 0.98$, has the particularity of corresponding to a spontaneous reversion of ICP without osmotherapy. These findings suggest the hypothesis that hypertonic saline not only decreases ICP but also affects the dynamics of HRV in terms of conditional mean and conditional variance. In fact, Dias et al. [11] have shown that the first effect of hypertonic saline administration is an increase of cerebral blood flow with reduction of cerebral vascular resistance while ABP is minimally affected. These effects, although small, may nevertheless be sufficient to induce variations in HRV dynamics.

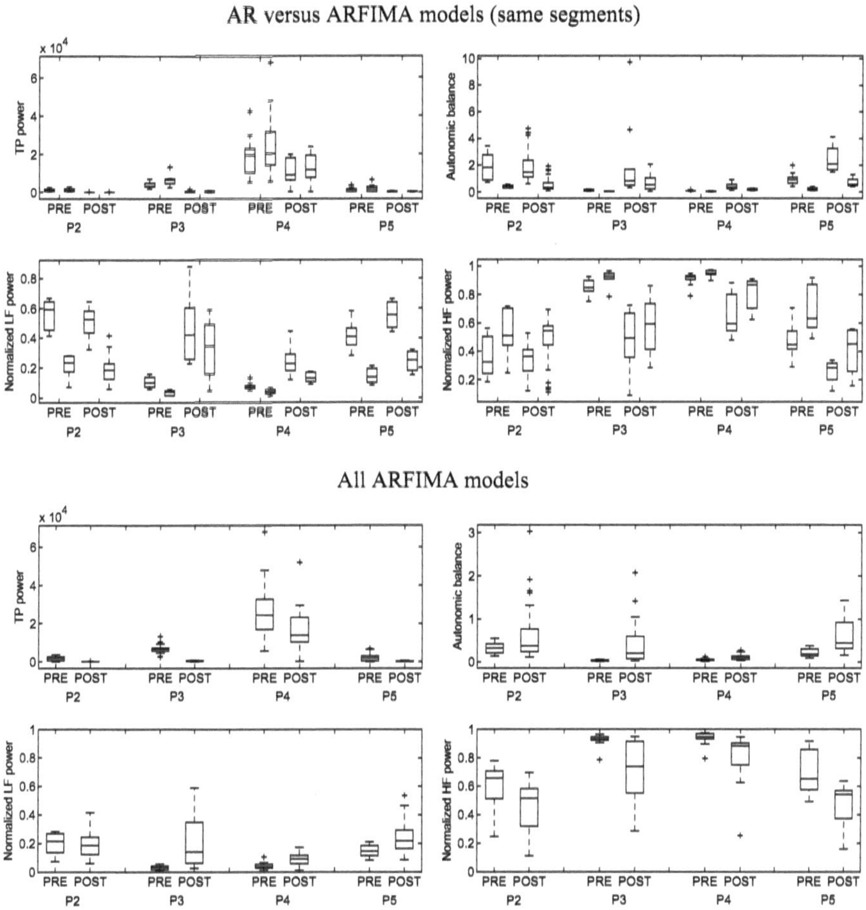

Fig. 17.7 Box plot of spectral measures over TP, B, LFn and HFn for each patient in pre-DC and post-DC: in the *upper panel* first box plot for each case correspond to AR models and the second to ARFIMA models

As mentioned above, the management of the refractory intracranial hypertension of patients P2 to P5 involved a surgical approach by DC. The immediate effect of the surgical procedure in the HRV dynamics was studied by comparing the parametric HRV measures over the admissible models in the periods of 24 h before (PRE) and after (POST) DC.

Figure 17.7 summarizes the distributions of TP, B, LFn and HFn for each patient in PRE and POST decompressive craniectomy, calculated over the segments described in Table 17.3. In the upper panels, the represented values were obtained from segments for which adequate AR (left boxplots) and ARFIMA-GARCH (right boxplots) models were found. Differences both in median values and dispersion are observable, in particular for LFn and B, both depending on LF. These findings

Table 17.3 Features resulting from modeling RR series segments of patients P2 to P5, in the periods of 24 h before (PRE) and after (POS) DC, with AR and ARFIMA-GARCH models

		Patients							
		P2		P3		P4		P5	
		PRE	POS	PRE	POS	PRE	POS	PRE	POS
Total number of segments		10	30	18	28	21	15	20	27
% Segments with valid models	AR	50%	77%	44%	37%	81%	33%	65%	15%
	ARFIMA	100%	100%	100%	86%	100%	100%	95%	96%
Median length (beats)	AR	1877	1653	2456.5	1553	1857	1634	2680	4096
	ARFIMA	3043.5	2277	3896	2994.5	2215	3000	3358	4096

indicate that the detrending strategies (smooth prior detrending and long-memory filtering) have impact in the spectral measures.

Note that the percentage of segments fitted with ARFIMA models is much higher and also the median length of those segments is larger than those fitted with AR models. This fact prompts the study of the spectral measures obtained for all segments for which ARFIMA model was estimated, which are represented in the lower panels of Fig. 17.7. Globally, TP decreases after DC while B stays at a comparable level. Regarding the two traditional bands, HFn decreases and LFn increases except in P2. As a matter of fact, the absolute measures HF and LF decrease for all patients except LF in P4. The less marked reduction in LF may be explained by the sustained need of hemodynamic support managed with noradrenaline infusion according to NCCU protocol [11].

The changes of these spectral measures may be related to the decrease of intracranial pressure expected after DC [12]. In spite of absence of relevant differences in the median of B with DC, the dispersion (evaluated by interquartile range) increased. In a recent study by Dias et al. [12] over nine patients submitted to DC (including the four patients P2–P5) the correlation between nonparametric based B and ICP was studied. A positive significant correlation of 0.4 was found in the 12 h immediately preceding DC, becoming negligible and non significant in the post-DC period.

For patients P4 and P5 who evolved to BD, the time of ZF, was documented c.f. Fig. 17.6. It is then possible to study retrospectively HRV dynamics in 24 h periods from the monitoring restore time after DC to ZF and after ZF. In addition to the expected reduced HRV, ARFIMA-GARCH modeling indicates also less heteroscedasticity in BD, measured by the percentage of segments in each 24 h with $u + v \geq 0.65$, Fig. 17.8. Note that there are the two values for patient P5 in day 6 resulting from the fact that ZF occurred within that 24 h period. The marked reduction between the two values reflects an immediate effect of ZF. Thus, ARFIMA-GARCH models allow to notice that heteroscedasticity decreases simultaneously with the irreversible compromising of cerebral perfusion pressure.

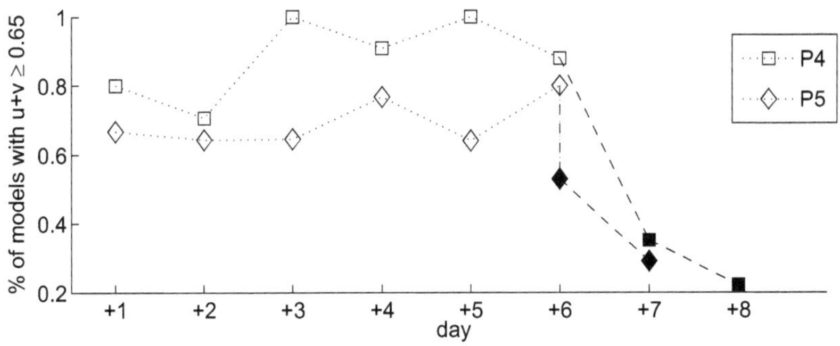

Fig. 17.8 Percentage of models with $u + v \geq 0.65$ for BD patients, P4 and P5. *Dark marks* correspond to time after zero flow (ZF)

In summary, ARFIMA-GARCH modeling allows the quantification of long range correlations, d, and volatility, u and v, which appear to reflect changes in HRV dynamics related both with the acute brain lesion and the medical treatments effects.

17.5 Final Comments

The last decade has witnessed an increase of interest in the characterization of complexity and nonlinearity in HRV with the aim of risk assessment. Here two specific complex/nonlinear characteristics of HRV, long memory and time-varying variance, are considered from a model based parametric point of view. Concerning long memory, several studies in the literature [23, 26], and references there included, indicate that the application of heuristic approaches such as DFA, lead to the same information on HRV. However, the model based ARFIMA approach has the advantage of allowing the design of individualized filters and simultaneously the obtention of clinically interpretable measures such as low and high frequency components in HRV.

The potential of HRV analysis as a tool to assess the sympathetic and parasympathetic dysfunction of the ANS after ABI has not been fully investigated yet, although Norris et al. [37] suggested that HRV could be used to predict the outcome in critically injured patients as one of a set of *new vital signs*. The parametric ARFIMA-GARCH approach advocated in this study provides new measures to describe the individual patient HRV dynamics, which may be particularly relevant to clinicians who manage critically ill patients with complex medical conditions that change rapidly and sometimes unpredictably. Further studies are warranted to fully appraise ARFIMA-GARCH modeling of HRV in acute brain injury.

Acknowledgements This work was supported in part by Portuguese funds through CIDMA UID/MAT/04106/2013 and CMUP UID/MAT/00144/2013, funded by the Portuguese Foundation for Science and Technology (FCT -Fundação para a Ciência e a Tecnologia), projects TEC2013-42140-R and TIN2014-53567-R from the MINECO with European Regional Development Fund (FEDER), Spain, and by Grupo Consolidado BSICoS (T-96) from DGA (Aragón, Spain) and European Social Fund (EU).

References

1. Almeida, R., Gouveia, S., Rocha, A.P., Pueyo, E., Martínez, J.P., Laguna, P.: Exploring QT variability dependence from heart rate in coma and brain death on pediatric patients. IEEE Trans. Biomed. Eng. **53**, 1317–1329 (2006)
2. Almeida, R., Silva, M.J., Rocha, A.P.: Exploring QT variability dependence from heart rate in coma and brain death on pediatric patients. In: Proceedings of Computing in Cardiology Conference (CinC), 2013, pp. 61–64 (2013)
3. Baillie, R.T., Cecen, A.A., Erkal, C.: Normal heartbeat series are nonchaotic, nonlinear, and multifractal: new evidence from semiparametric and parametric tests. Chaos **19**, 028503 (2009)
4. Beran, J., Feng, Y., Ghosh, S., Kulik, R.: Statistics for Long-Memory Processes: Probabilistic Properties and Statistical Methods. Springer, New York (2012)
5. Bianchi, A.M., Mainardi, L.T., Cerutti, S.: Time-frequency analysis of biomedical signals. Trans. Inst. Meas. Control **22**, 215–230 (2000)
6. Biswas, A.K., Scott, W.A., Sommerauer, J.F., Luckett, P.M.: Heart rate variability after acute traumatic brain injury in children. Crit. Care Med. **28**(12), 3907–3912 (2000)
7. Bolea, J., Almeida, R., Laguna, P., Sörnmo, L., Martínez, J.P.: BioSigBrowser, biosignal processing interface. In: Final Program and Abstract Book 9th International Conference on Information Technology and Applications in Biomedicine, ITAB 2009, Art. No. 5394301 (2009)
8. Bollerslev, T.: Generalized autoregressive conditional heteroscedasticity. J. Econ. **31**, 307–327 (1986)
9. Buchman, T.G., Stein, P.K., Goldstein, B.: Heart rate variability in critical illness and critical care. Curr. Opin. Crit. Care **8**(4), 311–315 (2002)
10. Dias, C., Maia, I., Cerejo, A., Varsos, G., Smielewski, P., Paiva, J.A., Czosnyka, M.: Pressures, flow, and brain oxygenation during plateau waves of intracranial pressure. Neurocrit. Care **21**(1), 124–32 (2014)
11. Dias, C., Silva, M.J., Pereira, E., Monteiro, E., Maia, I., Barbosa, S., Silva, S., Honrado, T., Cerejo, A., Aries, M.J.H., Smielewski, P., Paiva, J.A., Czosnyka, M.: Optimal cerebral perfusion pressure management at bedside: a single-center pilot study. Neurocrit. Care **23**(1), 92–102 (2015)
12. Dias, C., Almeida, R., Vaz Ferreira, A., Silva, J., Monteiro, E., Cerejo, A., Rocha, A.P.: Heart rate variability and multimodal brain monitoring before and after decompressive craniectomy in traumatic brain injury. Intensive Care Med. Exp. **4**(Suppl 1), 30:209 (2016)
13. Gang, Y., Malik, M.L.: Heart rate variability in critical care medicine. Curr. Opin. Crit. Care **8**(5): 371–375 (2002)
14. Goldberger, A.L., Amaral, L.A.N., Glass, L., Hausdorff, J.M., Ivanov, P.C., Mark, R.G., Mietus, J.E., Moody, G.B., Peng, C.K., Stanley, H.E.: Physiobank, physiotoolkit and physionet: components of a new research resource for complex physiologic signals. Circulation **101**, e215–e220 (2000)
15. Haji-Michael, P.G., Vincent, J.L., Degaute, J.P., van de Borne, P.: Power spectral analysis of cardiovascular variability in critically ill neurosurgical patients. Crit. Care Med. **28**(7), 2578–2583 (2000)
16. Hosking, J.R.M.: Fractional differencing. Biometrika **68**, 165–176 (1981)

17. Hu, Y.M., Tsoukalas, C.: Conditional volatility properties of sleep-disordered breathing. Comput. Biol. Med. **36**(3), 303–312 (2006)
18. Hurvich, C.M., Ray, B.K.: Estimation of the memory parameter for nonstationary or noninvertible fractionally integrated processes. J. Time Ser. Anal. **16**, 17–41 (1995)
19. Kahraman, S., Dutton, R.P., Hu, P., Stansbury, L., Xiao, Y., Stein, D.M., Scalea, T.M.: Heart rate and pulse pressure variability are associated with intractable intracranial hypertension after severe traumatic brain injury. J. Neurosurg. Anesthesiol. **22**(4), 296–302 (2010)
20. Kakar, V., Nagaria, J., John Kirkpatrick P.: The current status of decompressive craniectomy. Br. J. Neurosurg. **23**(2), 147–157 (2009)
21. Kamath, M.V., Watanabe, M., Upton, A. (eds.): Heart Rate Variability (HRV) Signal Analysis: Clinical Applications. CRC Press, Boca Raton (2013)
22. Leite, A., Rocha, A.P., Silva, M.E., Costa, O.: Modelling long-term heart rate variability: an ARFIMA approach. Biomed. Tech. **51**, 215–219 (2006)
23. Leite, A., Rocha, A.P., Silva, M.E., Gouveia, S., Carvalho, J., Costa, O.: Long-range dependence in heart rate variability data: ARFIMA modelling vs detrended fluctuation analysis. Proc. Comput. Cardiol. **34**, 21–24 (2007)
24. Leite, A., Rocha, A.P., Silva, M.E.: Long memory and volatility in HRV: an ARFIMA-GARCH approach. Proc. Comput. Cardiol. **36**, 165–168 (2009)
25. Leite, A., Rocha, A.P., Silva, M.E.: Beyond long memory in heart rate variability: an approach based on fractionally integrated autoregressive moving average time series models with conditional heteroscedasticity. Chaos **23**(2), 023103 (2013)
26. Leite, A., Silva, M.E., Rocha, A.P.: Scaling exponents in heart rate variability. In: Lita da Silva, J., Caeiro, F., Natário, I., Braumann, C.A. (eds.) Advances in Regression, Survival Analysis, Extreme Values, Markov Processes and Other Statistical Applications, pp. 259–267. Springer, Berlin (2013)
27. Leite, A., Rocha, A.P., Silva, M.E.: Modeling volatility in heat rate variability. In: 2016 Annual International Conference of the IEEE Engineering in Medicine and Biology Society (EMBC). IEEE Engineering in Medicine and Biology Society Conference Proceedings, pp. 3582–3585 (2016)
28. Ljung, G.M., Box, G.E.P.: On a measure of lack of fit in time series models. Biometrika **65**, 297–303 (1978)
29. Luis, A., Santos, A.S., Dias, C., Almeida, R., Rocha, A.P.: Heart rate variability during plateau waves of intracranial pressure: a pilot descriptive study. In: 2015 Annual International Conference of the IEEE Engineering in Medicine and Biology Society (EMBC). IEEE Engineering in Medicine and Biology Society Conference Proceedings, pp. 6142–6145 (2015)
30. Maddala, G.S., Kim, I.M.: Unit Roots, Cointegration, and Structural Change (Themes in Modern Econometrics). Cambridge University Press, Cambridge (1999)
31. Mainardi, L.T., Bianchi, A.M., Cerutti, S.: Time-frequency and time-varying analysis for assessing the dynamic responses of cardiovascular control. Crit. Rev. Biomed. Eng. **30**, 175–217 (2002)
32. Martínez, J.P., Almeida, R., Olmos, S., Rocha, A.P., Laguna, P.: Wavelet-based ECG delineator: evaluation on standard databases. IEEE Trans. Biomed. Eng. **51**, 570–581 (2004)
33. Mazzeo, A.T., LaMonaca, E., DiLeo, R., Vita, G., Santamaria, L.B.: Heart rate variability: a diagnostic and prognostic tool in anesthesia and intensive care. Acta Anaesthesiol. Scand. **55**(7), 797–811 (2011)
34. Mowery, N.T., Norris, P.R., Riordan, W., Jenkins, J.M., Williams, A.E., Morris Jr., J.A.: Cardiac uncoupling and heart rate variability are associated with intracranial hypertension and mortality: a study of 145 trauma patients with continuous monitoring. J. Trauma **65**(3), 621–627 (2008)
35. Nakagawa, T., Ashwal, S., Mathur, M., Mysore, M.: Guidelines for the determination of brain death in infants and children: an update of the 1987 task force recommendations-executive summary. Ann. Neurol. **71**(4), 573–585 (2012)
36. Neidzwiecki, M.: Identification of Time-Varying Processes. Wiley, New York (2000)

37. Norris, P.R., Morris, J.A., Ozdas, A., Groga, E.L., Williams, A.E.: Heart rate variability predicts trauma patient outcome as early as 12 h: implications for military and civilian triage. J. Surg. Res. **129**(1), 122–128 (2005)
38. Rajendra Acharya, U., Paul Joseph, K., Kannathal, N., Lim, C.M., Suri, J.S.: Heart rate variability: a review. Med. Biol. Eng. Comput. **44**(12), 1031–1051 (2006)
39. Robinson, P.M.: Gaussian semiparametric estimation of long range dependence. Ann. Stat. **23**(5), 1630–1661 (1995)
40. Rocha, A.P., Leite, A., Gouveia, S., Lago, P., Costa, O., Freitas, A.F.: Spectral characterization of long-term ambulatory heart rate variability signals. In: 5th IMA Conference on Mathematics in Signal Processing (2000)
41. Rocha, A.P., Almeida, R., Leite, A., Silva, M.J., Silva, M.E.: Long-term HRV in critically ill pediatric patients: comma versus brain death. Comput. Cardiol. **41**, 89–92 (2014)
42. Rocha, A.P., Leite, A., Silva, M.E.: Volatility leveraging in heart rate: health vs disease. Proc. Comput. Cardiol. **43**, 25–28 (2016)
43. Ryan, M.L., Thorson, C.M., Otero, C.A., Vu, T., Proctor, K.G.: Clinical applications of heart rate variability in the triage and assessment of traumatically injured patients. Anesthesiol. Res. Pract. **2011**, 416590 (2011)
44. Sassi, R., Cerutti, S., Lombardi, F., Malik, M., Huikuri, H., Peng, C.-K., Schmidt, G., Yamamoto, Y.: Advances in heart rate variability signal analysis: joint position statement by the e-Cardiology ESC Working Group and the European Heart Rhythm Association co-endorsed by the Asian Pacific Heart Rhythm Society. Europace **17**(9), 1341–1353 (2015)
45. Signorini, M.G., Sassi, R., Cerutti, S.: Working on the NOLTISALIS database: measurement of nonlinear properties in heart rate variability signals. In: Proceedings of IEEE- EMBS International Conference, 2001, pp. 547–550 (2001)
46. Stein, P.K., Bosner, M.S., Kleiger, R.E., Conger, B.M.: Heart rate variability: a measure of cardiac autonomic tone. Am. Heart J. **127**(5), 1376–1381 (1994)
47. Sykora, M., Czosnyka, M., Liu, X., Donnelly, J., Nasr, N., Diedler, J., et al.: Autonomic impairment in severe traumatic brain injury: a multimodal neuromonitoring study. Crit. Care Med. **44**(6), 1173–1181 (2016)
48. Tarvainen, M.P., Ranta-Aho, P.O., Karjalainen, P.A.: An advanced detrending method with application to HRV analysis. IEEE Trans. Biomed. Eng. **49**(2):172–175 (2002)
49. Task Force of the European Society of Cardiology and the North American Society of Pacing and Electrophysiology: Heart rate variability. Standards of measurement, physiological interpretation, and clinical use. Eur. Heart J. **17**(3), 354–381 (1996)
50. Vallbona, C., Cardus, D., Spencer, W.A., Hoff, H.E.: Patterns of sinus arrhythmia in patients with lesions of the central nervous system. Am. J. Cardiol. **16**(3), 379–89 (1965)
51. Vanderlei, L.C., Pastre, C.M., Hoshi, R.A., Carvalho, T.D., Godoy, M.F.: Basic notions of heart rate variability and its clinical applicability. Rev. Bras. Cir. Cardiovasc. **24**(2), 205–217 (2009)
52. Velasco, C.: Gaussian semiparametric estimation of nonstationary time series. J. Time Ser. Anal. **20**, 87–127 (1999)
53. Wijdicks, E.F.M.: The diagnosis of brain death. N. Engl. J. Med. **344**, 1215–1221 (2001)
54. Wilcox, R.R.: Fundamentals of Modern Statistical Methods, 2nd edn. Academic Press, New York (2010)
55. Winchell, R.J., Hoyt, D.B.: Spectral analysis of heart rate variability in the ICU: a measure of autonomic function. J. Surg. Res. **63**(1), 11–16 (1996)

Chapter 18
Age and Gender Dependency of Complexity Measures of Short-Term Heart Rate Time Series

Rico Schroeder and Andreas Voss

Abstract In the recent years, nonlinear complexity analysis of heart rate dynamics was increasingly applied improving the predictability of different diseases and related severity levels and/or risks. A limitation of short-term heart rate variability analysis is the sparse availability of studies providing statistically relevant values for nonlinear complexity indices from healthy subjects. Furthermore, most studies in this field neglected to consider influencing effects of age and gender on complexity indices. In this chapter, both the change of heart rate complexity with aging and the gender influence on complexity measures of short-term heart rate time series are presented essentially based on the results of two representative studies–the KORA S4 study (healthy subjects) and the MUSIC2 study (heart failure patients).

18.1 Introduction

The cardiac autonomic nervous system (ANS) with its two branches, sympathetic and parasympathetic nervous system, was proven to be a crucial component in physiological and pathological responses of the cardiovascular system [1]. ANS dysregulation triggered amongst others by aging, anxiety and stress, obesity and organic causes contributes to cardiovascular pathology, including hypertension, ischemic heart disease, metabolic syndrome, arrhythmias and congestive heart failure, and often to disease-related fatal outcomes [1–4].

An optimal, pronounced level of individual HRV indicates a healthy function, inherent self-regulatory capacity, adaptability, or resilience whereas less variability of the heart rate relates to system depletion, chronic stress, pathology, or inadequate functioning in various levels of self-regulatory control systems [5–8].

Over the last decades, the importance of heart rate variability (HRV) as a tool for assessing the autonomic nervous system activity in many different diseases and conditions has steadily increased. In addition to frequently used long-term HRV analysis, short-term HRV analysis has progressively been applied in recent

R. Schroeder • A. Voss (✉)
Institute of Innovative Health Technologies IGHT, Ernst-Abbe-Hochschule, Jena, Germany
e-mail: andreas.voss@eah-jena.de

R. Barbieri et al. (eds.), *Complexity and Nonlinearity in Cardiovascular Signals*,
DOI 10.1007/978-3-319-58709-7_18

years due to its suitability for ambulatory care and short-term patient monitoring and to receive test results almost immediately. A limitation of short-term HRV analysis is the sparse availability of studies providing statistically relevant values for the age and gender dependencies of nonlinear complexity HRV indices in healthy subjects. Age is one of the major independent risk factors for cardiovascular diseases [9]. Autonomic imbalance toward sympathetic preponderance has been shown to facilitate the occurrence of heart disease. Reduced vagally modulated HRV has demonstrated to be associated with increased cardiovascular morbidity and mortality in the elderly [10]. Careful examination of the changes occurring in the cardiovascular system during ageing reveals an increased stiffness of the vasculature, degenerative processes in the heart [11] and a significant shift in autonomic modulation of heart rate toward a sympathetic predominance [12] might be promising for an additional risk stratification in patients with diseases of the cardiovascular system [13].

Several studies demonstrated particularly age-related but also gender-related variation in long-term HRV as serious influence on the majority of linear and nonlinear HRV indices [14, 15]. Methods of nonlinear dynamics (NLD) providing a considerably extended and complex analysis of long-term HRV gained considerably in importance in recent years [16–18] and were also partially used in the field of short-term HRV analysis [19, 20]. However, valid information for short-term complexity HRV indices of nonlinear dynamics is rarely available because mostly only relatively small numbers of subjects were enrolled [17, 21].

18.2 Age- and Gender-Specific Dependency of HRV: A Short Survey

This section is a brief review of key observations about sex and gender differences of autonomic regulation in humans. It is known that heart diseases rather often have a dramatic influence on the autonomic regulation. Therefore, we investigate age and gender effects on autonomic regulation separately for healthy subjects and cardiovascular disease patients.

18.2.1 Age- and Gender-Specific Dependency of Autonomic Control and HRV in Healthy Subjects

Research on animals and humans [2, 22] has shown a relative dominance of the vagal tone in the autonomic control of the female heart and in contrast a preponderance of sympathetic mediated responses in males. Dart et al. (2002) [2] assume specific gender-related effects of hormones and possibly a divergent distribution of gender-roles during human evolution behind the different functioning of the ANS according to gender resulting in e.g. a higher systolic blood pressure rise on aerobic exercise, a higher hemodynamic response to isometric exercise and higher tolerances to cold temperature and repeated hypoglycemia in males compared to females. In general, a higher vagal activity known as cardio-protective is associated with a self-reported better health and wellbeing, and longer life [23, 24].

Performing a meta-analysis about sex differences in the healthy human heart rate variability, Koenig et al. (2016) [24] describe the paradoxical situation that females in some studies have an increased resting heart rate, which was found out to be a general independent CVD risk factor [25], but had no increased risk for a cardiovascular disease (CVD) compared to males [26]. Studies were able to demonstrate that the power of heart rate for prediction of mortality in CVD is different regarding to the gender [27, 28]. One possible explanation for this finding might be gender-related differences in the autonomic control of the heart, indirectly assessable by analysis of the heart rate variability.

Generally, aging is associated with increasing human sympathetic nervous activity (elevated muscle sympathetic nerve activity and norepinephrine spillover) and decreasing parasympathetic activity (reduced HRV and cardiovagal baroreflex sensitivity) at rest, the exact mechanisms are still under discussion [29–32].

With aging, the variation of heart rate increases during childhood due to maturation of the ANS and other control systems and decreases in aging adults linked to a loss of complexity of physiological dynamics caused by structural factors (e.g. loss of sinoatrial pacemaker cells, loss of neurons in the brain and spinal cord or loss of arterial distensibility) and/or functional changes (e.g. altered coupling between regulatory components components) [5, 33–37].

By analysis of high-resolution fetal magnetocardiograms reflecting the fetal cardiac activity, investigations of the development of HRV with aging are possible from 15th week of gestation onwards. In doing so, a substantial increase of the fetal HRV after the 30th week of gestation was proven for both genders [38, 39]. Additionally, an increase of complexity (higher approximate entropy) in fetal heart rate from the 20th week with progression of gestation was demonstrated possibly reflecting a changing parasympathetic influence during maturation of the ANS or/and an increased system coupling and feedback due to the maturation of organ systems [40, 41]. Gender-specific effects on the fetal HRV are contradictory published describing a different as well as a comparable HRV characteristic related to the gender [42]. Studies found gender-specific differences in the fetal HRV report an increased fetal HRV and ANS activity in male than female fetuses [43, 44]. Additionally, an increasing fetal heart rate from the 6th week of gestation (138 bmp) to the 9th week (177 bmp) following a decreasing fetal heart rate from the 9th week of gestation to the end of normal pregnancy (140 bpm) was proven [41, 45, 46]. The decreasing behavior of heart rate is assumed to be probably linked with changes in vagal function or with a rising parasympathetic activity, which overrides the accelerating influence of the sympathetic tone [41].

High complexity and fractal correlation properties of heart rate dynamics were found to be similar comparing children and young healthy subjects suggesting a highly adaptive cardiovascular regulatory system whereas a progressive complexity loss and alterations of fractal heart rate dynamics were ascertained in healthy subjects from middle age (>30 years) to old age [36]. Interestingly, Cysarz et al. (2011) [47] found higher complexity and long-term fractal scaling properties in heart rate dynamics in children aged between 7 and 13 years compared to both young preschool children (<7 years) and adolescents (>13 years). In case of the maturation from child to adolescence, they assume complex changes probably due to hormonal changes (e.g. increased production of androgens at adrenarche typically occurring between 6 and 8 years of age for both genders [48]) during pre-puberty and puberty behind this behavior.

From the analysis of the power law behavior (1/f fluctuation) it is known that the power spectrum of the nonharmonic component of HRV in the log-log scaled power spectrum show a linear slope with increasing frequency [49, 50]. Lipsitz et al. (1990) [51] found an increase of the spectral exponent of short-term HRV with aging, indicating complexity loss in the regulation of the heartbeat in elderly subjects. Sakata et al. (1999) [52] investigated 24-h HRV during daily life, suggesting that the nonharmonic component of 24-h HRV in normal subjects include at least two 1/f fluctuations (β-slopes) that differ in dynamics and age dependency and could differ also in physiological origins.

However, numerous studies have explicitly addressed the gender dependency of autonomic regulation and were published on this topic, but yielded inconsistent findings.

18.2.2 Age- and Gender-Specific Dependency of Autonomic Control and HRV in Cardiovascular Disease Patients

Age was demonstrated as highly significant determinant for a CVD in both genders, whereby CVD occur earlier in males than females. Life-long risk for CVD and both morbidity and mortality caused by CVD were shown to be are higher for males compared to females [24, 53–55]. In a study, analyzing over one million deaths of the Finnish "Causes of Death Register", the authors' [54] report of an age-standardized risk of death from CVD, that was 80% higher for males than for females. Considering the age-adjusted CVD mortality, this gender difference was most pronounced at the age 40–44 years and occurs in general in subjects <45 years which is largely attributable to circulating estrogens in women before the menopause. Furthermore, in males older than 55 years they found an increase of the CVD mortality rate in parallel to the non-vascular mortality, whereas in women older than 60 years the CVD mortality increased significantly more rapidly than the non-vascular mortality.

Explanations for a higher CVD incidence and mortality in young- to middle-aged males compared to females are related to the complex interaction among genetic, hormonal and environmental factors. They are manifold and comprise e.g. estrogen's protective effects in women before the menopause and sex-specific gene regulation causing gender differences in e.g. blood pressure, body fat distribution, insulin resistance, and inflammation predispose preferably men to a cardiovascular risk. Additionally, various factors according to the life-style, which occur more frequently in the male population, as excessive alcohol consumption, smoking and poor eating habits contribute to a higher risk for cardiovascular deaths and disease in males. [53, 56–58].

Effects of age and gender on the autonomic function using relatively novel nonlinear complexity measures as shown in this book chapter are only rarely been described in the literature (see Table 18.1, without guarantee for completeness). In summary, it was demonstrated a loss of complexity of cardiovascular time series and altered fractal scaling of heart rate (most scaling measures indicate loss of fractal scaling) dynamics with aging especially prominent during daytime at standing position. With regard to the gender, the authors report almost exclusively that the majority of nonlinear fractal scaling and complexity measures did not significantly differ (only in trend) between males and females. Additionally, it was found that age dependency of nonlinear indices is more pronounced in females than in males.

18.3 Investigating Age- and Gender-Specific Dependencies in Healthy Subjects (KORA Study) and in Heart Failure Patients (MUSIC2 Study)

In this chapter, we provide information about age- and gender-related short-term complexity HRV indices for healthy subjects and patients suffering from heart failure. Short-term 5 min HRV complexity indices will be determined from 782 healthy females and 1124 healthy males of the KORA S4 database (Cooperative Health Research in the Region Augsburg), and from 302 patients with ischemic heart failure (118 males, 25 females) or nonischemic heart failure (110 males, 49 females) of the MUSIC2 database (Muerte Subita en Insuficiencia Cardiaca or Sudden Death in Heart Failure). We apply various complexity measures and investigate the influence of age and gender in terms of five age decades (25–34, 35–44, 45–54, 55–64 and 65–74 years).

Due to the limited available space for this chapter, but the enormous range of factors influencing the HRV and the large amount of several HRV analysis methods, we decided to present only age- and gender-related effects on some complexity indices from short-term HRV analysis in detail.

Table 18.1 Results of literature research about published age- and/or gender effects on nonlinear heart rate complexity measures

Authors Year/Reference	Subjects (N) (♂/♀)/Age	Length of analyzed time series Conditions/postures	Nonlinear analysis methods and indices	Age and/or gender effects Analyzed subgroups	Results of nonlinear HRV analysis
Acharya et al. 2004 [21]	150 healthy subjects (75/75)/5–70 years	20 min Relaxed sitting and lying position	ApEn, LE, DFA ($\alpha 1$,$\alpha 2$)	Age Age groups: 10 ± 5, 25 ± 10, 40 ± 15, and 60 ± 5 years	– Complexity decreased with aging (ApEn ↓, LLE ↓) – Long range correlation ($\alpha 2$ ↓) decreased with aging
Beckers et al. 2006 [17]	276 healthy subjects (141/135)/18–71 years	13 h daytime 7 h nighttime posture is not explained	CD, LE, ApEn, PL (1/f slope), FD, DFA ($\alpha 1$,$\alpha 2$)	Age and gender Age groups: <30, 30–39, 40–49, 50–59, and >60 years	– All analyzed nonlinear indices were significantly correlated with age – Age dependency of NLD indices was especially prominent during daytime – Complexity decreased with aging (ApEn ↓, LE ↓, CD ↓) – FD, 1/f slope (↓) and short range correlation ($\alpha 1$ ↓) decreased with aging – Long range correlation ($\alpha 2$ ↑) increased with aging up to 60 years – Stabilization in the age decline of FD and ApEn at the age of $\geq$40 years – Age dependency of NLD indices was more pronounced in the females – Analyzed nonlinear indices did not significantly differ concerning the gender – ApEn was higher in females compared with males – LE and DFA $\alpha 1$ were lower in females compared with males

Boettger et al. 2010 [13]	131 healthy subjects (75/56)/20–90 years	30 min (daytime) Supine resting position, normal breathing	Hc, SD (plvar, phvar), FD	Age and gender Age groups: all ages, <30, 30–60, >60 years Gender groups: 75♂ (42.6 ± 17 years), 56♀ (42.8 ± 19 years)	– Complexity significantly decreased with aging (Hc ↓, phvar ↓) – No effects of gender on complexity indices
Cysarz et al. 2011 [47]	409 healthy subjects (189/220)/1–22 years	>4 h Regular school days activities	ApEn PL (ß), DFA ($\alpha1$,$\alpha2$)	Age and gender Preschool children (<7 years) Primary school children ($7 \leq$ age < 10 years) Pre-adolescence ($10 \leq$ age < 14 years) Adolescence (age ≥ 14 years)	– Complexity and long-term fractal scaling property were higher during age 7–13 years compared to age <7 years and age >13 years – DFA indices ($\alpha1$ ↑,$\alpha2$ ↑) increased with aging in the investigated young cohort – From 10 years of age, higher fractal and complexity indices (ApEn, ß) in males than females were found
Fukusaki et al. 2000 [78]	373 healthy subjects (124/249)/16–69 years	5 min (daytime) Supine resting position, normal breathing	PL (integrated power, ß)	Age and gender Age groups: 16–29, 30–39, 40–49, 50–59, 60– years	– ß significantly increased with aging up to 59 years – ß was significantly lower in women (more complexity) than in men
Iyengar et al. 1996 [84]	20 healthy subjects (10/10)/21–81 years	2 h Supine resting position, normal breathing	DFA ($\alpha1$,$\alpha2$)	Age Gender matched age groups: 21–34 and 68–81 years	– Short range correlation ($\alpha1$ ↑) was significantly increased in elderly subjects – Long range correlation ($\alpha2$ ↓) was significantly decreased in elderly subjects

(continued)

Table 18.1 (continued)

Authors Year/Reference	Subjects (N) (♂/♀)/Age	Length of analyzed time series Conditions/postures	Nonlinear analysis methods and indices	Age and/or gender effects Analyzed subgroups	Results of nonlinear HRV analysis
Kapidzic et al. 2014 [85]	56 healthy subjects (28/28)/19–44 years	512 RR intervals (daytime) Supine position at paced and spontaneous breathing	SampEn, cross-SampEn	Age and gender Sex matched age groups 19–25 and 35–44 years	– SampEn significantly decreased with age only in men under paced breathing – Pattern of dependencies of SampEn and cross-SampEn on paced breathing frequency were significantly different in men and in middle-aged subjects – Lower entropy measures in middle-aged males indicating reduced/less complex partial cardiac and respiratory control, and central cardio-respiratory control
Koenig et al. 2016 [24]	63,612 participants (31,642/31,970)/various	Meta-analysis, various length Several (supine, semi-supine, auto-nomic test, seated, lying, standing)	SampEn, ApEn, DFA (α1,α2)	Gender Meta-analysis	– Complexity (SampEn) was higher in females than males – Short range correlation (α1) was lower in females than males – ApEn showed no significant findings for both fixed-/random-effect models
Nemati et al. 2013 [86]	40 healthy subjects (20/20)/21–85 years	2500 RR,SBP,RESP values Supine resting position, normal breathing	CE, TE from RESP to RR, TE from SBP to RR	Age and gender Young males and females (both 21–34 years), Old males and females (both 68–85 years)	– Complexity was decreased in the old group, particularly in males (CE ↓) – TE from RESP to RR intervals was significantly reduced with aging – No significant changes in TE from SBP to RR with aging

Perseguini et al. 2011 [87]	24 healthy subjects (14/10)/60–75 years	250 RR values (daytime) Supine resting and standing, normal breathing	SE, SD (0V,1V,2LV,2ULV)	Gender Two age-matched gender groups	– Complexity (SE) in standing position was higher in males than females – SD pattern 1V is increased in females than males in supine position – SE was higher while standing than in supine position in males – 2LV was lower while standing than in supine position in females
Pikkujämsä et al. 1999 [36]	114 healthy subjects (66/48)/1–82 years	24 h Usual everyday activities	ApEn PL (ß), DFA ($\alpha1,\alpha2$)	Age and gender Children (<15 years), Young adults (<40 years), Middle-age (40–60 years), Old-age (>60 years)	– Similar complexity (ApEn) and fractal correlation properties ($\alpha1,\alpha2$, ß) in young adults and children – Progressive loss of complexity (ApEn ↓) and alterations of fractal correlation properties ($\alpha1$ ↓,$\alpha2$ ↓, ß ↓) in subjects of middle-age towards old-age – Women had a significantly lower $\alpha1$ than men
Pikkujämsä et al. 2001 [88]	389 healthy subjects (200/189)/40–59 years	13 min (daytime) Supine and sitting position, normal breathing	ApEn, DFA ($\alpha1$)	Gender Two age-matched gender groups	– Complexity (ApEn) was higher in women than men in sitting position, but not different in supine position – Low $\alpha1$ in women than men in both supine and sitting position – Interindividual variation in fractal/complexity measures of heart rate behavior was relatively small compared to the variation of traditional linear indices

(continued)

Table 18.1 (continued)

Authors Year/Reference	Subjects (N) (♂/♀)/Age	Length of analyzed time series Conditions/postures	Nonlinear analysis methods and indices	Age and/or gender effects Analyzed subgroups	Results of nonlinear HRV analysis
Porta et al. 2014 [89]	100 healthy subjects (54/46)/21–70 years	~13 min (daytime) Supine resting and standing, normal breathing	MB (NCI^{MB}) LP (NCI^{LP}) CE (NCI^{CE})	Age Gender matched age groups: 21–30, 31–40, 41–50, 51–60, and 61–70 years	– Complexity loss of heart period dynamics with aging ($NCI^{MB,LP,CE}$ ↓) at supine – Complexity loss of systolic arterial pressure series with age ($NCI^{LP,CE}$ ↓) at supine – Complexity of heart period increased with aging ($NCI^{MB,LP}$ ↑) during standing – Complexity of respiration dynamics was unaffected by aging
Reulecke et al. 2016 [90]	24 healthy subjects (12/12)/26 ± 5 years	5 min (daytime) Head-up tilt test, supine/orthostatic 70° phase, normal breathing	SD (Renyi entropy), Hc, DFA (α1,α2), ATI, IA (Porta index and Ehlers index)	Gender Two age-matched gender groups	– Increased α1 during orthostatic phase in males compared to baseline – Increased time irreversibility indices in men than women in orthostatic phase – Renyi entropy and CE revealed no gender differences during head-up tilt test
Ryan et al. 1994 [81]	67 healthy subjects (40/27)/20–90 years	8 min (daytime) Supine resting position, metronomic and normal breathing	ApEn	Age and gender Age groups: 20–39, 40–64, and 65–90 years	– Complexity decreased with age (ApEn ↓) – Complexity (ApEn) was higher in women than men
Snieder et al. 2007 [91]	203 twin pairs (196/210)/34–63 years	8.5 min (daytime) Seated in reclined position, normal breathing	ApEn	Age and gender Multiple regression analysis	– Complexity decreased significantly with aging (ApEn ↓) in both genders – Complexity (ApEn) was higher in females compared with males
Stein et al. 2009 [92]	585 elderly adults (?/?)/65–79 years	24 h Posture is not explained	DFA (α1), PL (1/f-slope)	Age Age groups: 65–69, 70–74, 75–79, and <80 years	– Short range correlation decreased with aging (α1 ↓) up to ≥75 years old – Power-law slope (1/f slope ↓) decreased with aging

Takahashi et al. 2012 [93]	44 healthy subjects (44/0)/20–70 years	200 RR intervals Supine resting position, normal breathing	SE, CE (CI, NCI), SD (0V,1V,2LV,2ULV)	Age Male age groups: 20–30 years and 60–70 years	– SE in old subjects remain similar to young subjects – Complexity decreased with aging (CI ↓ and NCI ↓) – SD pattern 0V is increased and 2LV and 2ULV are reduced in the old group
Vandeput et al. 2012 [18]	276 healthy subjects (141/135)/18–71 years	13 h daytime 7 h nighttime Posture is not explained	CD, ApEn, LE, PL (1/f slope), FD, DFA (α1,α2)	Age and gender Age groups: <30, 30–39, 40–49, 50–59, and >60 years	– All analyzed nonlinear indices were significantly correlated with age – Long range correlation (α2 ↑) increased with aging – Complexity decreased with aging (ApEn ↓, LE ↓, CD ↓) – FD, 1/f slope, and short range correlation (α1 ↓) decreased with aging – Strongest relation with age could be found for FD – Nonlinear indices showed a stronger correlation with age in females than males
Voss et al. 2012 [59]	1906 healthy subjects (1124/782)/25–74 years	5 min (daytime) Supine resting position, normal breathing	RE, SE, Hc, SPPA_Entropy, SD (forbword, plvar, phvar, RE, SE, wpsum02, wpsum13, wsdvar, 0V, 1V, 2V, 2LV, 2ULV, ASC, DESC,MP, PEAK, VAL), DFA (α1,α2), IA (Porta index, Ehlers index and Guzik index), ATI	Age Two age groups: 25–49 (young), 50–74 (elderly) years, Five age decade groups: 25–34, 35–44, 45–54, 55–64, and 65–74 years	– For almost all short-term HRV indices age dependence was found – Complexity decreased significantly with aging (RE ↓, SE ↓, Hc ↓, MSE ↓, change of several SD indices) up to approx. 55 years – Short and long range correlation (α1 ↑, α2 ↑) increased up to approx. 45 years – IA showed no significant age-dependent differences – Predictability increased (reduced complexity) significantly with aging (decay values of ATI function ↓) up to approx. 45 years

(continued)

Table 18.1 (continued)

Authors Year/Reference	Subjects (N) (♂/♀)/Age	Length of analyzed time series Conditions/postures	Nonlinear analysis methods and indices	Age and/or gender effects Analyzed subgroups	Results of nonlinear HRV analysis
Voss et al. 2013 [60]	1906 healthy subjects (1124/782)/25–74 years	5 min (daytime) Supine resting position, normal breathing	RE, SE, Hc, MSE, SPPA_Entropy, SD (forbword, plvar, phvar, RE, SE, wpsum02, wsdvar), DFA (α1,α2)	Age and gender Five age decade groups per gender: 25–34, 35–44, 45–54, 55–64, and 65–74 years	– General decrease of complexity over all age groups irrespective of the gender (RE ↓, SE ↓, Hc ↓, change of several SD indices) – No meaningful age and gender influence on SPPA_entropy – General increase of short- and long-term fractal correlation (α1 ↑, α2 ↑) with aging irrespective of the gender – Significant differences of many nonlinear HRV indices up to 55 years in both genders – Gender dependencies nearly disappear above the age of 55 years – Irrespective of the gender, highest age-related differences are present between 35–44 vs. 45–54 years – Age groups 45–54 vs. 55–64 years reveal higher differences in females than in males

Voss et al. 2015 [37]	1906 healthy subjects (1124/782)/25–74 years	5 min (daytime) Supine resting position, normal breathing	RE, SE, Hc, SPPA_Entropy, SD (forbword, plvar, phvar, RE, SE, wpsum02, wpsum13, wsdvar, 0V, 1V, 2V, 2LV, 2ULV, ASC, DESC,MP, PEAK, VAL), DFA ($\alpha 1$,$\alpha 2$), IA (Porta index, Ehlers index and Guzik index), ATI	Age and gender Two age groups per gender: 25–49 years (young) and 50–74 years (elderly) Five age decade groups per gender: 25–34, 35–44, 45–54, 55–64, and 65–74 years	– General significant decrease of complexity with aging, irrespectively of gender (RE ↓, SE ↓, Hc ↓, change of several SD indices, ATI decay values ↑) – General increase of short- and long-term fractal correlation ($\alpha 1$ ↑, $\alpha 2$ ↑) with aging irrespective of the gender – Complexity decrease was most pronounced comparing males between 35–44 and 45–54 years – IA indices were significantly increased in young males compared to females – ATI decay was significantly different comparing young females and males – Gender dependencies disappear with an age older than 55 years, 10 years earlier than the disappearance of age dependencies – Gender influences were considerably weaker than age influences – Behavior of age dependencies is more pronounced in men than in women

Complexity analysis: *ApEn* approximate entropy, *CD* correlation dimension, *Hc* compression entropy, *CE* conditional entropy (*CI* complexity index, *NCI* or NCI^{CE} normalized CI), *LE* lyapunov exponent, *LP* local predictability approach utilized in the model-free class (NCI^{LP} normalized CI), *MB* linear model-based approach (NCI^{MB} normalized CI), *MSE* multiscale entropy, *RE* renyi entropy, *SampEn* sample entropy, *SD* symbolic dynamics (*0V* symbol patterns with no variation, *1V* symbol patterns with one variation, *2V* symbol patterns with two variations, *2LV* symbol patterns with two like variations, *2ULV* symbol patterns with two unlike variations, *forbword* forbidden words=number of seldom ($p < 0.001$) or never occurring word types, *MP* portion of missing patterns, *plvar* portion of low-variability patterns in NN time series, *phvar* portion of high-variability patterns in NN time series, *wpsum02* portion of words consisting only of the symbols '0' and '2', *wpsum13* portion of words consisting only of the symbols '1' and '3', *wsdvar* standard deviation of word sequence, *ASC* portion of three successive symbols form an ascending ramp, *DESC* portion of three successive symbols form a descending ramp, *PEAK* portion of three successive symbols form a peak, *VAL* portion of three successive symbols form a valley), *SE* shannon entropy, *SPPA_Entropy* shannon entropy from segmented Poincaré plot analysis, *TE* transfer entropy. *Scaling Analysis*: *DFA* detrended fluctuation analysis ($\alpha 1$ short-term fractal exponent, $\alpha 2$ long-term fractal exponent), *FD* fractal dimension, *PL* power-law slope analysis (*1/f-slope* slope of the log-transformed cardiac interval power spectrum, β long-term scaling exponent beta of the 1/f slope). *Irregularity analysis*: *ATI* auto-transinformation. *Irreversibility analysis* (IA)/*Time series*: *RR* beat-to-beat interval duration, *SBP* systolic blood pressure, *RESP* respiration

18.3.1 Patients and Patients Data

To illustrate the influence of age and gender on complexity indices derived by short-term analysis of beat-to-beat time series, exemplary results of two different studies will be demonstrated in the following sections. In the first study [37, 59, 60], 5-min ECG recordings (500 Hz sample rate, lead II) from a cohort of 1906 healthy subjects from the KORA S4 database were investigated. In the second study, 5-min ECG recordings (200 Hz sample rate, lead X) of 208 patients with heart failure (HF) from the MUSIC2 database [61] were analyzed. In both studies, the recording of short-term ECGs was performed in supine resting position after 5–10 min resting phase allowing the person examined a relaxation and an adaptation to the measurement. Both studies fulfilled the Helsinki Declaration [62] and were approved by the Ethical Committees of the Bavarian Medical Association and Bavarian commissioner for data protection and privacy (KORA) and by the investigation committees of the MUSIC2 study participating institutions. All participations were informed about the respective study in detail and signed a consent form before participating in the study.

18.3.1.1 KORA S4 Database

KORA denotes a regional research platform for follow-upped population-based surveys dealing with epidemiology, health economics, and health care research [63]. The survey S4 was initiated in order to assess health indicators and health care, to quantify prevalence of risk factors for cardiovascular and chronic diseases, and to study the impact of lifestyle, metabolic and genetic factors on cardiovascular and other chronic diseases (http://www.helmholtz-muenchen.de/en/kora/about-kora/aims-main-research-areas/index.html; http://www.bioshare.eu/content/survey-s4-2000). The enrolled KORA survey S4 random-sampled population, acquired between 1999 and 2001 in Augsburg and surrounding districts, consists of 4107 participants aged between 25 and 74 years. During the survey period, among others, standard 12-lead ECG examinations (Hörmann Bioset 9000, Hörmann Medizinelektronik, Germany), laboratory tests, glucose tolerance tests, and investigations about the impact of environmental stressors were conducted. Furthermore, each participant was personal interviewed by means of a detailed standardized questionnaire about e.g. the individual health state, general living conditions, use of medical services and intake of medication. Based on exclusion criteria and evaluation of the questionnaire, 1906 subjects (782 females, 1124 males) from the S4 cohort allocated as healthy controls were selected for the investigation of age- and gender-dependencies on complexity measures in a healthy population [37]. 2201 subjects were excluded from the further analyses fulfilling exclusion criteria as documented diseases, intake of drugs affecting heart rate regulation or established cardiac arrhythmia [59].

18.3.1.2 MUSIC2 Database

MUSIC2, an abbreviation for a multicenter longitudinal study in Spain, was initiated to assess risk factors of all cause of mortality and sudden cardiac death in patients with HF with a NYHA (New York Heart Association) class II and III [64]. The enrolled MUSIC2 population, acquired between 2003 and 2004 in eight university hospitals, included 509 HF patients whose HF arises from different etiologies [65]. The diagnosis of HF by experienced cardiologists was confirmed by several medical examinations as radiography and/or echocardiography, short- and long-term ECG (three orthogonal leads, ELA Medical Syneflash™ MMC, Plymouth, Minnesota, USA) and stress ECG. The inclusion of the HF patients based on carefully selected inclusion as sinus rhythm, NYHA II or III, left ventricular ejection fraction LVEF <40%, left ventricular diastolic diameter LVDD > 60 mm and left ventricular hypertrophy LVH > 14 mm. Exclusion criteria were amongst others patients with an active pacemaker or defibrillator, severe valvular diseases, severe hepatic, pulmonary or renal diseases and other criteria influencing the autonomic regulation. The HF patients were optimally treated with e.g. ACE inhibitors (74%), beta-blockers (70%), diuretics (65%) and digitalis (21%). According to the KORA study [37] investigating healthy subjects aged between 25 and 74 years, 302 of 509 patients with ischemic HF (118 males, 25 females) or nonischemic HF (110 males, 49 females) from the MUSIC2 cohort were located in the same age range and thus selected for a comparable analysis of age- and gender-dependencies on complexity measures in HF.

18.3.2 Data Preprocessing

Before HRV complexity analysis were performed, data preprocessing of the acquired 5-min ECG recordings from the KORA study and MUSIC study was necessary in order to transform the ECGs in beat-to-beat time series suitable for HRV analysis and to reduce or avoid influences of e.g. noise, artefacts and ectopic beats on complexity results. First, in the ECG recordings, R peaks marking the electric excitation of heart muscle were detected automatically using either in-house developed software (KORA study) or ELA Medical Synetec™ software (MUSIC2 study) and then visual inspected for correctness and abnormalities (arrhythmia, ectopic beats, artefacts or disturbances) [60]. Time series with abnormalities >10% with regard to the total number of heartbeats were excluded from the further analysis. During the inspection, missing or misplaced R peaks were set manually to the correct position and erroneously detected R peaks were removed. From the locations of the R waves, heart rate time series (tachograms) consisting of successive beat-to-beat (RR) interval durations were extracted. Secondly, artefacts, disturbances and ectopic beats, which occur rarely, in the resulting RR interval time series were detected by a special adaptive variance estimation algorithm [66] and replaced automatically by interpolated "normal" beats. As result, "normal-to-normal RR interval" (NN) time series suitable for complexity analysis were obtained. Examples of typical NN time series of a young healthy male and female (both selected from the age decade 35–44 years) as well as an elderly healthy male and female (age decade 65–74 years) are shown in Fig. 18.1.

18.3.3 Short-Term Analysis of HRV

18.3.3.1 Time Domain Analysis

From traditional time domain (TD) analysis, two typical representative indices which describe the linear characteristics of time series were calculated according to the Task Force of the European Society of Cardiology and the North American Society of Pacing and Electrophysiology [67]. These are the indices *meanNN* [ms] and *sdNN* [ms] describing the mean value and standard deviation of NN time series. Additionally, the complexity index *shannon_h* [bit] describing Shannon entropy was estimated on the basis of the NN density distribution using a class width of 8 ms [59].

18.3.3.2 Nonlinear Complexity Analysis

Due to the inability of linear HRV indices to quantify complex signal properties of time series, an increased application of analysis methods investigating nonlinear dynamics (NLD) of short-term HRV was observed in the recent years. As explained in this book in chapter "Symbolic dynamics, Poincare plot analyses and compression entropy method estimate complexity in biological time series" by Schulz et al. and in [37, 59, 60], following nonlinear methods and indices were included in this study.

From symbolic dynamics (SD) method [68, 69], seven indices describing the dynamics of coarse-grained NN time series were estimated on the basis of word sequences (symbol alphabet {0,1,2,3} [18] and word length of three successive symbols).

- *shannon_SD* [bit] and *fwrenyi*4 [bit]: Shannon and Renyi entropy (weighting coefficient of $\alpha = 4$) of the word distribution of 64 possible word types, complexity measure
- *forbword*: forbidden words quantify the number of seldom ($p < 0.001$) or never occurring word types
- *wpsum*02: relative portion of words consisting only of the symbols '0' and '2', measure for decreased HRV
- *wsdvar*: standard deviation of word sequence
- *phvar*5: portion of high-variability patterns in the NN time series (>5 ms)

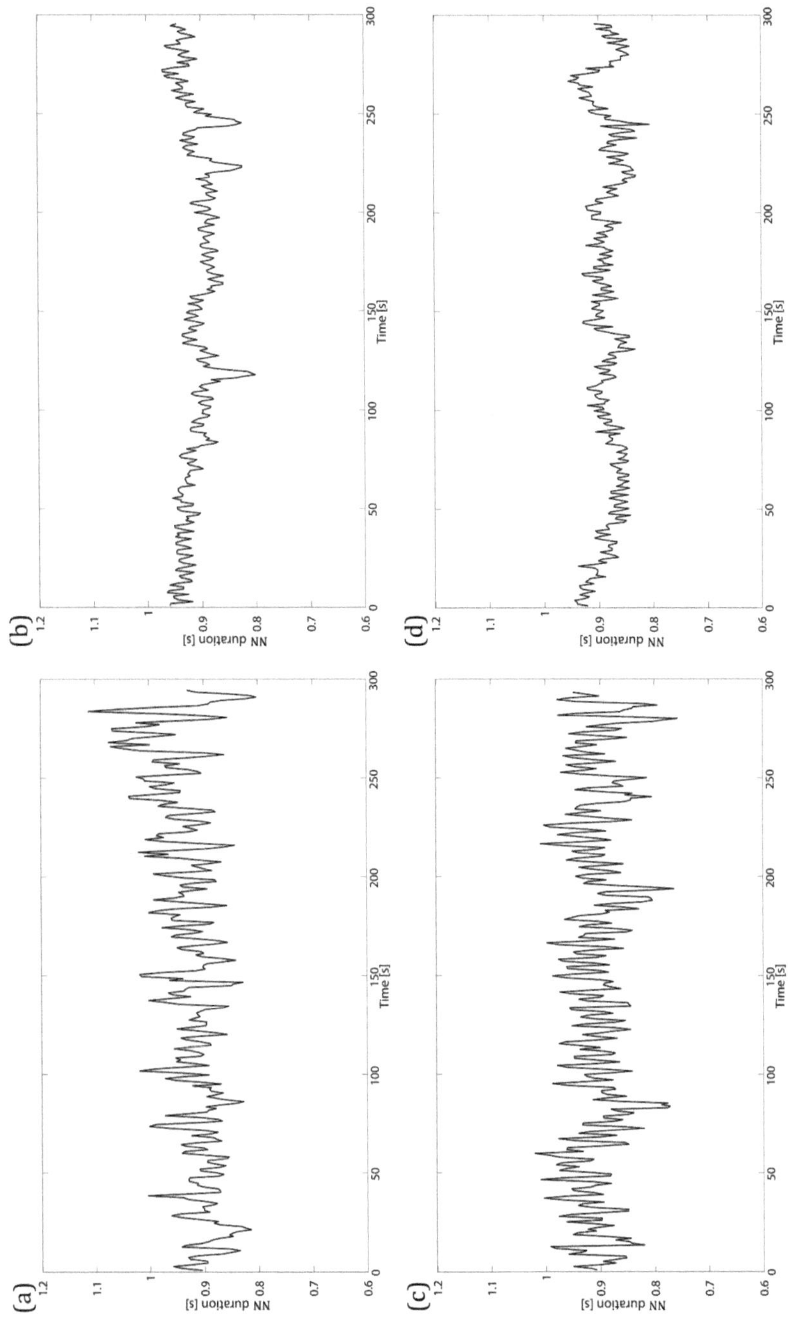

Fig. 18.1 Beat-to-beat interval (NN) time series of **(a)** a young healthy male aged 40 years, **(b)** an elderly healthy male aged 66 years, **(c)** a young healthy female aged 39 years, and **(d)** an elderly healthy female aged 70 years

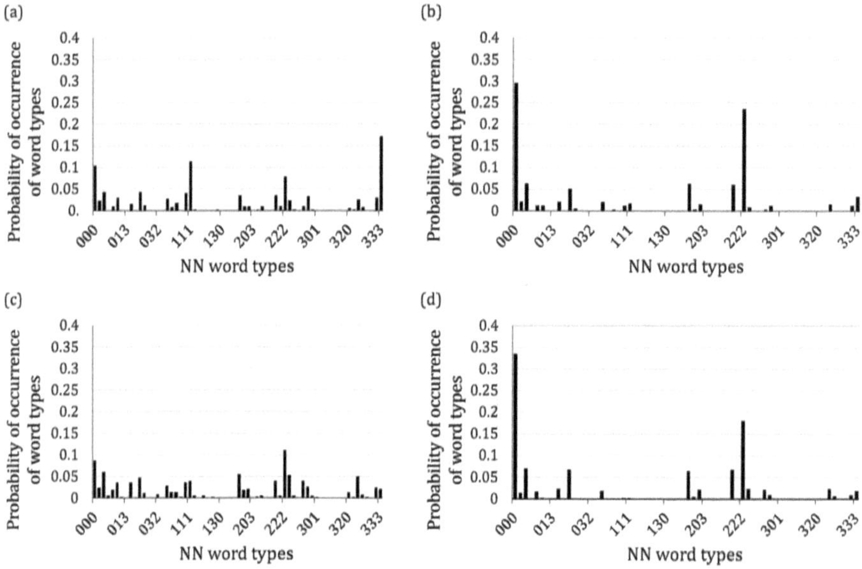

Fig. 18.2 Word type distributions (histograms) obtained from beat-to-beat interval (NN) time series applying symbolic dynamic analysis of (**a**) a young healthy male aged 40 years, (**b**) an elderly healthy male aged 66 years, (**c**) a young healthy female aged 39 years, and (**d**) an elderly healthy female aged 70 years

– *plvar*20: portion of low-variability patterns in the NN time series (< 20 ms).

Exemplary word type distributions (histograms) obtained from NN time series of a young healthy male and female as well as an elderly healthy male and female are illustrated in Fig. 18.2.

The compression entropy (Hc) index $H_c^{3,3}$ quantifies the complexity of NN time series by analyzing the lossless compressibility of the time series using a window length and a look-ahead buffer length of 3 [70]. Figure 18.3 presents the compressibility of NN time series, by means of $H_c^{3,3}$ estimated for a young and elderly healthy male and a young and elderly healthy female.

In contrast to indices (*SD1*, *SD2*, *SD1/SD2*) from traditional Poincaré plot analysis [71, 72] describing linear properties of a time series, indices from the enhanced segmented Poincaré plot analysis (SPPA) retain nonlinear features of time series [73]. As result of SPPA, a 12 × 12 probability matrix (each rectangle of the matrix has a size of *SD1* × *SD2*) is determined describing the coarse-grained distribution of the occurrence of points in the 45° rotated cloud of points of a Poincaré plot. In this study, two row probabilities, *SPPA_r_6* and *SPPA_r_7*, situated close to the center of the cloud of points, and the Shannon entropy *SPPA_entropy* [bit] were estimated on basis of the 12 × 12 probability matrices. Examples of segmented Poincaré plots obtained from NN time series of young and elderly males and females are given in Fig. 18.4.

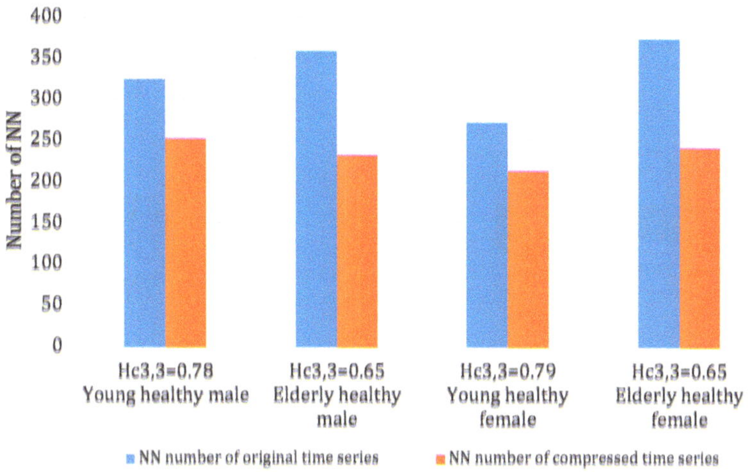

Fig. 18.3 Analysis of the compressibility of beat-to-beat interval (NN) time series by estimation of the compression entropy index $H_c{}^{3,3}$ (window length w = 3, buffer length b = 3). Illustrative comparison of the length of the original (*blue*) vs. compressed (*orange*) NN time series for a young healthy male aged 40 years ($H_c{}^{3,3} = 0.78$), an elderly healthy male aged 66 years ($H_c{}^{3,3} = 0.65$), a young healthy female aged 39 years ($H_c{}^{3,3} = 0.79$), and an elderly healthy female aged 70 years ($H_c{}^{3,3} = 0.65$)

Additionally, indices from the detrended fluctuation analysis (DFA) [37, 59] and multiscale entropy (MSE) [60] were included in this study. According to Peng et al. [74], the scaling exponents α from DFA quantifies fractal and complex scaling properties of time series. The short-term fractal scaling exponent α_1 considering the range $n = 4$–16 beats was calculated. Figure 18.5 presents exemplary DFA plots obtained from NN time series of young and elderly males and females.

MSE analysis introduced by Costa et al. [15] investigates the complexity of a physiologic time series based on the quantification of the sample entropy (SampEn) of the time series over a temporal scale range. For this, from a one-dimensional time series of length N, several successive coarse-grained time series representing the dynamics of the system at different timescales τ are generated. For each coarse-grained time series, the SampEn is estimated. SampEn considers the conditional probability that two sequences similar for m successive data points (probability p_m) remain similar at the next point $m + 1$ (probability p_{m+1}) within a given tolerance r, where self-matches are excluded from the probability calculation [14, 75].

Low SampEn values reflect a high degree of regularity and predictability of a time series whereas high SampEn values indicate less self-similarity and predictability [75]. In the present studies, two MSE values [bit] (Entropy-scale1, −scale 2) for coarse-grained time series over the scales $\tau = 1$–2 were determined using $m = 2$ and $\tau = 0.15$ [76]. MSE analysis of higher scales requires longer time series. For $\tau = 1$, the value of MSE corresponds to the SampEn value due to the fact that coarse-grained time series and original time series are identical.

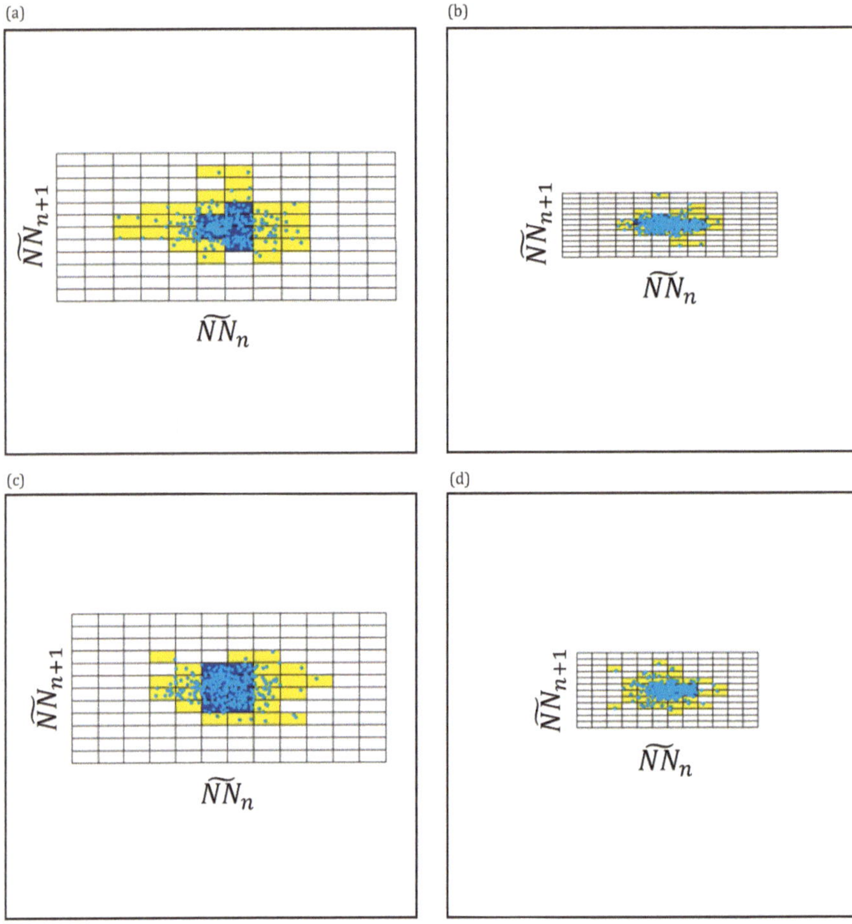

Fig. 18.4 Segmented Poincaré plots (segmentation into 12 × 12 rectangles each adapted to a size of *SD1* × *SD2*) of beat-to-beat interval (NN) time series of (**a**) a young healthy male aged 40 years, (**b**) an elderly healthy male aged 66 years, (**c**) a young healthy female aged 39 years, and (**d**) an elderly healthy female aged 70 years

18.3.4 Statistical Analyses

All statistical analyses were performed using IBM SPSS Statistics (version 19.0). In compliance with [37, 60], age- and gender-dependencies of the calculated linear HRV and nonlinear complexity indices were investigated performing two tests. For the tests, both the healthy KORA cohort and the HF patients of the MUSIC2 study were divided into the five age decade groups (1) 25–34 years, (2) 35–44 years, (3) 45–54 years, (4) 55–64 years, and (5) 65–74 years, separately for females and males (Table 18.2). Test I was performed to analyze gender-specific age dependencies

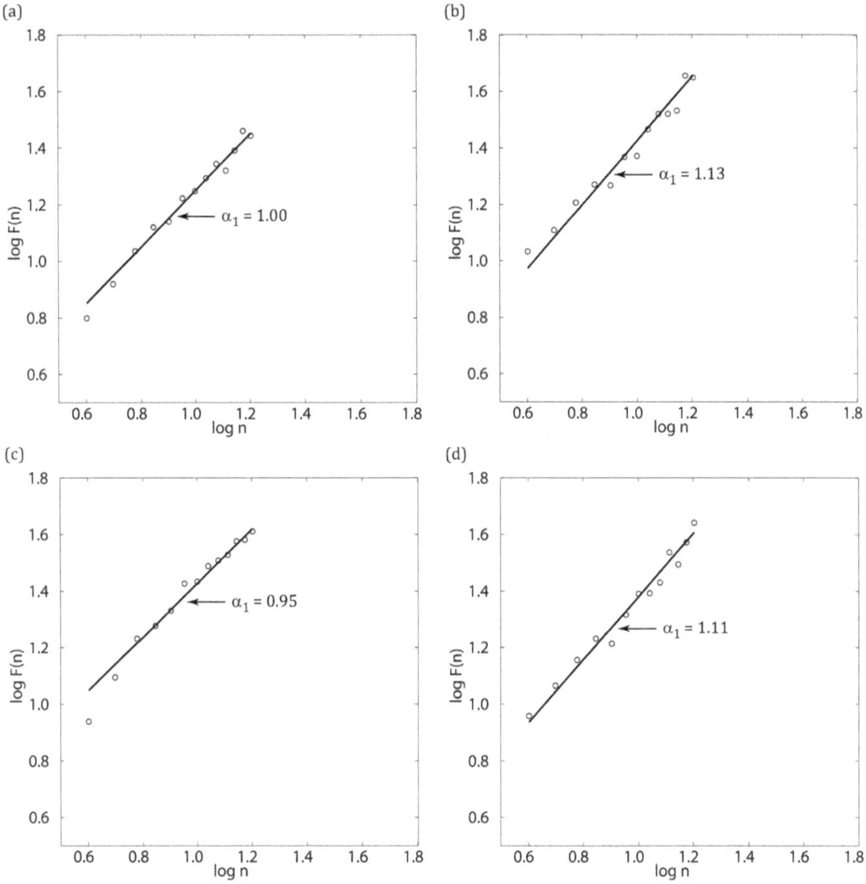

Fig. 18.5 Detrended fluctuation analysis plots calculated from beat-to-beat interval (NN) time series of (**a**) a young healthy male aged 40 years, (**b**) an elderly healthy male aged 66 years, (**c**) a young healthy female aged 39 years, and (**d**) an elderly healthy female aged 70. Scaling exponent $\alpha 1$ is estimated as slope of averaged fluctuation values $F(n)$ in a double logarithmic plot of $F(n)$ vs. segment size n (here $n = 4$–16 beats) and quantifies the strength of short-term fractal correlations within NN time series

of HRV indices within the five age groups using the Kruskal-Wallis test followed by Mann-Whitney U tests between adjacent age decades. In Test II, the gender dependency of HRV indices in each age decade was analyzed by the Mann-Whitney U test, to assess whether differences of HRV indices between females and males change with the ageing process. Test I and Test II were performed separately for healthy subjects (KORA study) and HF patients (MUSIC2 study). In HF patients, for Test I and II the age groups (1) and (2) were not considered for statistical analysis due to the insufficient number of available subjects in each group that might otherwise lead to unreliable test results. Following significance levels were considered: $p \geq 0.01$—NS means not significant, $p < 0.01$—“o” lowest significance

Table 18.2 Number of female and male subjects in five successive age decade groups (1–5) divided into healthy subjects from the KORA study cohort and patients with heart failure from the MUSIC2 study

	Healthy subjects		Heart failure patients	
Age decade groups	Number of females	Number of males	Number of females	Number of males
(1) 25—34 years	208	330		
(2) 35—44 years	259	292		
(3) 45—54 years	158	235	16	66
(4) 55—64 years	95	183	23	70
(5) 65—74 years	62	84	27	72

level, $p < 0.0007$—"oo" significance level using Bonferroni correction, $p < 10^{-10}$—"ooo" high significance level, and $p < 10^{-20}$—"oooo" highest significance level.

The Kolmogorov-Smirnov test on normality of a distribution revealed a non-normal distribution for most of the calculated HRV indices. For this reason, median values and interquartile ranges of all HRV indices were estimated for each age- and gender-specific group. In accordance with the vast majority of scientific publications focused on HRV analysis, presenting almost always mean values of HRV indices in its results, the following result section demonstrates also only mean values and associated standard deviations to facilitate comparisons to other study results. Median values and interquartile ranges of the HRV indices can be found as supplementary data in Voss et al. [37].

18.3.5 Results and Discussion

18.3.5.1 Age and Gender Influences in Healthy Subjects

Statistical analysis results quantifying mainly the age effects on HRV indices in healthy subjects (Test I) are presented separately for males in the Tables 18.3 and 18.4, and for females in the Tables 18.5 and 18.6. Table 18.7 primarily presents the degree of the gender influence on HRV indices comparing healthy females and males in the same age decade group (Test II).

In relation to TD indices, Test I revealed that ageing did not seem to have any influence on *meanNN* (Fig. 18.6a). Irrespective of the gender, *sdNN* as a marker for the overall heart rate variability decreased very considerably with ageing (Kruskal-Wallis test, and Fig. 18.6b) which can be clearly seen in Fig. 18.1 where the variance of the tachograms is reduced in the elderly man and woman compared to the younger ones. Whereas *sdNN* in the first two age decade groups (25–34 years vs. 35–44 years) differed only marginally, high significant differences of *sdNN* were found comparing middle-aged adults (35–44 years vs. 45–54 years). From an age of 55 onward, *sdNN* differences had started to disappear. In Test II, *meanNN* in the youngest age decade (25–34 years) was considerably lower in females compared

Table 18.3 Univariate statistical analysis results of Mann-Whitney U tests and a Kruskal-Wallis test quantifying changes of time domain and nonlinear heart rate variability (HRV) indices over successive age decades in healthy male subjects

Test I: Males		Mann-Whitney U tests Age decade comparisons				Kruskal-Wallis test
		1	2	3	4	
Domain	HRV-index	*2*	*3*	*4*	*5*	1–5
TD	meanNN	–	–	–	–	–
	sdNN	o	oo	o	–	oooo
	shannon_h	o	oo	o	–	oooo
SD	shannon_SD	oo	oooo	oo	–	oooo
	forbword	oo	oooo	o	–	oooo
	wpsum02	oo	oo	o	–	oooo
	wsdvar	o	oo	o	o	oooo
	phvar5	oo	oo	o	–	oooo
	plvar20	oo	oooo	oo	–	oooo
	fwrenyi4	oo	oooo	oo	–	oooo
DFA	$\alpha 1$	oo	oo	–	–	oooo
Hc	$H_c^{3,3}$	oo	oooo	o	–	oooo
SPPA	SPPA_r_6	–	–	–	–	oo
	SPPA_r_7	–	–	–	–	–
	SPPA_ entropy	–	–	–	–	o
MSE	Entropy–scale1	oo	o	–	–	ooo
	Entropy–scale2	o	o	–	–	ooo

TD time domain, *SD* symbolic dynamics, *DFA* detrended fluctuation analysis, *Hc* compression entropy, *SPPA* segmented Poincaré plot analysis, *MSE* multiscale entropy; age decade groups—1 = 25–34 years, 2 = 35–44 years, 3 = 45–54 years, 4 = 55–64 years, and 5 = 65–74 years; level of significance—"–" no significance ($p \geq 0.01$), "o" $p < 0.01$, "oo" $p < 0.0007$, "ooo" $p < 10^{-10}$, and "oooo" $p < 10^{-20}$

to males. Apart from that, Test II revealed no further significant gender-specific differences of TD indices when comparing females and males of the same age decade.

For HRV indices quantifying the nonlinear complexity of NN time series, Test I revealed loss of complexity and uncertainty accompanied with diminishing of heart rate variability with ageing, irrespectively of gender. Test II showed gender-specific differences of some nonlinear HRV indices in the young- and middle-aged decades that disappeared completely considering the age decade 65–74 years. The influence of the gender on nonlinear HRV indices, quantified by the significance level and the number of significant indices, seemed to be much lower than the influence of age on these indices. In more detail, with exception of SPPA_entropy, all entropy-based indices (*shannon_h* of TD, *shannon_SD* and *fwrenyi*4 of SD, $Hc^{3,3}$ of Hc, and Entropy –scale1 and –scale2 of MSE) decreased significantly ($p \geq 10^{-10}$) with ageing for both males and females indicating complexity loss (Test I–Kruskal-Wallis test, and Fig. 18.7). The decrease of entropy with aging is

Table 18.4 Results of the descriptive statistics considering healthy male subjects of five successive age decade groups

Test I: Males		Mean ± standard deviation of five age decade groups				
Domain	*HRV-index*	*1*	*2*	*3*	*4*	*5*
TD	meanNN	939 ± 129	925 ± 138	923 ± 134	904 ± 123	906 ± 123
	sdNN	50.0 ± 20.9	44.6 ± 16.8	36.8 ± 14.6	32.8 ± 14.7	29.6 ± 13.2
	shannon_h	4.45 ± 0.52	4.30 ± 0.51	4.02 ± 0.55	3.83 ± 0.56	3.68 ± 0.54
SD	shannon_SD	3.15 ± 0.42	2.95 ± 0.44	2.62 ± 0.47	2.42 ± 0.49	2.29 ± 0.47
	Forbword	25.5 ± 9.5	30.1 ± 9.3	36.2 ± 8.5	38.9 ± 8.6	39.8 ± 9.1
	wpsum02	0.45 ± 0.22	0.53 ± 0.22	0.67 ± 0.20	0.73 ± 0.20	0.79 ± 0.18
	Wsdvar	1.32 ± 0.42	1.23 ± 0.43	1.02 ± 0.46	0.88 ± 0.48	0.75 ± 0.43
	phvar5	0.48 ± 0.19	0.39 ± 0.19	0.28 ± 0.17	0.22 ± 0.16	0.19 ± 0.16
	plvar20	0.07 ± 0.16	0.12 ± 0.22	0.25 ± 0.28	0.36 ± 0.32	0.43 ± 0.31
	fwrenyi4	2.55 ± 0.54	2.31 ± 0.54	1.93 ± 0.51	1.72 ± 0.48	1.59 ± 0.46
DFA	α1	0.93 ± 0.20	0.99 ± 0.23	1.10 ± 0.22	1.15 ± 0.22	1.13 ± 0.26
Hc	$H_c^{3,3}$	0.80 ± 0.08	0.77 ± 0.08	0.71 ± 0.09	0.67 ± 0.09	0.65 ± 0.08
SPPA	SPPA_r_6	34.2 ± 4.4	34.8 ± 4.2	35.5 ± 4.0	35.9 ± 4.2	36.7 ± 4.8
	SPPA_r_7	34.5 ± 5.5	34.7 ± 5.1	34.7 ± 4.8	35.3 ± 4.6	35.9 ± 5.1
	SPPA_ entropy	4.02 ± 0.12	4.01 ± 0.10	4.01 ± 0.13	3.99 ± 0.14	3.94 ± 0.22
MSE	Entropy–scale1	1.92 ± 0.33	1.80 ± 0.32	1.72 ± 0.31	1.67 ± 0.31	1.65 ± 0.37
	Entropy–scale2	2.10 ± 0.38	2.00 ± 0.31	1.91 ± 0.35	1.85 ± 0.34	1.77 ± 0.38

TD time domain, *SD* symbolic dynamics, *DFA* detrended fluctuation analysis, *Hc* compression entropy, *SPPA* segmented Poincaré plot analysis, *MSE* multiscale entropy; age decade groups—1 = 25–34 years, 2 = 35–44 years, 3 = 45–54 years, 4 = 55–64 years, and 5 = 65–74 years

exemplary illustrated in Fig. 18.3 where the compression of the NN time series is higher in the elderly male and female compared to the young male and female. The compression of NN time series is comparable for males and females of the same age decade. Additionally, loss of both complexity and variability of heart rate time series with ageing, irrespective of the gender, was underpinned by the high significant (Kruskal-Wallis test, $p < 10^{-20}$) decreasing SD indices *wsdvar* and *phvar5* and the considerably increasing SD indices *forbword*, *wpsum02*, and *plvar20*. Figure 18.2 shows that the variability of occurring word types from the SD analysis is considerably decreased (lower number of occurring word types and higher probability of dominant word types) in elderly subjects compared to young subjects. From DFA, α_1 increased significantly over all age decade groups indicating a general increase in the properties of short-term fractal correlations of heart rate dynamics with ageing in healthy females and males (Fig. 5). From SPPA analysis, *SPPA_r_6* in males and *SPPA_r_7* (Fig. 18.7c) in females were considerably increased (Kruskal-Wallis test, at least $p < 0.0007$) over all ages. Additionally, in Fig. 18.4, a decreased variance of NN time series that leads to smaller dimensions of the SPPA plots due to the adaption on *SD1* and *SD2*, can be clearly seen in elderly subjects compared to the younger ones. However, comparing successive age decades, an age dependency of the two SPPA indices was not identified by

Table 18.5 Univariate statistical analysis results of Mann-Whitney U tests and a Kruskal-Wallis test quantifying changes of time domain and nonlinear heart rate variability (HRV) indices over successive age decades in healthy female subjects

Test I: Females		Test I: Females				
		Mann-Whitney U tests Age decade comparisons				Kruskal-Wallis test
		1	*2*	*3*	*4*	
Domain	*HRV-index*	*2*	*3*	*4*	*5*	*1-5*
TD	meanNN	–	–	o	–	–
	sdNN	–	oo	oo	–	oooo
	shannon_h	o	oo	oo	–	oooo
SD	shannon_SD	oo	oo	oo	–	oooo
	forbword	o	oo	oo	–	oooo
	wpsum02	o	oo	o	–	oooo
	wsdvar	o	oo	oo	–	oooo
	phvar5	oo	oo	o	–	oooo
	plvar20	oo	oo	oo	–	oooo
	fwrenyi4	oo	oo	oo	–	oooo
DFA	$\alpha 1$	oo	o	–	–	ooo
Hc	$H_c^{3,3}$	o	oo	o	–	oooo
SPPA	SPPA_r_6	–	–	–	–	–
	SPPA_r_7	–	–	–	–	ooo
	SPPA_ entropy	–	–	–	–	–
MSE	Entropy–scale1	o	–	–	–	ooo
	Entropy–scale2	–	–	–	–	oo

TD time domain, *SD* symbolic dynamics, *DFA* detrended fluctuation analysis, *Hc* compression entropy, *SPPA* segmented Poincaré plot analysis, *MSE* multiscale entropy; age decade groups—1 = 25–34 years, 2 = 35–44 years, 3 = 45–54 years, 4 = 55–64 years, and 5 = 65–74 years; level of significance—"–" no significance ($p \geq 0.01$), "o" $p < 0.01$, "oo" $p < 0.0007$, "ooo" $p < 10^{-10}$, and "oooo" $p < 10^{-20}$

use of the Mann-Whitney U-test. Considering the results of the Mann-Whitney U tests comparing two successive age decades (Test I), highest significant changes ($p < 10^{-20}$) of complexity occurred in males between the age groups 2 and 3 (age range 35–54 years). In females, the changes of NLD indices between the age groups 2 and 3 were not as pronounced as they were in males ($p < 0.01$ vs. $p < 10^{-20}$). From an age of 45 onward, changes of NLD indices between the age groups 3 and 4 began to decline in males and disappeared almost completely between the age groups 4 and 5 (55–64 years vs. 65–74 years) in both females and males. According to Test II, only the nonlinear index *SPPA_r_7* of SPPA pointed out a significant gender dependency in the age decades 25–34 years and 35–44 years. Few additional indices from NLD showed gender differences at trend level ($p < 0.01$) in dependency of the age decade under consideration. With ageing, the number of HRV indices influenced by the gender was decreasing. In the oldest investigated age decade (65–74 years), gender differences of the considered HRV indices didn't exist any longer.

Table 18.6 Results of the descriptive statistics considering healthy female subjects of five successive age decade groups

Test I: Females		Mean ± standard deviation of five age decade groups				
Domain	*HRV-index*	*1*	*2*	*3*	*4*	*5*
TD	meanNN	900 ± 116	903 ± 122	903 ± 109	868 ± 118	873 ± 110
	sdNN	48.7 ± 19.0	45.4 ± 20.5	36.9 ± 13.8	30.6 ± 12.4	27.8 ± 11.8
	shannon_h	4.43 ± 0.50	4.31 ± 0.53	4.04 ± 0.49	3.76 ± 0.54	3.62 ± 0.53
SD	shannon_SD	3.26 ± 0.44	3.05 ± 0.46	2.76 ± 0.48	2.51 ± 0.49	2.3 ± 0.53
	Forbword	22.7 ± 10.8	27.5 ± 10.3	33.8 ± 9.5	36.9 ± 10.1	39.6 ± 10.0
	wpsum02	0.41 ± 0.24	0.49 ± 0.24	0.64 ± 0.21	0.74 ± 0.18	0.79 ± 0.19
	Wsdvar	1.35 ± 0.43	1.25 ± 0.44	1.02 ± 0.42	0.84 ± 0.40	0.71 ± 0.43
	phvar5	0.51 ± 0.19	0.44 ± 0.19	0.33 ± 0.19	0.25 ± 0.18	0.19 ± 0.17
	plvar20	0.05 ± 0.13	0.09 ± 0.18	0.16 ± 0.23	0.32 ± 0.31	0.42 ± 0.35
	fwrenyi4	2.71 ± 0.58	2.46 ± 0.58	2.11 ± 0.56	1.81 ± 0.52	1.62 ± 0.51
DFA	$\alpha 1$	0.85 ± 0.21	0.93 ± 0.22	1.02 ± 0.24	1.04 ± 0.23	1.10 ± 0.26
Hc	$H_c^{3,3}$	0.80 ± 0.07	0.78 ± 0.08	0.73 ± 0.08	0.68 ± 0.09	0.65 ± 0.10
SPPA	SPPA_r_6	35.5 ± 4.4	35.5 ± 3.8	35.4 ± 4.0	34.9 ± 4.2	35.5 ± 4.3
	SPPA_r_7	31.2 ± 5.1	32.3 ± 5.2	33.2 ± 4.8	35.0 ± 4.9	36.0 ± 5.7
	SPPA_ entropy	4.04 ± 0.07	4.02 ± 0.08	4.01 ± 0.16	3.99 ± 0.15	3.99 ± 0.14
MSE	Entropy–scale1	1.96 ± 0.29	1.89 ± 0.30	1.83 ± 0.27	1.75 ± 0.28	1.71 ± 0.30
	Entropy–scale2	2.02 ± 0.27	2.03 ± 0.33	1.96 ± 0.29	1.87 ± 0.33	1.81 ± 0.32

TD time domain, *SD* symbolic dynamics, *DFA* detrended fluctuation analysis, *Hc* compression entropy, *SPPA* segmented Poincaré plot analysis, *MSE* multiscale entropy; age decade groups— 1 = 25–34 years, 2 = 35–44 years, 3 = 45–54 years, 4 = 55–64 years, and 5 = 65–74 years

The results of the KORA study are supported by several further studies on this field, e.g. [13, 17, 77–79] and Table 18.1. Amongst others, structural (loss of sinoatrial pacemaker cells, arterial distensibility, slight degree of left ventricular hypertrophy) and functional (changed interaction between regulatory components, decrease of the early filling rate) modifications of the cardiovascular system caused by the natural aging process lead to a reduction of HRV with aging [77]. In comparison to heart rate alterations caused by sympathetic influences, Fukusaki et al. [78] found a predominance of age-related change in the vagal modulation of heart rate primarily mediated by the aging process per se and not by physiologic changes characteristic of normal aging. Boettger et al. [13] investigated 131 healthy, medication-free participants in the age range between 20 and 90 years, and found that short-term HRV and its complexity significantly decreased with age. They assumed a shift of autonomic balance towards sympathetic predominance in higher age groups, limiting the reactiveness of the cardiovascular system to adjust to different demands and increasing the risk for developing tachyarrhythmias. In addition, they found that the decrease of complexity further corroborated with a shift toward a sympathetically dominated modulation of heart rate time series. This shift appears to start from an age of 30 onwards, when the correlation between loss in complexity/variability and age becomes linear. Beckers et al. [17] investigated

Table 18.7 Univariate statistical analysis results of Mann-Whitney U tests quantifying gender-depending changes of time domain and nonlinear heart rate variability (HRV) indices separately for each age decade

Test II: Male vs. Female		Mann-Whitney U tests				
		Age decades				
Domain	*HRV-index*	*1*	*2*	*3*	*4*	*5*
TD	meanNN	oo	–	–	–	–
	sdNN	–	–	–	–	–
	shannon_h	–	–	–	–	–
SD	shannon_SD	o	o	–	–	–
	forbword	o	o	–	–	–
	wpsum02	–	–	–	–	–
	wsdvar	–	–	–	–	–
	phvar5	–	–	–	–	–
	plvar20	–	o	o	–	–
	fwrenyi4	o	o	o	–	–
DFA	$\alpha 1$	o	o	o	o	–
Hc	$H_c^{3,3}$	–	–	–	–	–
SPPA	SPPA_r_6	o	–	–	–	–
	SPPA_r_7	oooo	oo	o	–	–
	SPPA_ entropy	–	–	–	–	–
MSE	Entropy–scale1	–	o	o	–	–
	Entropy–scale2	–	–	–	–	–

TD time domain, *SD* symbolic dynamics, *DFA* detrended fluctuation analysis, *Hc* compression entropy, *SPPA* segmented Poincaré plot analysis, *MSE* multiscale entropy; age decade groups—1 = 25–34 years, 2 = 35–44 years, 3 = 45–54 years, 4 = 55–64 years, and 5 = 65–74 years; level of significance—"–" no significance ($p \geq 0.01$), "o" $p < 0.01$, "oo" $p < 0.0007$, "ooo" $p < 10^{-10}$, and "oooo" $p < 10^{-20}$

24-h ECG recordings from 135 women and 141 men aged between 18 and 71 years, and found that nonlinear indices were significantly negative correlated with age. An age decade analysis showed stabilization in the age decline of the fractal dimension and approximate entropy at the age of ≥40 years. They suggested that the vagal pathways seemed to be more involved in the generation of nonlinear fluctuations. In addition, they could show that higher nonlinear behavior was evident during the night. They also did not found clear differences comparing nonlinear indices of males and females. They assumed their findings can be related to the concept of decreasing autonomic modulation with advancing age.

Moreover, we found that above the age of 55 years, after the menopause in females (and males), gender differences on complexity HRV indices are marginal or absent. The influence of female sex hormones on autonomic regulation and estrogen and its effect on cardiac vagal function could be demonstrated [80, 81]. The underlying causes for gender related differences in age-associated HRV complexity changes are still unclear. However, it can be assumed, that these differences might

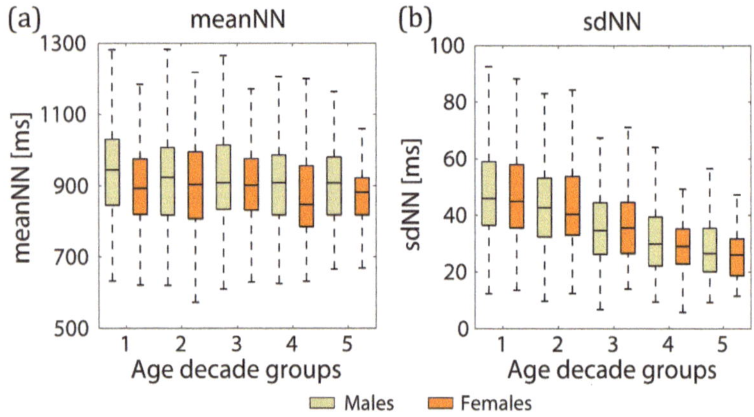

Fig. 18.6 Graphical box plot presentation of the linear time domain (TD) indices (**a**) *meanNN* and (**b**) *sdNN* calculated for five age decade groups separately for males (*yellow-green*) and females (*orange*)

be due to differences in afferent receptor stimulation, in central reflex transmission, in the efferent nervous system, and in postsynaptic signaling [37].

18.3.5.2 Age and Gender Influences in Heart Failure Patients

Considering age and gender effects on both linear and nonlinear HRV indices of patients with HF (MUSIC 2 study) did not reveal significant differences. This is partly in accordance with Stein et al. [82] who found no age, gender or race-related influences on linear TD indices (meanNN, sdNN, rmssd) investigating Holter recordings of 30 males and 40 females with HF (with a recent acute exacerbation) separated in two age groups (41–64 years vs. 65–93 years).

Aronson et al. [83] demonstrated gender-related differences in linear TD and frequency domain (FD) indices from 24-h HRV analysis only in a subset of patients with nonischemic HF (55 males and 39 females, age range 21–87 years). Thereby, from TD sdNN, rmssd, and pNN50 were higher in females than in males. From FD the total power, low-frequency power, and high-frequency power were higher in females compared to males. They concluded that females with nonischemic HF have an attenuated sympathetic activation and vagal withdrawal compared to males.

These results were not confirmed by the result of the MUSIC 2 study. Following the study of Aronson et al., we also investigated ischemic and nonischemic HF patients separately but could not found any age- and gender- related effect on neither

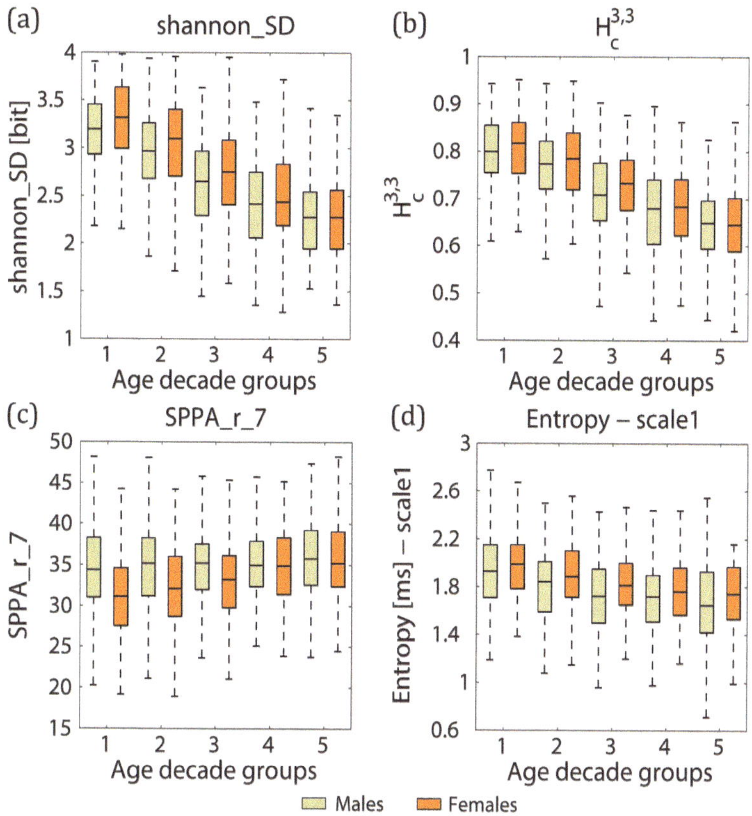

Fig. 18.7 Graphical box plot presentation of the nonlinear indices (**a**) *Shannon_SD* of the symbolic dynamics (SD) analysis, (**b**) $H_c^{3,3}$ of the compression entropy (Hc) analysis, (**c**) *SPPA_r_7* of the segmented Poincaré plot analysis (SPPA), and (d) *Entropy–scale1* of the multiscale entropy (MSE) calculated for five age decade groups separately for males (*yellow-green*) and females (*orange*)

linear nor nonlinear HRV indices between the age groups 45–54, 55–64 and 65–74 years. Possible causes of the different study results might be related to a different length of investigated times series (long-term vs. short-term analysis), the relative small number of enrolled HF patients in the MUSIC 2 study, and the not available age groups 25–34 and 35–44 years, comorbidities of HF patients, and different medication of the patients.

The loss of age- and gender-influences on HRV indices found in the MUSIC 2 study might be caused predominant by the HF disease itself and/or by overlaying effects of the medication or comorbidities of HF patients masking HRV age and gender effects.

18.3.6 Conclusion

Analyses of short-term nonlinear complexity indices revealed considerable age and gender dependencies in the KORA study. Especially, a high age-associated decline of HRV complexity was demonstrated in healthy subjects whereas the effect of age on the complexity indices was highest in middle-aged adults (35–44 years vs. 45–54 years) and started to disappear from an age of 55 onward. The influence of gender on complexity indices of HRV was less pronounced compared to the age influence. Moreover, the behavior of age dependencies is different between males and females, and much more pronounced in men than in women. The age dependency of HRV could mainly be caused by significant cardiovascular modifications, both structural as loss of sinoatrial pacemaker cells or of arterial distensibility and functional as altered coupling between regulatory components.

Possible causes for the gender differences in the younger ages could be related to different hormonal situations leading to changed sympathovagal balance accompanied with a higher sympathetic tone and a lower parasympathetic tone in men and vice versus in women. The disappearing effect of gender differences with aging can be presumably explained by hormonal restructuring especially caused by the menopause in women but also in men leading to more comparable hormonal stages.

The results, particularly of the KORA S4 study, are important for future research activities analyzing short-term HRV analysis especially in case of major age and/or gender differences of the enrolled subjects and when investigating subjects of younger age decades where the age effect is significantly more pronounced than in the subjects of elderly age decades.

Additionally, further studies should investigate how far the capability of short-term complexity HRV indices as indicators for cardiovascular diseases or as tool for preliminary diagnosis of certain autonomous regulation influencing diseases is dependent on age and gender influences.

Acknowledgments We gratefully acknowledge the cooperation of the Hospital Santa Creu i Sant Pau—ICCC and UPC Barcelona (MUSIC2 study; Prof. Antonio Bayes de Luna, Prof. Iwona Cygankiewicz, and Prof. Pere Caminal) and the cooperation of the Helmholtz Zentrum München–German Research Center for Environmental Health (KORA S4; Prof. Annette Peters and Siegfried Perz).

References

1. Bairey Merz, C.N., Elboudwarej, O., Mehta, P.: The autonomic nervous system and cardiovascular health and disease: a complex balancing act. JACC Heart Fail. **3**(5), 383–385 (2015)
2. Dart, A.M., Du, X.J., Kingwell, B.A.: Gender, sex hormones and autonomic nervous control of the cardiovascular system. Cardiovasc. Res. **53**(3), 678–687 (2002)
3. Dennis, P.A., et al.: Posttraumatic stress, heart rate variability, and the mediating role of behavioral health risks. Psychosom. Med. **76**(8), 629–637 (2014)

4. Lindmark, S., et al.: Dysregulation of the autonomic nervous system can be a link between visceral adiposity and insulin resistance. Obes. Res. **13**(4), 717–728 (2005)
5. McCraty, R., Shaffer, F.: Heart rate variability: new perspectives on physiological mechanisms, assessment of self-regulatory capacity, and health risk. Glob. Adv. Health Med. **4**(1), 46–61 (2015)
6. Segerstrom, S.C., Nes, L.S.: Heart rate variability reflects self-regulatory strength, effort, and fatigue. Psychol. Sci. **18**(3), 275–281 (2007)
7. Singer, D.H., et al.: Low heart rate variability and sudden cardiac death. J. Electrocardiol. **21**, S46–S55 (1988)
8. Thayer, J.F., et al.: Heart rate variability, prefrontal neural function, and cognitive performance: the neurovisceral integration perspective on self-regulation, adaptation, and health. Ann. Behav. Med. **37**(2), 141–153 (2009)
9. Pearson, T.A.: New tools for coronary risk assessment: what are their advantages and limitations? Circulation. **105**(7), 886–892 (2002)
10. Tsuji, H., et al.: Reduced heart rate variability and mortality risk in an elderly cohort. The Framingham heart study. Circulation. **90**(2), 878–883 (1994)
11. Folkow, B., Svanborg, A.: Physiology of cardiovascular aging. Physiol. Rev. **73**(4), 725–764 (1993)
12. De Meersman, R.E., Stein, P.K.: Vagal modulation and aging. Biol. Psychol. **74**(2), 165–173 (2007)
13. Boettger, M.K., et al.: Influence of age on linear and nonlinear measures of autonomic cardiovascular modulation. Ann. Noninvasive Electrocardiol. **15**(2), 165–174 (2010)
14. Voss, A., et al.: Methods derived from nonlinear dynamics for analysing heart rate variability. Philos. Trans. A: Math. Phys. Eng. Sci. **367**(1887), 277–296 (2009)
15. Costa, M., Goldberger, A.L., Peng, C.K.: Multiscale entropy analysis of complex physiologic time series. Phys. Rev. Lett. **89**(6), 068102 (2002)
16. Stein, P.K., et al.: Traditional and nonlinear heart rate variability are each independently associated with mortality after myocardial infarction. J. Cardiovasc. Electrophysiol. **16**(1), 13–20 (2005)
17. Beckers, F., Verheyden, B., Aubert, A.E.: Aging and nonlinear heart rate control in a healthy population. Am. J. Physiol. Heart Circ. Physiol. **290**(6), H2560–H2570 (2006)
18. Vandeput, S., et al.: Nonlinear heart rate dynamics: circadian profile and influence of age and gender. Med. Eng. Phys. **34**(1), 108–117 (2012)
19. Guzzetti, S., et al.: Symbolic dynamics of heart rate variability: a probe to investigate cardiac autonomic modulation. Circulation. **112**(4), 465–470 (2005)
20. Valencia, J.F., et al.: Complexity of the short-term heart-rate variability. IEEE Eng. Med. Biol. Mag. **28**(6), 72–78 (2009)
21. Acharya, U.R., et al.: Heart rate analysis in normal subjects of various age groups. Biomed. Eng. Online. **3**(1), 24 (2004)
22. Du, X.J., Fang, L., Kiriazis, H.: Sex dimorphism in cardiac pathophysiology: experimental findings, hormonal mechanisms, and molecular mechanisms. Pharmacol. Ther. **111**(2), 434–475 (2006)
23. Jarczok, M.N., et al.: Investigating the associations of self-rated health: heart rate variability is more strongly associated than inflammatory and other frequently used biomarkers in a cross sectional occupational sample. PLoS One. **10**(2), e0117196 (2015)
24. Koenig, J., Thayer, J.F.: Sex differences in healthy human heart rate variability: a meta-analysis. Neurosci. Biobehav. Rev. **64**, 288–310 (2016)
25. Cooney, M.T., et al.: Elevated resting heart rate is an independent risk factor for cardiovascular disease in healthy men and women. Am. Heart J. **159**(4), 612 (2010.) -619 e3.
26. Cordero, A., Alegria, E.: Sex differences and cardiovascular risk. Heart. **92**(2), 145–146 (2006)
27. Palatini, P.: Heart rate as a cardiovascular risk factor: do women differ from men? Ann. Med. **33**(4), 213–221 (2001)
28. Sacha, J.: Interaction between heart rate and heart rate variability. Ann. Noninvasive Electrocardiol. **19**(3), 207–216 (2014)

29. Antelmi, I., et al.: Influence of age, gender, body mass index, and functional capacity on heart rate variability in a cohort of subjects without heart disease. Am. J. Cardiol. **93**(3), 381–385 (2004)
30. Kingwell, B.A., et al.: Heart rate spectral analysis, cardiac norepinephrine spillover, and muscle sympathetic nerve activity during human sympathetic nervous activation and failure. Circulation. **90**(1), 234–240 (1994)
31. Laitinen, T., et al.: Age and gender dependency of baroreflex sensitivity in healthy subjects. J. Appl. Physiol. (1985). **84**(2), 576–583 (1998)
32. Laitinen, T., et al.: Age dependency of cardiovascular autonomic responses to head-up tilt in healthy subjects. J. Appl. Physiol. (1985). **96**(6), 2333–2340 (2004)
33. Finley, J.P., Nugent, S.T.: Heart rate variability in infants, children and young adults. J. Auton. Nerv. Syst. **51**(2), 103–108 (1995)
34. Goldberger, A.L.: Non-linear dynamics for clinicians: chaos theory, fractals, and complexity at the bedside. Lancet. **347**(9011), 1312–1314 (1996)
35. Kaplan, D.T., et al.: Aging and the complexity of cardiovascular dynamics. Biophys. J. **59**(4), 945–949 (1991)
36. Pikkujamsa, S.M., et al.: Cardiac interbeat interval dynamics from childhood to senescence: comparison of conventional and new measures based on fractals and chaos theory. Circulation. **100**(4), 393–399 (1999)
37. Voss, A., et al.: Short-term heart rate variability–influence of gender and age in healthy subjects. PLoS One. **10**(3), e0118308 (2015)
38. Lange, S., et al.: Influence of gestational age, heart rate, gender and time of day on fetal heart rate variability. Med. Biol. Eng. Comput. **43**(4), 481–486 (2005)
39. Van Leeuwen, P., et al.: Changes in the frequency power spectrum of fetal heart rate in the course of pregnancy. Prenat. Diagn. **23**(11), 909–916 (2003)
40. Fleisher, L.A., et al.: Complementary and non-coincident increases in heart rate variability and irregularity during fetal development. Clin. Sci. (Lond.) **92**(4), 345–349 (1997)
41. Van Leeuwen, P., et al.: Fetal heart rate variability and complexity in the course of pregnancy. Early Hum. Dev. **54**(3), 259–269 (1999)
42. Tendais, I., et al.: Sex differences in the fetal heart rate variability indices of twins. J. Perinat. Med. **43**(2), 221–225 (2015)
43. Bernardes, J., et al.: Linear and complex heart rate dynamics vary with sex in relation to fetal behavioural states. Early Hum. Dev. **84**(7), 433–439 (2008)
44. DiPietro, J.A., et al.: Fetal neurobehavioral development: associations with socioeconomic class and fetal sex. Dev. Psychobiol. **33**(1), 79–91 (1998)
45. Pillai, M., James, D.: The development of fetal heart rate patterns during normal pregnancy. Obstet. Gynecol. **76**(5 Pt 1), 812–816 (1990)
46. Van Lith, J.M., et al.: Fetal heart rate in early pregnancy and chromosomal disorders. Br. J. Obstet. Gynaecol. **99**(9), 741–744 (1992)
47. Cysarz, D., et al.: Unexpected course of nonlinear cardiac interbeat interval dynamics during childhood and adolescence. PLoS One. **6**(5), e19400 (2011)
48. Bjorklund, D.F., Blasi, C.H.: Child and Adolescent Development: An Integrated Approach. Wadsworth Cengage Learning, Belmont, CA (2011)
49. Kobayashi, M., Musha, T.: 1/f fluctuation of heartbeat period. I.E.E.E. Trans. Biomed. Eng. **29**(6), 456–457 (1982)
50. Saul, J.P., et al.: Analysis of long term heart rate variability: methods, 1/f scaling and implications. Comput. Cardiol. **14**, 419–422 (1988)
51. Lipsitz, L.A., et al.: Spectral characteristics of heart rate variability before and during postural tilt. Relations to aging and risk of syncope. Circulation. **81**(6), 1803–1810 (1990)
52. Sakata, S., et al.: Aging and spectral characteristics of the nonharmonic component of 24-h heart rate variability. Am. J. Phys. **276**(6 Pt 2), R1724–R1731 (1999)
53. Berry, J.D., et al.: Lifetime risks of cardiovascular disease. N. Engl. J. Med. **366**(4), 321–329 (2012)

54. Mikkola, T.S., et al.: Sex differences in age-related cardiovascular mortality. PLoS One. **8**(5), e63347 (2013)
55. Mosca, L., Barrett-Connor, E., Wenger, N.K.: Sex/gender differences in cardiovascular disease prevention: what a difference a decade makes. Circulation. **124**(19), 2145–2154 (2011)
56. McCarthy, J.J.: Gene by sex interaction in the etiology of coronary heart disease and the preceding metabolic syndrome. Nutr. Metab. Cardiovasc. Dis. **17**(2), 153–161 (2007)
57. Vartiainen, E., et al.: Thirty-five-year trends in cardiovascular risk factors in Finland. Int. J. Epidemiol. **39**(2), 504–518 (2010)
58. Winham, S.J., de Andrade, M., Miller, V.M.: Genetics of cardiovascular disease: importance of sex and ethnicity. Atherosclerosis. **241**(1), 219–228 (2015)
59. Voss, A., et al.: Short-term heart rate variability–age dependence in healthy subjects. Physiol. Meas. **33**(8), 1289–1311 (2012)
60. Voss, A., et al.: Influence of age and gender on complexity measures for short term heart rate variability analysis in healthy subjects. Conf. Proc. IEEE Eng. Med. Biol. Soc. **2013**, 5574–5577 (2013)
61. Voss, A., et al.: Short-term vs. long-term heart rate variability in ischemic cardiomyopathy risk stratification. Front. Physiol. **4**, 364 (2013)
62. World Medical Association Declaration of Helsinki: Ethical principles for medical research involving human subjects. JAMA. **284**(23), 3043–3045 (2000)
63. Holle, R., et al.: KORA—a research platform for population based health research. Gesundheitswesen. **67**(Suppl. 1), S19–S25 (2005)
64. Voss, A., et al.: Risk stratification in ischemic heart failure patients with linear and nonlinear methods of heart rate variability analysis. In: Vander Sloten, J., et al. (eds.) 4th European Conference of the International Federation for Medical and Biological Engineering: ECIFMBE 2008 23–27 November 2008 Antwerp, Belgium, pp. 1–4. Springer Berlin Heidelberg, Berlin (2009)
65. Voss, A., et al.: Linear and nonlinear heart rate variability risk stratification in heart failure patients. In: 2008 Computers in Cardiology. IEEE, Bologna (2008)
66. Wessel, N., et al.: Nonlinear analysis of complex phenomena in cardiological data. Herzschrittmachert. Elektrophysiol. **11**(3), 159–173 (2000)
67. Task Force: Heart rate variability: standards of measurement, physiological interpretation and clinical use. Task Force of the European Society of Cardiology and the north American Society of Pacing and Electrophysiology. Circulation. **93**(5), 1043–1065 (1996)
68. Kurths, J., et al.: Quantitative analysis of heart rate variability. Chaos. **5**(1), 88–94 (1995)
69. Voss, A., et al.: The application of methods of non-linear dynamics for the improved and predictive recognition of patients threatened by sudden cardiac death. Cardiovasc. Res. **31**(3), 419–433 (1996)
70. Baumert, M., et al.: Forecasting of life threatening arrhythmias using the compression entropy of heart rate. Methods Inf. Med. **43**(2), 202–206 (2004)
71. Brennan, M., Palaniswami, M., Kamen, P.: Do existing measures of Poincare plot geometry reflect nonlinear features of heart rate variability? I.E.E.E. Trans. Biomed. Eng. **48**(11), 1342–1347 (2001)
72. Kamen, P.W., Krum, H., Tonkin, A.M.: Poincare plot of heart rate variability allows quantitative display of parasympathetic nervous activity in humans. Clin. Sci. (Lond.) **91**(2), 201–208 (1996)
73. Voss, A., et al.: Segmented Poincare plot analysis for risk stratification in patients with dilated cardiomyopathy. Methods Inf. Med. **49**(5), 511–515 (2010)
74. Peng, C.K., et al.: Quantification of scaling exponents and crossover phenomena in nonstationary heartbeat time series. Chaos. **5**(1), 82–87 (1995)
75. Richman, J.S., Moorman, J.R.: Physiological time-series analysis using approximate entropy and sample entropy. Am. J. Physiol. Heart Circ. Physiol. **278**(6), H2039–H2049 (2000)
76. Kuntamalla, S., Lekkala Ram Gopal, R.: Reduced data Dualscale entropy analysis of HRV signals for improved congestive heart failure detection. Meas. Sci. Rev. **14**, 294 (2014)

77. Ferrari, A.U.: Modifications of the cardiovascular system with aging. Am. J. Geriatr. Cardiol. **11**(1), 30–33 (2002)
78. Fukusaki, C., Kawakubo, K., Yamamoto, Y.: Assessment of the primary effect of aging on heart rate variability in humans. Clin. Auton. Res. **10**(3), 123–130 (2000)
79. Liu, C.C., Kuo, T.B., Yang, C.C.: Effects of estrogen on gender-related autonomic differences in humans. Am. J. Physiol. Heart Circ. Physiol. **285**(5), H2188–H2193 (2003)
80. Moodithaya, S., Avadhany, S.T.: Gender differences in age-related changes in cardiac autonomic nervous function. J. Aging Res. **2012**, 679345 (2012)
81. Ryan, S.M., et al.: Gender- and age-related differences in heart rate dynamics: are women more complex than men? J. Am. Coll. Cardiol. **24**(7), 1700–1707 (1994)
82. Stein, P.K., et al.: Heart rate variability is independent of age, gender, and race in congestive heart failure with a recent acute exacerbation. Am. J. Cardiol. **79**(4), 511–512 (1997)
83. Aronson, D., Burger, A.J.: Gender-related differences in modulation of heart rate in patients with congestive heart failure. J. Cardiovasc. Electrophysiol. **11**(10), 1071–1077 (2000)
84. Iyengar, N., et al.: Age-related alterations in the fractal scaling of cardiac interbeat interval dynamics. Am. J. Phys. **271**(4 Pt 2), R1078–R1084 (1996)
85. Kapidzic, A., et al.: Nonlinear properties of cardiac rhythm and respiratory signal under paced breathing in young and middle-aged healthy subjects. Med. Eng. Phys. **36**(12), 1577–1584 (2014)
86. Nemati, S., et al.: Respiration and heart rate complexity: effects of age and gender assessed by band-limited transfer entropy. Respir. Physiol. Neurobiol. **189**(1), 27–33 (2013)
87. Perseguini, N.M., et al.: Spectral and symbolic analysis of the effect of gender and postural change on cardiac autonomic modulation in healthy elderly subjects. Braz. J. Med. Biol. Res. **44**(1), 29–37 (2011)
88. Pikkujamsa, S.M., et al.: Determinants and interindividual variation of R–R interval dynamics in healthy middle-aged subjects. Am. J. Physiol. Heart Circ. Physiol. **280**(3), H1400–H1406 (2001)
89. Porta, A., et al.: Effect of age on complexity and causality of the cardiovascular control: comparison between model-based and model-free approaches. PLoS One. **9**(2), e89463 (2014)
90. Reulecke, S., et al.: Men and women should be separately investigated in studies of orthostatic challenge due to different gender-related dynamics of autonomic response. Physiol. Meas. **37**(3), 314–332 (2016)
91. Snieder, H., et al.: Sex differences and heritability of two indices of heart rate dynamics: a twin study. Twin Res. Hum. Genet. **10**(2), 364–372 (2007)
92. Stein, P.K., et al.: Heart rate variability and its changes over 5 years in older adults. Age Ageing. **38**(2), 212–218 (2009)
93. Takahashi, A.C., et al.: Aging reduces complexity of heart rate variability assessed by conditional entropy and symbolic analysis. Intern. Emerg. Med. **7**(3), 229–235 (2012)

Chapter 19
Complexity and Nonlinearities in Cardiorespiratory Signals in Sleep and Sleep Apnea

Carolina Varon and Sabine Van Huffel

Abstract Sleep is a complex physiological process that plays a key role in maintaining homeostasis, well-being and overall health. It has an internal structure characterized by sleep stages, which is often affected by either the high demands of the current 24-h society or by different sleep disorders. In fact, these disturbances to the regular sleep structure have been strongly associated with reductions of cognitive and behavioral performance, depression, memory loss, and cardiovascular diseases. With this in mind, multiple studies have analyzed different physiological variables during each sleep stage, and how their dynamics are affected by sleep disorders such as sleep apnea. This chapter discusses the main findings reported in literature with special focus on the dynamics of heart rate and respiration.

The first part of this chapter focuses on the description and quantification of the different cardiac and respiratory dynamics during healthy sleep. Linear and nonlinear approaches that have been used to analyze the complexity of the cardiorespiratory system during different sleep stages are presented. For instance, the application of methods such as detrended fluctuation analysis, dynamic warping, correlation dimension, and entropy measures, on cardiac and respiratory signals during sleep are described. In addition, an overview of the cardiorespiratory interactions during sleep is provided, and different algorithms for sleep staging are compared.

The second part discusses the effect that sleep apnea has on the cardiorespiratory dynamics and interactions, and special focus is on how this effect has been quantified using linear and nonlinear techniques. The problem of sleep apnea detection is presented together with a comparison between the performances obtained using linear and nonlinear techniques.

Finally, a summary of the publicly available datasets that can be used for the analysis of heart rate and/or respiration during sleep and sleep apnea are presented.

C. Varon (✉) • S. Van Huffel
Department of Electrical Engineering (ESAT), KU Leuven, STADIUS Centre for Dynamical Systems, Signal Processing and Data Analytics, Leuven, Belgium

IMEC, Leuven, Belgium
e-mail: carolina.varon@esat.kuleuven.be

R. Barbieri et al. (eds.), *Complexity and Nonlinearity in Cardiovascular Signals*,
DOI 10.1007/978-3-319-58709-7_19

19.1 Introduction

This chapter is divided in two main parts. The first part focuses on the description and quantification of the different cardiac and respiratory dynamics during healthy sleep. Linear and nonlinear approaches that have been used to analyse the complexity of the cardiorespiratory system during different sleep stages are presented. For instance, the application of methods such as detrended fluctuation analysis, dynamic warping, correlation dimension, and entropy measures, on cardiac and respiratory signals during sleep are described. In addition, an overview of the cardiorespiratory interactions during sleep is provided, and different algorithms for sleep staging are compared.

The second part of this chapter discusses the effect that sleep apnea has on the cardiorespiratory dynamics and interactions, and special focus is put on how this effect has been quantified using linear and nonlinear techniques. The problem of sleep apnea detection is presented together with a comparison between the performances obtained using linear and nonlinear techniques.

19.2 Sleep Monitoring

Sleep is a complex process that plays a key role in maintaining homeostasis, well-being and overall health [1]. During a lifetime, humans normally spend up to one-third of the time sleeping, however, the current 24-h society is reducing this time and keeping people from getting the necessary amount of sleep. In fact, sleep deprivation has been strongly associated with reductions of cognitive and behavioral performance, depression, memory loss, and cardiovascular diseases [2]. Not only society but also sleep disorders play a key role in the reduction of sleep quality. These disorders have been grouped in seven major categories: insomnia; sleep-related breathing disorders; central disorders of hypersomnolence; circadian rhythm sleep-wake disorders; sleep-related movement disorders; parasomnias; and other sleep disorders [2, 3], and they all have different effects on the human body. For instance, cardiorespiratory irregularities have been observed in patients suffering from circadian rhythm sleep disorder such as delayed sleep-phase syndrome, advanced sleep-phase syndrome and non-24-h sleep-wake disorder, and in patients with obstructive sleep apnea (OSA) [1, 2, 4].

The gold standard in sleep monitoring and in the diagnosis of sleep disorders is the polysomnography (PSG), which is a sleep test usually recorded in a sleep laboratory [5, 6]. This test includes the overnight recording of, amongst others, the electrical activity of the brain using electroencephalography (EEG), the electrical activity of the heart using the electrocardiogram (ECG), airflow, respiratory effort measured on the chest and abdomen, and blood oxygen saturation (SaO_2). These signals are visually inspected by sleep experts in order to annotate the occurrence of different sleep stages [6], or events related to sleep disorders [5]. Although PSG is the most powerful tool in sleep medicine, it requires overnight hospitalization, costly sleep center facilities and sleep experts, and the use of instrumentation that might intervene with the normal sleep pattern. In order to overcome these limitations, many studies have focused on the development of non-intrusive technologies that can be used at home and that can monitor sleep during more than one single night. As a result, it is expected that the clinical practice of sleep medicine can be improved, and that early diagnosis of sleep disorders can be achieved with a non-intrusive screening tool [4]. One technology that has been considered for the development of such a tool is actigraphy. It can be recorded using wrist-worn devices and it has been proven to be useful in the monitoring of sleep in healthy adults [7]. However, its accuracy in the diagnosis of sleep disorders remains a challenge [8].

Apart from actigraphy, the changes in the cardiorespiratory signals during sleep have also been investigated and considered for the development of non-intrusive screening tools [4, 9–13]. The fact that cardiorespiratory activity changes with the sleep pattern (i.e. sleep stages) [14] has motivated many researchers to derive linear and nonlinear features from both the ECG and the respiratory signals. These features have then been used to detect either sleep stages, or events related to sleep disorders in particular to obstructive sleep apnea (OSA), or both simultaneously [11, 12, 15, 16]. Studies where both tasks are performed separately, namely sleep staging and OSA detection, have shown that promising results can be obtained using cardiorespiratory-based algorithms. However, their performance can be compromised when sleep staging is performed in the presence of OSA events [17]. All in all, a lot of improvements are still needed in the development of automatic sleep analysis before they can be accepted in sleep medicine [18].

19.3 Sleep Staging

Sleep is not just a *steady state*, instead it is a complex physiological process with an internal structure characterized by sleep stages [5]. These sleep stages were originally standardised by Rechtschaffen and Kales in 1968 [19] and later in 2007, they were redefined by the American Association of Sleep Medicine (AASM) [5]. The following five different sleep stages have been identified:

- Stage W (Wakefulness)
- Stage R (Rapid eye movement—REM)
- Stage N1 (non REM—NREM 1)
- Stage N2 (NREM 2)
- Stage N3 (NREM 3)

The REM sleep stage can be subdivided into tonic REM and phasic REM [20]. In order to differentiate between them, wave characteristics in the electroencephalogram (EEG), electrooculogram (EOG) and electromyogram (EMG) have been clearly identified for each stage [5]. This differentiation is typically done using consecutive windows of 30 s, where one sleep stage is assigned to each window. As a result, a sleep profile, or so-called *hypnogram*, can be obtained for the entire night.

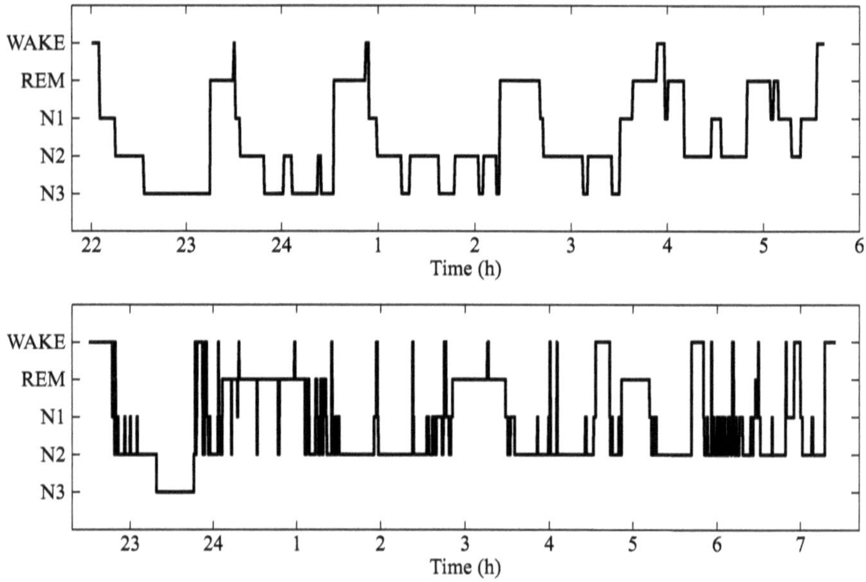

Fig. 19.1 Exemplary hypnogram for a healthy sleep stage distribution (*upper plot*), and real hypnogram recorded from a patient suffering from sleep apnea (*lower plot*)

Figure 19.1 shows an exemplary hypnogram of a healthy sleep stage distribution and a real hypnogram recorded from a patient suffering from sleep apnea.

A series of sleep cycles can be observed in the healthy distribution of sleep stages depicted in Fig. 19.1. One sleep cycle typically lasts between 90 and 110 min, and it ideally follows a sequence that starts with wake followed by light sleep, namely N1 and N2, then a period of stage N3, which is also called deep sleep, and ends with REM sleep. From the lower plot of Fig. 19.1 it is possible to observe that this sequence is destroyed in patients suffering from apnea. This particular example reveals an increased amount of time in light sleep (N1 and N2) and Wake, and it shows that deep sleep is only reached once during the entire night. These interruptions of the sleep cycle (i.e. sleep fragmentation) disrupt the restorative function of sleep and result in the multiple symptoms associated with sleep apnea. These symptoms will be discussed later in this chapter.

Sleep stages have a profound effect on the autonomic nervous system (ANS) [1]. During NREM, the activity of the parasympathetic nervous system (PNS) increases with respect to the one of the sympathetic nervous system (SNS), which is relatively low and stable. In tonic REM sleep, the latter is further reduced, while the PNS activity remains more or less unchanged. A different pattern is observed during phasic REM, where both the PNS and SNS activities increase with respect to the previous stages. At this point the activity of the SNS becomes variable and much higher than the PNS [14, 21]. Since the ANS is in charged of the unconscious control of both the heart and lungs, these variations are also reflected in the cardiac and respiratory activities.

19.3.1 Respiratory Changes During Sleep

In general, healthy adults tend to breath faster and shallower during sleep than during wakefulness, causing hypoventilation and hypoxia [22]. Nevertheless, clear differences in respiratory dynamics have been observed between sleep stages [22, 23]. In NREM sleep, the ventilation is reduced when compared to wakefulness, and the regularity of the respiration increases as the sleep goes from stage N1 to stage N3 [22]. On the other hand, during REM sleep the respiratory rate and depth tend to be very irregular [4, 14, 22], and the air volume tends to be similar [23] or even lower [22] than during NREM. In fact, the minute ventilation ($\dot{V}_E$)[1] can be reduced up to 84% of the values during wakefulness. This hypoventilation can be explained by a reduction in the ventilatory drive typical of REM [22]. Examples of these variations in the regularity and depth of the respiration are depicted in Fig. 19.2. These segments of respiratory effort show that indeed there are morphological differences between sleep stages and that the bandwidth of the respiratory signal during deep sleep reveals a more regular breathing. All these changes suggest that the respiratory control during sleep can go from being entirely automatic during NREMS, to behavioral-based during REMS [14], where the metabolic control of respiratory drive is disturbed [23].

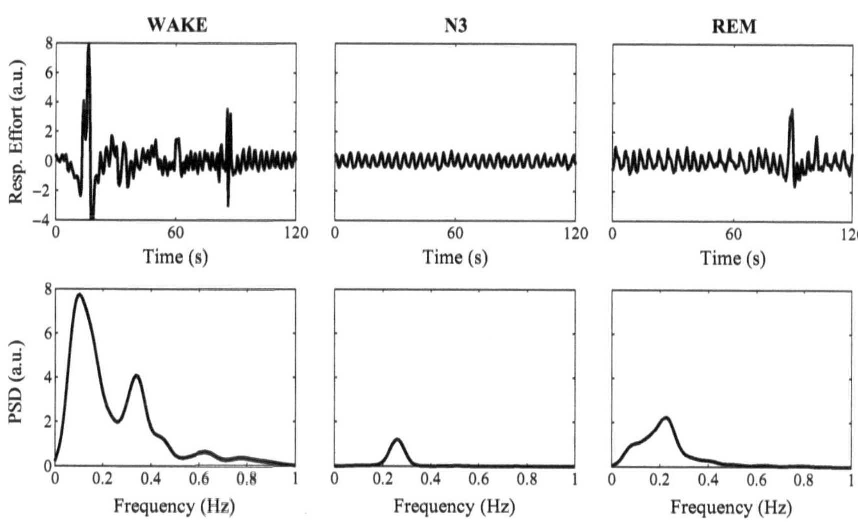

Fig. 19.2 Two-minute segments of respiratory effort measured on the thorax (*upper plots*) and their corresponding PSD (*lower plots*). Note the clear regularity during deep sleep (N3), where a dominant respiratory frequency can be identified. In contrast, during both REM and Wake a wider respiratory bandwidth is observed

[1] Minute ventilation is the total air volume entering the lungs per minute.

Different linear and nonlinear techniques have been used for the analysis of respiratory dynamics during sleep stages. From previous observations it is clear that simple features like the respiratory frequency, the respiratory amplitude [22], and the power of respiration in the different frequency bands of HRV [4] clearly differ between sleep and wake. Although these features have a certain discriminative power, their accuracy is limited by the presence of artefacts and nonstationarities in physiological signals like the respiration. In addition, it is generally accepted that the respiratory signal is a nonlinear deterministic process, and that mathematical tools for the description of complex and chaotic processes are of great use in the analysis of respiratory signals during sleep [24].

19.3.1.1 Detrended Fluctuation Analysis on Respiratory Variability During Sleep

Respiration can be seen as an integrative process, where each present breath is correlated with previous breaths [23]. Therefore, short- and long-term correlations are expected, and studies have shown that this is indeed the case. Long-term correlations were unveiled during REM sleep in healthy adults using detrended fluctuation analysis (DFA) [25]. This was done after applying DFA to breath-to-breath (BB) signals, which are computed as the time differences between consecutive peaks in the respiratory signal during deep sleep, REM sleep and Wake. In order to avoid transient influences caused by the transitions between sleep stages, 45 s epochs at the beginning and at the end of each stage are often removed from the analysis [23, 25]. Linear and higher order polynomials were used in the fitting procedure of the DFA, resulting in DFA1, DF2, and DFAn, for linear, quadratic and order n polynomials, respectively. Figure 19.3 shows the log-log plot of the fluctuation function $F(L)$ against the time scale L, computed for a healthy adult during WAKE, N3, and REM sleep, using DFA1. This method allows to identify long-term correlations in the data when the scaling exponent is larger than 0.5 (i.e., $\alpha > 0.5$). In this particular example the results reported in [25] for sleep and in [26] for wakefulness were reproduced, where $\alpha = 0.94$ during REM sleep and $\alpha = 0.91$ during wakefulness confirm the long-term correlations in the breath-to-breath signals. In contrast, during deep sleep and light sleep, a loss of long-term correlations were observed. In [25] it was also shown that similar results can be found for different trends, namely using DFA2, DFA3, and higher polynomials. These results suggest that there is a difference between the autonomic regulation of respiration during REM and NREM sleep, and that the breath intervals are not completely independent from each other during REM sleep.

Apart from the long-term correlations discovered using DFA on the BB signal, short-term correlations have been found after applying DFA on drive and timing respiratory components [23]. Examples of these components are the minute ventilation ($\dot{V}_E$), tidal volume (V_T), respiratory drive ($[V_I/T_I]$) that is computed as the inspiratory volume (V_I) divided by the inspiratory time (T_I), and the respiratory time ($[T_I/T_{Tot}]$), which results after dividing T_I by the total time of the respiratory

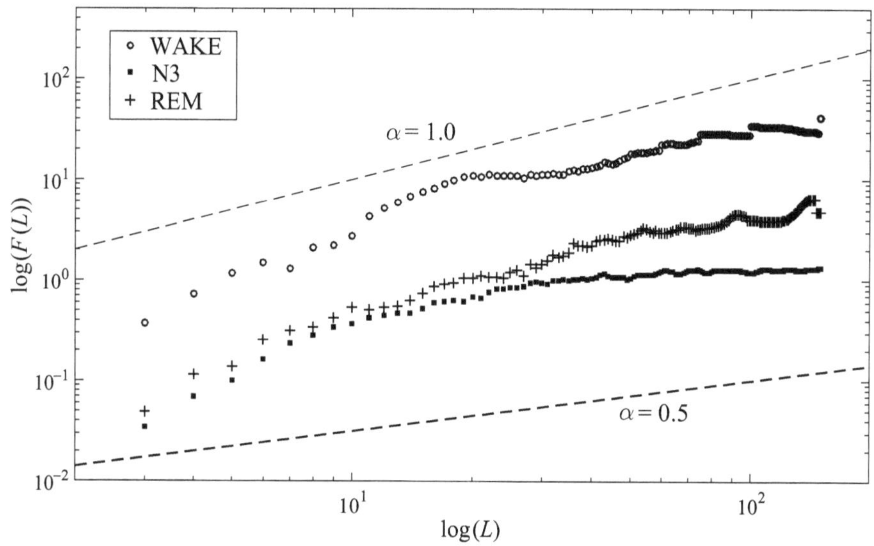

Fig. 19.3 Detrended fluctuation analysis (DFA1) applied to breath-to-breath intervals of a healthy adult. Representative segments of 5 min were taken for each sleep stage in order to illustrate the findings reported in [25]. The scaling exponent α is equal to 0.91 in Wake, 0.55 during N3, and 0.94 during REM

cycle (T_{Tot}). Two main scales were defined in [23] to analyze the correlations within the data. Scales between 7 and 14 breaths were taken for the analysis of short-term correlations (STC), and scales between 16 and (possibly larger than) 35 breaths were considered for long-term (LTC) correlations. After applying DFA2 on the different drive and timing respiratory components, the correlations indicated in Table 19.1 were found to be significant. The LTC indicate that even though the respiration becomes more irregular during REM, the respiratory pattern is not random, and the breath intervals are not independent from each other. These results are in agreement with those reported in [25]. The STC, on the other hand, suggest that the regulation of respiratory timing and drive are different, and that the latter remains unchanged during NREM and REM sleep. In other words, since the respiratory timing does not display STC, while the respiratory drive does, it is clear that timing and drive are regulated in a different way. One hypothesis is that the control of respiratory drive is done automatically and purely metabolic, in order to maintain blood gas homeostasis, while respiratory timing is more weakly and behavioural-based controlled to allow voluntary respiratory acts [24, 27].

Even though the mechanisms responsible for the respiratory regulation during sleep have not yet been fully understood, DFA has proven to be a powerful tool to identify regulatory differences between REM and NREM.

Table 19.1 Short- (STC) and long-term (LTC) correlations during REM and NREM sleep obtained using DFA2 in [23]

	STC	LTC
NREM	V_I/T_I	NONE
	V_T	
	$\dot{V}_E$	
REM	V_I/T_I	T_I/T_{Tot}
	V_T	V_I/T_I
	$\dot{V}_E$	$\dot{V}_E$
		V_T

19.3.1.2 Dynamic Warping on Respiratory Effort During Sleep

Based on the increased regularity of the respiratory signal during sleep, a self-similarity function can be used to discriminate between wake and sleep segments. In [28, 29], this similarity was measured using dynamic warping (DW) in time and frequency domain [30]. Dynamic time warping (DTW) allows to assess the similarity between two time series by means of a non-linear mapping of one onto the other.

Given the time series $A = \{a_i\}_{i=1}^{n}$ and $B = \{b_i\}_{j=1}^{m}$, a *warping matrix* $D \in \mathbb{R}^{n \times m}$ is constructed with entries d_{ij} defined as

$$d_{ij} = (a_i - b_j)^2. \tag{19.1}$$

In order to find the best alignment between the time series, an optimal path in D must be retrieved such that the cumulative distance between the series is minimized. This cumulative distance (DW) is defined as

$$DTW(A, B) = \left[\frac{1}{K} \sqrt{\sum_{k=1}^{K} w_k} \right], \tag{19.2}$$

with $w_k = (i,j)_k$ the kth element of the path $W = \{w_k\}_{k=1}^{K}$, and K the total length of the path. This procedure is illustrated in Fig. 19.4, where two real respiratory segments during sleep are used.

In principle, all the distance values in D could be taken into account for the computation of the optimal path. However, the computational complexity of the algorithm increases with the length of the series, hence, in order to reduce the space search, five conditions can be used:

- *Monotonicity:* Guarantees that the indexes i and j are always increasing or remaining the same during the search, $i_l \geq i_{l-1}$ and $j_l \geq j_{l-1}$
- *Continuity:* The path takes indexes with one step at the time, $i_l - i_{l-1} \leq 1$ and $j_l - j_{l-1} \leq 1$
- *Boundary constraints:* $i_1 = 1$, $i_K = n$, $j_1 = 1$, and $j_K = m$

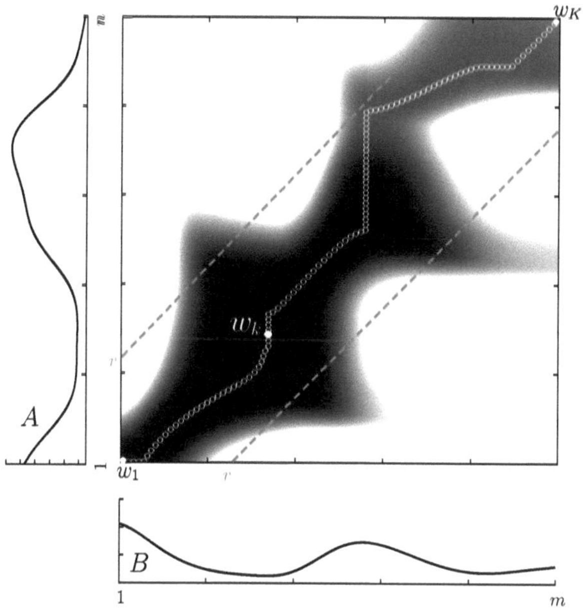

Fig. 19.4 Dynamic time warping (DTW) between the time series *A* and *B*, which correspond to two respiratory segments of 5 s during deep sleep. The warping matrix *D* contains the distances between the series, with white indicating higher distances. The *white circles* depict the optimal path retrieved between the boundaries defined by *r*

- *Warping window:* It is unlikely to find an optimal path far from the diagonal of D. This condition restricts the changes in i and j to a band of width r, such that $r > 0$ and $|i_l - j_l| \leq r$
- *Slope constraints:* $i_{l_p} - i_{l_0}/j_{l_p} - j_{l_0} \leq q$ and $j_{l_p} - j_{l_0}/i_{l_p} - i_{l_0} \leq p$, with $p, q \geq 0$, and p and q the number of steps in the horizontal direction and vertical direction, respectively.

This DTW procedure finds the minimum distance value between two time series, while dynamic frequency warping (DFW) computes the minimum distance between two PSDs.

Since respiration tends to be more regular during sleep, respiratory segments recorded during sleep are expected to be more self-similar than to segments recorded during wakefulness. In addition, the respiratory effort signal is known to be influenced by respiratory movements, as can be concluded from a correlation coefficient $r = 0.56$ with $p < 0.0001$ obtained between DTW features and parameters derived from actigraphy [29]. This suggests that body activity can be estimated from the respiratory effort. With this in mind, respiratory effort signals were recorded on the thorax of healthy volunteers, and different features such as time domain, frequency domain, and sample entropy were computed. A linear discriminant classifier was used and an accuracy of about 76% was reached for the discrimination between wake, REM, and NREM. A lower accuracy ($\approx 64\%$) was obtained when a discrimination between light and deep sleep was attempted. These performances can be improved by adding features computed from actigraphy as in [28]. In this case, DTW and DFW features were used together with actigraphy and a discrimination between wake and sleep was achieved with an accuracy of about 95%. Despite this improved performance, the discriminatory power of the respiratory effort and actigraphy is very limited for the classification of the different sleep stages. Nevertheless, this limitation can be overcome by the inclusion of cardiac and cardiorespiratory information into the classifier. This will be discussed further in this chapter.

19.3.1.3 Other Nonlinear Techniques Applied to Respiratory Signals During Sleep

Correlation dimension (CD) has also been used to analyze the complexity and nonlinearities of the respiratory signal during sleep stages. In [31], a population of healthy volunteers was used, and a differentiation between the correlation dimension during REM and NREM sleep was made. Correlation dimension was lower during deep sleep than during REM, which indicates once again that the respiration behaves in a more regular way during deep sleep, and that the complexity of the respiratory signal is higher during REM [24, 31]. In order to test for the presence of nonlinearities in the respiratory signal, a comparison between the CD calculated from respiratory movements on the chest and the CD from surrogate data was performed. Significant differences were found between both CDs during REM and NREM, with much higher values during REM. This suggests that respiratory signals have nonlinear properties during sleep [24]. Similar results were also obtained in pre-term infants despite the fact that sleep dynamics are known to be completely different between both age groups [31].

A different measure of complexity that has been used for the analysis of respiratory signals during sleep is the *approximate entropy* (ApEn). In [24], the use of ApEn confirmed once again that an increased regularity in the respiratory signal can be found in deep sleep with respect to REM. Similarly, a surrogate data analysis was also performed using the ApEn, and again nonlinear dynamics in the respiratory signals were found during sleep. Finally, a comparison between the fundamental frequency of the respiratory signal obtained from the PSD, and the ApEn was performed for each sleep stage. A weak linear relationship between both parameters was found. However, no differentiation could be made between sleep stages using only the respiratory frequency despite the significant differences in ApEn. These results suggest that ApEn allows to extract more information about sleep stages than the classical spectral analysis.

19.3.2 Cardiac Changes During Sleep

Sleep stages deeply influence both the blood pressure (BP) and the heart rate (HR). For instance, NREMS are accompanied by bradycardia, hypotension and a reduction in cardiac output [1, 14, 21], effects which are more pronounced during deep sleep (N3). Bradycardias have been associated with an increase in parasympathetic activation, while hypotension results from vasodilation caused by a reduced sympathetic modulation [21]. During REMS, on the other hand, HR and BP can increase to levels similar to those recorded during wakefulness [21]. Nevertheless, it is important to differentiate between phasic REM, where there is rapid eye movement, and tonic REM, where there is not. Phasic REM is characterized by unstable HR and BP, and great transients in the cerebral blood flow [14]. Instead, during tonic REM, HR and BP can be reduced to values even lower than those recorded during NREMS. These variations in BP and HR during REM sleep, have been associated with potential cardiovascular risk for patients suffering from coronary and cerebrovascular diseases [1], and with myocardial ischemia [21].

It is clear that cardiac activity is influenced by the sleep stages, and this can be observed not only from the different values of HR but also from the *Heart Rate Variability* (HRV) [32]. This variability can be studied using the RR-interval time series, which is computed using the time differences between consecutive R-waves in the ECG. When looking at the RR-interval time series computed for a full night (see Fig. 19.5), it is possible to observe changes in both the absolute value of the HR and the HRV during the different sleep stages. In fact, each sleep stage is characterized by a different pattern in the RR series. Figure 19.6 shows the RR series during 2 min of Wake, 2 min of deep sleep (N3), and 2 min of REM. As shown in the figure, the HR displays more regular oscillations during N3 than during Wake and REM. These oscillations are mainly modulated by the respiration through the well-known *respiratory sinus arrhythmia* (RSA). During Wake and REM, the RSA is attenuated by other sympathetic and parasympathetic modulators, which cause significant transient increases in HR, as can be seen in Fig. 19.6. Spectral analysis of the RR-interval time series has shown an increase in the HF accompanied by a reduced LF component during NREM, and an opposite effect during wakefulness and REM sleep [33]. These results are in agreement with a higher sympathetic activation during REM and wakefulness. All these changes in autonomic control are influenced not only by the current sleep stage, but also by the preceding one and by the point in the sleep cycle [34]. For example, it has been observed that a period of REM sleep at the end of the night is accompanied by a larger sympathetic activation than a REM segment at the beginning of the night [35]. This again, can be associated with an increased risk of cardiovascular events during the early morning [35].

Linear techniques such as the mean, standard deviation, autocorrelation, and spectral analysis have been used to study the HRV during sleep. They have shown that the HRV is, indeed, significantly higher during REM sleep than during NREM [32], however, they are not able to describe instantaneous changes in the control of the HR. Therefore, the use of nonlinear techniques is of paramount importance to assess the time-dependent behavior of the HR.

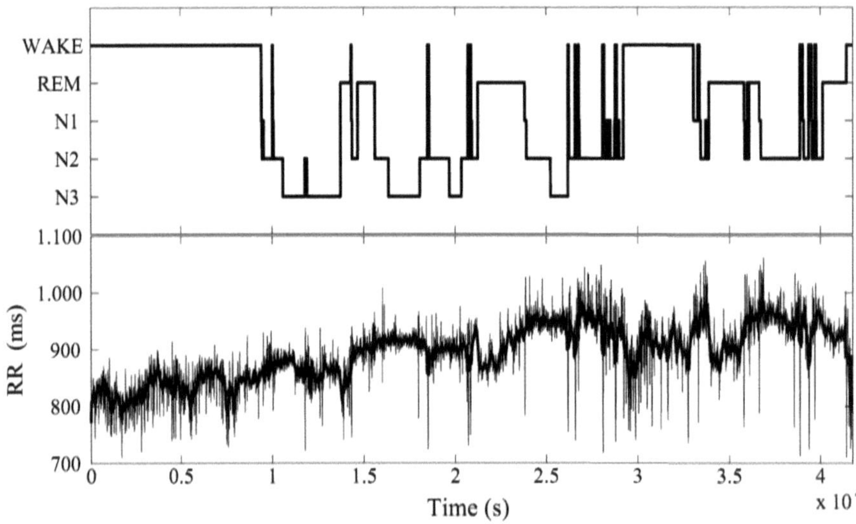

Fig. 19.5 Hypnogram of a healthy subject (*upper plot*) and the progression of the RR-interval time series throughout the night (*lower plot*). Note the general trend towards a decreased HR (i.e. increase in RR-interval) during sleep, and the apparent changes in HRV during the different sleep stages

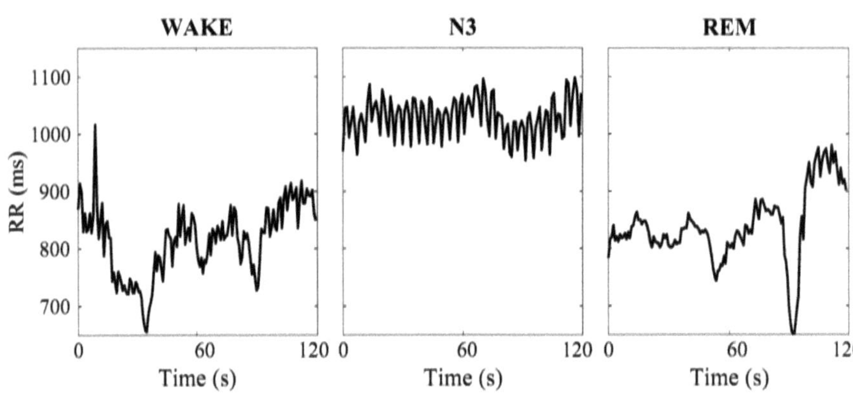

Fig. 19.6 Segments of RR-interval time series of 2 min during WAKE, deep sleep (N3), and REM

One way of analyzing the short- and long-term variability of the HR is by means of the Poincaré plot, where each RR-interval RR_n is plotted against the next interval RR_{n+1}. The scattering and the trajectory of the points describe the dynamics (depending on the extracted indices either linear and/ or nonlinear) of the HRV. The dispersion of the points along the y-axis for a given x indicates the beat-to-beat variability, while the total dispersion of the points in the x-axis describes the overall variation of the HR. Figure 19.7 shows three Poincaré plots obtained from RR-

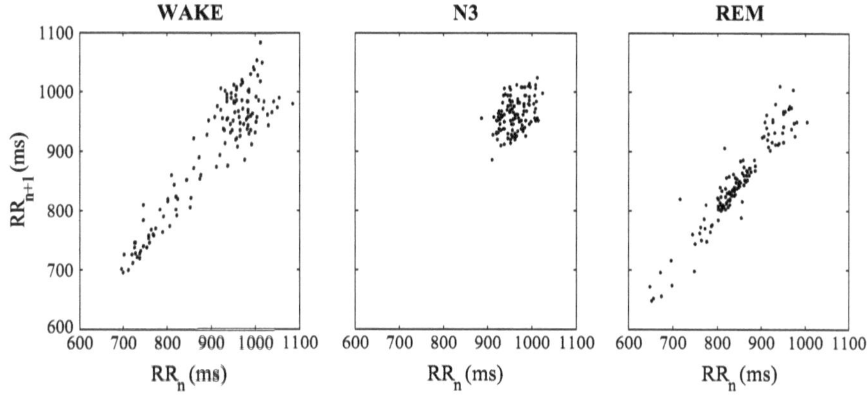

Fig. 19.7 Poincaré plots of three RR-interval time series of 2 min during WAKE, deep sleep (N3), and REM

interval segments during wakefulness, deep sleep (N3), and REM sleep of a healthy adult. Here, the results presented in [32] are reproduced, and it is clear that high correlations exist between consecutive RR-intervals for the different sleep stages. The plots show that during N3 sleep, the overall variation of the HR is lower while the beat-to-beat variability is higher. This confirms that during this sleep stage the HR is more regular and mainly modulated by the RSA. In fact, these results can also be associated with a more regular respiratory signal during N3, as discussed in Sect. 19.3.1.

The reduced beat-to-beat variability during REM can be explained by a higher regulation of the short-term variability of the HR during REM [36]. In addition, the plots of Fig. 19.7 confirm that the overall variability of the HR during REM is similar to the one recorded during wakefulness.

Different features can be extracted from the Poincaré plots. For instance, an ellipse can be fitted to the plot and the minor axis (SD1) can be associated with the beat-to-beat variability, and the major axis (SD2) with the long-term HRV. The ratio between the axes (SD1/SD2) indicates the relationship between short- and long-term variation of the HR [37]. These features were compared against typical linear and nonlinear measures of HRV in [38, 39], and high correlations were found between SD1 and HF and RMSSD. This is not a surprise, since the short-term variability is mainly vagally mediated. Similarly, SD2 was found to be highly correlated with SDNN, which is an indicator of the global HRV [38, 39]. A more interesting result was the high correlation obtained between SD1/SD2 and the short-term scaling exponent (α_1) computed using DFA [39]. More specifically, α_1 defined within the range of 4 and 11 heart beats. These findings suggest that nonlinear information of HRV might be described by the ratio SD1/SD2. However, these results need to be interpreted carefully, since other studies have shown that this ratio is only correlated with linear indices of HRV [38].

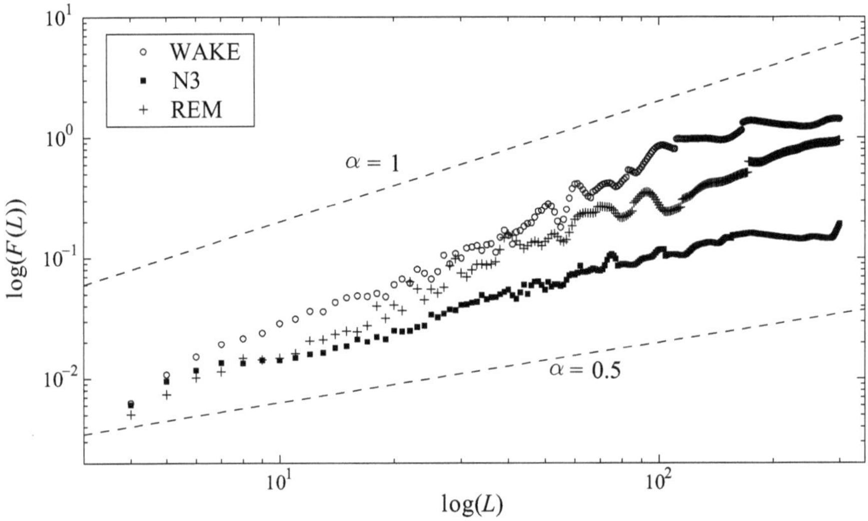

Fig. 19.8 Second order detrended fluctuation analysis (DFA2) of 2 min segments of RR-interval time series during wakefulness, REM, and deep sleep. For comparison, the *dashed lines* indicate uncorrelated ($\alpha = 0.5$) and correlated ($\alpha = 1$) behaviour

19.3.2.1 HRV Analysis During Sleep Using Detrended Fluctuation Analysis

Detrended fluctuation analysis (DFA) has been extensively used for the analysis of heart rate variability during sleep. Figure 19.8 shows the fluctuation curves ($F(L)$) for three different RR-interval segments recorded during wakefulness, deep sleep, and REM of a healthy adult. These curves were obtained using DFA2, where a second order polynomial was used in the detrending phase. The line with $\alpha = 0.5$ indicates uncorrelated behavior of the RR-intervals, while the line with $\alpha = 1$ indicates high correlations in the data. These results are consistent with the ones reported in [9, 40, 41], where quadratic and cubic polynomials were used in the analysis. Clear differences were found between sleep stages. In particular, the slope of the fluctuation function during N3 sleep is close to 1/2, which suggests a *loss* of long-term correlations, whereas some short-term correlations might be taking place during this stage. The latter are noticeable in the very small values of L, where it is clear that only after few heart beats in N3, the correlations are lost. In contrast, during REM and wakefulness clear long-term correlations can been identified. In [40], different degree polynomials were used for the detrending phase in order to evaluate the dependency of the results to the trends in the data. Again, the same results were found using a polynomial of up to a fourth degree. Furthermore, this analysis provided evidence that during deep sleep, the short-term correlations vanish after a time shorter than the breathing cycle time, which comprises few RR-intervals. This is, again, a strong indication of a higher cardiorespiratory coupling during N3.

The study of the different dynamics of the HRV during sleep stages is regularly performed using epochs of few minutes, where the transition periods between stages are eliminated [9, 40, 41]. Often, these periods are taken as the first and last 45 s of each sleep stage. In [40], however, the RR-interval time series computed from the whole night were analyzed using DFA, without differentiating between sleep stages. As a result, a scaling exponent $\alpha = 0.85$ was obtained, which suggests the presence of long-term correlations in the beat-to-beat variability during the entire night. Moreover, the authors compared these results with those obtained after eliminating the transition periods and concatenating all the sleep stages together. The resultant fluctuation curve ($F(L)$) was similar to the one obtained during light sleep, with a lower scaling exponent. These results indicate that the long-term correlations identified during the entire night are caused by the transition periods between sleep stages.

One technique that can be used to analyze the whole night of recording without overestimating the long-term correlations, and without removing the transition periods, is *progressive* detrended fluctuation analysis (PDFA) [42]. The main difference between PDFA and DFA is that the integrated time series is no longer divided into non-overlapping windows of size L. Instead, PDFA analyzes partial sums of the total time series separately. The length p of these partial sums is increased progressively from 1 to the length of the series, i.e., $p = 1 \ldots N$. Next, each partial sum is divided into non-overlapping epochs of fixed length L, which are then detrended as in DFA. As a result, a plot of the new fluctuation function against time can be obtained. This curve allows to identify the transition periods where the scaling exponent changes between one sleep stage to the other. In fact, the PDFA curves can be used to detect *ascending* transitions, namely, from deep sleep to either lighter sleep or wakefulness [42, 43], with more pronounced PDFA changes during the transitions to wakefulness. Conversely, *descending* transitions are not clearly identified from the PDFA curves. One possible explanation for this is the increased sympathetic tone experienced during the ascending transitions, and the fact that PDFA seems to be more sensitive to autonomic arousals [43]. Consequently, PDFA can be used to detect autonomic arousals and transitions to wake as demonstrated in [44].

Different studies have shown that DFA can be used to discriminate between sleep stages. For instance, Penzel et al. [41] compared the discrimination power of the scaling exponents against features derived from spectral analysis of HRV. Two different scaling parameters were computed, namely, α_1 defined in the range between 10 and 40 beats, and α_2 defined as the slope of the fluctuation curve between 70 and 300 beats. Furthermore, a discrimination between light sleep, deep sleep, REM, and wake was performed. The study showed that 78.4% of the RR-interval segments were correctly classified using only the DFA features, and after including the mean and standard deviation (SDNN) of the RR-intervals, this performance increased to 85%. In contrast, after using the classical spectral parameters of HRV, only 51.4% of the segments were assigned to the correct sleep stage. Finally, a combination of all DFA, spectral features, mean, and SDNN, resulted in 84% of correctly classified sleep segments. It is clear that the best discrimination could be done using DFA and time domain parameters of HRV.

The performance achieved using only spectral features of HRV is limited by the use of fixed LF and HF frequency bands. This is due to the time varying modulation of the ANS. In order to overcome this limitation, the boundaries of the HRV analysis can be adapted by means of time-frequency analysis. As a result, an accuracy of about 93% can be achieved in the classification of sleep and wake using only spectral features of HRV [45]. Note, however, that no differentiation can be made between the sleep stages using this approach.

19.3.2.2 Entropy Measures of Cardiac Activity in Healthy Sleep

During sleep, several physiological mechanisms such as the baroreflex, central oscillations, sympathovagal balance, hormonal regulators, amongst others, play a key role in the regulation of the heart rate. This regulation, however, is strongly modified during sleep [21], and this can be clearly observed from the complexity of the HRV [46]. To measure this complexity during sleep, different entropy measures like Shannon entropy (SE), approximate entropy (ApE), sample entropy (SampEn), and conditional entropy (CE) [47] have been used, and the main findings will be described next.

Since REM sleep is characterized by an irregular HR and a higher autonomic activation that reaches values similar or even higher than those recorded during wakefulness, the complexity of the HRV during this sleep stage is expected to be higher [46]. However, opposite results have been reported on the behavior of entropy measures during REM sleep. In [48], for instance, SE and corrected conditional entropy (CCE) were studied in both young and old healthy subjects. No significant differences were found between the entropy values of different sleep stages in the young population. In contrast, older subjects displayed a significant reduction in complexity during wakefulness, light sleep and REM, with a more pronounced effect during the latter. This suggests that ageing causes a reduction in the complexity of the cardiovascular system, which in a long-term can be associated with less flexibility of the system to respond to stressful situations. Furthermore, this reduction in complexity is larger during REM, which again confirms that REM can be linked with a higher cardiovascular risk.

In a different study on healthy subjects, the SampEn was computed for different frequency components of the HRV, and then compared between sleep stages [49]. Five minute segments of RR-interval time series were first decomposed into three main components using a five-level wavelet decomposition. These components correspond to the very low frequency band (VLF: 0.0.0375 Hz), low frequency band (LF: 0.0375–0.15 Hz), and high frequency band (HF: >0.15 Hz). Then, mean, standard deviation (SD), and SampEn were computed for each component and for each segment, after which a comparison between sleep stages was performed. The main results of this study are the increased SampEn values during deep sleep in the VLF band, and the increased SD during REM in all frequency bands. In fact, the SD during REM is significantly higher than during wakefulness, while no clear differences were found in SampEn values between these two stages. Even though the mechanisms responsible for the VLF variations of HRV are not yet fully understood, the results obtained during deep sleep can be associated with thermoregulatory changes, as suggested in [49]. The behavior observed during REM, on the other hand, is consistent to the one reported in [40], where a similar nonlinear behavior was observed during REM and Wake. A possible explanation for these results is that nonlinear HRV features describe the stability and the energy expenditure during each sleep stage. In particular, deep sleep is characterized by a regular HRV with a strong cardiorespiratory coupling, which results in a low energy expenditure, while REM sleep is accompanied by irregular HR and a high energy expenditure similar to wakefulness [49].

19.3.2.3 Other Nonlinear Measures of Cardiac Activity During Sleep

A different way of quantifying the complexity of the cardiac activity during sleep was adopted in [50], and it used the entire ECG signal, instead of only the RR-interval time series. ECG segments of 2 min 44 s were selected from a dataset recorded from 12 healthy adults during sleep, and different nonlinear features such as correlation dimension (D2), largest Lyapunov exponent (L1), and Kolmogorov entropy (K2), were computed. The main goal of the study was to identify possible differences between the ECG morphology during different sleep stages. Results indicate that D2 and K2 are the most discriminant features for the task of sleep staging, followed by L1. Moreover, L1 was found to be significantly larger and D2 significantly lower during REM sleep than during deep sleep. These findings reveal a lower complexity and a more chaotic behavior of the ECG during REM sleep.

With these results and those obtained using DFA and entropy-based features, it is possible to conclude that the increased HRV during REM sleep is associated with a less predictable ECG dynamics, and with a reduction of the RSA as a main modulator of the HR caused by a reduction of the parasympathetic modulation.

19.3.3 Cardiorespiratory Changes During Sleep

The analysis of linearities and nonlinearities of the heart rate and respiration showed that the effect of sleep stages was very similar in both [25]. As a result, it can be hypothesized that either both the cardiac and respiratory systems are affected in the same way, or that the cardiorespiratory coupling is modified during sleep.

Three different forms of cardiorespiratory coupling have been identified during sleep [51, 52], namely, respiratory sinus arrhythmia (RSA), cardiorespiratory phase synchronization (CRPS), and time delay stability (TDS). These different forms of coupling act in different time scales, and they all reflect different regulatory mechanisms. In [52], the HR and the respiratory signal recorded from 189 healthy adults were analysed, and it was shown that the RSA amplitude is highest during deep sleep, and lowest during REM and wakefulness. No clear differences between REM and wake were observed, but a tendency to lower RSA during wake was reported. These results are consistent with the different respiratory and cardiac findings discussed in previous sections. More interesting results were the clear differences in the strengths of the other two forms of cardiorespiratory coupling during sleep stages. The degree of CRPS, on the one hand, was highest during deep sleep and it significantly decreased during REM sleep to values even lower than those during wakefulness. An important finding is that compared to RSA, CRPS exhibits more pronounced changes between sleep stages. This behaviour of CRPS is consistent among all age groups as reported in [53], however, the degree of synchronization is reduced with age. This indicates that the effect of sleep on this form of cardiorespiratory coupling is maintained regardless of the age.

A comparison between RSA and CRPS revealed that they are two completely independent forms of coupling [52]. For example, it is known that RSA is a frequency dependent phenomenon, with a nonlinear response to changes in the respiratory frequency [54]. In contrast, the CRPS does not display any dependency on the breathing frequency, in fact, different degrees of CRPS can be observed in a period of constant RSA.

The strength of the TDS, on the other hand, displays a completely different behaviour during REM and deep sleep than the one of RSA and CRPS [52]. TDS was much higher during wakefulness and light sleep, and lowest during REM and deep sleep. Furthermore, a remarkable difference was revealed between light and deep sleep, despite their similarities in the linear and nonlinear features of HRV. These results suggest that TDS reveals the effect of different regulation mechanisms of heart rate and respiration during sleep. In addition, TDS acts on a different time scale, namely, between 150 and 220 s, while CRPS is often observed within 20–45 s, and RSA comprises only few heart beats. The study of these three forms of coupling during sleep stages reveal that the autonomic regulation manifests itself in different ways and at different time scales. In effect, this is a step forward towards the understanding of the underlying autonomic control mechanisms of the cardiorespiratory system.

The results obtained with TDS can be explained by a different brain-heart connectivity during sleep stages. In [52], the coupling between different EEG channels with both the HR and the respiration was quantified using the TDS. This quantification revealed a different topology in the network during sleep stages. A higher connectivity was identified during light sleep and wake, a lower connectivity during REM sleep, and a loss of connectivity during deep sleep. Indeed, these results are very similar to those obtained using TDS on the HR and respiration, which suggests that TDS offers information about connectivity and network topology that the other forms of coupling cannot describe. Moreover, these results are also consistent with those reported in [55], where information dynamics was used to quantify the interactions between the frequency components of the EEG and the HRV in the HF band, during sleep. It was shown that information flow is bidirectional between brain and heart systems during sleep. In fact, the EEG β

rhythms are the ones that appear to be sending most of the information to the heart. Furthermore, the links between brain and heart are stronger during light sleep, reduced during deep sleep, and partially or completely lost during REM. In other words, the brain components seem to be decoupled from the heart system during deep sleep and REM. This effect can be possibly explained by the changes in the dynamics of the heart occurring during sleep, which on their turn affect the coupling mechanisms with the brain and the respiration. Besides, it is also possible that the definition of the time scale limits the description of the network topology. Therefore, studies with different representations of the cardiac, respiratory, and brain activity need to be performed in order to enhance the understanding of this complex network. Methods that can be considered for this include the point process analysis presented in [56] and the instantaneous transfer entropy proposed in [57]. The advantage of these methodologies is that they can be used to analyze the whole night of recording without overestimating the long-term correlations, and without removing the transition periods between the different sleep stages.

19.3.4 Algorithms for Sleep Staging

It is clear that sleep stages affect the morphology of the ECG, the heart rate, the respiration, and the cardiorespiratory coupling. For this reason, heart rate and respiration have been extensively used in the development of algorithms for sleep staging. In fact, multiple algorithms have been proposed and they can be separated into three main groups according to the type of classification they perform, namely, sleep-wake, wake-REM-NREM (WRN), and wake-REM-light sleep-deep sleep (WRLD). Typically, sleep staging algorithms attempt to generate a hypnogram with a 30 s resolution [4, 17]. Nevertheless, in [58] a classification was made on 1 min segments in order to match with the classical sleep apnea detection, which is often performed on a minute-by-minute basis.

Algorithms based solely on the respiratory signal have achieved accuracies of up to 76% in WRN classification, and 63.8% in WRLD classification [29]. The algorithm proposed in [29] derives different features from the respiratory effort, such as, time and frequency parameters, non-linear features using dynamic time and frequency warping, and sample entropy. When only the features derived using dynamic warping were used, an accuracy of about 94% was achieved for sleep-wake classification. Algorithms based on the HR and HRV consist of either very simple rules as in [59], or of features derived using empirical mode decomposition, discrete wavelet transform, and linear and nonlinear dynamics features as in [60]. Even though these algorithms have proven to be good discriminators between sleep stages, mainly between sleep and wake stages, it is known that the fusion of features derived from the ECG and actigraphy allow for a better accuracy in sleep-wake classification [28]. In addition, the fusion between features derived from each modality separately and those quantifying cardiorespiratory coupling can achieve higher performances for the different classification tasks. Tables 19.2, 19.3, and 19.4

Table 19.2 Performance of sleep-wake classification algorithms

Reference	Signals	Features	Kappa	Acc. (%)
Redmond et al. [4]	HR, Rs, EDR	Time and frequency domain	0.6	89
Domingues et al. [61]	RR, Rs, Act	Frequency domain, CRPS	–	80.2
Long et al. [45]	HR, Act	Frequency domain	0.64	95.8
Long et al. [28]	Rs, Act	DTW, DFW	0.66	95.7
Willemen et al. [13]	HR, Rs	Time and frequency domain	0.69	92

Table 19.3 Performance of WRN classification algorithms

Reference	Signals	Features	Kappa	Acc. (%)
Redmond et al. [4]	HR, Rs, EDR	Time and frequency domain	0.46	76.1
Ebrahimi et al. [62]	HR, Rs	Time and frequency domain, EMD, SampEn, DFA, SE, ApE	–	89.32
Ebrahimi et al. [60]	HR	Time domain, Wavelet, DFA, ApE, SampEn SE, EMD	–	80.67
Domingues et al. [61]	RR, Rs, Act	Frequency domain, CRPS	–	66.4
Mendez et al. [63]	HR	Time-varying spectra	–	79.3
Long et al. [29]	Rs	Time and frequency domain, DTW, DFW, SampEn	0.45	76.2
Fonseca et al. [10]	HR, Rs	Time and frequency domain, DTW	0.56	80
Willemen et al. [13]	HR, Rs	Time and frequency domain	0.62	81
Kortelainen et al. [64]	HR, Act	Time-varying spectra	0.44	79
Xiao et al. [65]	HR	Time and frequency domain, DFA, Multiscale entropy, Mutual Information	0.46	72.58

Table 19.4 Performance of WRLD classification algorithms

Reference	Signals	Features	Kappa	Acc. (%)
Isa et al. [66]	HR	Time and frequency domain	0.26	60
Long et al. [29]	Rs	Time and frequency domain, DTW, DFW, SampEn	0.38	63.8
Fonseca et al. [10]	HR, Rs	Time and frequency domain, DTW	0.49	69
Willemen et al. [13]	HR, Rs	Time and frequency domain	0.56	69

summarize the results obtained with different algorithms, modalities, and for the three different classification tasks on healthy subjects. In the tables, Rs stands for respiratory signal, EDR is the ECG-derived respiration, Act stands for actigraphy, and EMD stands for empirical mode decomposition. After comparing the results, it is clear that nonlinear and complexity features have an added value in the different classification tasks.

The performance of the different classifiers is determined using the annotations provided by the sleep experts. However, disagreement between scorers is known to play a key role in the reduction of this performance. In [67], a sleep stage agreement of 82.6% was found between more than 2500 scorers, with a higher agreement for the REM stage followed by N2 and wake. The agreement for N1 and N3 was 63% and 67.4%, respectively. These results suggest that there is room for improvement in the definition of the rules for sleep staging, and that an agreement of about 83% needs to be taken into account when comparing classification results.

19.4 Sleep Apnea

Sleep apnea is an under-diagnosed sleep-related breathing disorder that affects up to 10% of middle-aged adults [68]. The reason for its under-diagnosis relies on the unawareness of the symptoms among the general population. Usually, people wait too long before seeking professional help, and they only do it when serious symptoms appear or when the disorder has reached an advanced stage. Clear symptoms of sleep apnea include excessive daytime sleepiness, fatigue, nocturia, irritability, depression, amongst others. In addition, sleep apnea is considered a risk factor for morbidity and mortality due to its long-term effect on the cardiovascular system [69]. This effect is linked with systemic hypertension and increased sympathetic modulation, which in a long-term compromise the well-functioning of the heart [70]. Not to mention the fact that hypertension is recognized as a common cause of cardiovascular and cerebrovascular diseases.

Sleep apnea is currently diagnosed using polysomnography (PSG), which is a sleep test that monitors different physiological signals such as heart rate and respiration. This test is typically performed in a hospital setting and it requires the supervision of a clinical expert, factors that make PSG an uncomfortable and costly procedure. For this reason, multiple studies have focussed on the development of less invasive devices for the diagnosis of sleep apnea. These devices are often designed using the heart rate and respiratory signals due to the clear effect that apnea has on the cardiorespiratory system [11, 12, 15, 71, 72].

Apneas and hypopneas are two types of respiratory events that are respectively characterized by a complete absence or reduction of airflow for at least 10 s [73]. These events may occur up to hundreds of times per night, and they often last for 30–60 s [9]. Apneas can be subdivided into three categories according to the respiratory effort: obstructive sleep apneas (OSA), central sleep apneas (CSA), and mixed apneas (MSA). During OSA events, the interruption of the respiratory flow is caused by an obstruction of the upper airways despite the respiratory effort, while in CSA events, this is caused by an absence of respiratory effort. MSA events are a combination of OSA and CSA and they typically start with a reduction of the respiratory effort that leads to an upper airway obstruction. Hypopneas, on the other hand, are characterized by a reduction of airflow, caused by either an obstruction or a reduced respiratory effort, that results in a drop of oxygen saturation. All apneas and hypopneas end with an arousal, which is a physiological state where the ANS restores the breathing regulation. The sympathetic tone increases as a consequence of these arousals and this on its turn causes an increase in heart rate. Figure 19.9 shows a segment of heart rate and respiratory effort before, during and after an OSA

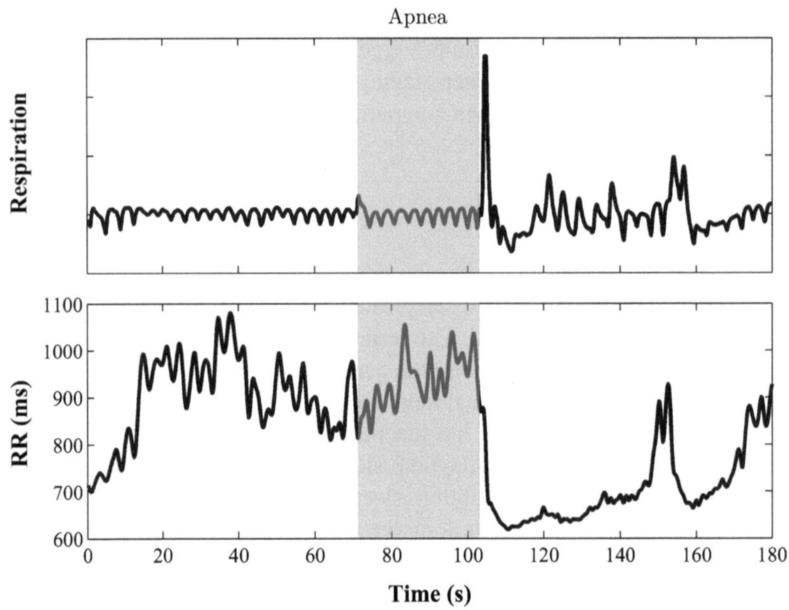

Fig. 19.9 Respiration and heart rate during an OSA episode, which is indicated by the *gray area*

episode. Note the increase in respiratory effort followed by an increase in heart rate at the end of the apnea event. These arousals are responsible for the disruption of healthy sleep patterns, as can be seen in Fig. 19.1, which results in severe sleepiness during the day. The effects of sleep apnea on the dynamics of the cardiorespiratory system and on the "normal" sleep pattern will be described next.

19.4.1 Effect on Respiratory Dynamics

Sleep apnea is characterized by frequent discontinuations in the respiratory pattern, which result in breath-to-breath (BB) intervals larger than 10 s [74]. This is, however, not the only effect that apnea has on the respiratory dynamics. The analysis of the distribution of BB signals recorded from patients suffering from sleep apnea, showed that the variability and regularity of the breathing pattern is affected during nonapneic periods [74]. In fact, an increased variability was reported in OSA patients when compared to the normal population, during all sleep stages. These changes in breathing variability could be associated with a higher upper-airway resistance typical in OSA patients.

Apart from BB variability, the respiratory movements appear to be random in OSA patients, according to the analysis of the largest Lyapunov exponent and correlation dimension performed in [75]. These results were later confirmed in

[76], where the phase-space plots computed using respiratory effort signals during sleep periods of OSA patients, reveal the presence of random attractors. The latter study also attempted to analyze the correlation dimension, but its computation was not possible during sleep segments with apnea due to the lack of a plateau in the correlation integral curves. These results suggest an increase in complexity during these segments, however, further investigation is still needed to support these findings. In addition to segments with apnea, periods during wakefulness with eyes closed, and sleep with nasal continuous positive airway pressure (nCPAP) were also analyzed in [76]. The values of correlation dimension derived from these segments were compared against surrogate data, and results indicate that respiratory movements display a nonlinear deterministic behavior. Furthermore, the values of correlation dimension were different for two pressure values in nCPAP, namely, 3 cm H_2O and 8 cm H_2O. As a result, correlation dimension could be considered as an index for titration of the CPAP treatment [76].

19.4.2 Effect on Cardiac Dynamics

The cardiac effects of apnea have been widely investigated using several approaches, such as linear and nonlinear techniques of HRV and morphological changes in the ECG [11, 12, 41, 77–80]. For instance, spectral HRV analysis performed on apnea patients during wakefulness has revealed a possible autonomic dysfunction caused by sleep apnea [77]. This was concluded after observing lower HF values in patients than in controls, during free and controlled breathing periods. Similarly, a reduced vagal activation was also observed in apnea patients, during all sleep stages, and an increased sympathovagal balance was observed during wakefulness and N2 [41].

The complexity of the HRV signal in apnea patients was investigated in [78] by means of sample entropy. It was concluded that, similarly to the respiratory signal, the HRV pattern seems to be more regular in apnea patients. Therefore, a lower complexity of the HRV can be associated with the cyclic pattern consisting of bradycardia and tachycardia characteristic of apnea events.

The dynamics of the HRV signal during sleep stages in apnea patients have also been investigated using DFA [9, 41, 81]. In this context, a comparison between the long-term correlations described in Sect. 19.3.1 and those obtained in sleep apnea has been performed. This comparison is illustrated in Fig. 19.10, where the results presented in [41] were reproduced using the HRV signal recorded from one sleep apnea patient of the sleep laboratory of the University Hospital Leuven (UZ Leuven), Belgium. Here, the DFA3 was used and the scaling parameter α was computed in the range $70 \leq t \leq 300$. The slopes of the fluctuation curves depicted in Fig. 19.10, suggest that in the apnea patient the long-term correlations observed during REM sleep and wakefulness are maintained, while no long-term correlations are observed neither in deep nor in light sleep. These results are very similar to those obtained for the healthy population, which indicate that the long-term control mechanisms of the heart rate are only slightly affected by sleep apnea [41, 81].

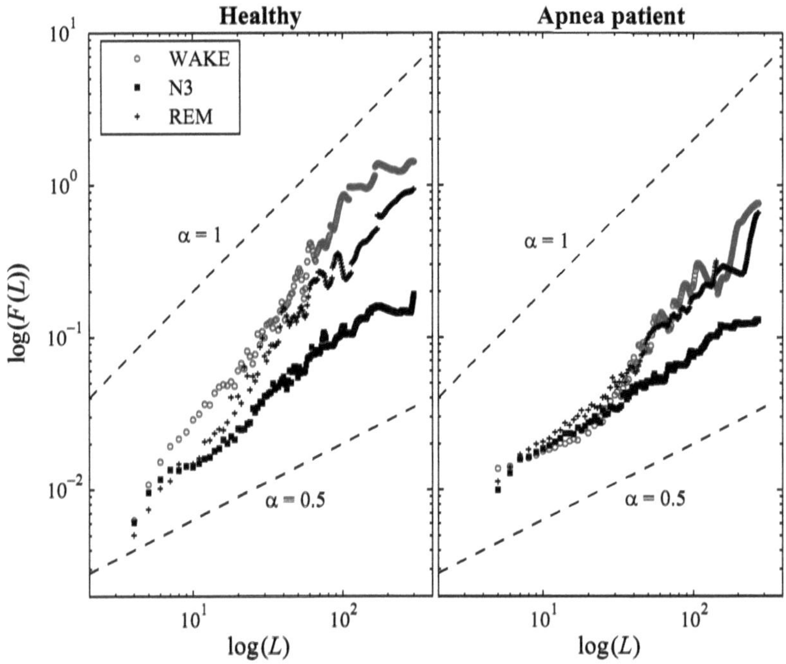

Fig. 19.10 Third order detrended fluctuation analysis (DFA3) of 5 min segments of RR-interval time series during wakefulness, REM and deep sleep of a healthy adult (same as in Fig. 19.8) and an apnea patient. For comparison, the *dashed lines* indicate uncorrelated ($\alpha = 0.5$) and correlated ($\alpha = 1$) behaviour. The scaling parameters for the healthy subject and apnea patient were, respectively, 1.14 and 1.25 during wakefulness, 0.57 and 0.55 during N3, and 1.23 and 0.96 during REM

19.4.3 Effect on Cardiorespiratory Interactions

Sleep apnea is associated with a reduced vagal activation during sleep, which might lead to changes in the way the respiration interacts with the heart rate through the RSA. This has been studied in [82], where the HR control during inhalation was found to be different than during exhalation. The increase in HR during inhalation is significantly affected during OSA events, while the HR decrease during exhalation seems to remain unchanged. The latter, however, disappears at the end of the event, possibly due to the arousal. These results suggest an inhibition of the vagal activation after the apnea event, which agrees with the increased sympathovagal balance used by many in sleep apnea detectors [79].

In [83], the cardiorespiratory coupling was analyzed using the cross-spectrum between the RR-interval time series and respiratory signal, in the LF (0.01–0.1 Hz) and HF (0.1–0.4 Hz) frequency bands. The study was based on the fact that there is a high-frequency coupling (HFC) related to the RSA, and a low-frequency coupling (LFC) associated with cyclic patterns in respiration of about 25–50 s. The LFC appears to be more accentuated in apnea patients, and it is directly associated with apnea episodes. Furthermore, it was demonstrated that OSA events display a broad band LFC, while CSA a narrow band LFC. These differences in the cardiorespiratory coupling could be used to differentiate between CSA and OSA, as proposed in [83].

19.4.3.1 Assessment of Cardiorespiratory Coupling Using Orthogonal Subspace Projections

The cardiorespiratory coupling was also investigated in [12], where orthogonal subspace projections (OSP) on the RR-interval time series and the ECG-derived respiratory signal (EDR) were used. This approach decomposes the HR signal into two different components, one containing all variations related to respiration, and one where all other modulators different from respiration are described. The general procedure to perform this decomposition will be described next.

Given are two physiological signals, namely the RR-interval time series denoted by $\mathbf{X}$ and the EDR signal denoted by $\mathbf{Y}$. All the information contained in $\mathbf{X}$ that is linearly related to $\mathbf{Y}$, can be computed by projecting $\mathbf{X}$ onto a subspace $\mathbb{V}$ defined by variations in $\mathbf{Y}$. Once the basis $\mathbf{V}$ for the subspace $\mathbb{V}$ is computed, any signal $\mathbf{X}$ can be projected onto the subspace, by means of $\mathbf{X_Y} = \mathbf{PX}$, with $\mathbf{P}$ a projection matrix defined as $\mathbf{P} = \mathbf{V}(\mathbf{V}^T\mathbf{V})^{-1}\mathbf{V}$. Note that $\mathbf{X_Y}$ describes the dynamics of $\mathbf{X}$ related to $\mathbf{Y}$, and the part of $\mathbf{X}$ that is related to other mechanisms can be computed as $\mathbf{X_{Y\perp}} = \mathbf{QX}$, with $\mathbf{Q} = I - \mathbf{P}$.

It is clear that the main result of this algorithm is the computation of a component of heart rate related to respiration denoted by $\mathbf{X_Y}$, and a component related to mechanisms other than respiration denoted by $\mathbf{X_{Y\perp}}$.

The construction of the subspace $\mathbb{V}$ can be done using wavelet decomposition of the respiratory signal and selecting only the coefficients in a certain frequency band. In [12], this approach was used, and two main frequency bands were used, namely LF: <0.07 Hz and HF: 0.07–0.6 Hz. As a result, the subspace related to respiration was constructed using the wavelet coefficients in those frequency bands and their delayed versions, using delays from 1 to m s. The relative power of each component was then computed as $F = (\mathbf{X}_\mathbf{Y}^T\mathbf{X_Y})/(\mathbf{X}^T\mathbf{X})$ for each subspace, or each frequency band. Figure 19.11 shows these power values for two different datasets: the Physionet Apnea-ECG dataset and a dataset collected in the sleep laboratory of UZ Leuven. For the latter, annotations of the different types of apnea were available, which allowed a differentiation between the cardiorespiratory interactions for each type. These results are again in agreement with the previous findings, where reduced interactions in the HF band were observed for apnea patients.

This approach takes into account only the linear cardiorespiratory interactions. In order to describe the nonlinear interactions, a nonlinear transformation is applied to both the HR signal and the respiratory signal as in [84]. Similar results were obtained, however, a better discrimination between normal and OSA events was

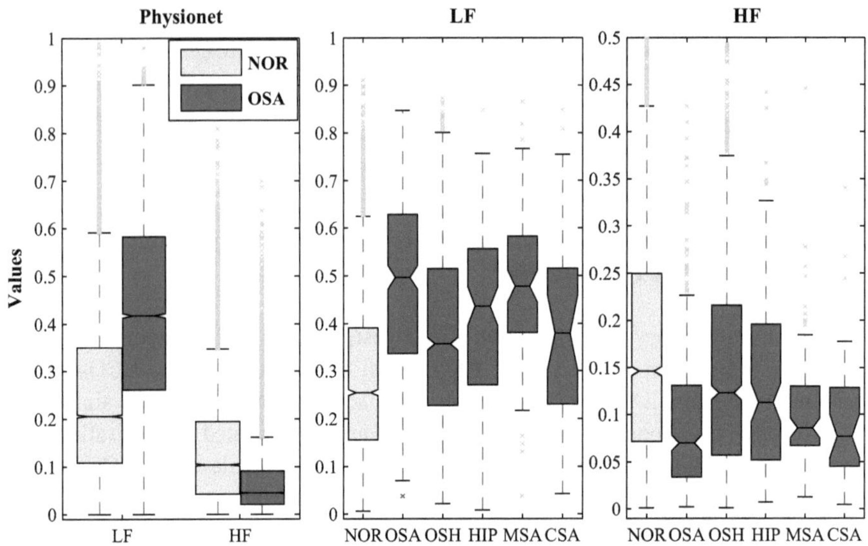

Fig. 19.11 Relative power of HR projected onto the respiration in the LF and HF bands. The *middle* and *right plot* show the values for the dataset recorded in UZ Leuven. Normal segments: NOR; Obstructive apneas: OSA; obstructive hypopneas: OSH; hypopneas: HPA; central apneas: CSA; and mixed apneas: MSA, are indicated

achieved after taking nonlinearities into account. In particular, higher nonlinearities were found in the LF band, which lead to an improvement in the classification performance.

A different form of cardiorespiratory coupling, namely, CRPS also appears to be affected by sleep apnea, in particular by OSA [85]. This has been studied in [85, 86], where the phase of the heart rate and respiratory signals were calculated using the Hilbert transform. Then, the phase difference was used to estimate the phase-locking periods, which correspond to times where the phase difference remains the same or below a certain threshold. Results of these studies indicate that the phase-locking periods are reduced in OSA, and their reduction is more pronounced in severe OSA patients. These results can be associated with the reduced vagal activation and the elevated sympathovagal balance observed in apnea patients as discussed before. In other words, the cardiorespiratory control might also be influenced by the ANS, which on its turn is affected by the repetitive occurrence of apnea episodes.

Multiple techniques have been used to quantify the changes in the cardiorespiratory coupling, and they have been used for the classification of apnea episodes from the ECG. Some of these techniques will be discussed next.

19.4.4 Sleep Apnea Detection

The clear changes in respiration, heart rate, and cardiorespiratory coupling during apnea episodes have motivated the development of detection algorithms using as few signals as possible. For example, [11, 12, 15, 71, 87] proposed to derive several features from the electrocardiographic (ECG) signal to detect sleep apnea. Moreover, instead of measuring the respiratory signal, those algorithms first compute the so-called ECG-derived respiration (EDR), and then derive a set of features that can later be used for the discrimination of apneas. Algorithms to compute the EDR include the R-peak amplitude [88], principal component analysis (PCA) [89], and kernel PCA [90].

The performance of the sleep apnea algorithms is typically evaluated on the *Apnea-ECG* dataset, publicly available in Physionet. However, many algorithms are tested on private datasets. Hence, the interscorer disagreement reported in [91] must be considered when comparing algorithms. An overall agreement of 93.9% for scoring respiratory events was reported, with an agreement for OSA segments, hypopnea segments, and CSA segments of 77.1%, 65.4%, and 52.4%, respectively. These results reinforce the fact that the detection of hypopneas and central apneas constitute a big challenge in sleep apnea detection. Hypopneas, for instance, appear very similar to normal segments in the ECG, which makes them very difficult to identify using only this signal. Obstructive sleep apnea, on the other hand, is much easier to detect from the ECG, and many algorithms have been proposed to solve this problem. In [79] a summary of the results of the Physionet challenge of 2000 is presented, and the algorithm with the best performance that uses a fully automated approach is described in [11]. This algorithm uses the time and frequency features of HRV, and it achieves an accuracy of 89.4% for the detection of sleep apnea on a minute-by-minute basis. A different fully automated algorithm was recently proposed in [12], and it uses features derived by means of OSP and features describing the changes in morphology of the ECG. An accuracy of 84.74% was reached by this algorithm on the Physionet dataset. Mendez et al. reported accuracies of 89% and 85% using empirical mode decomposition and wavelet analysis of the HRV, respectively, on the same dataset [92]. However, these last results were achieved after manually correcting the position of the Rpeaks in the ECG.

On a different (private) dataset, the spectral features of HRV achieved an accuracy of 70.4%, and together with values of sample entropy, an accuracy of 72.9% was reached [78]. In a different study, the time domain features, namely, the mean and the standard deviation of the RR-interval series, together with the scaling parameters derived using DFA2, achieved an accuracy of 74.4% in the discrimination of apnea cases [41].

The detection of CSA has been tackled in different studies. For instance, in [87] features derived from the recurrent plots obtained from the EDR signal achieved a positive predictive value of 60% and a sensitivity of 84%. In [83], on the other hand, a PPV of 58% and a sensitivity of 81% were obtained using features derived from the cross spectral power between the RR-interval time series and the EDR.

As can be seen, the sleep apnea algorithms that use linear techniques have outperformed those based on nonlinear approaches for the detection of sleep apnea. Nevertheless, complexity measures might be useful for the detection of central sleep apnea. The detection of hypopneas, however, from the cardiac and respiratory signals remains an open problem.

The aforementioned methodologies focused on the minute-by-minute classification, which might be later used to assess the severity of the disease in each particular patient. Other methodologies have focused on the assessment of the severity of OSA for each patient, without looking into each particular event. An example of such a methodology is the one presented in [93], where symbolic dynamics applied to HRV was used together with clinical data, such as age, neck circumference, daytime sleepiness, and intensity of snoring, to classify patients with high probability of OSA. This algorithm reached an area under the ROC curve of 94.1% in a private dataset. Even though this algorithm integrates clinical information with physiological variables, other algorithms, like those described in [94–96], which are based solely on the linear dynamics of HRV still reach an accuracy of 100% on the detection of OSA patients. These results suggest that linear dynamics of HRV might be enough to detect patients suffering from obstructive sleep apnea. However, it is important to note that these studies have not included patients with heart disorders, which are known to be closely related to the sleep apnea syndrome [69]. Hence, the question of whether nonlinear dynamics can improve the classification of sleep apnea in the general population remains open.

19.5 Public Datasets for Sleep Staging and Sleep Apnea Detection

The following list summarizes the public datasets that are available for the analysis of the dynamics of HRV and/or respiration during sleep and sleep apnea.

- The Apnea-ECG database [71] is commonly used to test ECG-based algorithms for sleep apnea detection. It contains 70 ECG recordings, of which eight are accompanied by signals of respiratory effort, oro-nasal airflow, and blood oxygen saturation.
- The MIT-BIH Polysomnographic Database [97] is regularly used for the analysis of sleep stages, and it might be used for the study of chronic obstructive sleep apnea syndrome. The dataset includes, amongst other signals, 16 ECG recordings accompanied by respiratory signals, which were either measured with a nasal thermistor or derived from inductance plethysmography.
- Sleep recordings and hypnograms in European data format (EDF) [98]. This dataset contains 61 PSGs, which contain oro-nasal respiratory signals that can be used for the analysis of respiration during different sleep stages.

- The CAP Sleep Database [99] consists of 108 PSGs that include the ECG signal, airflow, and respiratory effort recorded on both the abdomen and the thorax. This dataset is useful for the analysis of HRV and respiratory dynamics during sleep stages. The data includes recordings from healthy subjects and patients diagnosed with either nocturnal frontal lobe epilepsy, REM behavior disorder, periodic leg movements, insomnia, narcolepsy, sleep-disordered breathing, or bruxism.
- The Sleep Heart Health Study (SHHS) [100–102] can be used to analyse the impact of sleep-disordered breathing on the cardiovascular system. In this dataset more than 9000 PSGs are included, which contain, among other signals, the ECG, the respiratory effort on both the abdomen and thorax, and the airflow.
- The DREAMS Apnea Database [103] consists of 12 PSGs of patients suffering from sleep apnea. It contains ECG recordings, respiratory effort on both the abdomen and thorax, and nasal airflow. Annotations of sleep apnea events and sleep stages are available, which allow to use this dataset for both tasks described in this chapter.
- The St Vincent's University Hospital/University College Dublin sleep apnea database [104] contains 25 PSGs from adults suffering from sleep-disordered breathing. The recordings include the ECG signals, respiratory efforts, and oro-nasal airflow, among others. Both sleep stages and respiratory events are annotated in the dataset.

19.6 Conclusions

This chapter focused on the cardiorespiratory changes during sleep, and on how they are affected by sleep apnea. Special attention was on the complexity of the heart rate and respiratory signals during different sleep stages and on how their complexity is reduced due to the occurrence of sleep apnea events. In addition, the use of different nonlinear techniques for the assessment of cardiorespiratory dynamics was discussed in the framework of sleep staging and sleep apnea detection. For the sleep staging problem it was shown that a slight increase in performance could be obtained by including nonlinear features of HRV. However, for sleep apnea detection this has not been the case yet. The best algorithms in literature are based on features derived from the HRV and EDR signal using linear techniques, in particular, the classical time and frequency domain features of HRV. This can be seen as an advantage, since the clinical interpretability of linear methods is often simpler than that of the nonlinear counterparts. As a consequence, linear methods might be more easily accepted in a clinical setting.

Despite the good performance of the algorithms describing linear dynamics, it is important to keep in mind two main considerations. On the one hand, most of the public and private datasets on sleep apnea that have been used in the literature, do not include patients suffering from a cardiovascular disease. This can be seen as a limitation, since it is well-known that sleep apnea has a long-term effect on the cardiovascular system. Therefore, the effect of the nonlinear behaviour of HRV on the classification performance is still unknown in this type of patients.

On the other hand, it is essential to consider the assumptions and limitations of the nonlinear methodologies that have been used for the detection of sleep apnea before concluding that nonlinearities are not useful for this task. For instance, the accurate computation of some nonlinear features (e.g. DFA and Lyapunov exponent) can only be done on long time series. Moreover, some approaches assume either stationarity (e.g. Sample entropy, orthogonal subspace projections) or a certain joint probability distribution between heart rate and respiration, not to mention the bias induced by artefacts in the time series. As can be seen, non-linear methods, with different assumptions, might still contribute to the improvement of sleep apnea detection in the general population.

All in all, the influence of nonlinearities of heart rate and respiration during sleep and sleep apnea is not yet fully understood. Further research is needed to assess how nonlinear approaches can improve the understanding of the physiology during sleep.

Acknowledgements Bijzonder Onderzoeksfonds KU Leuven (BOF): Center of Excellence (CoE) #: PFV/10/002 (OPTEC); SPARKLE—Sensor-based Platform for the Accurate and Remote monitoring of Kinematics Linked to E-health #: IDO-13-0358; The effect of perinatal stress on the later outcome in preterm babies #: C24/15/036; TARGID—Development of a novel diagnostic medical device to assess gastric motility #: C32-16-00364. Fonds voor Wetenschappelijk Onderzoek-Vlaanderen (FWO): Project # G.0A55.13N (Deep brain stimulation). Agentschap Innoveren & Ondernemen (VLAIO): STW 150466 OSA +; O&O HBC 2016 0184 eWatch. iMinds Medical Information Technologies: Dotatie-Strategisch basis onderzoek (SBO-2016); ICON: HBC.2016.0167 SeizeIT. Belgian Federal Science Policy Office: IUAP #P7/19/ (DYSCO, "Dynamical systems, control and optimization", 2012–2017). Belgian Foreign Affairs-Development Cooperation: VLIR UOS programs (2013–2019). EU: European Union's Seventh Framework Programme (FP7/2007-2013): EU MC ITN TRANSACT 2012, #316679; The HIP Trial: #260777. Erasmus +: INGDIVS 2016-1-SE01-KA203-022114. European Research Council. The research leading to these results has received funding from the European Research Council under the European Union's Seventh Framework Programme (FP7/2007-2013) / ERC Advanced Grant: BIOTENSORS (no 339804). This manuscript reflects only the authors' views and the Union is not liable for any use that may be made of the contained information. Carolina Varon is a postdoctoral fellow of the Research Foundation—Flanders (FWO).

References

1. Mancia, G.: Autonomic modulation of the cardiovascular system during sleep. New Engl. J. Med. **328**(5), 347–349 (1993)
2. AASM: International Classification of Sleep Disorders: Diagnostic and Coding Manual. Revised edition. American Academy of Sleep Medicine, Chicago (2001)
3. Sateia, M.J.: International classification of sleep disorders: highlights and modifications. CHEST J. **146**(5), 1387–1394 (2014)
4. Redmond, S.J., de Chazal, P., O'Brien, C., Ryan, S., McNicholas, W.T., Heneghan, C.: Sleep staging using cardiorespiratory signals. Somnologie-Schlafforschung und Schlafmedizin **11**(4), 245–256 (2007)
5. AASM: The AASM Manual for the Scoring of Sleep and Associated Events: Rules, Terminology and Technical Specifications. American Academy of Sleep Medicine, Westchester, IL (2007)

6. Berry, R.B., Budhiraja, R., Gottlieb, D.J., Gozal, D., Iber, C., Kapur, V.K., Marcus, C.L., Mehra, R., Parthasarathy, S., Quan, S.F., et al.: Rules for scoring respiratory events in sleep: update of the 2007 AASM manual for the scoring of sleep and associated events. J. Clin. Sleep Med. **8**(5), 597–619 (2012)
7. AASM: Practice parameters for the role of actigraphy in the study of sleep and circadian rhythms: an update for 2002. Sleep **26**(3), 337–341 (2003)
8. Sadeh, A.: The role and validity of actigraphy in sleep medicine: an update. Sleep Med. Rev. **15**(4), 259–267 (2011)
9. Penzel, T., Kantelhardt, J.W., Lo, C.-C., Voigt, K., Vogelmeier, C.: Dynamics of heart rate and sleep stages in normals and patients with sleep apnea. Neuropsychopharmacology **28**(1), S48 (2003)
10. Fonseca, P., Long, X., Radha, M., Haakma, R., Aarts, R.M., Rolink, J.: Sleep stage classification with ecg and respiratory effort. Physiol. Meas. **36**(10), 2027 (2015)
11. De Chazal, P., Heneghan, C., Sheridan, E., Reilly, R., Nolan, P., O'Malley, M.: Automated processing of the single-lead electrocardiogram for the detection of obstructive sleep apnoea. IEEE Trans. Biomed. Eng. **50**(6), 686–696 (2003)
12. Varon, C., Caicedo, A., Testelmans, D., Buyse, B., Van Huffel, S.: A novel algorithm for the automatic detection of sleep apnea from single-lead ECG. IEEE Trans. Biomed. Eng. **62**(9), 2269–2278 (2015)
13. Willemen, T., Van Deun, D., Verhaert, V., Vandekerckhove, M., Exadaktylos, V., Verbraecken, J., Van Huffel, S., Haex, B., Vander Sloten, J.: An evaluation of cardiorespiratory and movement features with respect to sleep-stage classification. IEEE J. Biomed. Health Informatics **18**(2), 661–669 (2014)
14. Moorcroft, W.H., Belcher, P.: The body during sleep. In: Understanding Sleep and Dreaming, pp. 117–140. Springer, New York (2005)
15. Kesper, K., et al.: ECG signal analysis for the assessment of sleep-disordered breathing and sleep pattern. Med. Biol. Eng. Comput. **50**(2), 135–144 (2012)
16. de Chazal, P., et al.: Automated detection of obstructive sleep apnoea at different time scales using the electrocardiogram. Physiol. Meas. **25**(4), 967 (2004)
17. Redmond, S.J., Heneghan, C.: Cardiorespiratory-based sleep staging in subjects with obstructive sleep apnea. IEEE Trans. Biomed. Eng. **53**(3), 485–496 (2006)
18. Penzel, T., Conradt, R.: Computer based sleep recording and analysis. Sleep Med. Rev. **4**(2), 131–148 (2000)
19. Kales, A., Rechtschaffen, A., University of California Los Angeles Brain Information Service, NINDB Neurological Information Network (US): A manual of standardized terminology, techniques and scoring system for sleep stages of human subjects. National Institutes of Health publication, vol. 204. U.S. National Institute of Neurological Diseases and Blindness, Neurological Information Network, Bethesda (1968)
20. Orem, J.: Physiology in Sleep. Elsevier, Philadelphia (2012)
21. Somers, V.K., Dyken, M.E., Mark, A.L., Abboud, F.M.: Sympathetic-nerve activity during sleep in normal subjects. New Engl. J. Med. **328**(5), 303–307 (1993)
22. Douglas, N., White, D., Pickett, C.K., Weil, J., Zwillich, C.: Respiration during sleep in normal man. Thorax **37**(11), 840–844 (1982)
23. Rostig, S., Kantelhardt, J.W., Penzel, T., Cassel, W., Peter, J.H., Vogelmeier, C., Becker, H.F., Jerrentrup, A.: Nonrandom variability of respiration during sleep in healthy humans. Sleep **28**(4), 411–417 (2005)
24. Burioka, N., Cornélissen, G., Halberg, F., Kaplan, D.T., Suyama, H., Sako, T., Shimizu, E.: Approximate entropy of human respiratory movement during eye-closed waking and different sleep stages. CHEST J. **123**(1), 80–86 (2003)
25. Kantelhardt, J.W., Penzel, T., Rostig, S., Becker, H.F., Havlin, S., Bunde, A.: Breathing during rem and non-rem sleep: correlated versus uncorrelated behaviour. Phys. A: Stat. Mech. Appl. **319**, 447–457 (2003)
26. Peng, C.-K., Mietus, J.E., Liu, Y., Lee, C., Hausdorff, J.M., Stanley, H.E., Goldberger, A.L., Lipsitz, L.A.: Quantifying fractal dynamics of human respiration: age and gender effects. Ann. Biomed. Eng. **30**(5), 683–692 (2002)

27. Rafferty, G., Gardner, W.: Control of the respiratory cycle in conscious humans. J. Appl. Physiol. **81**(4), 1744–1753 (1996)
28. Long, X., Fonseca, P., Foussier, J., Haakma, R., Aarts, R.M.: Sleep and wake classification with actigraphy and respiratory effort using dynamic warping. IEEE J. Biomed. Health Informatics **18**(4), 1272–1284 (2014)
29. Long, X., Foussier, J., Fonseca, P., Haakma, R., Aarts, R.M.: Analyzing respiratory effort amplitude for automated sleep stage classification. Biomed. Signal Process. Control **14**, 197–205 (2014)
30. Sakoe, H., Chiba, S.: Dynamic programming algorithm optimization for spoken word recognition. IEEE Trans. Acoust. Speech Signal Process. **26**(1), 43–49 (1978)
31. Sako, T., Burioka, N., Suyama, H., Nomura, T., Takeshima, T., Shimizu, E.: Nonlinear behavior of human respiratory movement during different sleep stages. Chronobiol. Int. **18**(1), 71–83 (2001)
32. Raetz, S., Richard, C., Garfinkel, A., Harper, R.: Dynamic characteristics of cardiac RR intervals during sleep and waking states. Sleep **14**(6), 526–533 (1991)
33. M. Bonnet, D. Arand, Heart rate variability: sleep stage, time of night, and arousal influences. Electroencephalogr. Clin. Neurophysiol. **102**(5), 390–396 (1997)
34. Busek, P., Vankova, J., Opavský, J., Salinger, J., Nevsimalova, S.: Spectral analysis of heart rate variability in sleep. Physiol. Res. **54**(4), 369 (2005)
35. Verrier, R.L., Josephson, M.E.: Impact of sleep on arrhythmogenesis. Circ. Arrhythm. Electrophysiol. **2**(4), 450–459 (2009)
36. Haddad, G.G., Epstein, R.A., Epstein, M.A.E., Leistner, H.L., Mellins, R.B.: The RR interval and RR variability in normal infants during sleep. Pediatr. Res. **14**(6), 809–811 (1980)
37. Tulppo, M.P., Makikallio, T., Takala, T., Seppanen, T., Huikuri, H.V.: Quantitative beat-to-beat analysis of heart rate dynamics during exercise. Am. J. Physiol.-Heart Circ. Physiol. **271**(1), H244–H252 (1996)
38. Brennan, M., Palaniswami, M., Kamen, P.: Do existing measures of poincare plot geometry reflect nonlinear features of heart rate variability? IEEE Trans. Biomed. Eng. **48**(11), 1342–1347 (2001)
39. Hoshi, R.A., Pastre, C.M., Vanderlei, L.C.M., Godoy, M.F.: Poincaré plot indexes of heart rate variability: relationships with other nonlinear variables. Auton. Neurosci. **177**(2), 271–274 (2013)
40. Bunde, A., Havlin, S., Kantelhardt, J.W., Penzel, T., Peter, J.-H., Voigt, K.: Correlated and uncorrelated regions in heart-rate fluctuations during sleep. Phys. Rev. Lett. **85**(17), 3736 (2000)
41. Penzel, T., Kantelhardt, J.W., Grote, L., Peter, J.-H., Bunde, A.: Comparison of detrended fluctuation analysis and spectral analysis for heart rate variability in sleep and sleep apnea. IEEE Trans. Biomed. Eng. 50(10), 1143–1151 (2003)
42. Staudacher, M., Telser, S., Amann, A., Hinterhuber, H., Ritsch-Marte, M.: A new method for change-point detection developed for on-line analysis of the heart beat variability during sleep. Phys. A: Stat. Mech. Appl. **349**(3), 582–596 (2005)
43. Telser, S., Staudacher, M., Ploner, Y., Amann, A., Hinterhuber, H., Ritsch-Marte, M.: Can one detect sleep stage transitions for on-line sleep scoring by monitoring the heart rate variability? Somnologie **8**(2), 33–41 (2004)
44. Telser, S., Staudacher, M., Hennig, B., Ploner, Y., Amann, A., Hinterhuber, H., Ritsch-Marte, M.: Temporally resolved fluctuation analysis of sleep ECG. J. Biol. Phys. **33**(1), 19–33 (2007)
45. Long, X., Fonseca, P., Haakma, R., Aarts, R.M., Foussier, J.: Spectral boundary adaptation on heart rate variability for sleep and wake classification. Int. J. Artif. Intell. Tools **23**(03), 1460002 (2014)
46. Tobaldini, E., Nobili, L., Strada, S., Casali, K., Braghiroli, A., Montano, N.: Heart rate variability in normal and pathological sleep. Front. Physiol. **4**, 294 (2013). doi:10.3389/fphys.2013.00294,ISSN:1664-042X, http://journal.frontiersin.org/article/10.3389/fphys.2013.00294

47. Porta, A., Guzzetti, S., Montano, N., Furlan, R., Pagani, M., Malliani, A., Cerutti, S.: Entropy, entropy rate, and pattern classification as tools to typify complexity in short heart period variability series. IEEE Trans. Biomed. Eng. **48**(11), 1282–1291 (2001)
48. Viola, A.U., Tobaldini, E., Chellappa, S.L., Casali, K.R., Porta, A., Montano, N.: Short-term complexity of cardiac autonomic control during sleep: REM as a potential risk factor for cardiovascular system in aging. PLoS ONE **6**(4), e19002 (2011)
49. Vigo, D.E., Dominguez, J., Guinjoan, S.M., Scaramal, M., Ruffa, E., Solernó, J., Siri, L.N., Cardinali, D.P.: Nonlinear analysis of heart rate variability within independent frequency components during the sleep–wake cycle. Auton. Neurosci. **154**(1), 84–88 (2010)
50. Fell, J., Mann, K., Röschke, J., Gopinathan, M.: Nonlinear analysis of continuous ECG during sleep II. dynamical measures. Biol. Cybern. **82**(6), 485–491 (2000)
51. Bartsch, R.P., Liu, K.K., Ma, Q.D., Ivanov, P.C.: Three independent forms of cardio-respiratory coupling: transitions across sleep stages. In: Computing in Cardiology Conference (CinC), 2014, pp. 781–784. IEEE, Cambridge (2014)
52. Bartsch, R.P., Ivanov, P.C.: Coexisting forms of coupling and phase-transitions in physiological networks. In: Nonlinear Dynamics of Electronic Systems, pp. 270–287. Springer, Cham (2014)
53. Bartsch, R.P., Schumann, A.Y., Kantelhardt, J.W., Penzel, T., Ivanov, P.C.: Phase transitions in physiologic coupling. Proc. Natl. Acad. Sci. **109**(26), 10181–10186 (2012)
54. Angelone, A., Coulter, N.A.: Respiratory sinus arrhythmia: a frequency dependent phenomenon. J. Appl. Physiol. **19**(3), 479–482 (1964)
55. Faes, L., Nollo, G., Jurysta, F., Marinazzo, D.: Information dynamics of brain–heart physiological networks during sleep. New J. Phys. **16**(10), 105005 (2014)
56. Barbieri, R., Matten, E.C., Alabi, A.A., Brown, E.N.: A point-process model of human heartbeat intervals: new definitions of heart rate and heart rate variability. Am. J. Physiol.-Heart Circ. Physiol. **288**(1), H424–H435 (2005)
57. Valenza, G., Faes, L., Citi, L., Orini, M., Barbieri, R.: Instantaneous transfer entropy for the study of cardio-respiratory dynamics. In: Engineering in Medicine and Biology Society (EMBC), 2015 37th Annual International Conference of the IEEE, pp. 7885–7888. IEEE, Milano (2015)
58. Willemen, T., Van Deun, D., Verhaert, V., Pirrera, S., Exadaktylos, V., Verbraecken, J., Haex, B., Sloten, J.V.: Automatic sleep stage classification based on easy to register signals as a validation tool for ergonomic steering in smart bedding systems. Work **41**(supplement 1), 1985–1989 (2012)
59. Bulckaert, A., Exadaktylos, V., De Bruyne, G., Haex, B., De Valck, E., Wuyts, J., Verbraecken, J., Berckmans, D.: Heart rate-based nighttime awakening detection. Eur. J. Appl. Physiol. **109**(2), 317–322 (2010)
60. Ebrahimi, F., Setarehdan, S.-K., Ayala-Moyeda, J., Nazeran, H.: Automatic sleep staging using empirical mode decomposition, discrete wavelet transform, time-domain, and nonlinear dynamics features of heart rate variability signals. Comput. Methods Progr. Biomed. **112**(1), 47–57 (2013)
61. Domingues, A., Paiva, T., Sanches, J.M.: Hypnogram and sleep parameter computation from activity and cardiovascular data. IEEE Trans. Biomed. Eng. **61**(6), 1711–1719 (2014)
62. Ebrahimi, F., Setarehdan, S.-K., Nazeran, H.: Automatic sleep staging by simultaneous analysis of ecg and respiratory signals in long epochs. Biomed. Signal Process. Control **18**, 69–79 (2015)
63. Mendez, M.O., Matteucci, M., Castronovo, V., Ferini-Strambi, L., Cerutti, S., Bianchi, A.: Sleep staging from heart rate variability: time-varying spectral features and hidden markov models. Int. J. Biomed. Eng. Technol. **3**(3–4), 246–263 (2010)
64. Kortelainen, J.M., Mendez, M.O., Bianchi, A.M., Matteucci, M., Cerutti, S.: Sleep staging based on signals acquired through bed sensor. IEEE Trans. Inf. Technol. Biomed. **14**(3), 776–785 (2010)
65. Xiao, M., Yan, H., Song, J., Yang, Y., Yang, X.: Sleep stages classification based on heart rate variability and random forest. Biomed. Signal Process. Control **8**(6), 624–633 (2013)

66. Isa, S.M., Wasito, I., Arymurthy, A.M., et al.: Kernel dimensionality reduction on sleep stage classification using ECG signal. Int. J. Comput. Sci. Issues **8**, 115–123 (2011)
67. Rosenberg, R.S., Van Hout, S., et al.: The american academy of sleep medicine inter-scorer reliability program: sleep stage scoring. J. Clin. Sleep Med. **9**(1), 81–87 (2013)
68. Peppard, P.E., Young, T., Barnet, J.H., Palta, M., Hagen, E.W., Hla, K.M.: Increased prevalence of sleep-disordered breathing in adults. Am. J. Epidemiol. 177(9), 1006–1014 (2013)
69. Caples, S.M., Garcia-Touchard, A., Somers, V.K.: Sleep-disordered breathing and cardiovascular risk. Sleep **30**(3), 291–303 (2007)
70. Fletcher, E.C.: The relationship between systemic hypertension and obstructive sleep apnea: facts and theory. Am. J. Med. **98**(2), 118–128 (1995)
71. Penzel, T., Moody, G.B., Mark, R.G., Goldberger, A.L., Peter, J.H.: The apnea-ECG database. In: Proceedings of the Computers in Cardiology, pp. 255–258 (2000)
72. Willemen, T., Varon, C., Dorado, A.C., Haex, B., Vander Sloten, J., Van Huffel, S.: Probabilistic cardiac and respiratory based classification of sleep and apneic events in subjects with sleep apnea. Physiol. Meas. **36**(10), 2103 (2015)
73. Task Force AASM: Sleep-related breathing disorders in adults: recommendations for syndrome definition and measurement techniques in clinical research. Sleep **22**(5), 667–689 (1999)
74. Kowallik, P., Jacobi, I., Jirmann, A., Meesmann, M., Schmidt, M., Wirtz, H.: Breath-to-breath variability correlates with apnea-hypopnea index in obstructive sleep apnea. CHEST J. **119**(2), 451–459 (2001)
75. Bock, J., Gough, D.A.: Toward prediction of physiological state signals in sleep apnea. IEEE Trans. Biomed. Eng. **45**(11), 1332–1341 (1998)
76. Miyata, M., Burioka, N., Suyama, H., Sako, T., Nomura, T., Takeshima, T., Higami, S., Shimizu, E.: Non-linear behaviour of respiratory movement in obstructive sleep apnoea syndrome. Clin. Physiol. Funct. Imag. **22**(5), 320–327 (2002)
77. Wiklund, U., Olofsson, B., Franklin, K., Blom, H., Bjerle, P., Niklasson, U.: Autonomic cardiovascular regulation in patients with obstructive sleep apnoea: a study based on spectral analysis of heart rate variability. Clin. Physiol. **20**(3), 234–241 (2000)
78. Al-Angari, H.M., Sahakian, A.V.: Use of sample entropy approach to study heart rate variability in obstructive sleep apnea syndrome. IEEE Trans. Biomed. Eng. **54**(10), 1900–1904 (2007)
79. Penzel, T., et al.: Systematic comparison of different algorithms for apnoea detection based on electrocardiogram recordings. Med. Biol. Eng. Comput. **40**(4), 402–407 (2002)
80. Boudaoud, S., et al.: Corrected integral shape averaging applied to obstructive sleep apnea detection from the electrocardiogram. EURASIP J. Adv. Signal Process. **2007**(1), 032570 (2007)
81. Penzel, T., et al.: Cardiovascular and respiratory dynamics during normal and pathological sleep. Chaos **17**(1), 015116 (2007)
82. Bonsignore, M.R., Romano, S., Marrone, O., Insalaco, G.: Respiratory sinus arrhythmia during obstructive sleep apnoeas in humans. J. Sleep Res. **4**(s1), 68–70 (1995)
83. Thomas, R.J., et al.: Differentiating obstructive from central and complex sleep apnea using an automated electrocardiogram-based method. Sleep **30**(12), 1756 (2007)
84. Caicedo, A., Varon, C., Van Huffel, S.: Analysis of non-linear respiratory influences on sleep apnea classification. In: Computing in Cardiology Conference (CinC), 2014, pp. 593–596. IEEE, Cambridge (2014)
85. Kabir, M.M., Dimitri, H., Sanders, P., Antic, R., Nalivaiko, E., Abbott, D., Baumert, M.: Cardiorespiratory phase-coupling is reduced in patients with obstructive sleep apnea. PLoS ONE **5**(5), e10602 (2010)
86. Sola-Soler, J., Giraldo, B.F., Fiz, J.A., Jane, R.: Cardiorespiratory phase synchronization in osa subjects during wake and sleep states. In: Engineering in Medicine and Biology Society (EMBC), 2015 37th Annual International Conference of the IEEE, pp. 7708–7711. IEEE, Milano (2015)

87. Maier, C., Dickhaus, H., et al.: Central sleep apnea detection from ECG-derived respiratory signals. Methods Inf. Med. **49**(5), 462–466 (2010)
88. Moody, G.B., et al.: Derivation of respiratory signals from multi-lead ECGs. Proc. Comput. Cardiol. **12**(1985), 113–116 (1985)
89. Langley, P., et al.: Principal component analysis as a tool for analyzing beat-to-beat changes in ecg features: application to ECG-derived respiration. IEEE Trans. Biomed. Eng. **57**(4), 821–829 (2010)
90. Widjaja, D., et al.: Application of kernel principal component analysis for single-lead-ECG-derived respiration. IEEE Trans. Biomed. Eng. **59**(4), 1169–1176 (2012)
91. Rosenberg, R.S., Van Hout, S.: The american academy of sleep medicine inter-scorer reliability program: respiratory events. J. Clin. Sleep Med. **10**(4), 447 (2014)
92. Mendez, M., Corthout, J., Van Huffel, S., Matteucci, M., Penzel, T., Cerutti, S., Bianchi, A.: Automatic screening of obstructive sleep apnea from the ECG based on empirical mode decomposition and wavelet analysis. Physiol. Meas. **31**(3), 273 (2010)
93. Ravelo-García, A., Saavedra-Santana, P., Juliá-Serdá, G., Navarro-Mesa, J., Navarro-Esteva, J., Álvarez-López, X., Gapelyuk, A., Penzel, T., Wessel, N.: Symbolic dynamics marker of heart rate variability combined with clinical variables enhance obstructive sleep apnea screening. Chaos: Interdiscip. J. Nonlinear Sci. **24**(2), 024404 (2014)
94. Jarvis, M., Mitra, P.: Apnea patients characterized by 0.02 Hz peak in the multitaper spectrogram of electrocardiogram signals. In: Computers in Cardiology 2000, pp. 769–772. IEEE, Cambridge (2000)
95. Raymond, B., Cayton, R., Bates, R., Chappell, M.: Screening for obstructive sleep apnoea based on the electrocardiogram-the computers in cardiology challenge. In: Computers in Cardiology 2000, pp. 267–270. IEEE, Cambridge (2000)
96. De Chazal, P., Heneghan, C., Sheridan, E., Reilly, R., Nolan, P., O'Malley, M.: Automatic classification of sleep apnea epochs using the electrocardiogram. In: Computers in Cardiology 2000, pp. 745–748. IEEE, Cambridge (2000)
97. Ichimaru,Y., Moody, G.: Development of the polysomnographic database on CD-ROM. Psychiatr. Clin. Neurosci. **53**(2), 175–177 (1999)
98. Kemp, B., Zwinderman, A.H., Tuk, B., Kamphuisen, H.A., Oberye, J.J.: Analysis of a sleep-dependent neuronal feedback loop: the slow-wave microcontinuity of the EEG. IEEE Trans. Biomed. Eng. **47**(9), 1185–1194 (2000)
99. Terzano, M.G., Parrino, L., Sherieri, A., Chervin, R., Chokroverty, S., Guilleminault, C., Hirshkowitz, M., Mahowald, M., Moldofsky, H., Rosa, A., et al.: Atlas, rules, and recording techniques for the scoring of cyclic alternating pattern (cap) in human sleep. Sleep Med. **2**(6), 537–553 (2001)
100. Dean 2nd, D., Goldberger, A.L., Mueller, R., Kim, M., Rueschman, M., Mobley, D., Sahoo, S.S., Jayapandian, C.P., Cui, L., Morrical, M.G., et al.: Scaling up scientific discovery in sleep medicine: the national sleep research resource. Sleep **39**(5), 1151–1164 (2016)
101. Nieto, E.J., O'Connor, G.T., Rapoport, D.M., Redline, S.: The sleep heart health study: design, rationale, and methods. Sleep **20**(12), 1077–1085 (1997)
102. Redline, S., Sanders, M.H., Lind, B.K., Quan, S.F., Iber, C., Gottlieb, D.J., Bonekat, W.H., Rapoport, D.M., Smith, P.L., Kiley, J.P., et al.: Methods for obtaining and analyzing unattended polysomnography data for a multicenter study. Sleep **21**(7), 759–768 (1998)
103. Université de Mons: The dreams apnea database. http://www.tcts.fpms.ac.be/~devuyst/Databases/DatabaseApnea/. Accessed: 2017-02-10
104. St. Vincent's University Hospital/University College Dublin: Sleep apnea database. http://www.physionet.org/pn3/ucddb/. Accessed: 2017-02-10

MIX
Papier aus verantwortungsvollen Quellen
Paper from responsible sources
FSC® C105338

FSC
www.fsc.org

If you have any concerns about our products,
you can contact us on
ProductSafety@springernature.com

In case Publisher is established outside the EU,
the EU authorized representative is:
Springer Nature Customer Service Center GmbH
Europaplatz 3, 69115 Heidelberg, Germany

Printed by Libri Plureos GmbH
in Hamburg, Germany